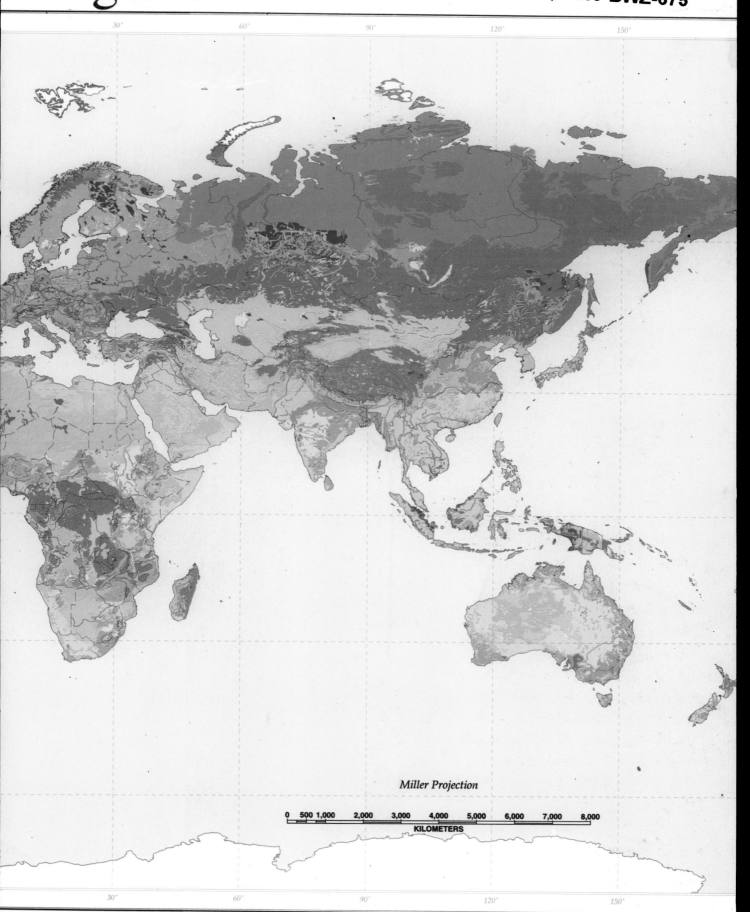

Miller Projection

KILOMETERS

0 500 1,000 2,000 3,000 4,000 5,000 6,000 7,000 8,000

ELEMENTS OF THE NATURE AND PROPERTIES OF SOILS

THIRD EDITION

Nyle C. Brady
Cornell University, Emeritus

Ray R. Weil
University of Maryland at College Park

Prentice Hall

Boston   Columbus   Indianapolis   New York   San Francisco   Upper Saddle River   Amsterdam
Cape Town   Dubai   London   Madrid   Milan   Munich   Paris   Montreal   Toronto   Delhi
Mexico City   Sao Paulo   Sydney   Hong Kong   Seoul   Singapore   Taipei   Tokyo

Vice President and Executive Editor: Vernon R. Anthony
Acquisitions Editor: William Lawrensen
Editorial Assistant: Lara Dimmick
Project Manager: Alicia Ritchy
Associate Managing Editor: Alexandrina Benedicto Wolf
Senior Operations Supervisor: Pat Tonneman
Operations Specialist: Laura Weaver
Art Director: Candace Rowley
Cover Designer: Anne DeMarinis
Cover photos: Ray Weil/Soil, Rock, Vegetation, and Water by Nile River.
Director of Marketing: David Gessell
Senior Marketing Coordinator: Alicia Wozniak

Campaign Marketing Manager: Leigh Ann Sims
Curriculum Marketing Manager: Thomas Hayward
Marketing Assistant: Les Roberts
Full-Service Project Management: Kelly Keeler; GGS Higher Education Resources, A Division of PreMedia Global, Inc.
Copyeditor: Kitty Wilson
Composition: GGS Higher Education Resources, A Division of PreMedia Global, Inc.
Printer/Binder: Edwards Brothers Malloy
Cover Printer: Lehigh-Phoenix Color
Text Font: AGaramond

Credits and acknowledgments borrowed from other sources and reproduced, with permission, in this textbook appear on appropriate page within text. Unless otherwise stated, all figures and tables belong to the authors.

SUSTAINABLE FORESTRY INITIATIVE

Certified Chain of Custody
Promoting Sustainable Forest Management
www.sfiprogram.org

SGS-SFI-COC-183

Library of Congress Cataloging-in-Publication Data

Brady, Nyle C.
    Elements of the nature and properties of soils / Nyle C. Brady, Ray R. Weil — 3rd ed.
       p. cm.
    Includes bibliographical references and index.
    ISBN 978-0-13-501433-2
  1. Soil science. 2. Soils. I. Weil, Ray R. II. Title.
    S591.B792 2010
    631.4—dc22

2008052697

15 14 13 12 11 10

Prentice Hall
is an imprint of

www.pearsonhighered.com

Paper bound   ISBN 10:   0-13-501433-6
                ISBN 13: 978-0-13-501433-2

Loose leaf   ISBN 10:   0-13-505195-9
                ISBN 13: 978-0-13-505195-5

Preface

Soils are at the heart of terrestrial ecosystems. An understanding of the soil system is therefore key to the success and environmental harmony of any human endeavor on the land. The importance of soils and the soil system is increasingly recognized by business and political leaders, by the scientific community, and by those who work with the land. Scientists and managers well versed in soil science are in short supply and becoming increasingly sought after.

This book is designed to help make your study of soils both fascinating and intellectually satisfying. Much of what you learn from these pages will be of enormous practical value in equipping you to meet the many natural-resource challenges of the 21st century. You will soon find that the soil system provides many opportunities to see practical applications for principles from such sciences as biology, chemistry, physics, and geology.

As is the case for its parent book, *The Nature and Properties of Soils*, 14th edition, this newest edition of *Elements of the Nature and Properties of Soils* strives to explain the fundamental principles of soil science in a manner that you will find relevant to your interests. Throughout, the text emphasizes the soil as a natural resource and soils as ecosystems. It highlights the many interactions between soils and other components of the larger forest, range, agricultural, wetland, and constructed ecosystems. This book is designed to serve you well, whether you expect this to be your only formal exposure to soil science or you are embarking on a comprehensive soil science education. It is meant to provide both an exciting, accessible introduction to the world of soils and a reliable, comprehensive reference that you will want to keep for your professional bookshelf.

Every chapter has been thoroughly updated with the latest advances, concepts, and applications. This edition includes new discussions on the pedosphere concept, ethnopedology, geophagy, soils and human health, organic farming, engineering properties of soils, nonsilicate colloids, inner- and outer-sphere complexes, effective CEC, the proton-balance approach to soil acidity, cation saturation, acid sulfate soils, acid precipitation, arid region soils, irrigation techniques, biomolecule binding, soil food-web ecology, disease suppressive soils, soil archaea, forest nutrient management, lead contamination, nutrient management, indicators of soil quality, soil ecosystem services, plant production for biofuels, global climate change, and many other topics of current interest in soil science. At the same time, this abridgement of the original book omits or simplifies some of the more technical details, presents fewer chemical equations and calculations, and focuses the text more clearly on the basics of soil science such that a survey of the field is be accomplished in 15 instead of 20 chapters, comprising about 600 instead of nearly 1000 pages. Among the major changes, all the nutrient cycles are now covered in a single chapter, allowing room for the addition of a new chapter on soil pollution and contamination.

In response to their popularity in recent editions, there are many new boxes that present either fascinating examples and applications or technical details and calculations. These boxes both *highlight* material of special interest and allow the logical thread of the regular text to flow smoothly without digression or interruption. Examples of applications boxes include case studies of how soils influenced the Hurricane Katrina levee failures, amelioration of selenium pollution in wetlands, capillary barriers for nuclear waste, the invasion of North American forests by earthworms, runoff farming in ancient deserts, and the debate over nitrate toxicity.

New boxes have also been added to provide detailed calculations for soil water content, profile water holding capacity, cation exchange capacity, and many other numerical problems.

Two extremely popular features of recent editions have been the high quality figures that help soils come alive, as well as the many World Wide Web universal resource locators (URLs) set in the margins of the relevant sections. These link websites developed by colleagues and organizations around the world and expand and elaborate on certain topics in ways that would not be possible in a printed book. These features are made even more effective and convenient on the new and expanded *Companion Website* (available at http://www.pearsonhighered.com/bradyweil/), where you will find beautiful full-color versions of most of the black-and-white figures printed in the book, as well as clickable links to all the URLs given in each chapter. The website also offers practice quizzes for each chapter that provide interactive feedback. These quizzes will both challenge you and build confidence in your growing understanding of the topics and their interrelationships.

Dr. Nyle Brady, having been in retirement for a number of years, decided to stand aside for the writing of the 14th edition of the full book and the 3rd edition of *Elements*, making these the first editions since 1952 not to see his direct participation. However, he remains as first author in recognition of the fact that his vision, wisdom, and inspiration continue to permeate both books. Although the responsibility for writing this edition was solely mine, I certainly could not have made all the improvements just mentioned without the many valuable suggestions, ideas, and corrections sent in by soil scientists, instructors, and students from around the world. I want to especially thank the professors who reviewed major parts of the book: Steve Thien, Kansas State University; Jan-Marie Traynor, County College of Morris; Iin Handayani, Murray State University; William C. Lindemann, New Mexico State University; and Eric Brevik, Dickinson State University. This new edition, like preceding editions, has greatly benefited from such contributions. The high level of professional devotion and camaraderie shared by so many students, teachers, and practitioners of soil science never ceases to amaze and inspire me.

Last, but not least, I wish to express my deep appreciation to Trish, my wife and spiritual partner, for her understanding, patience, encouragement, and sense of humor (exemplified by her joking reference to this book as *Dirts I Have Known*). Her support throughout the process enabled me to complete this labor of love.

About the Authors

Nyle C. Brady, a native of Manassa, Colorado, graduated from Brigham Young University in 1941 with a Bachelor's degree in Chemistry. In 1947 he received his PhD degree in Soil Science from the University of North Carolina. He served on the Cornell faculty from 1947 to 1973, and in 1952 became co-author of the world's most widely used college textbook on Soil Science. He was Head of the Department of Agronomy from July 1955 to December 1963 and served as the Director of the Cornell University Agricultural Experiment Station from September 1965 to July 1973. He was Associate Dean of the New York State College of Agriculture and Life Sciences from October 1970 to July 1973.

Dr. Brady served as the Director of Science and Education for the U.S. Department of Agriculture in Washington, D.C. from December 1963 to September 1965. From July 1973 to July 1981, he was Director General of the International Rice Research Institute in the Philippines. From 1981 to 1989, he served as Senior Assistant Administrator for Science and Technology of the United States Agency for International Development in Washington, D.C. From 1990 to 1994 he was a full time Senior Consultant for collaborative research and development programs of the World Bank in Washington, D.C. and the United Nations Development Program in New York. He is the recipient of four Honorary Doctoral Degrees from Brigham Young University (1979), Ohio State University (1991), University of the Philippines (1991), and N. C. State University (1992).

Ray R. Weil is Professor of Soil Science. He has earned degrees at Michigan State University, Purdue University, and Virginia Tech. Before coming to Maryland, he served in the Peace Corps in Ethiopia, managed a 500 acre organic farm in North Carolina, and was a Lecturer at the University of Malawi. He has become an international leader in sustainable agricultural systems in both developed and developing countries. Published in over 60 scientific journal articles and 6 books, his research focuses on cover crops and organic matter management for enhanced soil quality and nutrient cycling for water quality and sustainability. His research lab developed analytical methods for soil microbial biomass and active soil C that have been adopted by the USDA/NRCS and are used in ecosystem studies world wide. His contributions to improved cropping systems and soil management have been put into practice on farms large and small.

As a University of Maryland professor, Dr. Weil has taught over 5,000 undergraduate and graduate students, addressed over 3,000 farmers and farm advisors at meetings and field days, and helped train hundreds of researchers and managers in various companies and organizations. He has been the major advisor for 38 MS and PhD students. Weil is a Fellow of both the Soil Science Society of America and the American Society of Agronomy. He has twice been awarded a Fulbright Fellowship to support his work in developing countries. The synergism between Dr. Weil's teaching and research, and his ecological approach to soil science have found expression in various editions of this textbook since 1995.

contents

Earth, unique for soil and water. (NASA)

1

The Soils Around Us

*For in the end we will conserve
only what we love.
We will love only what we understand.
And we will understand only what
we are taught.*
—Baba Dioum, *African Conservationist*

Soils are crucial to life on Earth. From ozone depletion and global warming to rain forest destruction and water pollution, the world's ecosystems are impacted in far-reaching ways by processes carried out in the soil. To a great degree, the quality of the soil determines the nature of plant ecosystems and the capacity of land to support animal life and society. As human societies become increasingly urbanized, fewer people have intimate contact with the soil, and individuals tend to lose sight of the many ways in which they depend upon soils for their prosperity and survival. Indeed, the degree to which we are dependent on soils is likely to increase, not decrease, in the future.

Soils will continue to supply us with nearly all of our food (except for what can be harvested from the oceans). How many of us remember, as we eat a slice of pizza, that the pizza's crust began in a field of wheat, and its cheese began with grass, clover, and corn rooted in the soils of a dairy farm? Most of the fiber we use for lumber, paper, and clothing has its roots in the soils of forests and farmland. Although we sometimes use plastics and fiber synthesized from fossil petroleum as substitutes, in the long term we will continue to depend on terrestrial ecosystems for these needs.

In addition, biomass grown on soils is likely to become an increasingly important feedstock for fuels and manufacturing, as the world's finite supplies of petroleum are depleted during the course of this century. The early marketplace signs of this trend can be seen in the form of biofuels made from plant products, printers' inks made from soybean oil, and biodegradable plastics synthesized from cornstarch (Figure 1.1).

A stark reality of the 21st century is that the human population that demands all of these products will increase by several billion, while

Figure 1.1

Biofuels (left) produced from crops are far less polluting and have less impact on global warming than petroleum-based fuel. Soybean and other crops can substitute for petroleum to produce nontoxic inks (bottom), plastics, and other products. Cornstarch can be made into biodegradable plastics for such products as plastic bags and foam packing "peanuts" (upper right). (Photos courtesy of R. Weil)

the resource base available to provide them is actually *shrinking* because of soil degradation and urbanization. It is clear that we must greatly improve our understanding and management of the soil resource if we as a species are to survive and if we are to leave enough habitat for the survival of the other creatures that share this planet with us.

The Earth, our unique home in the vastness of the universe, is covered with life-sustaining air, water, and soil. However, we live in an age when human activities are changing the very nature of all three. Depletion of the ozone layer in the stratosphere is threatening to overload us with ultraviolet radiation. Increasing concentrations of carbon dioxide and methane gases are warming the planet and destabilizing the global climate. Tropical rain forests, and the incredible array of plant and animal species they contain, are disappearing at an unprecedented rate. Groundwater supplies are being contaminated in many areas and depleted in others. In parts of the world, the capacity of soils to produce food is being degraded, even as the number of people needing food is increasing. Bringing the global environment back into balance is a defining challenge of our times.

New understandings and new technologies will be needed to protect the environment and, at the same time, produce food and biomass to support society. The study of soil science has never been more important for foresters, farmers, engineers, natural resource managers, and ecologists alike.

1.1   SOILS AS MEDIA FOR PLANT GROWTH

Plant germination video: http://plantsinmotion.bio. indiana.edu/plantmotion/ earlygrowth/germination/ germ.html

In any ecosystem, whether your backyard, a farm, a forest, or a regional watershed, soils play six key roles (Figure 1.2). First among these is the support of plant growth. Soils provide a medium for plant roots and supply nutrient elements that are essential to the entire plant. Properties of the soil often determine the nature of the vegetation present and, indirectly, the number and types of animals (including people) that the vegetation can support.

When we think of the forests, prairies, lawns, and crop fields that surround us, we usually envision the **shoots**—the plant leaves, flowers, stems, and limbs—forgetting that half of the plant world, the **roots**, exists belowground. Because plant roots are

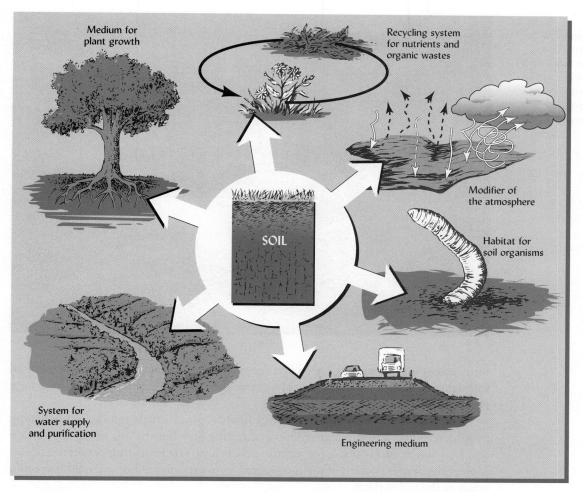

Figure 1.2
The many functions of soil can be grouped into six crucial ecological roles.

usually hidden from our view and difficult to study, we know much less about plant–environment interactions belowground than aboveground, but we must understand both to truly understand either. To begin with, let's list and then briefly discuss what a plant obtains from the soil in which its roots proliferate:

- Physical support
- Air
- Water
- Temperature moderation
- Protection from toxins
- Nutrient elements

The soil mass provides physical support, anchoring the root system so that the plant does not fall over or blow away. Occasionally, strong wind or heavy snow does topple a plant whose root system has been restricted by shallow or inhospitable soil conditions (Figure 1.3).

Plant roots depend on the process of respiration to obtain energy. Because root respiration, like our own respiration, produces carbon dioxide (CO_2) and uses oxygen (O_2), an important function of the soil is *ventilation*—allowing CO_2 to escape and fresh O_2 to enter the root zone. This ventilation is accomplished via networks of soil pores.

An equally important function of soil pores is to absorb rainwater and hold it where it can be used by plant roots. As long as plant leaves are exposed to sunlight, the plant requires a continuous stream of water to use in cooling, nutrient transport, turgor maintenance, and photosynthesis. Because plants use water continuously, but in

Figure 1.3
This wet, shallow soil failed to allow sufficiently deep roots to develop to prevent this tree from blowing over when snow-laden branches made it top-heavy during a winter storm. (Photo courtesy of R. Weil)

most places it rains only occasionally, the water-holding capacity of soils is essential for plant survival. A deep soil may store enough water to allow plants to survive long periods without rain (see Figure 1.4).

The soil also moderates temperature fluctuations. Perhaps you can recall digging in garden soil on a summer afternoon and feeling how hot the soil was at the surface and how much cooler just a few centimeters below. The insulating properties of soil protect the deeper portion of the root system from extremes of hot and cold that often occur at the soil surface.

Phytotoxic substances in soils may result from human activity, or they may be produced by plant roots, by microorganisms, or by natural chemical reactions. A good

Figure 1.4
A family of African elephants finds welcome shade under the leafy canopy of a huge acacia tree in this East African savanna. The photo was taken in the middle of a long dry season; no rain had fallen for almost five months. The tree roots are still using water from the previous rainy season stored several meters deep in the soil. The light-colored grasses are more shallow-rooted and have either set seed and died or gone into a dried-up, dormant condition. (Photo courtesy of R. Weil)

soil will protect plants from toxic concentrations of such substances by ventilating gases, by decomposing or adsorbing organic toxins, or by suppressing toxin-producing organisms. On the other hand, some microorganisms in soil produce growth-stimulating compounds that may improve plant vigor.

Soils supply plants with **mineral nutrients**. A fertile soil will provide a continuing supply of dissolved mineral nutrients in amounts and relative proportions appropriate for optimal plant growth. The nutrients include such metallic elements as potassium, calcium, iron, and copper, as well as such nonmetallic elements as nitrogen, sulfur, phosphorus, and boron. The plant takes these elements out of the soil solution and incorporates most of them into the thousands of different organic compounds that constitute plant tissue. Animals usually obtain their mineral nutrients indirectly from the soil by eating plants. Under some circumstances, animals (including humans) satisfy their craving for minerals by ingesting soil directly (Box 1.1).

Of the 92 naturally occurring chemical elements, 17 have been shown to be **essential elements**, meaning that plants cannot grow and complete their life cycles without them (Table 1.1). Essential elements used by plants in relatively large

Interactive periodic table— look up the essential elements: www.webelements.com

BOX 1.1
DIRT FOR DINNER?[a]

You are probably thinking, "dirt (excuse me, *soil*) for dinner? Yuck!" Of course, various birds, reptiles, and mammals are well known to consume soil at special "licks," and involuntary, inadvertent ingestion of soil by humans (especially children) is widely recognized as a pathway for exposure to environmental toxins (see Chapter 15), but most sophisticated residents of industrial countries, anthropologists and nutritionists included, find it hard to believe that anyone would *purposefully* ingest soil. Yet a long history of documented research on the subject shows that many people do routinely eat soil, often in amounts of 20 to 100 g (up to 1/4 pound) daily. Geophagy (deliberate "soil eating") is practiced in societies as disparate as those in Thailand, Turkey, rural Alabama, and urban Uganda (Figure 1.5). Immigrants from south Asia in the United Kingdom have brought the practice of soil eating to such cities as London and Birmingham. In fact, scientists studying the practice suggest that geophagy is a widespread and normal human behavior. Children and women (especially when pregnant) appear more likely than men to be geophagists. Poor people eat soil more commonly than the relatively well-to-do.

People usually do not eat just any soil but seek out a particular soil, be it the hardened clay of a termite nest, the soft, white soil in a particular riverbank, or the dark clay from a certain deep soil layer. People in different places and circumstances seek to consume different types of soils—some seek calcium-rich soils, others soil with high amounts of certain clays; still others seek red soils rich in iron. Interestingly, unlike many other animals, humans rarely appear to eat soil to obtain salt. Possible benefits from eating soil also vary and may include mineral nutrient supplementation (especially iron), detoxification of

Figure 1.5
Bars of clay soil sold for human consumption in a shop in Kampala, Uganda. (Photo courtesy of Peter W. Abrahams, University of Wales)

ingested poisons (by adsorption to clay—see Chapter 8), relief from stomachaches, survival in times of famine, and psychological comfort. Geophagists have been known to go to great lengths to satisfy their cravings for soil. But before you run out and add some local soil to your menu, consider the potential downsides to geophagy. Aside from the possibly difficult task of developing a taste for the stuff, the drawbacks to eating soil (especially surface soils) can include parasitic worm infection, lead poisoning, and mineral nutrient imbalances (because of adsorption of some mineral nutrients and release of others)—as well as premature tooth wear!

[a]This box is largely based on a fascinating book chapter by Abrahams (2005) and a review article by Stokes (2006).

Table 1.1

ELEMENTS ESSENTIAL FOR PLANT GROWTH AND THEIR SOURCES[a]

The chemical forms most commonly taken in by plants are shown in parentheses, with the chemical symbol for the element in bold type.

Macronutrients: Used in relatively large amounts (>0.1% of dry plant tissue)		Micronutrients: Used in relatively small amounts (<0.1% of dry plant tissue)
Mostly from air and water	Mostly from soil solids	From soil solids
Carbon (CO_2)	*Cations:*	*Cations:*
Hydrogen (H_2O)	Calcium (Ca^{2+})	Copper (Cu^{2+})
Oxygen (O_2, H_2O)	Magnesium (Mg^{2+})	Iron (Fe^{2+})
	Nitrogen (NH_4^+)	Manganese (Mn^{2+})
	Potassium (K^+)	Nickel (Ni^{2+})
		Zinc (Zn^{2+})
	Anions:	*Anions:*
	Nitrogen (NO_3^-)	Boron (H_3BO_3, $H_4BO_4^-$)
	Phosphorus ($H_2PO_4^-$, HPO_4^{2-})	Chlorine (Cl^-)
	Sulfur (SO_4^{2-})	Molybdenum (MoO_4^{2-})

[a] Many other elements are taken up from soils by plants but are not *essential* for plant growth (see Epstein and Bloom, 2005).

amounts are called **macronutrients**; those used in smaller amounts are known as **micronutrients**. To remember the 17 essential elements, try this mnemonic device:

C.B. HOPKiNS CaFé
Closed Monday Morning and Night
See You Zoon, the Mg

The bold letters indicate the chemical elements in this phrase; finding copper (Cu) and zinc (Zn) may require some imagination.

In addition to the mineral nutrients just listed, plants may also use minute quantities of organic compounds from soils. However, uptake of these substances is not necessary for normal plant growth. The organic metabolites, enzymes, and structural compounds making up a plant's dry matter consist mainly of carbon, hydrogen, and oxygen, which the plant obtains by photosynthesis from air and water, not from the soil.

Plants *can* be grown in nutrient solutions without any soil (a method termed **hydroponics**), but then the plant-support functions of soils must be engineered into the system and maintained at a high cost of time, energy, and management. Although hydroponic production on a small scale for a few high-value plants is feasible, production of the world's food and fiber and maintenance of natural ecosystems will always depend on millions of square kilometers of productive soils.

1.2 SOIL AS REGULATOR OF WATER SUPPLIES

For progress to be made in improving water quality, we must recognize that most of the water in our rivers, lakes, estuaries, and aquifers has either traveled through the soil or flowed over its surface. Imagine, for example, a heavy rain falling on the hills surrounding a river. If the soil allows the rain to soak in, some of the water may be stored in the soil and used by the trees and other plants, while some may seep slowly down through the soil layers to the groundwater, eventually entering the river over a period of months or years as base flow. If the water is contaminated, as it soaks through the upper layers of soil, it is purified and cleansed by soil processes that remove many impurities and kill potential disease organisms.

Contrast the preceding scenario with what would occur if the soil were so shallow or impermeable that most of the rain could not penetrate the soil, but ran off the hillsides on the soil surface, scouring surface soil and debris as it picked up speed, and entering the river rapidly and nearly all at once. The result would be a destructive flash flood of muddy water. The nature and management of soils has a major influence on the *purity* as well as the amount of water finding its way to aquatic systems. For those who live in a rural home, the purifying action of the soil in a septic drain field (see Section 6.8) is the main barrier that stands between what flushes down the toilet and the water running into the kitchen sink!

1.3 SOIL AS RECYCLER OF RAW MATERIALS

What would a world be like without the recycling functions performed by soils? Without reuse of nutrients, plants and animals would have run out of nourishment long ago. The world would be covered with a layer, possibly hundreds of meters high, of plant and animal wastes and corpses. Obviously, recycling must be a vital process in ecosystems, whether forests, farms, or cities. The soil system plays a pivotal role in the major geochemical cycles. Soils have the capacity to assimilate great quantities of organic waste, turning it into beneficial **humus**, converting the mineral nutrients in the wastes to forms that can be utilized by plants and animals, and returning the carbon to the atmosphere as carbon dioxide, where it again will become a part of living organisms through plant photosynthesis. Some soils can accumulate large amounts of carbon as soil organic matter, thus having a major impact on such global changes as the much-discussed *greenhouse effect* (see Sections 11.1 and 11.9).

1.4 SOIL AS MODIFIER OF THE ATMOSPHERE

The soil interacts in many ways with the Earth's blanket of air. In places where the soil is dry, poorly structured, and unvegetated, soil particles can be picked up by winds and contribute great quantities of dust to the atmosphere, reducing visibility, increasing human health hazards from breathing dirty air, and altering the temperature of the air and the planet. Moist, well-vegetated, and structured soil can prevent such dust-laden air. The evaporation of soil moisture is a major source of water vapor in the atmosphere, altering air temperature, composition, and weather patterns. Soils also breathe in and out. That is, they absorb oxygen and other gases such as methane, while they release gases such as carbon dioxide and nitrous oxide. These gas exchanges between the soil and the atmosphere have a significant influence on atmospheric composition and global warming.

Chronology of climate change science: www.aip.org/history/climate/timeline.htm

1.5 SOIL AS HABITAT FOR SOIL ORGANISMS

When we speak of protecting ecosystems, most people envision a stand of old-growth forest with its abundant wildlife, or perhaps an estuary with oyster beds and fisheries. Yet the most complex and diverse ecosystems on Earth are actually belowground! Soil is not a mere pile of broken rock and dead debris. A handful of soil may be home to *billions* of organisms, belonging to thousands of species. In even this small quantity of soil, there are likely to exist predators, prey, producers, consumers, and parasites (Figure 1.6).

How is it possible for such a diversity of organisms to live and interact in such a small space? One explanation is the tremendous range of niches and habitats in even a uniform-appearing soil. Some pores of the soil will be filled with water in which

Rangeland soil communities: www.blm.gov/nstc/soil/index.html

Figure 1.6
The soil is home to a wide variety of organisms, both large and very small. Here, a relatively large predator, a centipede (shown at about actual size), hunts for its next meal—which is likely to be one of the many smaller animals that feed on dead plant debris. (Photo courtesy of R. Weil)

swim organisms such as roundworms, diatoms, and rotifers. Tiny insects and mites may be crawling about in other larger pores filled with moist air. Micro-zones of good aeration may be only millimeters from areas of **anoxic** conditions. Different areas may be enriched with decaying organic materials; some places may be highly acidic, some more basic. Temperature, too, may vary widely.

Soils harbor much of the Earth's genetic diversity. Soils, like air and water, are important components of the larger ecosystem. Yet only now is soil quality taking its place, with air quality and water quality, in discussions of environmental protection.

1.6   SOIL AS ENGINEERING MEDIUM

Soil is probably the earliest and certainly one of the most widely used building materials. Nearly half the people in the world live in houses constructed from soil. Soil buildings vary from traditional African mud huts (Plate 79) to modern, environmentally friendly buildings built with cement-stabilized, hydraulically compacted "rammed-earth" walls (see Web link in margin).

Modern and historic buildings made of soil: www.eartharchitecture.org

"Terra firma, solid ground." We usually think of the soil as being firm and solid, a good base on which to build roads and all kinds of structures. Indeed, most

Figure 1.7
Better knowledge of the soils on which this road was built may have allowed its engineers to develop a more stable design, thus avoiding this costly and dangerous situation. (Photo courtesy of R. Weil)

structures rest on the soil, and many construction projects require excavation into the soil. Unfortunately, as can be seen in Figure 1.7, some soils are not as stable as others. Reliable construction on soils, and with soil materials, requires knowledge of the diversity of soil properties, as discussed later in this chapter. Designs for roadbeds or building foundations that work well in one location on one type of soil may be inadequate for another location with different soils.

Working with natural soils or excavated soil materials is not like working with concrete or steel. Properties such as bearing strength, compressibility, shear strength, and stability are much more variable and difficult to predict for soils than for manufactured building materials. Chapter 4 provides an introduction to some engineering properties of soils. Many other physical properties discussed will have direct application to engineering uses of soil. For example, Chapter 8 discusses the swelling properties of certain types of clays in soils. The engineer should be aware that when soils with swelling clays are wetted, they expand with sufficient force to crack foundations and buckle pavements. Much of the information on soil properties and soil classification discussed in later chapters will be of great value to people planning land uses that involve construction or excavation.

1.7   PEDOSPHERE AS ENVIRONMENTAL INTERFACE

The importance of soil as a natural body derives in large part from its role as an **interface** between the worlds of rock (the **lithosphere**), air (the **atmosphere**), water (the **hydrosphere**), and living things (the **biosphere**). Environments where all four of these worlds interact are often the most complex and productive on Earth. An estuary, where shallow waters meet the land and air, is an example of such an environment. Its productivity and ecological complexity far surpass those of a deep ocean trench, for example (where the hydrosphere is rather isolated), or the upper atmosphere (where rocks and water have little influence). The soil, or **pedosphere**, is another example of such an environment (Figure 1.8).

The concept of the soil as interface means different things at different scales. At the scale of kilometers, soils channel water from rain to rivers and transfer mineral elements from bed rocks to the oceans. They also remove and supply vast amounts of atmospheric gases, substantially influencing the global balance of methane and carbon dioxide. At a scale of a few meters (Figure 1.8b), soil forms the transition zone between hard rock and air, holding both liquid water and oxygen gas for use by plant roots. It transfers mineral elements from the Earth's rock crust to its vegetation. It processes or stores the organic remains of terrestrial plants and animals. At a scale of a few millimeters (Figure 1.8c), soil provides diverse microhabitats for air-breathing and aquatic organisms, channels water and nutrients to plant roots, and provides surfaces and solution vessels for thousands of biochemical reactions. Finally, at the scale of a few micrometers and smaller (less than one-millionth of a meter), soil provides ordered and complex surfaces, both mineral and organic, that act as templates for chemical reactions and interact with water and solutes. Its tiniest mineral particles form micro-zones of electromagnetic charge that attract everything from bacterial cell walls to proteins to conglomerates of water molecules. As you read the entirety of this book, the frequent cross-referencing between one chapter and another will remind you of the importance of scale and interfacing to the story of soil.

1.8   SOIL AS A NATURAL BODY

You may notice that this book sometimes refers to "soil," sometimes to "the soil," sometimes to "a soil," and sometimes to "soils." These variations of the word "soil" refer to two distinct concepts—*soil* as a material or *soils* as natural bodies. *Soil* is a

Figure 1.8

The pedosphere—interface of the worlds of rock (the lithosphere), air (the atmosphere), water (the hydrosphere), and life (the biosphere)—can be understood at many different scales. At the kilometer scale (a), soil participates in global cycles and the life of terrestrial ecosystems. At the meter scale (b), soil forms a transition zone between the hard rock below and the atmosphere above—a zone through which surface water and groundwater flow and in which plants and other living organisms thrive. At the millimeter scale (c), mineral particles form the skeleton of the soil that defines pore spaces, some filled with air and some with water, in which tiny creatures lead their lives. Finally, at the micro- and nanometer scales (d), soil minerals (lithosphere) provide charges, reactive surfaces that adsorb water and cations dissolved in water (hydrosphere), gases (atmosphere), and bacteria and complex humus macromolecules (biosphere). (Diagram courtesy of R. Weil)

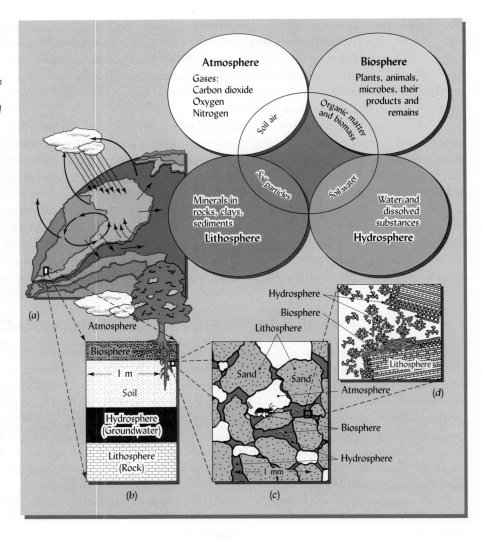

My friend, the soil; Hans Jenny—interview: http://findarticles.com/p/articles/mi_m0GER/is_1999_Spring/ai_54321347

material composed of minerals, gases, water, organic substances, and microorganisms. Some people (usually *not* soil scientists!) also refer to this material as *dirt*, especially when it is found where it is not welcome (e.g., in your clothes or under your fingernails).

A soil is a three-dimensional natural body in the same sense that a mountain, lake, or valley is. *The soil* is a collection of individually different soil bodies, often said to cover the land as the peel covers an orange. However, while the peel is relatively uniform around the orange, the soil is highly variable from place to place on Earth. One of the individual bodies, *a soil*, is to *the soil* as an individual tree is to the Earth's vegetation. Just as one may find sugar maples, oaks, hemlocks, and many other species of trees in a particular forest, so, too, might one find Christiana clay loams, Sunnyside sandy loams, Elkton silt loams, and other kinds of soils in a particular landscape.

Soils are natural bodies composed of soil (the material just described) *plus* roots, animals, rocks, artifacts, and so forth. By dipping a bucket into a lake, you may sample some of its water. In the same way, by digging or augering a hole into a soil, you may retrieve some soil. Thus, you can take a sample of soil or water into a laboratory and analyze its contents, but you must go out into the field to study a soil or a lake.

In most places, the rock exposed at the Earth's surface has crumbled and decayed to produce a layer of unconsolidated debris overlying the hard, unweathered rock. This unconsolidated layer is called the **regolith** and varies in thickness from virtually nonexistent in some places (i.e., exposed bare rock) to tens of meters in other places. The regolith material, in many instances, has been transported many kilometers from the site of its initial formation and then deposited over the bedrock which it now covers. Thus, all or part of the regolith may or may not be related to the rock now found below it. Where the underlying rock has weathered in place to the degree that it is loose enough to be dug with a spade, the term **saprolite** is used (see Plate 11).

Through their biochemical and physical effects, living organisms such as bacteria, fungi, and plant roots have altered the upper part—and, in many cases, the entire depth—of the regolith. Here, at the interface between the worlds of rock, air, water, and living things, soil is born. The transformation of inorganic rock and debris into a living soil is one of nature's most fascinating displays. Although generally hidden from everyday view, the soil and regolith can often be seen in road cuts and other excavations.

A soil is the product of both destructive and creative (synthetic) processes. Weathering of rock and microbial decay of organic residues are examples of destructive processes, whereas the formation of new minerals, such as certain clays, and of new stable organic compounds are examples of synthesis. Perhaps the most striking result of synthetic processes is the formation of contrasting layers called **soil horizons**. The development of these horizons in the upper regolith is a unique characteristic of soil that sets it apart from the deeper regolith materials (Figure 1.9).

Soil scientists specializing in **pedology** (*pedologists*) study soils as natural bodies, the properties of soil horizons, and the relationships among soils within a landscape. Other soil scientists, called **edaphologists**, focus on the soil as habitat for living things, especially plants. For both types of study it is essential to examine soils at all scales and in all three dimensions (especially the vertical dimension).

1.9   THE SOIL PROFILE AND ITS LAYERS (HORIZONS)

Soil scientists often dig a large hole, called a *soil pit*, usually several meters deep and about a meter wide, to expose soil horizons for study. The vertical section exposing a set of horizons in the wall of such a pit is termed a **soil profile**. Road cuts and other ready-made excavations can expose soil profiles and serve as windows to the soil. In an excavation open for some time, horizons are often obscured by soil material that has been washed by rain from upper horizons to cover the exposed face of lower horizons. For this reason, horizons may be more clearly seen if a fresh face is exposed by scraping off a layer of material several centimeters thick from the pit wall. Observing how soils exposed in road cuts vary from place to place can add a fascinating new dimension to travel. Once you have learned to interpret the different horizons (see Chapter 2), soil profiles can warn you about potential problems in using the land, as well as tell you much about the environment and history of a region. For example, soils developed in a dry region will have very different horizons from those developed in a humid region.

Horizons within a soil may vary in thickness and have somewhat irregular boundaries, but generally they parallel the land surface. This alignment is expected because the differentiation of the regolith into distinct horizons is largely the result of influences, such as air, water, solar radiation, and plant material, originating at the soil–atmosphere interface. Since the weathering of the regolith occurs first at the surface and works its way down, the uppermost layers have been changed the most, while the deepest layers are most similar to the original regolith, which is referred to as the

Google "soil profile" then click on "Image results."

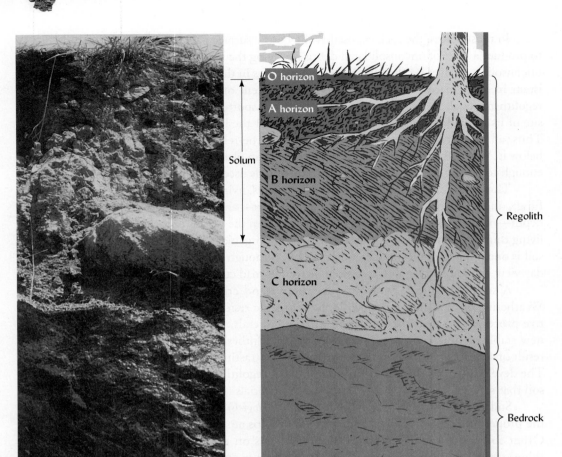

Figure 1.9

*Relative positions of the regolith, its soil, and the underlying bedrock. Note that the soil is a part of the regolith and that the A and B horizons are part of the **solum** (Latin for "soil" or "land"). The C horizon is the part of the regolith that underlies the solum, but may be slowly changing into soil in its upper parts. Sometimes the regolith is so thin that it has been changed entirely to soil; in such a case, soil rests directly on the bedrock.* (Photo courtesy of R. Weil)

soil's **parent material**. In places where the regolith was originally rather uniform in composition, the material below the soil may have a similar composition to the parent material from which the soil formed. In other cases, the regolith material has been transported long distances by wind, water, or glaciers and deposited on top of dissimilar material. In such a case, the regolith material found below a soil may be quite different from the upper layer of regolith in which the soil formed.

In undisturbed ecosystems, especially forests, organic materials formed from fallen leaves and other plant and animal remains tend to accumulate on the surface. There they undergo varying degrees of physical and biochemical breakdown and transformation, so that layers of older, partially decomposed materials may underlie the freshly added debris. Together, these organic layers at the soil surface are designated the **O horizons**.

Soil animals and percolating water move some of these organic materials downward to intermingle with the mineral grains of the regolith. These join the decomposing remains of plant roots to form organic materials that darken the upper mineral layers. Also, because weathering tends to be most intense nearest the soil surface, in many soils the upper layers lose some of their clay or other weathering products by leaching to the horizons below. **A horizons** are the layers nearest the surface that are

dominated by mineral particles but have been darkened by the accumulation of organic matter.

The organically enriched A horizon at the soil surface is sometimes referred to as **topsoil**. Plowing and cultivating a soil homogenizes and modifies the upper 12 to 25 cm (5 to 10 inches) of the soil to form a **plow layer**. In many soils, the majority of fine plant feeder roots can be found in the topsoil or plow layer. Sometimes contractors remove the plow layer from a site and sell or stockpile this topsoil for later use in establishing lawns and shrubs around newly constructed buildings (see Plate 44).

In some soils, intensely weathered and leached horizons that have not accumulated organic matter occur in the upper part of the profile, usually just below the A horizons. These horizons are designated **E horizons** (Figure 1.10).

The layers underlying the A and O horizons contain comparatively less organic matter than the horizons nearer the surface. Varying amounts of silicate clays, iron and aluminum oxides, gypsum, or calcium carbonate may accumulate in the underlying horizons. The accumulated materials may have been washed down from the horizons above, or they may have been formed in place through the weathering process. These underlying layers (commonly referred to as *subsoil*) are **B horizons** (Figures 1.10).

Plant roots and microorganisms often extend below the B horizon, especially in humid regions, causing chemical changes in the soil water, some biochemical weathering of the regolith, and the formation of **C horizons**. The C horizons are the least weathered part of the soil profile.

In some soil profiles, the component horizons are very distinct in color, with sharp boundaries that can be seen easily by even novice observers. In other soils, the

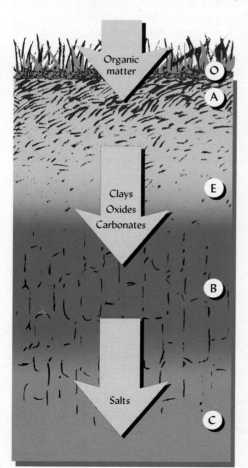

Figure 1.10
Horizons begin to differentiate as materials are added to the upper part of the profile and other materials are translocated to deeper zones. Under certain conditions, usually associated with forest vegetation and high rainfall, a leached E horizon forms between the organic-matter-rich A and B horizons. If sufficient rainfall occurs, soluble salts will be carried below the soil profile, perhaps all the way to the groundwater. Many soils lack one or more of the five horizons shown here.

color changes between horizons may be very gradual, and the boundaries more difficult to locate. Delineation of the horizons present in a soil profile often requires a careful examination, using all the senses. In addition to seeing the colors in a profile, a soil scientist may feel, smell, and listen to the soil (as in Box 4.1), as well as conduct chemical tests, to distinguish the horizons present. The importance of the various soil layers is highlighted in Box 1.2.

BOX 1.2
USING INFORMATION FROM THE ENTIRE SOIL PROFILE

Soils are three-dimensional bodies that carry out important ecosystem processes at all depths in their profiles. Depending on the particular application, the information needed to make proper land management decisions may come from soil layers as shallow as the upper 1 or 2 cm or as deep as the lowest layers of saprolite (Figure 1.11).

For example, the upper few centimeters of soil often hold the keys to plant growth and biological diversity, as well as to certain hydrologic processes. Here, at the interface between the soil and the atmosphere, living things are most numerous and diverse. Forest trees largely depend for nutrient uptake on a dense mat of fine roots growing in this zone. The physical condition of this thin surface layer may also determine whether rain will soak in or run downhill on the land surface. Certain pollutants, such as lead from highway exhaust, are also concentrated in this zone. For many types of soil investigations it will be necessary to sample the upper few centimeters separately so that important conditions are not overlooked.

On the other hand, it is equally important not to confine one's attention to the easily accessible "topsoil," for many soil properties are to be discovered only in the deeper layers. Plant-growth problems are often related to inhospitable conditions in the B or C horizons that restrict the penetration of roots. Similarly, the great volume of these deeper layers may control the amount of plant-available water held by a soil. For the purposes of recognizing or mapping different types of soils, the properties of the B horizons are often paramount. Not only is this the zone of major accumulations of minerals and clays, but the layers nearer the soil surface also are too quickly altered by management and soil erosion to be a reliable source of information for the classification of soils.

In deeply weathered regoliths, the lower C horizons and saprolite play important roles. These layers, generally at depths below 1 or 2 m, and often as deep as 5 to 10 m, greatly affect the suitability of soils for most urban uses that involve construction or excavation. The proper functioning of on-site sewage disposal systems and the stability of building foundations are often determined by regolith properties at these depths. Likewise, processes that control the movement of pollutants to groundwater or the weathering of geologic materials may occur at depths of many meters. These deep layers also have major ecological influences because, although the intensity of biological activity and plant rooting may be quite low, the total impact can be great as a result of the enormous volume of soil that may be involved. This is especially true of forest systems in warm climates.

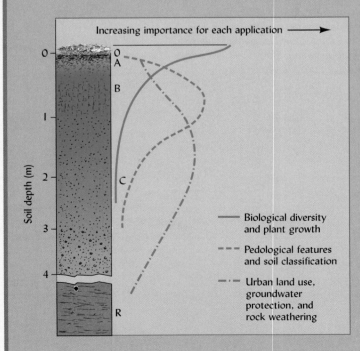

Figure 1.11
Information important to different soil functions and applications is most likely to be obtained by studying different layers of the soil profile. (Diagram courtesy of R. Weil)

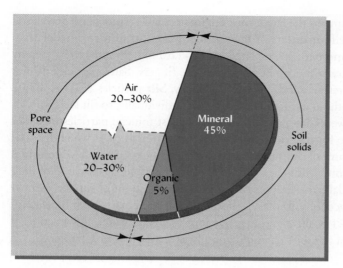

Figure 1.12
Volume composition of a loam surface soil when conditions are good for plant growth. The broken line between water and air indicates that the proportions of these two components fluctuate as the soil becomes wetter or drier. Nonetheless, a nearly equal proportion of air and water is generally ideal for plant growth.

1.10 SOIL: THE INTERFACE OF AIR, MINERALS, WATER, AND LIFE

We stated that where the regolith meets the atmosphere, the worlds of air, rock, water, and living things are intermingled. In fact, the four major components of soil are air, water, mineral matter, and organic matter. The relative proportions of these four components greatly influence the behavior and productivity of soils. In a soil, the four components are mixed in complex patterns; however, the proportion of soil volume occupied by each component can be represented in a simple pie chart. Figure 1.12 shows the approximate proportions (by volume) of the components found in a loam surface soil in good condition for plant growth. Although a handful of soil may at first seem to be a solid thing, it should be noted that only about half the soil volume consists of solid material (mineral and organic); the other half consists of pore spaces filled with air or water. Of the solid material, typically most is mineral matter derived from the rocks of the Earth's crust. Only about 5% of the *volume* in this ideal soil consists of organic matter. However, the influence of the organic component on soil properties is often far greater than its small proportion would suggest. Since it is far less dense than mineral matter, the organic matter accounts for only about 2% of the *weight* of this soil.

The spaces between the particles of solid material are just as important to the nature of a soil as are the solids themselves. It is in these pore spaces that air and water circulate, roots grow, and microscopic creatures live. Plant roots need both air and water. In an optimum condition for most plants, the pore space will be divided roughly equally among the two, with 25% of the soil volume consisting of water and 25% consisting of air. If there is much more water than this, the soil will be waterlogged. If much less water is present, plants will suffer from drought. The relative proportions of water and air in a soil typically fluctuate greatly as water is added or lost. Soils with much more than 50% of their volume in solids are likely to be too compacted for good plant growth. Compared to surface soil layers, subsoils tend to contain less organic matter, less total pore space, and a larger proportion of small pores (*micropores*), which tend to be filled with water rather than with air.

1.11 MINERAL (INORGANIC) CONSTITUENTS OF SOILS

Except in organic soils, most of the soil's solid framework consists of **mineral** particles. The larger soil particles (stones, gravel, and coarse sands) are generally rock fragments consisting of several different minerals. Smaller particles tend to be made of a single mineral.

Electron micrographs and other clay images:
www.minersoc.org/pages/gallery/claypix/index.html

Excluding, for the moment, the larger rock fragments such as stones and gravel, soil particles range in size over four orders of magnitude: from 2.0 mm to smaller than 0.0002 mm in diameter. **Sand** particles are large enough (2.0 to 0.05 mm) to be seen by the naked eye and feel gritty when rubbed between the fingers. Sand particles do not adhere to one another; therefore, sands do not feel sticky. **Silt** particles (0.05 to 0.002 mm) are too small to see without a microscope or to feel individually, so silt feels smooth but not sticky, even when wet. **Clay** particles are the smallest mineral particles ($<$0.002 mm) and adhere together to form a sticky mass when wet and hard clods when dry. The smaller particles ($<$0.001 mm) of clay (and similar-sized organic particles) have **colloidal** properties and can be seen only with the aid of an electron microscope. Because of their extremely small size, colloidal particles possess a tremendous amount of surface area per unit of mass. Since the surfaces of soil colloids (both mineral and organic) exhibit electro-magnetic charges that attract positive and negative ions as well as water, this fraction of the soil is the seat of most of the soil's chemical and physical activity (see Chapter 8).

Soil Texture

The proportion of particles in these different size ranges is described by **soil texture**. Terms such as *sandy loam, silty clay*, and *clay loam* are used to identify the soil texture. Texture has a profound influence on many soil properties, and it affects the suitability of a soil for most uses. To understand the degree to which soil properties can be influenced by texture, imagine sunbathing first on a sandy beach (loose sand), and then on a clayey beach (sticky mud).

To anticipate the effect of clay on the way a soil will behave, it is necessary to know the *kinds* of clays as well as the *amount* present. As home builders and highway engineers know all too well, soils containing certain **high-activity clays** make very unstable material on which to build because they swell when wet and shrink when dry. This shrink-and-swell action can easily crack foundations and cause retaining walls to collapse. These clays also become extremely sticky and difficult to work with when they are wet. In contrast, **low-activity clays**, formed under different conditions, can be very stable and easy to work with. Learning about the different types of clay minerals will help us understand many of the physical and chemical differences among soils in various parts of the world (see Box 1.3).

Soil Structure

Sand, silt, and clay particles can be thought of as the building blocks from which soil is constructed. The way these building blocks are put together is called **soil structure**. The particles may remain relatively independent of each other, but more commonly they are associated together in aggregates of different-size particles. These aggregates may take the form of roundish granules, cubelike blocks, flat plates, or other shapes. Soil structure (the way particles are arranged together) is just as important as soil texture (the relative amounts of different sizes of particles) in governing how water and air move in soils. Both structure and texture fundamentally influence many processes in soil, including the growth of plant roots.

1.12   SOIL ORGANIC MATTER

Soil organic matter consists of a wide range of organic (carbonaceous) substances, including living organisms (the soil **biomass**), carbonaceous remains of organisms that once occupied the soil, and organic compounds produced by current and past metabolism in the soil. The remains of plants, animals, and microorganisms are continuously broken down in the soil, and new substances are synthesized by other microorganisms. Over time, organic matter is lost from the soil as carbon dioxide produced by microbial

BOX 1.3
OBSERVING SOILS IN DAILY LIFE

Your study of soils can be enriched if you make an effort to become aware of the many daily encounters with soils and their influences that go unnoticed by most people. When you dig a hole to plant a tree or set a fence post, note the different layers encountered and note how the soil from each layer looks and feels. If you pass a construction site, take a moment to observe the horizons exposed by the excavations. An airplane trip is a great opportunity to observe how soils vary across landscapes and climatic zones. If you are flying during daylight hours, ask for a window seat. Look for the shapes of individual soils in plowed fields if you are flying in spring or fall (Figure 1.13).

Soils can give you clues to understanding the natural processes going on around you. Down by the stream, use a magnifying glass to examine the sand deposited on the banks or bottom. It may contain minerals not found in local rocks and soils, but originating many kilometers upstream. When you wash your car, see if the mud clinging to the tires and fenders is of a different color or consistency than the soils near your home. Does the "dirt" on your car tell you where you have been driving? Forensic investigators have been known to consult with soil scientists to locate crime victims or establish guilt by matching

soil clinging to shoes, tires, or tools with the soils at a crime scene.

Other examples of soil hints can be found even closer to home. The next time you bring home celery or leaf lettuce from the supermarket, look carefully for bits of soil clinging to the bottom of the stalk or leaves (Figure 1.14). Rub the soil between your thumb and fingers. Smooth, very black soil may indicate that the lettuce was grown in mucky soils, such as those in New York state or southern Florida. Light brown, smooth-feeling soil with only a very fine grittiness is more typical of California-grown produce, while light-colored, gritty soil is common on produce from the southern Georgia–northern Florida vegetable-growing region. In a bag of dry pinto beans, you may come across a few lumps of soil that escaped removal in the cleaning process because of being the same size as the beans. Often this soil is dark colored and very sticky, coming from the "thumb" area of Michigan, where a large portion of the U.S. dry bean crop is grown.

Opportunities to observe soils in daily life range from the remote and large-scale to the close-up and intimate. As you learn more about soils, you will undoubtedly be able to see more examples of their influence in your surroundings.

Figure 1.13
The light- and dark-colored soil bodies, as seen from an airliner flying over central Texas, reflect differences in drainage and topography in the landscape. (Photo courtesy of R. Weil)

Figure 1.14
The black, mucky soil clinging to the base of this celery stalk indicates that it was grown on organic soils, probably in New York state. (Photo courtesy of R. Weil)

respiration. Because of such loss, repeated additions of new plant and/or animal residues are necessary to maintain soil organic matter.

Under conditions that favor plant production more than microbial decay, large quantities of atmospheric carbon dioxide used by plants in photosynthesis are sequestered in the abundant plant tissues that eventually become part of the soil organic matter. Since carbon dioxide is a major cause of the greenhouse effect, which is warming Earth's climate, the balance between accumulation of soil organic matter and its loss through microbial respiration has global implications. In fact, more carbon is stored in the world's soils than in the world's plant biomass and atmosphere combined.

Figure 1.15

Abundant organic matter, including plant roots, helps create physical conditions favorable for the growth of higher plants as well as microbes (left). In contrast, soils low in organic matter, especially if they are high in silt and clay, are often cloddy (right) and not suitable for optimum plant growth. (Photos courtesy of N. C. Brady)

Even so, organic matter comprises only a small fraction of the mass of a typical soil. By weight, typical well-drained mineral surface soils contain from 1 to 6% organic matter. The organic matter content of subsoils is even smaller. However, the influence of organic matter on soil properties, and consequently on plant growth, is far greater than the low percentage would indicate (see also Chapter 11).

Organic matter binds mineral particles into a granular soil structure that is largely responsible for the loose, easily managed condition of productive soils. Part of the soil organic matter that is especially effective in stabilizing these granules consists of certain gluelike substances produced by various soil organisms, including plant roots (Figure 1.15).

Organic matter also increases the amount of water a soil can hold and the proportion of water available for plant growth (Figure 1.16). In addition, organic matter is a major source of the plant nutrients phosphorus and sulfur and the primary source of nitrogen for most plants. As soil organic matter decays, these nutrient elements, which are present in organic combinations, are released as soluble ions that can be taken up by plant roots. Finally, organic matter, including plant and animal residues, is the main

Figure 1.16

Soils higher in organic matter have greater water-holding capacities than soils low in organic matter. The soil in each container has the same texture, but the one on the right is lower in organic matter. The same amount of water was applied to each container. The depth of water penetration was less in the high organic matter soil (left) because of its greater water-holding capacity. It required a greater volume of the low organic matter soil to hold the same amount of water. (Photo courtesy of N. C. Brady)

food that supplies carbon and energy to soil organisms. Without it, biochemical activity so essential for ecosystem functioning would come to a near standstill.

Humus, usually black or brown in color, is a collection of very complex organic compounds that accumulate in soil because they are relatively resistant to decay. Just as clay is the colloidal fraction of soil mineral matter, so humus is the colloidal fraction of soil organic matter. Because of their charged surfaces, both humus and clay act as contact bridges between larger soil particles; thus, both play an important role in the formation of soil structure. The surface charges of humus, like those of clay, attract and hold both nutrient ions and water molecules. However, gram for gram, the capacity of humus to hold nutrients and water is far greater than that of clay. All in all, small amounts of humus may remarkably increase the soil's capacity to promote plant growth.

1.13 SOIL WATER: A DYNAMIC SOLUTION

Water is of vital importance in the ecological functioning of soils. The presence of water in soils is essential for the survival and growth of plants and other soil organisms. The soil moisture regime, often reflective of climatic factors, is a major determinant of the productivity of terrestrial ecosystems, including agricultural systems. Movement of water, and substances dissolved in it, through the soil profile is of great consequence to the quality and quantity of local and regional water resources. Water moving through the regolith is also a major driving force in soil formation.

Water is held within soil pores with varying degrees of tenacity depending on the amount of water present and the size of the pores. The attraction between water and the surfaces of soil particles greatly restricts the ability of water to flow.

When the soil moisture content is optimal for plant growth (Figure 1.12), the water in the large- and intermediate-sized pores can move about in the soil and can easily be used by plants. As a plant grows, however, its roots remove water from the largest pores first. Soon the larger pores hold only air, and the remaining water is found only in the intermediate- and smallest-sized pores. The water in the smallest pores is so strongly held on particle surfaces that plant roots cannot pull it away. Consequently, not all soil water is *available* to plants.

Soil Solution

Because soil water is never pure water, but contains hundreds of dissolved organic and inorganic substances, it may be more accurately called the **soil solution**. The soil solids, particularly the fine organic and inorganic colloidal particles (clay and humus), release nutrient elements to the soil solution from which they are taken up by plant roots. The soil solution tends to resist changes in its composition even when compounds are added or removed from the soil. This ability to resist change is termed the soil **buffering capacity** and is dependent on many chemical and biological reactions, including the attraction and release of substances by colloidal particles (see Chapter 8).

Many chemical and biological reactions are dependent on the relative levels of hydrogen ions (H^+) and hydroxyl ions (OH^-) in the soil solution, which are commonly determined by measuring the **pH** of the soil. The pH is a logarithmic scale used to express the degree of soil acidity or alkalinity (Figures 1.17 and 1.18). The pH is considered a master variable of soil chemistry and is of great significance to nearly all aspects of soil science.

1.14 SOIL AIR: A CHANGING MIXTURE OF GASES

Approximately half of the volume of the soil consists of pore spaces of varying sizes (refer to Figures 1.12), which are filled with either water or air. When water enters the soil, it displaces air from some of the pores; the air content of a soil is therefore

Figure 1.17
At neutrality (pH 7) the H^+ and OH^- ions of a solution are balanced, their respective numbers being the same. At pH 6, the H^+ ions are 10 times greater, whereas the OH^- ions are only one-tenth as numerous. The solution therefore is acid at pH 6, there being 100 times more H^+ ions than OH^- ions present. At pH 8, the exact reverse is true; the OH^- ions are 100 times more numerous than the H^+ ions. Hence, the pH 8 solution is alkaline.

inversely related to its water content. If we think of the network of soil pores as the ventilation system of the soil connecting airspaces to the atmosphere, we can understand that when too many pores are filled with water the ventilation system becomes clogged. Think how stuffy the air would become if the ventilation ducts of a classroom became clogged. Because oxygen could not enter the room, nor carbon dioxide leave it, the air in the room would soon become depleted of oxygen and enriched in carbon dioxide and water vapor by the respiration (breathing) of the people in it. In an air-filled soil pore surrounded by water-filled smaller pores, the metabolic activities of plant roots and microorganisms have a similar effect.

The composition of soil air varies greatly from place to place in the soil. In local pockets, some gases are consumed by plant roots and by microbial reactions, and others are released, thereby greatly modifying the composition of the soil air. Soil air generally has a higher moisture content than the atmosphere; the relative humidity of soil air is nearly 100% unless the soil is very dry. The content of carbon dioxide (CO_2) is usually much higher, and that of oxygen (O_2) somewhat lower, than contents of these gases found in the atmosphere.

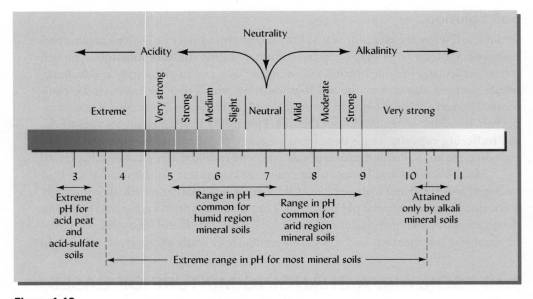

Figure 1.18
Extreme range in pH for most mineral soils and the ranges commonly found in humid region and arid region soils. Also indicated are the maximum alkalinity for alkali soils and the minimum pH likely to be encountered in very acid peat soils.

The amount and composition of air in a soil are determined to a large degree by the water content of the soil. The air occupies those soil pores not filled with water. As the soil drains from a heavy rain or irrigation, large pores are the first to be filled with air, followed by medium-sized pores, and finally the small pores, as water is removed by evaporation and plant use. This explains the tendency for soils with a high proportion of tiny pores to be poorly aerated. In extreme cases, lack of oxygen both in the soil air and dissolved in the soil water may fundamentally alter the chemical reactions that take place in the soil solution. This is of particular importance to understanding the functions of wetland soils (Chapter 7).

1.15   INTERACTION OF FOUR COMPONENTS TO SUPPLY PLANT NUTRIENTS

As you read our discussion of each of the four major soil components, you may have noticed that the impact of one component on soil properties is seldom expressed independently from that of the others. Rather, the four components interact with each other to determine the nature of a soil. For example, organic matter, because of its physical binding power, influences the arrangement of the mineral particles into clusters and, in so doing, increases the number of large soil pores, thereby influencing the water and air relationships.

Essential Element Availability

Perhaps the most important interactive process involving the four soil components is the provision of essential nutrient elements to plants. Plants absorb essential nutrients, along with water, directly from one of these components: the soil solution. However, the amount of essential nutrients in the soil solution at any one time is sufficient to supply the needs of growing vegetation for only a few hours or days. Consequently, the soil solution nutrient levels must be constantly replenished.

Fortunately, relatively large quantities of these nutrients are associated with both inorganic and organic soil solids. By a series of chemical and biochemical processes, nutrients are released from these solid forms to replenish those in the soil solution. For example, the tiniest colloidal-sized particles—both clay and humus—exhibit negative and positive charges. These charges tend to attract or **adsorb**[1] oppositely charged ions from the soil solution and hold them as **exchangeable ions**. Through ion exchange, elements such as Ca^{2+} and K^+ are released from this state of electrostatic adsorption on colloidal surfaces and escape into the soil solution. In the example below, a H^+ ion in the soil solution is shown to exchange places with an adsorbed K^+ ion on the colloidal surface:

$$\boxed{\text{colloid}}\ K^+ + H^+\text{ion} \longrightarrow \boxed{\text{colloid}}\ H^+ + K^+\text{ion} \tag{1.1}$$

Adsorbed   Soil   Adsorbed   Soil
                solution   solution

The K^+ ion thus released can be readily taken up (absorbed) by plants. Some scientists consider that this ion exchange process is among the most important of chemical reactions in nature.

Nutrient ions are also released to the soil solution as soil microorganisms decompose organic tissues. Plant roots can readily absorb all of these nutrients from the soil solution, provided there is enough O_2 in the soil air to support root metabolism.

Cation exchange in action, University of New England: www.une.edu.au/agss/ozsoils/images/SSCATXCH.dcr

Managing plant nutrients in Africa: http://www.fao.org/ag/magazine/spot3.htm

---

[1]*Adsorption* refers to the attraction of ions to the surface of particles, in contrast to *absorption*, the process by which ions are taken *into* plant roots. The adsorbed ions are exchangeable with ions in the soil solution.

Table 1.2
QUANTITIES OF SIX ESSENTIAL ELEMENTS FOUND IN UPPER 15 CM OF REPRESENTATIVE SOILS IN TEMPERATE REGIONS

Essential element	Humid region soil			Arid region soil		
	In solid framework, kg/ha	Exchangeable, kg/ha	In soil solution, kg/ha	In solid framework, kg/ha	Exchangeable, kg/ha	In soil solution, kg/ha
Ca	8,000	2,250	60–120	20,000	5,625	140–280
Mg	6,000	450	10–20	14,000	900	25–40
K	38,000	190	10–30	45,000	250	15–40
P	900	—	0.05–0.15	1,600	—	0.1–0.2
S	700	—	2–10	1,800	—	6–30
N	3,500	—	7–25	2,500	—	5–20

Most soils contain large amounts of plant nutrients relative to the annual needs of growing vegetation. However, the bulk of most nutrient elements is held in the structural framework of primary and secondary minerals and organic matter. Only a small fraction of the nutrient content of a soil is present in forms that are readily available to plants. Table 1.2 will give you some idea of the quantities of various essential elements present in different forms in typical soils of humid and arid regions.

Figure 1.19 illustrates how the two solid soil components interact with the liquid component (soil solution) to provide essential elements to plants. Plant roots do not ingest soil particles, no matter how fine, but are able to absorb only nutrients that are dissolved in the soil solution. Because elements in the coarser soil framework of the soil are only slowly released into the soil solution over long periods of time, the bulk of most nutrients in a soil is not readily available for plant use. Nutrient elements in the framework of colloid particles are somewhat more readily available to plants, as these particles break down much faster because of their greater surface area. Thus, the structural framework is the major storehouse and, to some extent, a significant source of essential elements in many soils.

1.16   NUTRIENT UPTAKE BY PLANT ROOTS

To be taken up by a plant, a nutrient element must be in a soluble form and must be located *at the root surface*. Often, parts of a root are in such intimate contact with soil particles (see Figure 1.20) that a direct exchange may take place between nutrient ions adsorbed on the surface of soil colloids and H^+ ions from the surface of root cell walls. In any case, the supply of nutrients in contact with the root will soon be depleted. This fact raises the question of how a root can obtain additional supplies once the nutrient ions at the root surface have all been taken up into the root. There are three basic mechanisms by which the concentration of nutrient ions at the root surface is maintained (Figure 1.21).

First, **root interception** comes into play as roots continually grow into new, undepleted soil. For the most part, however, nutrient ions must travel some distance in the soil solution to reach the root surface. This movement can take place by **mass flow**, as when dissolved nutrients are carried along with the soil water flowing toward a root that is actively drawing water from the soil. In this type of movement, the nutrient ions are somewhat analogous to leaves floating down a stream. On the other hand, plants can continue to take up nutrients even at night, when water is only slowly absorbed into the roots. Nutrient ions continually move by **diffusion** from

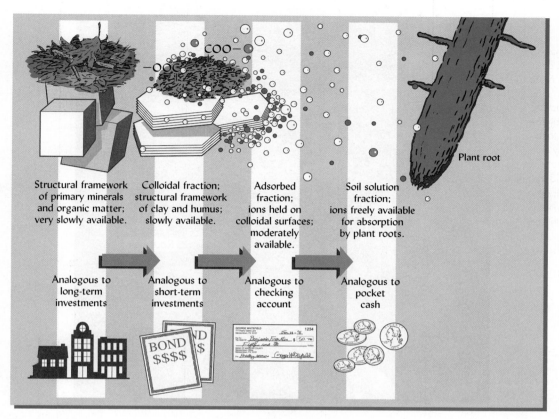

Figure 1.19

Nutrient elements exist in soils in various forms characterized by different accessibility to plant roots (shown here to be analogous to financial assets varying in degree of liquidity). The bulk of the nutrients is locked up in the structural framework of primary minerals, organic matter, clay, and humus. A smaller proportion of each nutrient is adsorbed in a swarm of ions near the surfaces of soil colloids (clay and organic matter). From the swarm of adsorbed ions, a still smaller amount is released into the bulk soil solution, where uptake by plant roots can take place. (Diagram courtesy of R. Weil)

areas of greater concentration toward the nutrient-depleted areas of lower concentration around the root surface.

Because nutrient uptake is an active metabolic process, conditions that inhibit root metabolism may also inhibit nutrient uptake. Examples of such conditions include excessive soil water content or soil compaction resulting in poor soil aeration, excessively hot or cold soil temperatures, and aboveground conditions that result in low translocation of sugars to plant roots. We can see that plant nutrition involves biological, physical, and chemical processes and interactions among many different components of soils and the environment.

1.17 SOIL QUALITY, DEGRADATION, AND RESILIENCE

Soil is a basic resource underpinning all terrestrial ecosystems. Managed carefully, soils are a *reusable* resource, but in the scale of human lifetimes, they cannot be considered a *renewable* resource. As we shall see in the next chapter, most soil profiles are thousands of years in the making. In all regions of the world, human activities are destroying some soils far faster than nature can rebuild them. As mentioned in the opening paragraphs of this chapter, growing numbers of people are demanding more and more from the Earth's fixed amount of land. Nearly all of the soils best suited for

Global population counter: www.ibiblio.org/lunarbin/worldpop

CHAPTER ONE

Figure 1.20
Scanning electron micrograph (SEM) cross section of a barley root growing in field soil. Note the intimate contact between the root and the soil, made more so by the long, thin root hairs that permeate the nearby soil and bind it to the root. The root itself is about 0.3 mm in diameter. (Photo courtesy of Margaret McCully, CSIRO, Plant Industry, Canberra, Australia)

Figure 1.21
Three principal mechanisms by which nutrient ions dissolved in the soil solution come into contact with plant roots. All three mechanisms operate simultaneously, but one mechanism or another may be most important for a particular nutrient. For example, in the case of calcium, which is generally plentiful in the soil solution, mass flow alone can usually bring sufficient amounts to the root surface. However, in the case of phosphorus, diffusion is needed to supplement mass flow because the soil solution is very low in this element in comparison to the amounts needed by plants. (Diagram courtesy of R. Weil)

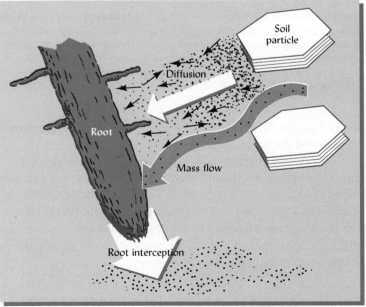

growing crops are already being farmed. Therefore, as each year brings millions more people to feed, the amount of cropland per person continuously declines. In addition, many of the world's major cities were originally located where excellent soils supported thriving agricultural communities, so now much of the very best farmland is being lost to suburban development as these cities thoughtlessly expand.

Finding more land on which to grow food is not easy. Most additional land brought under cultivation comes at the cost of clearing natural forests, savannas, and grasslands. Images of the Earth made from orbiting satellites show the resulting decline in land covered by forests and other natural ecosystems. Thus, as the human population struggles to feed itself, wildlife populations are deprived of vital habitat, and overall biodiversity suffers. Efforts to reduce and even reverse human population growth must be accelerated if our grandchildren are to inherit a livable world. In the meantime, if there is to be space for both people and wildlife, the best of our existing farmland soils will require improved and more intensive management. Soils completely washed away by erosion or excavated and paved over by urban sprawl are permanently lost, for all practical purposes. More often, soils are degraded in quality rather than totally destroyed.

Soil quality is a measure of the ability of a soil to carry out particular ecological functions, such as those described in Sections 1.1 to 1.6. Soil quality reflects a combination of *chemical, physical*, and *biological* properties. Some of these properties are relatively unchangeable, inherent properties that help define a particular type of soil. Soil texture and mineral makeup are examples. Other soil properties, such as structure and organic matter content, can be significantly changed by management. These more changeable soil properties can indicate the status of a soil's quality relative to its potential, in much the same way that water turbidity or oxygen content indicates the water-quality status of a river.

Mismanagement of forests, farms, and rangeland causes widespread degradation of soil quality by erosion that removes the topsoil, little by little (see Chapter 14). Another widespread cause of soil degradation is the accumulation of salts in improperly irrigated soils in arid regions (see Chapter 9). When people cultivate soils and harvest the crops without returning organic residues and mineral nutrients, the soil's supply of organic matter and nutrients becomes depleted (see Chapter 11). Such depletion is particularly widespread in sub-Saharan Africa, where degrading soil quality is reflected in diminished capacity to produce food. Contamination of a soil with toxic substances from industrial processes or chemical spills can degrade its capacity to provide habitat for soil organisms, to grow plants that are safe to eat, or to safely recharge ground and surface waters (see Chapter 15). Degradation of soil quality by pollution is usually localized, but the environmental impacts and costs involved are very large. **Sustainable** soil management means using soils in ways that will provide current benefits without jeopardizing capacity of the soil resource to satisfy the needs of future generations.

While sustainable soil use that protects soil quality must be the first priority, it is often necessary to attempt to restore the quality of soils that have already been degraded. Some soils have sufficient **resilience** to recover from minor degradation if left to revegetate on their own. In other cases, more effort is required to restore degraded soils (see Chapter 14). Organic and inorganic amendments may have to be applied, vegetation may have to be planted, physical alterations by tillage or grading may have to be made, or contaminants may have to be removed. As societies around the world assess the damage already done to their natural and agricultural ecosystems, the science of **restoration ecology** has rapidly evolved to guide managers in restoring plant and animal communities to their former levels of diversity and productivity. The job of **soil restoration**, an essential part of these efforts, requires in-depth knowledge of all aspects of the soil system.

1.18   CONCLUSION

The Earth's soil consists of numerous soil individuals, each of which is a three-dimensional natural body in the landscape. Each individual soil is characterized by a unique set of properties and soil horizons as expressed in its profile. The nature of the soil layers seen in a particular profile is closely related to the nature of the environmental conditions at a site.

Overview of soil quality concepts: http://soils.usda.gov/sqi/concepts/concepts.html

Proverbs about soil. Yoseph Araya. Scroll to p.40: http://www.iuss.org/Bulletins/IUSS%20Bulletin%20103.pdf

Soils perform six broad ecological functions. They 1) act as the principal medium for plant growth, 2) regulate water supplies, 3) modify the atmosphere, 4) recycle raw materials and waste products, 5) provide habitat for many kinds of organisms, and 6) serve as a major engineering medium for human-built structures. Soil is thus a major ecosystem in its own right. The soils of the world are extremely diverse, each type of soil being characterized by a unique set of soil horizons. A typical surface soil in good condition for plant growth consists of about half solid material (mostly mineral, but with a crucial organic component, too) and half pore spaces filled with varying proportions of water and air. These components interact to influence myriad complex soil functions, a good understanding of which is essential for wise management of our terrestrial resources.

If we take the time to learn the language of the land, the soil will speak to us.

STUDY QUESTIONS

1. As a society, is our reliance on soils likely to increase or decrease in the decades ahead? Explain.
2. Discuss how *a soil*, a natural body, differs from *soil*, a material that is used in building a roadbed.
3. What are the six main roles of soil in an ecosystem? For each of these ecological roles, suggest one way in which interactions occur with another of the six roles.
4. Think back over your activities during the past week. List as many incidents as you can in which you came into direct or indirect contact with soil.
5. Figure 1.12 shows the volume composition of a loam surface soil in ideal condition for plant growth. To help you understand the relationships among the four components, redraw this pie chart to represent what the situation might be after the soil has been compacted by heavy traffic. Then draw another pie chart to show how the four components would be related on a mass (weight) basis

rather than on a volume basis. Hint: How much does the air weigh?
6. Explain in your own words how the soil's nutrient supply is held in different forms, much the way that a person's financial assets might be held in different forms.
7. List the essential nutrient elements that plants derive mainly from the soil.
8. Are all elements contained in plants essential nutrients? Explain.
9. Define these terms: *soil texture, soil structure, soil pH, humus, soil profile, B horizon, soil quality, solum,* and *saprolite.*
10. Describe four processes that commonly lead to degradation of soil quality.
11. Compare the pedological and edaphological approaches to the study of soils. Which is more closely aligned with geology and which with ecology?

REFERENCES

Abrahams, P. W. 2005. "Geophagy and the involuntary ingestion of soil," pp. 435–457, in O. Selinus (ed.), *Essentials of medical geology.* Elsevier, The Hague.

Epstein, E., and A. J. Bloom. 2005. *Mineral nutrition of plants: Principles and perspectives,* 2nd ed. Sinauer Associates, Sunderland, MA.

Food and Agriculture Organization of the United Nations. 2005. *Global forest resources assessment 2005.* www.fao.org/forestry/site/fra2005/en (verified 17 November 2008).

Stokes, T. 2006. The earth-eaters. *Nature* 444:543–554.

2
Formation of Soils from Parent Materials

It is a poem of existence . . . not a lyric but a slow epic whose beat has been set by eons of the world's experience. . . .
—JAMES MICHENER, *CENTENNIAL*

The first astronauts to explore the moon labored in their clumsy pressurized suits to collect samples of rocks and dust from the lunar surface. These they carried back to Earth for analysis. It turned out that moon rocks are similar in composition to those found deep in the Earth—so similar that scientists concluded that the moon itself began when a stupendous collision between a Mars-sized object and the young planet Earth spewed molten material into orbit around the planet. The force of gravity eventually pulled this material together to form the moon. On the moon, this rock remained unchanged or crumbled into dust with the impact of meteors. On Earth, the rock at the surface, eventually coming in contact with water, air, and living things, was transformed into something new, into many different kinds of living soils. This chapter reveals the story of how rock and dust become "the ecstatic skin of the Earth."[1]

We will study the processes of soil formation that transform the lifeless regolith into the variegated layers of the soil profile. We will also learn about the environmental factors that influence these processes to produce soils in Belgium so different from those in Brazil, soils on limestone so different from those on sandstone, and soils in the valley bottoms so different from those on the hills.

---

[1] The apt description of soil as "ecstatic skin" is from a delightfully readable account of soils by Logan (1995). Earth may not be the only planet with a skin of soil. Data from Mars (see Kerr, 2005, for details) suggest erosion and formation of secondary minerals such as gypsum from the movement and evaporation of surface water but almost no weathering and no clays. The Mars Reconnaissance Orbiter discovered soil-covered glaciers on Mars and in 2008 the robotic Phoenix Mars Lander scraped away some of the polar soil to reveal a white slab of ice. Scientists have concluded that the Martian surface was briefly flooded with water billions of years ago but has been too cold for flowing water since.

Every landscape is comprised of a suite of different soils, each influencing ecological processes in its own way. Whether we intend to modify, exploit, preserve, or simply understand the landscape, our success will depend on our knowing how soil properties relate to the environment on each site and to the landscape as a whole.

2.1   WEATHERING OF ROCKS AND MINERALS

The influence of weathering, the physical and chemical breakdown of particles, is evident everywhere. Nothing escapes it. Weathering breaks up rocks and minerals, modifies or destroys their physical and chemical characteristics, and carries away the finer fragments and soluble products. However, weathering also synthesizes new minerals of great significance in soils. The nature of the rocks and minerals being weathered determines the rates and results of the breakdown and synthesis (Figure 2.1).

Characteristics of Rocks and Minerals

Geologists classify Earth's rocks as igneous, sedimentary, and metamorphic. Igneous rocks are those formed from molten magma and include such common rocks as granite and basalt (Figure 2.2).

Igneous rock is composed of such primary minerals[2] as light-colored quartz, muscovite, and feldspars and dark-colored biotite, augite, and hornblende. The mineral grains in igneous rocks interlock and are randomly dispersed, sometimes giving a salt-and-pepper appearance. In general, dark-colored minerals contain iron and magnesium and are more easily weathered.

Sedimentary rocks form when weathering products released from other, older rocks collect under water as sediment and eventually reconsolidate into new rock. For example, quartz sand weathered from granite and deposited near the shore of a prehistoric sea may become cemented by calcium or iron in the water to

Animations of rock formation (click on Chapter 6):
www.classzone.com/books/earth_science/terc/navigation/visualization.cfm

Figure 2.1
Two stone markers, photographed on the same day in the same cemetery, illustrate the effect of rock type on weathering rates. The date and initials carved in the slate marker in 1798 are still sharp and clear, while the date and figure of a lamb carved in the marble marker in 1875 have weathered almost beyond recognition. The slate rock consists largely of resistant silicate clay minerals, while the marble consists mainly of calcite, which is much more easily attacked by acids in rainwater. (Photos courtesy of R. Weil)

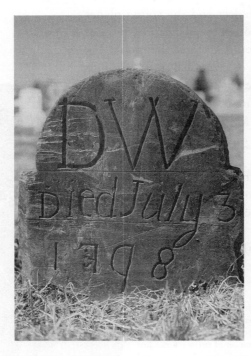

---

[2] Primary minerals have not been altered chemically since they formed as molten lava solidified. *Secondary minerals* are recrystallized products of the chemical breakdown and/or alteration of primary minerals.

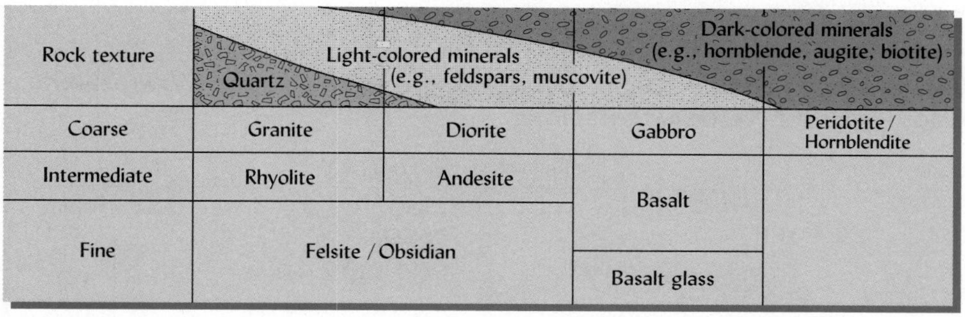

Rock texture	Quartz	Light-colored minerals (e.g., feldspars, muscovite)		Dark-colored minerals (e.g., hornblende, augite, biotite)
Coarse	Granite	Diorite	Gabbro	Peridotite / Hornblendite
Intermediate	Rhyolite	Andesite	Basalt	
Fine	Felsite / Obsidian		Basalt glass	

Figure 2.2

Classification of some igneous rocks in relation to mineralogical composition and the size of mineral grains in the rock (rock texture). Worldwide, light-colored minerals and quartz are generally more prominent than are the dark-colored minerals.

become a solid mass called sandstone. Similarly, clays may be compacted into shale. The resistance of a given sedimentary rock to weathering is determined by its particular dominant minerals and by the cementing agent. Sedimentary rocks are the most common type of rock encountered, covering about 75% of the Earth's land surface.

Metamorphic rocks are formed from other rocks by a process of change termed *metamorphism.* As Earth's continental plates shift, and sometimes collide, forces are generated that can uplift great mountain ranges or cause huge layers of rock to be pushed deep into the crust. These movements subject igneous and sedimentary rock masses to tremendous heat and pressure. These forces may slowly compress and partially remelt and distort the rocks, as well as break the bonds holding together the original minerals. Igneous rocks such as granite may be modified to form gneiss, a metamorphic rock in which light and dark minerals have been reoriented into bands. Sedimentary rocks, such as limestone and shale, may be metamorphosed to marble and slate, respectively. Slate may be further metamorphosed into phyllite or schist, which typically features mica crystallized during metamorphism.

Metamorphic rocks are usually harder and more strongly crystalline than the sedimentary rocks from which they formed. The particular minerals that dominate a given metamorphic rock influence its resistance to chemical weathering (see Table 2.1 and Figure 2.1).

Metamorphic rocks and how they form, California State University: http://seis.natsci.csulb.edu/bperry/ROCKS.htm

Weathering: A General Case

Weathering is a biochemical process that involves both destruction and synthesis. Moving from left to right in the weathering diagram (Figure 2.3), the original rocks and minerals are destroyed by both *physical disintegration* and *chemical decomposition.* Without appreciably affecting their composition, physical disintegration breaks down rock into smaller rocks and eventually into sand and silt particles that are commonly made up of individual minerals. Simultaneously, the minerals decompose chemically, releasing soluble materials and synthesizing new minerals, some of which are resistant end products. New minerals form either by minor chemical alterations or by complete chemical breakdown of the original mineral and resynthesis of new minerals. During the chemical changes, particle size continues to decrease, and constituents continue to dissolve in the aqueous weathering solution. The dissolved substances may recombine into new (secondary) minerals, may leave the profile in drainage water, or may be taken up by plant roots.

Animated overview of weathering: www.uky.edu/AS/Geology/howell/goodies/elearning/module07swf.swf

Table 2.1
SELECTED MINERALS FOUND IN SOILS, LISTED IN ORDER OF INCREASING RESISTANCE TO WEATHERING UNDER CONDITIONS COMMON IN HUMID TEMPERATE REGIONS

Primary minerals		Secondary minerals		
		Gypsum	$CaSO_4 \cdot 2H_2O$	Least resistant
		Calcite[a]	$CaCO_3$	
		Dolomite[a]	$CaCO_3 \cdot MgCO_3$	
Olivine	$Mg,FeSiO_4$			
Anorthite	$CaAl_2Si_2O_8$			
Augite[b]	$Ca_2(Al,Fe)_4 (Mg,Fe)_4Si_6O_{24}$			
Hornblende[b]	$Ca_2Al_2Mg_2Fe_3Si_6O_{22}(OH)_2$			
Albite	$NaAlSi_3O_8$			
Biotite	$KAl(Mg,Fe)_3Si_3O_{10}(OH)_2$			
Orthoclase	$KAlSi_3O_8$			
Microcline	$KAlSi_3O_8$			
Muscovite	$KAl_3Si_3O_{10}(OH)_2$			
		Clay minerals	Al silicates	
Quartz	SiO_2			
		Gibbsite	$Al_2O_3 \cdot 3H_2O$	
		Hematite	Fe_2O_3	
		Goethite	$FeOOH$	Most resistant

[a] In semiarid grasslands, dolomite and calcite are more resistant to weathering than suggested because of low rates of acid weathering.
[b] The given formula is only approximate because the mineral is so variable in composition.

Three groups of minerals that remain in well-weathered soils are shown on the right side of Figure 2.3: (1) silicate clays, (2) very resistant end products, including iron and aluminum oxide clays, and (3) very resistant primary minerals, such as quartz. In highly weathered soils of humid tropical and subtropical regions, the oxides of iron and aluminum, and certain silicate clays with low Si/Al ratios, predominate because most other constituents have been broken down and removed.

Physical Weathering (Disintegration)

Temperature   Heating of rocks by sunlight or fires causes expansion of their constituent minerals. As some minerals expand more than others, temperature changes set up differential stresses that eventually cause the rock to crack apart.

Because the outer surface of a rock is often warmer or colder than the more protected inner portions, some rocks may weather by **exfoliation**—the peeling away of outer layers (Figure 2.4). This process may be sharply accelerated if ice forms in the surface cracks. When water freezes, it expands with a force of about 1465 Mg/m², disintegrating huge rock masses and dislodging mineral grains from smaller fragments.

Abrasion by Water, Ice, and Wind   When loaded with sediment, water has tremendous cutting power, as is amply demonstrated by the gorges, ravines, and valleys around the world. The rounding of riverbed rocks and beach sand grains is further evidence of the abrasion that accompanies water movement.

Windblown dust and sand also can wear down rocks by abrasion, as can be seen in the many picturesque rounded rock formations in certain arid regions. In glacial areas, huge moving ice masses embedded with soil and rock fragments grind down rocks in their path and carry away large volumes of material.

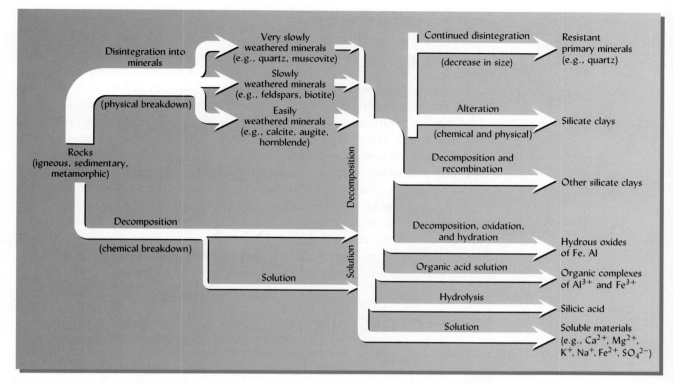

Figure 2.3

Pathways of weathering that occur under moderately acid conditions common in humid temperate regions. The disintegration of rocks into small individual mineral grains is a physical process, whereas decomposition, recombination, and solution are chemical processes. Alteration of minerals involves both physical and chemical processes. Note that resistant primary minerals, newly synthesized secondary minerals, and soluble materials are products of weathering. In arid regions the physical processes predominate, but in humid tropical areas decomposition and recombination are most prominent. (Diagram courtesy of N. C. Brady)

Plants and Animals   Plant roots sometimes enter cracks in rocks and pry them apart, resulting in some disintegration. Burrowing animals may also help disintegrate rock somewhat. However, such influences are of little importance in producing parent material when compared to the drastic physical effects of water, ice, wind, and temperature change.

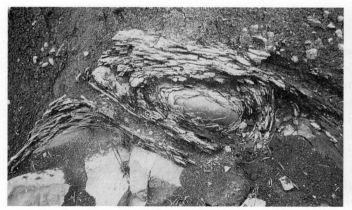

Figure 2.4

(Left) Concentric weathering called exfoliation. A combination of physical and chemical processes stimulate the mechanical breakdown, which produces layers that appear much like the leaves of a cabbage. (Right) Concentric bands of light and dark colors indicate that chemical weathering (oxidation and hydration) has occurred from the outside inward, producing iron compounds that differ in color. (Photos courtesy of N. C. Brady, left, and R. Weil, right)

Biogeochemical Weathering

While physical weathering is accentuated in very cold or very dry environments, chemical reactions are most intense where the climate is wet and hot. However, both types of weathering occur together, and each tends to accelerate the other. For example, physical abrasion (rubbing together) decreases the size of particles and therefore increases their surface area, making them more susceptible to rapid chemical reactions.

Chemical weathering is enhanced by such *geological* agents as the presence of water and oxygen, as well as by such *biological* agents as the acids produced by microbial and plant-root metabolism. That is why the term **biogeochemical weathering** is often used to describe the process. The various agents act in concert to convert primary minerals (e.g., feldspars and micas) to secondary minerals (e.g., clays and carbonates) and release plant nutrient elements in soluble forms (see Figure 2.3). Water and oxygen play important roles in the many chemical weathering reactions.

Microorganism activity is also key. Had there been no living organisms on Earth, chemical weathering processes would probably have proceeded 1000 times more slowly, with the result that little, if any, soil would have developed on our planet.

2.2   FACTORS INFLUENCING SOIL FORMATION[3]

We learned in Chapter 1 that *the soil* is a collection of *individual soils*, each with distinctive profile characteristics. This concept of soils as organized natural bodies derived initially from late-19th-century field studies by a brilliant Russian team of soil scientists led by V. V. Dukochaev. They noted similar profile layering in soils hundreds of kilometers apart, provided that the climate and vegetation were similar at the two locations. Such observations and much careful subsequent field and laboratory research has led to the recognition of five major factors that control the formation of soils:

1. *Parent materials* (geological or organic precursors to the soil)
2. *Climate* (primarily precipitation and temperature)
3. *Biota*, including people (living organisms, especially native vegetation, microbes, soil animals, and, increasingly, human beings)
4. *Topography* (slope, aspect, and landscape position)
5. *Time* (the period of time since the parent materials began to undergo soil formation)

The five soil-forming factors: www.soils.umn.edu/academics/classes/soil2125/doc/slab2sff.htm

Soils are often defined in terms of these factors as *dynamic natural bodies having properties derived from the combined effects of climate and biotic activities, as modified by topography, acting on parent materials over periods of time.*

We will now examine how each of these five factors affects the outcome of soil formation. However, as we do, we must keep in mind that these factors do not exert their influences independently. Indeed, interdependence is the rule. For example, contrasting climatic regimes are likely to be associated with contrasting types of vegetation, and perhaps differing topography and parent material as well. Nonetheless, in certain situations one of the factors has had the dominant influence in determining differences among a set of soils. Soil scientists refer to such a set of soils as a **lithosequence**, **climosequence**, **biosequence**, **toposequence**, or **chronosequence**.

---

[3]Many of our modern concepts concerning the factors of soil formation are derived from the work of Hans Jenny (1941 and 1980) and E. W. Hilgard (1921), American soil scientists whose books are considered classics in the field.

2.3 PARENT MATERIALS

Geological processes have brought to the Earth's surface numerous parent materials in which soils form (Figure 2.5). The nature of the parent material profoundly influences soil characteristics. For example, a soil might inherit a sandy texture (see Section 4.2) from a coarse-grained, quartz-rich parent material such as granite or sandstone. Soil texture, in turn, helps control the percolation of water through the soil profile, thereby affecting the translocation of fine soil particles and plant nutrients.

The chemical and mineralogical composition of parent material also influences both chemical weathering and the natural vegetation. For example, the presence of limestone in parent material will slow the development of acidity that typically occurs in humid climates.

Parent material deposition: http://sis.agr.gc.ca/cansis/taxa/genesis/pmdep/ontario.html

The nature of the parent material greatly influences the kinds of clays that can develop as the soil evolves (see Section 8.5). The parent material also may contain clay minerals, perhaps from a previous weathering cycle. In turn, the nature of the clay minerals present markedly affects the kind of soil that develops.

Classification of Parent Materials

Inorganic parent materials can either be formed in place as residual material weathered from rock, or they can be transported from one location and deposited at another (Figure 2.6). In wet environments (such as swamps and marshes), incomplete decomposition may allow organic parent materials to accumulate from the residues of many generations of vegetation. Although it is their chemical and physical properties that most influence soil development, parent materials are

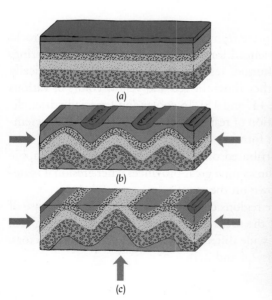

Figure 2.5

Diagrams showing how geological processes have brought different rock layers to the surface in a given area. (a) Unaltered layers of sedimentary rock with only the uppermost layer exposed. (b) Lateral geological pressures deform the rock layers through a process called crustal warping. At the same time, erosion removes much of the top layer, exposing part of the first underlying layer. (c) Localized upward pressures further reform the layers, thereby exposing two more underlying layers. As these four rock layers are weathered, they give rise to the parent materials on which different kinds of soils can form. (d) Crustal warping that lifted up the Appalachian Mountains tilted these sedimentary rock formations that were originally laid down horizontally. This deep road cut in Virginia illustrates the abrupt change in soil parent material (lithosequence) as one walks along the ground surface at the top of this photograph. (Photo courtesy of R. Weil)

Figure 2.6
How various kinds of parent material are formed, transported, and deposited.

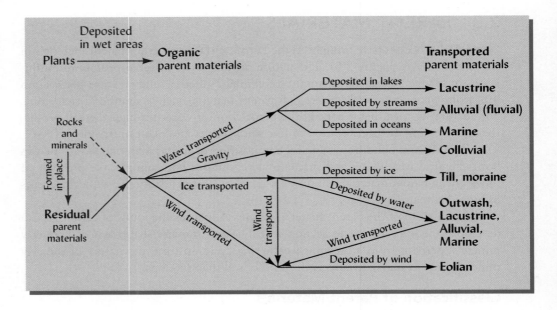

often classified with regard to the mode of placement in their current location, as seen on the right side of Figure 2.6.

Although these terms properly relate only to the placement of the parent materials, people sometimes refer to the soils that form from these deposits as *organic soils, glacial soils, alluvial soils,* and so forth. These terms are quite nonspecific because parent material properties vary widely within each group and because the effect of parent material is modified by the influence of climate, organisms, topography, and time.

Residual Parent Material

Residual parent material develops in place from weathering of the underlying rock. In stable landscapes it may have experienced long and possibly intense weathering. Where the climate is warm and very humid, residual parent materials are typically thoroughly leached and oxidized, and they show the red and yellow colors of various oxidized iron compounds (see Plates 9, 11, and 15). In cooler and especially drier climates, the color and chemical composition of residual parent material tends to resemble more closely the rock from which it formed.

Residual materials are widely distributed on all continents. The physiographic map of the United States (Figure 2.7) shows nine great provinces where residual materials are prominent (shades of olive/brown on the map).

A great variety of soils occupy the regions covered by residual debris because of the marked differences in the nature of the rocks from which these materials evolved. The varied soils are also a reflection of wide differences in other soil-forming factors, such as climate and vegetation (Sections 2.4 and 2.5).

Colluvial Debris

Colluvial debris, or **colluvium**, is made up of poorly sorted rock fragments detached from the heights above and carried downslope, mostly by gravity, assisted in some cases by frost action. Rock fragment (talus) slopes, cliff rock debris (detritus), and similar heterogeneous materials are good examples. Avalanches are made up largely of such accumulations.

Colluvial parent materials are frequently coarse and stony because physical rather than chemical weathering has been dominant. Stones, gravel, and fine materials are interspersed (not layered), and the coarse fragments are rather angular. Packing voids, spaces

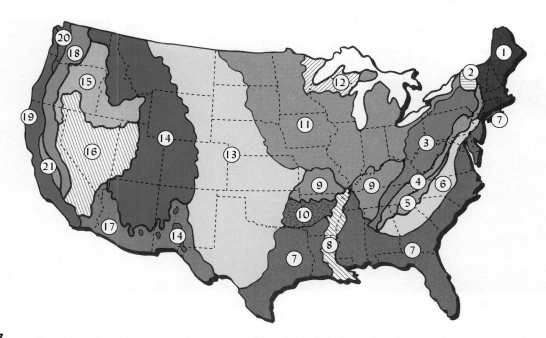

Figure 2.7

Generalized physiographic and regolith map of the United States. The regions are as follows (major areas of residual parent material are italicized and shown in shades of olive/brown on the map):

1. New England: mostly glaciated metamorphic rocks.
2. Adirondacks: glaciated metamorphic and sedimentary rocks.
3. *Appalachian Mountains and plateaus:* shales and sandstones.
4. *Limestone valleys and ridges:* mostly limestone.
5. Blue Ridge Mountains: sandstones and shales.
6. *Piedmont plateau:* metamorphic rocks.
7. Atlantic and Gulf coastal plain: unconsolidated sediments; sands, clays, and silts.
8. Mississippi floodplain and delta: alluvium.

9. *Limestone uplands:* mostly limestone and shale.
10. *Sandstone uplands:* mostly sandstone and shale.
11. Central lowlands: mostly glaciated sedimentary rocks with till and loess.
12. Superior uplands: glaciated metamorphic and sedimentary rocks.
13. *Great Plains region:* sedimentary rocks.
14. *Rocky Mountain region:* sedimentary, metamorphic, and igneous rocks.
15. Northwest intermountain: mostly igneous rocks; loess in river basins.

16. Great Basin: gravels, sands, alluvial fans; igneous and sedimentary rocks.
17. Southwest arid region: gravel, sand, and other debris of desert and mountain.
18. *Sierra Nevada and Cascade mountains:* igneous and volcanic rocks.
19. *Pacific Coast province:* mostly sedimentary rocks.
20. Puget Sound lowlands: glaciated sedimentary rocks.
21. California central valley: alluvium and outwash.

created when tumbling rocks come to rest against each other (sometimes at precarious angles), help account for the easy drainage of many colluvial deposits and also for their tendency to be unstable and prone to slumping and landslides, especially if disturbed by excavations.

Alluvial Deposits

Floodplains   Streams deposit three general classes of parent materials: *floodplains, alluvial fans,* and *deltas.* The part of a river valley that is inundated during floods is a floodplain. Sediment carried by the swollen stream is deposited during the flood, with the coarser materials being laid down near the river channel where the water is deeper and flowing with more turbulence and energy. Finer materials settle out in the calmer flood waters farther from the channel. Each major flooding episode lays down a distinctive layer of sediment, creating the stratification that characterizes alluvial soils (Figure 2.8).

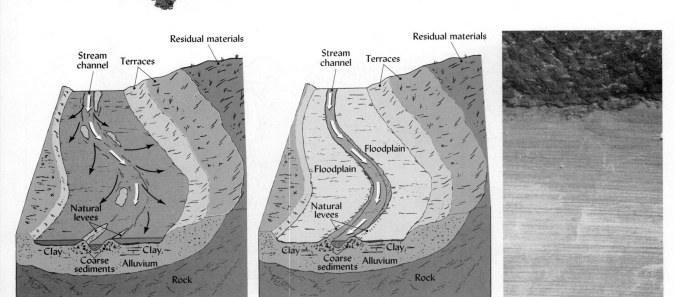

Figure 2.8

Floodplain development. (a) A stream at flood stage has overflowed its banks and is depositing sediment in the floodplain. The coarser particles are deposited nearest the stream channel where the water is flowing most rapidly, while the finer particles settle out where the water is moving more slowly. (b) After the flood the sediments are in place and vegetation is growing. (c) Profile of a soil on the Mississippi River floodplain showing contrasting thin layers of sand and silt sediments in the C horizon. Each layer resulted from a separate flooding episode. (Diagrams and photo courtesy of R. Weil)

If, over a period of time, there is a change in grade, a stream may cut down through its already well-formed alluvial deposits. This cutting action leaves **terraces** above the floodplain on one or both sides. Some river valleys feature two or more terraces at different elevations, each reflecting a past period of alluvial deposition and stream cutting.

Soils derived from alluvial sediments generally have characteristics seen as desirable for human settlement and agriculture. These characteristics include nearly level topography, proximity to water, high fertility, and high productivity. However, use of floodplain soils for home sites and urban development should generally be avoided. As the many disastrous floods of recent years have illustrated, building on a floodplain, no matter how great the investment in flood-control measures, all too often leads to tragic loss of life and property during serious flooding. In many areas, installation of systems for drainage and flood protection has proven costly and ineffective. Steps are therefore being taken to reestablish the wetland conditions of certain flood-prone agricultural areas that originally were natural wetlands. These and other alluvial soils can provide natural habitats, such as bottomland forests, which are very productive of timber and support a high diversity of birds and other wildlife.

Alluvial Fans   Streams that leave a narrow valley in an upland area and suddenly descend to a much broader valley below deposit sediment in the shape of a fan, as the water spreads out and slows down (see Figure 2.9). The rushing water tends to sort the sediment particles by size, first dropping the gravel and coarse sand, then depositing the finer materials toward the bottom of the alluvial fans. The soils derived from alluvial debris often prove very productive, although they may be quite coarse textured.

Delta Deposits   In a few river systems, considerable suspended material settles near the mouth of the river, forming a delta. A delta often is a continuation of a floodplain (its front, so to speak). It is clayey in nature and is likely to be poorly drained as well.

Figure 2.9
Characteristically shaped alluvial fan in a valley in central Nevada. Although alluvial fan areas are usually small and sloping, they can develop into productive, well-drained soils. Arrows indicate flow of water. (Photo courtesy of R. Weil)

Delta marshes are among the most extensive and biologically important of wetland habitats. Many of these habitats are today being protected or restored, but civilizations both ancient and modern have also developed important agricultural areas (often specializing in the production of rice) by creating drainage and flood-control systems on the deltas of such rivers as the Amazon, Euphrates, Ganges, Hwang Ho, Mississippi, Nile, Po, and Tigris.

Coastal Sediments

Streams eventually deposit much of their sediment loads in oceans, estuaries, and gulfs. The coarser fragments settle out near the shore and the finer particles at a distance (Figure 2.10). Over long periods of time, these underwater sediments build up, in some cases becoming hundreds of meters thick. Changes in the relative elevations

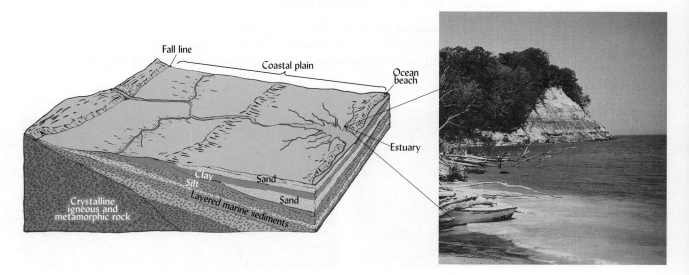

Figure 2.10
Diagram showing sediments laid down in marine waters and washed off the interior hills onto coastal areas. The diagram represents the coastal plain of the southeastern United States where such sediments cover older crystalline igneous and metamorphic rocks. Changes in the location of the shoreline and currents over time have resulted in sediment layers consisting alternately of fine clay, silts, coarse sands, and gravels. The photo shows such layering in coastal marine sediments along the Chesapeake Bay in Maryland.
(Diagram and photo courtesy of R. Weil)

of sea and land may later raise these marine deposits above sea level, creating a coastal plain. The deposits are then subject to a new cycle of weathering and soil formation.

A coastal plain usually has only moderate slopes, being more level in the low-lying parts nearer the coastline and more hilly farther inland, where streams and rivers flowing down the steeper grades have more deeply dissected the landscape. The land surface in the lower coastal portion may be only slightly above the water table during part of the year, so wetland forests and marshes often characterize such parent materials.

Marine and other coastal deposits are quite variable in texture. Some are sandy, as is the case in much of the Atlantic seaboard coastal plain. Others are high in clay, as are deposits found in the Atlantic and Gulf coastal flatwoods and in the interior pinelands of Alabama and Mississippi. Where streams have cut down through layers of marine sediments (as in the detailed block diagram in Figure 2.10), clays, silts, and sand may be encountered side by side. Because seawater is high in sulfur, many marine sediments are high in sulfur and go through a period of acid-forming sulfur oxidation at some stage of soil formation (see Section 9.6 and Plate 109).

Parent Materials Transported by Glacial Ice and Meltwaters

Debris transport by alpine glacier:
www.uwsp.edu/geO/faculty/lemke/glacial_processes/MoraineMovie.html

During the Pleistocene Epoch (about 10^4 to 10^7 years ago), up to 20% of the world's land surface—northern North America, northern and central Europe, and parts of northern Asia—was invaded by a succession of great ice sheets, some more than 1 km thick. Present-day glaciers in polar regions and high mountains cover about one-third as much area but are not nearly so thick as the glaciers of the Great Pleistocene Ice Age. Even so, if the current global warming trend continues, these present-day glaciers will largely melt, causing an increase in sea level and flooding of coastal areas around the world.

As the glacial ice pushed forward, the existing regolith with much of its mantle of soil was swept away, hills were rounded, valleys were filled, and, in some cases, the underlying rocks were severely ground and gouged. Thus, the glacier became filled with rock and all kinds of unconsolidated materials, carrying great masses of these materials as it pushed ahead (Figure 2.11). Finally, as the ice melted and the glacier retreated, a mantle of glacial debris or **drift** remained. This provided a new regolith and fresh parent material for soil formation.

Till and Associated Deposits   The name *drift* is applied to all material of glacial origin, whether deposited by ice or by associated waters. The materials deposited directly by the ice, called **till**, are heterogeneous (unstratified) mixtures of debris, which vary in size

Figure 2.11

(*Left*) Tongues of a modern-day glacier in Canada. Note the evidence of transport of materials by the ice and the "glowing" appearance of the major ice lobe. (*Right*) This U-shaped valley in the Rocky Mountains illustrates the work of glaciers in carving out land forms. The glacier left the valley floor covered with glacial till. Some of the material gouged out by the glacier was deposited many miles down the valley. (Left photo A-16817-102 courtesy of National Air Photo Library, Surveys and Mapping Branch, Canadian Department of Energy, Mines, and Resources; Right photo courtesy of R. Weil)

from boulders to clay. Till may therefore be somewhat similar in appearance to colluvial materials, except that the coarse fragments are more rounded from their grinding journey in the ice, and the deposits are often much more densely compacted because of the great weight of the overlying ice sheets. Figure 2.12 shows how glacial sheets deposited several types of soil parent materials, including ridges of till called **moraines**.

Glacial Outwash and Lacustrine Sediments   The torrents of water gushing forth from melting glaciers carried vast loads of sediment. In valleys and on plains where the glacial waters were able to flow away freely, the sediment formed **outwash plains** (Figure 2.12).

When the ice front came to a standstill, where there was no ready escape for the water, ponding began; ultimately, very large lakes were formed (Figure 2.12). The **lacustrine deposits** formed in these glacial lakes range from coarse delta materials and beach deposits near the shore to larger areas of fine silts and clay deposited from the deeper, more still waters at the center of the lake. Areas of inherently fertile (though not always well-drained) soils developed from these materials as the lakes dried.

Parent Materials Transported by Wind

Wind can most effectively pick up material from soil or regolith that is loose, dry, and unprotected by vegetation. Dry, barren landscapes have served, and continue to serve, as sources of parent material for soils forming as far away as the opposite side of the

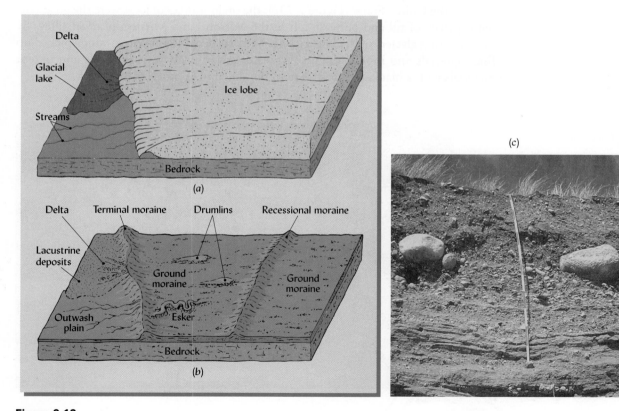

Figure 2.12

Illustration of how several glacial materials were deposited. (a) A glacier ice lobe moving to the left, feeding water and sediments into a glacial lake and streams and building up till near its front. (b) After the ice retreats, terminal, ground, and recessional moraines are uncovered along with cigar-shaped hills (drumlins), the beds of rivers that flowed under the glacier (eskers), and lacustrine, delta, and outwash deposits. (c) The stratified glacial outwash in the lower part of this soil profile in North Dakota is overlain by a layer of glacial till containing a random assortment of particles, ranging in size from small boulders to clays. Note the rounded edges of the rocks, evidence of the churning action within the glacier. Scale is marked every 10 cm. (Photo courtesy of R. Weil)

globe. The smaller the particles, the higher and farther the wind will carry them. Wind-transported (**eolian**) materials important as parent material for soil formation include, from largest to smallest particle size: **dune sand**, **loess** (pronounced "luss"), and **aerosolic dust**. Windblown **volcanic ash** from erupting volcanoes is a special case that is also worthy of mention.

Dune Sand   Along the beaches of the world's oceans and large lakes and over vast barren deserts, strong winds pick up medium and fine sand grains and pile them into hills of sand called *dunes*. Because most other minerals have been broken down and carried away by the waves, beach sand usually consists mainly of quartz, which is devoid of plant nutrients and highly resistant to weathering action. Nonetheless, over time dune grasses and other pioneering vegetation may take root, and soil formation may begin. Desert sands, too, are usually dominated by quartz, but they may also include substantial amounts of other minerals that could contribute more to the establishment of vegetation and the formation of soils, should sufficient rainfall occur. The pure-white dunes of sand-sized gypsum at White Sands, New Mexico, are a dramatic example of weatherable minerals in desert sands.

Loess   The windblown materials called *loess* are composed primarily of silt-sized particles. They cover wide areas in the central United States, eastern Europe, Argentina, and central China (Figure 2.13*a*). Loess may be blown for hundreds of kilometers. The deposits farthest from the source are thinnest and consist of the finest particles.

In the United States (Figure 2.13*b*), the main sources of loess were the great barren expanses of till and outwash left in the Missouri and Mississippi river valleys by the retreating glaciers of the last Ice Age. During the winter months, winds picked up fine materials and moved them southward, covering the existing soils and parent materials with a blanket of loess that accumulated to as much as 8 m thick.

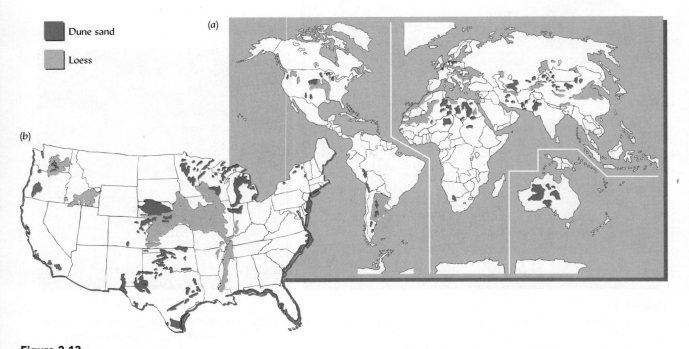

Dune sand

Loess

Figure 2.13

(*a*) Major eolian deposits of the world include the loess deposits in Argentina, eastern Europe, northern China, and the large areas of dune sands in north Africa and Australia. (*b*) Approximate distribution of loess and dune sand in the United States. The soils that have developed from loess are generally silt loams, often quite high in fine sands.

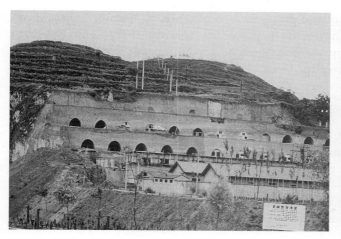

Figure 2.14
Villagers carve houses out of thick loess deposits in Xian, China. The loess consists mainly of silty materials bound together by small amounts of clay. The clay binder helps stabilize the loess when excavated, but only if the material is protected from rain, as in these vertical walls. Sloping excavations of this material would quickly slump and wash away when saturated with rain. Vertical road cuts are therefore a common feature of loessial landscapes around the world. (Photos courtesy of Raymond Miller, University of Maryland)

In central and western China, loess deposits reaching 30 to 100 m in depth cover some 800,000 km² (Figure 2.14). These materials have been windblown from the deserts of central Asia and are generally not associated directly with glaciers. These and other loess deposits tend to form silty soils of rather high fertility and potential productivity.

Aerosolic Dust Very fine particles (about 1 to 10 μm) carried high into the air may travel for thousands of kilometers before being deposited, usually with rainfall. These fine particles are called *aerosolic dust* because they can remain suspended in air, due to their very small size. Although this dust has not blanketed the receiving landscapes as thickly as is typical for loess, it does accumulate at rates that make significant contributions to soil formation. Much of the calcium carbonate in soils of the western United States probably originated as windblown dust. Recent studies have shown that dust, originating in the Sahara Desert of northern Africa and transported over the Atlantic Ocean in the upper atmosphere, is the source of much of the calcium and other nutrients found in the highly leached soils of the Amazon basin in South America. Likewise, in the springtime, dust from wind storms in the loess region of China blows across the Pacific Ocean to add soil parent materials (and air pollution) to the western part of North America.

Dust across the oceans, NASA. Click on "China during April of 1998": http://toms.gsfc.nasa.gov/aerosols/dust01.html

Volcanic Ash During volcanic eruptions cinders fall in the immediate vicinity of the volcano, while fine, often glassy, ash particles may blanket extensive areas downwind. Soils developed from volcanic ash are most prominent within a few hundred kilometers of the volcanoes that ring the Pacific Ocean. Important areas of volcanic ash parent materials occur in Japan, Indonesia, New Zealand, western United States (in Hawaii, Montana, Oregon, Washington, and Idaho), Mexico, Central America, and Chile. The soils formed are uniquely light and porous and tend to accumulate organic matter more rapidly than other soils in the area (Section 3.7). The volcanic ash tends to weather rapidly into allophane, a type of clay with unusual properties (see Section 8.5).

Organic Deposits

Organic material accumulates in marshes, swamps, and other wet places where plant growth exceeds the rate of residue decomposition. In such areas residues accumulate over the centuries from wetland plants such as pondweeds, cattails, sedges, reeds, mosses, shrubs, and certain trees. These residues sink into the water, where their decomposition is limited by lack of oxygen. As a result, organic deposits often accumulate up to several meters in depth (Figure 2.15). Collectively, these organic deposits are called **peat**.

Types of Peat Materials   Based on the nature of the parent materials, four kinds of peat are recognized:

1. Moss peat, the remains of mosses such as sphagnum
2. Herbaceous peat, residues of herbaceous plants such as sedges, reeds, and cattails
3. Woody peat, from the remains of woody plants, including trees and shrubs
4. Sedimentary peat, remains of aquatic plants (e.g., algae) and of fecal material of aquatic animals

In cases where a wetland area has been drained, woody peats tend to make very productive agricultural soils that are especially well suited for vegetable production. While moss peats have high water-holding capacities, they tend to be quite acid. Sedimentary peat is generally undesirable as an agricultural soil. This material is

Disappearing marshes. Cornelia Dean, *New York Times*. Also click on the A Marsh Mess video: www.nytimes.com/2005/11/ 15/science/earth/15marsh. html?ex=1289710800&en= debebd7482392dcc&ei= 5088&partner=rssnyt& emc=rss

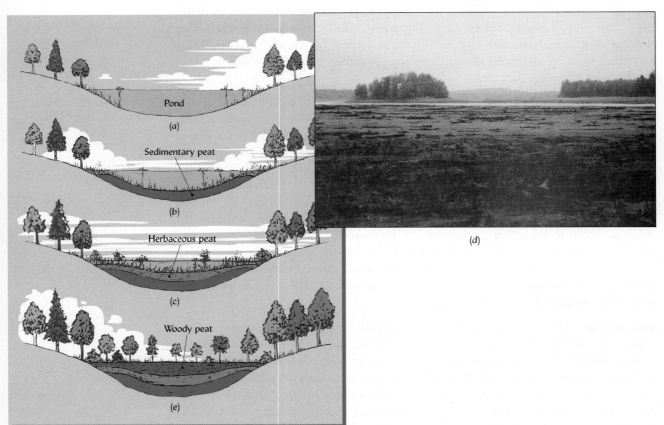

Figure 2.15

Stages in the formation and use of a typical woody peat bog. (a) A pond, typically formed by glacial action, receives nutrients and sediments running off the surrounding uplands. These encourage aquatic plant growth, especially in the shallow water around the pond edges. (b–d) Organic debris fills the bottom of the pond as increasingly rooted, emergent vegetation invades. (e) Eventually shrubs and trees take root in the peat and cover the area. Many such bogs have been cleared of trees and drained by ditches to remove some of the water, exposing an organic muck soil that is often highly productive for vegetable crops. The bog in the photo is in central Michigan. (Photo courtesy of R. Weil)

highly colloidal and compact and is rubbery when wet. Herbaceous peat is typical of coastal marshes.

The organic material is called **peat**, or **fibric**, if the residues are sufficiently intact to permit the plant fibers to be identified. If most of the material has decomposed sufficiently so that little fiber remains, the term **muck** or **sapric** is used. In mucky peats (hemic materials) only some of the plant fibers can be recognized.

Recognizing that the effects of **parent materials** on soil properties are modified by the combined influences of **climate**, **biotic activities**, **topography**, and **time**, we will now turn to these other four factors of soil formation, starting with climate.

2.4   CLIMATE

Climate is often the most influential of the four factors acting on parent material because it determines the nature and intensity of the weathering that occurs over large geographic areas. The principal climatic variables influencing soil formation are **effective precipitation** and *temperature*, both of which affect the rates of chemical, physical, and biological processes.

Effective Precipitation

Water is essential for all the major chemical weathering reactions. To be effective in soil formation, water must penetrate into the regolith. The seasonal rainfall distribution, evaporative demand, site topography, and soil permeability interact to determine how effectively precipitation can influence soil formation. The greater the depth of water penetration, the greater the depth of weathering soil and development. Surplus water percolating through the soil profile transports soluble and suspended materials from the upper to the lower layers. It may also carry away soluble materials in the drainage waters. Thus, percolating water stimulates weathering reactions and helps differentiate soil horizons.

Likewise, a deficiency of water is a major factor in determining the characteristics of soils of dry regions. Soluble salts are not leached from these soils, and in some cases they build up to levels that curtail plant growth. Soil profiles in arid and semiarid regions are also apt to accumulate carbonates and certain types of cracking clays.

Temperature

For every 10 °C rise in temperature, the rates of biochemical reactions more than double. Temperature and moisture both influence the organic matter content of soil through their effects on the balance between plant growth and microbial decomposition. If warm temperatures and abundant water are present in the profile at the same time, the processes of weathering, leaching, and plant growth will be maximized. The very modest profile development characteristic of cold areas contrasts sharply with the deeply weathered profiles of the humid tropics.

Considering soils with similar temperature regime, parent material, topography, and age, increasing effective annual precipitation generally leads to increasing clay and organic matter contents, greater acidity, and lower ratio of Si/Al (an indication of more highly weathered minerals). However, many places have experienced climates in past geologic epochs that were not at all similar to the climate evident today. This fact is illustrated in certain old landscapes in arid regions, where highly leached and weathered soils stand as relics of the humid tropical climate that prevailed there many thousands of years ago.

Climate also influences the natural vegetation. Humid climates favor the growth of trees. In contrast, grasses are the dominant native vegetation in subhumid and semiarid regions, while shrubs and brush of various kinds dominate in arid areas.

Thus, climate exerts its influence partly through a second soil-forming factor, the living organisms.

2.5   BIOTA: LIVING ORGANISMS—INCLUDING HUMANS

Humus synthesis, biochemical weathering, profile mixing, nutrient cycling, and aggregate formation and slope stability are all enhanced by the activities of organisms in the soil. Microbes, plants, and animals—including people—all play a role. Often the greatest influence is that of the natural vegetation.

Role of Natural Vegetation

Organic Matter Accumulation   The effect of vegetation on soil formation can be seen by comparing properties of soils formed under grassland and forest vegetation near the boundary between these two ecosystems (Figure 2.16). In the grassland, much of the organic matter added to the soil is from the deep, fibrous, grass root systems. By contrast, tree leaves falling on the forest floor are the principal source of soil organic matter in the forest. Another difference is the frequent occurrence in the grasslands of fires that destroy large amounts of aboveground material but stimulate even greater contributions from roots. Also, the much greater acidity under many forests inhibits the action of certain soil organisms that otherwise would mix much of the surface litter into the mineral soil. As a result, the soils under grasslands generally develop a thicker A horizon with a deeper distribution of organic matter than in comparable soils under forests, which characteristically store most of their organic matter in the forest floor (O horizons) and a thin A horizon. The microbial community in a

Figure 2.16

Natural vegetation influences the type of soil eventually formed from a given parent material (calcareous till, in this example). The forested soil exhibits surface layers (O horizons) of leaves and twigs in various stages of decomposition, along with a thin, mineral A horizon, into which some of the surface litter has been mixed. In contrast, most of the organic matter in the grassland is added as fine roots distributed throughout the upper 1 m or so, creating a thick, mineral A horizon. Also note that calcium carbonate has been solubilized and has moved down to the lower horizons (Ck) in the grassland soils, while it has been completely removed from the profile in the more acidic, leached forested soil. Under both types of vegetation, clay and iron oxides move downward from the A horizon and accumulate in the B horizon, encouraging the formation of characteristic soil structure. In the forested soil, the zone above the B horizon usually becomes a distinctly bleached E horizon, partly because most of the organic matter is restricted to the near-surface layers, and partly because decomposition of the forest litter generates organic acids that remove the brownish iron oxide coatings. Compare these mature profiles to the changes over time discussed in Sections 2.7 and 2.8.

(Diagrams courtesy of R. Weil)

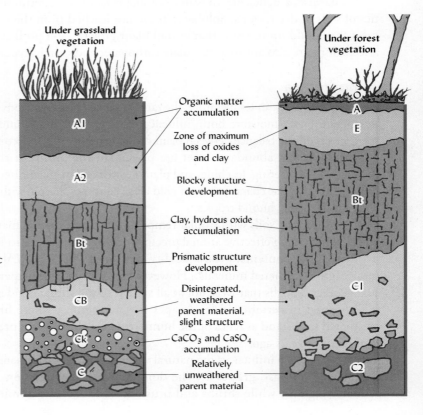

typical grassland soil is dominated by bacteria, while that of the forest soil is dominated by fungi (see Chapter 10 for details). Differences in microbial action affect the aggregation of the mineral particles into stable granules and the rate of nutrient cycling. The light-colored, leached E horizon typically found under the O or A horizon of a forested soil results from the action of organic acids generated mainly by fungi in the acidic forest litter. An E horizon is generally not found in a grassland soil.

Cation Cycling by Trees   The ability of natural vegetation to accelerate the release of nutrient elements from minerals by biogeochemical weathering, and to take up these elements from the soil, strongly influences the characteristics of the soils that develop. Soil acidity is especially affected. Differences occur not only between grassland and forest vegetation, but also between different species of forest trees. Litter falling from coniferous trees (e.g., pines, firs, spruces, and hemlocks) will recycle only small quantities of calcium, magnesium, and potassium compared to those recycled by litter from some deciduous trees (e.g., yellow poplar, beech, oaks, and maples) that take up and store much larger amounts of these cations (Figure 2.17).

Heterogeneity in Rangelands   In arid and semiarid rangelands, competition for limited soil water does not permit vegetation dense enough to completely cover the soil surface. Scattered shrubs or bunch grasses are interspersed with openings in the plant canopy where the soil is bare or partially covered with plant litter. The widely scattered vegetation alters soil properties in several ways. Plant canopies trap windblown dust that is often relatively rich in silt and clay. Roots scavenge nutrients such as nitrogen, phosphorus, potassium, and sulfur from the interplant areas. These nutrients are then deposited with the leaf litter under the plant canopies. The decaying litter adds organic

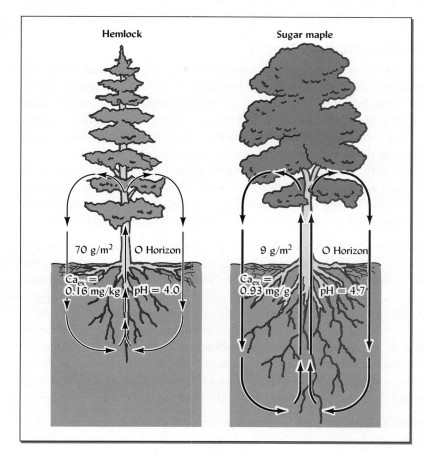

Figure 2.17

Nutrient cycling is an important process by which plants affect the soil in which they grow, altering the course of soil development and the suitability of the soil environment for future generations of vegetation. For example, hemlock (a conifer) and sugar maple (a deciduous hardwood) differ markedly in their ability to accelerate mineral weathering, mobilize nutrient cations, and recycle them to the upper soil horizons. Sugar maple roots are efficient at taking up Ca from soil minerals, and the maple leaves produced contain high concentrations of Ca. When these leaves fall to the ground, they decompose rapidly and release large amounts of Ca^{2+} ions that become adsorbed as exchangeable Ca^{2+} on humus and clay in the O and A horizons. This influx of Ca^{2+} ions may somewhat retard acidification of the surface layers. However, the maple roots' efficient extraction of Ca from minerals in the parent material may accelerate acidification and weathering in deeper soil horizons. In contrast, hemlock needles are Ca-poor, much slower to decompose, and therefore result in a thicker O horizon, greater acidity in the O and upper mineral horizons, but possibly less rapid weathering of minerals in the underlying parent material. [Data for a Connecticut forest reported by van Breemen and Finzi (1998)]

Figure 2.18
The scattered bunch grasses of this semiarid rangeland in the Patagonia region of Argentina have created "islands" of soil with enhanced fertility and thicker A horizons. A lens cap placed at the edge of one of these islands provides scale and highlights the increased soil thickness under the plant canopy. Such small-scale soil heterogeneity associated with plants is common where soil water limitations prevent complete plant ground cover. (Photos courtesy of Ingrid C. Burke, Short-Grass Steppe Long-Term Ecological Research Program, Colorado State University)

acids, which lower the soil pH and stimulate mineral weathering. As time goes on, the relatively bare soil areas between plants decline in fertility and may increase in size as they become impoverished and even less inviting for the establishment of plants. Simultaneously, the vegetation creates "islands" of enhanced fertility, thicker A horizons, and often more deeply leached calcium carbonate (see Figure 2.18).

Role of Animals, Including People

The role of animals in soil-formation processes must not be overlooked. Large animals such as gophers, moles, and prairie dogs bore into the lower soil horizons, bringing materials to the surface. Their tunnels are often open to the surface, encouraging movement of water and air into the subsurface layers. In localized areas, they enhance mixing of the lower and upper horizons by creating, and later refilling, underground tunnels. For example, dense populations of prairie dogs may completely turn over the upper meter of soil in the course of several thousand years. Old animal burrows in the lower horizons often become filled with soil material from the overlying A horizon, creating profile features known as *crotovinas* (Figure 2.19). In certain situations, animal activity may arrest soil development by increasing the loss of soil by erosion.

Earthworms, Ants, and Termites   Earthworms, ants, and termites mix the soil as they burrow, significantly affecting soil formation. Earthworms ingest soil particles and organic residues, enhancing the availability of plant nutrients in the material that passes through their bodies. They aerate and stir the soil and increase the stability of soil aggregates, thereby assuring ready infiltration of water. Ants and termites, as they build mounds, also transport soil materials from one horizon to another. In general, the mixing activities of animals, sometimes called **pedoturbation**, tends to undo or counteract the tendency of other soil-forming processes to accentuate the differences among soil horizons. Termites and ants may also retard soil profile development by denuding large areas of soil around their nests, leading to increased loss of soil by erosion.

Human Influences and Urban Soils   Human activities widely influence soil formation. For example, it is believed that Native Americans regularly set fires to maintain

Figure 2.19
Abandoned animal burrows in one horizon filled with soil material from another horizon are called crotovinas. In this Illinois prairie soil, dark, organic-matter-rich material from the A horizon has filled in old prairie dog burrows that extend into the B horizon. The dark circular shapes in the subsoil mark where the pit excavation cut through these burrows. Scale marked every 10 cm. (Photo courtesy of R. Weil)

several large areas of prairie grasslands in Indiana and Michigan. In more recent times, human destruction of natural vegetation (trees and grass) and subsequent tillage of the soil for crop production has abruptly modified soil formation. Likewise, irrigating an arid region soil drastically influences the soil-forming factors, as does adding fertilizer and lime to soils of low fertility. In surface mining and urbanizing areas today, bulldozers may have an effect on soils almost akin to that of the ancient glaciers; they level and mix soil horizons and set the clock of soil formation back to zero.

In other situations, people actually engineer new soils (see Plate 78), such as those used in most golf greens and certain athletic fields, the cover material used to vegetate and seal completed landfills (see Plate 64), and the plant media on rooftop gardens. Humans may even reverse the processes of erosion and sedimentation that normally destroy soils and counteract soil formation (see Section 2.6 and Chapter 14). For example, in a recent project called *Mud to Parks* (see Web link in margin), calcareous sediment dredged from the bottom of the Illinois River was placed as a thick layer of muddy parent material on barren, highly disturbed land. Within a year, the new parent material had dried out, was supporting lush vegetation, and began to develop such soil characteristics as granular and prismatic structure.

Illinois river muck becomes topsoil for new Chicago park, NPR, David Schaper: www.npr.org/templates/story/story.php?storyId=1919840

2.6 TOPOGRAPHY

Topography relates to the configuration of the land surface and is described in terms of differences in elevation, slope, and landscape position—in other words, the lay of the land. The topographical setting may either hasten or retard the work of climatic forces. Steep slopes generally encourage rapid soil loss by erosion and allow less rainfall to enter the soil before running off. In semiarid regions, the lower effective rainfall on steeper slopes also results in less complete vegetative cover, so there is less plant contribution to soil formation. For all of these reasons, steep slopes prevent the formation of soil from getting very far ahead of soil destruction. Therefore, soils on steep terrain tend to have rather shallow, poorly developed profiles in comparison to soils on nearby, more level sites (Figure 2.20).

In swales and depressions where runoff water tends to concentrate, the regolith is usually more deeply weathered and soil profile development is more advanced.

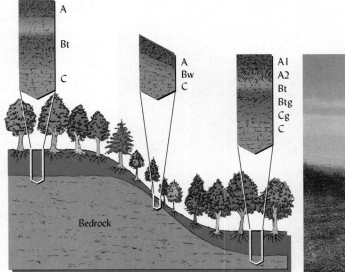

Figure 2.20

Topography influences soil properties, including soil depth. The diagram on the left shows the effect of slope on the profile characteristics and the depth of a soil on which forest trees are the natural vegetation. The photo on the right illustrates the same principle under grassland vegetation. Often a relatively small change in slope can have a great effect on soil development. See Section 2.9 for explanation of horizon symbols. (Photo courtesy of R. Weil)

However, in the lowest landscape positions, water may saturate the regolith to such a degree that drainage and aeration are restricted. Here the weathering of some minerals and the decomposition of organic matter are retarded, while the loss of iron and manganese is accelerated. In such low-lying topography, special profile features characteristic of wetland soils may develop (see Section 7.7 on the soils of wetlands).

Soils commonly occur together in the landscape in sequence called a **catena** (from the Latin meaning "chain"—visualize a length of chain suspended from two adjacent hills with each link in the chain representing a soil). Each member of the catena occupies a characteristic topographic position. Soils in a catena generally exhibit properties that reflect the influence of topography on water movement and drainage. A **toposequence** is a type of catena, in which the differences among the soils result almost entirely from the influence of topography because the soils in the sequence all share the same parent material and have similar conditions regarding climate, vegetation, and time (see, for example, Figure 2.20 and Plate 16).

Interaction with Vegetation   Topography often interacts with vegetation to influence soil formation. In grassland–forest transition zones, trees are commonly confined to the depressions where soil is generally wetter than in upland positions. As would be expected, the nature of the soil in the depressions is quite different from that in the uplands. As an example, if water stands for part or all of the year, low-lying areas may give rise to peat bogs and, in turn, to organic soils.

Slope Aspect   Topography affects the absorbance of solar energy in a given landscape. In the northern hemisphere, south-facing slopes are more perpendicular to the sun's rays and are generally warmer and thereby commonly lower in moisture than their north-facing counterparts (see also Figure 7.20). Consequently, soils on the south slopes tend to be lower in organic matter and are not so deeply weathered.

Salt Buildup   In arid and semiarid regions, topography influences the buildup of soluble salts. Dissolved salts from surrounding upland soils move on the surface and through the underground water table to the lower-lying areas (see Section 9.12).

Figure 2.21

An interaction of topography and parent material as factors of soil formation. The soils on the summit, toeslope, and floodplain in this idealized landscape have formed from residual, colluvial, and alluvial parent materials, respectively.

There they rise to the surface as the water evaporates, often accumulating to plant-toxic levels.

Parent Material Interactions   Topography can also interact with parent material. For example, in areas of tilted beds of sedimentary rock, the ridges often consist of resistant sandstone, while the valleys are underlaid by more weatherable limestone. In many landscapes, topography reflects the distribution of residual, colluvial, and alluvial parent materials, with residual materials on the upper slopes, colluvium covering the lower slopes, and alluvium filling the valley bottom (Figure 2.21).

2.7   TIME

Soil-forming processes take time to show their effects. The clock of soil formation starts ticking when a landslide exposes new rock to the weathering environment at the surface, when a flooding river deposits a new layer of sediment on its floodplain, when a glacier melts and dumps its load of mineral debris, or when a bulldozer cuts and fills a landscape to level a construction or mine-reclamation site.

Rates of Weathering   When we speak of a "young" or a "mature" soil, we are not so much referring to the age of the soil in years as to the degree of weathering and profile development. Time interacts with the other factors of soil formation. For example, on a level site in a warm climate, with much rain falling on permeable parent material rich in reactive minerals, weathering and soil profile differentiation will proceed far more rapidly than on a site with steep slopes and resistant parent material in a cold, dry climate.

In a few instances, soils form so rapidly that the effect of time on the process can be measured in a human life span. For example, dramatic mineralogical, structural, and color changes occur within a few months to a few years when certain sulfide-containing materials are first exposed to air by excavation, wetland drainage, or sediment dredging (see Plate 109 and Section 9.6). Under favorable conditions, organic matter may accumulate to form a darkened A horizon in freshly deposited, fertile alluvium in a mere decade or two. In some cases, incipient B horizons become discernible on humid-region mine spoils in as few as 40 years. Structural alteration and coloring by accumulated iron may form a simple B horizon within a few centuries if the parent material is sandy and the climate is humid. The same degree of B-horizon development would take much longer under conditions less favorable for weathering and leaching. The accumulation of silicate clays and the formation of blocky structure in B horizons usually become noticeable only after several thousand years. Developing in resistant rock, a mature, deeply weathered soil may be many thousands of years in the making (Figure 2.22).

Example of Soil Genesis Over Time   Figure 2.22 is worth studying carefully, as it illustrates changes that typically take place during soil development on residual rock in a warm, humid climate. During the first 100 years, lichens and mosses establish themselves on the bare exposed rock and begin to accelerate its breakdown and the accumulation of dust and organic matter. Within a few hundred years, grasses, shrubs, and stunted trees have taken root in a deepening layer of disintegrated rock and soil, adding greatly to the accumulation of organic materials and to the formation of the A and C horizons. During the next 10,000 years or so, successions of forest trees establish themselves and the activities of a multitude of tiny soil organisms transform the surface plant litter into a distinct O horizon. The A horizon thickens somewhat, becomes darker in color, and develops a stable granular structure. Soon, a bleached zone appears just below the A horizon as soluble weathering products, iron oxides, and clays are moved with the water and organic acids percolating down from the litter layer. These transported materials begin to accumulate in a deeper layer, forming a B horizon. The process continues with more silicate clay accumulating and blocky structure forming as the B horizon thickens and becomes more distinct. Eventually, the silicate clays themselves break down, some silica is leached away, and new clays containing less silica form in the B horizon. As the entire profile continues

Figure 2.22

Progressive stages of soil profile development over time for a residual igneous rock in a warm, humid climate that is conducive to forest vegetation. The time scale increases logarithmically from left to right, covering more than 100,000 years. Note that the mature profile (right side of this figure) expresses the full influence of the forest vegetation as illustrated in Figure 2.16. (Diagram courtesy of R. Weil)

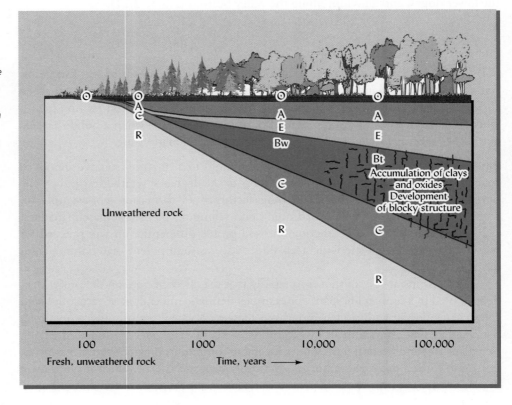

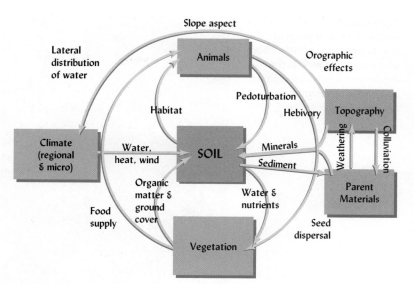

Figure 2.23
Parent material, topography, climate, and organisms (vegetation and animals) do not act independently. Rather, they are linked in many ways and influence the formation of soils in concert. The influence of each factor shown is modified by the length of time it has been acting, although time as a soil-forming factor is not shown here. [Adapted from Monger et al. (2005)]

to deepen over time, the zone of weathered, unconsolidated rock may become many meters thick.

The five soil factors of soil formation act simultaneously and interdependently to influence the nature of soils that develop at a site. Figure 2.23 illustrates some of the complex interactions that can help us predict what soil properties are likely to be encountered in a given environment. We will now turn our attention to the *processes* that cause parent materials to change into soils under the influence of these interacting soil-forming factors.

2.8 FOUR BASIC PROCESSES OF SOIL FORMATION[4]

The accumulation of regolith from the breakdown of bedrock or the deposition (by wind, water, ice, etc.) of unconsolidated geologic materials may precede or, more commonly, occur simultaneously with the development of the distinctive horizons of a soil profile. During the formation (**genesis**) of a soil from parent material, the regolith undergoes many profound changes brought about by four broad soil-forming processes (Figure 2.24) considered next. These four basic soil-forming, or **pedogenic**, processes help define what distinguishes soils from layers of sediment deposited by geologic processes.

Transformations occur when soil constituents are chemically or physically modified or destroyed and others are synthesized from the precursor materials. **Translocations** involve the movement of inorganic and organic materials laterally within a horizon or vertically from one horizon up or down to another. Water, either percolating down with gravity or rising up by capillary action, is the most common translocation agent. Inputs of materials to the developing soil profile from outside sources are considered **additions**. A very common example is the input of organic matter from fallen plant leaves and sloughed-off roots (the carbon having originated in the atmosphere). **Losses** from the soil profile occur by leaching to groundwater, erosion of surface materials, or other forms of removal. Erosion, a major loss agent,

---

[4]For the classic presentation of the processes of soil formation, see Simonson (1959). In addition, detailed discussion of these basic processes and their specific manifestations can be found in Birkeland (1999), Fanning and Fanning (1989), Buol et al. (2005), and Schaetzl and Anderson (2005).

Soil-forming processes animated:
www.environment.ualberta.ca/soa/process2.cfm

Figure 2.24

A schematic illustration of additions, losses, translocations, and transformations as the fundamental processes driving soil-profile development.
(Diagram courtesy of R. Weil)

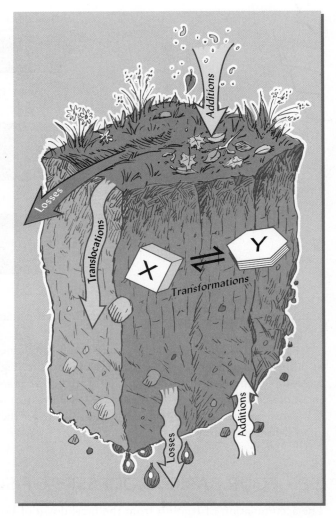

often removes the finer particles (humus, clay, and silt), leaving the surface horizon relatively sandier and less rich in organic matter than before.

These processes of soil genesis, operating under the influence of the environmental factors discussed previously, give us a logical framework for understanding the relationships between particular soils and the landscapes and ecosystems in which they function. In analyzing these relationships for a given site, ask yourself: What are the materials being added to this soil? What transformations and translocations are taking place in this profile? What materials are being removed? And how have the climate, organisms, topography, and parent material at this site affected these processes over time?

2.9   THE SOIL PROFILE

At each location on the land, the Earth's surface has experienced a particular combination of influences from the five soil-forming factors, causing a different set of layers (horizons) to form in each part of the landscape, thus slowly giving rise to the natural bodies we call **soils**. Each soil is characterized by a given sequence of these horizons. A vertical exposure of this sequence is termed a **soil profile**. We will now consider the major horizons making up soil profiles and the terminology used to describe them.

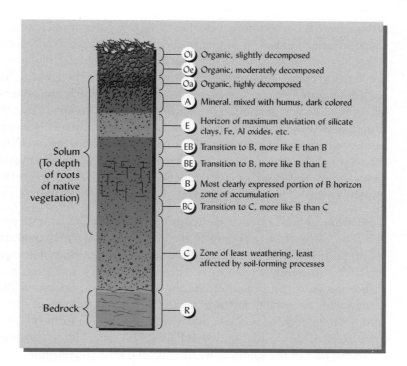

Figure 2.25
Hypothetical mineral soil profile showing the major horizons that may be present in a well-drained soil in the temperate humid region. Any particular profile may exhibit only some of these horizons, and the relative depths vary. A soil profile may exhibit more detailed subhorizons than indicated here. The solum normally includes the A, E, and B horizons plus some cemented layers of the C horizon.

Within the figure:
- **Oi** Organic, slightly decomposed
- **Oe** Organic, moderately decomposed
- **Oa** Organic, highly decomposed
- **A** Mineral, mixed with humus, dark colored
- **E** Horizon of maximum eluviation of silicate clays, Fe, Al oxides, etc.
- **EB** Transition to B, more like E than B
- **BE** Transition to B, more like B than E
- **B** Most clearly expressed portion of B horizon zone of accumulation
- **BC** Transition to C, more like B than C
- **C** Zone of least weathering, least affected by soil-forming processes
- **R**

Solum (To depth of roots of native vegetation) brackets the O through BC region.
Bedrock brackets the R region.

The Master Horizons and Layers[5]

Six **master** soil horizons are commonly recognized and are designated using the capital letters O, A, E, B, C, and R (Figure 2.25). Subordinate horizons may occur within a master horizon and these are designated by lowercase letters following the capital master horizon letter (e.g., Bt, Ap, or Oi).

O Horizons   The O horizons generally form above the mineral soil or occur in an organic soil profile. They derive from dead plant and animal residues. Generally absent in grassland regions, O horizons usually occur in forested areas and are commonly referred to as the **forest floor** (see Plates 7, 19, and 70). Often three subordinate O horizons can be distinguished: O*i*, O*e,* and O*a* (Figure 2.25).

A Horizons   The topmost mineral horizons, designated A horizons, generally contain enough partially decomposed (humified) organic matter to give the soil a color darker than that of the lower horizons (Plates 4, 7, and 20). The A horizons are often coarser in texture, having lost some of the finer materials by translocation to lower horizons and by erosion.

E Horizons   These are zones of maximum leaching, or **eluviation** (from Latin *ex* or *e*, "out," and *lavere*, "to wash") of clay, iron, and aluminum oxides, which leaves a concentration of resistant minerals, such as quartz, in the sand and silt sizes. An E horizon is usually found underneath the A horizon and is generally lighter in color than either the A horizon above it or the horizon below. Such E horizons are quite common in soils developed under forests, but they rarely occur in soils developed under grassland. Distinct E horizons can be seen in the soils in Plates 10, 19, and 31.

---

[5] In addition to the six master horizons described in this section, L (for limnic, from the Greek *limne-*, "marsh") and W (water) are also considered to be master horizons or layers. The L horizon occurs only in certain organic soils and includes layers of organic and mineral materials deposited in water or by aquatic organisms (e.g., diatomaceous earth, sedimentary peat, and marl). Layers of water (frozen or liquid) found within (not overlying) certain soil profiles are designated as W master horizons.

B Horizons   B horizons form below an O, A, or E horizon and have undergone sufficient changes during soil genesis so that the original parent material structure is no longer discernable. In many B horizons, materials have accumulated, typically by washing *in* from the horizons above, a process termed **illuviation** (from the Latin *il*, "in," and *lavere*, "to wash"). In humid regions, B horizons are the layers of maximum accumulation of materials such as iron and aluminum oxides (Bo or Bs horizons—see Plates 9, 10, and 31) and silicate clays (Bt horizons), some of which may have illuviated from upper horizons and some of which may have formed in place. Such Bt horizons can be clearly seen in the middle depths of the profiles shown in Plates 1 and 11. In arid and semiarid regions, calcium carbonate or calcium sulfate may accumulate in the B horizon (giving Bk and By horizons, respectively). See Plates 3, 8, and 13.

C Horizon   The C horizon is the unconsolidated material underlying the solum (A and B horizons). It may or may not be the same as the parent material from which the solum formed. The C horizon is below the zones of greatest biological activity and has not been sufficiently altered by soil genesis to qualify as a B horizon. In dry regions, carbonates and gypsum may be concentrated in the C horizon. While loose enough to be dug with a shovel, C horizon material often retains some of the structural features of the parent rock or geologic deposits from which it formed (see, for example, the lower third of the profiles shown in Plates 7, 11, and 31). Its upper layers may in time become a part of the solum as weathering and erosion continue.

R Layers   These are consolidated rock, with little evidence of weathering.

Subdivisions Within Master Horizons

Often distinctive layers exist *within* a given master horizon, and these are indicated by a numeral *following* the letter designation. For example, if three different combinations of structure and colors can be seen in the B horizon, then the profile may include a sequence such as B1–B2–B3 (see Plates 4, 7, and 9).

If two different geologic parent materials (e.g., loess over glacial till) are present within the soil profile, the numeral 2 is placed in front of the master horizon symbols for horizons developed in the second layer of parent material. For example, a soil would have a sequence of horizons designated O–A–B–2C if the C horizon developed in glacial till while the upper horizons developed in loess.

Where a layer of mineral soil material was transported by *humans* (usually using machinery) from a source outside the pedon, the caret symbol (^) is inserted before the master horizon designation. For example, suppose a landscaping contractor hauls in and spreads a layer of sandy fill material over an existing soil in order to level a site. The resulting soil might eventually (after enough organic matter had accumulated to form an A horizon) have the following sequence of horizons: ^A–^C–2Ab–2Btb, where the first two horizons formed in the human-transported fill (hence the ^ prefixes), and the last two horizons were part of the underlying, now buried soil (hence the lowercase "b" designations).

Transition Horizons

Transitional layers between the master horizons (O, A, E, B, C, and R) may be dominated by properties of one horizon but also have characteristics of another. The two applicable capital letters are used to designate the transition horizons (e.g., AE, EB, BE, and BC), the dominant horizon being listed before the subordinate one (e.g., Plate 1). Letter combinations with a slash, such as E/B, are used to designate transition horizons where distinct parts of the horizon have properties of E while other parts have properties of B.

Table 2.2
LOWERCASE LETTER SYMBOLS TO DESIGNATE SUBORDINATE DISTINCTIONS WITHIN MASTER HORIZONS

Letter	Distinction	Letter	Distinction
a	Organic matter, highly decomposed	n	Accumulation of sodium
b	Buried soil horizon	o	Accumulation of Fe and Al oxides
c	Concretions or nodules	p	Plowing or other disturbance
d	Dense unconsolidated materials	q	Accumulation of silica
e	Organic matter, intermediate decomposition	r	Weathered or soft bedrock
f	Frozen soil	s	Illuvial organic matter and Fe and Al oxides
ff	Dry permafrost	ss	Slickensides (shiny clay wedges)
g	Strong gleying (mottling)	t	Accumulation of silicate clays
h	Illuvial accumulation of organic matter	u	Presence of human-manufactured materials (artifacts)
i	Organic matter, slightly decomposed	v	Plinthite (high iron, red material)
j	Jarosite (yellow sulfate mineral)	w	Distinctive color or structure without clay accumulation
jj	Cryoturbation (frost churning)	x	Fragipan (high bulk density, brittle)
k	Accumulation of carbonates	y	Accumulation of gypsum
m	Cementation or induration	z	Accumulation of soluble salts

Subordinate Distinctions

Because the capital letter designates the nature of a master horizon in only a very general way, specific horizon characteristics may be indicated by a lowercase letter following the master horizon designation. Subordinate distinctions include special physical properties and the accumulation of particular materials such as clays and salts (Table 2.2). By way of illustration, a Bt horizon is a B horizon characterized by clay accumulation (t from the German *ton*, meaning clay). The significance of several other subordinate horizon designations will be discussed in the next chapter.

Horizons in a Given Profile

It is not likely that the profile of any one soil will show all of the horizons that collectively are shown in Figure 2.25. The ones most commonly found in well-drained soils are Oi and Oe (or Oa) if the land is forested; A or E (or both, depending on circumstances); Bt or Bw; and C. Conditions of soil genesis will determine which others are present and their clarity of definition.

When a virgin (never-cultivated) soil is plowed for the first time, the upper 15 to 20 cm becomes the plow layer or Ap horizon (Figure 2.26 and Plate 4). Cultivation, of course, obliterates the original layered condition of the upper portion of the profile, and the Ap horizon becomes more or less homogeneous. In some cultivated land, serious erosion produces a **truncated profile** (Plate 65). Another, sometimes perplexing, profile feature is the presence of a buried soil resulting from natural or human action. In profile study and description, such a situation requires careful analysis.

Soil Genesis in Nature

Not every contrasting layer of material found in soil profiles is a **genetic horizon** that developed as a result of the processes of soil genesis such as those just described. The parent materials from which many soils develop contained contrasting layers *before* soil genesis started. For example, such parent materials as glacial outwash, marine deposits, or recent alluvium may consist of various layers of fine and coarse particles laid down by separate episodes of sedimentation. Consequently, in characterizing soils, we must recognize not only the genetic horizons and properties that come into

Figure 2.26
Generalized profile of the Miami silt loam, one of the Alfisols of the eastern United States, before and after land is plowed and cultivated. The surface layers (O, A, and E) are mixed by tillage and are termed the Ap (plowed) horizon. If erosion occurs, they may disappear, at least in part, and some of the B horizon will be included in the furrow slice.

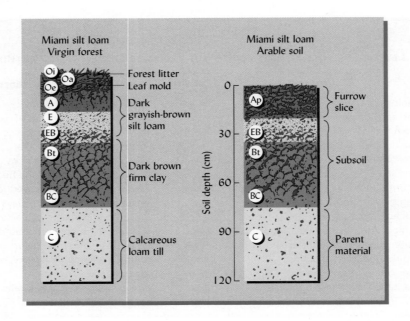

being during soil genesis, but also those layers or properties that may have been inherited from the parent material.

2.10   CONCLUSION

The parent materials from which soils develop vary widely around the world and from one location to another only a few meters apart. Knowledge of these materials, their sources or origins, mechanisms for their weathering, and means of transport and deposition are essential to understanding soil genesis.

Soil formation is stimulated by *climate* and living *organisms* acting on *parent materials* over periods of *time* and under the modifying influence of *topography*. These five major factors of soil formation determine the kinds of soil that will develop at a given site. With their rapidly expanding populations and technologies, people play an even-increasing role as organisms that influence soil formation. When all of these factors are the same at two locations, the kind of soil at these locations should also be the same.

Soil genesis starts when layers or horizons not present in the parent material begin to appear in the soil profile. Organic matter accumulation in the upper horizons, the downward movement of soluble ions, the synthesis and downward movement of clays, and the development of specific soil particle groupings (structure) in both the upper and lower horizons are signs that the process of soil formation is under way. As we have learned, soil bodies are dynamic in nature. Their genetic horizons continue to develop and change. Consequently, in some soils horizon differentiation has only begun, while in others it is well advanced.

The four general processes of soil formation (gains, losses, transformations, and translocations) and the five major factors influencing these processes provide us with an invaluable logical framework in selecting a site for a particular purpose and in predicting the nature of soil bodies likely to be found on a particular site. Conversely, analysis of the horizon properties of a soil profile can tell us much about the nature of the climatic, biological, and geological conditions (past and present) at the site.

Characterization of the horizons in the profile leads to the identity of a soil individual, which is then subject to classification—the topic of the next chapter.

STUDY QUESTIONS

1. What is meant by the statement *weathering combines the processes of destruction and synthesis*? Give an example of these two processes in the weathering of a primary mineral.

2. How is water involved in the main types of chemical weathering reactions?

3. Explain the weathering significance of the ratio of silicon to aluminum in soil minerals.

4. Give an example of how parent material may vary across large geographic regions on one hand, but may also vary within a small parcel of land on the other.

5. Name the five factors affecting soil formation. With regard to each of these factors of soil formation, compare a forested Rocky Mountain slope to the semiarid grassland plains far below.

6. How do *colluvium, till,* and *alluvium* differ in appearance and agency of transport?

7. What is *loess,* and what are some of its properties as a parent material?

8. Give two specific examples for each of the four broad processes of soil formation.

9. Assuming a level area of granite rock was the parent material in both cases, describe in general terms how you would expect two soil profiles to differ, one in a warm, semiarid grassland and the other in a cool, humid pine forest.

10. For the two soils described in question 5, make a profile sketch using master horizon symbols and subordinate suffixes to show the approximate depths, sequence, and nature of the horizons you would expect to find in each soil.

11. Visualize a slope in the landscape near where you live. Discuss how specific soil properties (such as colors, horizon thickness, types of horizons present, etc.) would likely change along the toposequence of soils on this slope.

REFERENCES

Binkley, D., and O. Menyailo (eds.). 2005. *Tree species effects on soils: Implications for global change.* Kluwer Academic Publishers, Dordrecht.

Birkeland, P. W. 1999. *Soils and geomorphology,* 3rd ed. Oxford University Press, New York.

Buol, S. W., R. J. Southard, R. C. Graham, and P. A. McDaniel. 2005. *Soil genesis and classification,* 5th ed. Iowa State University Press, Ames, IA.

Fanning, D. S., and C. B. Fanning. 1989. *Soil: Morphology, genesis, and classification.* John Wiley & Sons, New York.

Hilgard, E. W. 1921. *Soils: Their formation, properties, composition, and plant growth in the humid and arid regions.* Macmillan, London.

Jenny, H. 1941. *Factors of soil formation: A system of quantitative pedology.* Originally published by McGraw-Hill; Dover, Mineola, NY.

Jenny, H. 1980. *The soil resource—Origins and behavior.* Ecological Studies, Vol. 37. Springer-Verlag, New York.

Kerr, R. A. 2005. "And now, the younger, dry side of Mars is coming out." *Science* **307**:1025–1026.

Likens, G. E., and F. H. Bormann. 1995. *Biogeochemistry of a forested ecosystem,* 2nd ed. Springer-Verlag, New York.

Logan, W. B. 1995. *Dirt: The ecstatic skin of the Earth.* Riverhead Books, New York.

Marlin, J. C., and R. G. Darmody. 2005. "Returning the soil to the land: The mud to parks projects." *The Illinois Steward,* Spring. Available online at: http://www.istc.illinois.edu/special_projects/il_river/IL-steward.pdf. (verified 22 November 2008).

Monger, H. C., J. J. Martinez-Rios, and S. A. Khresat. 2005. "Arid and semiarid soils." In D. Hillel (ed.), *Encyclopedia of soils in the environment,* pp. 182–187. Elsevier, Oxford.

Richter, D. D., and D. Markewitz. 2001. *Understanding soil change.* Cambridge University Press, Cambridge, UK.

Schaetzl, R., and S. Anderson. 2005. *Soils—Genesis and geomorphology.* Cambridge University Press, Cambridge, UK.

Simonson, R. W. 1959. "Outline of a generalized theory of soil genesis," *Soil Sci. Soc. Amer. Proc.* **23**:152–156.

Soil Survey Division Staff. 1993. *Soil survey manual.* U.S. Department of Agriculture Handbook 18. Soil Conservation Service. http://soils.usda.gov/technical/manual/ (verified 18 February 2007).

van Breemen, N., and A. C. Finzi. 1998. "Plant–soil interactions: Ecological aspects and evolutionary implications," *Biogeochemistry* **42**:1–19.

3 Soil Classification

It is embarrassing not to be able to agree on what soil is. In this the pedologists are not alone. Biologists cannot agree on a definition of life and philosophers on philosophy.
—HANS JENNY, *THE SOIL RESOURCE: ORIGIN AND BEHAVIOR*

Roy Simonson investigates a soil (R. Weil)

We classify things in order to make sense of our world. We do it whenever we call things by group names, based on their important properties. Imagine a world without classifications. Imagine surviving in the woods knowing only that each plant was a plant, not which are edible by people, which attract wildlife, or which are poisonous. So, too, our understanding and management of soils and terrestrial systems would be hobbled if we knew only that a soil was a soil. How could we organize our information about soils, learn from others' experience, or communicate our knowledge to clients, colleagues, and students?

In this chapter we will learn how soils are classified as natural bodies on the basis of their profile characteristics. Soil classification allows us to take advantage of research and experience at one location to predict the behavior of similarly classified soils at another location. Soil names such as Histosols or Vertisols conjure up similar mental images in the minds of soil scientists everywhere, whether they live in the United States, Europe, Japan, developing countries, or elsewhere. A goal of the classification system is to create a universal language of soils that enhances communication among users of soils around the world.

In order to make practical use of soil information, land managers must know not only the "what" and "why" of soils but also the "where." If builders of an airport runway are to avoid the hazards of swelling clay soils, they must know *where* these troublesome soils are located. An irrigation expert probably must know *where* soils with suitable properties can be found. Almost any project involving soils, from planning a game park to fertilizing a farm field, will benefit from classification and geographic mapping of soils. This chapter provides an introduction to some of the tools that tell us *what is where*.

3.1   CONCEPT OF INDIVIDUAL SOILS

Compared to most other sciences, the organized study of soils is rather young, having begun in the 1870s when the Russian scientist V. V. Dokuchaev and his associates first conceived the idea that soils exist as natural bodies in nature. Russian soil scientists soon developed a system for classifying natural soil bodies, but poor international communications and the reluctance of some scientists to acknowledge such radical ideas delayed the universal acceptance of the natural bodies concept. In the United States, it was not until the late 1920s that C. F. Marbut of the U.S. Department of Agriculture (USDA), one of the few scientists who grasped the concept of soils as natural bodies, developed a soil classification scheme based on these principles.

We recognize the existence of individual entities, each of which we call *a soil*. Just as human individuals differ from one another, soil individuals have characteristics distinguishing each from the others. The gradation in soil properties from one soil individual to an adjacent one can be compared to the gradation in the wavelengths of light as you move from one color to another in a rainbow. The change is gradual, and yet we identify a boundary differentiating between what we call green and what we call blue.

Pedon, Polypedon[1], and Series

Soils in the field are heterogeneous; that is, the profile characteristics are not exactly the same in any two points within the soil individual you may choose to examine. Consequently, it is necessary to characterize a soil individual in terms of an imaginary three-dimensional unit called a **pedon** (rhymes with "head on," from the Greek *pedon,* ground; see Figure 3.1). It is the smallest sampling unit that displays the full range of properties characteristic of a particular soil.

Pedons occupy from about 1 to 10 m² of land area. Because it is what is actually examined during field investigation of soils, the pedon serves as the fundamental unit of soil classification. However, a soil unit in a landscape usually consists of a group of very similar pedons, closely associated together in the field. Such a group of similar

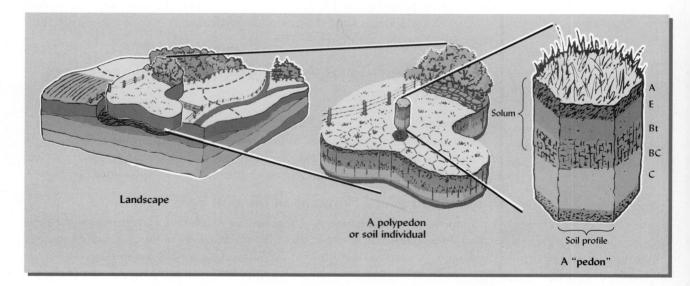

Figure 3.1

A schematic diagram to illustrate the concept of pedon and of the soil profile that characterizes it. Note that several contiguous pedons with similar characteristics are grouped together in a larger area (outlined by broken lines) called a polypedon *or soil individual. Several soil individuals are present in the landscape on the left.* (Diagram courtesy of R. Weil)

[1]The polypedon concept is no longer used by the USDA for reasons outlined by Ditzler (2005).

pedons, or a **polypedon**, is of sufficient size to be recognized as a landscape component termed a **soil individual**.

All the soil individuals in the world that have in common a suite of soil profile properties and horizons that fall within a particular range are said to belong to the same **soil series**. A soil series, then, is a class of soils, not a soil individual, in the same way that *Pinus sylvestrus* is a species of tree, not a particular individual tree. There are more than 20,000 soil series in the United States alone. They are the basic units used to classify soils. Units delineated on detailed soil maps are not purely one soil but are usually named for the soil series to which *most* of the pedons within the unit belong.

Groupings of Soil Individuals

In the concept of soils, the most specific extreme is that of a natural body called *a soil,* characterized by a three-dimensional sampling unit (pedon), related groups of which (polypedons) are included in a soil individual. At the most general extreme is *the soil,* a collection of all these natural bodies that is distinct from water, solid rock, and other natural parts of the Earth's crust. Hierarchical soil classification schemes generally group soils into classes at increasing levels of generality between these two extremes.

Links to online information about soil classification systems from around the world: www.itc.nl/~rossiter/research/rsrch_ss_class.html#National

Many cultures have traditional names for various classes of soils that help convey the people's collective knowledge about their soil resources (see Box 3.1). Scientific classification of soils began in the late 1800s stemming from the work of Dokuchaev in Russia (see Section 2.2). Many countries have developed and continue to use their own national soil classification systems.[2] To provide a global vocabulary for communicating about soils and a reference by which various national soil classification systems can be compared and correlated, scientists working through the Food and Agriculture Organization of the United Nations have developed a three-tier classification system known as the World Reference Base for Soils (see Table A.1 in Appendix A).

The Soil Survey Staff of the U.S. Department of Agriculture began in 1951 to collaborate with soil scientists from many countries to devise a classification system comprehensive enough to address all soils in the world, not just those in the United States. Finally published in 1975, and revised in 1999, the resulting system, *Soil Taxonomy,* is used in the United States and approximately 50 other countries. This system will be employed throughout this text.

3.2 COMPREHENSIVE CLASSIFICATION SYSTEM: SOIL TAXONOMY[3]

Soil Taxonomy[4] provides a hierarchical grouping of natural soil bodies. The system is based on *soil properties* that can be objectively observed or measured rather than on presumed mechanisms of soil formation. Its unique international nomenclature gives a definite connotation of the major characteristics of the soils in question.

Diagnostic Surface Horizons of Mineral Soils

Precise measurements are used to define certain **diagnostic soil horizons**, the presence or absence of which help determine the place of a soil in the classification system. The diagnostic horizons that occur at the soil surface are called **epipedons** (from the Greek *epi,* "over," and *pedon,* "soil"). The epipedon includes the upper part of the soil

---

[2]See Appendix A for summaries of the World Reference Base for Soils and the Canadian and Australian Systems of Soil Classification. For more information on other national systems and their interrelationships, see Eswaran et al. (2003).
[3]For a complete description of *Soil Taxonomy,* see Soil Survey Staff (1999). The first edition of *Soil Taxonomy* was published as Soil Survey Staff (1975). For an explanation of the earlier U.S. classification system, see USDA (1938).
[4]Taxonomy is the science of the principles of classification. For a review of the achievements and challenges of *Soil Taxonomy,* see SSSA (1984).

BOX 3.1
ETHNOPEDOLOGY: LOCAL SOIL KNOWLEDGE

For thousands of years, most societies were primarily agricultural, and almost everyone worked with soils on a daily basis. Through trial and error, people learned which soils were best suited to various crops and which responded best to different kinds of management. As farmers passed their observations and traditions from one generation to the next, they summarized their knowledge about soils by developing unique systems of soil classification. In some regions this local knowledge about soils helped shape agricultural systems that were sustainable for centuries. For example, formal Chinese soil classification goes back two millennia. In Beijing one may still visit the most recent (built in 1421) of a series of large sacrificial altars covered with five differently colored types of soils representing five regions of China (listed here with modern names in parentheses): (1) whitish saline soils (Salids) from the western deserts; (2) black organic-rich soils (Mollisols) from the north; (3) blue-gray waterlogged soils (e.g., Aquepts) from the east; (4) reddish iron-rich soils (Ultisols) from the south; and (5) yellow soils (Inceptisols) from the central loess plateau.

Local languages often reflect a sophisticated and detailed knowledge of how soils differ from one another. *Ethnopedological* studies carried out by anthropologists with an interest in soils (or soil scientists with an interest in anthropology) have documented many hitherto underappreciated indigenous classification systems for local soils. These systems most commonly classify soils using soil color, texture, hardness, moisture, organic matter, and topography, as well as other soil properties (Figure 3.2). Most of these properties are those observable in the surface horizon, the part of the soil with which farmers come into daily contact. In

this respect, the local classifications differ from most scientific classification schemes, which tend to focus on the subsurface horizons (as indicated in Figure 1.11). Rather than view this as a weakness, we can see the two approaches as complementary.

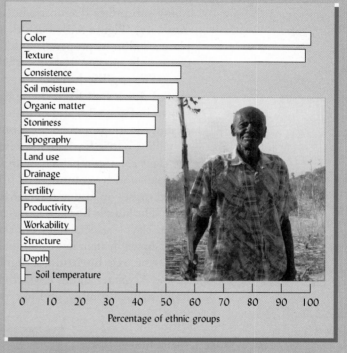

Figure 3.2
Soil characteristics used in soil classification among 62 ethnic groups around the world. [Data from Barrera-Bassols et al. (2006); photo courtesy of R. Weil]

darkened by organic matter, the upper eluvial horizons, or both. It may include part of the B horizon if the latter is significantly darkened by organic matter. Eight are recognized, but only five occur naturally over wide areas.

The **mollic epipedon** (Latin *mollis,* "soft") is a mineral surface horizon noted for its dark color (see Plates 8 and 20) associated with its accumulated organic matter (>0.6% organic C throughout), for its thickness (generally >25 cm), and for its softness even when dry. It has a high base saturation[5] greater than 50%.

The **umbric epipedon** (Latin *umbra,* "shade"; hence, dark) has the same general characteristics as the mollic epipedon except the percentage base saturation is lower. This mineral horizon commonly develops in areas with somewhat higher rainfall and where the parent material has lower content of calcium and magnesium.

The **ochric epipedon** (Greek *ochros,* "pale") is a mineral horizon that is either too thin, too light in color, or too low in organic matter to be either a mollic or umbric horizon. It is usually not as deep as the mollic or umbric epipedons (see Plates 1, 4, 7, and 11).

---

[5]The percentage base saturation is the percentage of the soil's negatively charged sites (cation exchange capacity) that are satisfied by attracting nonacid (or *base*) cations (such as Ca^{2+}, Mg^{2+}, and K^+) (see Section 9.3).

The **melanic epipedon** (Greek *melas,* "melan," "black") is a mineral horizon that is very black in color due to its high organic matter content (organic carbon >6%). It is characteristic of soils high in such minerals as allophane, developed from volcanic ash. It is extremely fluffy for a mineral soil (see Plate 2).

The **histic epipedon** (Greek *histos,* "tissue") is a 20 to 60 cm thick layer of **organic soil materials** overlying a mineral soil. Formed in wet areas, the histic epipedon is a layer of peat or muck with a black to dark brown color and a very low density.

Diagnostic Subsurface Horizons

Many subsurface diagnostic horizons are used to characterize different soils in *Soil Taxonomy* (Figure 3.3). Each diagnostic horizon provides a characteristic that helps place a soil in its proper class in the system. We will briefly discuss a few of the more commonly encountered subsurface diagnostic horizons.

The **argillic horizon** is a subsurface accumulation of silicate clays that have moved downward from the upper horizons or have formed in place. Examples are shown in Figure 3.4 and in Plate 1 between 50 and 90 cm. The clays often are found as shiny coatings termed *argillans* or *clay skins* (see Plates 18 and 24).

The **natric horizon** likewise has silicate clay accumulation (with clay skins), but the clays are accompanied by more than 15% exchangeable sodium on the colloidal complex and by columnar or prismatic soil structural units. The natric horizon is found mostly in arid and semiarid areas.

The **kandic horizon** has an accumulation of Fe and Al oxides as well as low-activity silicate clays (e.g., kaolinite), but clay skins need not be evident. The clays are low in activity as shown by their low cation-holding capacities (<16 cmol$_c$/kg clay) (see Figure 3.5).

The **oxic horizon** is a highly weathered subsurface horizon that is very high in Fe and Al oxides and in low-activity silicate clays (e.g., kaolinite). It is generally physically stable, crumbly, and not very sticky, despite its high clay content. It is found

Brief overview (with pictures) of diagnostic epipedons and subsurface horizons: http://soils.umn. edu/academics/classes/ soil2125/doc/s5chp1.htm

Figure 3.3

The mollic epipedon (a diagnostic horizon) in this soil includes genetic horizons designated Ap, A2, and Bt1, all darkened by the accumulation of organic matter. A subsurface diagnostic horizon, the argillic horizon, overlaps the mollic epipedon. The argillic horizon is the zone of illuvial clay accumulation (Bt1 and Bt2 horizons in this profile). Scale marked every 10 cm.

(Photo courtesy of R. Weil)

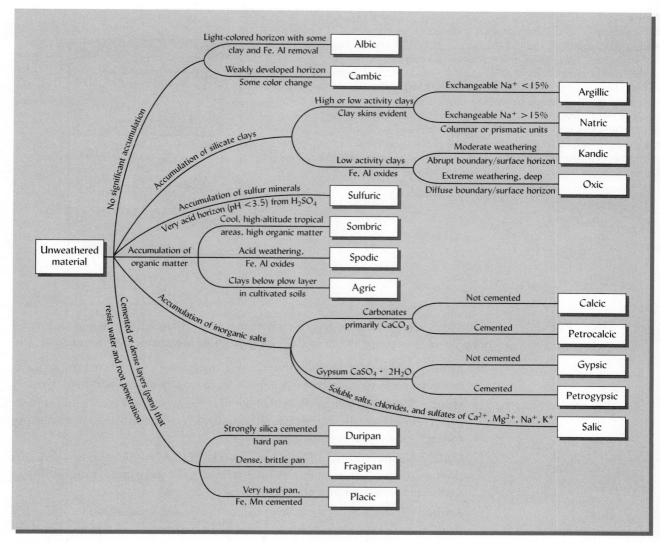

Figure 3.4

Names and major distinguishing characteristics of subsurface diagnostic horizons. Among the characteristics emphasized is the accumulation of silicate clays, organic matter, Fe and Al oxides, calcium compounds, and soluble salts, as well as materials that become cemented or highly acidified, thereby constraining root growth. The presence or absence of these horizons plays a major role in determining in which class a soil falls in Soil Taxonomy. See Chapter 8 for a discussion of low- and high-activity clays.

mostly in humid tropical and subtropical regions (see Plate 9, between about 1 and 3 feet on the scale and Plate 25 at 70 to 170 cm).

The **spodic horizon** is an illuvial horizon that is characterized by the accumulation of colloidal organic matter and aluminum oxide (with or without iron oxide). It is commonly found in highly leached forest soils of cool humid climates, typically on sandy-textured parent materials (see Plate 10, reddish-brown and black layers below the whitish layer).

The **albic horizon** is a light-colored eluvial horizon that is low in clay and oxides of Fe and Al. These materials have largely been moved downward from this horizon (see Plate 10, starting at about 10 cm depth).

Soil Moisture Regimes (SMR)

Soil moisture regime (*SMR*) refers to the presence or absence of either water-saturated conditions (usually groundwater) or plant-available soil water during specified periods

Global soil moisture regimes map: http://soils. usda.gov/use/worldsoils/ mapindex/smr.html

Figure 3.5

Vertical variation in clay content and cation exchange capacity (CEC) in a soil with thick albic (E1-E2-E3) and kandic (Bt1-Bt2) horizons. Note the well-expressed "clay bulge" that marks the kandic horizon. Similar clay enrichment (plus clay skins or other visual evidence of clay illuviation) characterizes an argillic horizon. This is a kandic rather than an argillic horizon because there was no visible evidence of clay illuviation and because the accumulated clay is of low-activity types, meaning the CEC of the clay is less than 16 cmol$_c$/kg of clay. This is a very old, highly mature soil that formed under humid, subtropical conditions in sandy sediments in the upper coastal plain of Georgia. It is classified in the Kandiudults great group in Soil Taxonomy. *(Data from Shaw et al., 2000)*

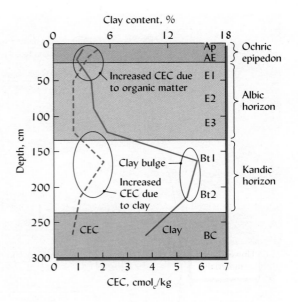

in the year. Several moisture regime classes are used to characterize soils and are helpful not only in classifying soils but in suggesting the most sustainable long-term use of soils:

Aquic.   Soil is saturated with water and virtually free of gaseous oxygen for sufficient periods of time for evidence of poor aeration (gleying and mottling) to occur.

Udic.   Soil moisture is sufficiently high year-round in most years to meet plant needs. An extremely wet moisture regime with excess moisture for leaching throughout the year is termed **perudic**.

Ustic.   Soil moisture is intermediate between Udic and Aridic regimes—generally there is some plant-available moisture during the growing season, although significant periods of drought may occur.

Aridic.   The soil is dry for at least half of the growing season and moist for less than 90 consecutive days. This regime is characteristic of arid regions. The term *torric* is used to indicate the same moisture condition in certain soils that are both hot and dry in summer, though they may not be hot in winter.

Xeric.   This soil moisture regime is found in typical Mediterranean-type climates, with cool, moist winters and warm, dry summers. Like the Ustic regime, it is characterized by having long periods of drought in the summer.

Soil Temperature Regimes

Soil temperature regimes, such as frigid, mesic, and thermic, are used to classify soils at some of the lower levels in *Soil Taxonomy.* The cryic (Greek *kryos,* "very cold") temperature regime distinguishes some higher-level groups. These regimes are based on mean annual soil temperature, mean summer temperature, and the difference between mean summer and winter temperatures, all at 50 cm depth.

3.3   CATEGORIES AND NOMENCLATURE OF SOIL TAXONOMY

There are six hierarchical categories of classification in *Soil Taxonomy*: (1) *order,* the highest (broadest) category, (2) *suborder,* (3) *great group,* (4) *subgroup,* (5) *family,* and (6) *series* (the most specific category). The lower categories fit within the higher categories. Thus, each order has several suborders, each suborder has several great groups, and so forth.

Nomenclature of Soil Taxonomy

Although it may seem strange at first sight, the nomenclature system is easy to learn after a bit of study. It has a logical construction and conveys a great deal of information about the nature of the soils named. The nomenclature is used throughout this book, especially to identify the kinds of soils shown in illustrations. If you make a conscious effort to identify the parts of each soil class mentioned in the text and figure captions and recognize the level of category indicated, the system will become second nature.

The names of the classification units are combinations of syllables, most of which are derived from Latin or Greek, and are root words in several modern languages. Because each part of a soil name conveys a concept of soil character or genesis, the name automatically describes the general kind of soil being classified. For example, soils of the order **Aridisols** (from the Latin *aridus,* dry, and *solum,* soil) are characteristically dry soils in arid regions. Thus, the names of orders are combinations of (1) formative elements, which generally define the characteristics of the soils, and (2) the ending *sols.*

The names of **suborders** automatically identify the order of which they are a part. For example, soils of the suborder **Aquolls** are the wetter soils (from the Latin *aqua,* "water") of the Mollisols order. Likewise, the name of the **great group** identifies the suborder and order of which it is a part. **Argiaquolls** are Aquolls with clay or argillic (Latin *argilla,* "white clay") horizons. In the following illustration, note that the three letters *oll* identify each of the lower categories as being in the M**oll**isols order:

M**oll**isols   Order
Aqu**olls**   Suborder
Argiaqu**olls**   Great group
Typic Argiaqu**olls**   Subgroup

If one is given only the subgroup name, the great group, suborder, and order to which the soil belongs are automatically known.

Family names in general identify subsets of the subgroup that are similar in texture, mineral composition, and mean soil temperature at a depth of 50 cm. Thus the name **fine, mixed, mesic, active Typic Argiaquolls** identifies a family in the Typic Argiaquolls subgroup with a fine texture, mixed clay mineral content, mesic (8 to 15 °C) soil temperature, and clays active in cation exchange.

Soil series are specific types of soils named after a geographic feature (town, river, etc.) near where they were first recognized. In detailed field soil surveying, soil series are sometimes further differentiated on the basis of surface soil texture, degree of erosion, slope, or other characteristics. These practical subunits are called soil **phases**, however, soil phases are *not* a category in the *Soil Taxonomy* system.

In the U.S., "official state soils" share the same level of distinction as official state flowers and birds: http://soils.usda.gov/gallery/state_soils/

3.4   SOIL ORDERS

Each of the world's soils is assigned to one of 12 **orders**, largely on the basis of soil properties that reflect a major course of development, with considerable emphasis placed on the presence or absence of major diagnostic horizons (Table 3.1). As an example, many soils that developed under grassland vegetation have the same general sequence of horizons and are characterized by a mollic epipedon—a thick, dark, surface horizon that is high in non-acid cations. They are included in the same order: Mollisols. Note that all order names have a common ending, *sols* (from the Latin *solum,* "soil").

The general conditions that enhance the formation of soils in the different orders are shown in Figure 3.6. From soil profile characteristics, soil scientists can ascertain the relative degree of soil development in the different orders, as shown in this figure. Note that soils with essentially no profile layering (Entisols) have the least

Maps and photos of each soil order: http://soils.ag.uidaho.edu/soilorders/index.htm

Table 3.1
NAMES OF SOIL ORDERS IN *SOIL TAXONOMY* WITH THEIR DERIVATION AND MAJOR CHARACTERISTICS
The bold letters in the order names indicate the formative element used as the ending for suborders and lower taxa within that order.

Name	Formative element	Derivation	Pronunciation	Major characteristics
Alfisols	alf	Nonsense symbol, Aluminum Al, iron Fe	Ped**alf**er	Argillic, natric, or kandic horizon; high-to-medium base saturation
Andisols	and	Jap. *ando,* "black soil"	**And**esite	From volcanic ejecta, dominated by allophane or Al-humic complexes
Aridisols	id	L. *aridus,* "dry"	Ar**id**	Dry soil, ochric epipedon, sometimes argillic or natric horizon
Entisols	ent	Nonsense symbol	Rec**ent**	Little profile development, ochric epipedon common
Gelisols	el	Gk. *gelid,* "very cold"	J**el**ly	Permafrost, often with cryoturbation (frost churning)
Histosols	ist	Gk. *histos,* "tissue"	**Hist**ology	Peat or bog; >20% organic matter
Inceptisols	ept	L. *inceptum,* "beginning"	Inc**ept**ion	Embryonic soils with few diagnostic features, ochric or umbric epipedon, cambic horizon
Mollisols	oll	L. *mollis,* "soft"	M**oll**ify	Mollic epipedon, high base saturation, dark soils, some with argillic or natric horizons
Oxisols	ox	Fr. *oxide,* "oxide"	**Ox**ide	Oxic horizon, no argillic horizon, highly weathered
Sp**od**osols	od	Gk. *spodos,* "wood ash"	P**od**zol; odd	Spodic horizon commonly with Fe, Al oxides and humus accumulation
Ultisols	ult	L. *ultimus,* "last"	**Ult**imate	Argillic or kandic horizon, low base saturation
Vertisols	ert	L. *verto,* "turn"	Inv**ert**	High in swelling clays; deep cracks when soil is dry

development, while the deeply weathered soils of the humid tropics (Oxisols and Ultisols) show the greatest soil development. The effect of climate (temperature and moisture) and of vegetation (forests or grasslands) on the kinds of soils that develop is also indicated in Figure 3.6. Study Table 3.1 and Figure 3.6 to better understand the relationship between soil properties and the terminology used in *Soil Taxonomy.*

To some degree, most of the soil orders occur in climatic regions that can be described by moisture and temperature regimes. Figure 3.7 illustrates some of the relationships among the soil orders with regard to these climatic factors. While only the Gelisols and Aridisols orders are defined directly in relation to climate, Figure 3.7 indicates that orders with the most highly weathered soils tend to be associated with the warmer and wetter climates.

Profiles for each soil order are shown in color on Plates 1 through 12. A color-coded soil map for the United States, showing the major areas dominated by each order, can be found on the back endpaper of this book; a general world map of the 12 soil orders is printed on the front endpaper. In studying these maps, try to confirm that the distribution of the soil orders is in accordance with what you know about the climate in various regions of the world.

Although a detailed description of all the lower levels of soil categories is far beyond the scope of this (or any other) book, a general knowledge of the 12 soil orders is essential for understanding the nature and function of soils in different environments. The simplified key given in Figure 3.8 helps illustrate how *Soil Taxonomy* can be used to key out the order of any soil based on observable and measurable properties of the soil profile. Because certain diagnostic properties take precedence over others, the key must always be used starting at the top, and working down. It will be useful to review this key after reading about the general characteristics, nature, and occurrence of each soil order.

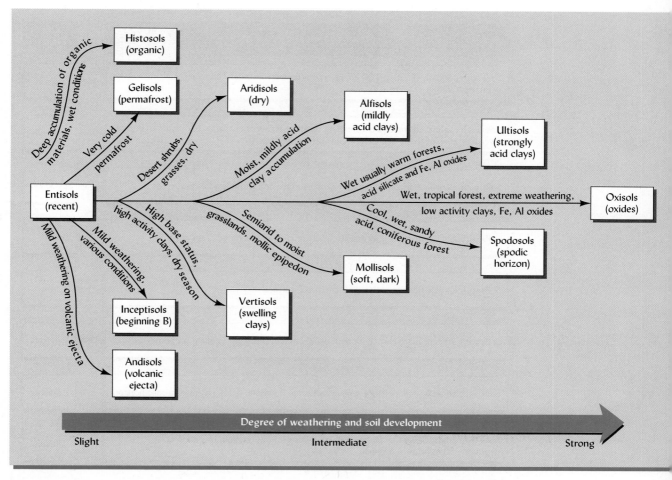

Figure 3.6

Diagram showing general degree of weathering and soil development in the different soil orders classified in Soil Taxonomy. *Also shown are the general climatic and vegetative conditions under which soils in each order are formed.*

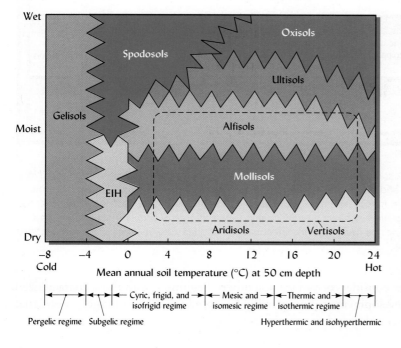

Figure 3.7

Diagram showing the general soil moisture and soil temperature regimes that characterize the most extensive soils in each of eight soil orders. Soils of the other four orders (Andisols, Entisols, Inceptisols, and Histosols) may be found under any of the soil moisture and temperature conditions (including the area marked EIH). Major areas of Vertisols are found only where clayey materials are in abundance and are most extensive where the soil moisture and temperature conditions approximate those shown inside the box with broken lines. Note that these relationships are only approximate and that less extensive areas of soils in each order may be found outside the indicated ranges. For example, some Ultisols (Ustults) and Oxisols (Ustox) have soil moisture levels for at least part of the year that are much lower than this graph would indicate. (The terms used at the bottom to describe the soil temperature regimes are those used in helping to identify soil families.)

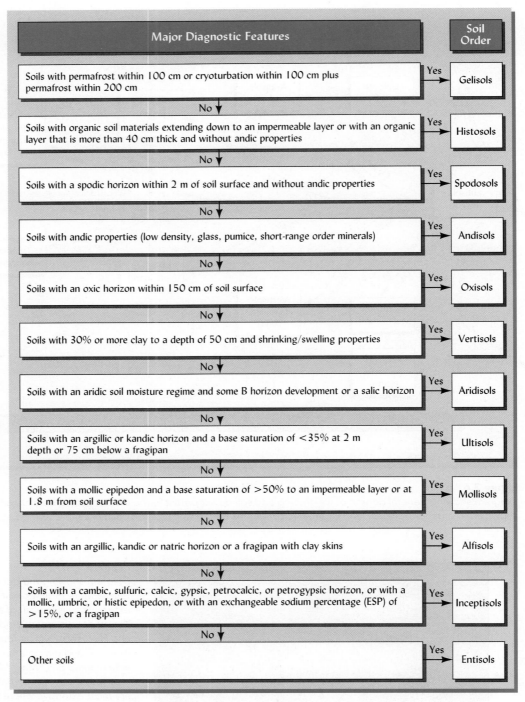

Major Diagnostic Features		Soil Order
Soils with permafrost within 100 cm or cryoturbation within 100 cm plus permafrost within 200 cm	Yes →	Gelisols
No ↓		
Soils with organic soil materials extending down to an impermeable layer or with an organic layer that is more than 40 cm thick and without andic properties	Yes →	Histosols
No ↓		
Soils with a spodic horizon within 2 m of soil surface and without andic properties	Yes →	Spodosols
No ↓		
Soils with andic properties (low density, glass, pumice, short-range order minerals)	Yes →	Andisols
No ↓		
Soils with an oxic horizon within 150 cm of soil surface	Yes →	Oxisols
No ↓		
Soils with 30% or more clay to a depth of 50 cm and shrinking/swelling properties	Yes →	Vertisols
No ↓		
Soils with an aridic soil moisture regime and some B horizon development or a salic horizon	Yes →	Aridisols
No ↓		
Soils with an argillic or kandic horizon and a base saturation of <35% at 2 m depth or 75 cm below a fragipan	Yes →	Ultisols
No ↓		
Soils with a mollic epipedon and a base saturation of >50% to an impermeable layer or at 1.8 m from soil surface	Yes →	Mollisols
No ↓		
Soils with an argillic, kandic or natric horizon or a fragipan with clay skins	Yes →	Alfisols
No ↓		
Soils with a cambic, sulfuric, calcic, gypsic, petrocalcic, or petrogypsic horizon, or with a mollic, umbric, or histic epipedon, or with an exchangeable sodium percentage (ESP) of >15%, or a fragipan	Yes →	Inceptisols
No ↓		
Other soils	Yes →	Entisols

Figure 3.8

A simplified key to the 12 soil orders in Soil Taxonomy. In using the key, always begin at the top. Note how diagnostic horizons and other profile features are used to distinguish each soil order from the remaining orders. Entisols, having no such special diagnostic features, key out last. Also note that the sequence of soil orders in this key bears no relationship to the degree of profile development and adjacent soil orders may not be more similar than nonadjacent ones. See Section 3.2 for explanations of the diagnostic horizons.

We will now consider each of the soil orders, beginning with those characterized by little profile development and progressing to those with the most highly weathered profiles (as represented from left to right in Figure 3.6).

3.5 ENTISOLS (RECENT: LITTLE IF ANY PROFILE DEVELOPMENT)

Weakly developed mineral soils without natural genetic B horizons (see Plate 4) belong to the Entisols order. Most have an ochric epipedon and a few have human-made anthropic or agric epipedons. Some have albic subsurface horizons. Soil productivity ranges from very high for certain Entisols formed in recent alluvium to very low for those forming in shifting sand or on steep rocky slopes.

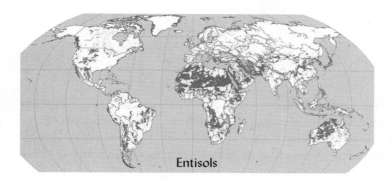

Entisols

16.3% of global and 12.2% of U.S. ice-free land

Suborders are:
Aquents (wet)
Arents (mixed horizons)
Fluvents (alluvial deposits)
Orthents (typical)
Psamments (sandy)

This is an extremely diverse group of soils with little in common, other than the lack of evidence for all but the earliest stages of soil formation. Entisols are either young in years or their parent materials have not reacted to soil-forming factors. On some parent materials, such as fresh lava flows or recent alluvium (Fluvents), there has been too little time for much soil formation. In extremely dry areas, scarcity of water and vegetation may inhibit soil formation. Likewise, frequent saturation with water (Aquents) may delay soil formation. Some Entisols occur on steep slopes, where the rates of erosion may exceed the rates of soil formation, preventing horizon development. Others occur on construction sites where bulldozers destroy or mix together the soil horizons, causing the existing soils to become Entisols (some have suggested that these be called *urbents,* or urban Entisols).

A

C

Figure 3.9
Profile of a Psamment formed on sandy alluvium in Virginia. Note the accumulation of organic matter in the A horizon but no other evidence of profile development. The A horizon is 30 cm thick. (Photo courtesy of R. Weil)

3.6   INCEPTISOLS (FEW DIAGNOSTIC FEATURES: INCEPTION OF B HORIZON)

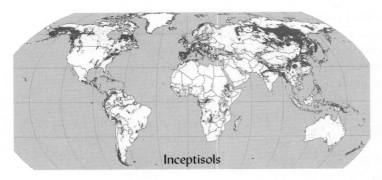

9.9% of global and 9.1% of U.S. ice-free land

Suborders are:
   Anthrepts (human-made, high phosphorus, dark surface)
   Aquepts (wet)
   Cryepts (very cold)
   Gelepts (permafrost)
   Udepts (humid climate)
   Ustepts (semiarid)
   Xerepts (dry summers, wet winters)

In Inceptisols, the beginning, or *inception*, of a B horizon is evident, and some diagnostic features are present. However, the well-defined profile characteristics of soils thought to be more mature have not yet developed. For example, a cambic horizon showing some color or structural change is common in Inceptisols (Plate 7), but a more mature illuvial B horizon such as an argillic cannot be present. Other subsurface diagnostic horizons that may be present in Inceptisols include duripans, fragipans, and calcic, gypsic, and sulfuric horizons. The epipedon in most Inceptisols is an ochric, although a plaggen or weakly expressed mollic or umbric epipedon may be present. Inceptisols show more significant profile development than Entisols, but are defined to exclude soils with diagnostic horizons or properties that characterize certain other soil orders. Thus, soils with only slight profile development occurring in arid regions or containing permafrost or andic properties are excluded from the Inceptisols. They fall, instead, in the soil orders Aridisols, Gelisols, or Andisols, as discussed in later sections. Inceptisols are widely distributed throughout the world. As with Entisols, Inceptisols are found in most climatic and physiographic conditions.

3.7   ANDISOLS (VOLCANIC ASH SOILS)

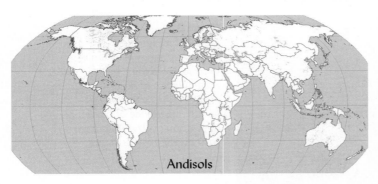

0.7% of global and 1.7% of U.S. ice-free land

Suborders are:
   Aquands (wet)
   Cryands (cold)
   Gelands (very cold)
   Torrands (hot, dry)
   Udands (humid)
   Ustands (moist/dry)
   Vitrands (volcanic glass)
   Xerands (dry summers, moist winters)

Andisols are usually formed on volcanic ash and cinders deposited in recent geological times (Figure 3.10). They are commonly found near the volcano source or in areas downwind from the volcano, where a sufficiently thick layer of ash has been deposited during eruptions. The principal soil-forming process has been the rapid transformation

Figure 3.10
An Andisol developed in layers of volcanic ash and pumice in central Africa.
(Photo courtesy of R. Weil)

Melanic Epipedon

Pumice layer

Weathered layers of volcanic ash and pumice

Buried A horizon

Oldest layers of volcanic pumice

Underlying layer of expanding clay

of volcanic ash to produce amorphous or poorly crystallized silicate minerals such as **allophane** and **imogolite** and the iron oxy-hydroxide **ferrihydrite**. Some Andisols have a melanic epipedon, a surface diagnostic horizon that has a high organic matter content and dark color (see Plate 2). The accumulation of organic matter is quite rapid due largely to its protection in aluminum–humus complexes. Little downward translocation of the colloids, or other profile development, has taken place. Like the Entisols and Inceptisols, Andisols are young soils, usually having developed for only 5000 to 10,000 years.

Andisols have a unique set of **andic properties** characterized by a high content of volcanic glass and/or a high content of amorphous or poorly crystalline iron and aluminum minerals. The combination of these minerals and the high organic matter results in light, fluffy soils that are easily tilled, yet have a high water-holding capacity and resist erosion by water. They are mostly found in regions where rainfall keeps them from being susceptible to erosion by wind. Andisols are usually of high natural fertility, except that phosphorus availability is severely limited by the extremely high phosphorus retention capacity of the andic materials (see Section 12.3). Fortunately, proper management of plant residues and fertilizers can usually overcome this difficulty.

In the United States, the area of Andisols is not extensive since recent volcanic action is not widespread. However, Andisols do occur in some very productive wheat- and timber-producing areas of Washington, Idaho, Montana, and Oregon. Likewise, this soil order represents some of the best farmland found in Chile, Ecuador, Colombia, and much of Central America.

3.8   GELISOLS (PERMAFROST AND FROST CHURNING)

8.6% of global and 7.5% of U.S. ice-free land

Suborders are:
   Histels (organic)
   Orthels (no special features)
   Turbels (cryoturbation)

Gelisols are young soils with little profile development. Cold temperatures and frozen conditions for much of the year slow the process of soil formation. The principal defining feature of these soils is the presence of a **permafrost** layer (see Plates 5 and 14). Permafrost is a layer of material that remains at temperatures below 0 °C for more than two consecutive years. It may be a hard, ice-cemented layer of soil material (e.g., designated Cfm in profile descriptions), or, if dry, it may be uncemented (e.g., designated Cff). In Gelisols, the permafrost layer lies within 100 cm of the soil surface, unless **cryoturbation** is evident within the upper 100 cm, in which case the permafrost may begin as deep as 200 cm from the soil surface.

Cryoturbation, or *frost churning*, moves the soil material to form broken, convoluted horizons (e.g., designated Cjj) at the top of the permafrost in some Gelisols. The frost churning also may form patterns on the ground surface, such as hummocks and ice-rich polygons that may be several meters across. In some cases rocks forced to the surface form rings or netlike patterns.

The permafrost in the southern part of the Gelisols region is only 1 or 2°C below freezing, so even small changes can cause melting and cause the soil to completely lose its bearing strength (see Figure 3.11, *right*).

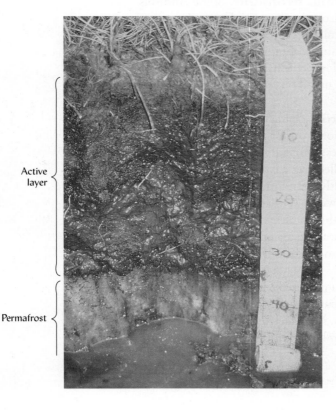

Active layer

Permafrost

Figure 3.11

Gelisols in Alaska. (Left) The soil is in the suborder Histels and has a histic epipedon and permafrost. This soil was photographed in Alaska in July. Scale in cm. (Right) Melting of the permafrost under this section of the Alaska Highway caused the soil to lose all bearing strength and collapse. Scale in cm. (Left photo courtesy of James G. Bockheim, University of Wisconsin; right photo courtesy of John Moore, USDA/NRCS)

Scientists have observed regional permafrost melting in much of the Arctic. This response of Gelisols is thought to be an early symptom of global climate change caused by greenhouse gas emissions (see Sections 11.1 and 11.9). Unfortunately, the melting of permafrost and deepening of the active layer in Gelisols are expected to accelerate this trend as the enormous pools of organic carbon once locked away in the permafrost become exposed to decay, thus releasing yet more greenhouse gases to the atmosphere.

3.9   HISTOSOLS (ORGANIC SOILS WITHOUT PERMAFROST)

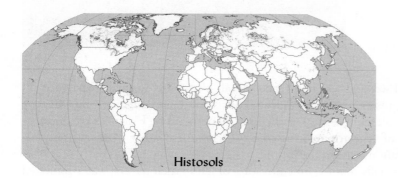

1.2% of global and 1.3% of U.S. ice-free land

Suborders are:
Fibrists (fibers of plants obvious)
Folists (leaf mat accumulations)
Hemists (fibers partly decomposed)
Saprists (fibers not recognizable)

Histosols consist of one or more thick layers of *organic soil material.* They have undergone little profile development because of the anaerobic environment in which they form.

The organic matter in Histosols ranges from peat to muck. *Peat* is comprised of the brownish, only partially decomposed, fibrous remains of plant tissues (see Figure 3.12, Plate 26). *Muck,* on the other hand, is a black material in which decomposition is much more complete and the organic matter is highly humified (Figure 3.12). Muck is like a black ooze when wet and powdery when dry.

Figure 3.12

A tidal marsh Histosol. The inset shows the fibric (peaty) organic material that contains recognizable roots and rhizomes of marsh grasses that died perhaps centuries ago, the anaerobic conditions having preserved the tissues from extensive decay. The soil core (held horizontally for the photograph) gives some idea of the soil profile, the surface layer being at the right and the deepest layer at the left. The water level is usually at or possibly above the soil surface. (Photos courtesy of R. Weil)

Figure 3.13

Soil subsidence due to rapid organic matter decomposition after artificial drainage of Histosols in the Florida Everglades. The house was built at ground level, with the septic tank buried about 1 m below the soil surface. Over a period of about 60 years, more than 1.2 m of the organic soil has "disappeared." The loss has been especially rapid because of Florida's warm climate, but artificial drainage that lowers the water table and continually dries out the upper horizons is an unsustainable practice on any Histosol. (Photo courtesy of George H. Snyder, Everglades Research and Education Center, Belle Glade, FL)

While not all wetlands contain Histosols, all Histosols (except Folists) occur in wetland environments. They can form from equatorial to arctic regions, but they are most prevalent in cold climates, up to the limit of permafrost. Horizons are differentiated by the type of vegetation contributing the residues, rather than by translocations and accumulations within the profile.

Whether artificially drained for cultivation or left in their natural water-saturated state, Histosols possess unique properties resulting from their high organic matter contents. Histosols are generally black to dark brown in color. They are extremely lightweight (0.15 to 0.4 Mg/m^3) when dry, being only about 10 to 20% as dense as most mineral soils. Histosols also have high water-holding capacities on a mass basis. While a mineral soil will absorb and hold from 20 to 40% of its weight of water, a cultivated Histosol may hold a mass of water equal to 200 to 400% of its dry weight. These soils also possess very high cation exchange capacities (typically 150 to 300 $cmol_c/kg$) that increase with increasing soil pH.

Some Histosols make very productive farmlands, but the organic nature of the materials requires liming, fertilization, tillage, and drainage practices quite different from those applied to soils in the other 11 orders. If other than wetland plants are to be grown, the water table is usually lowered to provide an aerated zone for root growth. This practice, of course, alters the soil environment and causes the organic material to oxidize, resulting in the disappearance of as much as 5 cm of soil per year in warm climates (Figure 3.13).

3.10   ARIDISOLS (DRY SOILS)

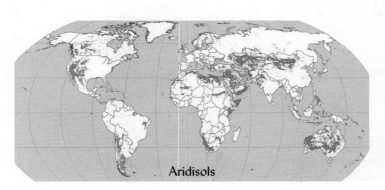

Aridisols

12.7% of global and 8.8% of U.S. ice-free land

Suborders are:
Argids (clay)
Calcids (carbonate)
Cambids (typical)
Cryids (cold)
Durids (duripan)
Gypsids (gypsum)
Salids (salty)

Figure 3.14

Two features characteristic of some Aridisols. (Left) Wind-rounded pebbles have given rise to a desert pavement. (Right) A petrocalcic horizon of cemented calcium carbonate. (Photos courtesy of R. Weil)

Aridisols occupy a larger area globally than any other soil order except Entisols. Water deficiency is a major characteristic of these soils. The soil moisture level is sufficiently high to support plant growth for no longer than 90 consecutive days. The natural vegetation consists mainly of scattered desert shrubs and short bunchgrasses. Soil properties, especially in the surface horizons, may differ substantially between interspersed bare and vegetated areas (see Section 2.5).

Aridisols are characterized by an ochric epipedon that is generally light in color and low in organic matter (see Plate 3). The processes of soil formation have brought about a redistribution of soluble materials, but there is generally not enough water to leach these materials completely out of the profile. Therefore, they often accumulate at a lower level in the profile. These soils may have a horizon of accumulation of calcium carbonate (calcic), gypsum (gypsic), soluble salts (salic), or exchangeable sodium (natric). Under certain circumstances, carbonates may cement together the soil particles and coarse fragments in the layer of accumulation, producing hard layers known as **petrocalcic** horizons (Figure 3.14). These hard layers act as impediments to plant root growth and also greatly increase the cost of excavations for buildings.

Some Aridisols (the *Argids*) have an argillic horizon, most probably formed under a wetter climate that long ago prevailed in many areas that are deserts today. With time and the addition of carbonates from calcareous dust and other sources, many argillic horizons become engulfed by carbonates (*Calcids*).

In stony or gravelly soils, erosion may remove all the fine particles from the surface layers, leaving behind a layer of wind-rounded pebbles that is called **desert pavement** (see Figure 3.14 and Plate 55). Pebbles in desert pavement often have a shiny coating called **desert varnish** (Plate 57).

Without irrigation, Aridisols are not suitable for growing cultivated crops. Even with irrigation, soluble salts may accumulate in the upper horizons to levels that most crop plants cannot tolerate. Some areas are used for low-intensity grazing. Overgrazing of Aridisols causes the once-grassy areas to become increasingly bare and the soils between the scattered shrubs to succumb to erosion by the desert winds and occasional thunderstorms. Plentiful sunshine makes areas of Aridisols attractive for energy production via the installation of large arrays to solar voltaic collectors or the

Type "desert pavement" into Google and click on "Image results" to see a wide variety of photos. Solving the mystery of desert varnish: http://sciencenow.sciencemag.org/cgi/content/full/2006/707/1

planting of biofuel crops such as *Jatropha,* a bushy, oil-seed producing plant capable of growing in very dry conditions.

3.11   VERTISOLS (DARK, SWELLING, AND CRACKING CLAYS)[6]

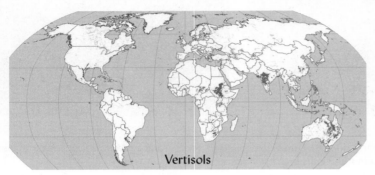

Vertisols

2.4% of global and 1.7% of U.S. ice-free land

Suborders are:
   Aquerts (wet)
   Cryerts (cold)
   Torrerts (hot summer, very dry)
   Uderts (humid)
   Usterts (moist/dry)
   Xererts (dry summers, moist winters)

The main soil-forming process affecting Vertisols is the shrinking and swelling of clay as these soils go through periods of drying and wetting. Vertisols have a high content (>30%) of sticky, swelling, and shrinking-type clays. Most Vertisols are dark, even blackish in color, to a depth of 1 m or more (Plate 12). However, unlike for most other soils, the dark color of Vertisols is not necessarily indicative of a high organic matter content. The organic matter content of dark Vertisols typically ranges from as much as 5 or 6% to as little as 1%.

Vertisols typically develop from limestone, basalt, or other calcium- and magnesium-rich parent materials, which encourage the formation of swelling-type, high-activity clays.

Vertisols are found mostly in subhumid to semiarid environments where the native vegetation is usually grassland. The climate features dry periods of several months during which clay shrinks, causing the soils to develop deep, wide cracks that are diagnostic for this order (Figure 3.15*a*). The surface soil granules may slough off into the cracks, giving rise to a partial inversion of the soil (Figure 3.16*a*). This accounts for the association with the term *invert,* from which this order derives its name.

When the rains come, water entering the large cracks moistens the clay in the subsoils, causing it to swell. The repeated shrinking and swelling of the subsoil clay results in a kind of imperceptibly slow "rocking" movement of great masses of soil. As the subsoil swells, blocks of soil shear off from the mass and rub past each other under pressure, giving rise in the subsoil to shiny, grooved, tilted surfaces called **slickensides** (Figure 3.16*c*). Eventually, this back-and-forth motion may form bowl-shaped depressions with relatively deep profiles surrounded by slightly raised areas in which little soil development has occurred and in which the parent material remains close to the surface (see Figure 3.16*b*). The resulting pattern of micro-highs and micro-lows on the land surface, called **gilgai**, is usually discernable only where the soil is untilled (Figure 3.15*b*).

The high shrink–swell potential of Vertisols makes them extremely problematic for any kind of highway or building construction (Plate 43). This property also makes agricultural management very difficult.

---

[6]See Coulombe et al. (1996) for a detailed review of the properties and mode of formation of Vertisols.

(a)

(b)

Figure 3.15

(a) Wide cracks formed during the dry season in the surface layers of this Vertisol in India. Surface debris can slough off into these cracks and move to subsoil. When the rains come, water can move quickly to the lower horizons, but the cracks are soon sealed, making the soils relatively impervious to the water. (b) Once the cracks have sealed, water may collect in the "microlows," making the gilgai relief easily visible as in this Texas vertisol.

(Photo (a) courtesy of N. C. Brady; (b) courtesy of K. N. Potter, USDA/ARS, Temple, Texas)

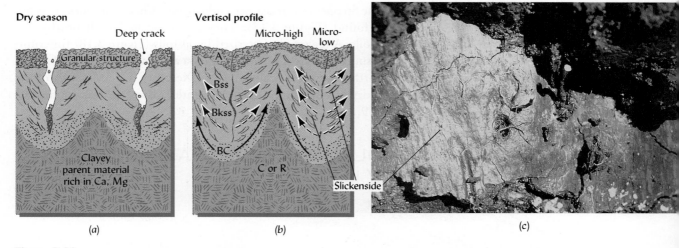

(a)

(b)

(c)

Figure 3.16

Vertisols are high in swelling-type clay and have wedgelike structures in the subsoil. (a) During the dry season, large cracks appear as the clay shrinks upon drying. Some of the surface soil granules fall into cracks under the influence of wind and animals. This action causes a partial mixing, or inversion, of the horizons. (b) During the wet season, rainwater pours down the cracks, wetting the soil near the bottom of the cracks first, and then the entire profile. As the clay absorbs water, it swells the cracks shut, entrapping the collected granular soil. The increased soil volume causes lateral and upward movement of the soil mass. The soil is pushed up between the cracked areas. As the subsoil mass shears from the strain, smooth surfaces or slickensides form at oblique angles. (c) An example of a slickenside in a Vertisol. Note the grooved, shiny surface. The white spots in the lower right of the photo are calcium carbonate concretions that often accumulate in a Bkss horizon. (Diagrams and photo courtesy of R. Weil)

3.12   MOLLISOLS (DARK, SOFT SOILS OF GRASSLANDS)

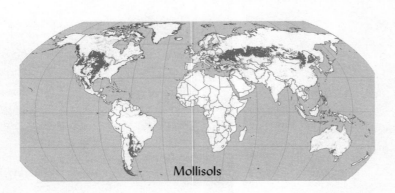
Mollisols

6.9% of global and 22.4% of U.S. ice-free land

Suborders are:
   Albolls (albic horizon)
   Aquolls (wet)
   Cryolls (cold)
   Gelolls (very cold)
   Rendolls (calcareous)
   Udolls (humid)
   Ustolls (moist/dry)
   Xerolls (dry summers, moist winters)

The principal process in the formation of Mollisols is the accumulation of calcium-rich organic matter, largely from the dense root systems of prairie grasses, to form the thick, soft Mollic epipedon that characterizes soils in this order (Plates 8, 13, and 20). This humus-rich surface horizon may extend 60 to 80 cm deep. Its cation exchange capacity is more than 50% saturated with base cations (Ca^{2+}, Mg^{2+}, etc.). Mollisols in humid regions generally have higher organic matter and darker, thicker mollic epipedons than their lower-moisture-regime counterparts (see Section 11.8).

The surface horizon generally has granular or crumb structures, largely resulting from an abundance of organic matter and swelling-type clays. In many cases, the highly aggregated soil is not hard when dry, hence the name *Mollisol,* implying softness (Table 3.1). In addition to the mollic epipedon, Mollisols may have an argillic (clay), natric, albic, or cambic subsurface horizon, but not an oxic or spodic horizon.

Mollisols are dominant in the Great Plains of North America, the Pampas of South America, and the Steppes of Eurasia (see endpapers). Where soil moisture is not limiting, Udolls are found (Figure 3.17). They are associated with nearby wet Mollisols termed Aquolls. A region characterized by Ustolls (Figure 3.18) (intermittently dry during the summer) extends from Manitoba and Saskatchewan in Canada to southern Texas. Farther west are sizable areas of Xerolls (with a Xeric moisture regime, which is very dry in summer but moist in winter). Two Mollisols profiles are included in Figure 3.19.

3.13   ALFISOLS (ARGILLIC OR NATRIC HORIZON, MODERATELY LEACHED)

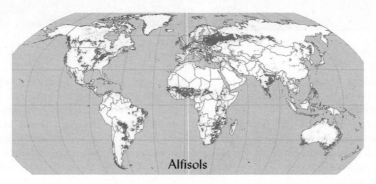
Alfisols

9.6% of global and 14.5% of U.S. ice-free land

Suborders are:
   Aqualfs (wet)
   Cryalfs (cold)
   Udalfs (humid)
   Ustalfs (moist/dry)
   Xeralfs (dry summers, moist winters)

The Alfisols are more strongly weathered than soils in the orders just discussed, but less so than Spodosols and Ultisols (see following). They are found in cool to hot humid areas (see Figure 3.7) as well as in the semiarid tropics and Mediterranean

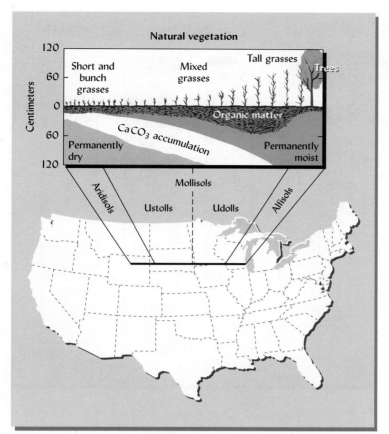

Figure 3.17

Correlation between natural grassland vegetation and certain soil orders is graphically shown for a transect across north central United States. The controlling factor, of course, is climate. Note the deeper organic matter and deeper zone of calcium accumulation, sometimes underlain by gypsum, as one proceeds from the drier areas in the west toward the more humid region where prairie soils are found. Alfisols may develop under grassland vegetation, but more commonly occur under forests and have lighter-colored surface horizons.

Figure 3.18

Typical landscape dominated by Ustolls (Montana). These productive soils produce much of the food and feed in the United States. (Photo courtesy of R. Weil)

climates. Most often, Alfisols develop under native deciduous forests, although in some cases, as in California and parts of Africa, savanna (mixed trees and grass) is the native vegetation.

Alfisols are characterized by a subsurface diagnostic horizon in which silicate clay has accumulated by illuviation (see Plate 1). Clay skins or other signs of clay movement are present in such a B horizon (see Plates 18 and 24). In Alfisols, this

Figure 3.19

Monoliths of profiles representing three soil orders. The suborder names are in parentheses. Genetic (not diagnostic) horizon designations are also shown. Note the spodic horizons in the Spodosol characterized by humus (Bh) and iron (Bs) accumulation. In the Alfisol is found the illuvial clay horizon (Bt), and the structural B horizon (Bw) is indicated in the Mollisols. The thick dark surface horizon (mollic epipedon) characterizes both Mollisols. Note that the zone of calcium carbonate accumulation (Bk) is near the surface in the Ustoll, which has developed in a dry climate. The E/B horizon in the Alfisol has characteristics of both E and B horizons.

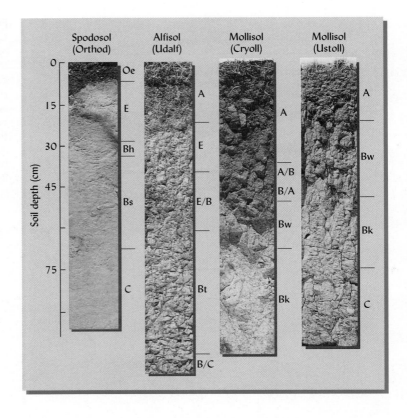

clay-rich horizon is only moderately leached, and its cation exchange capacity is more than 35% saturated with non-acid cations (Ca^{2+}, Mg^{2+}, etc.). In most Alfisols this horizon is termed *argillic* because of its accumulation of silicate clays. The horizon is termed *natric* if, in addition to having an accumulation of clay, it is more than 15% saturated with sodium and has prismatic or columnar structure (see Figures 4.8 and 9.26). In some Alfisols in subhumid tropical regions, the accumulation is termed a *kandic* horizon (from the mineral kandite) because the clays have a low cation exchange capacity.

Alfisols very rarely have a mollic epipedon, for such soils would be classified in the Argiudolls or other suborder of Mollisols with an argillic horizon. Instead, Alfisols typically have a relatively thin, gray to brown ochric epipedon (Plate 1 shows an example) or an umbric epipedon. Those formed under deciduous temperate forests commonly have a light-colored, leached *albic* E horizon immediately under the A horizon (see Plate 21).

3.14 ULTISOLS (ARGILLIC HORIZON, HIGHLY LEACHED)

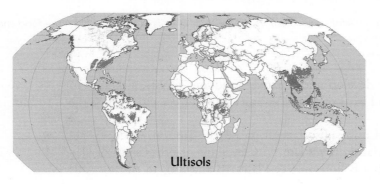

Ultisols

8.5% of global and 9.6% of U.S. ice-free land

Suborders are:
Aquults (wet)
Humults (high humus)
Udults (humid)
Ustults (moist/dry)
Xerults (dry summers, moist winters)

The principal processes involved in forming Ultisols are clay mineral weathering, translocation of clays to accumulate in an argillic or kandic horizon, and leaching of non-acid cations from the profile. Most Ultisols have developed under moist conditions in warm to tropical climates. Ultisols are formed on old land surfaces, usually under forest vegetation, although savanna or even swamp vegetation is also common. They often have an ochric or umbric epipedon, but are characterized by a relatively acidic B horizon that has less than 35% of the exchange capacity satisfied with non-acid cations. The clay accumulation may be either an argillic horizon or, if the clay is of low activity, a kandic horizon. Ultisols commonly have both an epipedon and a subsoil that is quite acid and low in plant nutrients.

Ultisols are more highly weathered and acidic than Alfisols but less acid than Spodosols and less highly weathered than the Oxisols. Except for the wetter members of the order, their subsurface horizons are commonly red or yellow in color, evidence of accumulations of oxides of iron (see Plate 11). Certain Ultisols that formed under fluctuating wetness conditions have horizons of iron-rich mottled material called **plinthite** (see Plates 15 and 37). This material is soft and can be easily dug from the profile so long as it remains moist. When dried in the air, however, plinthite hardens irreversibly into a kind of ironstone that is virtually useless for cultivation (Plates 29 and 33), but can be used to make durable bricks for building (Plate 48).

Although Ultisols are not naturally as fertile as Alfisols or Mollisols, they respond well to good management. They are located mostly in regions of long growing seasons and of ample moisture for good crop production. The silicate clays of Ultisols are usually of the nonsticky, low-activity type, which, along with the presence of iron oxides and aluminum, promotes ready workability.

3.15   SPODOSOLS (ACID, SANDY, FOREST SOILS, HIGHLY LEACHED)

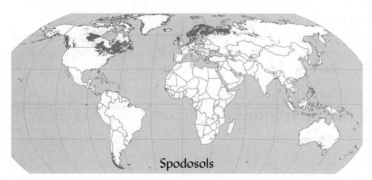

Spodosols

2.6% of global and 3.3% of U.S. ice-free land

Suborders are:
   Aquods (wet)
   Cryods (cold)
   Gelods (very cold)
   Humods (humus)
   Orthods (typical)

Spodosols occur mostly on coarse-textured, acid parent materials subject to ready leaching. They occur only in moist to wet areas, commonly where it is cold or temperate (see Figure 3.7), and also in some tropical and subtropical areas. Intensive acid leaching is the principal soil-forming process. They are mineral soils with a *spodic* horizon, a subsurface accumulation of illuviated organic matter, and an accumulation of aluminum oxides with or without iron oxides (see Plates 10 and 31). This usually thin, dark, illuvial horizon typically underlies a light, ash-colored, eluvial *albic* horizon.

Spodosols form under forest vegetation, especially under coniferous species whose needles are low in non-acid cations like calcium and high in acid resins. As this litter decomposes, strongly acid organic compounds are released and carried down into the permeable profile by percolating waters. Some of the leaching organic compounds may precipitate and form a black-colored Bh horizon. Leaching organic acids

Figure 3.20
(Left) A Spodosol in northern Michigan exhibits discontinuous, wavy Bh and Bs genetic horizons, which comprise the relatively deep spodic diagnostic horizon. The nearly white, discontinuous eluvial horizon (E) consists mainly of uncoated quartz sand particles and is an albic diagnostic horizon. The dark, organic-enriched surface horizon (an ochric diagnostic horizon) shows the smooth lower boundary and uniform thickness characteristic of an Ap horizon formed by plowing after the original coniferous forest was cleared. (Right) A much shallower Spodosol in Scotland also exhibits a spodic diagnostic horizon comprised of Bh and Bs genetic horizons. Scale at left marked every 10 cm, knife at right has a 12 cm long handle. (Photos courtesy of R. Weil)

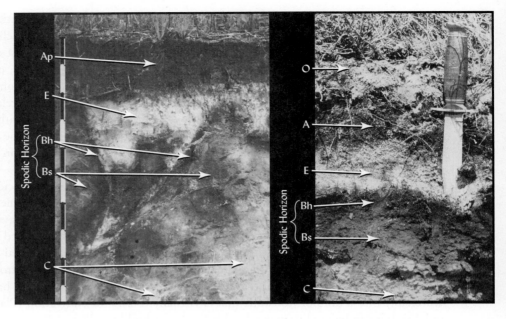

bind with iron and aluminum, removing these metals from the A and E horizons and carrying them downward. This iron and aluminum eventually precipitates in a reddish-brown-colored Bs horizon, usually just below the black-colored Bh horizon. Together, the Bh and Bs horizons constitute the spodic diagnostic horizon that defines the Spodosols. The depth at which the spodic horizon forms can vary from less than 20 cm to several meters. As iron oxides (and most other minerals except quartz) are stripped from the E horizon by the organic leaching process, this horizon may become a nearly white albic diagnostic horizon that consists mainly of clean quartz sand. The leaching and precipitation often occur along wavy wetting fronts, thus yielding the striking profiles seen in Spodosols (Figure 3.20).

Spodosols are not naturally fertile. Because they are already quite acid and poorly buffered, many Spodosols and the lakes in watersheds dominated by soils of this order are susceptible to damage from acid rain (see Section 9.6).

3.16 OXISOLS (OXIC HORIZON, HIGHLY WEATHERED)

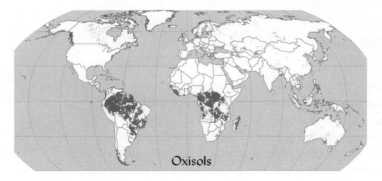

7.6% of global and <0.01% of U.S. ice-free land

Suborders are:
  Aquox (wet)
  Perox (very humid)
  Torrox (hot, dry)
  Udox (humid)
  Ustox (moist/dry)

The Oxisols are the most highly weathered soils in the classification system (see Figure 3.6). They form in hot climates with nearly year-round moist conditions; hence, the native vegetation is generally thought to be tropical rain forest. However, some Oxisols (Ustox) are found in areas that are today much drier than was the case when the soils were forming their oxic characteristics. Their most important diagnostic feature is a deep oxic subsurface horizon. Weathering and intense leaching have removed a large part of the silica from the silicate materials in this horizon. Some quartz and 1:1-type silicate clay minerals remain, but the hydrous oxides of iron and aluminum are dominant (see Chapter 8 for information on the various clay minerals). The epipedon in most Oxisols is either ochric or umbric. Usually the boundaries between subsurface horizons are indistinct, giving the subsoil a relatively uniform appearance with depth (Plates 9 and 25).

The clay content of Oxisols is generally high, but the clays are of the low-activity, nonsticky type. Consequently, when the clay dries out it is not hard and cloddy, but is easily worked. Also, Oxisols are resistant to compaction, so water moves freely through the profile. The depth of weathering in Oxisols is typically much greater than for most of the other soils, 20 m or more having been observed. The low-activity clays have a very limited capacity to hold nutrient cations such as Ca^{2+}, Mg^{2+}, and K^+, so they are typically of low natural fertility and moderately acid. The high concentration of iron and aluminum oxides also gives these soils a capacity to bind so tightly with what little phosphorus is present that phosphorus deficiency often limits plant growth once the natural vegetation is disturbed.

Roads and buildings are relatively easily constructed on most Oxisols because these soils are easily excavated, do not shrink and swell, and are physically very stable on slopes. The very stable aggregation of the clays, stimulated largely by iron compounds, makes these soils quite resistant to erosion.

3.17   LOWER-LEVEL CATEGORIES IN SOIL TAXONOMY

Suborders

As indicated next to the global distribution maps in previous sections, soils within each order are grouped into suborders on the basis of soil properties that reflect major environmental controls on current soil-forming processes. Many suborders are indicative of the moisture regime or, less frequently, the temperature regime under which the soils are found. Thus, soils formed under wet conditions generally are identified under separate suborders (e.g., Aquents, Aquerts, Aquepts), as being wet soils.

Great Groups

The great groups are subdivisions of suborders. More than 400 great groups are recognized. They are defined largely by the presence or absence of diagnostic horizons (such as umbric, argillic, and natric), which supply formative elements for the names of great groups.

Remember that the great group names are made up of these formative elements attached as prefixes to the names of suborders in which the great groups occur. Thus, Ustolls with a natric horizon (high in sodium) belong to the Natrustolls great group. As can be seen in the example discussed in Box 3.2, soil descriptions at the great group level can provide important information not indicated at the higher, more general levels of classification.

The names of selected great groups from two orders are given in Table 3.2. This list illustrates again the usefulness of *Soil Taxonomy*, especially the nomenclature it employs. Note from Table 3.2 that not all possible combinations of great group prefixes and suborders are used. In some cases a particular combination does not exist. For example, Aquolls occur in lowland areas but not on very old landscapes. Hence, there are no "Paleaquolls." Also, because *all* Ultisols contain an argillic horizon, the use of terms such as "Argiudults" would be redundant.

BOX 3.2
GREAT GROUPS, FRAGIPANS, AND ARCHAEOLOGIC DIGS

Soil Taxonomy is a communications tool that helps scientists and land managers share information. In this box we will see how misclassification, even at a lower level in *Soil Taxonomy*, such as the great group, can have costly ramifications.

In order to preserve our historical and prehistorical heritage, laws require that an archaeological impact statement be prepared prior to starting major construction work on the land. The archaeological impact is usually assessed in three phases. Selected sites are then studied by archaeologists, with the hope that at least some of the artifacts can be preserved and interpreted before construction activities obliterate them forever. Only a few relatively small sites can be subjected to actual archaeological digs because of the expensive skilled hand labor involved (Figure 3.21).

Such an archaeological impact study was ordered as a precursor to construction of a new highway in a mid-Atlantic state. In the first phase, a consulting company gathered soils and other information from maps, aerial photographs, and field investigations to determine where Neolithic people may have occupied sites. Then the consultants identified about 12 ha of land where artifacts indicated significant neolithic activities. The soils in one area were mapped mainly as Typic Dystrudepts. These soils formed in old colluvial and alluvial materials that, many thousands of years ago, had been along a river bank. Several representative soil profiles were examined by digging pits with a backhoe. The different horizons were described, and it was determined in which horizons artifacts were most likely to be found. What was not noted

was the presence in these soils of a fragipan, a dense, brittle layer that is extremely difficult to excavate using hand tools.

A fragipan is a subsurface diagnostic horizon used to classify soils, usually at the great group or subgroup level. Its presence would distinguish Fragiudepts from Dystrudepts.

When it came time for the actual hand excavation of sites to recover artifacts, a second consulting company was awarded the contract. Unfortunately, their bid on the contract was based on soil descriptions that did not specifically classify the soils as Fragiudepts—soils with very dense, brittle, hard fragipans in the layer that would need to be excavated by hand. So difficult was this layer to excavate and sift through by hand that it nearly doubled the cost of the excavation—an additional expense of about $1 million. Needless to say, there ensued a controversy as to whether this cost would be borne by the consulting firm that bid with faulty soils data, the original consulting firm that failed to adequately describe the presence of the fragipan, or the highway construction company that was paying for the survey.

This episode gives us an example of the practical importance of soil classification. The formative element *Fragi* in a soil great group name warns of the presence of a dense, impermeable layer that will be very difficult to excavate, will restrict root growth (often causing trees to topple in the wind or become severely stunted), may cause a perched water table (epiaquic conditions), and will interfere with proper percolation in a septic drain field.

Figure 3.21
An archaeological dig. (Photo courtesy of Antonio Segovia, University of Maryland)

Table 3.2

EXAMPLES OF GREAT GROUP NAMES FOR SELECTED SUBORDERS IN THE MOLLISOL AND ULTISOL ORDERS

	Dominant feature of great group			
	Argillic horizon	Central concept with no distinguishing features	Old land surfaces	Fragipan
Mollisols				
1. Aquolls (wet)	Argiaquolls	Haplaquolls	—	—
2. Udolls (moist)	Argiudolls	Hapludolls	Paleudolls	—
3. Ustolls (dry)	Argiustolls	Haplustolls	Paleustolls	—
4. Xerolls (Med.)[a]	Argixerolls	Haploxerolls	Palexerolls	—
Ultisols				
1. Aquults (wet)	—	—	Paleaquults	Fragiaquults
2. Udults (moist)	—	Hapludults	Paleudults	Fragiudults
3. Ustults (dry)	—	Haplustults	Paleustults	—
4. Xerults (Med.)[a]	—	Haploxerults	Palexerults	—

[a]Med. = Mediterranean climate; distinct dry period in summer.

Subgroups

Subgroups are subdivisions of the great groups. More than 2500 subgroups are recognized. The central concept of a great group makes up one subgroup, termed *Typic*. Thus, the Typic Hapludolls subgroup typifies the Hapludolls great group. Other subgroups may have characteristics that intergrade between those of the central concept and soils of other orders, suborders, or great groups. A Hapludoll with restricted drainage would be classified as an Aquic Hapludoll. One with evidence of intense earthworm activity would fall in the Vermic Hapludolls subgroup. Some intergrades may have properties in common with other orders or with other great groups. Thus, soils in the Entic Hapludolls subgroup are very weakly developed Mollisols, close to being in the Entisols order. The subgroup concept illustrates very well the flexibility of this classification system.

Families

Within a subgroup, soils fall into a particular family if, at a specified depth, they have similar physical and chemical properties affecting the growth of plant roots. About 8000 families have been identified. The criteria used include broad classes of particle size, mineralogy, cation exchange activity of the clay, temperature, and depth of the soil penetrable by roots. Terms such as *loamy, sandy*, and *clayey* are used to identify the broad particle size classes. Terms used to describe the mineralogical classes include *smectitic, kaolinitic, siliceous, carbonatic*, and *mixed*. The clays are described as *superactive, active, semiactive*, or *subactive* with regard to their capacity to hold cations. For temperature classes, terms such as *cryic, mesic,* and *thermic* are used. The terms *shallow* and *micro* are sometimes used at the family level to indicate unusual soil depths.

Thus, a Typic Argiudoll from Iowa, loamy in texture, having a mixture of moderately active clay minerals and with annual soil temperatures (at 50 cm depth) between 8 and 15 °C, is classed in the *loamy, mixed, active, mesic Typic Argiudolls* family. In contrast, a sandy-textured Typic Haplorthod, high in quartz and located in a cold area in eastern Canada, is classed in the *sandy, siliceous, frigid Typic Haplorthods* family. (Note that clay activity classes are not used for soils in sandy textural classes.)

Series

The series category is the most specific unit of the classification system. It is a subdivision of the family, and each series is defined by a specific range of soil properties involving primarily the kind, thickness, and arrangement of horizons. Features such as a hard pan within a certain distance below the surface, a distinct zone of calcium carbonate accumulation at a certain depth, or striking color characteristics may aid in series identification.

In the United States and many other countries, each series is given a name, usually from some town, river, or lake such as Fargo, Muscatine, Cecil, Mohave, or Ontario. There are about 23,000 soil series in the United States alone. For practical reasons, soil series are sometimes subdivided into **phases** based on certain properties of importance to land use and management (e.g., texture of the surface horizon, stoniness, slope, degree of erosion, or soluble salt content). Thus, "Pinole loam, 2 to 9% slopes" and "Hagerstown silt loam, stony phase" are examples of phases of soil series. Although not technically a category in *Soil Taxonomy*, soil phases are commonly represented on soil maps and used as management units.

The complete classification of a Mollisol, the Kokomo series, is given in Figure 3.22. This figure illustrates how *Soil Taxonomy* can be used to show the relationship between *the soil*, a comprehensive term covering all soils, and a specific soil series. The figure deserves study because it reveals much about the structure and use of *Soil Taxonomy*. If a soil series name is known, the complete *Soil Taxonomy* classification of the soil may be found on the Internet at the URL in the

Official *Soil Taxonomy* Soil Series Names and Descriptions. Click on "Soil Series Name Search." http://soils.usda.gov/technical/classification/scfile/index.html

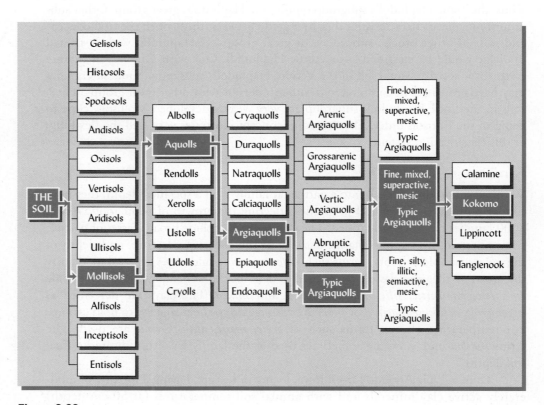

Figure 3.22

Diagram illustrating how one soil (Kokomo) keys out in the overall classification scheme. The shaded boxes show that this soil is in the Mollisols order, Aquolls suborder, Argiaquolls great group, and so on. In each category, other classification units are shown in the order in which they key out in Soil Taxonomy. *Many more families exist than are shown.*

Plate 1 *Alfisols—a Glossic Hapludalf in New York. Scale in cm.*

Plate 2 *Andisols—a Typic Melanudand from western Tanzania. Scale in 10 cm.*

Plate 3 *Aridisols—a skeletal Ustollic Calcicambid from Nevada. Shovel handle is 60 cm long.*

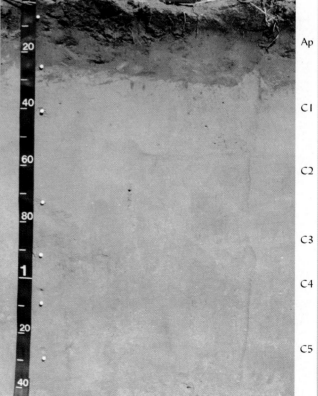

Plate 4 *Entisols—a Typic Udipsamment on a river flood plain in North Carolina.*

Plate 5 *Gelisols—a Typic Aquaturbel from Alaska. Permafrost below 32 cm on scale.*

Plate 6 *Histosols—a Limnic Haplosaprist from southern Michigan. Buried mineral soil at bottom of scale. Scale in feet.*

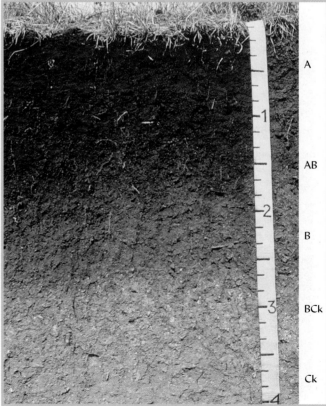

Plate 7 *Inceptisols—a Typic Eutrudept from Vermont.*

Plate 8 *Mollisols—a Typic Hapludoll from central Iowa. Mollic epipedon to 1.8 ft. Scale in feet.*

Plate 9 *Oxisols—a Udeptic Hapludox from central Puerto Rico. Scale in feet and inches.*

Plate 10 *Spodosols—a Typic Haplorthod in New Jersey. Scale in 10 cm increments.*

Plate 11 *Ultisols—a Typic Hapludult from central Virginia showing metamorphic rock structure in the saprolite below the 60-cm-long shovel.*

Plate 12 *Vertisols—a Typic Haplustert from Queensland, Australia, during wet season. Scale in meters.*

Plate 13 *Typic Argiustolls in eastern Montana with a chalky white calcic horizon (Bk and Ck) overlain by a Mollic epipedon (Ap, A2, and Bt).*

Plate 14 *Ice wedge and permafrost underlying Gelisols in the Seward Peninsula of Alaska. Shovel is 1 m long.*

Plate 15 *A Typic Plinthudult in central Sri Lanka. Mottled zone is plinthite, in which ferric iron concentrations will harden irreversibly if allowed to dry.*

Plate 16 *A soil catena or toposequence in central Zimbabwe. Redder colors indicate better internal drainage. Inset: B-horizon clods from each soil in the catena.*

Plate 17 *Redox concentrations (red) and depletions (gray) in Btg horizon from an Aquic Paleudalf.*

Plate 18 *Clay skins (argillans) appear as dark, shiny coatings in this Ultisol Bt horizon. Bar = 1 cm.*

Plate 19 *The boundary between the Oe and the E horizons of a forested Ultisol.*

Plate 20 *The effect of moisture on soil color. Right side of this Mollisol profile was sprayed with water.*

Plate 21 *Effect of poor drainage on soil color. Gray matrix colors and red redox concentrations in the B horizons of a Plinthaquic Paleudalf.*

Plate 22 *10YR hue page in Munsell color book with standard notation for color of hue 10YR, value 5, and chroma 6.*

Plate 23 *Roadcut in southern Brazil, exposing the profile of an Udalf with a sombric horizon. This dark, humus-rich subsurface horizon typically forms in humid, high-altitude tropical and subtropical mountains.*

Plate 24 *Thick clay skins (argillans) in an argillic (Bt) horizon. Image made from a very thin, polished slice of soil, magnified with a petrographic microscope using plain polarized (left) and cross-polarized light (right). Note the thin layers of illuvial clay.*

Plate 25 *Oxisol profile in Central Brazil. Scale in 10 cm.*

Plate 26 *Histosols—a Fibrist in central Scotland. Knife handle 11.5 cm.*

Ap

A2

C

2Ab

2Bb

Plate 27 *Sloping Entisol formed in shaley colluvium deposited over buried soil in Pennsylvania.*

Plate 28 *Green slickenside in glauconitic Marlton soil.*

Plate 29 *Hard plinthite concretions in tropical Alfisol Bv horizon.*

5 cm

Plate 30 *Yellow Jarosite formed by sulfidization.*

Plate 31 *Spodosol formed in glacial outwash in Michigan.*

1 cm

Plate 32 *Dark organic matter coatings (arrows) on blocky peds in a Kansas Mollisol.*

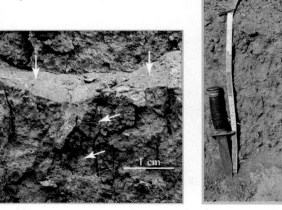

Plate 33 *Shallow laterite (hard plinthite) layer (knife point) in tropical Alfisol. Scale marked in 10 cm.*

Plate 34 *Traditional Sri Lankan slash-and-burn system. Farmers chop down patches of forest and burn the dead vegetation, returning many nutrients in the ash.*

Plate 35 *Gleyed colors along root channel in ped from C horizon.*

Plate 36 *Erosion of convex sites by tillage and water has exposed red B horizon material. Ultisols in Virginia.*

Plate 37 *Iron depletions and concentrations in C horizon of Ultisol in Alabama.*

Plate 38 *Oxidized (red) root zones in the A and E horizons indicate a hydric soil. They result from oxygen diffusion out from roots of wetland plants having aerenchyma tissues (air passages).*

Plate 39 *Dark (black) humic accumulation and gray humus depletion spots in the A horizon are indicators of a hydric soil. Water table is 30 cm below the soil surface.*

Plate 40   *Urban soils (referred by some as Urbents) often hold surprises in their profiles. Here a tree planting hole reveals a buried A horizon (top) and buried asphalt (lower).*

Plate 41   *Soil saturated beyond its liquid limit by torrential rains caused this landslide and mudflow that pushed huge tropical cloud forest trees downslope and demolished a village at the foot of this mountain in Honduras.*

Plate 42   *Mass wasting of clayey soils on steep slopes may occur when saturated with water, as in this rotational block slide in East Africa. Note man in center for scale.*

Plate 43   *The red, kaolinitic soil in the foreground was hauled in to build up a stable roadbed across this low-lying landscape. The black soils are rich in expansive clays, which would break up the pavement if used for the sub-base. South Central Tanzania.*

Plate 44   *Two large soil stockpiles on a construction site. The brown A horizon was set aside for landscaping topsoil, while the redder B horizon (back) was stockpiled for use as fill and road base.*

Plate 45   *An uneven layer of silt blankets over a glacial deposit of coarse sand and gravel in this Rhode Island Inceptisol. The profile water-holding capacity varies with the thickness of the silt, causing an irregular pattern of drought-stricken turfgrass (inset).*

Plate 46 *Bent trees indicate soil creep.*

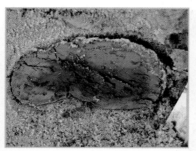

Plate 47 *Reduced interior of clay inclusions in sandy sediment.*

Plate 49 *Tillage research plots with rails for implements to avoid wheel effects. Auburn, Alabama.*

Plate 48 *Bricks made from hardened plinthite.*

Plate 50 *Adult 17-year cicada and emergence holes.*

Plate 51 *Dark green clover in N-deficient lawn.*

Plate 52 *Morrow plots at University of Illinois, Urbana.*

Plate 53 *Grapevines growing in Jory soil (Ultisol with low P availability) are dramatically stimulated (right) but they are little affected if growing in Chehalis soil (Mollisol with high P availability).*

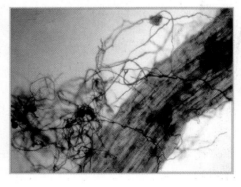

Plate 54 *Sorghum root colonized by AM fungi showing hyphae growing outside the root (extra radicular).*

Plate 55 *Desert pavement with a pebble removed to reveal vesicular pores in soil.*

Plate 56 *Evaporating basins for sea salt. Bloom of Halobacterium, an archaean with red photosynthesizing pigment.*

Plate 57 *Ancient carving in desert varnish, a Mn oxide rock coating deposited by bacteria. Nevada.*

Plate 58 *Alluvium deposited 1 m thick in 8 years since riprap installed on stream bank in an urban watershed.*

Plate 59 *Iron concentrations (orange lepidocrocite) on ped faces in Btg horizon of Alfisol in Chad.*

Plate 60 *Connecticut River valley in western Massachusetts. Note variable alluvial soils and presence of riparian forest buffer along the river bank.*

BOX 3.3
USING SOIL TAXONOMY TO UNDERSTAND A LANDSCAPE

In real-world landscapes, different soils exist alongside each other, often in complex patterns. Adjacent soils on a tract of land may belong to different families, subgroups, great groups, or even different soil orders. Figure 3.23 depicts a glaciated landscape in a humid temperate region (Iowa) where 2 to 7 m of loess overlies leached till and the native vegetation was principally tall grass prairie interspersed with small areas of trees. This landscape demonstrates how diagnostic horizons and other features of soil taxonomy are used to organize soils information. It also highlights the relationships among soils that allow us to make soil maps and interpret geographic soils information to help in planning projects on the land.

These relationships are reflected in the soil taxonomy names. The formative element *aqu* appears in the taxonomic name of the three wetter soils. It appears at the suborder level (Endoaquolls) for the very wet, poorly drained soils, but only at the subgroup level for the less wet, somewhat poorly drained soil (Aquic Hapludoll). The formative element *argi* is used in the name of two soils (Argiudolls) to indicate that enough clay has accumulated in the B horizon of these Mollisols to develop into an argillic diagnostic horizon. Argi does not appear in the names of the two Alfisols, because an accumulation of clay (argillic or similar horizon) is a required feature of all Alfisols. Subgroup modifiers also provide important information about the interrelationships of these soils in the landscape. For example, the modifier Cumulic indicates that the Wabash soil has an unusually thick mollic epipedon because soil material washing off the uplands and carried by local streams has accumulated in the low-lying floodplains where this soil is found. The modifier Mollic used for the Downs soil indicates that this soil is transitional between the Alfisols and Mollisols, the A horizon in the Downs soils being slightly too thin to classify as a mollic epipedon.

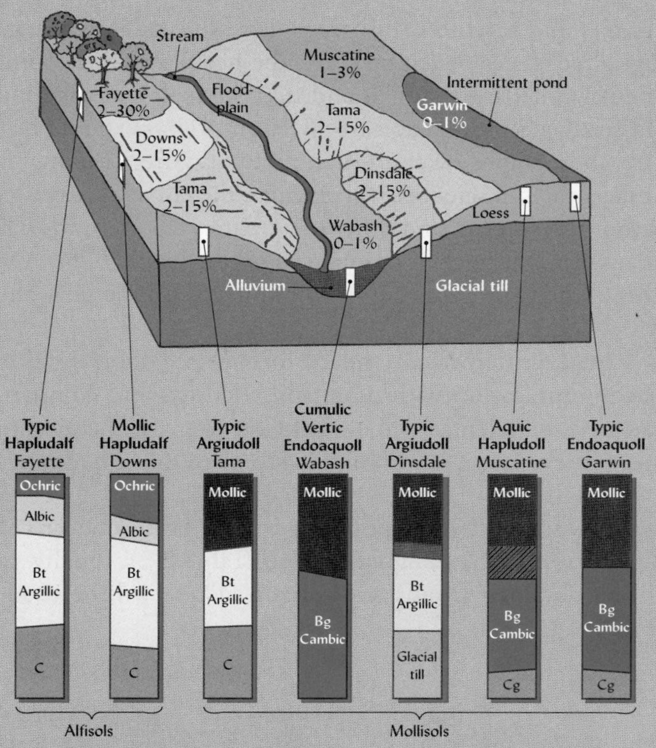

Figure 3.23

Soil taxonomy reflects soil–landscape relationships. Diagram courtesy of Ray Weil based on Riecken and Smith (1949).

margin. Box 3.3 illustrates how soil taxonomic information can assist in understanding the nature of a landscape.

3.18 TECHNIQUES FOR MAPPING SOILS[7]

Geographic information about soils is often best communicated to land managers by means of a soil map. Soil maps are in great demand as tools for practical land planning and management. Many soil scientists therefore specialize in mapping soils. Before beginning the actual mapping process, a soil scientist must learn as much as possible about the soils, landforms, and vegetation in the survey area. Therefore, the first step in mapping soils is to collect and study older or smaller-scale soil maps, geological and

---

[7]For an internationally oriented guide to all aspects of the process of making soil maps, see Legros (2006). For official procedures for making soil surveys in the United States, see USDA-NRCS (2006).

topographic maps, previous soil descriptions, and any other information available on the area. Once the soil survey begins, the soil scientist's task is threefold: (1) to define each soil unit to be mapped; (2) to compile information about the nature and classification of each soil; and (3) to delineate the boundaries where each soil unit occurs in the landscape. We will now discuss some of the procedures and tools that soil scientists use to delineate soils in the field.

Soil Description

USDA Soil Survey division field book for describing and sampling soils: http://soils.usda.gov/technical/fieldbook/

Soil scientists use computers and satellites, but they also use spades and augers. Despite all the technological advances of recent years, the heart of soil mapping is still the soil pit, a rectangular hole large enough and deep enough to allow one or more people to enter and study a typical pedon (see Section 3.1) as exposed on the pit face. Plates 1 to 12 are photographs taken of such pit faces. After cleaning away loose debris from the pit face, the soil scientist will examine the colors, texture, consistency, structure, plant rooting patterns, and other soil features to determine which horizons are present and at what depths their boundaries occur (Figure 3.24). Often the horizon boundaries are etched with a trowel or soil knife, as can be seen on the right side of Plate 9.

A soil description is then written in a standard format (see Table 3.3 for an example) that facilitates communication with other soil scientists and comparison with other soils. Sometimes the soil scientist will use field kits to make chemical tests, for pH and free carbonates (effervescence of carbon dioxide when dilute hydrochloric acid is added) (see Plate 91).

As far as possible at this stage, the soil horizons will be given master (A, E, B, etc.) and subordinate (2Bt, Ap, etc.) designations (see Table 2.2). Finally, samples of soil material will be obtained from each horizon. These will be used for

Figure 3.24

A soil pit allows detailed observations to be made of the soil in place. Here, several Natural Resources Conservation Service soil scientists describe a typical pedon for a mapping unit (Thorndale series) as revealed on the wall of a soil pit. The soil scientist standing in the pit is comparing the colors and textures of soil samples he has removed from several horizons while another soil scientist (upper right), records the observations in a notebook to make a soil profile description (such as the one in Table 3.3). Later, when these soil scientists are boring transects of auger holes to map soils in a landscape (see Figures 3.25 and 3.26), they can determine the identity of the soils they encounter by comparing the properties of samples brought up in their augers to the properties listed in detailed written descriptions of the soil pit. (Photo courtesy of R. Weil)

Table 3.3
SOIL PROFILE DESCRIPTION (WITH SOIL TAXONOMY DIAGNOSTIC HORIZONS) FOR THE THORNDALE SOIL SERIES[a]

Figure 3.24 shows soil scientists describing a Thorndale soil in Pennsylvania.

Horizon designation	Diagnostic horizon	Horizon boundaries	Description of horizon in typical pedon
Ap	Ochric epipedon	0–20 cm	Dark grayish brown (2.5Y 4/2) silt loam; weak medium granular structure; friable, slightly sticky, slightly plastic; many fine roots; neutral; clear smooth boundary.
Btg1		20–43 cm	Light olive gray (5Y 6/2) silty clay loam; moderate coarse subangular blocky structure; slightly firm, sticky, plastic; few medium and fine roots; many prominent dark grayish brown (10YR 4/2) clay films on faces of peds; common medium prominent reddish brown (5YR 4/3) masses of iron accumulations; slightly acid; clear smooth boundary.
Btg2	Argillic horizon	43–65 cm	Light olive gray (5Y 6/2) silty clay loam; weak coarse prismatic structure parting to moderate medium subangular blocky; firm, sticky, plastic; few fine roots; many prominent dark grayish brown (10YR 4/2) clay films on prisms and faces of peds; common medium and fine prominent reddish brown (5YR 4/4) masses of iron accumulations in the matrix; slightly acid; gradual smooth boundary.
Btxg	Fragipan	65–103 cm	Grayish brown (10YR 5/2) silty clay loam; weak very coarse prismatic structure parting to weak medium subangular blocky; firm, brittle, moderately sticky, moderately plastic; few fine roots; many faint dark grayish brown (10YR 4/2) clay films on prism faces and few faint dark grayish brown (10YR 4/2) clay films on faces of peds; common fine to medium prominent reddish brown (5YR 4/4) and brown (7.5YR 5/4) masses of iron accumulations in the matrix; slightly acid; abrupt smooth boundary.
C		103–163 cm	Strong brown (7.5YR 5/6) and reddish yellow (7.5YR 7/6) silt loam; massive; friable, slightly sticky, slightly plastic; common medium prominent grayish brown (2.5Y 5/2) iron depletions in the matrix; slightly acid.

[a]Thorndale soils are very deep, poorly drained soils formed in medium-textured colluvium derived from limestone, calcareous shale, and Siltstone. Slopes are 0 to 8%. Permeability is slow. Mean annual precipitation and temperature are about 100 cm and 12 °C. Taxonomic class is Fine-silty, mixed, active, mesic Typic Fragiaqualfs.
Adapted from USDA-NRCS (2002).

detailed laboratory analyses and for archiving. The laboratory analyses will provide information for the chemical, physical, and mineralogical characterization of each soil.

Using these techniques, the soil scientists assigned to map an area will familiarize themselves with the soils they expect to find, learning certain unique characteristics that they can look for to quickly identify each soil and distinguish it from other soils in the area.

Delineating Soil Boundaries

For obvious reasons, a soil scientist cannot dig pits at many locations on the landscape to determine which soils are present and their boundaries. Instead, he or she will bring up soil material from numerous small boreholes made with a hand auger or

(a)   (b)

Figure 3.25

Soil maps are prepared by soil scientists who examine the soils in the field using such tools as a hand-powered soil auger (a) or a truck-mounted hydraulic soil probe (b). (Photos courtesy of USDA/NRCS)

hydraulic probe (Figure 3.25). The texture, color, and other properties of the soil material from various depths can be compared mentally to characteristics of the known soils in the region.

With hundreds of different soils in many regions, this might seem to be a hopeless task. However, the job is not as daunting as one might suppose, for the soil scientist is not blindly or randomly boring holes. Rather, he or she is working from an understanding of the soil associations and how the five soil-forming factors determine which soils are likely to be found in which landscape positions. Usually there are only a few soils likely to occupy a particular location, so only a few characteristics must be checked. The soil auger is used primarily to confirm that the type of soil predicted to occur in a particular landscape position is the type actually there.

3.19   SOIL SURVEYS

NRCS National Cooperative Soil Survey: http://soils.usda.gov/partnerships/ncss/

A **soil survey** is more than simply a soil map. The glossary describes a *soil survey* as "a systematic examination, description, classification, and mapping of the soils in a given area." Once the natural bodies are delineated and their properties are described, the soil survey can aid in making interpretations for all kinds of soil uses.

Mapping Units

Because local features and requirements will dictate the nature of the soil maps and, in turn, the specific soil units that are mapped, the field *mapping units* may be somewhat different from the *classification units* found in *Soil Taxonomy*. The mapping units may represent some further differentiation below the soil series level—namely, *phases* of soil series; or the soil mappers may choose to group together similar or associated soils into conglomerate mapping units. Examples of such soil mapping units follow.

Consociations The smallest practical mapping unit for most detailed soil surveys is an area that contains primarily one soil series and usually only one phase of that soil series. For example, a mapping unit may be labeled as the consociation "Saybrook silt loam, 2 to 5% slopes, moderately eroded." Quality-control standards may indicate that a consociation mapping unit should be 50% "pure" and that the "impurities" should be so similar to the named phase that the differences do not affect land management. Inclusions of *contrasting* soils should occupy less than 15% of the consociation.

Soil Complexes and Soil Associations Sometimes *contrasting* soils occur adjacent to each other in a pattern so intricate that the delineation of each kind of soil on a soil map becomes difficult, if not impossible. In such cases, a soil *complex* is indicated on a soil map, and an explanation of the soils present in the complex is contained in the soil survey report. A complex often contains two or three distinctly different soil series. Relatively small-scale soil maps may display only soil *associations*—general groupings of soils that typically occur together in a landscape and could be mapped separately if a larger scale were used.

Using Soil Surveys

In the United States, soil scientists have worked for over 100 years to complete a detailed soil survey of the entire country. This effort, known as the *National Cooperative Soil Survey*, is an ongoing collaboration of federal, state, and local governments. The principal federal role is played by the Natural Resources Conservation Service in the U.S. Department of Agriculture.

In 2006, the U.S. National Cooperative Soil Survey was reorganized to focus on 273 Major Land Resource Areas (MLRAs), zones defined by ecological characteristics, instead of the 3000+ politically defined counties formerly used to organize soil survey activities. At about the same time, a digital database called the National Soil Information System (NASIS) was established to facilitate the effort to replace static paper soil survey reports with a dynamic resource for soils information adapted to many types of users.

The U.S. National Soil Information System (NASIS): http://nasis.usda.gov/intro

Interpretive Information

Use of soil surveys for land use or site planning requires that the geographic information on the maps be integrated with the descriptive information, detailed soil profile descriptions, and interpretive rankings for the mapping units. Soil surveys offer interpretive information on yield potentials for various crops, on the suitability of soils for different irrigation methods, drainage requirements for soils, land capability classification of each mapping unit, and other interpretive information on many nonagricultural land uses, including wildlife habitat, forestry, landscaping, waste disposal, building construction, and sources of roadbed materials.

Online soil survey reports and data: http://soils.usda.gov/survey/

Web Soil Survey

The traditional printed and bound soil survey maps and reports (Figure 3.26) are being replaced by the interactive electronic "web soil survey" available on the Internet (see margin). The user first *defines* the *area of interest* using an address or by delineating a geographic area on a large-scale map. The second step is to *view* the

Use Web Soil Survey here: http://websoilsurvey.nrcs.usda.gov/app/

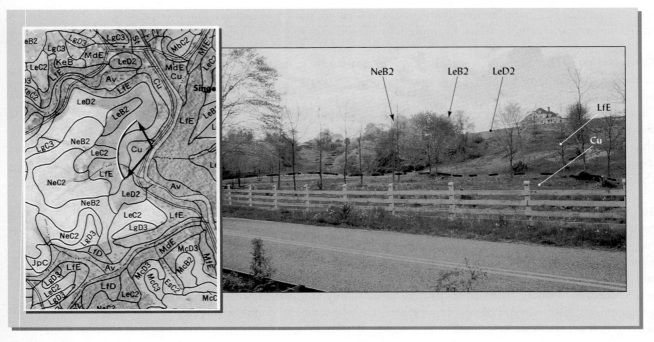

Figure 3.26
A small section of detailed soil survey map of Harford Country, Maryland (left), and a ground-level view of the part of the landscape represented on the map by the veiw arrows (right). The map is reproduced here at the original scale (1:15,840) and represents an area of about 1.1 km². The map symbols represent soil consociations such as LeD2, which is named for the soil phase "Legore silt loam, 12% to 15% moderately eroded," a soil in the Alfisols order. The level land in the foreground is an alluvial soil in the Inceptisols order, the Codorus silt loam. [Map from Smith and Matthews (1975); photo courtesy of R. Weil]

chosen area of interest at a larger scale on a display that includes the soil boundaries and map unit names over an air photo background. Figure 3.27 illustrates such a map, as displayed on the Web site. Finally, the Web site presents many options to *explore* the information by requesting maps that group soil series into suitability classes, slope classes, or other interpretive groupings. The user then can choose to save or print the map and related information or download the data for use in a geographic information system or GIS.

3.20   CONCLUSION

Rationale for concepts in Soil Taxonomy: http://soils .usda.gov/technical/ classification/taxonomy/ rationale/index.html

The soil that covers the Earth is actually composed of many individual soils, each with distinctive properties. Among the most important of these properties are those associated with the layers, or *horizons*, found in a soil profile. These horizons reflect the physical, chemical, and biological processes soils have undergone during their development. Horizon properties greatly influence how soils can and should be used.

Knowledge of the kinds and properties of soils around the world is critical to humanity's struggle for survival and well-being. A soil classification system based on these properties is equally critical if we expect to use knowledge gained at one location to solve problems at other locations where similarly classed soils are found. *Soil Taxonomy*, a classification system based on measurable soil properties, helps fill this need in more than 50 countries. Scientists constantly update the system as they learn more about the nature and properties of the world's soils and the relationships

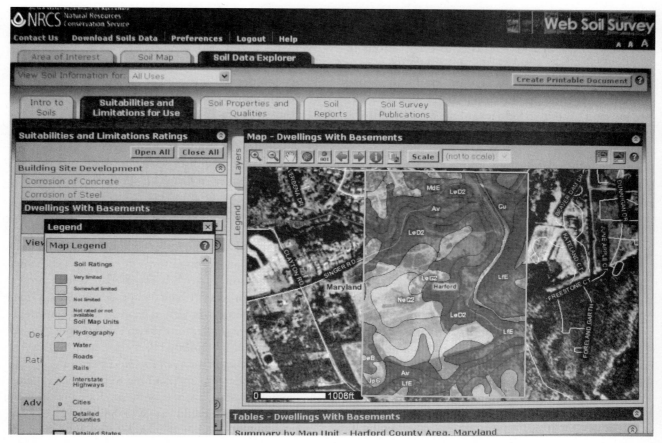

Figure 3.27

A screen shot from Web Soil Survey showing a defined area of interest (AOI, outlined rectangle in middle of map) in which soils information is superimposed over an air photo background. The map scale can be made large enough to show individual houses and driveways. In the screen shot the "soil data explorer" option has been used to color-code the soils with regard to their suitability rating as sites for dwellings with basements. A legend of soil units, suitability ratings, and geographic features is given in the panel at left. Tabular data (not shown) is also provided on the percent of the AOI occupied by each soil. The AOI shown here is the same area as in Figure 3.26 (note stream with a sharp bend in both figures). (Photo courtesy of R. Weil)

among them. In the remaining chapters of this book we will use taxonomic names whenever appropriate to indicate the kinds of soils to which a concept or illustration may apply.

Making soil surveys is both a science and an art by which many soil scientists apply their understanding of soils and landscapes to the real world. Mapping soils is not only a profession; many would say that it is a way of life. Working alone outdoors in all kinds of terrain and carrying all the necessary equipment, the soil scientist collects ground truth to be integrated with data from satellites and laboratories. The resulting soil maps and descriptive information in the soil survey reports and databases are used in countless practical ways by soil scientists and nonscientists alike. The soil survey, combined with powerful computerized geographic information systems, enables planners to make rational decisions about what should go where. The challenge for soil scientists and concerned citizens is to develop the foresight and fortitude to use criteria in the planning process that will help preserve our most valuable soils—not hasten their destruction under shopping malls and landfills.

STUDY QUESTIONS

1. Diagnostic horizons are used to classify soils in *Soil Taxonomy*. Explain the difference between a diagnostic horizon (such as an argillic horizon) and a genetic horizon designation (such as a Bt1 horizon). Give a field example of a diagnostic horizon that contains several genetic horizon designations.

2. Explain the relationships among a *soil individual*, a *polypedon*, a *pedon*, and a *landscape*.

3. Rearrange the following soil orders from the *least* to the *most* highly weathered: Oxisols, Alfisols, Mollisols, Entisols, and Inceptisols.

4. What is the principal soil property by which Ultisols differ from Alfisols? Inceptisols from Entisols?

5. Use the key given in Figure 3.8 to determine the soil order of a soil with the following characteristics: a spodic horizon at 30 cm depth, permafrost at 80 cm depth. Explain your choice of soil order.

6. Of the five soil-forming factors discussed in Chapter 2 (parent material, climate, organisms, topography, and time), choose *two* that have had the dominant influence on developing soil properties characterizing each of the following soil orders: Vertisols, Mollisols, Spodosols, and Oxisols.

7. To which soil order does each of the following belong: Psamments, Udolls, Argids, Udepts, Fragiudalfs, Haplustox, and Calciusterts.

8. What's in a name? Write a hypothetical soil profile description and land-use suitability interpretation for a hypothetical soil that is classified in the Aquic Argixerolls subgroup.

9. Explain why *Soil Taxonomy* is said to be a hierarchical classification system.

10. Name the soil taxonomy category and discuss the engineering implications of these soil taxonomy classes: Aquic Paleudults, Fragiudults, Haplusterts, Saprists, and Turbels.

11. Describe the kinds of information a soil mapper may use in deciding where to drill into the soil with an auger to bring up subsurface samples for study.

12. A soil mapper drew a boundary around an area in which he made six randomly located auger borings, two in soil *A*, with an argillic horizon more than 60 cm thick and strong brown in color, and the other four in soil *B*, with a somewhat lighter brown argillic horizon between 50 and 70 cm thick. Other soil properties, as well as management considerations, were similar for the two types of soils. Would the map unit delineated likely be a *soil association*, a *soil consociation*, or a *soil complex?* Explain.

13. Assume that you are planning to buy a 4-ha site on which to start a small orchard. Explain, step by step, how you could use Web Soil Survey to help determine whether the prospective site was suitable for your intended use.

14. Try to produce the map shown in Figure 3.27 by visiting http://websoilsurvey.nrcs.usda.gov/app/. Hints: The location is about 4 miles due south of the town of Bel Air, Maryland (enter these as the city and state under the "Navigate by Address" tab). Be sure to study the instructions available on the Web site.

REFERENCES

Barrera-Bassols, N., J. Alfred Zinck, and E. Van Ranst. 2006. "Symbolism, knowledge and management of soil and land resources in indigenous communities: Ethnopedology at global, regional and local scales," *Catena* **65**:118–137.

Coulombe, C. E., L. P. Wilding, and J. B. Dixon. 1996. "Overview of Vertisols: Characteristics and impacts on society," *Advances in Agronomy* **17**:289–375.

Ditzler, C. A. 2005. "Has the polypedon's time come and gone?" *HPSSS Newsletter*, February 2005, pp. 8–11. Commission on History, Philosophy and Sociology of Soil Science, International Union of Soil Sciences. www.iuss.org/Newsletter12C4-5.pdf (verified 23 February 2009).

Eswaran, H. 1993. "Assessment of global resources: Current status and future needs," *Pedologie* **43**(1):19–39.

Eswaran, H., T. Rice, R. Ahrens, and B. A. Stewart (eds.). 2003. *Soil classification: A global desk reference.* CRC Press, Boca Raton, FL.

Gong, Z., X. Zhang, J. Chen, and G. Zhang. 2003. "Origin and development of soil science in ancient China." *Geoderma* **115**:3–13.

Legros, J.-P. 2006. *Mapping of the Soil.* Science Publishers, Enfield, N. H.

PPI. 1996. "Site-specific nutrient management systems for the 1990's." Pamphlet. Potash and Phosphate Institute and Foundation for Agronomic Research, Norcross, Ga.

Riecken, F. F., and G. D. Smith. 1949. "Principal upland soils of Iowa, their occurrence and important properties," *Agron* **49** (revised). Iowa Agr. Exp. Sta.

Shaw, J. N., L. T. West, D. E. Radcliffe, and D. D. Bosch. 2000. "Preferential flow and pedotransfer functions for transport properties in sandy Kandiustults." *Soil Sci. Soc. Amer. J.* **64**:670–678.

Smith, H., and E. Matthews. 1975. *Soil Survey of Harford County Area, Maryland.* U.S. Soil Conservation Service, Washington, DC.

Soil Science Society of America. 1984. *Soil taxonomy, achievements and challenges.* SSSA Special Publication 14. Soil Sci. Soc. Amer., Madison, WI.

Soil Survey Staff. 1975. *Soil taxonomy: A basic system of soil classification for making and interpreting soil surveys.* Natural Resources Conservation Service, Washington, DC.

Soil Survey Staff. 1999. *Soil taxonomy: A basic system of soil classification for making and interpreting soil surveys,* 2nd ed. Natural Resources Conservation Service, Washington, DC.

Soil Survey Staff. 2006. *Keys to soil taxonomy.* U.S. Department of Agriculture, Natural Resources Conservation Service. http://soils.usda.gov/technical/classification/tax_keys/keysweb.pdf.

Talawar, S., and R. E. Rhoades. 1998. "Scientific and local classification and management of soils." *Agriculture and Human Values* **15**:3–14.

U.S. Department of Agriculture. 1938. *Soils and men.* USDA Yearbook. U.S. Government Printing Office, Washington, DC.

USDA-NRCS. 2002. *Official series description—Thorndale series.* National Cooperative Soil Survey. www2.ftw.nrcs.usda.gov/osd/dat/T/THORNDALE.html (posted December 2002; verified 04 October 2006).

USDA-NRCS. 2006. *National soil survey handbook,* title 430-vi. U.S. Department of Agriculture, Natural Resources Conservation Service. http://soils.usda.gov/technical/handbook/(posted 22 March 2006; verified 20 October 2006).

4

Soil Architecture and Physical Properties

And when that crop grew, and was harvested, no man had crumbled a hot clod in his fingers and let the earth sift past his fingertips.
—JOHN STEINBECK, *THE GRAPES OF WRATH*

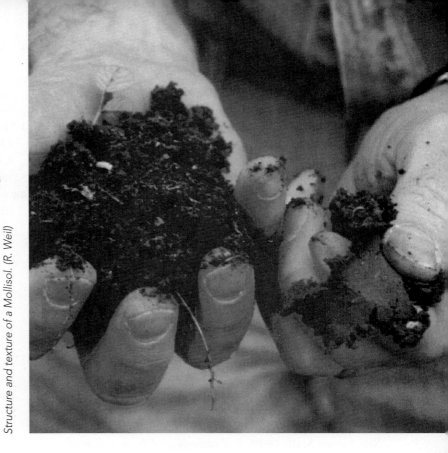

Structure and texture of a Mollisol. (R. Weil)

Soil physical properties profoundly influence how soils function in an ecosystem and how they can best be managed. Success or failure of both agricultural and engineering projects often hinges on the physical properties of the soil used. The occurrence and growth of many plant species are closely related to soil physical properties, as is the movement over and through soils of water and its dissolved nutrients and chemical pollutants.

Soil scientists use the color, texture, and other physical properties of soil horizons in classifying soil profiles and in making determinations about soil suitability for agricultural and environmental projects. Knowledge of basic soil physical properties is not only of great practical value in itself, but will also help in understanding many aspects of soils considered in later chapters.

The physical properties discussed in this chapter relate to the solid particles of the soil and the manner in which they are aggregated. If we think of the soil as a house, the primary particles in soil are the building blocks from which the house is constructed. **Soil texture** describes the sizes of the soil particles. The larger mineral particles usually are embedded in, and coated with, clay and other colloidal size materials. Where the larger mineral particles predominate, the soil is gravelly or sandy; where the mineral colloids are dominant, the soil is claylike. All gradations between these extremes are found in nature.

In building a house, the manner in which the building blocks are put together determines the nature of the walls, rooms, and passageways. Organic matter and other substances act as cement between individual particles, encouraging the formation of clumps or aggregates of soil.

Soil structure describes the manner in which soil particles are aggregated. This property, therefore, defines the nature of the system of pores and channels in a soil.

Together, soil texture and structure help determine the ability of the soil to hold and conduct the water and air necessary for sustaining life. These factors also determine how soils behave when used for highways and building foundations, or when manipulated by tillage. In fact, through their influence on the movement of water through and off soils, physical properties also exert considerable control over the destruction of the soil itself by erosion.

4.1 SOIL COLOR

We begin with soil color because color is often the most obvious characteristic of a soil. Although color itself has little effect on the behavior and use of soils, it does provide clues about other soil properties and conditions. To obtain the precise, repeatable description of colors needed for soil classification and interpretation, soil scientists compare a small piece of soil to standard color chips in special Munsell color charts. The Munsell charts use color chips arranged according to the three components of how people see color: the **hue** (in soils, usually redness or yellowness), the **value** (lightness or darkness, a value of 0 being black), and the **chroma** (intensity or brightness, a chroma of 0 being neutral gray). In a Munsell color book (carefully study Plate 22 in the color insert), color chips are arranged on pages with values increasing from the bottom to the top, and the chromas increasing from left to right, while hues change from one page to another.

Soils display a wide range of reds, browns, yellows, and even greens (see Plates 16 and 35). Some soils are nearly black, others nearly white. Some soil colors are very bright, others are dull grays. Soil colors may vary from place to place in the landscape (e.g., Plates 16 and 36) as well as with depth through the various layers (horizons) within a soil profile, or even within a single horizon or clod of soil (Plates 10, 17, and 37).

Color anarchy vs. system: www.urbanext.uiuc.edu/soil/less_pln/color/color.htm

Causes and Interpretation of Soil Colors

Three major factors influence soil colors: (1) organic matter content, (2) water content, and (3) the presence and oxidation states of iron and manganese oxides. Organic matter tends to coat mineral particles, darkening and masking the brighter colors of the minerals themselves (see Plates 8 and 32). Soils are generally darker (have low color value) when wet than when dry (Plate 20). Water content has a more profound indirect effect on soil colors. It influences the level of oxygen in the soil, and thereby the rate of organic matter accumulation which darkens the soil (Plate 39). Water also affects the oxidation state of iron and manganese (Plates 16, 17, and 21). In well-drained uplands, especially in warm climates, well-oxidized iron compounds impart bright (high chroma) reds and browns to the soil (Plates 9 and 11). Reduced iron compounds impart gray and bluish colors (low chroma) to poorly drained soil profiles (Plates 35 and 38). Under prolonged anaerobic conditions, reduced iron (which is far more soluble than oxided iron) is removed from particle coatings, often exposing the light gray colors of the underlying silicate minerals. Soil exhibiting gray colors from reduced iron and iron depletion is said to be **gleyed** (Plates 21 and 38).

4.2 SOIL TEXTURE (SIZE DISTRIBUTION OF SOIL PARTICLES)

Knowledge of the proportions of different-sized particles in a soil (i.e., the **soil texture**) is critical for understanding soil behavior and management. When investigating soils on a site, the texture of various soil horizons is often the first and most important property to determine, for one can draw many conclusions from this information. Furthermore, the texture of a soil in the field is not readily subject to change, so it is considered a basic property of a soil.

Nature of Soil Separates

Baseball mud, by Lesley Bannatyne: www.csmonitor .com/2005/1018/ p18s02-hfks.html?s5widep

Diameters of individual soil particles range over six orders of magnitude, from boulders (1 m) to submicroscopic clays ($<10^{-6}$ m). Scientists group these particles into soil **separates** according to several classification systems, as shown in Figure 4.1. The classification system established by the U.S. Department of Agriculture is used in this text. The size ranges for these separates are not purely arbitrary, but reflect major changes in how the particles behave and in the physical properties they impart to soils.

Gravels, cobbles, boulders, and other **coarse fragments** greater than 2 mm in diameter may affect the behavior of a soil, but they are not considered to be part of the **fine earth fraction** to which the term *soil texture* properly applies.

Sand   Particles smaller than 2 mm but larger than 0.05 mm are termed *sand*. Sand feels gritty between the fingers. The particles are generally visible to the naked eye and may be rounded or angular, depending on the degree of weathering and abrasion undergone. Coarse sand particles may be rock fragments containing several minerals, but most sand grains consist of a single mineral, usually quartz (SiO_2) or other primary silicate (Figure 4.2). The dominance of quartz means that the sand separate generally contains few plant nutrients.

As sand particles are relatively large, so, too, the pores between them are relatively large. These large pores cannot hold water against the pull of gravity (see Section 5.2) and so drain rapidly and promote entry of air into the soil. The relationship between particle size and **specific surface area** (the surface area for a given mass of particles) is illustrated in Figure 4.3. The large particles of sand have low specific surface areas, possess little capacity to hold water or nutrients, and do not stick together into a coherent mass (see Section 4.9). Due to the just described properties, most sandy soils are well aerated and loose but also infertile and prone to drought.

Silt   Particles smaller than 0.05 mm but larger than 0.002 mm in diameter are classified as *silt*. Although similar to sand in shape and mineral composition, individual

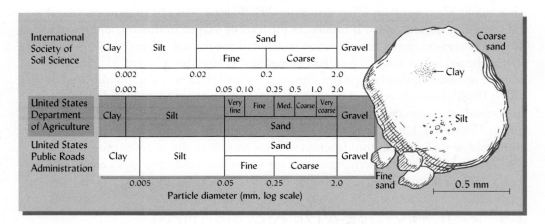

Figure 4.1

Classification of soil particles according to their size. The shaded scale in the center and the names on the drawings of particles follow the U.S. Department of Agriculture system, which is widely used throughout the world and in this book. The other two systems shown are also widely used by soil scientists and by civil engineers. The drawing illustrates the sizes of soil separates (note scale).

(Diagram courtesy of R. Weil)

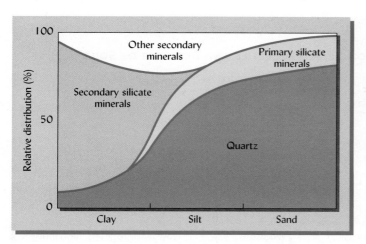

Figure 4.2

General relationship between particle size and kinds of minerals present. Quartz dominates the sand and coarse silt fractions. Primary silicates such as the feldspars, hornblende, and micas are present in the sands and, in decreasing amounts, in the silt fraction. Secondary silicates dominate the fine clay. Other secondary minerals, such as the oxides of iron and aluminum, are prominent in the fine silt and coarse clay fractions. (Diagram courtesy of N. C. Brady)

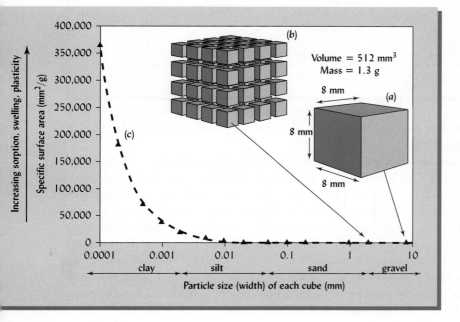

Figure 4.3

Surface area and particle size. Consider a single, gravel-sized cube 8 mm on a side and weighing 1.3 g (a). Each face has 64 mm^2 of surface area. The cube has six faces with a total of 384 mm^2 surface area (6 faces · 64 mm^2 per face) or a specific surface of 295 cm^2/g (384/1.3). If this cube were cut into smaller cubes so that each cube was only 2 mm on each side (b), then the same mass of material would now be present as 64 (4 · 4 · 4) smaller, sand-sized cubes. Each face of each small cube would have 4 mm^2 (2 mm · 2 mm) of surface area, giving 24 mm^2 of surface area for each cube (6 faces · 4 mm^2 per face). The total surface area would therefore be 1536 mm^2 (24 mm^2 per cube · 64 cubes), or a specific surface of 1182 mm^2/g (1536/1.3). This is four times as much surface area as the single large cube. (c) The specific surface area curve explains why nearly all of the adsorbing power, swelling, plasticity, heat of wetting, and other surface area related properties are associated with the clay fraction in mineral soils. (Diagram courtesy of R. Weil)

silt particles are so small as to be invisible to the unaided eye (see Figure 4.2). Rather than feel gritty when rubbed between the fingers, silt feels smooth or silky, like flour. Where silt is composed of weatherable minerals, the relatively small size (and large surface area) of the particles allows weathering rapid enough to release significant amounts of plant nutrients.

The pores between particles in silty material are much smaller (and much more numerous) than those in sand, so silt retains more water and lets less drain through. However, even when wet, silt itself does not exhibit much **stickiness** or **plasticity** (malleability). What little plasticity, cohesion, and adsorptive capacity some silt fractions exhibit is largely due to a film of adhering clay (see Figure 2.14). Because of their low stickiness and plasticity, soils high in silt and fine sand can be highly susceptible to erosion by both wind and water.

Clay Clay particles are smaller than 0.002 mm. They therefore have very large specific surface areas, giving them a tremendous capacity to adsorb water and other substances. A spoonful of clay may have a surface area the size of a football field (see Section 8.1). This large adsorptive surface causes clay particles to cohere

Table 4.1

GENERALIZED INFLUENCE OF SOIL SEPARATES ON SOME PROPERTIES AND BEHAVIOR OF SOILS[a]

Property/behavior	Rating associated with soil separates		
	Sand	Silt	Clay
Water-holding capacity	Low	Medium to high	High
Aeration	Good	Medium	Poor
Drainage rate	High	Slow to medium	Very slow
Soil organic matter level	Low	Medium to high	High to medium
Decomposition of organic matter	Rapid	Medium	Slow
Warm-up in spring	Rapid	Moderate	Slow
Compactability	Low	Medium	High
Susceptibility to wind erosion	Moderate (high if fine sand)	High	Low
Susceptibility to water erosion	Low (unless fine sand)	High	Low if aggregated, high if not
Shrink-swell potential	Very Low	Low	Moderate to very high
Sealing of ponds, dams, and landfills	Poor	Poor	Good
Suitability for tillage after rain	Good	Medium	Poor
Pollutant leaching potential	High	Medium	Low (unless cracked)
Ability to store plant nutrients	Poor	Medium to high	High
Resistance to pH change	Low	Medium	High

[a] Exceptions to these generalizations do occur, especially as a result of soil structure and clay mineralogy.

in a hard mass after drying. When wet, clay is sticky and can be easily molded (exhibits high plasticity).

Fine clay-size particles are so small that they behave as **colloids**—if suspended in water they do not readily settle out. Unlike most sand and silt particles, clay particles tend to be shaped like tiny flakes or flat platelets. The pores between clay particles are very small and convoluted, so movement of both water and air is very slow. In clayey soil the pores between particles are tiny in size, but huge in number, allowing the soil to hold a great deal of water, however much of it may be unavailable to plants (see Section 5.8). Each unique clay mineral (see Chapter 8) imparts different properties to the soils in which it is prominent. Therefore, soil properties such as shrink-swell behavior, plasticity, water-holding capacity, soil strength, and chemical adsorption depend on the *kind* of clay present as well as the *amount*.

Influence of Surface Area on Other Soil Properties

When particle size decreases, specific surface area and related properties increase greatly, as shown graphically in Figure 4.3. Fine colloidal clay has about 10,000 times as much surface area as the same weight of medium-sized sand. Soil texture influences many other soil properties in far-reaching ways (see Table 4.1) as a result of such fundamental surface phenomena as sorption of water films, chemicals, and nutrients; weathering of minerals; and growth of microbes.

4.3   SOIL TEXTURAL CLASSES

Beyond the three broad groups *sandy soils, clayey soils,* and *loamy soils,* the 12 specific **textural class** names shown in Figure 4.4 convey a more precise idea of the size distribution of particles and the general nature of soil physical properties. Most soils are some type of **loam**.

Interactive textural triangle:
http://courses.soil.ncsu.edu/
resources/physics/texture/
soiltexture.swf

Loams   The central concept of a **loam** may be defined as a mixture of sand, silt, and clay particles that exhibits the *properties* of those separates in about equal proportions.

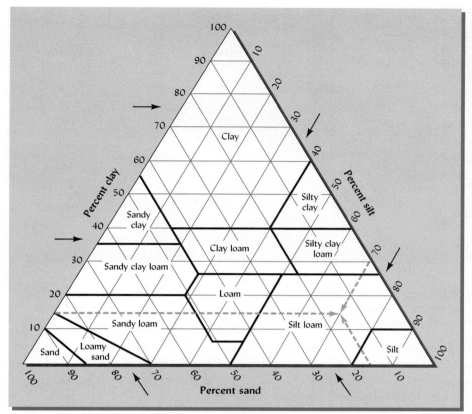

Figure 4.4
The major soil textural classes are defined by the percentages of sand, silt, and clay according to the heavy boundary lines shown on the textural triangle. To use the graph, first find the appropriate clay percentage along the left side of the triangle, then draw a horizontal line from that location across the graph. Next find the sand percentage along the base of the triangle, then draw a line inward going parallel to the triangle side labeled "Percent silt." The small arrows indicate the proper direction in which to draw the lines. The name of the compartment in which these two lines intersect indicates the textural class of the soil sample. Percentages for any two of the three soil separates is all that is required because the percentages for sand, silt, and clay add up to 100%. As an example, a soil that contains 15% sand, 15% clay, and 70% silt is indicated by the light dashed lines that intersect in the compartment labeled "Silt loam."

This definition does not mean that the three separates are present in equal *amounts* (that is why the loam class is not exactly in the middle of the triangle in Figure 4.4). This anomaly exists because a relatively small percentage of clay is required to engender clayey properties in a soil, whereas small amounts of sand and silt have a lesser influence on how a soil behaves. A loam in which sand is dominant is classified as a *sandy loam.* In the same way, some soils are classed as *silt loams, silty clay loams, sandy clay loams,* and *clay loams.* Note from Figure 4.4 that a *clay loam* may have as little as 26% clay, but to qualify as *sandy loam* or *silt loam,* a soil must have at least 45% sand or 50% silt, respectively.

Coarse Fragment Modifiers If a soil contains a significant proportion of particles larger than sand (termed **coarse fragments**), a qualifying adjective may be used as part of the textural class name. Coarse fragments that range from 2 to 75 mm along their greatest diameter are termed *gravel* or *pebbles,* those ranging from 75 to 250 mm are called *cobbles* (if round) or *channers* (if flat), and those more than 250 mm across are called *stones* or *boulders.* A *gravelly, fine sandy loam* is an example of such a modified textural class.

Alteration of Soil Textural Class

Over long periods of time, pedologic processes (see Chapter 2) such as illuviation, erosion, and mineral weathering can alter the textures of certain soil horizons. However, management practices generally do not alter the textural class of a soil on a field scale. Changing the texture of a given soil would require mixing it with another soil material of a different textural class. For example, the incorporation of large quantities of sand to change the physical properties of a clayey soil for use in greenhouse pots or for turfgrass areas would be considered to change the soil texture. However,

Figure 4.5
Altering soil texture can be a big job. Here, large amounts of sand (darker material) and gravel (whitish) have been imported and are being spread to improve the all-weather playability of an athletic field. Insets show laser-guided bulldozers evenly spreading the sand and gravel materials. (Photos courtesy of R. Weil)

adding peat or compost to a soil while mixing a potting medium does not constitute a change in texture, since this property refers only to the mineral particles. In fact the term *soil texture* is not relevant to artificial media that contain mainly perlite, peat, Styrofoam, or other nonsoil materials.

Great care must be exercised in attempting to ameliorate physical properties of fine-textured soils by adding sand. Where specifications (as for a landscape design) call for soil materials of a certain textural class, it is generally advisable to find a naturally occurring soil that meets the specification, rather than attempt to alter the textural class by mixing in sand or clay. Mixing in moderate amounts of fine sand or sand ranging widely in size may yield a product more akin to concrete than to a sandy soil. For some applications (such as golf putting greens and athletic fields; see Figure 4.5), the need for rapid drainage and resistance to compaction even when wet may justify the construction of an artificial soil from carefully selected uniform sands.

Determination of Textural Class by the "Feel" Method

Textural class determination is one of the most valuable field skills you will learn in soil science. Determining the textural class of a soil by its feel is of great practical value in soil survey, land classification, and any investigation in which soil texture may play a role.

The textural triangle (see Figure 4.4) should be kept in mind when determining the textural class by the feel method as explained in Box 4.1 and Figure 4.6.

Laboratory Particle-Size Analyses

The first and sometimes most difficult step in a particle-size analysis is the complete dispersion of a soil sample in water, so even the tiniest clumps are broken down into individual, primary particles. Separation into size groups can be accomplished by washing the suspended particles through standard sieves that are graded in size to separate the coarse fragments and sand separates, permitting the silt and clay fractions to pass through. A sedimentation procedure is usually used to determine the amounts of silt and clay. The principle involved is simple: Because soil particles are more dense than water, they tend to sink, settling at a velocity that is proportional to their *size.*

BOX 4.1
A METHOD FOR DETERMINING TEXTURE BY FEEL

The first and most critical step in the texture-by-feel method is to knead a walnut-sized sample of moist soil into a uniform puttylike consistency, slowly adding water if necessary. This step may take a few minutes, but a premature determination is likely to be in error as hard clumps of clay and silt may feel like sand grains. The soil should be moist, but not quite glistening. Try to do this with only one hand so as to keep your other hand clean for writing in a field notebook (and shaking hands with your client).

While squeezing and kneading the sample, note its malleability, stickiness, and stiffness, all properties associated with the clay content. A high silt content makes a sample feel smooth and silky, with little stickiness or resistance to deformation. A soil with a significant content of sand feels rough and gritty, and makes a grinding noise when rubbed near one's ear.

Get a feel for the amount of clay by attempting to squeeze a ball of properly moistened soil between your thumb and the side of your forefinger, making a ribbon of soil. Make the ribbon as long as possible until it breaks from its own weight (see Figure 4.6).

Interpret your observations as follows:

1. Soil will not cohere into a ball, falls apart: **sand**
2. Soil forms a ball, but will not form a ribbon: **loamy sand**
3. Soil ribbon is dull and breaks off when less than 2.5 cm long and
   a. Grinding noise is audible; grittiness is prominent feel: **sandy loam**
   b. Smooth, floury feel prominent; no grinding audible: **silt loam**
   c. Only slight grittiness and smoothness; grinding not clearly audible: **loam**
4. Soil exhibits moderate stickiness and firmness, forms ribbons 2.5 to 5 cm long, and
   a. Grinding noise is audible; grittiness is prominent feel: **sandy clay loam**
   b. Smooth, floury feel prominent; no grinding audible: **silty clay loam**
   c. Only slight grittiness and smoothness; grinding not clearly audible: **clay loam**
5. Soil exhibits dominant stickiness and firmness, forms shiny ribbons longer than 5 cm, and
   a. Grinding noise is audible; grittiness is dominant feel: **sandy clay**
   b. Smooth, floury feel prominent; no grinding audible: **silty clay**
   c. Only slight grittiness and smoothness; grinding not clearly audible: **clay**

A more precise estimate of sand content (and hence more accurate placement in the horizontal dimension of the textural class triangle) can be made by wetting a pea-sized clump of soil in the palm of your hand and smearing it around with your finger until your palm becomes coated with a souplike suspension of soil. The sand grains will stand out visibly and their volume as compared to the original "pea" can be estimated, as can their relative size (fine, medium, coarse, etc.).

It is best to learn the method using samples of known textural class. With practice, accurate textural class determinations can be made on the spot.

Figure 4.6
(*Top*) The gritty, noncohesive appearance and short ribbon of a sandy loam with about 15% clay. (*Middle*) The smooth, dull appearance and crumbly ribbon characteristic of a silt loam. (*Bottom*) The smooth, shiny appearance and long, flexible ribbon of a clay. (Photos courtesy of R. Weil)

In other words, "The bigger they are, the faster they fall." The equation that describes this relationship is referred to as *Stokes' law*:

$$V = kd^2$$

where *k* is a constant related to the force of gravity and density and viscosity of water, and *d* is the particle diameter.

4.4   STRUCTURE OF MINERAL SOILS

The term **soil structure** relates to the *arrangement* of sand, silt, clay, and organic particles in soils. The particles become aggregated together due to various forces and at different scales to form distinct structural units called **peds** or **aggregates**. When a mass of soil is excavated and gently broken apart, it tends to break into peds along natural zones of weakness. Although *aggregate* and *ped* can be used synonymously, the term *ped* is most commonly used to describe the large-scale structure evident when observing soil profiles and involving structural units which range in size from a few mm to about 1 m. At this scale, the attraction of soil particles to one another in patterns that define structural units is influenced mainly by such physical processes as freeze–thaw, wet–dry, shrink–swell, the penetration and swelling of plant roots, the burrowing of soil animals, and the activities of people and machines. Structural peds should not be confused with **clods**—the compressed, cohesive chunks of soil that can form artificially when wet soil is plowed or excavated. Most large peds are composed of, and can be broken into, smaller peds or aggregates (Figure 4.7). The network of **pores** within and between aggregates greatly influences the movement of air and water, the growth of plant roots, and the activities of soil organisms, including the accumulation and breakdown of organic matter.

Types of Soil Structure

Many types or shapes of peds occur in soils, often within different horizons of a particular soil profile. Some soils may exhibit a **single-grained** structural condition in which particles are not aggregated. At the opposite extreme, some soils (such as certain clay sediments) occur as large, cohesive masses of material and are described as

Figure 4.7

The hierarchical organization of soil structure. The larger structural units observable in a soil profile each contain many smaller units. The lower example shows how large prismatic peds typical of B horizons break down into smaller peds (and so on). The upper example illustrates how microaggregates smaller than 0.25 mm in diameter are contained within the granular macroaggregates of about 1 mm diameter that typify A horizons. The microaggregates often form around and occlude tiny particles of organic matter originally trapped in the macroaggregate. Note the two different scales for the prismatic and granular structures. (Diagram courtesy of R. Weil)

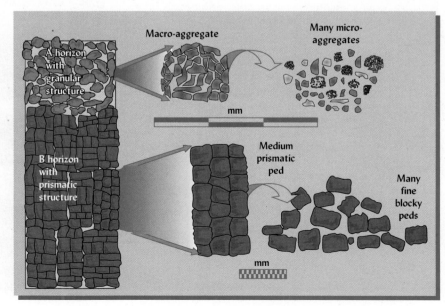

exhibiting a **massive** structural condition. However, most soils exhibit some type of aggregation and are composed of peds that can be characterized by their shape (or *type*), size, and distinctness (or *grade*). The four principal soil ped shapes are *spheroidal, platy, prismlike,* and *blocklike* (see Figure 4.8).

In describing soil structure (e.g., Table 3.3), soil scientists note not only the *type* (shape) of the structural peds present, but also the relative *size* (fine, medium, coarse) and degree of development or distinctness of the peds (*grades* such as strong, moderate, or weak). For example, the soil shown in Figure 4.8*d* might be described as having "weak, fine, subangular blocky structure." Generally, the structure of a soil is easier to observe when the soil is relatively dry. When wet, structural peds may swell and press closer together, making the individual peds less well defined. We will now turn our attention to the formation and stabilization of the granular aggregates that characterize surface horizons.

4.5 FORMATION AND STABILIZATION OF SOIL AGGREGATES

The granular aggregation of surface soils is a highly dynamic soil property. Some aggregates disintegrate and others form anew as soil conditions change. Generally, smaller aggregates are more stable than larger ones, so maintaining the much-prized larger aggregates requires great care. We will discuss practical means of managing soil structure and avoiding **soil compaction** after we consider the factors responsible for aggregate formation and stabilization.

Hierarchical Organization of Soil Aggregates[1]

Surface horizons are usually characterized by roundish granular structure that exhibits a hierarchy in which relatively large **macroaggregates** (0.25 to 5 mm in diameter) are comprised of smaller **microaggregates** (2 to 250 µm). The latter, in turn, are composed of tiny packets of clay and organic matter only a few µm in size. You may easily demonstrate the existence of this *hierarchy of aggregation* by selecting a few of the largest aggregates in a soil and gently crumbling them into many smaller-sized pieces. You will find that even the smallest specks of soil usually are not individual particles but can be rubbed into a smear of still smaller particles of silt, clay, and humus. At each level in the hierarchy of aggregates, different factors are responsible for binding together the subunits (Figure 4.9).

Processes Influencing Aggregate Formation and Stability in Soils

Both biological and physical–chemical (abiotic) processes are involved in the formation of soil aggregates. Physical–chemical processes tend to be most important at the smaller end of the scale, biological processes at the larger end. The physical–chemical processes of aggregate formation are associated mainly with clays and, hence, tend to be of greater importance in finer-textured soils. In sandy soils that have little clay, aggregation is almost entirely dependent on biological processes.

Most important among the physical–chemical processes are (1) flocculation, the mutual attraction among clay and organic molecules; and (2) the swelling and shrinking of clay masses.

---

[1]The hierarchical organization of soil aggregates and the role of soil organic matter in its formation were first put forward by Tisdall and Oades (1982). For a review of advances in this area during the two decades following, see Six et al. (2004).

Spheroidal

Characteristic of surface (A) horizons. Subject to wide and rapid changes.

Granular (porous)

Crumb (very porous)

Plate-like

Common in E horizons, may occur in any part of the profile. Often inherited from parent material of soil, or caused by compaction.

Block-like

Common in B horizons, particularly in humid regions. May occur in A horizons.

Angular blocky

Subangular blocky

Prism-like

Usually found in B horizons. Most common in soils of arid and semi-arid regions.

Columnar (rounded tops)

Prismatic (flat, angular tops)

Figure 4.8

The various structure types (shapes) found in mineral soils. Their typical location is suggested. The drawings illustrate their essential features and the photos indicate how they look in situ. For scale, note the 15-cm-long pencil in (e) and the 3-cm-wide knife blade in (d) and (f). [Photo (e) courtesy of J. L. Arndt, North Dakota State University; others courtesy of R. Weil]

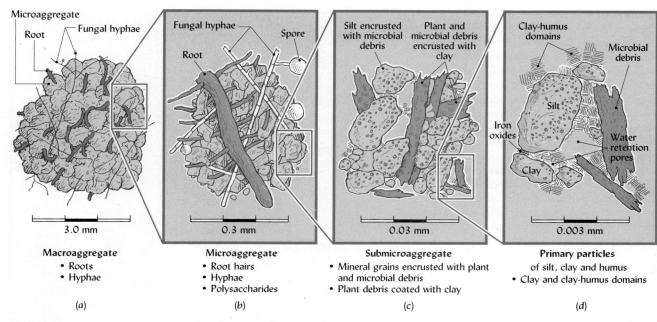

Figure 4.9

Larger aggregates are often composed of an agglomeration of smaller aggregates. This illustration shows four levels in this hierarchy of soil aggregates. The different factors important for aggregation at each level are indicated. (a) A macroaggregate composed of many microaggregates bound together mainly by a kind of sticky network formed from fungal hyphae and fine roots. (b) A microaggregate consisting mainly of fine sand grains and smaller clumps of silt grains, clay, and organic debris bound together by root hairs, fungal hyphae, and microbial gums. (c) A very small submicroaggregate consisting of fine silt particles encrusted with organic debris and tiny bits of plant and microbial debris (called particulate organic matter) *encrusted with even smaller packets of clay, humus, and Fe or Al oxides. (d) Clusters of parallel and random clay platelets interacting with Fe or Al oxides and organic polymers at the smallest scale. These organoclay clusters or domains bind to the surfaces of humus particles and the smallest of mineral grains.* (Diagram courtesy of R. Weil)

Flocculation of Clays and the Role of Adsorbed Cations   Except in very sandy soils that are almost devoid of clay, aggregation begins with the **flocculation** of clay particles into microscopic clumps, or *floccules* (Figure 4.10). If two clay platelets come close enough to each other, positively charged ions compressed in a layer between them will attract the negative charges on both platelets, thus serving as bridges to hold the platelets together. These processes lead to the formation of a small "stack" of parallel clay platelets, termed a *clay domain*. Polyvalent cations (e.g., Ca^{2+}, Fe^{2+}, Al^{3+}) also complex with hydrophobic humus molecules, allowing them to bind to clay surfaces, forming domains that are more random in orientation, resembling a house of cards. Clay/humus domains form bridges that bind to each other and to fine silt particles (mainly quartz), creating the smallest size groupings in the hierarchy of soil aggregates (Figure 4.9*d*). These domains, aided by the flocculating influence of polyvalent cations and humus, provide much of the long-term stability for the smaller (<0.25 mm) microaggregates. In certain highly weathered clayey soils (Ultisols and Oxisols), the cementing action of iron oxides and other inorganic compounds produces very stable small aggregates called **pseudosand**.

When monovalent cations, especially Na^+ (rather than polyvalent cations such as Ca^{2+} or Al^{3+}) are prominent, as in some soils of arid and semiarid areas, the attractive forces are not able to overcome the natural repulsion of one negatively charged clay platelet by another (Figure 4.10). The clay platelets cannot approach closely enough to flocculate, so remain in a dispersed, gel-like condition that causes the soil

Figure 4.10
The role of cations on the flocculation of clays. (a) Such di- and trivalent cations as Ca^{2+}, Fe^{3+}, and Al^{3+} are tightly adsorbed and can effectively neutralize the negative surface charges on the clay particles. These cations can also form bridges that bring clay particles together. (b) Monovalent ions, especially Na^+, with relatively large hydrated radii can cause clay particles to repel each other and create a dispersed condition. Three things contribute to such dispersion: (1) the large hydrated Na^+ ion does not get close enough to the clay to effectively neutralize the negative charges, (2) the single charge on Na^+ is not effective in forming a bridge between clay particles, and (3) compared to di- or trivalent ions, 2 or 3 times as many monovalent ions must crowd between clay particles in order to neutralize the charges on the clay surfaces. (Diagram courtesy of R. Weil)

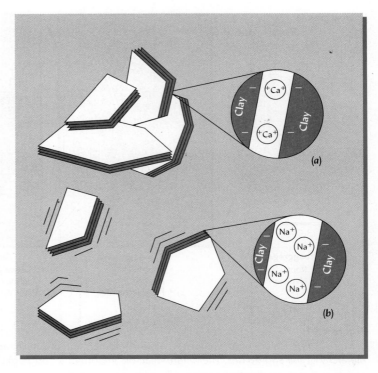

to become almost structureless, impervious to water and air, and very undesirable from the standpoint of plant growth (see Section 9.15).

Volume Changes in Clayey Materials   As a soil dries out and water is withdrawn, the platelets in clay domains move closer together, causing the domains and, hence, the soil mass to shrink in volume. As a soil mass shrinks, cracks will open up along zones of weakness. Over the course of many cycles (as occur between rain or irrigation events in the field) the network of cracks becomes better defined. Plant roots also have a distinct drying effect as they take up soil moisture in their immediate vicinity. Water uptake, especially by fibrous-rooted perennial grasses, accentuates the physical aggregation processes associated with wetting and drying. This effect is but one of many ways in which physical and biological soil processes interact.

Freezing and thawing cycles have a similar effect, since the formation of ice crystals is a drying process that also draws water out of clay domains. The swelling and shrinking actions that accompany freeze–thaw and wet–dry cycles in soils create fissures and pressures that alternately break apart large soil masses and compress soil particles into defined structural peds.

Glomalin: Where the carbon is?
www.ars.usda.gov/is/AR/
archive/sep02/soil0902.htm

Activities of Soil Organisms   Among the biological processes of aggregation, the most prominent are (1) the burrowing and molding activities of soil animals, (2) the enmeshment of particles by sticky networks of roots and fungal hyphae, and (3) the production of organic glues by microorganisms, especially bacteria and fungi. Earthworms (and termites) move soil particles about, often ingesting them and forming them into pellets or casts (see Chapter 11). Plant roots (particularly root hairs) and fungal hyphae exude sugarlike polysaccharides and other organic compounds, forming sticky networks that bind together individual soil particles and tiny microaggregates into larger macroaggregates (see Figure 4.9*a*). The threadlike fungi that associate with plant roots (called *mycorrhizae*; see Section 10.9)

Figure 4.11

The aggregates of soils high in organic matter are much more stable than are those low in this constituent. The low-organic-matter soil aggregates fall apart when they are wetted; those high in organic matter maintain their stability. (Photo courtesy of N. C. Brady)

produce a sticky sugar-protein called **glomalin**, which is thought to be an effective cementing agent. Bacteria also produce polysaccharides and other organic glues as they decompose plant residues.

Influence of Organic Matter   In most temperate zone soils, the formation and stabilization of granular aggregates is primarily influenced by soil organic matter (see Figure 4.11). Organic matter provides the energy substrate that makes possible the previously mentioned biological activities. During the aggregation process, soil mineral particles (silts and fine sands) become coated and encrusted with bits of decomposed plant residue and other organic materials. Complex organic polymers resulting from decay chemically interact with particles of silicate clays and iron and aluminum oxides to form bridges between individual soil particles, thereby binding them together in water-stable aggregates (see Figure 4.9*d*).

Influence of Tillage   Tillage can both promote and destroy aggregation. If the soil is not too wet or too dry, tillage can break large clods into natural aggregates, creating a temporarily loose, porous condition conducive to the easy growth of young roots and the emergence of tender seedlings. Tillage can also incorporate organic amendments into the soil and kill weeds.

Over longer periods, however, tillage greatly hastens the oxidative loss of soil organic matter from surface horizons, thus weakening soil aggregates. Tillage operations, especially if carried out when the soil is wet, also tend to crush or smear soil aggregates, resulting in loss of macroporosity and the creation of a ***puddled*** condition (see Figure 4.12).

Adventures in the rhizosphere, by Alex Blumberg: http://chicagowildernessmag .org/issues/spring1999/ underground.html

Compaction and its control: www.extension.umn.edu/ distribution/cropsystems/ DC3115.html

Figure 4.12

Puddled soil (left) and well-granulated soil (right). Plant roots and especially humus play the major role in soil granulation. Thus a sod tends to encourage development of a granular structure in the surface horizon of cultivated land. (Courtesy USDA Natural Resources Conservation Service)

4.6   TILLAGE AND STRUCTURAL MANAGEMENT OF SOILS

Tillage implements and what they do: www.WTAMU.EDU/ ~crobinson/TILLAGE/tillage .htm

When protected under dense vegetation and undisturbed by tillage, most soils (except perhaps some sparsely vegetated soils in arid regions) possess a surface structure sufficiently stable to allow rapid infiltration of water and to prevent crusting. However, for the manager of cultivated soils, the development and maintenance of stable surface soil structure and associated pore networks is a major challenge. Many studies have shown that aggregation and associated desirable soil properties such as water infiltration rate decline under long periods of tilled row-crop cultivation.

Tillage and Soil Tilth

Simply defined, **tilth** refers to the physical condition of the soil in relation to plant growth. Tilth depends not only on aggregate formation and stability, but also on such factors as bulk density (see Section 4.7), soil moisture content, degree of aeration, rate of water infiltration, drainage, and capillary water capacity. As might be expected, tilth often changes rapidly and markedly. For instance, the workability of fine-textured soils may be altered abruptly by a slight change in moisture.

A major aspect of tilth is soil **friability** (see also Section 4.9). Soils are said to be **friable** if their clods are not sticky or hard, but rather crumble easily, revealing their constituent aggregates. Generally, **soil friability** is enhanced when the **tensile strength** of individual aggregates (i.e., the force required to pull them apart) is relatively high compared to the tensile strength of the clods. This condition allows tillage or excavation forces to easily break down the large clods, while the resulting aggregates remain stable. As might be expected, friability can be markedly affected by changes in soil water content, especially for fine-textured soils. Each soil typically has an optimum water content for greatest friability.

Clayey soils are especially prone to puddling and compaction because of their high plasticity and cohesion. When puddled clayey soils dry, they usually become dense and hard. Proper timing of trafficking is more difficult for clayey than for sandy soils, because the former take much longer to dry to a suitable moisture content and may also become too dry to work easily. Increased soil organic matter content usually enhances soil friability and can partially alleviate the susceptibility of a clay soil to structural damage during tillage and traffic (Figure 4.13).

Figure 4.13

Sensitivity of fine-textured soils to mechanical damage by tillage is increased by wetness and decreased by soil organic matter. The sensitivity index on the vertical axis is based mainly on the tendency for clay dispersion that is associated with soil smearing and aggregate destruction. The soil is especially susceptible to this damage when excessively wet (wetter than the condition termed field capacity, as defined in Section 5.8). It is also evident that a high level of soil organic matter helps protect the aggregated clay during tillage. The relationships shown apply to clayey soils and would be much less extreme in sandy soils. [Graph based on data in Watts and Dexter (1997)]

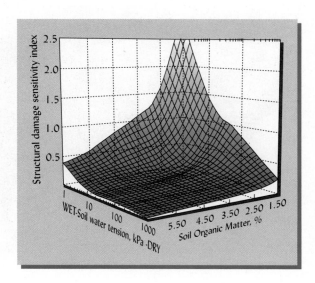

Some clayey soils of humid tropical regions are much more easily managed than those just described. The clay fraction of these soils is dominated by hydrous oxides of iron and aluminum, which are not as sticky, plastic, and difficult to work. These soils may have very favorable physical properties, since they hold large amounts of water but have such stable aggregates that they respond to tillage after rainfall much like sandy soils.

Farmers in temperate regions typically find their soils too wet for tillage just prior to planting time (early spring), while farmers in tropical regions may face the opposite problem of soils too dry for easy tillage just prior to planting (end of dry season). In tropical and subtropical regions with a long dry season, soil often must be tilled in a very dry state in order to prepare the land for planting with the onset of the first rains.

Conventional Tillage and Crop Production

Since the Middle Ages, the moldboard plow has been the primary tillage implement most used in the Western world.[2] Its purpose is to lift, twist, and invert the soil while incorporating crop residues and animal wastes into the plow layer (Figure 4.14). The moldboard plow is often supplemented by the disk plow, which is used to cut up residues and partially incorporate them into the soil. In conventional practice, such primary tillage is followed by a number of secondary tillage operations, such as harrowing to kill weeds and to break up clods, thereby preparing a suitable seedbed.

After the crop is planted, the soil may receive further secondary tillage to control weeds and to break up crusting of the immediate soil surface. Whether powered by humans, draft animals, or tractors, all these passes over the field can cause considerable compaction.

Conservation Tillage and Soil Tilth

In recent years, agricultural land-management systems have been developed that minimize the need for soil tillage. Since these systems also leave considerable plant residues on or near the soil surface, they protect the soil from erosion (see Section 14.6 for a detailed discussion). For this reason, the tillage practices followed in these systems are

Conservation tillage in vegetable production, by Mary Peet: www.cals.ncsu .edu/sustainable/peet/ tillage/c03tilla.html

Figure 4.14
While the action of the mold-board plow lifts, turns, and loosens the upper 15 to 20 cm of soil (the furrow slice), the counterbalancing downward force compacts the next lower layer of soil. This compacted zone can develop into a plowpan. Compactive action can be understood by imagining that you are lifting a heavy weight—as you lift the weight your feet press down on the floor below. (Photo courtesy of R. Weil)

---

[2] For an early but still valuable critique of the moldboard plow, see Faulkner (1943).

called *conservation tillage*. The U.S. Department of Agriculture defines *conservation tillage* as that which leaves at least 30% of the soil surface covered by residues. Under a system of **no-till**, one crop is planted in the residue of another, with virtually no tillage. Other minimum-tillage systems such as chisel plowing permit some stirring of the soil, but still leave a high proportion of the crop residues on the surface. These organic residues protect the soil from the beating action of raindrops and the abrasive action of the wind, thereby reducing water and wind erosion and maintaining soil structure.

Soil Crusting

Falling drops of water during heavy rains or sprinkler irrigation beat apart the aggregates exposed at the soil surface. In some soils the dilution of salts by this water stimulates the dispersion of clays. Once the aggregates are broken down, small particles and dispersed clay tend to wash into and clog the soil pores. Soon the soil surface is covered with a thin layer of fine, structureless material called a **surface seal**. The surface seal inhibits water infiltration and increases erosion losses.

As the surface seal dries, it forms a hard **crust** (Figure 4.15). In arid and semiarid regions, soil sealing and crusting can have disastrous consequences because high runoff losses leave little water available to support plant growth. Crusting can be minimized by keeping some vegetative or mulch cover on the land to reduce the impact of raindrops.

Soil Conditioners

Gypsum (calcium sulfate) can improve the physical condition of many types of soil. The more soluble gypsum products provide enough electrolytes (cations and anions) to promote flocculation and inhibit the dispersion of aggregates, thus preventing surface crusting. Replacement of sodium on clay surfaces by the calcium from gypsum may also promote flocculation. Field trials have shown that gypsum-treated soils permit greater water infiltration and are less subject to erosion than untreated soils. Similarly, gypsum can reduce the strength of hard subsurface layers, thereby allowing greater root penetration and subsequent plant uptake of water from the subsoil.

(c)

Figure 4.15

Scanning electron micrographs of the upper 1 mm of a soil with stable aggregation (a) compared to one with unstable aggregates (b). Note that the aggregates in the immediate surface have been destroyed and a surface crust has formed. The bean seedling (c) must break the soil crust as it emerges from the seedbed. [Photos (a) and (b) from O'Nofiok and Singer (1984), used with permission of Soil Science Society of America; photo (c) courtesy of R. Weil]

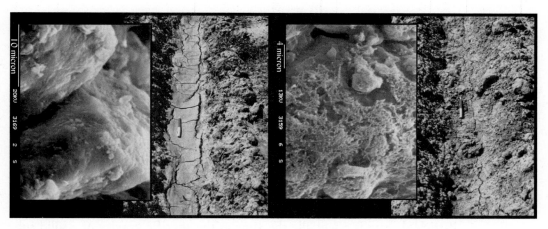

Figure 4.16
The dramatic stabilizing effect of synthetic polyacrylamide (PAM), which was used at rates of 1 to 2 kg/ha per irrigation in the furrow shown at right, but not in the furrow on the left (note the marker pens for scale). Scanning electron micrographs (scale on left edge) reveal the structure of the respective soils. Note the meshlike network of PAM enveloping the treated soil particles. [Photos courtesy of C. Ross from Ross et al. (2003)]

Certain synthetic organic polymers can stabilize soil structure in much the same way as do natural organic polymers such as bacterial polysaccharides. For example, polyacrylamide (PAM) is effective in stabilizing surface aggregates when applied at rates as low as 1 to 15 mg/L of irrigation water or sprayed on at rates as low as 1 to 4 kg/ha (Figure 4.16). Combining the use of PAM and gypsum can nearly eliminate irrigation-induced erosion.

Several species of algae that live near the soil surface are known to produce quite effective aggregate-stabilizing compounds. Application of small quantities of commercial preparations containing such algae may bring about a significant improvement in surface soil structure. The amount of amendment required is very small because the algae, once established in the soil, can multiply.

Various humic materials are marketed for their soil conditioning effects when incorporated at low rates (<500 kg/ha). However, carefully conducted research at many universities has failed to show that these materials have significantly affected aggregate stability or crop yield, as claimed.

General Guidelines for Managing Soil Tilth

The following guidelines are often relevant to managing soil tilth:

1. Minimizing tillage, especially moldboard plowing, disk harrowing, or rototilling, reduces the loss of aggregate-stabilizing organic matter.
2. Trafficking when the soil is dry and tilling when soil moisture is optimum will minimize destruction of soil structure.
3. Mulching with plant litter adds organic matter, encourages earthworm activity, and protects aggregates from beating rain and sunlight.
4. Adding organic materials can stimulate microbial supply of decomposition products that stabilize soil aggregates.
5. Sod crops in the rotation favor stable aggregation by maintaining soil organic matter, providing fine plant roots, and ensuring a period without tillage.
6. Cover crops and green manure crops can provide another good source of root action and organic matter for structural management.
7. Applying gypsum (or calcareous limestone if the soil is acidic) by itself or in combination with synthetic polymers can stabilizing surface aggregates.

The maintenance of a high degree of aggregation is one of the most important goals of soil management. This is so largely because the ability of soil to perform needed ecosystem functions is greatly influenced by soil porosity and density, properties which we shall now consider.

4.7   SOIL DENSITY AND COMPACTION

Particle Density

Soil **particle density**, D_p, is defined as the mass per unit volume of soil *solids* (in contrast to the volume of the *soil*, which would also include spaces between particles). Thus, if 1 cubic meter (m^3) of soil solids weighs 2.6 megagrams (Mg), the particle density is 2.6 Mg/m^3 (which can also be expressed as 2.6 g/cm^3).

Particle density is essentially the same as the **specific gravity** of a solid substance. The chemical composition and crystal structure of a mineral determines its particle density. Particle density is *not* affected by pore space, and therefore is not related to particle size or to the arrangement of particles (soil structure).

Particle densities for most mineral soils vary between the narrow limits of 2.60 to 2.75 Mg/m^3 because quartz, feldspar, micas, and the colloidal silicates that usually make up the major portion of mineral soils all have densities within this range. For general calculations concerning arable mineral surface soils (1 to 5% organic matter), a particle density of about 2.65 Mg/m^3 may be assumed if the actual particle density is not known. This number would be adjusted upward to 3.0 Mg/m^3 or higher when large amounts of high-density minerals such as magnetite, garnet, epidote, zircon, tourmaline, or hornblende are present. Likewise, it would be reduced for soils known to be high in organic matter, which has a particle density of only 0.9 to 1.4 Mg/m^3.

Bulk Density

A second important mass measurement of soils is **bulk density**, D_b, which is defined as the mass of a unit volume of dry soil. This volume includes both solids and pores. A careful study of Figure 4.17 should make clear the distinction between particle and

Figure 4.17

Bulk density, D_b, and particle density, D_p, of soil. Bulk density is the weight of the solid particles in a standard volume of field soil (solids plus pore space occupied by air and water). Particle density is the weight of solid particles in a standard volume of those solid particles. Follow the calculations through carefully and the terminology should be clear. In this particular case the bulk density is one-half the particle density, and the percent pore space is 50.

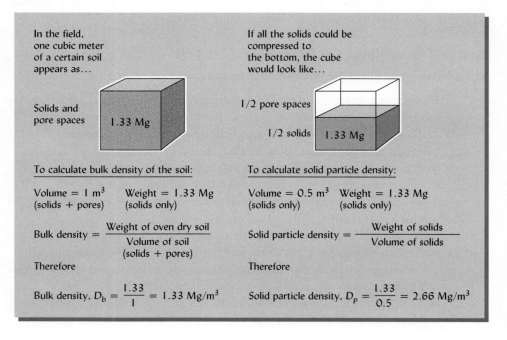

In the field, one cubic meter of a certain soil appears as...

Solids and pore spaces — 1.33 Mg

To calculate bulk density of the soil:

Volume = 1 m^3 Weight = 1.33 Mg
(solids + pores)   (solids only)

$$\text{Bulk density} = \frac{\text{Weight of oven dry soil}}{\text{Volume of soil (solids + pores)}}$$

Therefore

$$\text{Bulk density, } D_b = \frac{1.33}{1} = 1.33 \text{ Mg/m}^3$$

If all the solids could be compressed to the bottom, the cube would look like...

1/2 pore spaces

1/2 solids — 1.33 Mg

To calculate solid particle density:

Volume = 0.5 m^3 Weight = 1.33 Mg
(solids only)   (solids only)

$$\text{Solid particle density} = \frac{\text{Weight of solids}}{\text{Volume of solids}}$$

Therefore

$$\text{Solid particle density, } D_p = \frac{1.33}{0.5} = 2.66 \text{ Mg/m}^3$$

bulk density. Both expressions of density use only the mass of the solids in a soil; therefore, any water present is excluded from consideration.

There are several methods of determining soil bulk density by obtaining a known volume of soil, drying it to remove the water, and weighing the dry mass. A special coring instrument can obtain a sample of known volume without disturbing the natural soil structure. For surface soils, perhaps the simplest method is to dig a small hole, dry and weigh all the excavated soil, and then determine the soil volume by lining the hole with plastic film and filling it completely with a measured volume of water. This method is well adapted to stony soils in which it is difficult to use a core sampler.

Animation of particle vs. bulk density: www.landfood.ubc.ca/ soil200/components/ mineral.htm#114

Factors Affecting Bulk Density

Soils with a high proportion of pore space to solids have lower bulk densities than those that are more compact and have less pore space. Consequently, any factor that influences soil pore space will affect bulk density. Typical ranges of bulk density for various soil materials and conditions are illustrated in Figure 4.18. It would be worthwhile to study this figure until you have a good feel for these ranges of bulk density.

Effect of Soil Texture   As illustrated in Figure 4.18, fine-textured soils such as silt loams, clays, and clay loams generally have lower bulk densities than do sandy soils. This is true because the solid particles of the fine-textured soils tend to be organized in porous granules, especially if adequate organic matter is present. In these aggregated soils, pores exist both between *and* within the granules. This condition ensures high total pore space and a low bulk density. In sandy soil, however, organic matter contents generally are low, the solid particles are less likely to be aggregated, and the bulk densities are commonly higher than in the finer-textured soils. Similar amounts of large pores are present in both sandy and well-aggregated fine-textured soils, but sandy soils have few of the fine, within-ped pores, and so have less total porosity (Figure 4.19).

Depth in Soil Profile   Deeper in the soil profile, bulk densities are generally higher, probably as a result of lower organic matter contents, less aggregation, fewer roots and other soil-dwelling organisms, and compaction caused by the

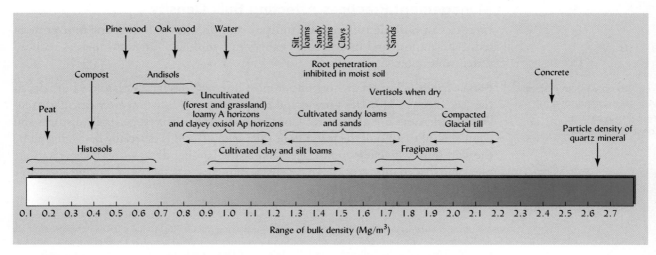

Figure 4.18
Bulk densities typical of a variety of soils and soil materials.

Figure 4.19
A schematic comparison of sandy and clayey soils showing the relative amounts of large (macro-) pores and small (micro-) pores in each. There is less total pore space in the sandy soils than in the clayey one because the clayey soil contains a large number of fine pores within each aggregate (a), but the sand particles (b), while similar in size to the clayey aggregates, are solid and contain no pore spaces within them. This is the reason why, among surface soils, those with coarse texture are usually more dense than those with finer textures. (Diagram courtesy of R. Weil)

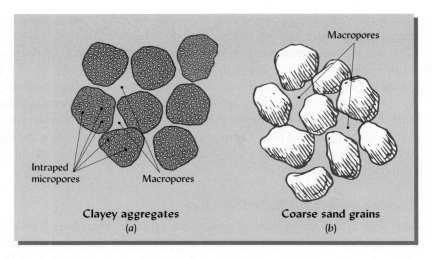

weight of the overlying layers. Very compact subsoils may have bulk densities of 2.0 Mg/m^3 or even greater. Many soils formed from till (see Section 2.3) have extremely dense subsoils as a result of past compaction by the enormous mass of glacial ice.

Useful Density Figures

Bulk densities—check "Earth" and "Sand": www.asiinstr.com/technical/ Material_Bulk_Density_ Chart_A.htm

For engineers involved with moving soil during construction, or for landscapers bringing in topsoil by the truckload, a knowledge of the bulk density of various soils is useful in estimating the weight of soil to be moved. A typical medium-textured mineral soil might have a bulk density of 1.25 Mg/m^3, or 1250 kilograms in a cubic meter.[3] A typical "half-ton" (1000 lb or 454 kg) load capacity pickup truck could carry less than 0.4 m^3 of this soil, even though the truck bed has room for about five times this volume of material.

The mass of soil in 1 ha to a depth of normal plowing (15 cm) can be calculated from soil bulk density. If we assume a bulk density of 1.3 Mg/m^3 for a typical arable surface soil, such a hectare-furrow slice 15 cm deep weighs about 2 million kg.[4]

Management Practices Affecting Bulk Density

Increases in bulk density usually indicate a poorer environment for root growth, reduced aeration, and undesirable changes in hydrologic function, such as reduced water infiltration.

Trees on construction sites: www.treesaregood.com/ treecare/avoiding_ construction.aspx

Forest Lands   The surface horizons of most forested soils have rather low bulk densities (see Figure 4.18). Tree growth and forest ecosystem function are particularly sensitive to increases in bulk density. Conventional timber harvest generally disturbs and compacts 20 to 40% of the forest floor (Figure 4.20) and is especially damaging along the skid trails where logs are dragged and at the landing decks—areas where logs are piled and loaded onto trucks. An expensive, but effective, means of moving logs while

---

[3] Most commercial landscapers and engineers in the United States still use English units. To convert values of bulk density given in units of Mg/m^3 into values of lb/yd^3, multiply by 1686. Therefore, 1 yd^3 of a typical medium-textured mineral soil with a bulk density of 1.25 Mg/m^3 would weigh over a ton (1686 × 1.25 = 2108 lb/yd^3).

[4] 10,000 m^2/ha × 1.3 Mg/m^3 × 0.15 m = 1950 Mg/ha, or about 2 million kg per ha to a depth of 15 cm. A comparable figure in the English system is 2 million lb per acre–furrow slice 6 to 7 in. deep.

Figure 4.20

Timber harvest with a conventional rubber-tired skidder in a boreal forest in western Alberta, Canada. Such practices cause significant soil compaction that can impair soil ecosystem functions for many years. Timber harvest practices that can reduce such damage to forest soils include selective cutting, use of flexible-track vehicles and overhead cable transport of logs, and abstaining from harvest during wet conditions. (Photo courtesy of Andrei Startsev, Alberta Environmental Center)

minimizing compactive degradation of forest lands is the use of cables strung between towers or hung from large balloons.

Intensive use of soils for access roads, trails, and campsites in forests and other areas with natural vegetation can also lead to increased bulk densities (Figure 4.21). An important consequence of increased bulk density is a diminished capacity of the soil to take in water, hence increased losses by surface runoff. Damage from hikers can be minimized by restricting foot traffic to well-designed, established trails that may include a thick layer of wood chips, or even a raised boardwalk in the case of heavily traveled paths over very fragile soils, such as in wetlands.

Figure 4.21

Impact of campers on the bulk density of forest soils, and the consequent effects on rainwater infiltration rates and runoff losses (see white arrows). At most campsites the high-impact area extends for about 10 m from the fire circle or tent pad. Managers of recreational land must carefully consider how to protect sensitive soils from compaction that may lead to death of vegetation and increased erosion. [Data from Vimmerstadt et al. (1982)]

Cherry trees, loved to death, by Adrian Higgins, *Washington Post*: www.washingtonpost.com/ wp-dyn/articles/A30233- 2005Apr6.html

Urban Soils   While it is usually not practical to modify the entire root zone of a tree planted in compacted urban soil, several practices can help (see also Section 7.6): (1) a planting hole that is as large as possible; (2) a thick layer of mulch spread out to the drip line (but not too near the trunk) to enhance root growth, at least near the surface; (3) a series of narrow trenches radiating out from the planting hole and back-filled with loose, enriched soil.

In some urban settings, it may be desirable to create an "artificial soil" that includes a skeleton of coarse angular gravel to provide strength and stability, and a mixture of loam-textured topsoil and organic matter to provide nutrient- and water-holding capacities. Also, large quantities of sand and organic materials are sometimes mixed into the upper few centimeters of a fine-textured soil on which putting green turf grass is to be grown.

Green Roofs   In the design of rooftop gardens, the mass of soil involved must be minimized. One might choose to grow only such shallow-rooted plants as sedums or turfgrasses so that a relatively thin layer of soil (say, 15 cm) could be used, keeping the total mass of soil from being too great. It may also be possible to reduce the cost of construction by selecting a natural soil having a relatively low bulk density, such as some well aggregated loams or peat soils. Often an artificial growing medium is created from such light-weight materials as perlite and peat. However, if anchoring trees and other plants is an important function of the soil in such an installation, very low-density materials would not be suitable. When very low-density materials are used, they may require a surface netting system to prevent wind from blowing them off the roof.

Agricultural Land   Although tillage may temporarily loosen the surface soil, in the long term intense tillage increases soil bulk density because it depletes soil organic matter and weakens soil structure (Table 4.2). The effect of cultivation can be minimized by adding crop residues or farm manure in large amounts and rotating cultivated crops with a grass sod.

Use of moldboard plows and disk harrows or repeated trips over the field by heavy machinery can form **plow pans**, or **traffic pans**, dense zones immediately below the plowed layer (Figure 4.22). Tillage, grazing, and off-road trafficking often compact soils quite deeply into the subsoil. It may take many years of restorative management for the subsoil to recover its natural degree of porosity and friability.

Large chisel-type plows (Figure 4.23) can be used in **subsoiling** to break up dense subsoil layers, thereby permitting root penetration (Figure 4.22). However, in some soils, the effects of subsoiling are quite temporary. Any tillage tends to reduce

Table 4.2
BULK DENSITY AND PORE SPACE OF SURFACE SOILS FROM CULTIVATED AND NEARBY UNCULTIVATED AREAS

Soil	Texture	Years cropped	Bulk density, Mg/m³		Pore space, %	
			Cultivated soil	Uncultivated soil	Cultivated soil	Uncultivated soil
Mean of 2 Udults (Maryland)	Sandy loam	50+	1.59	0.84	40.0	66.4
Mean of 2 Udults (Maryland)	Silt loam	50+	1.18	0.78	55.5	68.8
Mean of 3 Ustalfs (Zimbabwe)	Clay	20–50	1.44	1.20	54.1	62.6
Mean of 3 Ustalfs (Zimbabwe)	Sandy loam	20–50	1.54	1.43	42.9	47.2

For Maryland soils from Lucas and Weil (unpublished) and for Zimbabwe soils from Weil (unpublished).

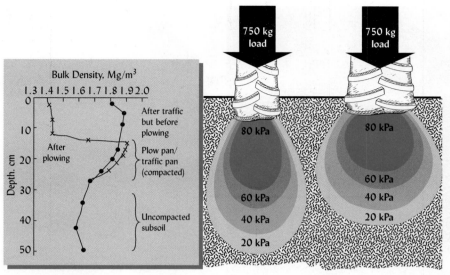

Figure 4.22
Vehicle tires compact soil to considerable depths. (Left) Representative bulk densities associated with traffic compaction on a sandy loam soil. Plowing can temporarily loosen the compacted surface soil (plow layer), but usually increases compaction just below the plow layer. (Right) Vehicle tires (750 kg load per tire) compact soil to about 50 cm. The more narrow the tire, the deeper it sinks and the deeper its compactive effect. The tire diagram shows the compactive pressure in kPa. For tire designs that reduce compaction, see Tijink and van der Linden (2000). (Diagrams courtesy of R. Weil)

soil strength, thus making the soil less resistant to subsequent compaction. To prevent compaction, which can result in yield reductions and loss of profitability, the number of tillage operations and heavy equipment trips over the field should be minimized and timed to avoid periods when the soil is wet.

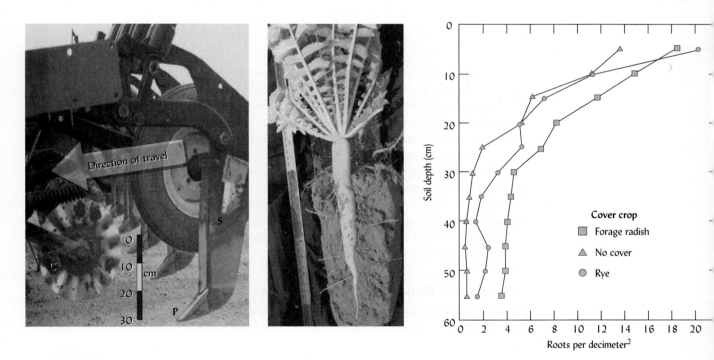

Figure 4.23
Alleviation of subsoil compaction. (Left) A heavy chisel plow, also known as a subsoiler or ripper, is pulled through the soil toward the left with the chisel point (P) about 40 cm deep. If the soil is relatively dry, the subsoiler will cause a network of cracks that enhance water, air, and root movement. However, subsoiling is a slow, energy-intensive operation, the benefits usually last for only a year or two, and the operation usually disturbs the soil surface, leaving it more susceptible to erosion. An alternative approach involves growing taprooted plants (such as the forage radish, Center) in the fall and spring, when the subsoil is relatively wet and easily penetrated by roots. The taproots then decay, leaving semipermanent channels in which roots of subsequent crops can grow, even when the soil is relatively dry and hard. (Right) Corn planted after forage radish had twice as many roots reaching the subsoil (below 30 cm) as corn planted after the fibrous rooted rye and nearly 10 times as many as corn planted in soil that had no cover crop during the winter. (Photos courtesy of R. Weil. Data from Chen and Weil, unpublished)

(a)   (b)

Figure 4.24
One approach to reducing soil compaction is to spread the applied weight over a larger area of the soil surface. Examples are extra-wide wheels on heavy vehicles used to apply soil amendments (left) and standing on a wooden board while preparing a garden seedbed in early spring (right). (Photos courtesy of R. Weil)

Another approach for minimizing compaction is to carefully restrict all wheel traffic to specific lanes, leaving the rest of the field (usually 90% or more of the area) free from compaction. Such **controlled traffic** systems are widely used in Europe, especially on clayey soils. Gardeners can practice controlled traffic by establishing permanent foot paths between planting beds, enhancing them with a thick mulch, sod grass, or stone paving.

An opposite strategy uses special wide tires fitted to heavy equipment so as to spread the weight over more soil surface, thus reducing the force applied per unit area (Figure 4.24*a*). Wider tires do lessen the compactive effect, but they also increase the percentage of the soil surface that is impacted. In an analogous practice, home gardeners can stand on wooden boards when preparing seedbeds in relatively wet soil (Figure 4.24*b*).

Influence of Bulk Density on Soil Strength and Root Growth

High bulk density may occur as a natural soil profile feature (for example, a fragipan), or it may be an indication of human-induced soil compaction. In any case, root growth is inhibited by excessively dense soils for a number of reasons, including the soil's resistance to penetration, poor aeration, slow movement of nutrients and water, and the buildup of toxic gases and root exudates.

Roots penetrate the soil by pushing their way into pores. If a pore is too small to accommodate the root cap, the root must push the soil particles aside and enlarge the pore. To some degree, the density *per se* restricts root growth, as the roots encounter fewer and smaller pores. However, root penetration is also limited by **soil strength**, the property of the soil that causes it to resist deformation. Soil strength increases with increasing bulk density and decreasing water content. Therefore, root growth is most restricted when compacted soils are relatively dry.

The more clay present in a soil, the smaller the average pore size, and the greater the resistance to penetration at a given bulk density. Therefore, if the bulk density is the same, roots more easily penetrate a moist sandy soil than a moist clayey one.

4.8   PORE SPACE OF MINERAL SOILS

One of the main reasons for measuring soil bulk density is that this value can be used to calculate pore space. (See Box 4.2.) For soils with the same particle density, the lower the bulk density, the higher the percent pore space (**total porosity**).

Total porosity varies widely among soils for the same reasons that bulk density varies. Values range from as low as 25% in compacted subsoils to more than 60% in well-aggregated, high-organic-matter surface soils. As is the case for bulk density, management can exert a decided influence on the pore space of soils (see Table 4.2). Cultivation tends to lower the total pore space compared to that of uncultivated soils because of decreased organic matter content and less aggregation.

Size of Pores

Bulk density values predict only *total* porosity. However, soil pores occur in a wide variety of sizes and shapes that largely determine what role the pore can play in the soil (Table 4.3). We will simplify our discussion at this point by referring only to **macropores** (larger than about 0.08 mm) and **micropores** (smaller than about 0.08 mm). Figure 4.25 shows that the decrease in organic matter and increase in clay that occur with depth in many profiles are associated with a shift from macropores to micropores.

Macropores   The macropores characteristically allow the ready movement of air and the drainage of water. They also are large enough to accommodate plant roots and the wide range of tiny animals that inhabit the soil (see Chapter 11). Macropores can occur as the spaces between individual sand grains in coarse-textured soils. Thus, even though a sandy soil has relatively low total porosity, the movement of air and water through such a soil is surprisingly rapid because of the dominance of the macropores.

In well-structured soils, the macropores are generally found between peds. These **interped pores** may occur as spaces between loosely packed granules or as the planar cracks between tight-fitting blocky and prismatic peds (see Plate 82).

BOX 4.2
CALCULATION OF PERCENT PORE SPACE IN SOILS

It is often desirable to calculate the pore space from data on bulk (D_b) and particle densities (D_p):

$$\% \text{ pore space} = 100\% - \left(\frac{D_b}{D_p} \times 100 \right)$$

Consider a cultivated clay soil with a bulk density determined to be 1.28 Mg/m³. If we have no information on the particle density, we assume that the particle density is approximately that of the common silicate minerals (i.e., 2.65 Mg/m³). We calculate the percent pore space using the formula above:

$$\% \text{ pore space} = 100\% - \left(\frac{1.28\,\text{Mg/m}^3}{2.65\,\text{Mg/m}^3} \times 100 \right)$$
$$= 100\% - 48.3 = 51.7$$

This value of pore space, 51.7%, is quite close to the typical percentage of air and water space described in Figure 1.12 for a well-granulated, medium- to fine-textured soil in good condition for plant growth. This simple calculation tells us nothing about the relative amounts of large and small pores, however, and so must be interpreted with caution.

Table 4.3
A SIZE CLASSIFICATION OF SOIL PORES AND SOME FUNCTIONS OF EACH SIZE CLASS
Pore sizes are actually a continuum and the boundaries between classes given here are inexact and somewhat arbitrary. The term micropore *is often broadened to refer to all the pores smaller than macropores.*

Simplified class	Class[a]	Effective diameter range (mm)	Characteristics and functions
Macropores	Macropores	0.08–5+	Generally found between soil peds (interped); water drains by gravity; effectively transmit air; large enough to accommodate plant roots, habitat for certain soil animals.
Micropores	Mesopores	0.03–0.08	Retain water after drainage; transmit water by capillary action; accommodate fungi and root hairs.
	Micropores	0.005–0.03	Generally found within peds (intraped); retain water that plants can use; accommodate most bacteria.
	Ultramicropores	0.0001–0.005	Found largely within clay groupings; retain water that plants cannot use; exclude most microorganisms.
	Cryptopores	< 0.0001	Exclude all microorganisms, too small for large molecules to enter.

[a] The pore size classes and boundary diameters are those cited in Soil Science Society of America (2001).

Figure 4.25
Volume distribution of organic matter, sand, silt, clay, and pores of macro- and microsizes in a representative medium-textured soil with good structure. Note that macropores are especially plentiful in the surface horizon (upper 30 cm). (Diagram courtesy of R. Weil)

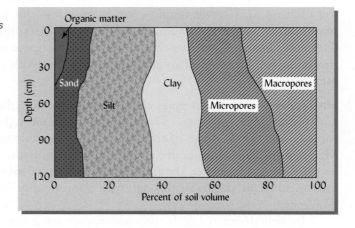

Macropores created by roots, earthworms, and other organisms constitute a very important type of pores termed **biopores**. These are usually tubular in shape and may be continuous for lengths of a meter or more (see Plate 81). In some clayey soils, biopores are the principal form of macropores, greatly facilitating the growth of plant roots (Table 4.4 and Plates 81 and 83). Perennial plants, such as forest trees and certain forage crops, are particularly effective at creating channels that serve as conduits for roots, long after the death and decay of the roots that originally created them. Two such old root channels, each about 8 mm in diameter, can be seen perforating the clay slickenside shown in Figure 3.16c.

Micropores   In contrast to macropores, micropores are usually filled with water in field soils. Even when not water-filled, they are too small to permit much air movement. Water movement in micropores is slow, and much of the water retained in these pores is not available to plants (see Chapter 5). Fine-textured soils, especially those without a stable granular structure, may have a preponderance of micropores, thus allowing relatively slow gas and water movement, despite the relatively large volume of total pore space. While the larger micropores accommodate plant root hairs and microorganisms, the smaller micropores (sometimes termed *ultramicropores* and

Table 4.4

DISTRIBUTION OF LOBLOLLY PINE ROOTS IN THE SOIL MATRIX AND IN OLD ROOT CHANNELS IN THE UPPERMOST METER OF AN ULTISOL IN SOUTH CAROLINA

The root channels were generally from 1 to 5 cm in diameter and filled with loose surface soil and decaying organic matter.

	Numbers of roots per m² of upper 1 m soil		
Root size, diameter	Soil matrix	Old root channels	Comparative increase in root density in the old channels, %
Fine roots, <4 mm	211	3617	94
Medium roots, 4–20 mm	20	361	95
Coarse roots, >20 mm	3	155	98

Calculated from Parker and Van Lear (1996).

cryptopores) are too small to permit the entrance of even the smallest bacteria or some decay-stimulating enzymes produced by the bacteria. These pores can act as hiding places for some adsorbed organic compounds (both naturally occurring and pollutants), thereby protecting them from breakdown for long periods of time, perhaps for centuries.

4.9 SOIL PROPERTIES RELEVANT TO ENGINEERING USES

Soil **consistence** is a term used by soil scientists to describe the ease with which a soil can be reshaped or ruptured. As a clod of soil is squeezed between the thumb and forefinger (or crushed underfoot, if necessary), observations are made on the amount of force needed to crush the clod and on the manner in which the soil responds to the force.

Moisture content greatly influences how a soil responds to stress; hence, moist and dry soils are given separate consistence ratings (Table 4.5). As described in Section 4.6, a moist clod that crumbles with only light pressure is said to be friable. Friable soils are easily excavated or tilled.

Engineers use the term **consistency** to describe how a soil resists *penetration* by an object, while the soil scientist's consistence describes resistance to *rupture*. Instead of crushing a clod of soil, the engineer attempts to penetrate it with either the blunt end of a pencil (some use their thumbs) or a thumbnail. For example, if the blunt end of a pencil makes only a slight indentation, but the thumbnail penetrates easily, the soil is rated as *very firm* (Table 4.5). Consistency, then, is a kind of simple field estimation of soil strength or penetration resistance (see Section 4.7).

Field observations of both consistence and consistency provide valuable information to guide decisions about loading and manipulating soils. For construction purposes, however, more precise measurements are needed of a number of related soil properties that help predict how a soil will respond to applied stress.

Soil Bearing Strength and Sudden Failure

Engineers define soil **bearing strength** as the capacity of a soil mass to withstand stresses without rupturing or becoming deformed. Failure of a soil to withstand stress can result in a building toppling over as its weight exceeds the soil's bearing strength.

Excavation safety animation of unstable soils: http://physics.uwstout.edu/geo/exca_s.avi

Table 4.5
SOME FIELD TESTS AND TERMS USED TO DESCRIBE THE CONSISTENCE AND CONSISTENCY OF SOILS
The consistency of cohesive materials is closely related to, but not exactly the same as, their consistence. Conditions of least coherence are represented by terms at the top of each column, those of greater coherence near the bottom.

	Soil consistence[a]			Soil consistency[b]	
Dry soil	Moist to wet soil	Soil dried then submerged in water	Field rupture (crushing) test	Soil at *in situ* moisture	Field penetration test
Loose	Loose	Not applicable	Specimen not obtainable	Soft	Blunt end of pencil penetrates deeply with ease
Soft	Very friable	Noncemented	Crumbles under very slight force between thumb and forefinger	Medium firm	Blunt end of pencil can penetrate about 1.25 cm with moderate effort
Slightly hard	Friable	Extremely weakly cemented	Crumbles under slight force between thumb and forefinger	Firm	Blunt end of pencil can penetrate about 0.5 cm
Hard	Firm	Weakly cemented	Crushes with difficulty between thumb and forefinger	Very firm	Blunt end of pencil makes slight indentation; thumbnail easily penetrates
Very hard	Extremely firm	Moderately cemented	Cannot be crushed between thumb and forefinger, but can be crushed slowly underfoot	Hard	Blunt end of pencil makes no indentation; thumbnail barely penetrates
Extremely hard	Slightly rigid	Strongly cemented	Cannot be crushed by full body weight underfoot		

[a] Abstracted from USDA-NRCS (2005).
[b] Modified from McCarthy (1993).

Similarly, an earthen dam or levee might give way under the pressure of impounded water, or pavements and structures might slide down unstable hillsides (Figure 1.7).

Cohesive Soils   Two components of strength apply to **cohesive soils** (essentially soils with a clay content of more than about 15%): (1) inherent electrostatic attractive forces (see *clay flocculation* in Section 4.5) and (2) frictional resistance. One laboratory test used to estimate soil strength is the direct **unconfined compression test** illustrated in Figure 4.26*a*. A cylindrical specimen of cohesive soil is placed vertically between two flat, porous stones (which allow water to escape from the compressed soil pores) and a slowly increasing downward force is applied. The soil column will first bulge out a bit and then fail—that is, give way suddenly and collapse—when the force exceeds the soil strength.

The strength of cohesive soils declines dramatically if the material is very wet and the pores are filled with water. Then the particles are forced apart so that neither the cohesive nor the frictional component is very strong, making the soil prone to failure, often with catastrophic results (such as mudslides, Figure 4.27, or levee failures, Box 4.3). On the other hand, if cohesive soils become more compacted or dry down, their strength increases as particles are forced into closer contact with one another—a result that has implications for plant root growth as well as for engineering (see Section 4.7).

Noncohesive Soils   The strength of dry, noncohesive soil materials such as loose sand depends entirely on frictional forces, including the interlocking of rough

Saturated soils lead to life in prison, by Adam Pitluk: www.riverfronttimes.com/content/printVersion/108522

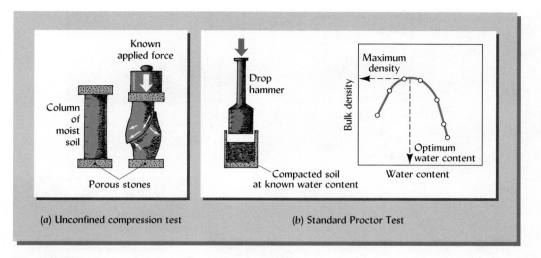

Figure 4.26

Two important tests to determine engineering properties of soil materials. (a) An unconfined compression test for soil strength. (b) The Proctor test for maximum density and optimum water content for compaction control.

Figure 4.27

Houses damaged by a mudslide that occurred when the soils of a steep hillside in Oregon became saturated with water after a period of heavy rains. The weight of the wet soil exceeded its shear strength, causing the slope to fail. Excavations for roads and houses near the foot of a slope can contribute to the lack of slope stability, as can removal of tree roots by large-scale clear-cutting on the slope itself. (Photo courtesy of John Griffith, Coos Bay, OR)

particle surfaces. One reflection of such interparticle friction is the **angle of repose**, the steepest angle to which a material can be piled without slumping. Smooth, rounded sand grains cannot be piled as steeply as can rough, interlocking sands. If a small amount of water bridges the gaps between particles, electrostatic attraction of the water for the mineral surfaces will increase the soil strength (as illustrated in Figure 4.28).

Collapsible Soils   Certain soils that exhibit considerable strength at low *in situ* water contents lose their strength suddenly if they become wet. Such soils may collapse without warning under a roadway or building foundation. A special case of soil collapse is **thixotropy**, the sudden liquification of a wet soil mass when subjected to vibrations, such as those accompanying earthquakes and blasting.

Soil liquefaction: www.ce.washington.edu/ ~liquefaction/html/content .html

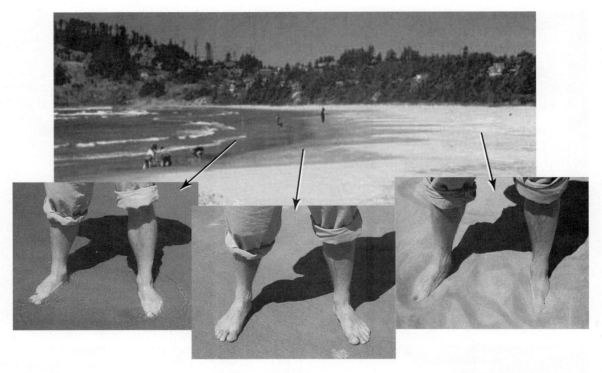

Figure 4.28

A beach in Oregon illustrates the concept of soil strength for sandy materials. The dry sand (lower right) has little strength and your feet easily mire into it as you walk along. There is nothing to hold the individual sand particles together. Closer to the ocean, where the soil has been thoroughly wetted by waves, but where there is no standing water (lower center), water films act as bridges between sand particles, holding them together and thereby resisting penetration by the feet. If you stand in shallow water along the edge of the ocean (lower left), once again your feet penetrate the surface sand because each sand particle is completely surrounded by water, which acts more as a lubricant than as a binding force. (Photos courtesy of R. Weil)

Settlement—Gradual Compression

Most foundation problems result from slow, often uneven, vertical subsidence or **settlement** of the soil. Soils to be used for a foundation or roadbed are compacted on purpose using heavy rollers (Figure 4.29) or vibrators. Compaction occurring after construction would result in uneven settlement and cracked pavements or foundations.

The **Proctor test** is used to guide efforts at compacting soil materials before construction. A specimen of soil is mixed to a given water content and placed in a holder, where it is compacted by a drop hammer. The bulk density (usually referred to as the *dry density* by engineers) is then measured. The process is repeated with increasing water contents until the data form a *Proctor curve* (Figure 4.26b), which indicates the soil water content that maximizes compactability. On construction sites, tank trucks may spray water to bring the soil water content to the determined optimum level before heavy equipment (such as that shown in Figure 4.29) compacts the soil to the desired density.

Compressibility   A **consolidation test** may be conducted on a soil specimen to determine its **compressibility**—how much its volume will be reduced by a given applied force. Because of the relatively low porosity and equidimensional shape of the individual mineral grains, very sandy soils resist compression once the particles have settled into a tight packing arrangement. They make excellent soils for foundations. The high porosity of clay floccules and the flakelike shape of clay particles give clayey

Figure 4.29
Compaction of soils used as foundations and roadbeds is accomplished by heavy equipment such as this sheepsfoot roller. The knobs ("sheepsfeet") concentrate the mass on a small impact area, punching and kneading the loose, freshly graded soil to optimum density.
(Photo courtesy of R. Weil)

soils much greater compressibility. Soils consisting mainly of organic matter (peats) have the highest compressibilities and generally are unsuitable for foundations. Perhaps the most famous example of uneven settlement due to slow compression is the Leaning Tower of Pisa in Italy. Unfortunately, most cases of uneven settlement result in headaches, not tourist attractions.

Expansive Soils

Damage caused by expansive soils in the United States rarely makes the evening news programs, although the total cost annually may exceed that caused by tornados, floods, and earthquakes. Expansive clays occur on about 20% of the land area in the United States and cause upwards of $6 billion in damages annually to pavements, foundations, and utility lines. The damages can be severe in certain sites in all parts of the country, but are most extensive in regions that have long dry periods alternating with periods of rain (see distribution of Vertisols, endpapers).

Some clays, particularly the smectites, swell when wet and shrink when dry (see Section 8.14). Expansive soils are rich in these types of clay. The electrostatic charges on clay surfaces attract water molecules from larger pores into the micropores within clay domains. The swelling and shrinkage cause sufficient movement of the soil to crack building foundations, burst pipelines, and buckle pavements.

Atterberg Limits

As a dry, clayey soil takes on increasing amounts of water, it undergoes dramatic and distinct changes in behavior and consistency. A hard, rigid solid in the dry state, it becomes a crumbly (friable) semisolid when a certain moisture content (termed the **shrinkage limit**) is reached. If it contains expansive clays, the soil also begins to swell in volume as this moisture content is exceeded. Increasing the water content beyond the **plastic limit** will transform the soil into a malleable, plastic mass and cause additional swelling. The soil will remain in this plastic state until its **liquid limit** is exceeded, causing it to transform into a viscous liquid that will flow when jarred. These critical water contents (measured in units of percent) are termed the **Atterberg limits**.

Determination of Atterberg limits, University of Texas at Arlington: http://geotech.uta.edu/lab/Main/atrbrg_lmts/

BOX 4.3
TRAGEDY IN THE BIG EASY—A LEVEE DOOMED TO FAIL[a]

In 2005, Hurricane Katrina was one of the worst natural disasters in American history, with some 100,000 homes flooded and over 1,000 people killed. Some of the worst flooding occurred when the 17th Street levee failed (Figure 4.30). Investigations later showed a faulty levee design did not properly deal with underlying layers of organic soils and sands. Poor design, combined with poor levee maintenance, allowed water to seep under the levee and weaken the soil at its base.

The levee, essentially a gently sloping mound of compacted clay soils, was covered on the landward side with a thin veneer of topsoil to support a protective grass mantle. To hold back floodwater and storm surges, engineers had constructed a concrete seawall along the crest of the levee. The seawall was attached to long steel pilings driven deep into the soil of the levee. The pilings were meant to both anchor the seawall and to prevent water from seeping through or under the levee. Out of sight, under the layers of clayey and loamy materials from which the levee was constructed, several layers of peat (buried Histosols) and sand provided for the weak link in the levee design.

Organic soils are highly compressible and have very low bearing and shear strengths. Both peats and sands are also highly permeable and conduct water readily. To perform their intended functions, the steel pilings attached to the seawall had to be long enough to penetrate through the peat/sand layers and into the more cohesive, higher-strength soil below. Unfortunately, the pilings in the 17th Street levee were too short and failed to penetrate through the peat layer (Figure 4.31). Whenever storms raised the water level in the canal, seepage under the levee would rise to saturate the soil at the foot of the levee. When saturated, the soil would lose most of its shear strength and resistance to compression. Apparently, the engineers designing the levee had data on the peat layers, but based their design on the *average* soil properties, rather than on the weakest soils present.

On 21 August 2005, with seepage water saturating the soil at the levee base and turning the peat layer into little more than "soup," Katrina's storm surge "snapped the chain" at the weakest link, toppling a 140 m long section of the seawall, pushing both it and the levee some 14 m inland. The storm-churned waters poured through the breach, inundating the city of New Orleans.

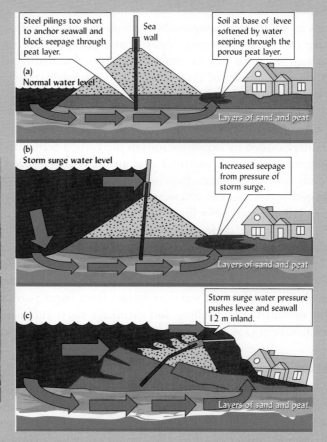

Figure 4.31
Ilustration (not to scale) of how the buried layers of low-strength organic soil (a) allowed seepage of storm surge water (b) dooming the levee and its seawall to failure (c).
(Diagram courtesy of R. Weil)

Figure 4.30
A large helicopter attempts emergency repairs to the 17th Street levee breach several days after Hurricane Katrina hit New Orleans and toppled this section of the levee and seawall. Large chunks of the levee can be seen some 14 m inland. (Photo courtesy of U.S. Army Corps of Engineers)

[a] Based on forensic engineering investigation information in Seed et al. (2005) and reporting by Marshall (2005) and Vartabedian and Braun (2006).

Smectite clays (see Section 8.3) generally have high liquid limits, especially if saturated with sodium. Kaolinite and other nonexpansive clays have low liquid limit values. The tendency of expansive clay soils to literally flow down steep slopes when the liquid limit is exceeded, producing mass wasting and landslides, is illustrated in Plates 41 and 42.

The expansiveness of a soil (and therefore the hazard of its destroying foundations and pavements) can be quantified as the *coefficient of linear extensibility* (COLE). Suppose a sample of soil is moistened to its plastic limit and molded into the shape of a bar with length L_M. If the bar of soil is allowed to air dry, it will shrink to length L_D. The COLE is the percent reduction in length of the soil bar upon shrinking.

Unified Classification System for Soil Materials

The U.S. Army Corps of Engineers and the U.S. Bureau of Reclamation have established a widely used system of classifying soil materials in order to aid in predicting the engineering behavior of different soils. The system first groups soils into coarse- and fine-grained soils. Each type of soil is then given a two-letter designation based primarily on its particle-size distribution (texture), Atterberg limits, and organic-matter content (e.g., GW for well-graded gravel, SP for poorly graded sands, CL for clay of low plasticity, and OH for organic-rich clays of high plasticity). This classification of soil materials helps engineers predict the soil strength, expansiveness, compressibility, and other properties so that appropriate engineering designs can be made for the soil at hand (Box 4.3).

4.10 CONCLUSION

Soils present an incredibly complex physical network of solid surfaces, pores, and interfaces that provides the setting for myriad chemical, biological, and physical processes. These in turn influence plant growth, hydrology, environmental management, and engineering uses of soil. The nature and properties of the individual particles, their size distribution, and their arrangement in soils determine the total volume of nonsolid pore space, as well as the pore sizes, thereby impacting on water and air relationships.

The properties of individual particles and their proportionate distribution (soil texture) are subject to little human control in field soils. However, it is possible to exert some control over the arrangement of these particles into aggregates (soil structure) and on the stability of these aggregates. Tillage and traffic must be carefully controlled to avoid undue damage to soil tilth, especially when soils are rather wet. Generally, nature takes good care of soil structure, and humans can learn much about soil management by studying natural systems. Vigorous and diverse plant growth, generous return of organic residues, and minimal physical disturbance are attributes of natural systems worthy of emulation. Proper plant species selection and management of chemical, physical, and biological factors can help ensure maintenance of soil physical quality. In recent years, these management goals in agriculture have been made more practical by the advent of conservation tillage systems that minimize soil manipulations while decreasing soil erosion and water runoff.

Particle size, moisture content, and plasticity of the colloidal fraction all help determine the stability of soil in response to loading forces from traffic, tillage, or building foundations. The physical properties presented in this chapter greatly influence nearly all other soil properties and uses, as discussed throughout this book.

STUDY QUESTIONS

1. If you were investigating a site for a proposed housing development, how could you use soil colors to help predict where problems might be encountered?

2. You are considering the purchase of some farmland in a region with variable soil textures. The soils on one farm are mostly sandy loams and loamy sands, while those on a second farm are mostly clay loams and clays. List the potential advantages and disadvantages of each farm as suggested by the texture of its soils.

3. Revisit your answer to question 2. Explain how soil structure in both the surface and subsurface horizons might modify your opinion of the merits of each farm.

4. Two different timber-harvest methods are being tested on adjacent forest plots with clay loam surface soils. Initially, the bulk density of the surface soil in both plots was 1.1 Mg/m^3. One year after the harvest operations, plot A soil had a bulk density of 1.48 Mg/m^3, while that in plot B was 1.29 Mg/m^3. Interpret these values with regard to the relative merits of systems A and B, and the likely effects on the soil's function in the forest ecosystem.

5. What are the textural classes of two soils, the first with 15% clay and 45% silt, and the second with 80% sand and 10% clay? (Hint: Use Figure 4.4.)

6. For the forest plot B in question 4, what was the change in percent pore space of the surface soil caused by timber harvest? Would you expect that most of this change was in the micropores or in the macropores? Explain.

7. Discuss the positive and negative impacts of tillage on soil structure. What is another physical consideration that you would have to take into account in deciding whether or not to change from a conventional to a conservation tillage system?

8. What would you, as a home gardener, consider to be the three best and three worst things that you could do with regard to managing the soil structure in your home garden?

9. What does the Proctor test tell an engineer about a soil, and why would this information be important?

10. In a humid region characterized by expansive soils, a homeowner experienced burst water pipes, doors that no longer closed properly, and large vertical cracks in the brick walls. The house had had no problems for over 20 years, and a consulting soil scientist blamed the problems on a large tree that was planted near the house some 10 years before the problems began to occur. Explain.

REFERENCES

Bigham, J. M., and E. J. Ciolkosz (eds.). 1993. *Soil Color.* SSSA Special Publication no. 31. Soil Science Society of America, Madison, WI.

Faulkner, E. H. 1943. *Plowman's Folly.* University of Oklahoma Press, Norman, Okla.

Marshall, B. 2005. "17th Street Canal levee was doomed—report blames Corps: Soil could never hold." *The Times-Picayune,* Wednesday, November 30, New Orleans.

McCarthy, D. F. 1993. *Essentials of Soil Mechanics and Foundations,* 4th ed. Prentice Hall, Englewood Cliffs, NJ.

Oades, J. M. 1993. "The role of biology in the formation, stabilization, and degradation of soil structure." *Geoderma* **56**:377–400.

O'Nofiok, O., and M. J. Singer. 1984. "Scanning electron microscope studies of surface crusts formed by simulated rainfall." *Soil Sci. Soc. Amer. J.* **48**:1137–1143.

Parker, M. M., and D. H. Van Lear. 1996. "Soil heterogeneity and root distribution of mature loblolly pine stands in Piedmont soils." *Soil Sci. Soc. Amer. J.* **60**:1920–1925.

Ross, C., R. E. Sojka, and J. A. Foerster. 2003. "Scanning electron micrographs of polyacrylamide-treated soil in irrigation furrows." *J. Soil Water Conserv.* **58**:327–331.

Seed, R. B., P. G. Nicholson, R. A. Dalrymple, J. Battjes, R. G. Bea, G. Boutwell, J. D. Bray, B. D. Collins, L. F. Harder, J. R. Headland, M. Inamine, R. E. Kayen, R. Kuhr, J. M. Pestana, R. Sanders, F. Silva-Tulla, R. Storesund, S. Tanaka, J. Wartman, T. F. Wolff, L. Wooten, and T. Zimmie. 2005. "Preliminary report on the performance of the New Orleans levee systems in hurricane Katrina on August 29, 2005—Preliminary findings from field investigations and associated studies shortly after the hurricane." *Report UCB/CITRIS – 05/01.* University of California at Berkeley and the American Society of Civil Engineers, Berkeley, CA.

Six, J., H. Bossuyt, S. Degryze, and K. Denef. 2004. "A history of research on the link between (micro)aggregates, soil biota, and soil organic matter dynamics." *Soil Tillage Res.* **79**:7–31.

Soil Science Society of America. 2001. *Glossary of Soil Science Terms 1996.* Soil Science Society of America, Madison, WI.

Tijink, F. G. J., and J. P. van der Linden. 2000. "Engineering approaches to prevent compaction in cropping systems with sugar beet. In R. Horn et al., eds., *Subsoil compaction: Distribution, processes, and consequences.* pp. 442–452. Catena Verlag, Reiskirchen, Germany.

Tisdall, J. M., and J. M. Oades. 1982. "Organic matter and water-stable aggregates in soils." *Soil Sci. Soc. Am. J.* **33**:141–163.

USDA-NRCS. 2005. *National soil survey handbook, title 430-vi.* U.S. Department of Agriculture, Natural Resources Conservation Service. http://soils.usda.gov/technical/handbook/ (posted September 2005; verified 12 December 2008).

Vartabedian, R., and S. Braun. 2006. "Fatal flaws: Why the walls tumbled in New Orleans." *Los Angeles Times,* Los Angeles, CA.

Vimmerstadt, J., F. Scoles, J. Brown, and M. Schmittgen. 1982. "Effects of use pattern, cover, soil drainage class, and overwinter changes on rain infiltration on campsites." *J. Environ. Qual.,* **11**:25–28.

Watts, C. W., and A. R. Dexter. 1997. "The influence of organic matter in reducing the destabilization of soil by simulated tillage." *Soil Tillage Res.* **42**:253–275.

5

Soil Water: Characteristics and Behavior

When the earth will . . . drink up the rain as fast as it falls.
—H. D. Thoreau, *The Journal*

Bringing water to arid valley soils. (R. Weil)

One of nature's simplest chemical compounds, water is a vital component of every living cell. Its unique properties promote a wide variety of physical, chemical, and biological processes. These processes greatly influence almost every aspect of soil development and behavior, from the weathering of minerals to the decomposition of organic matter, from the growth of plants to the pollution of groundwater.

We are all familiar with water. We drink it, wash with it, and swim in it. But water in the soil is something quite different from water in a drinking glass. In the soil, the intimate association between water and soil particles changes the behavior of both. Water causes soil particles to swell and shrink, to adhere to each other, and to form structural aggregates. Water participates in innumerable chemical reactions that release or tie up nutrients, create acidity, and wear down minerals so that their constituent elements eventually contribute to the saltiness of the oceans.

Certain soil water phenomena seem to contradict our intuition about how water ought to behave. Attraction to solid surfaces restricts some of the free movement of water molecules, making it less liquid and more solidlike in its behavior. In the soil, water can flow up as well as down. Plants may wilt and die in a soil whose profile contains a million kilograms of water per hectare. A layer of sand or gravel in a soil profile may actually inhibit drainage, rather than enhance it.

Soil–water interactions determine the rates of water loss by leaching, surface runoff and evapotranspiration, the balance between air and water in soil pores, the rate of change in soil temperature, the rate and kind of metabolism of soil organisms, and the capacity of soils to store and provide water for plant growth.

The characteristics and behavior of water in the soil comprise a common thread that interrelates nearly every chapter in this book. The principles contained in this chapter will help us understand why mudslides occur in water-saturated soils (Chapter 4), why earthworms may improve soil quality (Chapter 10), why wetlands contribute to global ozone depletion (Chapter 12), and why famine stalks humanity in certain regions of the world. Mastery of the principles presented in this chapter is fundamental to your working knowledge of the soil system.

5.1 STRUCTURE AND RELATED PROPERTIES OF WATER[1]

Water properties: www.biologylessons.sdsu.edu/classes/lab1/semnet/water.htm

The ability of water to influence so many soil processes is determined primarily by the structure of the water molecule. This structure also is responsible for the fact that water is mainly present on Earth as a liquid, not a gas. Water is, with the exception of mercury, the *only* inorganic (not carbon-based) liquid found on Earth at standard temperature and pressure. Water is a simple compound, its individual molecules containing one oxygen atom and two much smaller hydrogen atoms. The elements are bonded together covalently, each hydrogen atom sharing its single electron with the oxygen.

Polarity

Instead of lining up symmetrically on either side of the oxygen atom (H-O-H), the hydrogen atoms are attached to the oxygen in a V-shaped arrangement at an angle of only 105°. Water is therefore an asymmetrical molecule with its electrons spending more time nearer to the oxygen than to the hydrogen. Consequently, water molecules exhibit *polarity*; that is, the charges are not evenly distributed. Rather, the side on which the hydrogen atoms are located tends to be electropositive and the opposite side electronegative.

Polarity explains why water molecules are attracted to electrostatically charged ions and to colloidal surfaces. Cations such as H^+, Na^+, K^+, and Ca^{2+} become hydrated through their attraction to the oxygen (negative) end of water molecules. Likewise, negatively charged clay surfaces attract water, this time through the hydrogen (positive) end of the molecule. Polarity of water molecules also encourages the dissolution of salts in water since the ionic components have greater attraction for water molecules than for each other.

Hydrogen Bonding

Through a phenomenon called **hydrogen bonding**, a hydrogen atom of one water molecule is attracted to the oxygen end of a neighboring water molecule, thereby forming a low-energy bond between the two molecules. This type of bonding accounts for the polymerization of water.

Cohesion, Adhesion, and Surface Tension

Hydrogen bonding accounts for two basic forces responsible for water retention and movement in soils: the attraction of water molecules for each other (**cohesion**) and the attraction of water molecules for solid surfaces (**adhesion**). By adhesion (also called *adsorption*), some water molecules are held rigidly at the surfaces of soil solids. In turn, these tightly bound water molecules hold, by cohesion, other water molecules farther removed from the solid surfaces (Figure 5.1). The forces of adhesion and

---

[1]For more in-depth discussions of water–soil interactions, see Hillel (1998) or Warrick (2001).

Figure 5.1

The forces of cohesion (between water molecules) and adhesion (between water and solid surface) in a soil–water system. The forces are largely a result of H-bonding, shown as broken lines. The adhesive or adsorptive force diminishes rapidly with distance from the solid surface. The cohesion of one water molecule to another results in water molecules forming temporary clusters that are constantly changing in size and shape as individual water molecules break free or join up with others. The cohesion between water molecules also allows the solid to indirectly restrict the freedom of water for some distance beyond the solid–liquid interface.

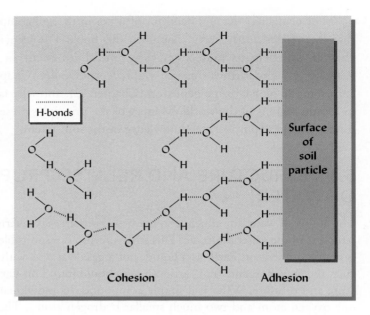

cohesion make it possible for the soil solids to retain water and control its movement and use. Adhesion and cohesion also make possible the property of plasticity possessed by clays (see Section 4.9).

Surface tension is another important property of water that markedly influences its behavior in soils. At liquid–air interfaces, surface tension results from the greater attraction of water molecules for each other (cohesion) than for the air. The net effect is an inward force at the surface that causes water to behave as if its surface were covered with a stretched elastic membrane (Figure 5.2). Because of the relatively high attraction of water molecules for each other, water has a high surface tension (72.8 millinewtons/m at 20 °C) compared to that of most other liquids (e.g., another low molecular weight compound, ethyl alcohol, 22.4 mN/m). As we shall see, surface tension is an important factor in the phenomenon of capillarity, which determines how water moves and is retained in soil.

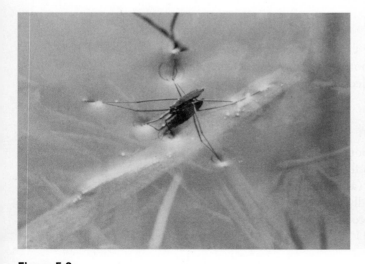
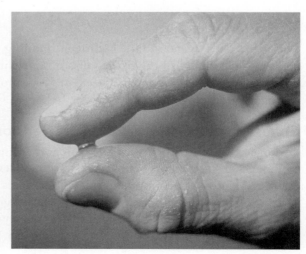

Figure 5.2

Everyday evidences of water's surface tension (left) as insects land on water and do not sink and of forces of cohesion and adhesion (right) as a drop of water is held between the fingers. (Photos courtesy of R. Weil)

5.2 CAPILLARY FUNDAMENTALS AND SOIL WATER

The movement of water up a wick typifies the phenomenon of capillarity. Two forces cause capillarity: (1) the attraction of water for the solid (adhesion or adsorption), and (2) the surface tension of water, which is due largely to the attraction of water molecules for each other (cohesion).

Change capillary radius (Kappilarradius) to see water rise. Universität Heidelberg: http://www.ito.ethz.ch:16080/filep/inhalt/seiten/exp1200/animation_1200.htm

Capillary Mechanism

Capillarity can be demonstrated by placing one end of a fine, clean glass tube in water. The water rises in the tube; the smaller the tube inside radius, the higher the water rises. The water molecules are attracted to the sides of the tube (adhesion) and start to spread out along the glass in response to this attraction. At the same time, the cohesive forces hold the water molecules together and create surface tension, causing a curved surface (called a *meniscus*) to form at the interface between water and air in the tube. Lower pressure under the meniscus in the glass tube allows the higher pressure on the free water to push water up the tube. The process continues until the water in the tube has risen high enough that its weight just balances the pressure differential across the meniscus.

The height of rise in a capillary tube is inversely proportional to the tube inside radius *r*. Capillary rise is also inversely proportional to the density of the liquid and is directly proportional to the liquid's surface tension and the degree of its adhesive attraction to the tube or soil surface. If we limit our consideration to water at a given temperature (e.g., 20 °C), then these factors can be combined into a single constant, and we can use a simple capillary equation to calculate the height of rise *h*:

$$h = \frac{0.15}{r} \tag{5.1}$$

where both *h* and *r* are expressed in centimeters. This equation tells us that the narrower the tube, the greater the capillary force and the higher the water rise in the tube (Figure 5.3*a*).

Height of Rise in Soils

Capillary forces are at work in all moist soils. However, the rate of movement and the rise in height are less than one would expect on the basis of soil pore size alone. One reason is that soil pores are not straight, uniform openings like glass tubes. Furthermore, some soil pores are filled with air, which may be entrapped, slowing down or preventing the movement of water by capillarity (see Figure 5.3*b*).

Since capillary movement is determined by pore size, it is the pore-size distribution discussed in Chapter 4 that largely determines the amount and rate of movement of capillary water in the soil. The abundance of medium- to large-sized capillary pores in sandy soils permits rapid initial capillary rise but limits the ultimate height of rise[2] (Figure 5.3*c*). Clays have a high proportion of very fine capillary pores, but frictional forces slow down the rate at which water moves through them. Consequently, in clays the capillary rise is slow, but in time it generally exceeds that of sands. Loams exhibit capillary properties between those of sands and clays.

---

[2] Note that if water rises by capillarity to a height of 37 cm above a free-water surface in a sand (as shown in the example in Figure 5.3*c*), then it can be estimated (by rearranging the capillary equation to $r = 0.15/h$) that the smallest continuous pores must have a radius of about 0.004 cm (0.15/37 = 0.004). This calculation gives an approximation of the minimum effective capillary pore radius in a soil.

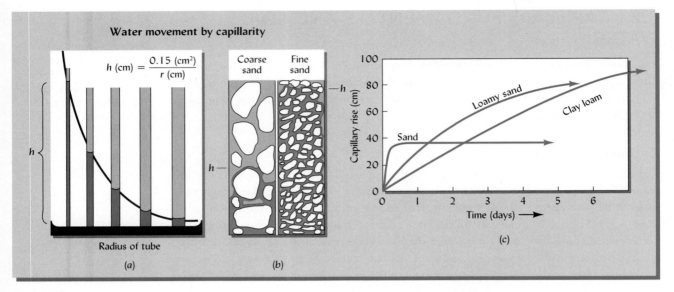

Figure 5.3

Upward capillary movement of water through tubes of different inside radius and soils with different pore sizes. (a) The capillary equation can be graphed to show that the height of rise h doubles when the tube inside radius is halved. This relationship can be demonstrated using glass tubes of different radius. (b) The same principle also relates pore sizes in a soil and height of capillary rise, but the rise of water in a soil is rather jerky and irregular because of the tortuous shape and variability in size of the soil pores (as well as because of pockets of trapped air). (c) The finer the soil texture, the greater the proportion of small-sized pores and, hence, the higher the ultimate rise of water above a free-water table. However, because of the much greater frictional forces in the smaller pores, the capillary rise is much slower in the finer-textured soil than in the sand. (Diagrams courtesy of R. Weil)

Capillarity is traditionally illustrated as an upward adjustment. But movement in any direction takes place, since the attractions between soil pores and water are as effective in forming a water meniscus in horizontal pores as in vertical ones (Figure 5.4). The significance of capillarity in controlling water movement in small pores will become evident as we turn to soil water energy concepts.

Figure 5.4

In this irrigated field in Arizona, water has moved up by capillarity from the irrigation furrow toward the top of the ridge (left), as well as horizontally to both sides and away from the irrigation furrow (right). (Photos courtesy of N. C. Brady)

5.3   SOIL WATER ENERGY CONCEPTS

Every day we can see that things tend toward a lower energy state (and that it takes an input of energy–work–to prevent them from doing so). Use your cell phone and its battery runs down from a fully-charged high potential energy state to a discharged, low energy state. If you should drop your phone, it would fall from its state of relatively high potential energy in your hand to a lower potential energy state on the floor (where it is closer to the source of gravitational pull). The difference in energy levels (that is how high off the floor you are holding the phone) determines how forcefully transition will occur. In this respect, soil water is no different–it tends to move from a higher to a lower energy state. Again, the *difference* in energy levels between water at adjacent points in the soil profile is what influences water movement.

Soil water energy and dynamics: http://faculty. washington .edu/slb/esc210/soils15.pdf

Forces Affecting Potential Energy

The discussion of the structure and properties of water in the previous section suggests three important forces affecting the energy level of soil water. First, adhesion, or the attraction of water to the soil solids (matrix), provides a **matric** force (responsible for adsorption and capillarity) that markedly reduces the energy state of water near particle surfaces. Second, the attraction of water to ions and other solutes results in **osmotic** forces and tends to reduce the energy state of water in the soil solution. Osmosis, movement of pure water across a semipermeable membrane into a solution, is evidence of the lower energy state of water in the solution. The third major force acting on soil water is **gravity**, which always pulls the water downward. The energy level of soil water at a given elevation in the profile is thus higher than that of water at some lower elevation. This difference in energy level causes water to flow downward.

Soil Water Potential

The *difference* in energy level of water from one site or one condition to another (e.g., between wet soil and dry soil) determines the direction and rate of water movement in soils and in plants. In a wet soil, most of the water is retained in large pores or thick water films around particles. Therefore, most of the water molecules in a wet soil are not very close to a particle surface and so are not held very tightly by the soil solids (the matrix). In this condition, the water molecules have considerable freedom of movement, so their energy level is near that of water molecules in a pool of pure water outside the soil. In a drier soil, however, the water that remains is located in small pores and thin water films and is therefore held tightly by the soil solids. Thus the water molecules in a drier soil have little freedom of movement, and their energy level is much lower than that of the water molecules in wet soil. If wet and dry soil samples are brought in touch with each other, water will move from the wet soil (higher energy state) to the drier soil (lower energy).

To evaluate the energy status of soil water in a particular location in the profile, its energy level is compared to that of pure water at standard pressure and temperature, unaffected by the soil and located at some reference elevation. The *difference* in energy levels between this pure water in the reference state and that of the soil water is termed soil **water potential** (Figure 5.5). The term *potential*, like the term *pressure*, implies a difference in energy status. Water will move from a soil zone having a high soil water potential to one having a lower soil water potential. This fact should always be kept in mind when thinking about the behavior of water in soils.

Several forces are implicated in soil water potential, each of which is a component of the **total soil water potential**, ψ_t. These components are due to differences in energy levels resulting from gravitational, matric, submerged hydrostatic, and osmotic forces and are termed **gravitational potential**, ψ_g; **matric potential**, ψ_m, **hydrostatic**

Figure 5.5

Relationship between the potential energy of pure water at a standard reference state (pressure, temperature, and elevation) and that of soil water. If the soil water contains salts and other solutes, the mutual attraction between water molecules and these chemicals reduces the potential energy of the water, the degree of the reduction being termed osmotic *potential. Similarly, the mutual attraction between soil solids (soil matrix) and soil water molecules also reduces the water's potential energy. In this case the reduction is called* matric potential. *Since both of these interactions reduce the water's potential energy level compared to that of pure water, the changes in energy level (osmotic potential and matric potential) are both considered to be negative. In contrast, differences in energy due to gravity (gravitational potential) are always positive because the reference elevation of the pure water is purposely designated at a site in the soil profile below that of the soil water. A plant root attempting to remove water from a moist soil would have to overcome all three forces simultaneously.*

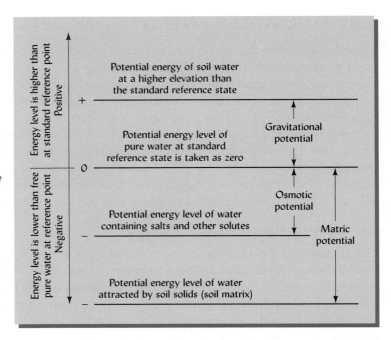

potential, ψ_h; and **osmotic potential**, ψ_o, respectively. All of these components act simultaneously to influence water behavior in soils. The general relationship of soil water potential to potential energy levels is shown in Figure 5.5 and can be expressed as:

$$\psi_t = \psi_g + \psi_m + \psi_o + \psi_h + \cdots \tag{5.2}$$

where the ellipsis ($\cdots$) indicates the possible contribution of additional potentials not yet mentioned.

Gravitational potential:
http://id.mind.net/~zona/
mstm/physics/
mechanics/ energy/
gravitationalPotentialEnergy/
gravitationalPotentialEnergy
.html

Gravitational Potential   The force of gravity pulls soil water toward the Earth's center. The gravitational potential, ψ_g, of soil water is the product of the acceleration due to gravity and the height of the soil water above a reference elevation. The reference elevation is usually chosen within the soil profile or at its lower boundary to ensure that the gravitational potential of soil water above the reference point will always be positive.

Following heavy precipitation, snow melt, or irrigation, gravity plays an important role in removing excess water from the upper horizons and in recharging groundwater below the soil profile (see Section 5.5).

Pressure Potential   The pressure potential component accounts for all other effects on soil water potential besides gravity and solute levels. Pressure potential most commonly includes (1) the positive hydrostatic pressure due to the weight of water in saturated soils and aquifers and (2) the negative pressure due to the attractive forces between the water and the soil solids or the soil matrix.

The **hydrostatic potential**, ψ_h, is a component that is operational only for water in saturated zones below the water table. Anyone who has dived to the bottom of a swimming pool has felt hydrostatic pressure on his or her eardrums.

The attraction of water to solid surfaces gives rise to the **matric potential** ψ_m, which is always negative because the water attracted by the soil matrix has an energy state lower than that of pure water. (These negative pressures are sometimes referred to as *suction* or *tension*. If these terms are used, their values are positive.) The matric potential is operational in unsaturated soil above the water table (Figure 5.6).

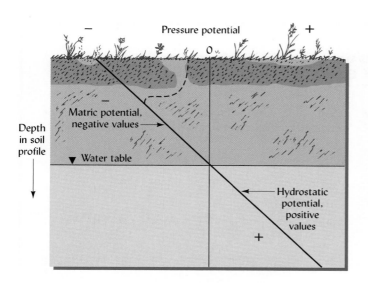

Figure 5.6
The matric potential and hydrostatic potential are both pressure potentials that may contribute to total water potential. The matric potential is always negative and the hydrostatic potential is positive. When water is in unsaturated soil above the water table (top of the saturated zone), it is subject to the influence of matric potentials. Water below the water table in saturated soil is subject to hydrostatic potentials. In the example shown here, the matric potential decreases linearly with elevation above the water table, signifying that water rising by capillary attraction up from the water table is the only source of water in this profile. Rainfall or irrigation (see dotted line) would alter or curve the straight line, but would not change the fundamental relationship described.

Matric potential, ψ_m, which results from adhesive forces and capillarity, influences both the retention and movement of soil water. Differences between the ψ_m of two adjoining soil zones encourage the movement of water from wetter (high energy state) areas to drier (low energy state) areas or from large pores to small pores. Although this movement may be slow, it is extremely important in supplying water to plant roots and in engineering applications.

Osmotic Potential   The osmotic potential, ψ_o, is attributable to the presence of both inorganic and organic solutes in the soil solution. As water molecules cluster around solute ions or molecules, the freedom of movement (and therefore the potential energy) of the water is reduced. The greater the concentration of solutes, the more osmotic potential is lowered. As always, water will tend to move to where its energy level will be lower, in this case to the zone of higher solute concentration. However, liquid water will move in response to differences in osmotic potential (the process termed **osmosis**) only if a *semipermeable membrane* exists between the zones of high and low osmotic potential, allowing water through but *preventing the movement of the solute*. If no membrane is present, movement of the solute, rather than of the water, largely equalizes concentrations.

Because soil zones are *not* generally separated by membranes, the osmotic potential, ψ_o, has little effect on the mass movement of water in soils. Its major effect is on the uptake of water by plant root cells that *are* isolated from the soil solution by their semipermeable cell membranes. In soils high in soluble salts, ψ_o may be lower (have a greater negative value) in the soil solution than in plant root cells. This leads to constraints in the uptake of water by plants. In very salty soil, the soil water osmotic potential may be low enough to cause cells in young seedlings to collapse (plasmolyze) as water moves from the cells to the lower osmotic potential zone in the soil.

Osmosis animation:
http://www.stolaf.edu/people/giannini/flashanimat/transport/osmosis.swf

Methods of Expressing Energy Levels

Several units can be used to express differences in energy levels of soil water. One is the *height of a water column* (usually in centimeters) whose weight just equals the potential under consideration. We have already encountered this means of expression since the *h* in the capillary equation (Section 5.2) tells us the matric potential of the water in a capillary pore. A second unit is the standard *atmosphere* pressure at sea level, which is 760 mm Hg or 1020 cm of water. Another unit termed *bar* approximates the pressure of a standard atmosphere. Energy may be expressed per unit of mass

Table 5.1
APPROXIMATE EQUIVALENTS AMONG EXPRESSIONS OF SOIL WATER POTENTIAL AND THE EQUIVALENT DIAMETER OF PORES EMPTIED OF WATER

Height of unit column of water, cm	Soil water potential, bars	Soil water potential, kPa[a]	Equivalent diameter of pores emptied, μm[b]
0	0	0	—
10.2	−0.01	−1	300
102	−0.1	−10	30
306	−0.3	−30	10
1,020	−1.0	−100	3
15,300	−15	−1,500	0.2
31,700	−31	−3,100	0.97
102,000	−100	−10,000	0.03

[a] The SI unit kilopascal (kPa) is equivalent to 0.01 bars.
[b] Smallest pore that can be emptied by equivalent tension as calculated using Eq. 5.1.

(**joules/kg**) or per unit of volume (**newtons/m²**). In the International System of Units (SI), 1 pascal (Pa) equals 1 newton (N) acting over an area of 1 m². In this textbook, we use Pa or kilopascals (kPa) to express soil water potential. Since other publications may use other units, Table 5.1 shows the equivalency among common means of expressing soil water potential.

5.4   SOIL WATER CONTENT AND SOIL WATER POTENTIAL

Types of soil water sensors: www.sowacs.com/sensors/index.html

The previous discussions suggest an inverse relationship between the water content of soils and the tenacity with which the water is held in soils. Many factors affect the relationship between soil water potential, ψ, and moisture content, θ. A few examples will illustrate this point.

Soil Water Versus Energy Curves

The relationship between soil water potential, ψ, and moisture content, θ, of three soils of different textures is shown in Figure 5.7. Such curves are sometimes termed *water release characteristic curves*, or simply *water characteristic curves*. The clay soil holds much more water at a given potential than does the loam or sand. Likewise, at a given moisture content, the water is held much more tenaciously (ψ_m is lower) in the clay than in the other two soils (note that soil water potential is plotted on a log scale).

Soil structure also influences soil water content–energy relationships. A well-granulated soil has more total pore space and greater overall water-holding capacity than one with poor granulation or one that has been compacted. Soil aggregation especially increases the relatively large inter-aggregate pores (Section 4.5) in which water is held with little tenacity (ψ_m is closer to zero). In contrast, a compacted soil will hold less total water, most of which will be held tightly in small- and medium-sized pores.

Measurement of Soil Water Status

The soil water characteristic curves just discussed highlight the importance of making two general kinds of soil water measurements: the *amount* of water present (water content) and the *energy status* of the water (soil water potential). In order to understand or

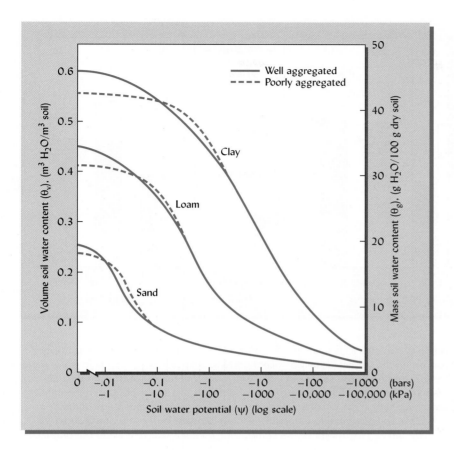

Figure 5.7

Soil water potential curves for three representative mineral soils. The curves show the relationship obtained by slowly drying completely saturated soils. The dashed lines show the effect of compaction or poor aggregation. The soil water potential, ψ(which is negative), is expressed in terms of bars (upper scale) and kilopascals (kPa) (lower scale). Note that the soil water potential is plotted on a log scale.

manage water supply and movement in soils, it is essential to have information (directly measured or inferred) on *both* types of measurements. For example, a soil water potential measurement might tell us whether water will move toward the groundwater, but without a corresponding measurement of the soil water content, we would not know the possible significance of the contribution to groundwater.

Generally, the behavior of soil water is most closely related to the energy status of the water, not to the amount of water in a soil. Thus, a clay loam and a loamy sand will both feel moist and will easily supply water to plants when the ψ_m is, say, −10 kPa. However, the amount of water held by the clay loam, and thus the length of time it could supply water to plants, would be far greater at this potential than would be the case for the loamy sand.

Water Content

The **volumetric water content**, θ_v, is defined as the volume of water associated with a given volume (usually 1 m³) of dry soil (see Figure 5.7). A comparable expression is the **mass water content**, θ_m, or the mass of water associated with a given mass (usually 1 kg) of dry soil. Both of these expressions have advantages for different uses. In most cases, we shall use the volumetric water content, θ_v, in this text.

As compaction reduces total porosity, it also increases θ_v (assuming a given θ_m), therefore often leaving too little air-filled pore space for optimal root activity. However, if a soil is initially very loose and highly aggregated (such as the forested A horizons described in Figure 5.8), moderate compaction may actually benefit plant growth by increasing the volume of pores that hold water between 10 and 1500 kPa of tension.

We think of plant root systems as exploring a certain depth of soil. We measure precipitation (and sometimes irrigation) as a depth of water (e.g., mm of rain). For

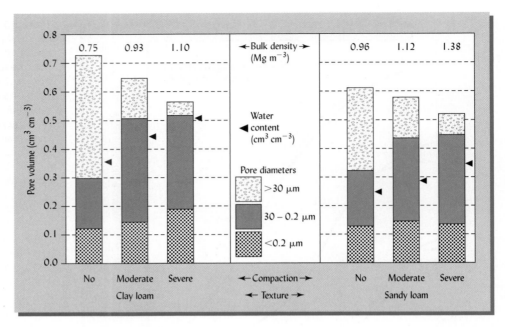

Figure 5.8

The compaction of two soils (clay loam on left, sandy loam at right) decreased total porosity mainly by converting the largest (usually air-filled) pores into smaller pores that hold water more tightly. These forested A horizon soils were initially so loose that the moderate compaction benefited plants by increasing the volume of water-holding 0.2 to 30 μm pores. On the other hand, the water originally in the uncompacted soil takes up a greater percentage of the pore volume (◄) when the soil is compacted, possibly leading to nearly water-saturated conditions. For example, here the clay loam with severe compaction contains 0.52 cm³ water but only 0.04 cm³ air per cm³ soil, less than the 0.10 cm³ air per cm³ soil (≈10% air porosity; see Section 7.2) thought to be required for good plant growth. [Adapted from Shestak and Busse (2005) with permission of The Soil Science Society of America]

such reasons, it is often convenient to express the volumetric water content as a *depth ratio* (depth of water per unit depth of soil). Conveniently, the numerical values for these two expressions are the same. For example, for a soil containing 0.1 m³ of water per m³ of soil (10% by volume), the depth ratio of water is 0.1 m of water per m of soil depth (see also Section 5.9).[3]

Gravimetric Method   The gravimetric method is a direct measurement of soil water content and is therefore the standard method by which all indirect methods are calibrated. The water associated with a given mass (and, if the bulk density of the soil is known, a given volume) of dry soil solids is determined. A sample of moist soil is weighed and then dried in an oven at a temperature of 105 °C for about 24 hours and finally weighed again. The weight loss represents the soil water. Box 5.1 provides examples of how θ_v and θ_m can be calculated. The gravimetric method is a *destructive* method (i.e., a soil sample must be removed for each measurement) and cannot easily be automated, thereby making it poorly suited for monitoring changes in soil moisture.

Electromagnetic Methods   Several non-destructive methods of measuring soil water content are based on the fact that the dielectric constant for water is very different from that of soil particles or air. The volumetric soil water content is calculated by a computer chip from signals sent to it by a probe that senses electrical properties of the

---

[3]When measuring amounts of water added to soil by irrigation, it is customary to use units of volume such as m³ and hectare-meter (the volume of water that would cover a hectare of land to a depth of 1 m). Generally, farmers and ranchers in the irrigated regions of the United States use the English units ft³ and acre-foot (the volume of water needed to cover an acre of land to a depth of 1 ft).

BOX 5.1
GRAVIMETRIC DETERMINATION OF SOIL WATER CONTENT

The gravimetric procedures for determining mass soil water content, θ_m, are relatively simple. Assume that you want to determine the water content of a 100-g sample of moist soil. You dry the sample in an oven kept at 105 °C and then weigh the soil again. Assume that the dried soil now weighs 70 g, which indicates that 30 g of water has been removed from the moist soil. Expressed in kilograms, this is 30 kg water associated with 70 kg dry soil.

Since the mass soil water content, θ_m, is commonly expressed in terms of kg water associated with 1 kg dry soil (not 1 kg of wet soil), it can be calculated as follows:

$$\frac{30 \text{ kg water}}{70 \text{ kg dry soil}} = \frac{X \text{ kg water}}{1 \text{ kg dry soil}}$$

$$X = \frac{30}{70} = 0.428 \text{ kg water/kg dry soil} = \theta_m$$

To calculate the volume soil water content, θ_v, we need to know the bulk density of the dried soil, which in this case we shall assume to be 1.3 Mg/m³. In other words, a cubic meter of this soil (*when dry*) has a mass of 1300 kg. From the above calculations, we know that the mass of water associated with this 1300 kg of dry soil is 0.428×1300, or 556 kg.

Since 1 m³ of water has a mass of 1000 kg, the 556 kg of water will occupy 556/1000, or 0.556 m³.

Thus, the volume water content is 0.556 m³/m³ of dry soil:

$$\frac{1300 \text{ kg soil}}{m^3 \text{ soil}} \times \frac{m^3 \text{ water}}{1000 \text{ kg water}} \times \frac{0.428 \text{ kg water}}{\text{kg soil}} = \frac{0.556 \text{ m}^3 \text{ water}}{m^3 \text{ soil}}$$

Assuming a soil that does not swell when wet, the relationship between the mass and volume water contents can be summarized as:

$$\boldsymbol{\theta_v = D_b \times \theta_m} \tag{5.3}$$

soil. The two most commonly used types of probes are *capacitance* and *TDR* (time domain reflectometry), both of which are easily integrated with data loggers and automated irrigation systems in the field (see Table 5.2 and Figure 5.9).

Water Potentials

Tensiometers   The tenacity with which water is attracted to soil particles is an expression of matric water potential, ψ_m. Field **tensiometers** (Figure 5.10) measure this attraction, or *tension*. The tensiometer is basically a water-filled tube closed at the bottom with a porous ceramic cup and at the top with an airtight seal. Once placed in the soil, water in the tensiometer moves through the porous cup into the adjacent soil until the water potential in the tensiometer is the same as the matric water potential in the soil. As the water is drawn out, a vacuum develops under the top seal, which can be measured by a vacuum gauge or an electronic transducer. If rain or irrigation rewets the soil, water will enter the tensiometer through the ceramic tip, reducing the vacuum or tension recorded by the gauge. Tensiometers are useful between 0 and –85 kPa potential, a range that includes half or more of the water stored in most soils. As the soil dries beyond 85 kPa, tensiometers fail because air is drawn in through the pores of the ceramic, relieving the vacuum. A solenoid switch can be fitted to a field tensiometer in order to automatically turn an irrigation system on and off.

Electrical Resistance Blocks   Electrical resistance blocks are made of porous gypsum, nylon, or fiberglass, suitably embedded with electrodes. When placed in moist soil, the porous block absorbs water in proportion to the soil water potential. The resistance to flow of electricity between the embedded electrodes decreases proportionately. It is possible to connect such blocks to data loggers or electronic switches so that irrigation systems can be turned on and off automatically at set soil moisture levels.

Table 5.2
SOME METHODS OF MEASURING SOIL WATER
More than one method may be needed to cover the entire range of soil moisture conditions.

| Method | Measures soil water | | | Used mainly in | | Comments |
	Content	Potential	Useful range, kPa	Field	Lab	
1. Gravimetric	X		0 to < −10,000		X	Destructive sampling; slow (1 to 2 days) unless microwave used. The standard for calibration.
2. Neutron scattering	X		0 to < −1500	X		Radiation permit needed; expensive equipment; not good in high-organic-matter soils; requires access tube.
3. Time domain reflectometry (TDR)	X		0 to < −10,000	X	X	Can be automated; accurate to ±1 to 2% volumetric water content; very sandy or salty soils need calibration; requires wave guides; expensive instrument.
4. Capacitance sensors	X		0 to < −1500	X	X	Can be automated; accurate to ±2 to 4% volumetric water content; sands or salty soils need calibration; simple, inexpensive sensors and recording instruments.
5. Resistance blocks		X	− 90 to < −1500	X	X	Can be automated; not sensitive near optimum plant water contents; may need calibration.
6. Tensiometer		X	0 to −85	X	X	Can be automated; accurate to ±0.1 to 1 kPa; limited range; inexpensive; needs periodic servicing to add water.
7. Thermocouple psychrometer		X	50 to < −10,000	X	X	Moderately expensive; wide range; accurate only to ±50 kPa.
8. Pressure membrane apparatus		X	50 to < −10,000		X	Used with gravimetric method to construct drier part of water characteristic curve.
9. Tension table		X	0 to −50		X	Used with gravimetric method to construct wetter part of water characteristic curve.

5.5   THE FLOW OF LIQUID WATER IN SOIL

All about groundwater flow: http://environment.uwe.ac .uk/geocal/SoilMech/water/ index.htm

Three types of water movement within the soil are recognized: (1) saturated flow, (2) unsaturated flow, and (3) vapor movement. In all cases, water flows in response to energy gradients, with water moving from a zone of higher to one of lower water potential. *Saturated flow* takes place when the soil pores are completely filled (or saturated) with water. *Unsaturated flow* occurs when the larger pores in the soil are filled with air, leaving only the smaller pores to hold and transmit water. *Vapor movement* occurs as vapor pressure differences develop in relatively dry soils.

Saturated Flow Through Soils

Under some conditions, at least part of a soil profile may be completely saturated; that is, all pores, large and small, are filled with water. The lower horizons of poorly drained soils are often saturated, as are portions of well-drained soils above stratified layers of clay. During and immediately following a heavy rain or irrigation, pores in the upper soil zones are often filled entirely with water.

Figure 5.9

Instrumental measurement of soil water content using time domain reflectometry (TDR). The instrument sends a pulse of electromagnetic energy down the two parallel metal rods of a waveguide that the soil scientist is pushing into the soil (inset photo). The TDR instrument makes precise picosecond measurements of the speed at which the pulse travels down the rods, a speed influenced by the nature of the surrounding soil. Microprocessors in the instrument analyze the wave patterns generated and calculate the apparent dielectric constant of the soil. Since the dielectric constant of a soil is mainly influenced by its water content, the instrument can accurately convert its measurements into volumetric water content of the soil. (Photos courtesy of R. Weil)

The quantity of water per unit of time, Q/t, that flows through a column of saturated soil (Figure 5.11) can be expressed by Darcy's law, as follows:

$$\frac{Q}{t} = AK_{\text{sat}}\frac{\Delta\psi}{L} \qquad (5.4)$$

where A is the cross-sectional area of the column through which the water flows, K_{sat} is the **saturated hydraulic conductivity**, $\Delta\psi$ is the change in water potential between the ends of the column (for example, $\psi_1 - \psi_2$), and L is the length of the column. For a given column, the rate of flow is determined by the ease with which the soil transmits water (K_{sat}) and the amount of force driving the water, namely the **water potential gradient** $\Delta\psi/L$. For saturated flow, this force may also be called the **hydraulic gradient**. By analogy, think of pumping water through a garden hose, with

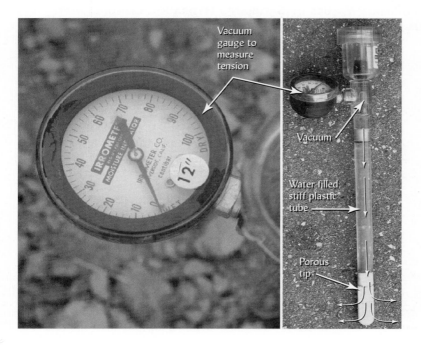

Figure 5.10

Tensiometer used to determine water potential in the field. The side view (right) shows the entire instrument. The tube is filled with water through the screw-off top. Once the instrument is tightly sealed, the white porous tip and the lower part of the plastic tube is inserted into a snug-fitting hole in the soil. The vacuum gauge (close up, left) will directly indicate the tension or negative potential generated as the soil draws the water out (curved arrows) through the porous tip. Note the scale goes up to only 100 centibars (= 100 kPa) tension at the driest. (Photos courtesy of R. Weil)

Figure 5.11

Saturated flow (percolation) in a column of soil with cross-sectional area A, cm². All soil pores are filled with water. At lower right, water is shown running off into a container to indicate that water is actually moving down the column. The force driving the water through the soil is the water potential gradient, $\psi_1 - \psi_2/L$, where both water potentials and length are expressed in cm (see Table 5.1). If we measure the quantity of water flowing out Q/t as cm³/s, we can rearrange Darcy's law (Eq. 5.4) to calculate the saturated hydraulic conductivity of the soil, K_{sat}, in cm/s as:

$$K_{sat} = \frac{Q}{A \cdot t} \frac{L}{\psi_1 - \psi_2} \quad (5.5)$$

Remember that the same principles apply where the water potential gradient moves the water in a horizontal direction.

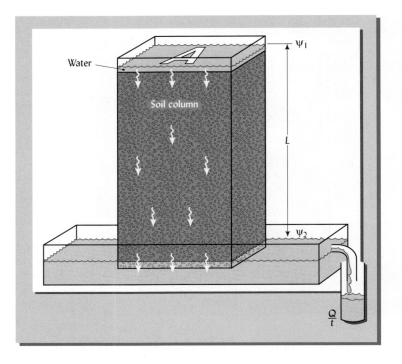

K_{sat} representing the size of the hose (water flows more readily through a larger hose) and $\Delta\psi/L$ representing the size of the pump that drives the water through the hose.

The units in which K_{sat} is measured are length/time, typically cm/s or cm/h. The K_{sat} is an important property that helps determine how well a soil or soil material will perform in such uses as irrigated cropland, sanitary landfill cover material, wastewater storage lagoon lining, and septic tank drain field (Table 5.3).

It should not be inferred from Figure 5.11 that saturated flow occurs only down the profile. The hydraulic force can also cause horizontal and even upward flow, as

Table 5.3
SOME APPROXIMATE VALUES OF SATURATED HYDRAULIC CONDUCTIVITY (IN VARIOUS UNITS) AND INTERPRETATIONS FOR SOIL USES

K_{sat}, cm/s	K_{sat}, cm/h	K_{sat}, in./h	Comments
1×10^{-2}	36	14	Typical of beach sand.
5×10^{-3}	18	7	Typical of very sandy soil, too rapid to effectively filter pollutants in wastewater.
5×10^{-4}	1.8	0.7	Typical of moderately permeable soils, K_{sat} between 1.0 and 15 cm/h considered suitable for most agricultural, recreational, and urban uses calling for good drainage.
5×10^{-5}	0.18	0.07	Typical of fine-textured, compacted, or poorly structured soils. Too slow for proper operation of septic tank drain fields, most types of irrigation, and many recreational uses such as playgrounds.
$<1 \times 10^{-8}$	$<3.6 \times 10^{-5}$	$<1.4 \times 10^{-5}$	Extremely slow; typical of compacted clay. K_{sat} of 10^{-5} to 10^{-8} cm/h may be required where nearly impermeable material is needed, as for wastewater lagoon lining or landfill cover material.

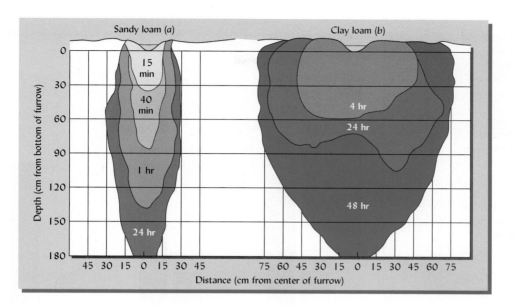

Figure 5.12

Comparative rates of irrigation water movement into a sandy loam and a clay loam. Note the much more rapid rate of movement in the sandy loam, especially in a downward direction.
[Redrawn from Cooney and Peterson (1955)]

occurs when groundwater wells up under a stream (see Section 6.6). Downward and horizontal flow is illustrated in Figure 5.12, which records the flow of water from an irrigation furrow into two soils, a sandy loam and a clay loam. The water moved down much more rapidly in the sandy loam than in the clay loam. On the other hand, horizontal movement (which would have been largely by *un*saturated flow) was much more evident in the clay loam.

Factors Influencing the Hydraulic Conductivity of Saturated Soils

Macropores   Anything affecting the size and configuration of soil pores will influence hydraulic conductivity. The total flow rate in soil pores is proportional to the fourth power of the radius. Thus, flow through a pore 1 mm in radius is equivalent to that in 10,000 pores with a radius of 0.1 mm. As a result, macropores (radius > 0.08 mm) account for nearly all water movement in saturated soils. However, air trapped in rapidly wetted soils can block pores and thereby reduce hydraulic conductivity. Similarly, the *interconnectedness* of pores is important as non-interconnected pores are like "dead-end streets" to flowing water. Vesicular pores in certain desert soils are examples (Plate 55).

The presence of biopores, such as root channels and earthworm burrows (typically >1 mm in radius), has a marked influence on the saturated hydraulic conductivity of different soil horizons (Plate 81). Because they usually have more macropore space, sandy soils generally have higher saturated conductivities than finer-textured soils. Likewise, soils with stable granular structure conduct water much more rapidly than do those with unstable structural units, which break down upon being wetted. Saturated conductivity of soils under perennial vegetation is commonly much higher than where annual plants are cultivated (Figure 5.13).

Preferential Flow   Scientists have been surprised to find more extensive pollution of groundwater from pesticides and other toxicants than would be predicted from traditional hydraulic conductivity measurements that assume uniform soil porosity. Apparently solutes (dissolved substances) are carried downward rapidly by water that moves through large macropores such as cracks and biopores, often before the bulk of the soil is thoroughly wetted. Mounting evidence suggests that this type of nonuniform water movement, referred to as **preferential flow**, greatly increases the chances of groundwater pollution (see Figure 5.14).

Figure 5.13

The effect of land management and soil texture on saturated conductivity (K_{sat}) of three soils in Canada. Soils under native woodlots had higher K_{sat} values, apparently due to higher organic matter contents and to preferential flow channels provided by decayed roots and burrowing animals. Tillage practices had little effect on conductivity in sand, but in loam and clay loam soils, conductivity was higher where no-tillage systems had been used, suggesting that no-till had increased the proportion of larger, water-conducting pores. [Drawn from averages of three methods of measuring K_{sat} in Reynolds et al. (2000)]

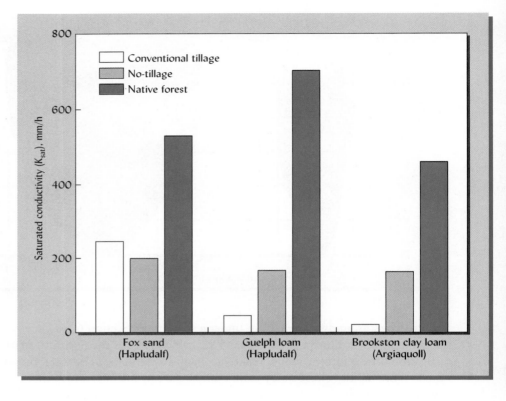

Macropores with continuity from the soil surface down through the profile encourage preferential flow. Such pores can result from animal burrowing, root channels, or clay shrinkage cracks. In very sandy soils, hydrophobic organic coatings on sand grains produce "fingers" of rapid wetting (see Plate 69). This "finger flow" probably is responsible for the finger-like shapes of the spodic horizon in some Spodosol profiles (e.g., Plate 10).

In some clay soils, water from the first rainstorm after a dry spell moves rapidly down shrinkage cracks, carrying with it soluble pesticides or nutrients that may be on the soil surface (Table 5.4). Chemicals and fecal bacteria leached by preferential flow can threaten human health, as well as environmental quality.

Unsaturated Flow in Soils

In unsaturated soils, most macropores are filled with air, leaving only the finer pores to accommodate water movement. The water content and, in turn, the tightness with which water is held (water potential) can be highly variable. In unsaturated soils, the

Figure 5.14

An illustration of preferential flow of water and pesticides downward to the water table. An herbicide (weed killer) was applied alongside a highway (right) with the expectation that downward movement into the water table would not be a serious problem since the surrounding soils were fine textured and would not be expected to readily permit infiltration of the chemical. As the deep rooted vegetation dried the soil, wide cracks formed in the swelling clay. Because of these cracks, the first heavy rain after a dry spell carried the chemicals into the groundwater before the soil could swell and shut the cracks. Through the groundwater the herbicide could move into nearby streams. [From DeMartinis and Cooper (1994), with permission of Lewis Publishers]

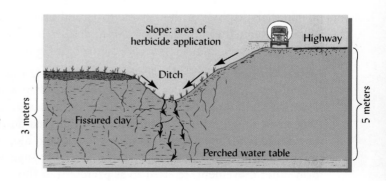

Table 5.4
LEACHING OF PESTICIDES BY PREFERENTIAL FLOW IN A SLOWLY PERMEABLE ALFISOL
Most of the spring leaching of three widely used pesticides took place following the first major storm of the year.

| | Leaching of pesticide, % of annual application (3-year average) | | |
Chemical	First storm	Spring season	First storm as % of season
Carbofuran	0.22	0.25	88
Atrazine	0.037	0.053	68
Cyanazine	0.02	0.02	100

Calculated from Kladivko et al. (1999).

primary driving force for water movement is the **matric potential gradient**, the difference in the matric potential of the moist soil areas and nearby drier areas into which the water is moving. Movement will be from a zone of thick moisture films (high matric potential, e.g., −1 kPa) to one of thin films (lower matric potential, e.g., −100 kPa).

Influence of Texture   Figure 5.15 shows the general relationship between matric potential, ψ_m (and, in turn, water content), and hydraulic conductivity of a sandy loam and clay soil. Note that at or near zero potential (which characterizes the saturated flow region), the hydraulic conductivity is thousands of times greater than at potentials that characterize typical unsaturated flow (−10 kPa and below).

How soil texture affects hydraulic properties: www.pedosphere.com/resources/texture/triangle_us.cfm

At high potential levels (high moisture contents), hydraulic conductivity is higher in the sand than in the clay. The opposite is true at low potential values (low moisture contents) when the clay soil has many more micropores that are still water-filled and can participate in unsaturated flow.

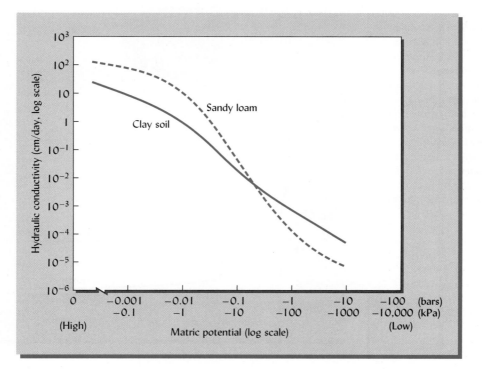

Figure 5.15
Generalized relationship between matric potential and hydraulic conductivity for a sandy soil and a clay soil (note log scales). Saturated flow takes place at or near zero potential, while much of the unsaturated flow occurs at a potential of −0.1 bar (−10 kPa) or below.

5.6   INFILTRATION AND PERCOLATION

A special case of water movement is the entry of free water (rainfall, snowmelt, or irrigation) into the soil at the soil–atmosphere interface. As we shall explain in Chapter 6, this is a pivotal process in landscape hydrology that greatly influences the moisture regime for plants and the potential for soil degradation, chemical runoff, and down-valley flooding.

Infiltration

The process by which water enters the soil pore spaces and becomes soil water is termed *infiltration*, and the rate at which water can enter the soil is termed the *infiltrability, i*:

$$i = \frac{Q}{A * t} \tag{5.6}$$

where Q is the volume quantity of water (m^3) infiltrating, A is the area of the soil surface (m^2) exposed to infiltration, and t is time (s). Since m^3 appears in the numerator and m^2 in the denominator, the units of infiltration can be simplified to m/s or, more commonly, cm/h. The infiltration rate is not constant over time, but generally decreases during an irrigation or rainfall episode. If the soil is quite dry when infiltration begins, all the macropores open to the surface will be available to conduct water into the soil. In soils with expanding types of clays, the initial infiltration rate may be particularly high as water pours into the network of shrinkage cracks. However, as infiltration proceeds, many macropores fill with water, and shrinkage cracks close up. The infiltration rate declines sharply at first and then tends to level off, remaining fairly constant thereafter (Figure 5.16).

Figure 5.16

The potential rate of water entry into the soil, or infiltration capacity, can be measured by recording the drop in water level in a double ring infiltrometer (top). Changes in the infiltration rate of several soils during a period of water application by rainfall or irrigation are shown (bottom). Generally, water enters a dry soil rapidly at first, but its infiltration rate slows as the soil becomes saturated. The decline is least for very sandy soils with macropores that do not depend on stable structure or clay shrinkage. In contrast, a soil high in expansive clays may have a very high initial infiltration rate when large cracks are open, but a very low infiltration rate once the clays swell with water and close the cracks. Most soils fall between these extremes, exhibiting a pattern similar to that shown for the silt loam soil. The dashed arrow indicates the level of K_{sat} for the silt loam illustrated. (Diagram courtesy of R. Weil)

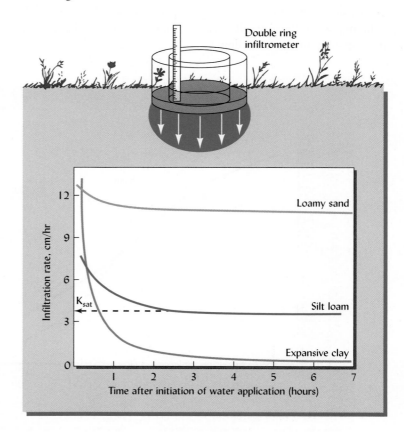

Percolation

Once it has infiltrated the soil, the water moves downward into the profile by the process termed **percolation**. Both saturated and unsaturated flow are involved in percolation of water down the profile, and rate of percolation is related to the soil's hydraulic conductivity. In the case of water that has infiltrated a relatively dry soil, the progress of water movement can be observed by the darkened color of the soil as it becomes wet (Figure 5.17). There usually appears to be a sharp boundary, termed a **wetting front**, between the dry underlying soil and the soil already wetted. During an intense rain or heavy irrigation, water movement near the soil surface occurs mainly by saturated flow in response to gravity. At the wetting front, however, water is moving into the underlying drier soil in response to matric potential gradients as well as gravity. During a light rain, both infiltration and percolation may occur mainly by unsaturated flow as water is drawn by matric forces into the fine pores without accumulating at the soil surface or in the macropores.

Water Movement in Stratified Soils

The fact that, at the wetting front, water is moving by unsaturated flow has important ramifications for how percolating water behaves when it encounters an abrupt change in pore sizes due to such layers as fragipans or claypans, or sand and gravel lenses. In some cases, such pore-size stratification may be created by soil managers,

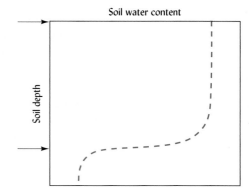

Figure 5.17

The wetting front 24 hours after a 5-cm rainfall. Water removal by plant roots had dried the upper 70 to 80 cm of this humid-region (Alabama) profile during a previous three-week dry spell. The clearly visible boundary results from the rather abrupt change in soil water content at the wetting front between the dry, lighter-colored soil and the soil darkened by the percolating water. The wavy nature of the wetting front in this natural field soil is evidence of the heterogeneity of pore sizes. The graph (right) indicates how soil water content decreases sharply at the wetting front. Scale in 10-cm intervals. (Photo courtesy of R. Weil)

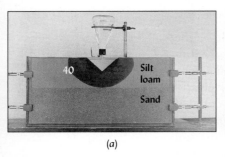
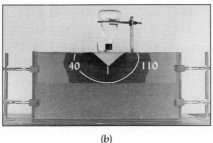
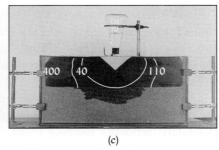

(a)   (b)   (c)

Figure 5.18

Downward water movement in soils with a stratified layer of coarse material. (a) Water is applied to the surface of a medium-textured topsoil. Note that after 40 min, downward movement is no greater than movement to the sides, indicating that in this case the gravitational force is insignificant compared to the matric potential gradient between dry and wet soil. (b) The downward movement stops when a coarse-textured layer is encountered. After 110 min, no movement into the sandy layer has occurred because the macropores of the sand provide less attraction for water than the finer-textured soil above. (c) After 400 min, the water content of the overlying layer becomes sufficiently high to give a water potential of about –1 kPa or more, and downward movement into the coarse material takes place. (Photos courtesy W. H. Gardner, Washington State University)

as when coarse plant residues are plowed under in a layer or a layer of gravel is placed under finer soil in a planting container. In all cases, the effect on water percolation is similar—that is, the downward movement is impeded—even though the causal mechanism may vary. The contrasting layer acts as a barrier to water flow and results in much higher field-moisture levels above the barrier than what would normally be encountered in freely drained soils. It is not surprising that percolating water should slow down markedly when it reaches a layer with finer pores, which therefore has a lower hydraulic conductivity. However, the fact that a layer of *coarser* pores will temporarily stop the movement of water may not be obvious (Figure 5.18 and Box 5.2).

The macropores of the sand offer less attraction for the water than do the finer pores of the overlying material. Since water always moves from higher to lower potential (to where it will be held more tightly), the wetting front cannot move readily into the sand. Eventually the downward-moving water will accumulate above the sand layer (if it cannot move laterally) and nearly saturate the pores at the soil–sand interface. The matric potential of the water at the wetting front will then fall to nearly zero or even become positive. Once this occurs, the water will be so loosely held by the fine-textured soil that gravity or hydrostatic pressure will force the water into the coarser layer.

Interestingly, a coarse sand layer in an otherwise fine-textured soil profile would also inhibit the *rise* of water from moist subsoil layers up to the surface soil, a situation that could be illustrated by turning Figure 5.18*b* upside down. The large pores in the coarse layer will not be able to support capillary movement up from the smaller pores in a finer layer. Consequently, water rises by capillarity up to the coarse-textured layer but cannot cross it to supply moisture to overlying layers. Thus, plants growing on some soils with buried gravel lenses are subject to drought since they are unable to exploit water in the lower soil layers. This principle also allows a layer of gravel to act as a capillary barrier under a concrete slab foundation to prevent water from soaking up from the soil and through the concrete floor of a home basement.

BOX 5.2
PRACTICAL APPLICATIONS OF UNSATURATED WATER FLOW IN CONTRASTING LAYERS

Unsaturated water flow always occurs from larger to smaller pores. Unsaturated flow is interrupted where soil texture abruptly changes from relatively fine to coarse because a larger pore cannot "pull" water from a smaller pore. If water is entering the system more rapidly than lateral capillarity can carry it away, a perched water table may develop above the interface between the two layers.

This phenomenon is applied in the design of golf course putting greens. The soil specified for the rooting zone consists almost entirely of sand in order to promote rapid infiltration of water and to resist compaction by foot traffic. However, water normally drains so fast through sand that too little is held to meet the needs of growing grass. This situation is remedied to some extent by constructing the putting green with a layer of gravel underneath the sand rooting zone. The large pores in the gravel temporarily stop the downward movement of water. The resulting perched water table (Figure 5.19) causes the sand layer to retain more water than it would otherwise, but still allows for rapid drainage of excess water.

The same principle is at the heart of a design proposed to keep nuclear wastes from contaminating groundwater during the many thousands of years required for the radionuclides to decay to harmless products. One plan is to store radioactive wastes in sealed containers kept in caverns deep inside Yucca Mountain, Nevada. Despite its desert location, the Yucca Mountain rock contains large amounts of water in pores and fractures, resulting in water dripping from the ceiling of the storage caverns. Although the containers for the heat-generating wastes are corrosion resistant, they would certainly corrode if exposed to moisture and air for thousands of years. To fend off the dripping water, a huge canopy made of special metal alloys was designed to cover the highly radioactive wastes. This is an extremely difficult and expensive approach, with no guarantee that the structure would not deteriorate over the millennia. Therefore, a far easier, less expensive, and more reliable alternative has been proposed. Burying the waste containers under mounds of first gravel and then sand (Figure 5.20) in a texture-stratified system would create a *capillary barrier* to protect the containers. Water dripping into the sand would be held by capillary forces in the relatively small pores between sand grains. As the water enters the sand, it moves by capillary flow along matric potential gradients.

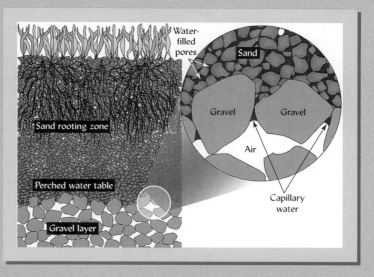

Figure 5.19
A gravel layer increases the water available to turfgrass roots in the sand rooting zone, while allowing rapid drainage if saturation occurs. (Diagram courtesy of R. Weil)

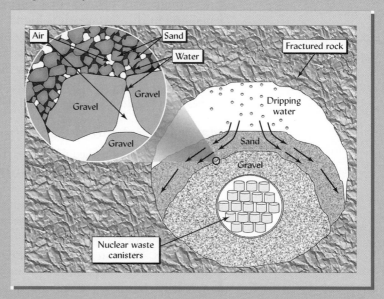

Figure 5.20
Proposed use of capillary principles to prevent dripping water from corroding toxic nuclear waste containers inside Yucca Mountain in Nevada. [Diagram courtesy of R. Weil, based on concepts in Carter and Pigford (2005)]

Its downward movement would be interrupted when it reached the much larger pores of the gravel layer. The containers would remain dry because both matric potential gradients and gravity would cause the water in the sand layer to move along the curved interface rather than into the gravel layer (arrows in Figure 5.20).

5.7   WATER VAPOR MOVEMENT IN SOILS

Water vapor moves from one point to another within the soil in response to differences in vapor pressure. Thus, water vapor will move from a moist soil where the soil air is nearly 100% saturated with water vapor (high vapor pressure) to a drier soil where the vapor pressure is somewhat lower. Also, water vapor will move from a zone of low salt content to one with a higher salt content (e.g., around a fertilizer granule). The salt lowers the vapor pressure of the water and encourages water movement from the surrounding soil.

If the temperature of one part of a uniformly moist soil is lowered, the vapor pressure will decrease and water vapor will tend to move toward this cooler part. Heating will have the opposite effect in that heating will increase the vapor pressure, and the water vapor will move away from the heated area.

Even though the amount of water vapor is small, seeds of some plants can absorb sufficient water vapor from the soil to stimulate germination. Likewise, water vapor movement may be of considerable significance to drought-resistant desert plants (*xerophytes*), many of which can exist at extremely low soil water contents. For instance, at night the surface horizon of a desert soil may cool sufficiently to cause vapor movement up from deeper layers. If cooled enough, the vapor may then condense as dewdrops in the soil pores, supplying certain shallow-rooted xerophytes with water for survival.

5.8 QUALITATIVE DESCRIPTION OF SOIL WETNESS

Feel and appearance of soils from field capacity to near wilting point: http://www.wy.nrcs.usda .gov/technical/soilmoisture/ soilmoisture.html

As an initially water-saturated soil dries down, both the soil as a whole and the soil water it contains undergo a series of gradual changes in physical behavior and in their relationships with plants. These changes are due mainly to the fact that the water remaining in the drying soil is found in smaller pores and thinner films where the water potential is lowered principally by the action of matric forces.

To study these changes and introduce the terms commonly used to describe varying degrees of soil wetness, we shall follow the moisture and energy status of soil during and after wetting. The terms to be introduced describe various stages along a continuum of soil wetness and should not be interpreted to imply that soil water exists in different "forms."

Maximum Retentive Capacity

When all soil pores are filled with water, the soil is said to be *water-saturated* (Figure 5.21) and at its **maximum retentive capacity**. The matric potential is close to zero, nearly the same as that of free water. The volumetric water content is essentially the same as the total porosity. The soil will remain at maximum retentive capacity only so long as water continues to infiltrate, for the water in the largest pores (sometimes termed **gravitational water**) will percolate downward, mainly under the influence of gravitational forces. Data on maximum retentive capacities and the average depth of soils in a watershed are useful in predicting how much rainwater can be stored in the soil temporarily, thus possibly avoiding downstream floods.

Field Capacity

Once the rain or irrigation has ceased, water in the largest soil pores will drain downward quite rapidly in response to the hydraulic gradient (mostly gravity). After one to three days, this rapid downward movement will become negligible as matric forces play a greater role in the movement of the remaining water (Figure 5.22). The soil then is said to be at its **field capacity**. In this condition, water has moved out of the macropores and air has moved in to take its place. The micropores or capillary pores are still filled with water and can supply plants with needed water. The matric potential will vary slightly from soil to soil but is generally in the range of -10 to -30 kPa, assuming

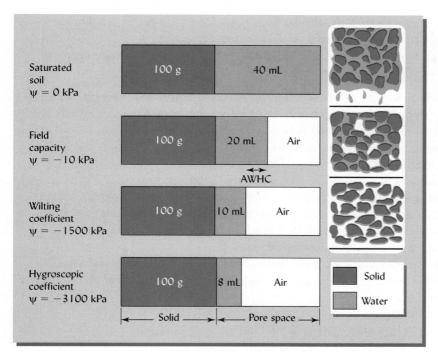

Volumes of water and air associated with 100 g of soil solids in a representative well-granulated silt loam. The top bar shows the situation when the soil is completely saturated with water. This situation will usually occur for short periods of time when water is being added. Water will soon drain out of the larger pores (macropores). The soil is then said to be at field capacity. Plants will remove water from the soil quite rapidly until they begin to wilt. When permanent wilting of the plants occurs, the soil water content is said to be at the wilting coefficient. There is still considerable water in the soil, but it is held too tightly to permit its absorption by plant roots. The water lost between field capacity and wilting coefficient is considered to be the soil's plant available water-holding capacity (AWHC). A further reduction in water content to the hygroscopic coefficient is illustrated in the bottom bar. At this point the water is held very tightly, mostly by the soil colloids.

Figure 5.21

drainage into a less-moist zone of similar porosity.[4] Water movement will continue to take place by unsaturated flow, but the rate of movement is very slow since it now is due primarily to capillary forces, which are effective only in micropores (Figure 5.21). The water found in pores small enough to retain it against rapid gravitational drainage, but large enough to allow capillary flow in response to matric potential gradients, is sometimes termed **capillary water.**

While all soil water is affected by gravity, the term *gravitational water* refers to the portion of soil water that readily drains away between the states of maximum retentive capacity and field capacity. Gravitational water includes much of the water

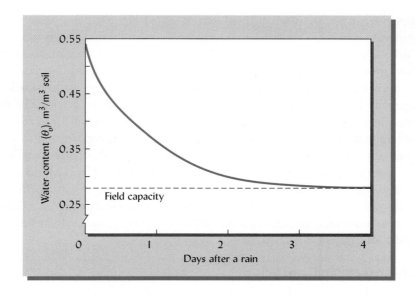

Figure 5.22
The water content of a soil drops quite rapidly by drainage following a period of saturation by rain or irrigation. After two or three days the rate of water drainage out of the soil is very slow and the soil is said to be at field capacity.
(Diagram courtesy of R. Weil)

[4]Note that because of the relationships pertaining to water movement in stratified soils (see Section 5.6), soil in a flower pot will cease drainage while much wetter than field capacity.

that transports chemicals such as nutrient ions, pesticides, and organic contaminants into the groundwater and, ultimately, into streams and rivers.

Field capacity is a very useful term because it refers to an approximate degree of soil wetness at which several important soil properties are in transition:

1. At field capacity, a soil is holding the maximal amount of water useful to plants. Additional water, while held with low energy of retention, would be of limited use to upland plants because it would remain in the soil for only a short time before draining, and, while in the soil, it would occupy the larger pores, thereby reducing soil aeration.

2. At field capacity, the soil is near its lower plastic limit—that is, the soil behaves as a crumbly semisolid at water contents below field capacity, and as a plastic putty-like material that easily turns to mud at water contents above field capacity (see Section 4.9). Therefore, field capacity approximates the optimal wetness for ease of tillage or excavation.

3. At field capacity, sufficient pore space is filled with air to allow good aeration for most aerobic microbial activity and for the growth of most plants (see Section 7.6).

Permanent Wilting Percentage, or Wilting Coefficient

Once an unvegetated soil has drained to its field capacity, further drying is quite slow, especially if the soil surface is covered to reduce evaporation. However, if plants are growing in the soil, they will remove water from their rooting zone, and the soil will continue to dry. The roots will remove water first from the largest water-filled pores, where the water potential is relatively high. As these pores are emptied, roots will draw their water from the progressively smaller pores and thinner water films in which the matric water potential is lower and the forces attracting water to the solid surfaces are greater. Hence, it will become progressively more difficult for plants to remove water from the soil at a rate sufficient to meet their needs.

As the soil dries, the rate of plant water removal may fail to keep up with plant needs, and herbaceous plant may begin to wilt during the daytime to conserve moisture. At first the plants will regain their turgor at night when water is not being lost through the leaves, and the roots can catch up with the plants' demand. Ultimately, however, an herbaceous plant will remain wilted night and day when its roots cannot generate water potentials low enough to coax the remaining water from the soil. Although they may not show wilting symtoms, most trees and other woody plants also have great difficulty obtaining any water from soil in this condition. The water content of the soil at this stage is called the **wilting coefficient**, or **permanent wilting percentage**, and by convention is taken to be the amount of water retained by the soil when the water potential is <1500 kPa (Figure 5.23). The soil will appear to be dusty dry, although some water remains in the smallest of the micropores and in very thin films (perhaps only 10 molecules thick) around individual soil particles (see Figure 5.21).

As illustrated in Figure 5.22, plant **available water** is considered to be that water retained in soils between the states of field capacity and wilting coefficient (between <10 to <30 kPa and <1500 kPa). The amount of capillary water remaining in the soil that is unavailable to higher plants can be substantial, especially in fine-textured soils and those high in organic matter.

Hygroscopic Coefficient

Although plant roots do not generally dry the soil much beyond the permanent wilting percentage, if the soil is exposed to the air, water will continue to be lost by evaporation. When soil moisture is lowered below the wilting point, the water molecules that

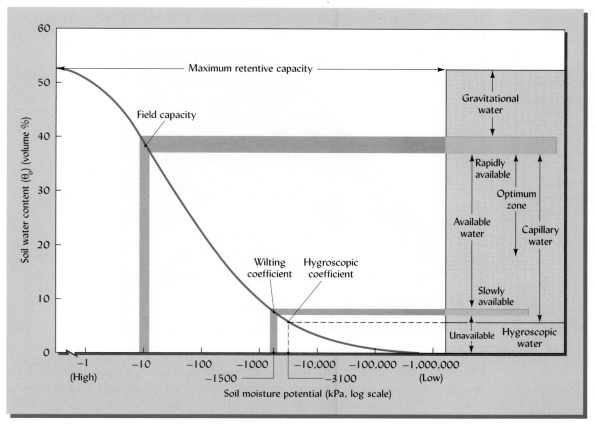

Figure 5.23

Water content–matric potential curve of a loam soil as related to different terms used to describe water in soils. The shaded bars in the diagram to the right suggest that measurements such as field capacity are only approximations. The gradual change in potential with soil moisture change discourages the concept of different "forms" of water in soils. At the same time, such terms as gravitational and available assist in the qualitative description of moisture utilization in soils.

remain are very tightly held, mostly being adsorbed by colloidal soil surfaces. This state is approximated when the atmosphere above a soil sample is essentially saturated with water vapor (98% relative humidity) and equilibrium is established at a water potential of <3100 kPa. The water is thought to be in films only 4 or 5 molecules thick and is held so rigidly that much of it is considered nonliquid and can move only in the vapor phase. The moisture content of the soil at this point is termed the **hygroscopic coefficient**.

5.9 FACTORS AFFECTING THE AMOUNT OF PLANT-AVAILABLE SOIL WATER

What is available water holding capacity? http://soils.usda.gov/sqi/publications/files/avwater.pdf

As illustrated in Figure 5.23, there is a relationship between the water potential of a given soil and the amount of water held at field capacity and at permanent wilting percentage, the two boundary properties determining the available water-holding capacity. This energy-controlling concept should be kept in mind as we consider the various soil properties that affect the amount of water a soil can hold for plant use.

The general influence of texture on field capacity, wilting coefficient, and **available water-holding capacity** is shown in Figure 5.24. Note that as fineness of texture increases, there is a general increase in available moisture storage from sands to

Figure 5.24

General relationship between soil water characteristics and soil texture. Note that the wilting coefficient increases as the texture becomes finer. The field capacity increases until we reach the silt loams, then levels off. Remember these are representative curves; individual soils would probably have values different from those shown.

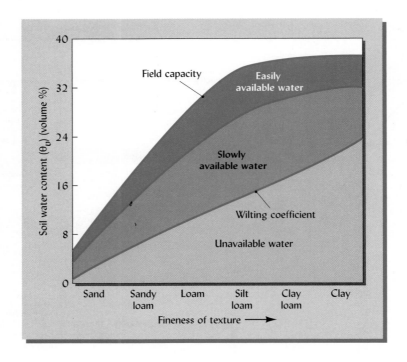

loams and silt loams. Plants growing on sandy soils are more apt to suffer from drought than are those growing on a silt loam in the same area (see Plate 45). However, clay soils frequently provide less available water than do well-granulated silt loams since the clays tend to have a high wilting coefficient.

The influence of organic matter deserves special attention. The available water-holding capacity of a well-drained mineral soil containing 5% organic matter is generally higher than that of a comparable soil with 3% organic matter. Evidence suggests that soil organic matter exerts both direct and indirect influences on soil water availability.

The direct effects are due to the very high water-holding capacity of organic matter, which, when the soil is at the field capacity, is much higher than that of an equal volume of mineral matter. Even though the water held by organic matter at the wilting point is also somewhat higher than that held by mineral matter, the amount of water available for plant uptake is still greater from the organic fraction (Figure 5.25).

Organic matter indirectly affects the amount of water available to plants because it helps stabilize soil structure and increase the total volume as well as the size of pores. This results in an increase in water infiltration and water-holding capacity with a simultaneous increase in the amount of water held at the wilting coefficient. Recognizing the beneficial effects of organic matter on plant-available water is essential for wise soil management.

Compaction Effects on Matric Potential, Aeration, and Root Growth

Soil compaction generally reduces the amount of water that plants can take up. First, as the clay particles are forced closer together, soil strength may increase beyond about 2000 kPa, the level considered to limit root penetration (Section 4.7). Second, compaction decreases the total pore space, which generally means that less water is retained at field capacity. Third, reduction in macropore size and numbers generally means less air pore space when the soil is near field capacity. Fourth, the creation of more very fine micropores will increase the permanent wilting coefficient and so decrease the available water content.

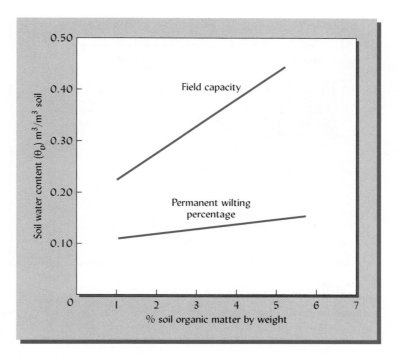

Figure 5.25

The effects of organic matter content on the field capacity and permanent wilting percentage of a number of silt loam soils. The differences between the two lines shown is the available soil moisture content, which was obviously greater in the soils with higher organic matter levels. [Redrawn from Hudson (1994); used with permission of the Soil & Water Conservation Society]

Least Limiting Water Range   We have already defined **plant-available water** as that held with matric potentials between field capacity (<10 to <30 kPa) and the permanent wilting point (<1500 kPa). Thus, plant-available water is that which is not held too tightly for roots to take up and yet is not held so loosely that it freely drains away by gravity. The **least limiting water range** is that range of water contents for which soil conditions do not severely restrict root growth. According to the least limiting water range concept, soils are too *wet* for normal root growth when so much of the soil pore space is filled with water that less than about 10% remains filled with air. At this water content, lack of oxygen for respiration limits root growth. In loose, well-aggregated soils, this water content corresponds quite closely to field capacity. However, in a compacted soil with very few large pores, oxygen supply may become limiting at lower water contents (and potentials) because some of the smaller pores will be needed for air.

The least limiting water range concept tells us that soils are too *dry* for normal root growth when the soil strength (measured as the pressure required to push a pointed rod through the soil) exceeds about 2000 kPa. This level of soil strength occurs at water contents near the wilting point in loose, well-aggregated soils, but may occur at considerably higher water contents if the soil is compacted (see Figure 5.26). To summarize, the least limiting water range concept suggests that root growth is limited by lack of oxygen at the wet end of the range and by the inability of roots to physically push through the soil at the dry end. Thus, compaction effects on root growth are most pronounced in dry soils (Figure 5.27).

Osmotic Potential

The presence of soluble salts, either from applied fertilizers or as naturally occurring compounds, can influence plant uptake of soil water. For soils high in salts, the osmotic potential tends to reduce available moisture because more water is retained in the soil at the permanent wilting coefficient than would be the case due to matric potential alone. In most humid region soils, these osmotic potential effects are insignificant, but they become of considerable importance for certain soils in dry regions that may accumulate soluble salts through irrigation or natural processes.

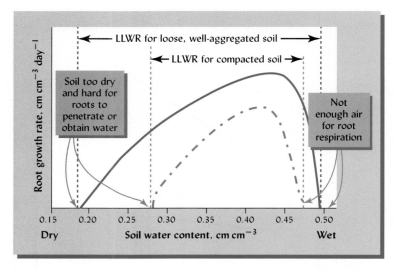

Figure 5.26

Compaction reduces the range of soil water contents suitable for plant growth (the least limiting water range, [LLWR]). Near the wet end of the soil water content scale (right), root growth (curved line) is limited by lack of air for root respiration. Once the soil dries a little, the largest pores drain and fill with air. Neither water nor air is then limiting and root growth rate becomes maximal. With further drying, low water potentials make it more difficult for roots to obtain moisture, and the soil increases its mechanical resistance to root penetration. Root growth declines until the soil is so dry that roots cannot grow at all (left). The lower (dashed) curve depicts the reduced rate of root growth that would pertain if the soil were compacted. Because compaction compresses the largest pores, it takes somewhat less water than before to create an oxygen-limited condition that reduces root growth. Toward the dry end of the scale, higher soil strength brings root growth to a halt at a water content that would still support considerable growth in an uncompacted soil. (Diagram courtesy of R. Weil, concepts from Da Silva and Kay, 1997)

Soil Depth and Layering

The total volume of available water will depend on the total volume of soil explored by plant roots (see Plate 85). This volume may be governed by the total depth of soil above root-restricting layers (see Figure 5.28), by the greatest rooting depth characteristic of a particular plant species, or even by the size of a flower pot chosen for containerized plants.

The capacity of soils to store available water determines to a great extent their usefulness for plant growth. To estimate the water-holding capacity of a soil, each soil horizon to which roots have access may be considered separately and then summed to give a total water-holding capacity for the profile (see Box 5.3).

Figure 5.27

Root growth of lodgepole pine tree seedlings in response to increased compaction at three soil water levels. Compaction affected root growth only when soil water was low, probably because root growth was limited by high soil strength. The tree seedlings were grown for 12 weeks in pots of mineral soil collected during timber harvest in British Columbia, Canada. The soil was compacted to three bulk density levels. Water was added as required to maintain volumetric water contents of 0.10–0.15 (low), 0.20–0.30 (medium), and 0.30–0.35 (high) cm³/cm³. [Drawn from data in Blouin et al. (2004)]

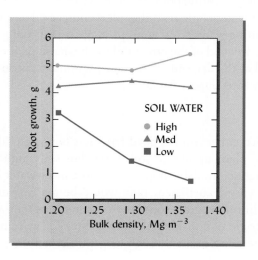

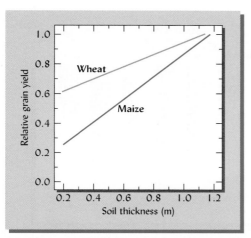

Figure 5.28

Relative grain yields for maize and wheat in relation to thickness of soil available for rooting. The crops were all grown with no-till management in Argiudolls and Paleudolls of the southeastern Pampas in Argentina. These soils held about 1.5 mm of available water per 1 cm of soil depth. Maize, which was grown during the hot, dry weather from spring to fall, was much more responsive to increased soil thickness than winter wheat, which was grown during the cooler, low-water-demand period from fall to spring. In this case, the soil thickness was limited by a root-restricting petrocalcic (cemented) horizon. [Redrawn from Sadras and Calvino (2001)]

BOX 5.3
TOTAL AVAILABLE WATER-HOLDING CAPACITY OF A SOIL PROFILE

The total amount of water available to a plant growing in a field soil can be estimated from the rooting depth of the plant and the amount of water held between field capacity and wilting percentage in each of the soil horizons explored by the roots. For each soil horizon, the mass available water-holding capacity (*AWHC*) is estimated as the difference between the mass water content at field capacity, θ_{mFC} (g water per 100 g soil at field capacity) and that at permanent wilting percentage, θ_{mWP}. We can convert this value into a volume water content, θ_v, by multiplying by the ratio of the bulk density, D_b, of the soil to the density of water, D_w. Finally, this volume ratio is multiplied by the thickness of the horizon to give the total centimeters of available water capacity *AWHC* in that horizon:

$$AWHC = (\theta_{mFC} - \theta_{mWP}) * D_b * D_w * L \tag{5.7}$$

For the first horizon described in Table 5.5, we can substitute values (with units) into Eq. 5.7:

$$AWHC = \left(\frac{22\,g}{100\,g} - \frac{8\,g}{100\,g}\right) * \frac{1.2\,Mg}{m^3} * \frac{1\,m^3}{1\,Mg} * 20\,cm = 3.36\,cm$$

Note that all units cancel out except cm, resulting in the depth of available water (cm) held by the horizon. In Table 5.5, the *AWHC* of all horizons within the rooting zone are summed to give a total *AWHC* for the soil–plant system. Since no roots penetrated to the last horizon (1.0 to 1.25 m), this horizon was not included in the calculation. We can conclude that for the soil–plant system illustrated, 14.13 cm of water could be stored for plant use. At a typical summertime water-use rate of 0.5 cm of water per day, this soil could hold about a four-week supply.

TABLE 5.5
CALCULATION OF ESTIMATED SOIL PROFILE AVAILABLE WATER-HOLDING CAPACITY

Soil depth, cm	Relative root length	Soil depth increment, cm	Soil bulk density, Mg/m³	Field capacity (FC), g/100 g	Wilting percentage (WP), g/100 g	Available water-holding capacity (AWHC), cm
0–20	xxxxxxxxx	20	1.2	22	8	$20 * 1.2\left(\frac{22}{100} - \frac{8}{100}\right) = 3.36\,cm$
20–40	xxxx	20	1.4	16	7	$20 * 1.4\left(\frac{16}{100} - \frac{7}{100}\right) = 2.52\,cm$
40–75	xx	35	1.5	20	10	$35 * 1.5\left(\frac{20}{100} - \frac{10}{100}\right) = 5.25\,cm$
75–100	xx	25	1.5	18	10	$25 * 1.5\left(\frac{18}{100} - \frac{10}{100}\right) = 3.00\,cm$
100–125 Total	—	25	1.6	15	11	No roots 3.36 + 2.52 + 5.25 + 3.00 = 14.13 cm

Figure 5.29

The soil depth above which 95% of all roots are located under different vegetation and soil types. The analysis used 475 root profiles reported from 209 geographic locations. The deepest rooting depths were found mainly in water-limited ecosystems. Within all but the wettest ecosystem, rooting was deeper in the sandier soils. Globally, 9 out of 10 profiles had at least 50% of all roots in the upper 30 cm of the soil profile and 95% of all roots in the upper 2 m (including any O horizons present). [Redrawn from Schenk and Jackson (2002)]

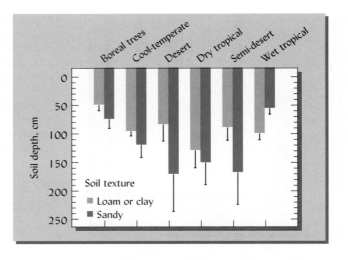

Root Distribution

The distribution of roots in the soil profile determines to a considerable degree the plant's ability to absorb soil water. Most plants, both annuals and perennials, have the bulk of their roots in the upper 25 to 30 cm of the profile. For most plants in limited-rainfall regions, roots explore relatively deep soil layers, but usually 95% of the entire root system is contained in the upper 2 m of soils. As Figure 5.29 shows, in more humid regions, rooting tends to be somewhat shallower. Perennial plants, both woody and herbaceous, produce some roots that grow very deeply (>3 m) and are able to absorb a considerable proportion of their moisture from deep subsoil layers. Even in these cases, however, it is likely that much of the root absorption is from the upper layers of the soil, provided these layers are well supplied with water. On the other hand, if the upper soil layers are moisture deficient, even annual row crops such as sunflower, corn, and soybeans will absorb much of their water from the lower horizons, provided that adverse physical or chemical conditions do not inhibit root exploration of these horizons.

5.10 CONCLUSION

Water impacts all life. The interactions and movement of this simple compound in soils help determine whether these impacts are positive or negative. An understanding of the principles that govern the attraction of water for soil solids and for dissolved ions can help maximize the positive impacts while minimizing the less desirable ones.

The water molecule has a polar structure that results in electrostatic attraction of water to both soluble cations and soil solids. These attractive forces tend to reduce the potential energy level of soil water below that of pure water. The extent of this reduction, called soil water potential, ψ, has a profound influence on a number of soil properties, but especially on the movement of soil water and its uptake by plants.

The water potential due to the attraction between soil solids and water (the matric potential, ψ_m) combines with the gravitational potential, ψ_g, to largely control water movement. This movement is relatively rapid in soils high in moisture and with an abundance of macropores. In drier soils, however, the adsorption of water on the soil solids is so strong that its movement in the soil and its uptake by plants are greatly reduced. As a consequence—even though there is still a significant quantity of water in the soil—plants may die for lack of water because that water is unavailable to them.

Water is supplied to plants by capillary movement toward the root surfaces and by growth of the roots into moist soil areas. In addition, vapor movement may be of significance in supplying water for drought-resistant desert species (xerophytes). The osmotic potential, ψ_o, becomes important in soils with high soluble salt levels that can impede plant uptake of water from the soil. Such conditions occur most often in soils with restricted drainage in areas of low rainfall and in potted indoor plants.

The characteristics and behavior of soil water are very complex. As we have gained more knowledge, however, it has become apparent that soil water is governed by relatively simple, basic physical principles. Furthermore, researchers are discovering the similarity between these principles and those governing the movement of groundwater and the uptake and use of soil moisture by plants—the subject of the next chapter.

STUDY QUESTIONS

1. What is the role of the *reference state of water* in defining soil water potential? Describe the properties of this reference state of water.

2. Imagine a root of a cotton plant growing in the upper horizon of an irrigated soil in California's Imperial Valley. As the root attempts to draw water molecules from this soil, what forces (potentials) must it overcome? If this soil were compacted by a heavy vehicle, which of these forces would be most affected? Explain.

3. Using the terms *adhesion, cohesion, meniscus, surface tension, atmospheric pressure*, and *hydrophilic surface*, write a brief essay to explain why water rises up from the water table in a mineral soil.

4. Suppose you were hired to design an automatic irrigating system for a wealthy homeowner's garden. You determine that the flower beds should be kept at a water potential above −60 kPa but not wetter than −10 kPa, as the annual flowers here are sensitive to both drought and lack of good aeration. The rough turf areas, however, can do well if the soil dries to as low as −300 kPa. Your budget allows either tensiometers or electrical resistance blocks to be hooked up to electronic switching valves. Which instruments would you use and where? Explain.

5. Suppose the homeowner referred to in question 4 increased your budget and asked to use the TDR method to measure soil water contents. What additional information about the soils, not necessary for using the tensiometer, would you have to obtain to use the TDR instrument? Explain.

6. A greenhouse operator was growing ornamental woody plants in 15-cm-tall plastic containers filled with a loamy sand. He watered the containers daily with a sprinkler system, letting any excess water drain out the bottom. His first batch of 1000 plants yellowed and died from too much water and not enough air. As an employee of the greenhouse, you suggest that he use 30-cm-tall pots for the next batch of plants. Explain your reasoning.

7. Suppose you measured the following data for a soil:

θ_m at different water tensions, kg water/kg dry soil

Horizon	Bulk density, Mg/m^3	−10 kPa	−100 kPa	−1500 kPa
A (0–30 cm)	1.28	0.28	0.20	0.08
B$_t$ (30–70 cm)	1.40	0.30	0.25	0.15
B$_x$ (70–120 cm)	1.95	0.20	0.15	0.05

Estimate the total available water-holding capacity (AWHC) of this soil in centimeters of water.

8. A forester obtained a cylindrical core ($L = 15$ cm, $r = 3.25$ cm) of soil from a field site. She placed all the soil in a metal can with a tight-fitting lid. The empty metal can weighed 300 g and when filled with the field-moist soil weighed 972 g. Back in the lab, she placed the can of soil, with lid removed, in an oven for several days until it ceased to lose weight. The weight of the dried can with soil (including the lid) was 870 g. Calculate both θ_m and θ_v.

9. Give four reasons why compacting a soil is likely to reduce the amount of water available to growing plants.

10. Since even rapidly growing, finely branched root systems rarely contact more than 1 or 2% of the soil particle surfaces, how is it that the roots can utilize much more than 1 or 2% of the water held on these surfaces?

11. For two soils subjected to "no," "moderate," or "severe" compaction, Figure 5.8 shows the volume fraction (cm^3 cm^{-3}) of pores in three size classes. The symbol ◄ indicates the volume fraction of water, θ_v (cm^3 cm^{-3}), in each soil. The figure

indicates that $\theta_v \approx 0.35$ for the uncompacted clay loam (bulk density = 0.75 g cm^{-3}). Show a complete calculation (with all units) to demonstrate that ◄ in the figure correctly indicates that $\theta_v \approx 0.36$ for the severely compacted sandy loam (bulk density = 1.10 g/cm^3).

12. Fill in the shaded cells of this table to show the cm of available water capacity in the entire 90-cm profile. Show complete calculations for the first (upper left) and last (lower right) of the shaded cells:

Soil depth, cm	Bulk density, D_b (Mg m^{-3})	Field capacity		Wilting point		Available water	
		θ_m, %[a]	θ_d, cm[b]	θ_m, %	θ_d, cm	θ_m, %	θ_d, cm
0–30	1.48	27.1		17.9		9.2	
30–60	1.51	27.5		18.1		9.4	
60–90	1.55	27.1		20.0		7.1	
0–90	—	—		—		—	

[a] θ_m is water content by mass and is reported here as percent (%), which is equivalent to g water/100 g dry soil.
[b] θ_d is water content by depth and is reported as cm of water held in the soil layer indicated.

REFERENCES

Blouin, V., M. Schmidt, C. Bulmer, and M. Krzic. 2004. "Soil compaction and water content effects on lodgepole pine seedling growth in British Columbia," in B. Singh (ed.), *Supersoil 2004.* Program and abstracts for the 3rd Australian–New Zealand soils conference. University of Sydney, Sydney, Australia, www.regional.org.au/au/asssi/supersoil2004/s14/oral/2036_blouinv. htm.

Carter, L. J., and T. H. Pigford. 2005. "Proof of safety at Yucca Mountain," *Science* **310**:447–448.

Cooney, J. J., and J. E. Peterson. 1955. *Avocado Irrigation.* Leaflet 50. California Agricultural Extension Service.

Da Silva, A. P., and B. D. Kay. 1997. "Estimating the least limiting water range of soil from properties and management," *Soil Sci. Soc. Amer. J.* **61**: 877–883.

DeMartinis, J. M., and S. C. Cooper. 1994. "Natural and man-made modes of entry in agronomic areas," in R. Honeycutt and D. Schabacker (eds.), *Mechanisms of Pesticide Movement in Ground Water.* Boca Raton, FL: Lewis Publishers 165–175.

Hillel, D. 1998. *Environmental Soil Physics.* Orlando, FL: Academic Press.

Hudson, B. D. 1994. "Soil organic matter and available water capacity," *J. Soil and Water Cons.* **49:** 189–194.

Kladivko, E. J., et al. 1999. "Pesticide and nitrate transport into subsurface tile drains of different spacing," *J. Environ. Qual.* **28**:997–1004.

Reynolds, W. D., et al. 2000. "Comparison of tension infiltrometer, pressure infiltrometer and soil core estimates of saturated conductivity," *Soil Sci. Soc. Amer. J.* **64**:478–484.

Sadras, V. O., and O. A. Calvino. 2001. "Quantification of grain yield response to soil depth in soybean, maize, sunflower and wheat," *Agron J.* **93**:577–583.

Schenk, H., and R. Jackson. 2002. "The global biogeography of roots," *Ecological Monographs* **72**:311–328.

Shestak, C. J., and M. D. Busse. 2005. "Compaction alters physical but not biological indices of soil health," *Soil Sci. Soc. Amer. J.* **69**:236–246.

Warrick, A. W. 2001. *Soil Physics Companion.* Boca Raton, FL: CRC Press.

Water cycles in the Teton Mountains. (R. Weil)

6

Soil and the Hydrologic Cycle

Both soil and water belong to the biosphere, to the order of nature, and—as one species among many, as one generation among many yet to come—we have no right to destroy them.
—DANIEL HILLEL, *OUT OF EARTH*

The world's water resources are *not* evenly distributed across space and time. The rain forests of the Amazon and Congo basins are drenched by more than 2000 mm of rain each year, while the deserts of North Africa and central Asia get by with less than 100 mm. In addition, the supply of water is not distributed evenly throughout the year. Rather, periods of high rainfall and flooding alternate with periods of drought.

Yet one could say that everywhere the supply of water is adequate to meet the needs of the plants and animals native to the natural communities of the area. Of course, this is so only because the plants and animals have adapted to the local availability of water. Early human populations, too, adapted to the local water supplies by settling where water was plentiful from rain or rivers, by developing techniques to harvest water for agriculture and store it in underground cisterns, and by adopting nomadic lifestyles that allowed them and their herds to follow the rains and the grass supply.

But "civilized" humans have not been willing to adapt their cultures to their environment. Rather, they have joined in organized efforts to adapt their environment to their desires. Hence, the ancients tamed the flows of the Tigris and Euphrates. We moderns dig wells in the Sahel, bottle up the mighty Nile at Aswan, pump out the aquifers under farms and suburbs, and create sprawling cities (with swimming pools and bluegrass lawns!) in the deserts of the American Southwest or on the sands of Arabia. Truly, cities like Las Vegas are gambling in more ways than one.

There is plenty of room for improvement in managing water resources, and many of the improvements are likely to come as a result of better management of soils. The soil plays a central role in

the cycling and use of water. For instance, by serving as a massive reservoir, soil helps moderate the adverse effects of excesses and deficiencies of water. The soil can help us treat and reuse wastewaters from animal, domestic, and industrial sources. The flow of water through the soil in these and other circumstances connects the chemical pollution of soils to the possible contamination of groundwater.

In Chapter 5 we considered the nature and movement of water in soils. In this chapter we will see how those characteristics apply to the practical management of water as it cycles between the soil, the atmosphere, and vegetation.

6.1   THE GLOBAL HYDROLOGIC CYCLE

Global Stocks of Water

There is enough water on Earth that if it were all aboveground at a uniform depth, it would cover the planet's surface to a depth of some 3 km. Most of this water, however, is relatively inaccessible and is not active in the annual cycling of water that supplies rivers, lakes, and living things.

The water that cycles more actively is in the surface layer of the oceans, in shallow groundwater, in lakes and rivers, in the atmosphere, and in the soil (see Figure 6.1). Although the combined volume is a tiny fraction of the water on Earth, these pools of water are accessible for movement in and out of the atmosphere and from one place on Earth's surface to another.

The Hydrologic Cycle

Primer on hydrologic cycle, Environment Canada: www.ec.gc.ca/water/en/nature/e_nature.htm

Solar energy drives the cycling of water from Earth's surface to the atmosphere and back again in what is termed the **hydrologic cycle** (Figure 6.2). About one-third of the solar energy that reaches Earth is absorbed by water, stimulating *evaporation*—the conversion of liquid water into water vapor. The water vapor moves up into the atmosphere, eventually forming clouds that can move from one region of the globe to another. Within an average of about 10 days, pressure and temperature differences in the atmosphere cause the water vapor to condense into liquid droplets or solid particles, which return to the Earth as rain or other precipitation.

The water cycle: http://micrometeorology.unl.edu/et/flash/watercycle_f.html

About 500,000 km^3 of water are evaporated from the Earth's surfaces and vegetation each year, some 110,000 km^3 of which falls as rain or snowfall on the continents. Some of the water falling on land runs off the surface of the soil, and some infiltrates the soil and drains into the groundwater, but most is evaporated back to the atmosphere. Both the surface runoff and groundwater seepage enter streams and

Figure 6.1

The sources of the Earth's water. The preponderance of water is found in the oceans, glaciers and ice caps, and deep groundwater (left), but most of these waters are inaccessible for rapid exchange with the atmosphere and the land. The mean residence times of a molecule of water in these pools are measured in thousands of years. The sources on the right, though much smaller in quantity, are actively involved in water movement through the hydrologic cycle with mean residence times measured in days. (Data from several sources)

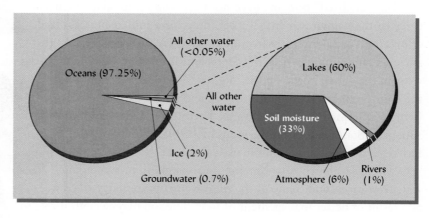

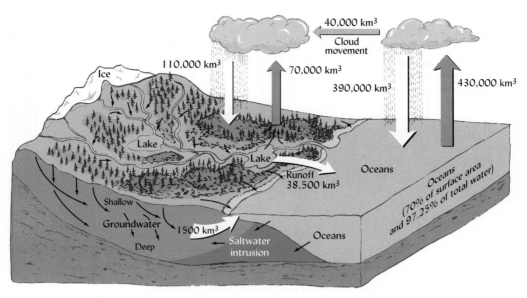

Figure 6.2

The hydrologic cycle upon which all life depends is very simple in principle. Water evaporates from the Earth's surface, both the oceans and continents, and returns in the form of rain or snowfall. The net movement of clouds brings some 40,000 km³ of water to the continents and an equal amount of water is returned through runoff and groundwater seepage that is channeled through rivers to the ocean. About 86% of the evaporation and 78% of the precipitation occurs in the ocean areas. However, the processes occurring on land areas where the soils are influential have impacts not only on humans but on all other forms of life, including those residing in the sea.

rivers that, in turn, flow into the oceans. The volume of water returned in this way is about 40,000 km³, which balances the same quantity of water that is transferred annually in clouds from the oceans to the continents.

A **watershed** is an area of land drained by a single system of streams and bounded by ridges that separate it from adjacent watersheds. All the precipitation falling on a watershed is either stored in the soil, returned to the atmosphere (see Section 6.2), or discharged from the watershed as surface or subsurface flow (runoff). Water is returned to the atmosphere either by **evaporation** from the land surface (vaporization of soil water) or, after plant uptake and use, by vaporization from the stomata on the surfaces of leaves (a process termed **transpiration**). Together, these two pathways of evaporative loss to the atmosphere are called **evapotranspiration**.

The disposition of water in a watershed is often expressed by the water-balance equation, which in its simplest form is:

$$P = ET + SS + D \qquad (6.1)$$

where P = precipitation, ET = evapotranspiration, SS = soil storage, and D = discharge.

Rearranging Equation 6.1 (D = P − ET − SS) shows that discharge can be increased only if ET and/or SS are decreased, changes that may or may not be desirable. For a forest watershed (sometimes termed a *catchment*), management may aim to maximize D so as to provide more water to downstream users. Clear-cutting the trees will almost certainly decrease ET and therefore increase discharge (Figure 6.3). In the case of a crop field, water applied in irrigation would be included on the left side of Equation 6.1. Irrigation managers may want to save water applied by minimizing unnecessary losses in D and allowing negative values for SS (withdrawals of soil storage water) during parts of the year.

Earth's water budget:
http://ww2010.atmos.uiuc.edu/(Gh)/guides/mtr/hyd/bdgt.rxml

Tools for watershed protection:
http://www.cwp.org/Resource_Library/Why_Watersheds/index.htm

Figure 6.3

Discharge (stream flow) from a southern Appalachian catchment before and after commercial clear-cutting of mixed hardwoods. Flow is expressed as cm of water above or below the long-term average flow (as measured prior to the timber harvest). Note the very high values for several years after the trees were felled and the site prepared for planting. The high flows resulted mainly from reduced transpiration water use in the absence of large trees. Dominated by Typic Hapludults and Typic Dystrochrepts, this 59 ha catchment is part of the Coweeta Hydrologic Laboratory, North Carolina, U.S. [Data from Swank et al. (2001)]

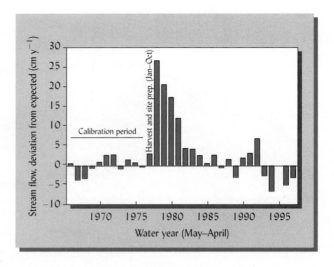

6.2   FATE OF PRECIPITATION AND IRRIGATION WATER

Some precipitation is intercepted by plant foliage and returned to the atmosphere by evaporation without ever reaching the soil. **Interception** and subsequent sublimation (vaporization directly from the solid state) of snow is especially important in coniferous forests, where 30 to 50% of the precipitation may never reach the soil. Water that does reach the ground may penetrate into the soil through the process of **infiltration**, especially if the soil surface structure is loose and open. If the rate of rainfall or snowmelt exceeds the infiltration capacity of the soil, some ponding may result, and considerable runoff and erosion may take place. In extreme cases, more than 50% of the precipitation may be lost as **surface runoff**. Runoff usually carries with it detached soil particles (**sediment**; see Chapter 14), as well as chemicals both attached to these particles and dissolved in the water.

Once water penetrates the soil, some of it is subject to downward percolation and eventual loss from the root zone by **drainage**. In humid areas and on irrigated land, up to 50% of the water input may be lost as drainage below the root zone. However, during subsequent periods of low rainfall, some of this water may move back up into the plant-root zone by rise of **capillary water**. Such movement is important to plants in areas with deep soils, especially in dry climates.

The water retained by the soil is referred to as **soil storage** water, some of which eventually moves upward by capillarity and is lost by evaporation from the soil surface. Much of the remainder is absorbed by plants and then moves through the roots and stems to the leaves, where it is lost by transpiration. The water thus lost to the atmosphere by evapotranspiration may later return to the soil as precipitation or irrigation water, and the cycle starts again.

Factors Affecting Infiltration

Timing of Precipitation   Heavy rainfall, even if of short duration, can supply water faster than most soils can absorb it. This accounts for the fact that in some arid regions, a rare cloudburst that brings 20 to 50 mm of water in a few minutes can result in flash flooding and gully erosion. A larger amount of precipitation spread over several days of gentle rain could move slowly into the soil, thereby increasing the stored water available for plant absorption, as well as replenishing the underlying groundwater. As illustrated in Figure 6.4, the timing of snowfall can affect the partitioning of spring snowmelt water between surface runoff and infiltration.

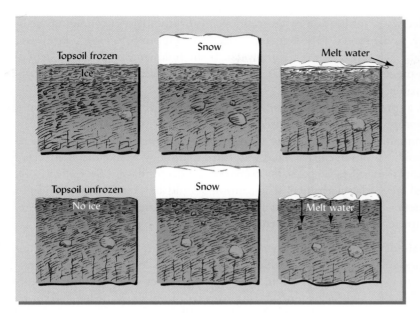

Figure 6.4
The relative timing of freezing temperatures and snowfall in the fall in some temperate regions drastically influences water runoff and infiltration into soils in the spring. The upper three diagrams illustrate what happens when the surface soil freezes before the first heavy snowfall. The snow insulates the soil so that it is still frozen and impermeable as the snow melts in the spring. The lower sequence of diagrams illustrates the situation when the soil is unfrozen in the fall when it is covered by the first deep snowfall.

Type of Vegetation   The vegetation and surface residues of perennial grasslands and dense forests protect the porous soil structure from the beating action of raindrops. Therefore, they encourage water infiltration and reduce the likelihood that soil will be carried off by any runoff that does occur. In general, very little runoff occurs from land under undisturbed forests or well-managed turfgrass.

Soil Management   Encouraging infiltration rather than runoff is usually a major objective of soil and water management. One approach is to allow more time for infiltration to take place by enhancing soil surface storage (Figure 6.5, *left*). A second approach is to maintain dense vegetation during periods of high rainfall. For example, **cover crops**, plants established between the principal crop-growing seasons, can

Figure 6.5
Managing soils to increase infiltration of rainwater. (left) Small furrow dikes on the right side of this field in Texas retain rainwater long enough for it to infiltrate rather than run off. (right) This saturated soil under a winter cover crop of hairy vetch is riddled with earthworm burrows that greatly increased the infiltration of water from a recent heavy rain. Scale in centimeters. (Photos courtesy of O. R. Jones, USDA Agricultural Research Service, Bushland, Texas, *left*, and R. Weil, *right*)

greatly enhance water infiltration by creating open root channels, encouraging earthworm activity, and protecting soil surface structure (Figure 6.5, *right*). However, remember that cover crops also transpire water. If the following crop will be dependent on soil storage water, care may be needed to kill the cover crop before it can dry out the soil profile.

A third approach is to maintain soil structure by minimizing compaction, whether by trampling or heavy equipment traffic. For example, soil compaction by forestry equipment used to skid logs or clear forestland for agricultural use can seriously impair soil infiltration capacity and watershed hydrology. While the ill effects of compaction can sometimes be partially overcome by deep tillage (see Section 4.7), environmental stewardship requires that caution should be exercised in the use of heavy equipment on any landscape and that disturbance of the natural soil and vegetation be completely avoided on as much of the land as possible.

Urban Watersheds   The use of heavy equipment to prepare land for urban development can severely curtail soil infiltration capacity and saturated hydraulic conductivity (Figure 6.6). Soil compaction by construction activities therefore results in drastically increased surface runoff during storms. The runoff burden carried by streams in urbanized watersheds is further increased because a large proportion of the land is covered with completely impermeable surfaces (rooftops, paved streets, and parking lots). Erosion of stream banks, toppling of trees, scouring of streambeds, and exposure of once-buried pipelines (Figure 6.7) are typical signs of the environmental degradation that ensues as streams become overwhelmed. Damage to streams in urban watersheds is accentuated because storm sewers and street gutters rush all this excess water off the land, requiring the stream to carry a huge volume of runoff concentrated in a short period of time. In recognition of the severe environmental problems caused by such excessive and concentrated runoff, urban planners and engineers are now working with soil scientists to reduce disruptions of the hydrologic cycle by urban development. Features such as permeable pavers that allow some infiltration even in parking lots (Figure 6.8, *left*) and rain gardens that catch runoff and release it slowly by infiltration (Figure 6.8, *right*) are part of what is termed *low-impact urban design*.

Soil Properties   Inherent soil properties affect the fate of precipitation. If the soil is loose and open (e.g., sands and well-granulated soils), a high proportion of the

How sprawl aggravates drought: www.smartgrowthamerica.org/waterandsprawl.html

Permeable paving vs. runoff pollution audio report: www.npr.org/templates/story/story.php?storyId=6165654

Figure 6.6

Grading and excavating land in preparation for urban/suburban development can greatly reduce soil permeability and hydraulic conductivity. The impairment results from compaction, damage to soil structure, and profile truncation (removal of A horizon). Both topsoil and subsoil horizons in these two New Zealand soils suffered three- to tenfold reductions in saturated hydraulic conductivity when the land was converted from a pasture to a housing development. The vertical lines indicate the variation among repeated measurements in the same soil, this variation being especially large in the pasture soils because of the random presence of earthworm channels.

[Data from Zanders (2001), used with permission from Landcare Research New Zealand, Ltd.; photo courtesy of R. Weil]

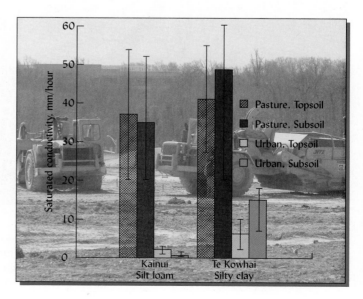

Figure 6.7

Exposure of a once-buried storm sewer manhole along a highly degraded stream that collects runoff from part of Baltimore, Maryland. The drastically increased storm runoff volume from the largely impermeable urban watershed has caused the stream to erode its banks and deeply cut into the hillside. The round manhole cover indicates the former land surface. (Photo courtesy of R. Weil)

incoming water will infiltrate the soil, and relatively little will run off. In contrast, heavy clay soils with unstable soil structures resist infiltration and encourage runoff. Other factors that influence the balance between infiltration and runoff include the slope of the land (steep slopes favoring runoff over infiltration) and impermeable layers within the soil profile. Such impermeable layers as fragipans and claypans (see Figure 3.5) can restrict infiltration and increase surface runoff once the upper horizons become saturated, even if the surface soil has intact structure and high porosity (Figure 6.9). All the soil and plant factors just discussed can result in some parts of a landscape contributing more runoff than others. Such spatial heterogeneity of infiltration and runoff is particularly important to ecosystem function in dry regions (see Section 9.11).

Figure 6.8

Two methods of increasing infiltration and slowing runoff in urbanized watersheds. Permeable pavers (left) allow grass to grow and water to infiltrate a parking lot while cars can still park without compacting soil or forming mud. Inlet to a small rain garden (right) that captures runoff from a suburban parking lot. The water is directed to a small depression. The pond that forms is ephemeral, holding the runoff water only temporarily. The permeable soil underlying the depression is designed to allow the water to infiltrate and seep away over a period of a few hours. The depression is planted with a variety of native plants that provide both beauty and wildlife habitat. (Photos courtesy of R. Weil)

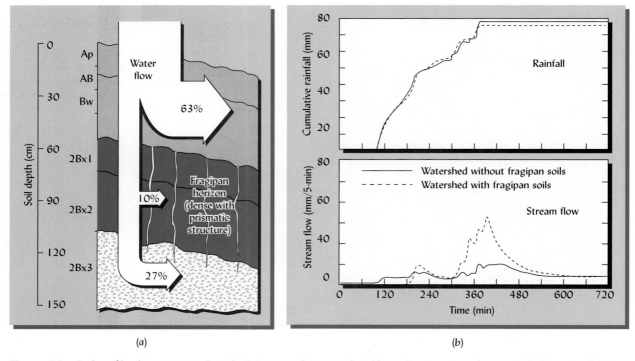

Figure 6.9 *Soil profile characteristics largely determine the vertical and lateral movement of water, including runoff from watersheds during rainstorms. In this example, a fragipan (left; also see Box 3.2) provides a barrier to both root growth and downward percolation of water. The zone above the fragipan may become saturated during wet weather, giving rise to a perched water table. As more rain falls, little can enter the already saturated surface horizons. Most of the water flows laterally, either through the soil above the fragipan or as surface runoff. The graphs (right) show the cumulative rainfall (upper) and volume of stream flow (lower) during and after a storm in which nearly 80 mm of rain fell in a 300-minute period. Of the two small (13 to 20 ha) watersheds represented, one has a large area of fragipan-containing soils (Fragiochrepts) formed in colluvium near the stream channel; the other watershed has no such fragipans. Although the rainfall was almost identical for the two nearby watersheds, stream flow (both the maximum and the overall volume) was much greater for the watershed with fragipans than the one without these impermeable layers.* [Soil profile based on data in Day et al. (1998); graphs from Gburek et al. (2006), with permission of Elsevier Science, Oxford, United Kingdom]

6.3   THE SOIL–PLANT–ATMOSPHERE CONTINUUM[1]

The flow of water through the soil–plant–atmosphere continuum (SPAC) ties together many of the processes we have just discussed: *interception, surface runoff, percolation, drainage, evaporation, plant water uptake, ascent of water to plant leaves,* and *transpiration* of water from the leaves back into the atmosphere (Figure 6.10).

Water Potentials

If a plant is to absorb water from the soil, the water potential must be lower (greater negative value) in the plant root than in the soil adjacent to the root. Likewise, movement up the stem to the leaf cells is in response to differences in water potential, as is the movement from leaf surfaces to the atmosphere. To illustrate the movement of water to sites of lower and lower water potential, Figure 6.10 shows that the water potential drops from –50 kPa in the soil, to –70 in the root, to –500 kPa at the leaf surfaces, and, finally, to –20,000 kPa in the atmosphere.

Two primary factors determine whether plants are well supplied with water: (1) the rate at which water is supplied by the soil to the absorbing roots and (2) the rate at which water is transpired from the plant leaves.

---

[1]The physics of water movement in plants has been quite controversial. For recent evidence that confirms that water moves up tall trees by the same capillary forces that control its movement in soil, see Tyree (2003).

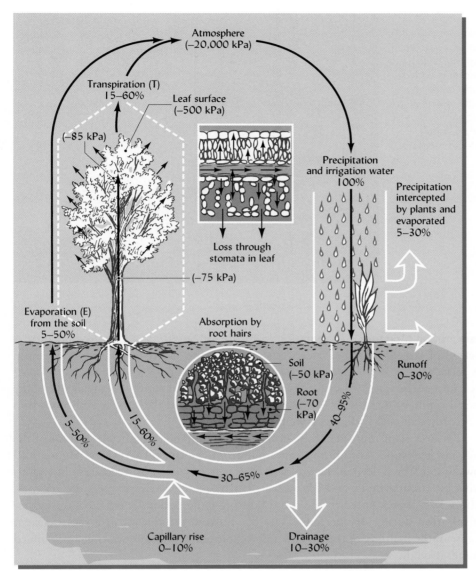

Figure 6.10

Soil–plant–atmosphere continuum (SPAC), showing water movement from soil to plants to the atmosphere and back to the soil in a humid to subhumid region. Water behavior through the continuum is subject to the same energy relations covering soil water that were discussed in Chapter 5. Note that the moisture potential in the soil is –50 kPa, dropping to –70 kPa in the root, declining still further as it moves upward in the stem and into the leaf, and is very low (–500 kPa) at the leaf–atmosphere interface, from whence it moves into the atmosphere, where the moisture potential is –20,000 kPa. Moisture moves from a higher to a lower moisture potential. Note the suggested ranges for partitioning of the precipitation and irrigation water as it moves through the continuum. Over 98% of the water absorbed by the roots of plants is transpired as water vapor over the course of the growing season.

Evapotranspiration

The evaporation (**E**) component of **evapotranspiration** (**ET**) may be viewed as a "waste" of water from the standpoint of plant productivity. However, much of the **transpiration** component (**T**) is essential for plant growth, providing the water that plants need for cooling, nutrient transport, photosynthesis, and turgor maintenance.

The **potential evapotranspiration** rate (PET) tells us how fast water vapor *would* be lost from a densely vegetated plant–soil system *if* soil water content were continuously maintained at an optimal level.

In practice, PET can be most easily estimated by applying a correction factor to the amount of water evaporated from an open pan of standard design (Figure 6.11). Water loss by transpiration from well-watered, dense vegetation is typically only about 65% as rapid as loss by evaporation from an open pan; hence, the correction factor for dense vegetation such as a lawn is typically 0.65 (it is lower for less dense vegetation):

$$\text{PET} = 0.65 \times \text{Pan evaporation} \qquad (6.2)$$

Values of PET range from more than 12 mm per day in hot, dry weather to as little as 1 mm per day in cold, damp weather.

Animated water use by plants:
http://micrometeorology
.unl.edu/et/flash/slide2.html

Evapotranspiration calculation spreadsheet:
http://biomet.ucdavis
.edu/irrigation_scheduling/
LIMP/LIMP.htm

Figure 6.11

A class A evaporation pan used to help estimate potential evapotranspiration (PET). Once a day, the water level is determined in the stilling well (small cylinder) and a measured amount of water is added to bring the level back up to the original mark. Evaporation from the pan integrates the effects of relative humidity, temperature, wind speed, and other climatic variables related to the vapor pressure gradient. Also shown is an anemometer to measure wind speed. (Photo courtesy of R. Weil)

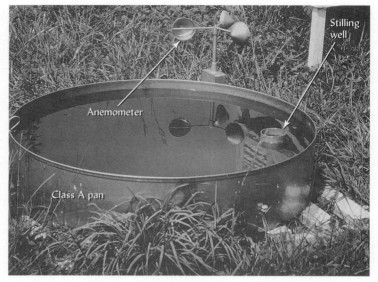

Effect of Soil Moisture   Evaporation from the soil surface at a given temperature is determined largely by soil surface wetness and by the ability of the soil to replenish this surface water as it evaporates. In most cases, the upper 15 to 25 cm of soil provides most of the water for surface evaporation. Unless a shallow water table exists, the upward capillary movement of water is very limited and the surface soil soon dries out, greatly reducing further evaporation loss.

Because plant roots penetrate deep into the profile, a significant portion of the water lost by evapotranspiration can come from the subsoil layers. As Figure 5.29 shows, water stored deep in the profile is especially important to vegetation in regions having alternating moist and dry seasons (such as Ustic or Xeric moisture regimes; Section 3.2). Water stored in the subsoil during rainy periods is available for evapotranspiration during dry periods. Plate 85 illustrates the death of a rooftop lawn because of the inability of the shallow soil to hold sufficient water during a prolonged summer drought.

Plant Water Stress   For dense vegetation growing in soil well supplied with water, ET will nearly equal PET. When soil water content is less than optimal, the plant will not be able to withdraw water from the soil fast enough to satisfy PET. The difference between PET and actual ET is termed the *water deficit*. A large deficit is indicative of high water stress.

Plant Characteristics   As the leaf area per unit land area (a ratio termed the **leaf area index, [LAI]**) increases, more radiation will be absorbed by the foliage to stimulate transpiration and less will reach the soil to promote evaporation. For annual crops, the LAI value typically varies from 0 at planting to a peak of perhaps 3 to 5 at flowering, then declines as the plant senesces, and finally drops back to 0 when the plant is removed at harvest (assuming that no weeds are allowed to grow). In contrast, perennial vegetation, such as pastures and forests, has very high leaf area indices both early and late in the growing season. Where leaf litter has accumulated on the forest floor, very little direct sunlight strikes the soil, and evaporation is very low throughout the year (Figure 6.12). Other plant characteristics, including rooting depth, length of life cycle, and leaf morphology, can influence the amount of water lost by evapotranspiration over a growing season. Plates 62 and 84 illustrate the competitiveness of mature trees for soil water supplies.

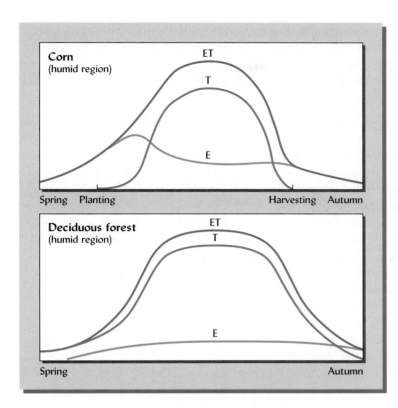

Figure 6.12

Relative rates of evaporation from the soil surface E, of transpiration from the plant leaves T, and the combined vapor loss ET for two field situations. (upper) A field of corn in a humid region. Until the plants are well established, most of the vapor loss is from the soil surface E, but as the plants grow, T soon dominates. The soil surface is shaded and E may actually decrease somewhat since most of the moisture is moving through the plants. As the plants reach maturity, T declines, as does the combined ET. For a nearby deciduous forested area (lower) the same general trend is illustrated, except there is relatively less evaporation from the soil surface and a higher proportion of the vapor loss is by transpiration. The soil is shaded by the leaf canopy most of the growing season in the forested area. It should be noted that these figures relate to a representative situation and that the actual field losses would be influenced by rainfall distribution, temperature fluctuations, and soil properties.

[Diagram courtesy of R. Weil and N. C. Brady]

Water Use Efficiency

Water use efficiency may be expressed in terms of dry matter produced per unit of water transpired (*T efficiency*) or the dry matter yield per unit of water lost by evapotranspiration (*ET efficiency*). In Figure 6.13, water use efficiency is expressed as kg of grain produced per m^3 of water used in evapotranspiration. For irrigation systems analysis, water use efficiency may consider additional aspects such as losses of water in storage reservoirs or leakage from canals (see Section 6.9).

Huge amounts of water are needed to produce the human food supply. Water use efficiency is largely driven by climatic factors. In arid regions, crops may use 1000 to more than 5000 kg of water to produce a single kg of grain. Ironically, in more

Conference on drought-resistant soils: www.fao.org/landandwater/agll/soilmoisture

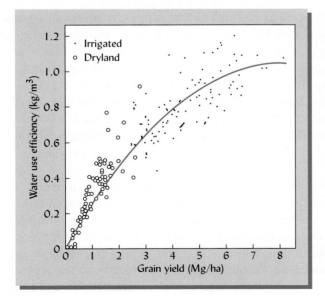

Figure 6.13

Evapotranspiration water-use efficiency, the dry matter produced for every unit of water used on the crop, generally increases as crop yield increases. This is true mainly because such yield-improving measures as fertilizer use, better cultivars, denser stands, and better pest control generally increase crop yield by a greater proportion than they do the use of water. This relationship generally applies where water is available to meet the demands of higher-yielding crops. Even so, the curve seems to level off at just over 1 kg grain per 1000 kg water (1 m^3 = 1000 kg water). The data shown are for irrigated and nonirrigated (dryland) wheat fields in the southern high plains region of the United States. [From Howell (2001), with permission of The American Society of Agronomy]

humid regions where water and rain are plentiful, much less water is needed for each kg of grain produced because the evaporative demand is far lower. When we take into account the water used to grow grains, fruits, vegetables, and feed for cattle, almost 7000 L (1700 gal) of water are required to grow a *single day's* food supply for one adult in the United States!

Evaporation and evapotranspiration from Florida's wetlands: http://aquat1.ifas.ufl.edu/guide/evaptran.html

ET Efficiency   Since evapotranspiration includes both transpiration from plants and evaporation from the soil surface, evapotranspiration efficiency is more subject to management than transpiration efficiency. Highest ET efficiency is attained where plant density and other growth factors minimize the proportion of ET attributable to evaporation from the soil. As long as the supply of water is not too limited, maintaining optimum conditions for plant growth (by closer plant spacing, fertilization, or selection of more vigorous varieties) increases the efficiency of water use by plants (Figure 6.13). However, if irrigation water is not available and the period of rainfall is very short, the increase in evapotranspiration by more vigorously growing plants may deplete stored soil water, resulting in serious water stress or even plant death before any harvestable yield has been produced.

In summary, water losses from the soil surface and from transpiration are determined by (1) climatic conditions, (2) plant cover in relation to soil surface (LAI), (3) efficiency of water use by different plants and management regimes, and (4) length and season of the plant-growing period.

6.4   CONTROL OF VAPOR LOSSES

Measures that can be taken to bring ET into closer balance with available soil water supplies, and thus reduce plant stress, include practices that limit the amount of leaf area exposed to solar radiation and thus limit transpiration. Such measures include wider plant spacing, limited fertilization in dry regions, and elimination of undesired plant leaf area (namely weeds). Other approaches, such as mulching and establishment of dense vegetation canopies, limit the exposure of soil to solar radiation and thus reduce the evaporation component of ET. Finally, where feasible, the supply of water in the soil may be increased by irrigating, usually with dramatically enhanced plant productivity.

Integrated weed management: http://ssca.usask.ca/conference/1997proceedings/Odonovan.html

One strategy widely practiced and once thought to be effective for semiarid region crop production is *summer fallow*. This farming system, which alternates bare **fallow** (an unvegetated period) one year with traditional cropping the next, has been used to conserve soil moisture in some low-rainfall environments. Transpiration water losses during the fallow year are minimized by controlling weeds with light tillage and/or with herbicides. It is hoped that some of the water saved during the fallow year remains in the profile the following year when a crop is planted, resulting in higher yields of that second-year crop than if the soil had been cropped every year. Early research on conventionally tilled soils showed that yields in the alternate years were often sufficiently high as to make up for the harvest foregone during the fallow year. This cropping practice is responsible for the checkerboard of dark fallow soils and golden ripening wheat that can be seen when flying over certain semiarid "breadbasket" regions.

Research now shows that summer fallow is probably not in the best interests of farmers or natural resource conservation. The main disadvantage of this system is the soil degradation that occurs, largely because of a negative soil organic matter balance (see Section 11.7) and wind erosion (see Section 14.11) during the fallow years. Conservation tillage practices (including "no-till") leave most of the crop residues on the soil surface where they reduce evaporation and protect the soil. From research at many sites we can now conclude that long-term productivity, profitability, and soil

Table 6.1

LONG-TERM AVERAGE GRAIN YIELDS IN THE SEMIARID NORTH AMERICAN GREAT PLAINS REGION FROM CROPPING SYSTEMS WITH OR WITHOUT SUMMER FALLOW

Note that in every case the system with fallow was less productive. Results like these are encouraging farmers to abandon summer fallow in favor of reduced tillage rotations that yield more and better maintain soil quality.

Location	Duration of experiment, years	Mean annual precipitation, cm	Rotation (and tillage)[a]	Long-term avg. yield, Mg ha^{-1} yr^{-1}
Akron, CO	10	41.8	WW–Fallow (CT)	1.1
			WW–C–M (NT)	1.5
Mandan, ND	12	42.7	SW–Fallow (CT)	1.1
			SW–WW–SF (NT)	1.7
Sidney, MT	13	34.5	SW–Fallow (CT)	1.2
			Cont. SW (NT)	1.9
Swift Current, SK	23	36.1	SW–Fallow (RT)	1.3
			Lentil–SW (RT)	1.6

[a] Abbreviations: C = corn, CT = Conventional till, M = millet, NT = no-till, RT = reduced till with herbicides, SF = sunflower, SW = spring wheat, WW = winter wheat.
Data from Varvel et al. (2006).

quality are likely to be optimized with cropping systems that use conservation tillage and keep the soil continuously vegetated with a diversity of crops (Table 6.1).

Control of Surface Evaporation (E)

More than half the precipitation in semiarid and subhumid areas is usually returned to the atmosphere by evaporation (E) directly from the soil surface. In natural rangeland systems, E is a large part of ET because plant communities tend to self-regulate toward configurations that minimize the deficit between PET and ET—generally low plant densities with large unvegetated areas between scattered patches of shrubs or bunchgrass (see Figure 2.18). In addition, plant residues on the soil surface are sparse. Evaporation losses are also high in arid-region irrigated soil, especially if inefficient practices are used (see Section 6.9). Even in humid-region rain-fed areas, E losses are significant during hot, rainless periods. Such moisture losses rob the plant community of much of its growth potential and reduce the water available for discharge to streams. Careful study of Figure 6.14 will clarify these relationships and the principles just discussed. Note that the gap between PET and ET represents the **soil water deficit**, which is a measure of how limiting water supply is for plant productivity (see also Plate 76).

For arable soils, the most effective practices aimed at controlling E are those that cover the soil. This cover can best be provided by mulches and by selected conservation tillage practices that leave plant residues on the soil surface, mimicking the soil cover of natural ecosystems.

Vegetative Mulches   A *mulch* is a material used to cover the soil surface primarily for the purpose of controlling evaporation, soil erosion, temperature, and/or weeds. Examples of *organic* mulches include straw, leaves, and crop residues. Mulches can be highly effective in checking evaporation, but they may be expensive and labor intensive. Mulching is therefore most practical for small areas (gardens and landscaping beds) and for high-value horticultural crops. It is much less labor intensive to produce a mulch "in place" by growing a cover crop or cash crop and letting the residues remain on the soil surface (see *Crop Residue and Conservation Tillage*, following).

Organic vs. plastic mulches:
http://jeq.scijournals.org/
cgi/content/full/30/5/1808

Figure 6.14

Partitioning of liquid water losses (discharge) and vapor losses (evaporation and transpiration) in regions varying from low (arid) to high (humid) levels of annual precipitation. The example shown assumes that temperatures are constant across the regions of differing rainfall. Potential evapotranspiration (PET) is somewhat higher in the low-rainfall zones because the lower relative humidity there increases the vapor pressure gradient at a given temperature. Evaporation (E) represents a much greater proportion of total vapor losses (ET) in the drier regions due to sparse plant cover caused by interplant competition for water. The greater the gap between PET and ET, the greater the deficit and the more serious the water stress to which plants are subject. (Diagram courtesy of R. Weil)

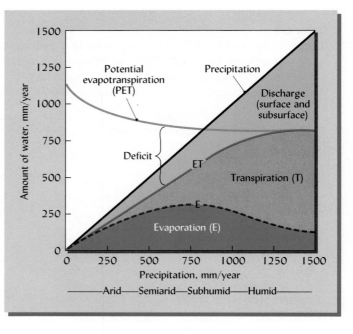

In addition to reducing evaporation, organic mulches may provide these benefits: (1) reduce soilborne diseases spread by splashing water; (2) provide a clean path for foot traffic; (3) reduce weed growth (if applied thickly); (4) increase water infiltration; (5) provide organic matter and, possibly, plant nutrients to the soil; (6) encourage earthworm populations; (7) reduce soil erosion; and (8) moderate soil temperatures, especially preventing overheating in summer months (see Section 7.11). This last effect may be detrimental if it slows soil warming in spring.

Plastic Mulches   Plastic films (and specially prepared paper) can be used as mulch to control evaporative water losses. The dark, opaque types also effectively control weeds. The mulch is often applied by machine, and plants grow through holes made in the film (Figure 6.15). In landscaping beds, they are often covered with a layer of tree-bark mulch or gravel for longer life and a more pleasing appearance.

Figure 6.15

For crops with high cash value, plastic mulches are commonly used. The plastic is installed by machine (left), and at the same time the plants are transplanted (right). Plastic mulches help control weeds, conserve moisture, encourage rapid early growth, and eliminate the need for cultivation. The high cost of plastic makes it practical only with the highest-value crops. (Photos courtesy of K. Q. Stephenson, Pennsylvania State University)

Figure 6.16

Conservation tillage leaves plant residues on the soil surface, reducing both evaporation losses and erosion. (Left) In a semiarid region (South Dakota) the straw from the previous year's wheat crop was only partially buried to anchor it against the wind while still allowing it to cover much of the soil surface. In the next year the left half of the field, now shown growing wheat, will be **stubble mulched,** *and the right half will be sown to wheat. (Right)* **No-till** *planted corn in a more humid region grows up through the straw left on the surface by a previous wheat crop. Note that with no-till, almost no soil is directly exposed to solar radiation, rain, or wind.* (Photos courtesy of R. Weil, *left,* and USDA Natural Resources Conservation Service, *right*)

Unlike their organic counterparts, plastic mulches can warm cool soils and are available in colors, such as red, that reflect light of special benefit to growing plants. However, most of the ancillary benefits listed above for organic mulches do not accrue from the use of plastic. Because plastic films interfere with water infiltration, permeable fabric mulches are often recommended instead, especially in landscaping beds. Another major problem with plastic mulches (even many of the permeable fabric types) is the difficulty of completely removing the plastic at the end of the growing season; after a number of years, scraps of plastic accumulate in the soil, causing an unsightly mess and interfering with water movement and cultivation. Some manufacturers now offer biodegradable and light-degradable plastic mulches (made from plant products) that remain intact for a month or so and then break down almost completely (see Figure 7.29).

Crop Residue and Conservation Tillage   Plant residues on the soil surface conserve soil moisture by reducing evaporation and increasing infiltration. **Conservation tillage** practices leave a high percentage of the residues from the previous crop on or near the surface (Figure 6.16, *left*). A conservation tillage practice widely used in subhumid and semiarid regions is **stubble mulch** tillage, which allows much of the wheat stubble or cornstalks from the previous crop to remain on or near the surface. Conservation tillage planters are capable of planting through the stubble and allow much of it to remain on the surface during the establishment of the next crop (see Figure 6.16). Unfortunately, plant growth in dry regions is usually insufficient to produce the large amount of residue mulches needed to greatly reduce evaporation and maximize water conservation.

Other conservation tillage systems that leave residues on the soil surface include no-till (see Figure 6.16, *right*), where the new crop is planted directly into the sod or residues of the previous crop, with almost no soil disturbance. The long-term soil water conserving effects of such tillage systems are shown in Box 6.1. Conservation tillage systems will receive further attention in Section 14.6.

BOX 6.1
WATER CONSERVATION PAYS OFF

Intriguing long-term research results show that the increasing use of conservation tillage can really contribute to increased yields in semiarid-region agriculture. In 1938, the average yield of sorghum in USDA research plots in Lubbock, Texas, was about 900 kg/ha. By 1997, this average had increased to 3830 kg/ha, an overall increase of some 325%. Further research suggests that about one-third of this yield increase could be attributed to improved varieties, the remaining two-thirds to other factors. Increased nutrient availability was ruled out as a yield-inducing factor since no fertilizers were applied to these plots.

This left increased water availability as the prime factor likely contributing to the higher yields. Analyses of rain and snowfall records showed no increase in precipitation that might have accounted for the increased yields.

However, soil moisture measurements made at planting time (Figure 6.17, *right*) were found to be much higher during the period from 1972 to 1997 than in the earlier period (1956 to 1972).

Records showed that tillage and residue management were the primary factors that accounted for this soil moisture difference. During the earlier period (1956 to 1972), the plots were tilled between crops to control weeds, and little residue remained on the soil. During the later period (1972 to 1997), there was a major shift to no-till crop production, which maintained crop residues on the soil surface. Together with improved weed control using herbicides, the no-till systems reduced evaporative losses, leaving extra moisture in the soil to accommodate the increased sorghum yields.

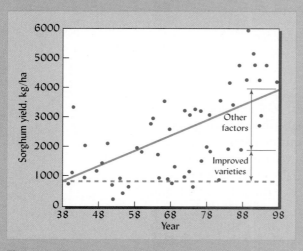

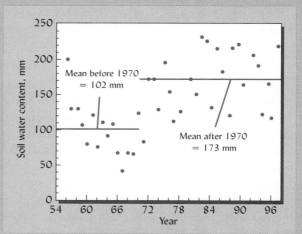

Figure 6.17
Sorghum yield increases 1938 to 1998 (left) and changes in soil water storage 1954 to 1997 (right). [From Unger and Baumhardt (1999)]

6.5   LIQUID LOSSES OF WATER FROM THE SOIL

In our discussion of the hydrologic cycle, we noted two types of liquid losses of water from soils: (1) percolation, or subsurface drainage water, and (2) surface runoff water (see Figure 6.2). *Percolation water* recharges the groundwater and moves chemicals out of the soil. *Runoff* contributes to storm flow in streams and moves both particles and chemicals. When the amount of rainfall entering a soil exceeds the water-holding capacity of the soil, losses by percolation will occur. Percolation losses are influenced by (1) the amount of rainfall and its distribution, (2) runoff from the soil, (3) evaporation, (4) the character of the soil, and (5) the nature of the vegetation.

Percolation–Evaporation Balance

In humid temperate regions, the rate of water infiltration into the soil (precipitation minus runoff) is greater, at least during certain seasons, than the rate of evapotranspiration. As soon as the soil field capacity is reached, percolation into the substrata

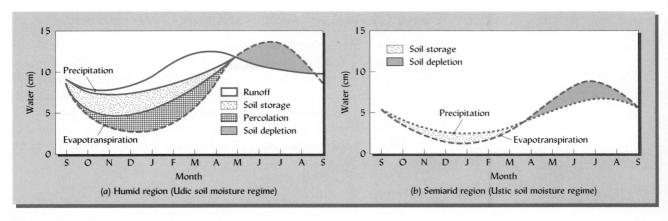

Figure 6.18

Seasonal water balance curves for two temperate zone regions: (a) a humid region and (b) a semiarid region. Note that actual ET shown is restrained by the water supply available. Potential evapotranspiration (PET), not shown, would be much higher, especially in (b). Percolation through the soil is absent in the semiarid region. In each case, water is stored in the soil and later released when ET exceeds precipitation.

occurs. In the example shown in Figure 6.18*a*, maximum percolation occurs during the winter and early spring, when evaporation is lowest. During the summer, little percolation occurs. In fact, evapotranspiration exceeds precipitation, resulting in a depletion of soil water.

In a semiarid region, as in a humid region, water is stored in the soil during the winter months and is used to meet the moisture deficit in the summer. But because of the low rainfall, little runoff and essentially no percolation out of the profile occur. Water may move to the lower horizons, but it is absorbed by plant roots and ultimately is lost by transpiration.

The comparative losses of water by evapotranspiration and percolation through soils found in different climatic regions are shown in Figure 6.19. These differences should be kept in mind while reading the following section on percolation and groundwaters.

Figure 6.19

Percentage of the water entering the soil that is lost by downward percolation and by evapotranspiration. Representative figures are shown for different climatic regions.

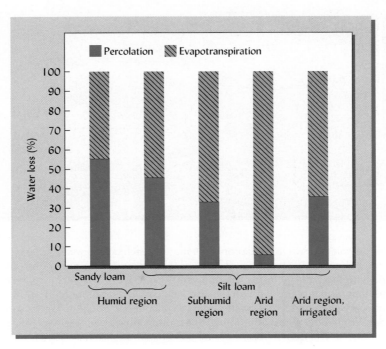

6.6   PERCOLATION AND GROUNDWATERS

Groundwater Basics:
www.groundwater.org/gi/
whatisgw.html

When drainage water moves downward through the soil and regolith, it eventually encounters a zone in which the pores are all saturated with water. Often this saturated zone lies above an impervious soil horizon (Figure 6.20) or a layer of impermeable rock or clay. The upper surface of this zone of saturation is known as the **water table**, and the water within the saturated zone is termed **groundwater**. The water table (Figure 6.21) is commonly only 1 to 10 m below the soil surface in humid regions but may be several hundred or even thousands of meters deep in arid regions. In swamps it is essentially at the land surface.

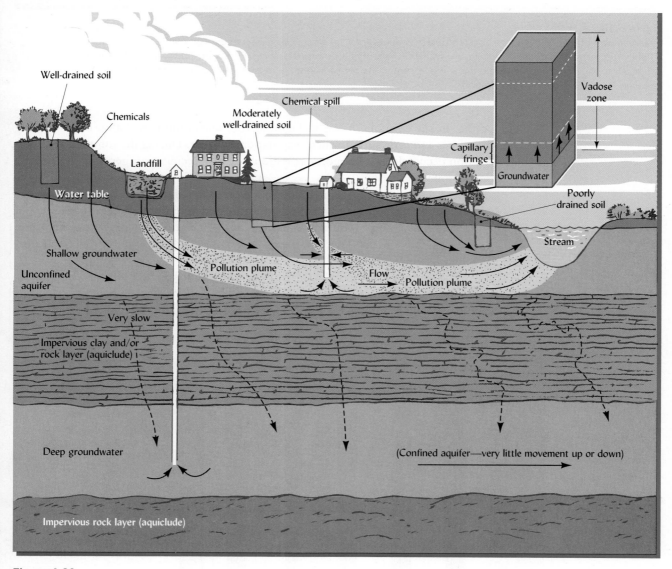

Figure 6.20

Precipitation and irrigation water percolate down the soil profile, ultimately reaching the water table and underlying shallow groundwater. The unsaturated zone above the water table is known as the vadose zone (upper right). Groundwater moves up from the water table by capillarity into the capillary fringe. Groundwater also moves horizontally down the slope toward a stream, carrying with it chemicals that have leached through the soil, including plant nutrients (N, P, Ca, etc.) as well as pesticides and other pollutants. A shallow well pumps groundwater from the unconfined aquifer near the surface. A deeper well exploits groundwater from a deep, confined aquifer. Two plumes of pollution are shown, one originating from landfill leachate, the other from a chemical spill. The former appears to be contaminating the shallow well. (Diagram courtesy of R. Weil)

Vadose
zone

Capillary
fringe

Water table

Ground-
water

Figure 6.21
The water table, capillary fringe, zone of unsaturated material above the water table (vadose zone), and groundwater are illustrated in this photograph. The groundwater can provide significant quantities of water for plant uptake.
(Photo courtesy of R. Weil)

The unsaturated zone above the water table is termed the **vadose zone** (see Figures 6.20 and 6.21). The vadose zone may include unsaturated materials underlying the soil profile, and so may be considerably deeper than the soil itself. In some cases, however, the saturated zone may be sufficiently near the surface to include the lower soil horizons, with the vadose zone confined to the upper soil horizons.

Shallow groundwater receives downward-percolating drainage water. Most of the groundwater, in turn, seeps laterally through porous geological materials (termed **aquifers**) until it is discharged into springs and streams. Groundwater may also be removed by pumping for domestic and irrigation uses. The water table will move up or down in response to the balance between the amount of drainage water coming in through the soil and the amount lost through pumped wells and natural seepage to springs and streams.

Shallow Groundwater

Groundwater that is near the surface can serve as a reciprocal water reservoir for the soil. As plants remove water from the soil, it may be replaced by upward capillary movement from a shallow water table. The zone of wetting by capillary movement is known as the **capillary fringe** (see Figure 6.21). Such movement can provide a steady and significant supply of water that enables plants to survive during periods of low rainfall. Capillary rise from shallow groundwater may also bring a steady supply of salts to the surface if the groundwater is brackish. (See Section 9.12 for details on this soil-degrading process.)

Movement of Chemicals in the Drainage Water[2]

Percolation of water through the soil to the water table not only replenishes the groundwater, it also dissolves and carries downward a variety of inorganic and organic chemicals found in the soil or on the land surface. Chemicals **leached** from the soil to the groundwater (and eventually to streams and rivers) in this manner include elements weathered from minerals, natural organic compounds resulting from the decay of plant residues, plant nutrients derived from natural and human sources, and various synthetic chemicals applied intentionally or inadvertently to soils.

Colorado farmers in crisis audio report:
www.npr.org/templates/story/story.php?storyId=5400947

Quality of U.S. waters:
www.epa.gov/305b/2000report/factsheet.pdf

---

[2]For a review of chemical transport through field soils, see Jury and Fluhler (1992).

Green roofs to counter
pollution audio report:
www.npr.org/templates/story/
story.php?storyId=5454152

Of great concern is the leaching of human pathogens and various highly toxic synthetic compounds, such as pesticides and their breakdown products or chemicals dissolved from waste disposal sites (see Chapter 15 for a detailed discussion of these pollution hazards). Figure 6.20 illustrates how the groundwater can be charged with these pathogens or chemicals, and how a plume of contamination spreads to downstream wells and bodies of water.

Chemical Movement Through Macropores

As shown in Figure 6.22, chemicals or pathogens may be washed from the soil surface into large pores, through which they can quickly move downward by **preferential flow** (see Section 5.5). Most of the water flowing through large macropores does not come into contact with the bulk of the soil. Such preferential flow is sometimes termed **bypass flow**, as it tends to move rapidly around, rather than through, the soil matrix. As a result, if chemicals have been incorporated into the upper few centimeters of soil, their movement into the larger pores is reduced, and downward leaching is greatly curtailed. On the other hand, leaching of chemicals is most serious if the chemicals are merely applied on the soil surface.

Figure 6.22

Preferential bypass flow in macropores transports soluble chemicals downward through a soil profile. Where the chemical is on the soil surface (left) and can dissolve in surface-ponded water when it rains, it may be transported rapidly down cracks, earthworm channels, and other macropores. Where the chemical is dispersed within the soil matrix in the upper horizon (right), most of the water moving down through the macropores will bypass the chemical, and thus little of the chemical will be carried downward. Note that channels not open all the way to the surface do not conduct water or contaminants by preferential flow.

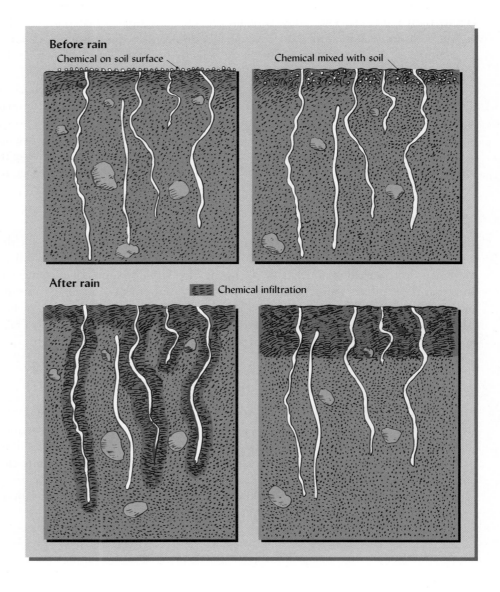

Table 6.2

INFLUENCE OF WATER APPLICATION INTENSITY ON THE LEACHING OF PESTICIDES THROUGH THE UPPER 50 CM OF A GRASS SOD–COVERED MOLLISOL

Metalaxyl is far more water soluble than Isazofos, but in each case, the heavy rain stimulated much more pesticide leaching through the macropores.

| | Pesticide leached as percentage of that applied to surface | |
Pesticide	10 cm of heavy rains	10 cm of light rains
Isazofos	8.8	3.4
Metalaxyl	23.8	13.9

Data from Starrett et al. (1996).

Intensity of Rain or Irrigation   During a high-intensity rainfall event following a few days of fair weather, water and its associated chemicals move downward rapidly through the macropores, bypassing the bulk of the soil. In contrast, a gentle rain that in time may provide as much water as the more intense event would likely thoroughly wet the upper soil aggregates, thereby minimizing the rapid downward percolation of both water and the chemicals and pathogens it carries. Table 6.2 illustrates the effects of rainfall intensity on the leaching of pesticides applied to turfgrass.

6.7   ENHANCING SOIL DRAINAGE[3]

Prolonged saturation of soils may be due to the low-lying landscape position in which the soil is found, such that the regional water table is at or near the soil surface for extended periods (**endoaquic**). In other soils, water may accumulate above an impermeable layer in the soil profile (**epiaquic**), creating a **perched water table** (see Figure 6.23). Soils with either type of saturation may be components of wetlands, transitional ecosystems between land and water that are characterized by anaerobic (no oxygen) conditions (see Section 7.7).

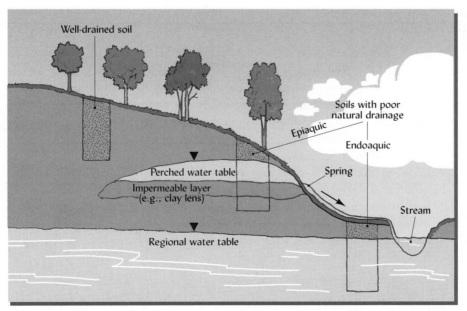

Figure 6.23

Cross section of a landscape showing the regional and perched water tables in relation to three soils, one well-drained and two with poor internal drainage. By convention, an inverted triangle (▼) identifies the level of the water table. The soil containing the perched water table is wet in the upper part, but unsaturated below the impermeable layer, and therefore is said to be epiaquic (Greek epi, "upper"), while the soil saturated by the regional water table is said to be endoaquic (Greek endo, "within"). Artificial drainage can help to lower both types of water tables. (Diagram courtesy of R. Weil)

---

[3] For a review of all aspects of artificial drainage, see Skaggs and van Schilfgaarde (1999).

Reasons for Enhancing Soil Drainage

Water-saturated, poorly aerated soil conditions are essential to the normal functioning of wetland ecosystems and to the survival of many wetland plant species. However, for most other land uses, these conditions are a distinct detriment.

Engineering Problems   During construction, the muddy, low-bearing-strength conditions of saturated soils make it very difficult to operate machinery (see Section 4.9). By the same token, soils used for recreation can withstand traffic much better if they are well drained. Houses built on poorly drained soils may suffer from uneven settlement and flooded basements during wet periods. Similarly, a high water table will result in capillary rise of water into roadbeds and around foundations, lowering the soil strength and leading to damage from frost heaving (see Section 7.8) if the water freezes in winter. Heavy trucks traveling over a paved road underlaid by a high water table create potholes and eventually destroy the pavement.

Plant Production   Water-saturated soils make the production of most upland crops and forest species difficult, if not impossible. In wet soil, farm equipment used for planting, tillage, or harvest operations may bog down. Except for a few specially adapted plant species (bald cypress trees, rice, cattails, etc.), most crop and forest species grow best in well-drained soils since their roots require adequate oxygen for respiration (see Section 7.1). Furthermore, a high water table early in the growing season will confine the plant roots to a shallow layer of partially aerated soil; the resulting restricted root system can lead to water stress later in the year, when the weather turns dry and the water table drops rapidly.

For these and other reasons, artificial drainage systems have been widely used to remove excess (gravitational) water and lower the water table in poorly drained soils. Land drainage is practiced in select areas in almost every climatic region but is most widely used to enhance the agricultural productivity of clayey alluvial and lacustrine soils. Drainage systems are also a vital, if sometimes neglected, component of arid region irrigation systems, where they are needed to remove excess salts and prevent waterlogging (see Section 9.18).

Artificial drainage is a major alteration of the soil system, and the following potential beneficial and detrimental effects should be carefully considered. In many instances laws designed to protect wetlands require that a special permit be obtained for the installation of a new artificial drainage system.

Benefits of Artificial Drainage

1. Increased bearing strength and improved soil workability, which allow more timely field operations and greater access to vehicular or foot traffic.
2. Less frost-heaving of foundations, pavements, and plants (e.g., see Figure 7.17).
3. Enhanced rooting depth, growth, and productivity of most upland plants due to improved oxygen supply and, in acid soils, lessened toxicity of manganese and iron (see Sections 7.3 and 7.5).
4. Reduced levels of fungal disease infestation in seeds and on young plants.
5. More rapid soil warming, resulting in earlier maturing crops (see Section 7.11).
6. Less production of methane and nitrogen gases that cause global environmental damages (see Chapters 11 and 12).
7. Removal of excess salts from irrigated soils and prevention of salt accumulation by capillary rise in areas of salty groundwater (see Section 10.3).

Detrimental Effects of Artificial Drainage

1. Loss of wildlife habitat, especially waterfowl breeding and overwintering sites.
2. Reduction in nutrient assimilation and other biochemical functions of wetlands (see Section 7.7).

3. Increased leaching of nitrates and other contaminants to groundwater.

4. Accelerated loss of soil organic matter, leading to subsidence of certain soils (see Sections 3.9 and 11.8).

5. Increased frequency and severity of flooding due to loss of runoff water retention capacity.

6. Greater cost of damages when flooding occurs on alluvial lands developed after drainage.

7. Increased of global warming as soil organic matter conversion to CO_2 is enhanced.

Surface Drainage Systems

Artificial drainage systems are designed to promote two general types of drainage: (1) *surface drainage* and (2) internal, or *subsurface, drainage*. The purpose of surface drainage is to remove water before it enters the soil.

Surface Drainage Ditches   Most surface drainage systems hasten the surface runoff of water by construction of shallow ditches with gentle side slopes that do not interfere with equipment traffic. If there is some slope on the land, the shallow ditches are usually oriented across the slope and across the direction of planting and cultivating, thereby permitting the interception of water as it runs off down the slope. These ditches can be made at low cost with simple equipment. For removing surface water from landscaped lawns, this system of drainage can be modified by constructing gently sloping swales rather than ditches.

Land Smoothing   Often, surface drainage ditches are combined with **land smoothing** to eliminate the ponding of water and facilitate its removal from the land. Small ridges are cut down and depressions are filled in using precision, laser-guided field-leveling equipment. The resulting land configuration permits excess water to move at a controlled rate over the soil surface to the outlet ditch and then on to a natural drainage channel. Land smoothing is also commonly used to prepare a field for flood irrigation (see Section 6.9).

Subsurface (Internal) Drainage

The purpose of subsurface drainage systems is to remove the groundwater from within the soil and to subsequently lower the water table. They require channels into which excess water can flow. Internal drainage occurs only when the pathway for drainage is located below the level of the water table (Figure 6.24). The flow of water from a saturated soil into a drainage outlet is in response to positive gravitational and submergence potentials (see Section 5.3). Box 6.2 provides an example of how knowledge of basic soil properties and water-movement principles can be applied in designing a system to alleviate drainage problems in an ornamental garden.

Deep Open-Ditch Drainage   If a ditch is excavated to a depth below the water table (Figures 6.24 and 6.26), water will seep from the saturated soil, where it is under a positive pressure, into the ditch, where its potential will be essentially zero. Once in the ditch, the water can flow rapidly off the field as it no longer must overcome the frictional forces that would delay its twisting journey through tiny soil pores. However, the ditches, being 1 m or more deep, present barriers to equipment. Therefore, deep-ditch drainage is generally practical only for sandy soils in which the ditches may be spaced quite far apart. Open ditches need regular maintenance to control vegetation and sediment buildup.

Buried Perforated Pipes   A network of perforated plastic pipe can be laid underground using specialized equipment (Figure 6.26, *right*). Water moves into the pipe through the perforations. The pipe should be laid with the slotted or perforated side *down*. This allows water to flow up into the pipe but protects against soil falling into and clogging

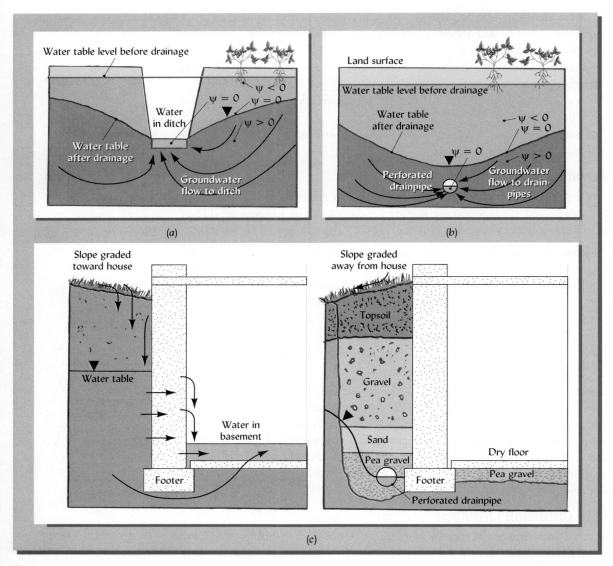

Figure 6.24

Three types of subsurface drainage systems. (a) Open ditches are used to lower the water table in a poorly drained soil. The wet season levels of the water table before and after ditch installation are shown. The water table is deepest next to the ditch, and the drainage effect diminishes with distance from the ditch. (b) Buried "tile lines" made of perforated plastic pipe act very much as the ditches in (a), but have two advantages: they are not visible after installation and they do not present any obstacle for surface equipment. Note the flow lines indicating the paths taken by water moving to the drainage ditches or pipes in response to the pressure potential gradients between the submerged water ($\psi > 0$) and free water in the drainage ditch and pipes ($\psi = 0$). (c) The water table around a building foundation before (left) and after (right) installation of a footer drain and correction of surface grading. The principles of water movement in soils are applied to keep the basement dry. (Diagrams courtesy of R. Weil)

the pipe. Sediment buildup will also be avoided if the pipe has the proper slope (usually a 0.5 to 1% drop) so water flows rapidly to the outlet ditch or stream.

Building Foundation Drains   Surplus water around building foundations can cause serious damage. The removal of this excess water is commonly accomplished using buried perforated pipe, placed alongside and slightly below the foundation or underneath the floor (Figure 6.24c). The perforated pipe must be sloped to allow water to move rapidly to an outlet ditch or sewer. Water will not seep into the basement for the same reason that it will not seep into a drainpipe placed above the water table.

SUCCESS OR FAILURE IN LANDSCAPE DRAINAGE DESIGN

A hedge of hemlock trees, all carefully pruned into ornamental shapes as part of an intricate landscape design, were dying again because of poor drainage. A very expensive effort to improve the drainage under the hemlocks had proved to be a failure, and the replanted hemlocks were again causing an unsightly blemish in an otherwise picture-perfect garden.

Finally, the landscape architect for the world-famous ornamental garden called in a soil scientist to assist her in finding a solution to the dying hemlock problem. Records showed that in the previous failed attempt to correct the drainage problem, contractors had removed all the hemlock trees in the hedge and had dug a trench under the hedge some 3 m deep (a). They had then backfilled the trench with gravel up to about 1 m from the soil surface, completing the backfill with a high-organic-matter, silt loam topsoil. It was into this silt loam that the new hemlock trees had been planted. Finally, a bark mulch had been applied to the surface.

The landscaper who had designed and installed the drainage system apparently had little understanding of the various soil horizons and their relation to the local hydrology. When the soil scientist examined the problem site, he found an impermeable claypan that was causing water from upslope areas to move laterally into the hemlock root zone. Basic principles of soil water movement also told him that water would not drain from the fine pores in the silt loam topsoil into the large pores of the gravel, and therefore the gravel in the trench would do no good in draining the silt loam topsoil (compare the situation to that in Figures 5.18 and 5.19). In fact, the water moving laterally over the impermeable layer created a perched water table that poured water into the gravel-filled trench, soon saturating both the gravel and the silt loam topsoil.

To cure the problem, the previous "solution" had to be undone (b). The dead hemlocks were removed, the ditches were reexcavated, and the gravel was removed from the trenches. Then the trench, except for the upper 0.5 m, was filled with a sandy loam subsoil to provide a suitable rooting medium for the replacement trees. The upper 0.5 m of the trench was filled with a sandy loam topsoil, which was also acid but contained a higher level of organic matter. The interface between the subsoil and surface soil was mixed so that there would be no abrupt change in pore configuration. This would allow an unsaturated wetting front to move down from the upper to the lower layers, drawing down any excess water.

About 1 m uphill from the trench, an *interceptor drain* was installed by digging a small trench through the impermeable clay layer that was guiding water to the area. A perforated drainage pipe surrounded by a layer of gravel was laid in the bottom of this trench with about a 1% slope to allow water to flow away from the area to a suitable outlet. The interceptor drain prevented the water moving laterally over the impermeable soil layer from reaching the evergreen hedge root zone.

Even though the replanting of the hedge was followed by an exceptionally rainy year, the new drainage system kept the soil well aerated, and the trees thrived. The principles of water movement explained in Sections 5.6 and 6.7 of this textbook were applied successfully in the field.

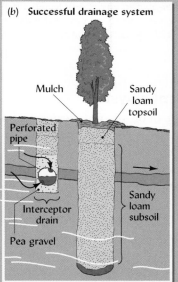

Figure 6.25
Failed and successful drainage designs.

6.8 SEPTIC TANK DRAIN FIELDS

Thousands of ordinary suburbanites get their first exposure to the importance of water movement through soils when they apply for a building permit for their new dream house. The local authorities will usually not allow a home to be built until arrangements are made for wastewater treatment. Typically, a soil scientist will come out to inspect the soils at the home site and judge their suitability for use as a septic tank drain field. If the soils are found unsuitable, the landowner may be denied the permit to build.

Septic tank drainage fields:
www.soil.ncsu.edu/
publications/Soilfacts/
AG-439-13/

(a)   (b)

Figure 6.26

(Left) An open-ditch drainage system designed to lower the water table during the wet season. The water flowing in the ditch has seeped in from the saturated soil. The capillary fringe above the water table can be seen as a dark band of soil. In order to protect water quality from chemicals and sediments that might reach the ditch via surface runoff, buffer strips of grass or forest vegetation should be established on both sides. Note in the background the piles of spoil, subsoil excavated to make the ditches. This material is usually spread thinly and mixed by tillage with the surface soil in the field, but it should first be tested for other detrimental properties. (Right) Specialized, laser-guided equipment laying a drain line made of corrugated plastic pipe. The pipe has perforations on the underside that allow water to seep in from a saturated soil. The trench will be backfilled with the soil shown piled on the left. The drain line must be laid deeper than the seasonal high water table if it is to remove any water. (Photos courtesy of R. Weil, left, and USDA/NRCS, right)

Operation of a Septic System

Septic systems and their maintenance:
http://extension.umd.edu/environment/Water/files/septic.html

The most common type of on-site wastewater treatment for homes not connected to municipal sewage systems is the *septic tank* and associated *drain field* (sometimes called *filter field* or *absorption field*). In essence, a septic drain field operates like artificial soil drainage in reverse. A network of perforated underground pipes is laid in trenches, but instead of draining water away from the soil, the pipes in a septic drain field carry wastewater *to* the soil, the water entering the soil via slits or perforations in the pipes (Figure 6.27). In a properly functioning septic drain field, the wastewater will enter the soil and percolate downward, undergoing several purifying processes before it reaches the groundwater. One of the advantages of this method of sewage treatment is that it has the potential to replenish local groundwater supplies for other uses.

The Drain Field   The water exiting the septic tank (see Figure 6.27) via a pipe near the top is termed the septic tank **effluent**. Although its load of suspended solids has been much reduced, it still carries organic particles, dissolved chemicals (including nitrogen), and microorganisms (including pathogens). The flow is directed to one or more buried pipes that constitute the **drain field**. Blanketed in gravel and buried in trenches about 0.6 to 2 m under the soil surface, these pipes are perforated on the bottom to allow the wastewater to seep out and enter the soil. It is at this point that soil properties play a crucial role. Septic systems depend on the soil in the drain field to (1) keep the effluent out of sight and out of contact with people, (2) treat or purify the effluent, and (3) conduct the purified effluent to the groundwater.

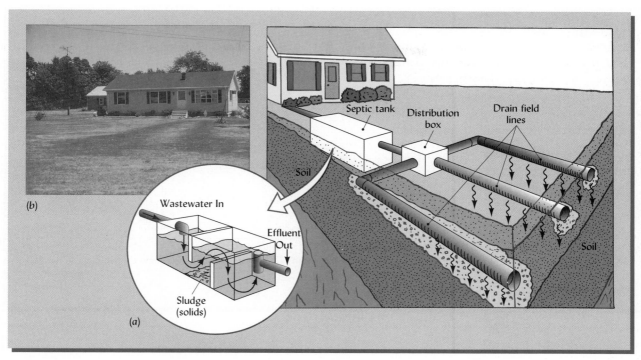

Figure 6.27

(a) *A septic tank and drain field constituting a standard system for on-site wastewater treatment. Most of the solids suspended in the household wastewater settle out in the concrete septic tank. Effluent from the tank flows to the drain field, where it seeps out of the perforated pipes and into the soil. In the soil, the effluent is purified by microbial, chemical, and physical processes as it percolates toward the groundwater.* (b) *Photo shows the telltale dark strips of lawn where poorly functioning septic tank drain lines are stimulating the grass with wastewater and nitrogen.* (Photo courtesy of R. Weil)

Soil Properties Influencing Suitability for a Septic Drain Field

The soil should have a *saturated hydraulic conductivity* (see Section 5.5) that will allow the wastewater to enter and pass through the soil profile rapidly enough to avoid backups that might saturate the surface soil with effluent, but slowly enough to allow the soil to purify the effluent before it reaches the groundwater. The soil should be sufficiently *well aerated* to encourage *microbial breakdown* of the wastes and *destruction of pathogens*. The soil should have some fine pores and clay or organic matter to adsorb and filter contaminants from the wastewater.

Soil properties that may disqualify a site for use as a septic drain field include impermeable layers such as a fragipan or a heavy claypan, gleying in the upper horizons, too steep a slope, or excessively drained sand and gravel. Septic tank drain fields installed where soil properties are not appropriate may result in extensive pollution of groundwater and in health hazards caused by seepage of untreated wastewater (Figure 6.27*b* and Plate 67).

Suitability Rating   The suitability of a site for septic drain field installation depends largely on soil properties that affect water movement and the ease of installation (see Table 6.3). For example, a septic drain field laid out on a slope greater than 15% may allow considerable lateral movement of the percolating water such that, at some point downslope, the wastewater will seep to the surface and present a potential health hazard (Plate 66).

The soil properties ideal for a septic drain field are nearly the opposite of those associated with the need for tile drainage. For example, instead of a high water table that requires lowering by drainage, septic drain field sites should have a low water table so that there is plenty of well-aerated soil to purify the wastewater before it

Drainage tips for hillside homeowners: www.wy.nrcs.usda.gov/ technical/ewpfactsheets/ homedrain.html

Table 6.3
SOIL PROPERTIES INFLUENCING SUITABILITY FOR A SEPTIC TANK DRAIN FIELD
Note that most of these soil properties pertain to the movement of water through the soil profile. Official design requirements vary among jurisdictions.

Soil property[a]	Limitations		
	Slight	Moderate	Severe
Flooding	—	—	Floods frequent to occasional
Depth to bedrock or impermeable pan, cm	>183	102–183	<102
Ponding of water	No	No	Yes
Depth to seasonal high water table, cm	>183	122–183	<122
Permeability (perc test) at 60 to 152 cm soil depth, mm/h	50–150	15–50	<15 or >150[b]
Slope of land, %	<8	8–15	>15
Stones >7.6 cm, % of dry soil by weight	<25	25–50	>50

[a]Assumes that soil does not contain permafrost and has not subsided more than 60 cm.
[b]Soil permeability (as determined by a perc test) greater than 150 mn/h is considered too fast to allow for sufficient filitering and treatment of wastes.
Adapted from Soil Survey Staff (1993), Table 620-17.

reaches the groundwater. Application of large quantities of wastewater through septic drain fields will actually raise the water table somewhat under the drain field.

Perc Test   A perc test determines the *percolation rate* (which is related to the saturated hydraulic conductivity described in Section 5.5), expressed in millimeters (or other unit of depth) of water entering the soil per hour. Some jurisdictions use the percolation rate to indicate whether the soil can accept wastewater rapidly enough to provide an acceptable disposal medium (Table 6.3). The test is simple to conduct (Figure 6.28) and should be carried out during the wettest season of the year.

To some degree, a low percolation rate can be compensated for by increasing the total length of drain field pipes, and hence increasing the area of land devoted to the

Figure 6.28
The perc test used in some places to help determine soil suitability for a septic tank drain field. On the site proposed for the drain field, a number of holes are augered, or excavated, to the depth where the perforated pipes are to be laid. The bottom of each hole is lined with gravel, and the holes are filled with water as a pretreatment to ensure that the soil is wet when the test is conducted. After the water has drained, the hole is refilled with water and a measuring rod is used to determine how long it takes for the water level to drops by 2.5 cm. This measurement is related to the saturated hydraulic conductivity. [Based on New York State Department of Health (2004); diagram courtesy of R. Weil]

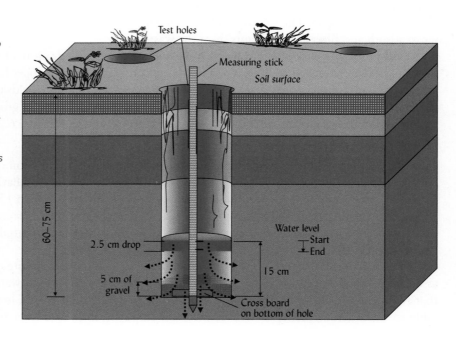

drain field. The size of the septic drain field is also influenced by the amount of waste-water that is likely to be generated (which may be estimated by the number of bedrooms in the house being served).

6.9   IRRIGATION PRINCIPLES AND PRACTICES[4]

In most regions of the world, insufficient water is the prime limitation to agricultural productivity. In semiarid and arid regions, intensive crop production is all but impossible without supplementing the meager rainfall provided by nature. However, if given supplemental water through irrigation, the sunny skies and fertile soils of some arid regions stimulate extremely high crop yields. The history of irrigation is nearly as old as the history of agriculture itself. Rice producers in Asia, wheat and barley producers in the Middle East, and corn producers in Central and South America were irrigating their crops well over 2000 years ago.

FAO's map/data of irrigated areas: www.fao.org/nr/water/aquastat/irrigationmap/index.stm

Irrigation Today

Food Production   During the past 50 years, the area of irrigated cropland has expanded greatly in many parts of the world, including the semiarid western United States (see Plate 111). The high productivity induced by irrigation is evident in the fact that irrigated land is responsible for some 40% of global crop production on only 15% of the world's cropland.

Expanded and improved irrigation, especially in Asia, has been a major factor in helping global food supplies keep up with, and even surpass, the global growth in population. As a result, irrigated agriculture remains the largest *consumptive* user of water resources, accounting for about 80% of all water consumed worldwide, in both developing and developed countries (Figure 6.29).

Landscaping   Irrigation is an integral part of such landscaping installations as golf courses, home lawns, and flower beds. In arid regions, the use of irrigation for landscaping is often predicated on the desire to maintain vegetation that conforms to an

Guide to desert landscaping: www.vvwater.org/guide/index.htm

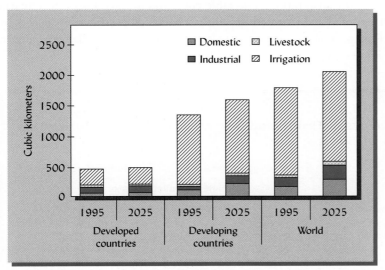

Figure 6.29
Consumptive use of water for domestic, industrial, livestock, and irrigation purposes in developing and developed countries and for the world. Consumptive use refers to water that is withdrawn by pumping from rivers or aquifers but not returned to the source after use. Most of it eventually evaporates. The values for 2025 are predicted. Note that irrigation is the dominant consumer of water in all cases. [From Rosegrant et al. (2002). Reproduced and adapted with permission from the International Food Policy Research Institute and the International Water Management Institute]

---

[4]For a fascinating account of water resources and irrigation management in the Middle East, see Hillel (1995). For a practical manual on small-scale irrigation with simple but efficient microtechnology, see Hillel (1997). For extensive information on irrigation of crops, see Lascano and Sojka (2007).

ideal notion of perpetual green lushness. In contrast, lawns, landscapes, and golf courses in many parts of the world now utilize only adapted vegetation that is capable of surviving the periods of dry weather and other adverse conditions characteristic of the local climate without irrigation. Increasing environmental awareness has engendered a growing trend toward more xerophytic landscaping (utilizing desert plants and rocks) in arid regions and generally toward more use of locally native vegetation that requires little or no irrigation.

Future Prospects   Among the problems facing irrigated agriculture in the future is the slowly dwindling availability of water because of (1) increased competition for water from a growing population of urban water users, (2) the overpumping of aquifers that has led to falling water tables, (3) reduction of storage capacity of existing reservoirs by siltation with eroded sediments (see Section 14.2), and (4) the need to allow a portion of river flows to maintain fish habitats downstream. Reducing waste and achieving greater efficiency of water use in irrigation are increasingly important aspects that will be emphasized in this section. Another major problem associated with irrigation, the salinization of soils and drainage waters, is considered in Chapter 9.

Water-Use Efficiency

Various measures of water-use efficiency are used to compare the relative benefits of different irrigation practices and systems. The most meaningful overall measure of efficiency would compare the output of a system (crop biomass or value of marketable product) to the amount of water allocated as an input into the system. There are many factors to consider (type of plants grown, reuse of "wasted" water by others downstream, etc.), so such comparisons must be made with caution.

Application Efficiency   A simple measure of water-use efficiency, sometimes termed the **water application efficiency**, compares the amount of water available or allocated to irrigate a field to the amount of water actually used in transpiration by the irrigated plants. In this regard, most irrigation systems are quite inefficient, with only about 10 to 30% of the water that is taken from the source transpired by the desired plants. Table 6.4 gives estimates of the various water losses that typically occur between allocation and transpiration. The table compares average losses in semiarid regions under both rainfed and irrigated agriculture. Much of the water loss occurs by evaporation and leakage in the reservoirs, canals, and ditches used to deliver water to irrigated fields (Figure 6.30).

Table 6.4

ESTIMATES OF WATER LOSSES IN TRADITIONAL IRRIGATED AND RAINFED AGRICULTURE IN SEMIARID AREAS OF THE WORLD

	Irrigated agriculture	Rainfed agriculture
	Percent of available water (%)[a]	
Storage and conveyance losses	30	0
Surface runoff and drainage losses[b]	44	40–50
Evaporation (from soil or water)	8–13	30–35
Transpiration by the crop	13–18	15–30

[a] Available water for irrigated agriculture = water stored in reservoirs or pumped from groundwater; available water for rainfed agriculture = rainfall.
[b] Some of the water lost by drainage and runoff may be reused by downstream irrigators.
From Wallace (2000).

Figure 6.30
Concrete-lined irrigation ditches (center) and standard-sized siphon pipes (right) can increase the efficiency of water delivery to the field. Unlined ditches (left) lose much of the water to adjacent soil areas or to the groundwater. Note the evidence of capillary movement above the water level in the unlined ditch. (Left and right photos courtesy of N. C. Brady; center photo courtesy of R. Weil)

Field Water Efficiency   Water-use efficiency *in the field* may be expressed as:

$$\text{Field water efficiency, \%} = \frac{\text{Water transpired by the crop}}{\text{Water applied to the field}} \times 100 \qquad (6.3)$$

Field water efficiency for traditional irrigation systems used in arid regions is often as low as 20 to 25%. For example, if transpiration uses 18% out of the total allocated water and 70% reaches the field, the field water efficiency = 100(18/70) = 25.7%. The water delivered to the field that is not transpired by the crop is lost as surface runoff, deep percolation below the root zone, and/or evaporation from the soil surface. Achieving a high level of field water efficiency (or a low level of water wastage) is very dependent on the skill of the irrigation manager and on the methods of irrigation used (Table 6.5).

Surface Irrigation

In surface irrigation systems, water is applied to the upper end of a field and allowed to distribute itself by gravity flow. Usually the land must be leveled and shaped so that the water will flow uniformly across the field. The water may be distributed in

Table 6.5
SOME CHARACTERISTICS OF THE THREE PRINCIPAL METHODS OF IRRIGATION

Methods and specific examples	Direct costs of installation, 2006 dollars/ha[a]	Labor requirements	Field water efficiency, %[b]	Suitable soils
Surface: basin, flood, furrow	600–900	High to low, depending on system	20–50	Nearly level land; not too sandy or rocky
Sprinkler: center pivot, movable pipe, solid set	900–1800	Medium to low	60–70	Level to moderately sloping; not too clayey
Microirrigation: drip, porous pipe, spitter, bubbler	1000–2000	Low	80–90	Steep to level slopes; any texture, including rocky or gravelly soils

[a]Average ranges from many sources. Costs of required drainage systems not included.
[b]Field water efficiency = 100 × (water transpired by crop/water applied to field).

furrows graded to a slight slope so that water applied to the upper end of the field will flow down the furrows at a controlled rate (Figure 6.30, *right*). In **border irrigation** systems, the land is shaped into broad strips 10 to 30 m wide, bordered by low dikes. Water is usually brought to surface-irrigated fields in supply ditches or gated pipes. The amount of water that enters the soil is determined by the permeability of the soil and by the length of time a given spot in the field is inundated with water. Achieving a uniform infiltration of exactly the required amount of water is very difficult and depends on controlling the slope and length of the irrigation runs across the field.

The **level basin** technique of surface irrigation, as is used for paddy rice and certain tree crops, alleviates these problems because each basin has no slope and is completely surrounded by dikes that allow water to stand on the area until infiltration is complete.

This method is not practical for highly permeable soils. On sloping land, terraces can be built in a modification of the level-basin method. Control over leaching and runoff losses is difficult, and the entire soil surface is wetted so that much water is lost by evaporation from the soil and by weed transpiration.

Sprinkler Systems

In sprinkler irrigation, water is sprayed through the air onto a field, simulating rainfall. Thus the entire soil surface, as well as plant foliage (if present), is wetted. This leads to evaporative losses similar to those described for surface systems. Furthermore, an additional 5 to 20% of the applied water may be lost by evaporation or windblown mist as the drops fly through the air. Plants often respond positively to the cooler, better-aerated sprinkler water, but wet leaves may increase the incidence of fungal diseases in some plants, such as grapes, fruit trees, and roses.

A sprinkler system should be designed to deliver water at a rate that is less than the infiltration capacity of the soil, so that runoff or excessive percolation will not occur. In practice, runoff and erosion may be problems. Overlapping of spray circles can help achieve a more even distribution of water. Because of better control over application rates, the field water-use efficiency is generally higher for sprinkler systems than for surface system.

Sprinkler irrigation is practical on a wider range of soil conditions than is the case for the surface systems. Various types of sprinkler systems are adapted to moderately sloping as well as level land. They can be used on soils with a wide range of textures, even those too sandy for surface irrigation systems.

The equipment costs for sprinkler systems are higher than those for surface-flow systems. Some types of sprinkler systems are set in place, others are moved by hand, and still others are self-propelled, either moving in large circles around a central pivot (Figure 6.31) or rolling slowly across a rectangular field. Most systems can be automated and adapted to deliver doses of pesticides or soluble fertilizers to plants.

Microirrigation

The most efficient irrigation systems in use today are those using microirrigation, whereby only a small portion of the soil is wetted (Figure 6.32, *left*) in contrast to the complete wetting accomplished by most surface and sprinkler systems.

Perhaps the best-established microirrigation system is *drip* (or *trickle*) *irrigation*, in which tiny emitters attached to plastic tubing apply water to the soil surface alongside individual plants (Plate 74). Water is applied at a low rate (sometimes drop by drop) but at a high frequency, with the objective of maintaining optimal soil water

Figure 6.31

Center pivot irrigation systems. The system at right is making a low-energy, precision application of water to a soybean crop and is rotating slowly towards the left. The photo at left shows another center pivot system with a heavy duty motor used to pump the water up from the groundwater. Also shown are a large tank of liquid fertilizer and the computer controls that allow nutrients to be injected into the water at precise rates during irrigation events. (Photos courtesy of R. Weil)

availability in the immediate root zone while leaving most of the soil volume dry (Figure 6.32).

Water is normally carried to the field in pipes, run through special filters to remove any grit or chemicals that might clog the tiny holes in the emitters, and then distributed throughout the field by means of a network of plastic pipes. Soluble fertilizers may be added to the water as needed.

If properly maintained and managed, microirrigation allows much more control over water application rates and spatial distribution than do either surface or sprinkler systems. Losses by supply-ditch seepage, sprinkler-drop evaporation, runoff, drainage (in excess of that needed to remove salts), soil evaporation, and weed transpiration can be greatly reduced or eliminated. Once a system is in place, the labor required for its operation is modest. Microirrigation systems are easily automated using either timers or, more effectively, computerized sensors that measure soil moisture and/or rainfall.

Figure 6.32

Two types of microirrigation. (Left) Drip or trickle irrigation with a single emitter for each seedling in a cabbage field. (Right) A microsprayer or spitter irrigating an individual tree in a home garden. In both cases, irrigation wets only the small portion of the soil in the immediate root zone. Small quantities of water applied at high frequency (such as once or twice a day) ensures that the root zone is kept almost continuously at an optimal moisture content. (Photos courtesy of R. Weil, *left*, and N. C. Brady, *right*)

Microirrigation often produces healthier plants and higher crop yields because the plant is never stressed by low water potentials or low aeration conditions that are associated with the feast-or-famine regime of infrequent, heavy water applications made by all surface irrigation systems and most sprinkler systems. A disadvantage or risk is that there is very little water stored in the soil at any time, so even a brief breakdown of the system could be disastrous in hot, dry weather. Because of its high water-use efficiency, microirrigation is most profitable where water supplies are scarce and expensive and where high-valued plants such as fruit trees are being grown.

6.10   CONCLUSION

The hydrologic cycle encompasses all movements of water on or near the Earth's surface. It is driven by solar energy, which evaporates water from the ocean, the soil, and vegetation. The water cycles into the atmosphere, returning elsewhere to the soil and the oceans in rain and snow.

The soil is an essential component of the hydrologic cycle. It receives precipitation from the atmosphere, rejecting some of it, which is then forced to run off into streams and rivers, and absorbing the remainder, which then moves downward to be either transmitted to the groundwater, taken up and later transpired by plants, or evaporated directly from soil surfaces and returned to the atmosphere.

The behavior and movement of water in soils and plants are governed by the same set of principles: water moves in response to differences in energy levels, moving from higher to lower water potential. These principles can be used to manage water more effectively and to increase the efficiency of its use.

Management practices should encourage movement of water into well-drained soils while minimizing evaporative (E) losses from the soil surface. These two objectives will provide as much water as possible for plant uptake and groundwater recharge. Water from the soil must satisfy the transpiration (T) requirements of healthy leaf surfaces; otherwise, plant growth will be limited by water stress. Practices that leave plant residues on the soil surface and that maximize plant shading of this surface will help achieve high efficiency of water use.

Extreme soil wetness, characterized by surface ponding and saturated conditions, is a natural and necessary condition for wetland ecosystems; however, for most other land uses, extreme wetness is detrimental. Drainage systems have therefore been developed to hasten the removal of excess water from soil and lower the water table so that upland plants can grow without aeration stress, and so the soil can better bear the weight of vehicular and foot traffic.

A septic tank drain field operates as a drainage system in reverse. Septic wastewaters can be disposed of and treated by soils if the soils are freely draining. Soils with low permeability or high water tables may indicate good conditions for wetland creation or appropriate sites for installation of artificial drainage for agricultural use, but they are not generally suited for septic tank drain fields.

Irrigation waters from streams or wells greatly enhance plant growth, especially in regions with scarce precipitation. With increasing competition for limited water resources, it is essential that irrigators manage water with maximal efficiency so that the greatest production can be achieved with the least waste of water resources. Such efficiency is encouraged by practices that favor transpiration over evaporation, such as mulching and the use of microirrigation.

As the operation of the hydrologic cycle causes constant changes in soil water content, other soil properties are also affected, most notably soil aeration and temperature, the subjects of the next chapter.

STUDY QUESTIONS

1. You know that the forest vegetation that covers a 120 km² wildland watershed uses an average of 4 mm of water per day during the summer. You also know that the soil averages 150 cm in depth and at field capacity can store 0.2 mm of water per mm of soil depth. However, at the beginning of the season the soil was quite dry, holding an average of only 0.1 mm/mm. As the watershed manager, you are asked to predict how much water will be carried by the streams draining the watershed during the 90-day summer period when 450 mm of precipitation falls on the area. Use the water balance equation to make a rough prediction of the stream discharge as a percentage of the precipitation and in cubic meters of water.

2. Draw a simple diagram of the hydrologic cycle using a separate arrow to represent these processes: *evaporation, transpiration, infiltration, interception, percolation, surface runoff,* and *soil storage.*

3. Describe and give an example of the *indirect* effects of plants on the hydrologic balance through their effects on the soil.

4. State the basic principle that governs how water moves through the SPAC. Give two examples, one at the soil–root interface and one at the leaf–atmosphere interface.

5. Define *potential evapotranspiration* and explain its significance to water management.

6. What is the role of evaporation from the soil (E) in determining water-use efficiency, and how does it affect ET? List three practices that can be used to control losses by E.

7. Weed control should reduce water losses by what process?

8. Comment on the relative advantages and disadvantages of organic versus plastic mulches.

9. What does conservation tillage conserve? How does it do it?

10. The small irrigation project you manage collects 2,000,000 m³ of water annually in a reservoir. Of this, 20% evaporates from the reservoir surface during the year. Of the remaining water, 25% is lost by evaporation and percolation into the soil during distribution via unlined canals before the water reaches the fields. The water is then applied by furrow irrigation, with averages 20% of the water applied percolating below the crop root zone, 20% running off into collection canals at the low end of the field, and 30% evaporating from the soil surface. By the time of crop harvest, ET has dried the soil to about the same water content it had before irrigation began. Average ET is 7 mm/day for a 180-day irrigation season and crop water-use efficiency averages 1.1 kg of dry matter/m³ water transpired. Show calculations (or make a spreadsheet) to estimate:
    (a) the overall water-use efficiency of the project (kg output/m³ water allocated),
    (b) the application water efficiency for the project,
    (c) the field water efficiency for the project, and
    (d) the number of hectares that can be irrigated in this project.

11. Explain under what circumstances earthworm channels might increase downward saturated water flow but not have much effect on the leaching of soluble chemicals applied to the soil.

12. What will be the effect of placing a perforated drainage pipe in the capillary fringe zone just above the water table in a wet soil? Explain in terms of water potentials.

13. What soil features may limit the use of a site for a septic tank drain field?

14. Which irrigation systems are likely to be used where: (a) water is expensive and the market value of crops produced per hectare is high, and (b) the cost of irrigation water is subsidized and the value of crop products that can be produced per hectare is low? Explain.

REFERENCES

Day, R. L., et al. 1998. "Water balance and flow patterns in a fragipan using in situ soil block," *Soil Sci.* **163:**517–528.

Gburek, W. J., B. A. Needelman, and M. S. Srinivasan. 2006. "Fragipan controls on runoff generation: Hydropedological implications at landscape and watershed scales," *Geoderma* **131:**330–344.

Hillel, D. 1995. *The Rivers of Eden.* New York: Oxford University Press.

Hillel, D. 1997. *Small-Scale Irrigation for Arid Zones.* FAO Development Series 2. (Rome: U.N. Food and Agriculture Organization).

Howell, T. A. 2001. "Enhancing water use efficiency in irrigated agriculture," *Agron. J.* **93:**281–289.

Jury, W. A., and H. Fluhler. 1992. "Transport of chemicals through soil: Mechanisms, models, and field applications," *Advances in Agronomy* **47**:141–201.

Lascano, R., and R. Sojka. 2007. *Irrigation of Agricultural Crops.* Agron. Monograph No. 30. 2nd ed. (Madison WI: Amer. Soc. Agronomy).

New York State Department of Health. 2004. *Individual residential wastewater treatment systems design handbook.* 10 New York Codes, Rules and Regulations Appendix 75-A. Oneida County Health Department, Albany, NY, www.oneidacounty .org/oneidacty/gov/dept/health/Sewage/75A/ 75ABooklet.pdf.

Rosegrant, M. W., X. Cai, and S. A. Cline. 2002. *Global water outlook to 2025: Averting an impending crisis.* International Food Policy Research Institute and the International Water Management Institute, Washington, DC, www.ifpri.org/pubs/fpr/ fprwater2025.pdf.

Skaggs, R. W., and J. van Schilfgaarde (eds.). 1999. *Agricultural Drainage.* Agronomy Series no. 38. (Madison, WI: Amer. Soc. Agron., Crop Sci. Soc. Amer., Soil Sci. Soc. Amer.).

Soil Survey Staff. 1993. *National Soil Survey Handbook.* Title 430-VI. (Washington, DC: USDA Natural Resources Conservation Service).

Starrett, S. K., N. E. Christians, and T. A. Austin. 1996. "Movement of pesticides under two irrigation regimes applied to turfgrass," *J. Environ. Qual.* **25**:566–571.

Swank, W. T., J. M. Vose, and K. J. Elliot. 2001. "Long-term hydrologic and water quality responses following commercial clear cutting of mixed hardwoods on a southern Appalachian catchment," *Forest Ecology & Management* **143**:163–178.

Tyree, M. T. 2003. "The ascent of water," *Nature* **423**:923.

Unger, P. W., and R. L. Baumhardt. 1999. "Factors related to dryland grain sorghum yield increases: 1939 through 1997," *Agron J.* **91**:870–875.

Varvel, G., W. Riedell, E. Deibert, B. McConkey, D. Tanaka, M. Vigil, and R. Schwartz. 2006. "Great Plains cropping system studies for soil quality assessment," *Renewable Agriculture and Food Systems* **21**:3–14.

Wallace, J. S. 2000. "Increasing agricultural water use efficiency to meet future food production," *Agric. Ecosyst. Environ.* **82**:105–119.

Zanders, J. 2001. "Urban development and soils," *Soil Horizons—A Newsletter of Landcare Research New Zealand, Ltd.* 6(Nov.):6.

7

Soil Aeration and Temperature

The naked earth is warm with Spring. . . .
—JULIAN GRENFELL,
INTO BATTLE

It is a central maxim of ecology that "everything is connected to everything else." This interconnectedness is one reason why soils are such fascinating (and challenging) objects of study. In this chapter we explore two aspects of the soil environment, aeration and temperature; they are not only closely connected to each other but are both also intimately influenced by many of the soil properties discussed in other chapters.

Since air and water share the pore space of soils, it is not surprising that much of what we learned about the texture, structure, and porosity of soils (Chapter 4) and the retention and movement of water in soils (Chapters 5 and 6) will have direct bearing on soil aeration. These are some of the physical parameters affecting aeration status, but chemical and biological processes also affect, and are affected by, soil aeration.

For the growth of plants and the activity of microorganisms, soil aeration status can be just as important as soil moisture status and can sometimes be even more difficult to manage. In most forest, range, agricultural, and ornamental applications, a major management objective is to maintain a high level of oxygen in the soil for root respiration. Yet it is also vital that we understand the chemical and biological changes that take place when the oxygen supply in the soil is depleted.

Soil temperatures affect plant and microorganism growth and also influence soil drying by evaporation. The movement and retention of heat energy in soils are often ignored, but they hold the key to understanding many important soil phenomena,

from frost-damaged pipelines and pavements to the Spring awakening of biological activity in soils. The unusually high soil temperatures that result from fires on forest-, range-, or croplands can markedly change critical physical and chemical soil properties.

We will see that increasing soil temperatures influence soil aeration largely through their stimulating effects on the growth of plants and soil organisms and on the rates of biochemical reactions. Nowhere are these interrelationships more critical than in the water-saturated soils of wetlands, ecosystems that will therefore receive special attention in this chapter.

7.1   SOIL AERATION—THE PROCESS

For plant roots and other soil organisms to readily carry on respiration, the soil must be well ventilated. Good ventilation allows the exchange of gases between the soil and the atmosphere to supply enough oxygen (O_2) while preventing the potentially toxic accumulation of gases such as carbon dioxide (CO_2), methane (CH_4), and ethylene (C_2H_6). Soil aeration status involves the rate of such ventilation, as well as the proportion of pore spaces filled with air, the composition of that soil air, and the resulting chemical oxidation or reduction potential in the soil environment.

Soil Aeration in the Field

Oxygen availability in field soils is regulated by three principal factors: (1) *soil macroporosity* (as affected by texture and structure), (2) *soil water content* (as it affects the proportion of porosity that is filled with air), and (3) *O_2 consumption* by respiring organisms (including plant roots and microorganisms). The term *poor soil aeration* refers to a condition in which the availability of O_2 in the root zone is insufficient to support optimal growth of upland plants and aerobic microorganisms. Typically, poor aeration becomes a serious impediment to plant growth when O_2 concentration drops below 0.1 L/L. This often occurs when more than 80 to 90% of the soil pore space is filled with water (leaving less than 10 to 20% of the pore space filled with air). The high soil water content not only leaves little pore space for air storage, but, more important, the water blocks the pathways by which gases could exchange with the atmosphere. Compaction can also cut off gas exchange, even if the soil is not very wet and has a large percentage of air-filled pores.

Excess Moisture

The extreme case of excess water occurs when all or nearly all of the soil pores are filled with water. The soil is then said to be **water saturated** or **waterlogged**. Waterlogged soil conditions are typical of wetlands and may also occur for short periods of time in well-drained soils when water is applied, or if wet soil has been compacted.

Plants adapted to life in waterlogged soils are termed **hydrophytes**. For example, a number of grass species, including rice, eastern gamagrass, and spartina marsh grasses transport oxygen for respiration down to their roots via hollow structures in their stems and roots known as **aerenchyma** tissues. Mangroves and other hydrophytic trees produce aerial roots and other structures that allow their roots to obtain O_2 while growing in water-saturated soils.

Soil aeration and plant growth:
www.uoguelph.ca/~mgoss/five/410_N06.html

Figure 7.1

Most plants depend on the soil to supply oxygen for root respiration and therefore are disastrously affected by even relatively brief periods of soil saturation during which oxygen becomes depleted. (Left) Sugar beets on a clay loam soil dying where the soil has become water-saturated in a compacted area. (Right) Pine trees dying in a sandy soil area that has become saturated as a result of flooding by beavers. A new community of plants better adapted to poorly aerated soil conditions is taking over the site. (Photos courtesy of R. Weil)

Most plants, however, are dependent on a supply of oxygen from the soil and suffer dramatically if good soil aeration is not maintained by drainage or other means (Figure 7.1). Some plants succumb to O_2 deficiency or toxicity of other gases within hours after the soil becomes saturated.

Gaseous Interchange

The more rapidly roots and microbes use up oxygen and release carbon dioxide, the greater is the need for the exchange of gases between the soil and the atmosphere. This exchange is facilitated by two mechanisms, **mass flow** and **diffusion**. Mass flow of air is much less important than diffusion in determining the total exchange that occurs. It is enhanced, however, by fluctuations in soil moisture content that force air in or out of the soil or by wind and changes in barometric pressure.

The great bulk of the gaseous interchange in soils occurs by *diffusion*. Through this process, each gas moves in a direction determined by its own *partial pressure*. The partial pressure of a gas in a mixture is simply the pressure this gas would exert if it alone were present in the volume occupied by the mixture. Thus, if the pressure of air is 1 atmosphere (~100 kPa), the partial pressure of oxygen, which makes up about 21% (0.21 L/L) of the air by volume, is approximately 21 kPa.

Because of diffusion along *partial pressure gradients*, the higher concentration of oxygen in the atmosphere will result in a net movement of this particular gas into the soil. Carbon dioxide and water vapor normally move in the opposite direction, since the partial pressures of these two gases are generally higher in the soil air than in the atmosphere. A representation of the principles involved in diffusion is given in Figure 7.2.

Figure 7.2

The process of diffusion between gases in a soil pore and in the atmosphere. The total gas pressure is the same on both sides of the boundary. The partial pressure of oxygen is greater, however, in the atmosphere. Therefore, oxygen tends to diffuse into the soil pore where fewer oxygen molecules per unit volume are found. The carbon dioxide molecules, on the other hand, move in the opposite direction owing to the higher partial pressure of this gas in the soil pore. This diffusion of O_2 into the soil pore and of CO_2 into the atmosphere will continue as long as the respiration by roots and microorganisms in the soil consumes O_2 and releases CO_2.

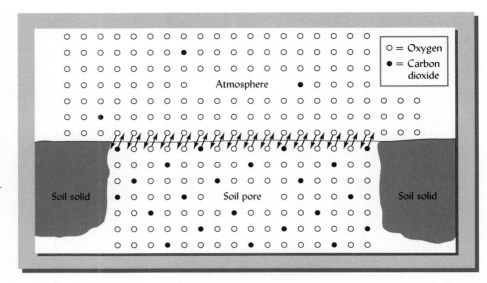

7.2 MEANS OF CHARACTERIZING SOIL AERATION

The aeration status of a soil can be characterized in several ways, including (1) the content of oxygen and other gases in the soil atmosphere, (2) the air-filled soil porosity, and (3) the chemical oxidation-reduction (redox) potential.

Gaseous Composition of the Soil Air

Oxygen   The atmosphere above the soil contains nearly 21% O_2, 0.035% CO_2, and more than 78% N_2. In comparison, soil air has about the same level of N_2 but is consistently lower in O_2 and higher in CO_2. The O_2 content may be only slightly below 20% in the upper layers of a soil with an abundance of macropores. It may drop to less than 5% or even to near zero in the lower horizons of a poorly drained soil with few macropores. Once the supply of O_2 is virtually exhausted, the soil environment is said to be **anaerobic**.

Carbon Dioxide and Other Gases   Since the N_2 content of soil air is relatively constant, there is a general inverse relationship between the contents of the other two major components of soil air—O_2 and CO_2—with O_2 decreasing as CO_2 increases. In cases where the CO_2 content becomes as high as 10%, it may be toxic to some plant processes. Soil air usually is much higher in water vapor than is the atmosphere, being essentially saturated except at or very near the surface of the soil (see Section 5.7). Also, under waterlogged conditions, the concentrations of gases such as methane (CH_4) and hydrogen sulfide (H_2S), which are formed as organic matter decomposes, are notably higher in soil air. Another gas produced by roots and microbes under anaerobic conditions is ethylene (C_2H_4). This gas is particularly toxic to plants, even in concentrations lower than 1 µL/L (0.0001%).

Air-Filled Porosity

Many researchers believe that microbiological activity and plant growth become severely inhibited in most soils when air-filled porosity falls below 20% of the pore space or 10% of the total soil volume (with correspondingly high water contents). One of the principal reasons that high water contents cause oxygen deficiencies for roots is that water-filled pores block the diffusion of oxygen into the soil to replace that used by respiration. In fact, oxygen diffuses 10,000 times faster through a pore filled with air than through a similar pore filled with water.

7.3 OXIDATION REDUCTION (REDOX) POTENTIAL[1]

Soil aeration markedly influences the reduction and oxidation states of the chemical elements. The reaction that takes place as the reduced state of an element is changed to the oxidized state may be illustrated by the oxidation of two-valent iron [Fe^{2+} or Fe(II)] in FeO to the trivalent form [Fe^{3+} or Fe(III)] in FeOOH:

Oxidation-reduction reactions: www.shodor.org/UNChem/advanced/redox/index.html

$$\underset{\text{Fe(II)}}{\overset{(2+)}{2FeO}} + 2H_2O \rightleftharpoons \underset{\text{Fe(III)}}{\overset{(3+)}{2FeOOH}} + 2H^+ + 2e^- \qquad (7.1)$$

As Reaction 7.1 proceeds to the *right*, each Fe(II) loses an electron (e^-) to become Fe(III) and forms H^+ ions by hydrolyzing H_2O. These H^+ ions lower the pH. When the reaction proceeds to the *left*, FeOOH acts as an **electron acceptor** and the pH rises as H^+ ions are consumed. The tendency or potential for electrons to be transferred from one substance to another in such reactions can be measured using a platinum electrode and is termed the **redox potential** (E_h).

The redox potential is usually measured in volts or millivolts as the reference state, the redox potential of the hydrogen couple $\frac{1}{2}H_2 \rightleftharpoons H^+ + e^-$ is arbitrarily taken as zero. If a substance will accept electrons easily, it is known as an *oxidizing agent*; if a substance supplies electrons easily, it is a *reducing agent*.

Role of Oxygen Gas

Oxygen gas (O_2) is an important example of a strong oxidizing agent, since it rapidly accepts electrons from many other elements. All aerobic respiration requires O_2 to serve as the electron acceptor as living organisms oxidize organic carbon to release energy for life.

Oxygen can oxidize both organic and inorganic substances. In a well-aerated soil with plenty of gaseous O_2, the E_h is in the range of 0.4 to 0.7 volt (V). As aeration is reduced and gaseous O_2 is depleted, the E_h declines to about 0.32 to 0.38 V. If organic-matter-rich soils are flooded under warm conditions, E_h values as low as −0.3 V can be found.

The relationship between changes in O_2 content and E_h of a wet soil is shown in Figure 7.3. Within a day or two after the warm soil became water saturated, aerobic

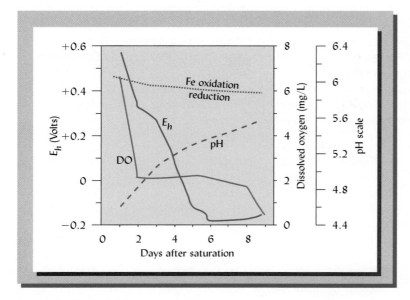

Figure 7.3

Changes in soil chemistry following water-saturation of a silt loam A horizon. In the first two days, aerobic and facultative microorganisms respired most of the dissolved O_2 (DO), thereby lowering the E_h of the soil solution. As DO was depleted and E_h dropped, conditions became suitable for anaerobic microorganisms. Substances such as the iron (Fe) in soil minerals are reduced by anaerobic microbes, which use them as the terminal electron acceptor for their metabolism. This process lowers the E_h further. The dotted line indicates that the E_h at which Fe^{3+} tends to be reduced to Fe^{2+} declines somewhat as the pH rises. The rise in pH depicted by the dashed line in the graph is in turn partly caused by the consumption of H^+ ions in reactions that reduce Fe^{3+} (see Reaction 7.1). Such reactions also change the color of the soil by dissolving certain minerals. [Adapted from Jenkinson and Franzmeier (2006) with permission of the Soil Science Society of America]

---

[1]For reviews of redox reactions in soils, see Bartlett and James (1993) and Bartlett and Ross (2005).

and facultative microorganisms oxidizing organic carbon in the soil (see Section 10.2) respired most of the O_2 initially present, thereby lowering the E_h of the soil solution.

Other Electron Acceptors

As O_2 becomes depleted and E_h drops, reducing conditions become established. With no O_2 available, only **anaerobic** microorganisms can survive. They must use substances other than O_2 as the terminal electron acceptors for their metabolism. For example, they may use iron in soil minerals. As they reduce iron, the E_h drops still further because electrons are consumed. The pH simultaneously rises because the reaction consumes H^+ ions (Reaction 7.1 goes to the left). When these reduction reactions proceed, soil colors change from the reds of iron oxides to the grays of reduced iron (see Plate 63 and Sections 4.1 and 7.7). Similar reactions involve the reduction or oxidation of C, N, Mn, S, and other elements from the solution, organic matter, and mineral components of the soil.

When the soil becomes essentially devoid of O_2, the soil redox potential falls below levels of about 0.38 to 0.32 V (at pH 6.5). After O_2 is used up, the next most easily reduced substance present is usually the N^{5+} in nitrate (NO_3^-). If the soil contains much NO_3^-, the E_h will remain near 0.28 to 0.22 V as the nitrate is reduced:

$$\underset{\text{N(V)}}{\overset{(5+)}{NO_3^-}} + 2e^- + 2H^+ \longleftrightarrow \underset{\text{N(III)}}{\overset{(3+)}{NO_2^-}} + H_2O \qquad (7.2)$$

Once all the N^{5+} in nitrate has been transformed into NO_2^-, N_2, and other N species, the E_h will drop further. At this point, organisms capable of reducing Mn will become active. Thus, as E_h values fall, the elements N, Mn, Fe, and S (in SO_4^{2-}) and C (in CO_2) accept electrons and become reduced, predominantly in the order shown in Table 7.1.

In other words, transformation of different elements requires different degrees of reducing conditions. The soil E_h must be lowered to -0.2 V before methane is produced, but reduction of NO_3^- to N_2 gas takes place when the E_h is as high as $+0.28$ V. We can conclude that soil aeration helps determine the specific chemical species present in soils and, in turn, the availability, mobility, and possible toxicity of many chemical elements.

Redoximorphic features in soils: http://nesoil.com/images/redox.htm

Table 7.1
OXIDIZED AND REDUCED FORMS AND CHARGES FOR SEVERAL ELEMENTS AND REDOX POTENTIALS E_h AT WHICH THE REDOX REACTIONS OCCURRED IN A SOIL AT pH 6.5

E_h values measured in soil are generally lower than the theoretical values for the reactions. At E_h levels lower than about 0.38 to 0.32 V, microorganisms utilize elements other than oxygen as their electron acceptor.

Element	Oxidized form	Charge on oxidized element	Reduced form	Charge on reduced element	E_h at which change of form occurs, V
Oxygen	O_2	0	H_2O	-2	0.38 to 0.32
Nitrogen	NO_3^-	$+5$	N_2	0	0.28 to 0.22
Manganese	Mn^{4+}	$+4$	Mn^{2+}	$+2$	0.22 to 0.18
Iron	Fe^{3+}	$+3$	Fe^{2+}	$+2$	0.11 to 0.08
Sulfur	SO_4^{2-}	$+6$	H_2S	-2	-0.14 to -0.17
Carbon	CO_2	$+4$	CH_4	-4	-0.20 to -0.28

E_h values from Patrick and Jugsujinda (1992).

7.4   FACTORS AFFECTING SOIL AERATION AND E_h

Drainage of Excess Water

Drainage of gravitational water out of the profile and concomitant diffusion of air into the soil takes place most readily in macropores. The most important factors influencing the aeration of well-drained soils are therefore those that determine the volume of the soil macropores. Soil texture, bulk density, aggregate stability, organic matter content, and biopore formation are among the soil properties that help determine macropore content and, in turn, soil aeration (see Section 4.8).

Rates of Respiration in the Soil

The concentrations of both O_2 and CO_2 are largely dependent on microbial activity, which in turn depends on the availability of organic carbon compounds as food. Incorporation of large quantities of manure, crop residues, or sewage sludge may alter the soil air composition appreciably. Likewise, the cycling of plant residues by leaf fall, root mass decay, and root excretion in natural ecosystems provides the substrate for microbial activity. Respiration by plant roots and enhanced respiration by soil organisms near the roots are also significant processes. All these processes are very much enhanced as soil temperature increases (see Section 7.8).

Soil Heterogeneity

Subsoils are usually more deficient in oxygen than are topsoils. Not only is the water content usually higher (in humid climates), but the total pore space, as well as the macropore space, is generally much lower in the deeper horizons. In addition, the pathway for diffusion of gases into and out of the soil is longer for deeper horizons. However, if organic substrates are in low supply, the subsoil may still be aerobic because O_2 can diffuse fast enough to replace that used by respiration. For this reason, certain recently flooded soils are anaerobic in the upper 50 to 100 cm and are aerobic below.

Tillage   One cause of soil heterogeneity is tillage, which has both short-term and long-term effects on soil aeration. In the short term, stirring the soil often allows it to dry out faster and also mixes in large quantities of air. These effects are especially evident on somewhat compacted, fine-textured soils, on which plant growth often responds immediately after a cultivation to control weeds or "knife in" fertilizer. In the long term, however, tillage may reduce macroporosity (see Section 4.6).

Pore Size   Oxygen gas will diffuse much more slowly through the small, largely water-filled pores within a soil aggregate than through the largely air-filled pores between aggregates. Therefore, anaerobic conditions may occur in the center of an aggregate, only a few mm from well-aerated conditions near the surface of the same aggregate (Figure 7.4).

In some upland soils, large subsoil pores such as cracks between peds and old root channels may periodically fill with water, causing localized zones of poor aeration. This condition is expressed by gray, reduced surfaces on peds with reddish, oxidized interiors (see Plate 82). In a normally saturated soil, such large pores may cause the opposite effect (peds with oxidized faces but reduced interiors), as they facilitate O_2 diffusion into the soil during dry periods.

Plant Roots   Respiration by the roots of upland plants usually depletes the O_2 in the soil just outside the root. The opposite can occur in wet soils growing hydrophytic plants. Aerenchyma tissues in these plants may transport surplus O_2 into the roots, allowing some to diffuse into the soil and produce an oxidized zone in an otherwise anaerobic soil (see, for example, Plate 38).

Figure 7.4

The oxygen content of soil air in a wet aggregate from an Aquic Hapludoll (Muscatine silty clay loam) from Iowa. The measurements were made with a unique microelectrode. Note that the oxygen content near the aggregate center was zero, while that near the edge of the aggregate was 21%. Thus pockets of oxygen deficiency can be found in a soil whose overall oxygen content may not be low. [From Sexstone et al. (1985)]

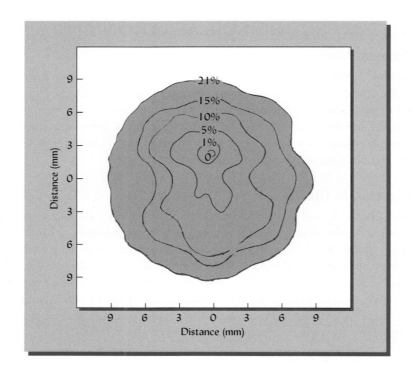

For all these reasons, aerobic and anaerobic processes may proceed simultaneously and in very close proximity to each other in the same soil. This heterogeneity of soil aeration should be kept in mind when considering the role that soils play in elemental cycling and ecosystem function.

7.5   ECOLOGICAL EFFECTS OF SOIL AERATION

Effects on Organic Residue Degradation

Poor aeration slows down the rate of decay, as evidenced by the relatively high levels of organic matter that accumulate in poorly drained soils. The *nature* as well as the *rate* of microbial activity is determined by the O_2 content of the soil. Where O_2 is present, aerobic organisms are active (see Section 11.2). In the absence of gaseous oxygen, anaerobic organisms take over, and decomposition is much slower. Poorly aerated soils therefore tend to contain a wide variety of only partially oxidized products such as ethylene gas (C_2H_4), alcohols, and organic acids, many of which can be toxic to higher plants and to many decomposer organisms. The latter effect helps account for the formation of Histosols in wet areas where inhibition of decomposition allows thick layers of organic matter to accumulate. In summary, the presence or absence of oxygen gas completely modifies the nature of the decay process and its effect on plant growth.

Oxidation–Reduction of Elements

Nutrients   Through its effects on the redox potential, the level of soil oxygen largely determines the forms of several inorganic elements, as shown in Table 7.2. The oxidized states of the nitrogen and sulfur are readily utilizable by higher plants. Reduced forms of iron and manganese may be so soluble that toxicities may occur. However, some reduction of iron may be beneficial as it will release phosphorus from insoluble iron-phosphate compounds. Such phosphorus release has implications for eutrophication (see Section 12.3) when it occurs in saturated soils or in underwater sediments.

Table 7.2
OXIDIZED AND REDUCED FORMS OF SEVERAL IMPORTANT ELEMENTS

Element	Normal form in well-oxidized soils	Reduced form found in waterlogged soils
Carbon	CO_2, $C_6H_{12}O_6$	CH_4, C_2H_4, CH_3CH_2OH
Nitrogen	NO_3^-	N_2, NH_4^+
Sulfur	SO_4^{2-}	H_2S, S^{2-}
Iron	Fe^{3+} [Fe(III) oxides]	Fe^{2+} [Fe(II) oxides]
Manganese	Mn^{4+} [Mn(IV) oxides]	Mn^{2+} [Mn(II) oxides]

In the neutral-to-alkaline soils of drier areas, oxidized forms of iron and manganese are tied up in highly insoluble compounds, resulting in deficiencies of these elements. Such differences illustrate the interaction of aeration and soil pH in supplying available nutrients to plants (see Chapter 12).

Toxic Elements   Redox potential determines the species of such potentially toxic elements as chromium, arsenic, and selenium, markedly affecting their impact on the environment and food chain (see Section 15.6). Reduced forms of arsenic are most mobile and toxic, giving rise to toxic levels of this element in drinking water. In contrast, it is the oxidized hexavalent form of chromium (Cr^{6+}) that is mobile and very toxic to humans. In neutral to acid soils, easily decomposed organic materials can be used to reduce the chromium to the less toxic Cr^{3+} form, which is not subject to ready reoxidation (Figure 7.5).

Greenhouse Gases   The production of nitrous oxide (N_2O) and methane (CH_4) in wet soils is of universal significance. These two gases, along with carbon dioxide (CO_2), are responsible for about 80% of the anthropogenic global warming (see Section 11.9). The atmospheric concentrations of these gases have been increasing at alarming rates each year for the past half century or more.

Methane gas is produced by the reduction of CO_2 when the E_h is -0.2 V, a condition common in natural wetlands and in rice paddies. Nitrous oxide is also produced in large quantities by wetland soils, as well as sporadically by upland soils. Because of the biological productivity and diversity of wetlands (Section 7.7), soil scientists are

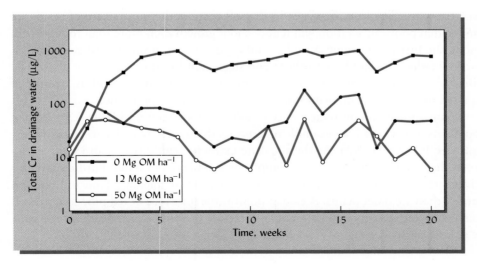

Figure 7.5
Effect of adding decomposable organic matter (OM) on the concentration of chromium in water draining from a chromium-contaminated soil. Here dried cattle manure was added as the decomposable OM. As the manure oxidized, it caused the reduction of the toxic, mobile Cr^{6+} to the relatively immobile, nontoxic Cr^{3+}. Note the log scale for the Cr in the water, indicating that the 50 Mg manure ha^{-1} addition caused the Cr level to be lowered approximately 100-fold. The coarse-textured soil was a Typic Torripsamment in California. [Data from Losi et al. (1994)]

Figure 7.6

The relationship between soil redox potentials (E_h) and emissions of three "greenhouse" gases from soils. Because the three gases differ widely in their potential for global warming per mole of gas, their emissions are expressed here as CO_2 equivalents. Note the low global warming potential from all three gases when E_h is between −0.15 and +0.18 V. Although this study used small containers of flooded soil in the lab, it may be desirable to manage the aeration status of flooded soils in rice paddies and wetlands with these results in mind. Manipulating organic matter additions, water table levels, water flow rates, and the duration of flooding might allow managers to maintain the soil within the "window" where E_h is too high to stimulate microbial methanogenesis (CH_4 production) but also too low to stimulate production of much N_2O or CO_2. Such management, if feasible, could potentially reduce the contribution of wet soils to global warming. [From Yu and Patrick (2004), with permission of the Soil Science Society of America]

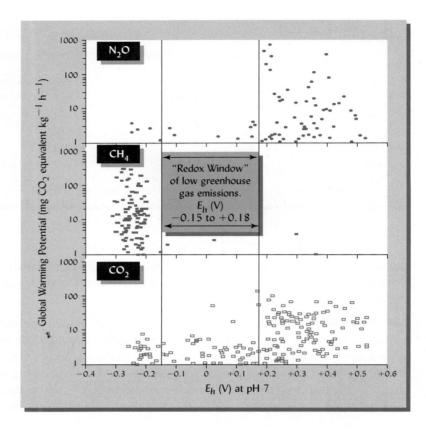

seeking means of managing greenhouse gases from these wetland soils without resorting to draining (i.e., destroying) them. Fortunately, it may be possible to minimize production of all three major greenhouse gases by maintaining soil E_h in a moderately low range that is feasible for many rice paddies and natural wetlands (Figure 7.6).

Effects on Activities of Higher Plants

It is the lack of oxygen in the root environment rather than the excessive wetness itself that impairs plant growth in overly wet soils. This fact explains why flooding a soil with stagnant water is generally much more damaging to plants—even for some hydrophytes—than flooding with flowing, oxygenated water.

Hemoglobin helps plants survive flooding: www.umanitoba.ca/afs/fiw/040729.html

Plant Growth   The lack of oxygen in the soil changes the metabolism of the entire plant. Often poor soil aeration will reduce shoot growth even more than root growth. Among the first plant responses to low soil oxygen is the closure of leaf stomata, followed by reductions in photosynthesis and sugar translocation within the plant. The ability of the root to take up water and nutrients is inhibited, and as a result of impaired root metabolism, plant hormones are thrown out of balance.

Plant species vary in their ability to tolerate poor aeration (Table 7.3). Among crop plants, sugar beet is an example of a species that is very sensitive to poor soil aeration (see Figure 7.1, *left*). At the opposite extreme, paddy rice is an example of a species that can grow with its roots completely submerged in water. Furthermore, for a given species of plant, the young seedlings may be more tolerant of low soil aeration porosity than are older plants. A case in point is the tolerance of red pine to restricted drainage during its early development and its poor growth or even death on the same site at later stages (see Figure 7.1, *right*).

Table 7.3
PLANT TOLERANCE TO A HIGH WATER TABLE AND RESTRICTED AERATION
The plants on the left thrive in wetlands. Those on the right are very sensitive to poor aeration.

Plants adapted to grow well with a water table at the stated depth

<10 cm	15 to 30 cm	40 to 60 cm	75 to 90 cm	>100 cm
Bald cypress	Alsike clover	Birdsfoot trefoil	Beech	Arborvitae
Black spruce	Bermuda grass	Black locust	Birch	Barley
Common cattail	Black willow	Bluegrass	Cabbage	Beans
Cranberries	Cottonwood	Linden	Corn	Cherry
Phragmites grass	Creeping bentgrass	Mulberry	Hairy vetch	Hemlock
Mangrove	Deer tongue	Mustard	Millet	Oats
Pitcher plant	Eastern gamagrass	Red maple	Peas	Peach
Reed canary grass	Ladino clover	Sorghum	Red oak	Sand lovegrass
Rice	Loblolly pine	Sycamore		Sugar beets
Skunk cabbage	Orchard grass	Weeping love grass		Walnut
Spartina grass	Redtop grass	Willow oak		Wheat
Swamp white oak	Tall fescue			White pine

Knowledge of plant tolerance to poor aeration is useful in choosing appropriate species to revegetate wet sites. The occurrence of plants specially adapted to anaerobic conditions is useful in identifying wetland sites (see Section 7.7).

Nutrients and Water Uptake　Low O_2 levels constrain root respiration and impair root function. The root cell membrane may become less permeable to water, so that plants may actually have difficulty taking up water and some species will wilt and desiccate in a waterlogged soil. Likewise, plants may exhibit nutrient deficiency symptoms on poorly drained soils even though the nutrients may be in good supply. Furthermore, toxic substances (e.g., ethylene gas) produced by anaerobic microorganisms may harm plant roots and adversely affect plant growth.

7.6　AERATION IN RELATION TO SOIL AND PLANT MANAGEMENT

In general, in the field, aeration may be enhanced by implementing the principles outlined in Sections 4.5–4.7 regarding the maintenance of soil aggregation and tilth. Equally important are systems to increase both surface and subsurface drainage (see Section 6.7) and encourage the production of vertically oriented biopores (e.g., earthworm and root channels) that are open to the surface (Sections 4.5 and 6.7). Here we will briefly consider special steps to avoid aeration problems for container-grown plants, landscape trees, and lawns.

Container-Grown Plants

Potted plants frequently suffer from waterlogging and poor aeration even though potting mixes are engineered to minimize waterlogging. Mineral soil generally makes up no more than one-third of the volume of most potting mixes, the remainder being composed of inert, lightweight and coarse-grained materials such as perlite (expanded volcanic glass), vermiculite (expanded mica—not soil vermiculite), or pumice (porous volcanic rock). In order to achieve maximum aeration and minimum weight, some potting mixes contain no mineral soil at all. Most mixes also contain some peat, shredded bark, wood chips, compost, or other stable organic material that adds macroporosity and holds water.

Despite the fact that most planting containers have holes to allow drainage of excess water, the bottom of the container still creates a perched water table. As is the case with stratified soils in the field (Section 5.6), water drains out of the holes at the bottom of the pot *only* when the soil at the bottom is saturated with water and the water potential is positive. The finer pores in the medium remain filled with water, leaving no room for air, and anaerobic conditions soon prevail. The situation is aggravated if the potting medium contains much mineral soil. In any case, the use of as tall a pot as possible will allow for better aeration in the upper part of the medium. Watering should be deferred until the soil *near the bottom of the container* has begun to dry.

Tree and Lawn Management

Planting city trees the right way:
http://www.forestry.iastate.edu/publications/b1047.pdf

In transplanting woody species, special caution must be taken to prevent poor aeration or waterlogging immediately around the young root system. Figure 7.7 illustrates the right and the wrong ways to transplant trees into a compacted soil.

The aeration of well-established, mature trees must also be safeguarded. If operators push surplus excavated soil around the base of a tree during landscape grading (Figure 7.8), the tree's feeder roots near the original soil surface become deficient in oxygen even if the overburden is only 5 to 10 cm in depth. One should build a protective wall (a *dry well*) or install a fence around the base of a valuable tree before grading operations begin in order to preserve the original soil surface for a radius of several meters from the trunk. This measure will allow the tree's roots to continue to access the O_2 they need. Failure to observe these precautions can easily kill a large, valuable tree, although it may take a year or two to do so.

Management systems for heavily trafficked lawns commonly include the installation of perforated drainage pipes. Another practice that enhances soil aeration in compacted lawn areas is **core cultivation**. This procedure removes thousands of small cores of soil from the surface horizon, thereby permitting gas exchange to take place more easily (Figure 7.9). Spikes that merely punch holes in the soil are much less effective than corers, since compaction is increased in the soil surrounding a spike.

Figure 7.7

Providing a good supply of air to tree roots can be a problem, especially when trees are planted in fine-textured, compacted soils of urban areas. A machine-dug hole with smooth sides will act as a "tea cup" and fill with water, suffocating tree roots. Breather tubes, a larger rough-surfaced hole, and a layer of surface mulch in which some fine tree roots can grow are all measures that can improve the aeration status of the root zone. (Diagram courtesy of R. Weil)

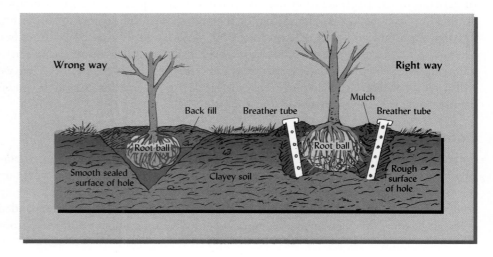

Figure 7.8

Protection of valuable trees during landscape grading operations. Even a thin layer of soil spread over a large tree's root system can suffocate the roots and kill the tree. (Inset) In order to preserve the original ground surface so that tree feeder roots can obtain sufficient oxygen, a dry well may be constructed of brick or any decorative material. The dry well may be incorporated into the final landscape design or filled in at a rate of a few centimeters per year. (Photos courtesy of R. Weil)

7.7 WETLANDS AND THEIR POORLY AERATED SOILS[2]

Poorly aerated areas called **wetlands** cover about 14% of the world's ice-free land, with the greatest areas occurring in the cold regions of Canada, Russia, and Alaska (Table 7.4). Once considered mere wastelands waiting to be drained, about half of

Figure 7.9

One way of increasing the aeration of compacted soil is by core cultivation. The machine removes small cores of soil, leaving holes about 2 cm in diameter and 5 to 8 cm deep. This method is commonly used on high-traffic turf areas. Note that the machine removes the cores and does not simply punch holes in the soil, a process that would increase compaction around the hole and impede air diffusion into the soil. (Photos courtesy of R. Weil)

---

[2]Two well-illustrated, nontechnical, yet informative publications on wetlands are Welsh et al. (1995) and CAST (1994). For a compilation of technical papers on hydric soils and wetlands, see Richardson and Vepraskas (2001).

Table 7.4
MAJOR TYPES OF WETLANDS AND THEIR GLOBAL AREAS

Wetland type	Global area, 1000s km²	Percent of ice-free land area	Percent of all wetland areas
Inland (swamps, bogs, etc.)	5415	3.9	28.8
Riparian or ephemeral	3102	2.3	16.5
Organic (Histosols)	1366	1.0	7.3
Salt-affected, including coastal	2230	1.6	11.9
Permafrost-affected (Histels)	6697	4.9	35.6

Data from Eswaren et al. (1996).

Wetland definition and function:
www.stemnet.nf.ca/CITE/ecowetlands.htm

the world's natural wetlands areas have been destroyed by human activities. The same is true for the continental United States, whose current 400,000 km² of wetlands is less than half of the area that existed when European settlement began. Wetlands are now highly prized for providing such ecosystem services as wildlife habitat, water purification, flood reduction, shoreline protection, recreational opportunities, natural products, and, perhaps most significantly, the potential to mitigate global warming (see Chapter 11). Most wetland losses occurred as farmers used artificial drainage (see Section 6.7) to convert them into cropland. In recent decades, filling and drainage for urban development has also taken its toll on wetland areas (see Box 7.1).

Defining Wetlands

While there are many different types of wetlands, they all share a key feature, namely *soils that are water-saturated near the surface for prolonged periods when soil temperatures and other conditions are such that plants and microbes can grow and remove the soil oxygen, thereby assuring anaerobic conditions.* It is largely the prevalence of anaerobic conditions that determines the kinds of plants, animals, and soils found in these areas.

BOX 7.1
IT'S THE LAW

Not only is it a bad idea ecologically to drain or fill wetlands, it's also against the law! In the United States and many other countries, knowingly destroying a wetland can bring severe penalties. The case reported here (Figure 7.10) reflects the change from a generation ago when most wetland destruction was caused by farmers installing agricultural drainage, to today when wetlands in industrial countries are most threatened by urban/suburban development. The newspaper reports that the developer had been informed (even warned) about the wetland area on his 1000-hectare development tract. He nonetheless filled the wetland areas so he could build hundreds of homes on these sites. To add further environmental insult, he installed septic drain fields (see Section 6.8) on these seasonally saturated soils. The judge sentenced the man to nine years in prison followed by three years of supervised release. His business partners were also imprisoned and fined. The severity of this punishment should "send a message" to other would-be violators.

Developers Sentenced in Wetlands Case

A federal judge sentenced three Mississippi real estate developers to prison yesterday for filling in wetlands and selling the property to low- and fixed-income families, marking the end of the nation's largest wetlands criminal prosecution.

Figure 7.10
Clipping of Washington Post *story by Eilperin (2005).* (Photo courtesy of R. Weil)

There is widespread agreement that the wetter end of a wetland occurs where the water is too deep for rooted, emergent vegetation to take hold. The difficulty is in precisely defining the so-called *drier end* of the wetland.

Since uses and management of wetlands are regulated by governments in the United States and in many other countries, billions of dollars are at stake in determining what is and what is not protected as a wetland.

Thousands of environmental professionals are employed in the process of **wetland delineation**—finding the exact drier-end boundaries of wetlands on the ground. Wetland delineation is *not* done in front of a computer screen but is a sweaty, muddy, tick- and mosquito-ridden business that those trained in soil science are uniquely qualified to carry out.

What do these scientists look for to indicate the existence of a wetland system? Most authorities agree that three characteristics can be found in any wetland: (1) a wetland hydrology or water regime, (2) hydric soils, and (3) hydrophytic plants.

Wetland Hydrology

The balance between inflows and outflows, as well as the water storage capacity of the wetland itself, determines how wet it will be and for how long. The temporal pattern of water table changes is termed the **hydroperiod**. For a coastal marsh, the hydroperiod may be daily, as the tides rise and fall. For inland swamps, bogs, or marshes, the hydroperiod is more likely to be seasonal. Some wetlands may never be flooded, although they are saturated within the upper soil horizons.

Wetlands function, regulation, and management: http://www.epa.gov/owow/wetlands/

If the period of saturation occurs when the soil is too cold for microbial or plant-root activity to take place, oxygen may be dissolved in the water or entrapped in aggregates within the soil. Consequently, true anaerobic conditions may not develop, even in flooded soils. Remember, it is the anaerobic condition, not just saturation, that makes a wetland a wetland.

The more slowly water moves through a wetland, the longer the *residence time* and the more likely that wetland functions and reactions will be carried out. For this reason, actions that speed water flow, such as creating ditches or straightening stream meanders, are generally considered degrading to wetlands and are to be avoided.

All wetlands are water saturated some of the time, but many are not saturated all the time. Systematic field observations, assisted by instruments to monitor the changing level of the water table, may be required to ascertain the frequency and duration of flooding or saturated conditions.

In the field, even during dry periods, there are many signs one can look for to indicate where saturated conditions frequently occur. Past periods of flooding will leave water stains on trees and rocks and a coating of sediment on the plant leaves and litter. Drift lines of once-floating branches, twigs, and other debris also suggest previous flooding. Trees with extensive root masses above ground indicate an adaptation to saturated conditions. But perhaps the best indicator of saturated conditions is the presence of **hydric soils**.

Hydric Soils[3]

In order to assist in delineating wetlands, soil scientists developed the concept of hydric soils. In *Soil Taxonomy* (see Chapter 3), these soils are mostly (but not exclusively) classified in the order Histosols, in Aquic suborders such as Aquents and

---

[3]The U.S. Department of Agriculture Natural Resources Conservation Service defines a hydric soil as one "that formed under conditions of saturation, flooding or ponding long enough during the growing season to develop anaerobic conditions in the upper part." For an illustrated field guide to features that indicate hydric soils, see Hurt et al. (1996). For a current list of soil series considered to be hydric soils, see http://soils.usda.gov/use/hydric/.

Aqualfs, or in Aquic subgroups. These soils generally have an Aquic or Peraquic moisture regime (see Section 3.2).

Three properties help define hydric soils. First, they are subject to *periods of saturation* that inhibit the diffusion of O_2 into the soil. Second, for substantial periods of time they undergo *reduced conditions* (see Section 7.3); that is, electron acceptors other than O_2 are reduced. Third, they exhibit certain features termed *hydric soil indicators*. Such indicators are discussed in Box 7.2.

Hydrophytic Vegetation

Photos of hydrophytic vegetation: www.bixby.org/parkside/multimedia/vegetation

Plants that have evolved special mechanisms to adapt to life in saturated, anaerobic soils comprise the **hydrophytic vegetation** that distinguishes wetlands from other systems. Typical adaptive features include hollow aerenchyma tissues that transport O_2 down to

---

BOX 7.2
HYDRIC SOIL INDICATORS

Hydric soil indicators are features associated (sometimes only in specific geographic regions) with the occurrence of saturation and reduction. Most of the indicators can be observed in the field by digging a small pit to a depth of about 50 cm. They principally involve the loss or accumulation of various forms of Fe, Mn, S, or C. Thick, dark surface layers can also be indicators of hydric conditions in which organic matter decomposition has been inhibited (see, for example, Plates 6 and 39).

Iron, when reduced to Fe(II), becomes sufficiently soluble that it migrates away from reduced zones and may precipitate as Fe(III) compounds in more aerobic zones. Zones where reduction has removed or depleted the iron coatings from mineral grains are termed **redox depletions**. They commonly exhibit the gray, low-chroma colors of the bare, underlying minerals (see Section 4.1 for an explanation of chroma). Also, iron itself turns gray to blue-green when reduced. The contrasting colors of redox depletions or reduced iron and zones of reddish oxidized iron result in unique mottled **redoximorphic features** (see, for example, Plates 15 and 17). Other redoximorphic features involve reduced Mn. These include the presence of hard black *nodules* that sometimes resemble shotgun pellets. Under severely reduced conditions the entire soil matrix may exhibit low-chroma colors, termed gley. Colors with a chroma of 1 or less quite reliably indicate reduced conditions (Figure 7.11).

Always keep in mind that redoximorphic features are indicative of hydric soils only when they occur in the upper horizons. Many soils of upland areas exhibit redoximorphic features only in their deeper horizons, due to the presence of a fluctuating water table at depth. Upland soils that are saturated or even flooded for short periods, especially if during cold weather, are not wetland (hydric) soils.

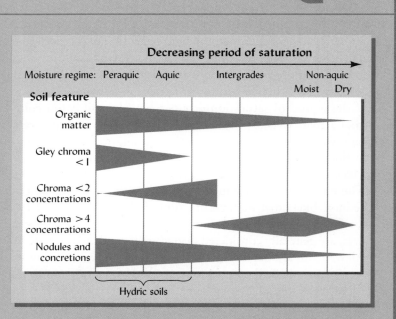

Figure 7.11
The relationship between the occurrence of some soil features and the annual duration of water-saturated conditions. The absence of iron concentrations (mottles) with colors of chroma >4 and the presence of strong expressions of the other features are indications that a soil may be hydric. Peraquic refers to a moisture regime in which soils are saturated with water throughout the year. For other moisture regimes, see Section 3.2. [Adapted from Veneman et al. (1999)]

A unique redoximorphic feature associated with certain wetland plants is the presence, in an otherwise gray matrix, of reddish oxidized iron around root channels where O_2 diffused out from the aerenchyma-fed roots of a hydrophyte (see Plate 38). These *oxidized root zones* exemplify the close relationship between hydric soils and hydrophytic vegetation.

the roots. Certain trees (such as bald cypress) produce adventitious roots, buttress roots, or knees. Other species spread their roots in a shallow mass on or just under the soil surface, where some O_2 can diffuse even under a layer of ponded water. The leftmost column in Table 7.3 lists a few common hydrophytes. Not all the plants in a wetland are likely to be hydrophytes, but the majority usually are.

Wetland Chemistry

Wetland chemistry is characterized by low redox potentials (see Section 7.3). Many wetland functions depend on *variations* in the redox potential; that is, in certain zones or for certain periods of time oxidizing conditions alternate with reducing ones. For example, even in a flooded wetland, O_2 will be able to diffuse from the atmosphere or from oxygenated water into the soil, creating a thin *oxidizing zone* (see Figure 7.12 and Plate 63). The diffusion of O_2 within the saturated soil is extremely limited, so that a few centimeters deeper into the profile, O_2 is eliminated and the redox potential becomes low enough for reactions such as nitrate reduction to take place. The close proximity of the oxidized and anaerobic zones allows water passing through wetlands to be stripped of N by the sequential oxidation of ammonium to nitrate and then the reduction of the nitrate to various nitrogen gases that escape into the atmosphere (see Section 12.1).

Redox   To be considered a wetland, redox potentials should become low enough for iron reduction to produce redoximorphic features. Still lower E_h will allow carbon or sulfate reduction to produce methane gas or rotten-egg-smelling hydrogen sulfide (H_2S) gas. Toxic elements such as chromium and selenium undergo redox reactions that may help remove them from the water before it leaves the wetland. Acids from industry or mine drainage may also be neutralized by reactions in hydric soils. This array of unique chemical reactions contributes greatly to the benefits that society and the environment gain from wetlands.

Constructed Wetlands

Scientists and engineers not only preserve natural wetlands but also construct artificial ones for specific purposes, such as wastewater treatment (see, for example, Box 12.2).

Certain regulations allow for the destruction of some natural wetland areas, provided that new wetlands are constructed or that previously degraded wetlands are restored. This process, termed **wetland mitigation**, has been only partially successful, as scientists still have much to learn about how wetlands work and the constructed wetlands are rarely monitored for proper functioning.

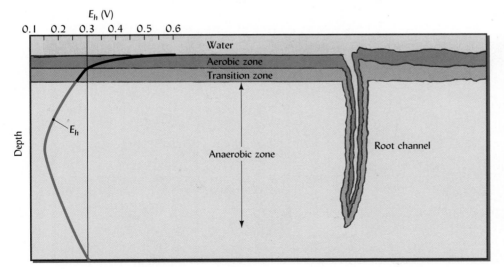

Figure 7.12

Representative redox potentials within the profile of an inundated hydric soil. Many of the biological and chemical functions of wetlands depend on the close proximity of reduced and oxidized zones in the soil. The changes in redox potential at the lower depths depend largely on the vertical distribution of organic matter. In some cases, low subsoil organic matter results in a second oxidized zone beneath the reduced zone. (Diagram courtesy of R. Weil)

We have seen how greatly soil aeration is influenced by soil water. We will now turn our attention to soil temperature, another physical soil property that is closely related to both soil water and aeration.

7.8 PROCESSES AFFECTED BY SOIL TEMPERATURE

The temperature of a soil greatly affects the physical, biological, and chemical processes occurring in that soil and in the plants growing on it (Figure 7.13).

Figure 7.13
Soil temperature ranges associated with a variety of soil processes.

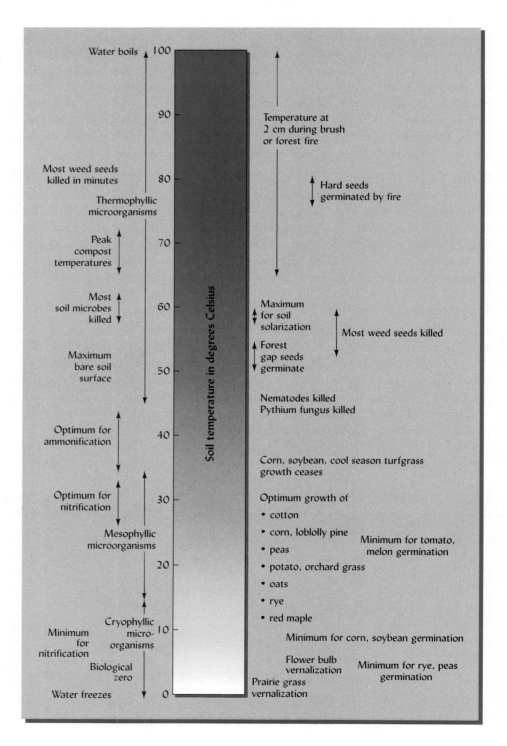

Plant Processes

Most plants are much more sensitive to *soil* temperature than to aboveground *air* temperature, but this is not often appreciated since air temperature is more commonly measured. Contrary to what one might expect, adverse soil temperature generally influences shoot growth and photosynthesis more than root growth (as was seen in the study described in Figure 7.14). Most plants have a rather narrow range of soil temperatures for optimal growth. For example, two species that evolved in warm regions, corn and loblolly pine, grow best when the soil temperature is about 25 to 30 °C. In contrast, the optimal soil temperature for cereal rye and red maple, two species that evolved in cool regions, is in the range of 12 to 18 °C.

In temperate regions, cold soil temperature often limits the productivity of crops and natural vegetation. The life cycles of plants are also greatly influenced by soil temperature. For example, tulip bulbs require chilling in early winter to develop flower buds, although flower development is suppressed until the soil warms up the following spring.

In warm regions, and in the summer in temperate regions, soil temperatures may be too high for optimal plant growth, especially in the upper few centimeters of soil. For example, bent grass, a cool-season grass prized for providing an excellent "playing surface" on golf greens, often suffers from heat stress when grown in warmer regions (Figure 7.14). Even plants of tropical origin, such as corn and tomato, are adversely affected by soil temperatures higher than 35 °C. Seed germination may also be reduced by high soil temperatures.

Seed Germination Many plants require specific soil temperatures to trigger seed germination, accounting for much of the difference in species between early- and late-season weeds in cultivated land. Likewise, the seeds of certain plants adapted to open gaps in a forest stand are stimulated to germinate by the greater fluctuations and maximum soil temperatures that occur where the forest canopy is disturbed by timber harvest or wind-thrown trees. The seeds of certain prairie grasses and grain crops require a period of cold soil temperatures (2 to 4 °C) to enable them to germinate the following spring, a process termed *vernalization*.

Root Functions Root functions such as nutrient uptake and water uptake are sluggish in cool soils with temperatures below the optimum for the particular species. One result is that nutrient deficiencies, especially of phosphorus, often occur in young plants in early spring, only to disappear when the soil warms later in the season. On bright, sunny days in winter and early spring when the soil is still cold, evergreen plants may become desiccated and even die because the slow water uptake by roots in the cold

Measuring soil temperature video: http://videogoogle .com/videoplay?docid =8773072375890666921& pr=goog-sl

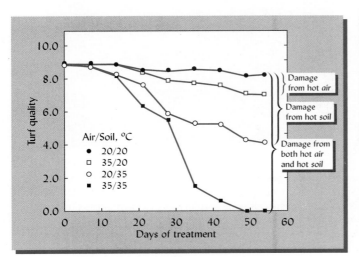

Figure 7.14

Effects of air and soil temperatures on turfgrass quality. Heat stress is a major problem for bent grass (Agrostis spp.) grown on golf greens in warmer climates. Researchers in this study grew bent grass for 60 days while controlling both soil and air temperatures. Turfgrass quality (color, vigor, etc.) was rated from 0 (dead) to 10 (best quality). Compare the small decline in turf quality caused by increasing air temperature from 20 to 35 °C to the much greater decline caused by the same increase in soil temperature. High temperature in both air and soil caused the worst effects. Among the plant parameters measured, photosynthesis was affected more than root growth by soil temperature. Other research has shown that a fine spray of water combined with a large fan can cool both air and soil on a golf green. [From Xu and Huang (2000)]

soils cannot keep up with the high evaporative demand of bright sun on the foliage. Such *winter burn* can be prevented by covering the shrubs with a shade cloth.

Microbial Processes

Microbial processes are influenced markedly by soil temperature changes (Figure 7.15). Although it is commonly assumed that microbial activity virtually ceases below 5 °C (a benchmark referred to as *biological zero*), low rates of soil microbial activity and organic matter decomposition have been measured in the permafrost layers of Gelisols at temperatures as low as −20 °C. In fact, given that some 80% of the Earth's biosphere is colder than 5 °C, it should not be surprising that microbes have widely adapted to life at cold temperatures.

Nonetheless, microbial activity is far greater at warm temperatures; the rates of microbial processes such as respiration typically more than double for every 10 °C rise in temperature (see Figure 7.16). The optimum temperature for microbial decomposition processes may be 35 to 40 °C, considerably higher than the optimum for plant growth. The dependence of microbial respiration on warm soil temperatures has important implications for soil aeration (see Section 7.7) and for the decomposition of plant residues and, hence, the cycling of the nutrients they contain.

In environments with hot, sunny summers (maximum daily air temperatures >35 °C), a controlled heating process called **soil solarization** can be used to suppress pests and diseases in some high-value crops. In this process, the ground is covered with a clear plastic film that traps enough heat to raise the temperature of the upper few centimeters of soil to as high as 50 to 60 °C.

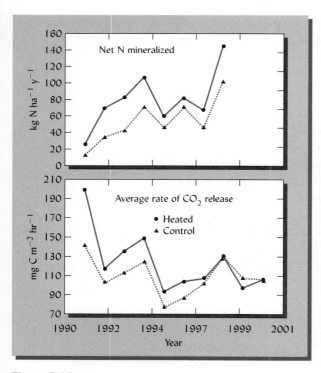

Figure 7.15

Elevated soil temperature accelerates biological processes, such as (upper graph) nitrogen release from organic matter (mineralization) and (lower graph) carbon dioxide release by soil respiration (about 80% due to microorganisms and 20% due to plant roots). In this mid-latitude hardwood forest, electric heating cables were buried in certain plots to maintain soil temperature at 5 °C above normal throughout the year. Cables were also installed in the control plots (photo) to assure an equal degree of physical disturbance, but no electricity was applied. The biological processes represented are discussed in detail in Chapters 11 and 12. [Graphs from data in Melillo et al. (2002); photo courtesy of R. Weil]

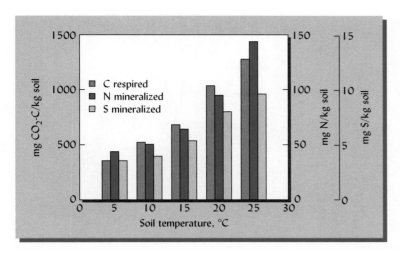

Figure 7.16
Effect of soil temperature on cumulative microbial respiration (CO_2 release) and net nitrogen and sulfur mineralization in surface soils from hardwood forest in Michigan. The soil water content was adequate for microbial growth throughout the 32-week study period. Note the near doubling of the microbial activity with a 10-degree change in soil temperature (compare findings for 15 vs. 25 °C). Means from four sites are shown.
[From MacDonald et al. (1995)]

As we shall see in Chapter 15, warm soil temperatures are also critical for pollution remediation technologies that utilize specialized microorganisms to degrade petroleum products, pesticides, and other organic contaminants in soils.

Freezing and Thawing

When soil temperatures fluctuate above and below 0 °C, the water in the soil undergoes cycles of freezing and thawing. During alternate freezing and thawing, zones of pure ice, called *ice lenses,* form within the soil, and ice crystals form and expand. Great pressure develops due mainly to ice lens growth rather than to the 9% increase in volume that water undergoes when it freezes. In a saturated soil with a puddled structure, the frost action breaks up the large masses and greatly improves granulation. In contrast, for soils with good aggregation to begin with, freeze–thaw action when the soil is very wet can lead to structural deterioration.

Alternate freezing and thawing can force objects upward in the soil, a process termed **frost heaving**. Objects subject to heaving include stones, fence posts, and perennial tap-rooted plants (Figure 7.17). This action, which is most severe where the soil is silty in texture, wet, and lacking a covering of snow or dense vegetation, can drastically reduce stands of alfalfa, some clovers, and trefoil.

Freezing can also heave shallow foundations, roads, and runways that have fine material as a base. Gravels and pure sands are normally resistant to frost damage, but silts and sandy soils with modest amounts of finer particles are particularly susceptible. Very clay-rich soils do not usually exhibit much frost heave, but ice-lens segregation can still occur and can lead to severe loss of strength when thawing occurs. To avoid damage by freezing soil temperatures, foundation footings (as well as water pipelines) should be set into the soil below the maximum depth to which the soil freezes—a depth that ranges from less than 10 cm in subtropical zones, such as south Texas and Florida, to more than 200 cm in very cold climates.

Permafrost

Perhaps the most significant global phenomenon involving soil temperatures is the thawing in recent years of some of the permafrost (permanently frozen ground) in arctic regions. Nearly 25% of the land areas of the Earth are underlain by permafrost. Rising temperatures since the late 1980s have caused some of the upper layers of permafrost to thaw. In parts of Alaska, for example, temperatures in top layers of permafrost have risen about 4 °C since the late 1980s, resulting in melting rates of about a meter in a decade. Such melting drastically affects the physical foundation of buildings and roads,

Ice lens formation:
www.aip.org/png/html/frost.htm

Images of permafrost and more:
www.earthscienceworld.org/images/search/results.html?Keyword=permafrost

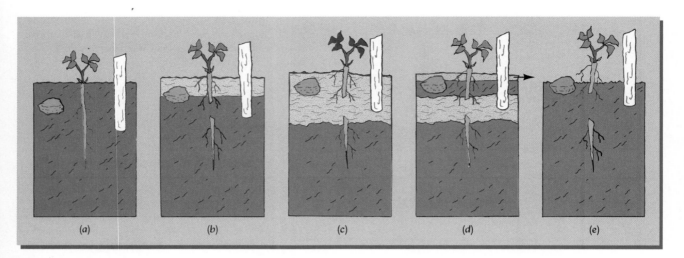

Figure 7.17

How frost heaving moves objects upward. (a) Position of the object (stone, plant, or fence post) before the soil freezes. (b) As lenses of pure ice form in the freezing soil by attraction of water from the unfrozen soil below, the frozen soil tightens around the upper part of the object, lifting it somewhat—enough to break the root in the case of the plant. (c) The objects are lifted upwards as ice-lens formation continues with deeper penetration of the freezing front. (d) As for freezing, thawing commences from the surface downward. Water from thawing ice lenses escapes to the surface because it cannot drain downward through the frozen soil. The soil surface subsides while the heaved objects are held in the "jacked-up" position by the still-frozen soil around their lower parts. (e) After complete thaw, the stone is closer to the surface than previously (although rarely at the surface unless erosion of the thawed soil has occurred), and the upper part of the broken root is exposed, so that the plant it is likely to die. (Diagram courtesy of R. Weil)

as well as the stability of root zones of forests and other such vegetation in the region. Trees fall and buildings collapse as the frozen layers melt. Worse, the thawing of arctic permafrost is expected to further accelerate global warming, as decomposition of organic materials long trapped in the frozen layers of Histels releases vast quantities of carbon dioxide into the atmosphere (see Figure 3.11).

Soil Heating by Fire

Fire is one of the most far-reaching ecosystem disturbances in nature. In addition to the obvious aboveground effects of forest, range, or crop-residue fires, the brief but sometimes dramatic changes in soil temperature also may have lasting impacts below ground. Unless the fire is artificially stoked with added fuel, the temperature rise itself is usually very brief and is limited to the upper few centimeters of soil. But the high temperatures (>125 °C being common) may essentially distill various fractions of the organic matter (Figure 7.18). As these volatilized hydrocarbon compounds reach cooler soil particles deeper in the soil, they condense (solidify) on the surface of the soil particles and fill some of the surrounding pore spaces with water-repellent (hydrophobic) hydrocarbons. Consequently, when rain comes, water infiltration in even a sandy soil is greatly reduced in comparison to unburned areas. This effect of soil temperature is quite common on chaparral lands in semiarid regions and may be responsible for the disastrous mudslides that occur when the layer of soil above the hydrophobic zone becomes saturated with rainwater (see Plate 71).

Fires also affect the germination of certain seeds that have hard coatings that prevent them from germinating until they are heated above 70 to 80 °C. On the other hand, burning of straw in wheat fields generates similar soil temperatures, but with the effect of killing most of the weed seeds near the surface and thus greatly reducing subsequent weed infestation. The heat and ash may also hasten the cycling

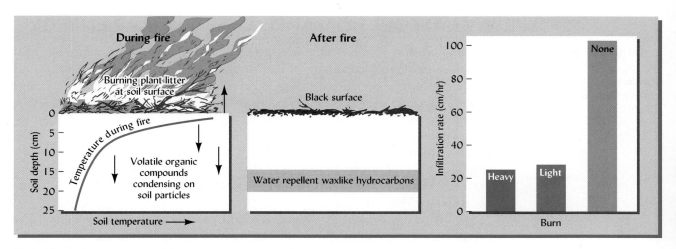

Figure 7.18

(Left) Wildfires of a lodgepole pine stand heat up the surface layers of this sandy soil (an Inceptisol) in Oregon. (Center) Note that the soil temperature is increased sufficiently near the surface to volatilize organic compounds, some of which then move down into the soil and condense (solidify) on the surface of cooler soil particles. These condensed compounds are waxlike hydrocarbons that are water repellent. As a consequence (right), the infiltration of water into the soil is drastically reduced and remains so for a period of at least six years. See also Plate 71. [From Dryness (1976)]

of plant nutrients. Fires set to clear land of timber slash may burn long and hot enough to seriously deplete soil organic matter and kill so many soil organisms that forest regrowth is inhibited.

Contaminant Removal

The removal of certain organic pollutants from contaminated soils can be accomplished by raising the soil temperature in place using electromagnetic radiation. The resulting temperatures are sufficiently high to vaporize some contaminants which can then be flushed from the soil by air (Figure 7.19).

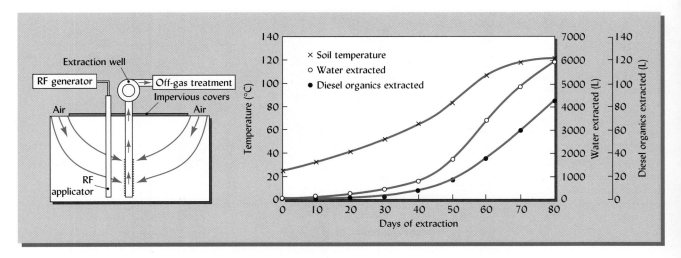

Figure 7.19

Increasing soil temperatures can extract organic pollutants from soils. (Left) Radio frequency (RF) electromagnetic radiation was used to gradually increase temperatures in a block of soil containing diesel fuel. At the higher temperatures, organic hydrocarbons were vaporized (along with water) and then removed from the soil using an extraction well. (Right) Soil temperatures increase with time near the RF applicator, along with the quantity of organic compounds and water extracted from the soil. While this procedure is rather expensive, it permits remediation of the soil without having to remove it from its natural setting or subject it to extremely high temperatures. [Modified from Lowe et al. (2000)]

7.9   ABSORPTION AND LOSS OF SOLAR ENERGY[4]

Green roofs vs. urban heat islands: www.artic.edu/webspaces/greeninitiatives/greenroofs/

The temperature of soils in the field is directly or indirectly dependent on at least three factors: (1) the net amount of heat energy the soil absorbs, (2) the heat energy required to bring about a given change in the temperature of a soil, and (3) the energy used for processes such as evaporation, which are constantly occurring at or near the surface of soils.

Solar radiation is the primary source of energy to heat soils. But clouds and dust particles intercept the sun's rays and absorb, scatter, or reflect much of the energy (Figure 7.20). Only about 35 to 40% of the solar radiation actually reaches the Earth in cloudy humid regions compared to 75% in cloud-free arid areas. The global average is about 50%.

Little of the solar energy reaching the Earth actually results in soil warming. The energy is used primarily to evaporate water from the soil or leaf surfaces or is radiated or reflected back to the sky. Only about 10% is absorbed by the soil and can be used to warm it. Even so, this energy is of critical importance to soil processes and to plants growing on the soils.

Albedo   The fraction of incident radiation that is reflected by the land surface is termed the **albedo** and ranges from as low as 0.1 to 0.2 for dark-colored, rough soil surfaces to as high as 0.5 or more for smooth, light-colored surfaces. Vegetation may affect the surface albedo either way, depending on whether it is dark green and growing or yellow and dormant.

Figure 7.20

Schematic representation of the radiation balance in daytime and nighttime in the spring or early summer in a temperate region. About half the solar radiation reaches the Earth, either directly or indirectly, from sky radiation. Most radiation that strikes the Earth in the daytime is used as energy for evapotranspiration or is radiated back to the atmosphere. Only a small portion, perhaps 10%, actually heats the soil. At night the soil loses some heat, and some evaporation and thermal radiation occur.

(Diagram courtesy of N.C. Brady and R. Weil)

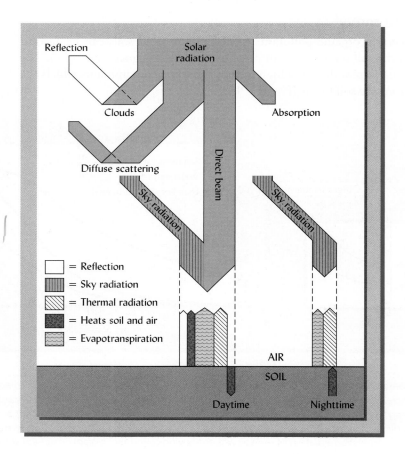

[4]For application of these principles to the role of soil moisture in models of global warming, see Lin et al. (2003).

The fact that dark-colored soils absorb more energy than lighter-colored ones does not necessarily imply, however, that dark soils are always warmer. In fact, the darkest soils are often the wettest and so slow to warm.

Aspect   The angle at which the sun's rays strike the soil influences soil temperature. If the land is sloped toward the sun, the incoming rays are perpendicular to the soil surface, and energy absorption (and soil temperature increase) is greatest (Figure 7.21). That is why south-facing slopes (in the northern hemisphere) are usually warmer and drier than north-facing slopes. Planting crops on the south side of soil ridges is one method of controlling the soil aspect on a microscale.

Rain   Mention should be made of the effect of rain or irrigation water on soil temperature. For example, in temperate zones, spring rains definitely warm the surface soil as the water moves into it. Conversely, in the summer, rainfall cools the soil, since it is often cooler than the soil it penetrates. However, spring rain, by increasing the amount of solar energy used in evaporating water from the soil, can accentuate low temperatures.

Soil Cover   Bare soils warm up more quickly and cool off more rapidly than those covered with vegetation, snow, or mulches. Frost penetration during the winter is considerably greater in bare, noninsulated land. Timber-harvest practices that leave less than about 50% shade will likely allow soil warming that could hasten the loss of soil organic matter or the onset of anaerobic conditions in wet soils.

Figure 7.21

(Inset) Effect of slope aspect on solar radiation received per unit land area. Slope (a) is north facing and receives solar radiation at an angle of 45° to the ground surface so only 5 units of solar radiation (arrows) hit the unit of land area. The same land area on the south-facing slope (b) receives 7 units of radiation at a 90° angle to the ground. In other words, if a given amount of radiation from the sun strikes the soil at right angles, the radiation is concentrated in a relatively small area, and the soil warms quite rapidly. This is one of the reasons why north slopes tend to have cooler soils than south slopes. It also accounts for the colder soils in winter than in summer. (Photo) A view looking eastward toward a forested mountain in Virginia illustrates the temperature effect. The main ridge (left to right) is running north and south and the smaller side ridges east and west (up and down). The dark patches are pine trees in this predominantly hardwood deciduous forest. The pines dominate the southern slopes on each east-west ridge. The soils on the southern slopes are warmer and therefore drier, less deeply weathered, and lower in organic matter. (Photo and diagram courtesy of R. Weil)

Table 7.5
MAXIMUM SURFACE TEMPERATURES FOR FOUR TYPES OF SURFACES ON A SUNNY AUGUST DAY IN COLLEGE STATION, TEXAS

Type of surface	Maximum temperature, °C	
	Day	Night
Green, growing turfgrass	31	24
Dry, bare soil	39	26
Brown, summer-dormant grass	52	27
Dry synthetic sports turf	70	29

Data from Beard and Green (1994).

Even low-growing vegetation such as turfgrass has a very noticeable influence on soil temperature and on the temperature of the surroundings (Table 7.5). Much of the cooling effect is due to heat dissipated by transpiration of water. To experience this effect, on a blistering hot day, try having a picnic on an asphalt parking lot instead of on a growing green lawn!

7.10   THERMAL PROPERTIES OF SOILS

Specific Heat of Soils

A dry soil is more easily heated than a wet one because the amount of energy required to raise the temperature of water by 1 °C is much higher than that required to warm soil solids by 1 °C. When this relationship is expressed per unit mass—for example, in calories per gram (cal/g)—it is called **specific heat**, or heat capacity, c. The specific heat of pure water is about 1.00 cal/g (or 4.18 joules per gram, J/g); that of dry soil is about 0.2 cal/g (0.8 J/g).

The specific heat largely controls the degree to which soils warm up in the spring, wetter soils warming more slowly than drier ones. Furthermore, if the water does not drain freely from the wet soil, it must be evaporated, a process that is very energy consuming, as the next section will show.

Energy-efficient geothermic temperature control systems can both warm and cool buildings via heat exchange between the soil and a network of pipes laid underground. Subsoils are generally warmer than the atmosphere in the winter and cooler than the atmosphere in the summer. Therefore, water circulating through the network of pipes absorbs heat from the soil during the winter and releases it to the soil in the summer. The high specific heat of soils permits a large exchange of energy to take place without greatly modifying the soil temperature.

Heat of Vaporization

The evaporation of water from soil surfaces requires a large amount of energy, 540 kilocalories (kcal) or 2.257 megajoules (mJ) for every kilogram of water vaporized. This energy comes from solar radiation or from the surrounding soil. In either case, evaporation can cool the soil, much the way it chills a person who comes out from the water after swimming on a windy day.

The low temperature of a wet soil is due partially to evaporation and partially to high specific heat. The temperature of the upper few centimeters of wet soil is commonly 3 to 6 °C lower than that of a moist or dry soil. This is a significant factor in the spring in a temperate zone, when a few degrees will make the difference between

Heat energy concepts: http://hyperphysics .phy-astr.gsu.edu/hbase/ thermo/heat.html

Soil heat capacity saves energy in buildings: www.geoexchange.org/ geothermal/videos.html? task=videodirectlink&id=3

the germination or lack of germination of seeds, or the microbial release or lack of release of nutrients from organic matter.

Thermal Conductivity of Soils

As shown in Section 7.9, some of the solar radiation that reaches the Earth slowly penetrates the profile largely by conduction; this is the same process by which heat moves to the handle of a cast-iron frying pan. The movement of heat in soil is analogous to the movement of water (see Section 5.5), the rate of flow being determined by a driving force and by the ease with which heat flows through the soil. This can be expressed as Fourier's Law:

$$Q_h = K * \frac{\Delta T}{x}$$ (7.3)

where Q_h is the *thermal flux,* the quantity of heat transferred across a unit cross-sectional area in a unit time; K is the **thermal conductivity** of the soil; and $\Delta T/x$ is the temperature gradient over distance x that serves as the driving force for the conduction of heat.

The thermal conductivity, K, of soil is influenced by a number of factors, the most important being the moisture content of the soil and the degree of compaction (see Figure 7.22). Heat passes through water many times faster than through air. Wet, compacted soil would be the poorest insulator or the best conductor of heat. Relatively dry, loose soil makes a good insulating material. Buildings built mostly underground need little heating or cooling because they take advantage of both the low thermal conductivity and relatively high heat capacity of large volumes of soil.

Heat conduction provides a means of temperature adjustment, but, because it is slow, changes in subsoil temperature lag behind those of the surface layers. Moreover, seasonal and daily temperature changes are always less in the subsoil. In temperate regions, surface soils in general are expected to be warmer in summer and cooler in winter than the subsoil. Soil thermal conductivity can also affect air temperature above the soil, as shown in Figure 7.23.

Variation with Time and Depth

Figure 7.24 illustrates the considerable seasonal variations of soil temperature that occur in temperate region soils. The surface layer temperatures vary more or less according to the temperature of the air, although these layers are generally warmer than the air throughout the year. In the subsoil, the seasonal temperature increases and decreases lag behind changes registered in the surface soil and in the air.

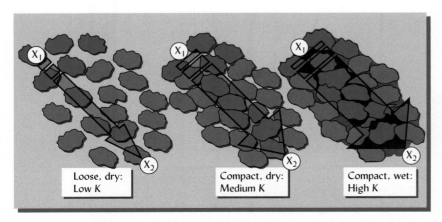

Loose, dry: Low K   Compact, dry: Medium K   Compact, wet: High K

Figure 7.22

Bulk density and water content affect heat transfer through soils from a warm zone (X_1) to a cooler zone (X_2). The rate of heat transfer is in proportion to the arrow thickness. Soil compaction increases particle-to-particle contact, which in turn hastens heat transfer because the thermal conductivity of mineral particles is much higher than that of air. If the remaining gaps between particles become filled with water instead of air, thermal conductivity increases still more because water also conducts heat better than air. Therefore wet, compacted soils transfer heat most rapidly. (Diagram courtesy of R. Weil)

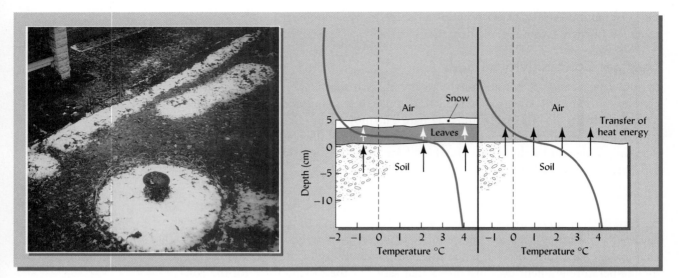

Figure 7.23

Transfer of heat energy from soil to air. The scene, looking down on a garden after an early fall snowstorm, shows snow on the leaf-mulched flower beds, but not on areas where the soil is bare or covered with thin turf. The reason for this uneven accumulation of snow can be seen in the temperature profiles. Having stored heat from the sun, the soil layers are often warmer than the air as temperatures drop in fall (this is also true at night during other seasons). On bare soil, heat energy is transferred rapidly from the deeper layers to the surface, the rate of transfer being enhanced by high moisture content or compaction, which increases the thermal conductivity of the soil. As a result, the soil surface and the air above it are warmed to above freezing, so snow melts and does not accumulate. The leaf mulch, which has a low thermal conductivity, acts as an insulating blanket that slows the transfer of stored heat energy from the soil to the air. The upper surface of the mulch is therefore hardly warmed by the soil, and the snow remains frozen and accumulates. A heavy covering of snow can itself act as an insulating blanket. (Photo and diagram courtesy of R. Weil)

Figure 7.24

Average monthly soil temperatures for 6 of the 12 months of the year at different soil depths at College Station, TX (1951–1955). Note the lag in soil temperature change at the lower depths. For example, surface soil temperatures in March respond to the warming of spring, while temperatures of the deep subsoil still reflect the cold of winter. Subsoil temperatures are less variable than air and surface soil temperatures, although there is some temperature fluctuation even at 300 cm deep.

[From Fluker (1958)]

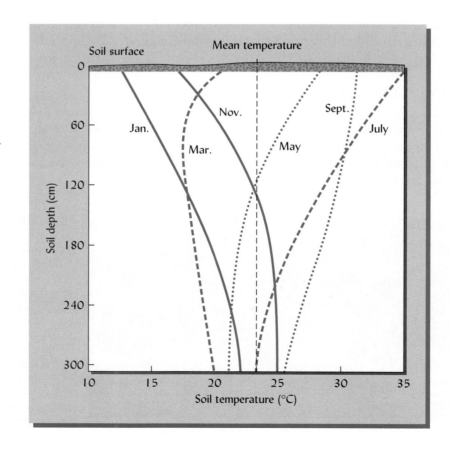

Compared to the surface soil and the air, deep soil layers are generally warmer in the late fall and winter and cooler in the spring and summer. Soil reaches its maximum daily temperature later in the afternoon or evening than does the air, the lag time being greater and the fluctuation less pronounced for greater depths. Deeper than 4 to 5 m, the temperature changes little and approximates the mean annual air temperature (a fact experienced by people visiting deep caverns).

7.11   SOIL TEMPERATURE CONTROL

Practical management of soil temperature mainly involves practices that affect the cover or mulch on the soil and those that reduce excess soil moisture (see Sections 6.7 and 7.10).

Organic Mulches and Plant-Residue Management

Mulching effects: www.ianrpubs.unl.edu/ epublic/pages/publicationD .jsp?publicationId=187

Soil temperatures are influenced by soil cover and especially by organic residues or other types of mulch on the soil surface. Figure 7.25 shows that mulches effectively buffer extremes in soil temperatures. In periods of hot weather, they keep the surface soil cooler than where no cover is used; in contrast, during cold weather they keep the soil warmer than it would be if bare.

The forest floor is a prime example of a natural temperature-modifying mulch. It is not surprising, therefore, that timber harvest practices can markedly affect forest soil temperature regimes (Figure 7.26). Disturbance of the leaf mulch, changes in water content due to reduced evapotranspiration, and compaction by machinery are all factors that influence soil temperatures through thermal conductivity. Reduced shading after tree removal also lets in more solar radiation.

Mulch from Conservation Tillage   The use of mulches has been extended to field-crop culture in areas that have adopted conservation tillage practices. Conservation tillage leaves most or all of the crop residues on the soil surface, thereby allowing farmers to grow mulch in place rather than having to transport it to the field.

Concerns in Cool Climates   While the mulch provided by no-till gives great control over erosion (see Section 14.6), it can also lower soil temperatures early in the season.

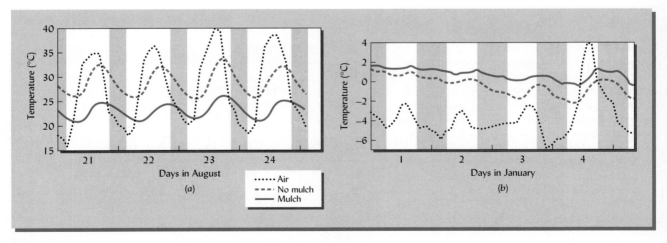

Figure 7.25

(a) Influence of straw mulch (8 tons/ha) on air temperature at a depth of 10 cm during an August hot spell in Bushland, TX. Note that the soil temperatures in the mulched area are consistently lower than where no mulch was applied. (b) During a cold period in January, the soil temperature was higher in the mulched than in the unmulched area. The shaded bars represent nighttime.

[Redrawn from Unger (1978); used with permission of American Society of Agronomy]

Figure 7.26

Soil temperature in an aspen-spruce boreal forest after two levels of harvest and soil compaction. One harvest procedure removed only the stems (tree trunks) with branches and foliage left on the soil, while a second procedure removed whole trees and stripped woody materials and litter to expose the mineral soil (this was done to simulate the kind of damage often inflicted by poorly managed harvest equipment). The soils were either left undisturbed during harvest, or they were severely compacted. The compacted treatment is shown only for the whole-tree removal procedure as compaction did not affect soil temperature where harvest removed only tree trunks. Exposure of the mineral soil A horizon resulted in much warmer temperatures in summer and somewhat colder soil in winter. Compaction of this soil mainly slowed warming in summer, partly because of a higher water content (and therefore a higher heat capacity). The Aquepts (Luvic Gleysols in the Canadian soil classification) at this site in British Columbia included about 20–30 cm of silt loam material over a clay loam. [From Tan et al. (2005)]

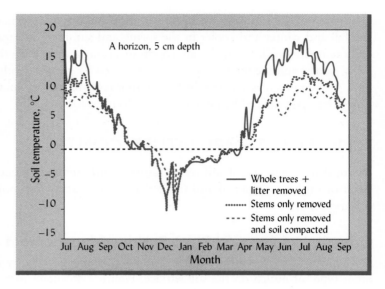

In cold regions, the lower soil temperatures can inhibit seed germination, seedling performance, and, often, crop yields. This effect is well illustrated by the data in Figure 7.27, which also features an innovative way to alleviate this problem by pushing aside the residues in just a narrow band over the seed row in the no-tillage system. Another approach to solving this problem is to ridge the soil, permit water to drain out of the ridge, and then plant on the drier, warmer ridgetop (or on the south side of the ridge—see Section 7.9).

Advantages in Warm Climates   In warm regions, delayed planting is not a problem. In fact, the cooler near-surface soil temperatures under a mulch may reduce heat stress on roots during summer. Plant-residue mulches also conserve soil moisture by decreasing evaporation. The resulting cooler, moist surface layer of soil is an important part of no-tillage systems because it allows roots to proliferate in this zone, where nutrient and aeration conditions are optimal.

Figure 7.27

Tillage effects on hourly temperature changes near the surface of a cold Alfisol in northern British Columbia. The soil had been managed to grow barley under no-tillage (NT) and conventional (clean surface without residues) tillage (CT) systems for the previous 14 years. In the clean-tilled soil, midafternoon temperatures peaked at 4 °C higher than those in the residue-mulch-covered no-tillage soil. A modification of the no-tillage system (MNT) that pushed aside the residues in a narrow (7.5-cm wide) band over the seeding row eliminated much of the temperature depression while keeping most of the surface covered by the soil- and water-conserving mulch. Note the daily temperature changes and the general warming trend during the seven days shown. [From Arshad and Azooz (1996)]

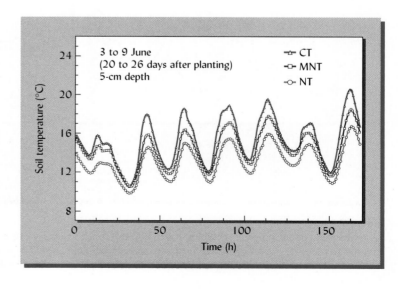

Figure 7.28

These winter-grown southern California strawberries will come to market when prices are still high because of the effect of the clear plastic mulch on soil temperature.

(Photo courtesy of R. Weil)

Plastic Mulches

One of the reasons for the popularity of plastic mulches for gardens and high-value specialty crops is their effect on soil temperature (see Section 6.4 for their effect on soil moisture). In contrast to organic mulches, plastic mulches generally increase soil temperature, clear plastic having a greater heating effect than black plastic. In temperate regions, this effect can be used to extend the growing season or to hasten production to take advantage of the higher prices offered by early-season markets (Figure 7.28).

Major disadvantages of clear, black, and colored plastic mulches are the nonrenewable fossil fuels used in their manufacture, the difficulty of removing the material from the field at the end of the season, and the problem of properly disposing of all that shredded plastic waste. One solution may be found in newer biodegradable plastic films that have been manufactured from such natural renewable raw materials as corn starch (Figure 7.29).

Figure 7.29

Laying down a special biodegradable clear plastic mulch to hasten soil warming for a no-till sweet corn crop on hilly cropland in Pennsylvania. The farmer's no-till system keeps the soil covered with plant residues that prevent erosion but also slow soil warming in spring. Each strip of plastic mulch covers two rows of corn seeds that have already been planted with a no-till planter. The plastic film acts like a greenhouse to trap solar energy, warm the soil, and hasten the corn germination and early growth. When the corn seedlings are about 20 cm tall the farmer will slit the plastic film, allowing the plants to grow unimpeded. By the time the corn canopy has closed, the plastic will have largely disappeared, having served its function of getting the corn off to an early start.

(Photo courtesy of R. Weil)

Table 7.6
SOIL TEMPERATURE AND TOMATO YIELD WITH STRAW OR BLACK PLASTIC MULCH[a]
The data are averages for two years of tomato production on a sandy loam Ultisol near Griffin, Georgia. The straw kept the surface soil from rising to detrimentally high temperatures, while it also increased infiltration of rainwater and reduced soil compaction. Daily drip irrigation supplied plenty of water, but could not overcome the temperature effects of the black plastic mulch.

	Not irrigated		Irrigated daily	
	Straw mulch	Plastic mulch	Straw mulch	Plastic mulch
Average soil temperature, °C	24	37	24	35
Tomato yield, Mg/ha	68	30	70	24

[a]Soil temperature measured at 5 cm below the soil surface, average of weeks 2–10 of the growing season.
Data calculated from Tindall et al. (1991).

In warmer climates, and during the summer months, the soil-heating effect of plastic mulches may be quite detrimental, inhibiting root growth in the upper soil layers and sometimes seriously decreasing crop yields (Table 7.6).

7.12   CONCLUSION

Soil aeration and soil temperature critically affect the quality of soils as habitats for plants and other organisms. Most plants have definite requirements for soil oxygen along with limited tolerance for carbon dioxide, methane, and other such gases found in poorly aerated soils. Some microbes, such as the nitrifiers and general-purpose decay organisms, are also constrained by low levels of soil oxygen. Through its effect on soil redox potential (E_h) and acidity (pH), soil aeration status helps determine the forms present, availability, mobility, and possible toxicity of such elements as nitrogen, sulfur, carbon, iron, manganese, chromium, and many others.

Soils with extremely wet moisture regimes are unique with respect to their morphology and chemistry and to the plant communities they support. Such hydric soils are characteristic of wetlands and help these ecosystems perform myriad valuable functions.

Plants as well as microbes are also quite sensitive to differences in soil temperature, particularly in temperate climates where low temperatures can limit essential biological processes. Soil temperature also impacts the use of soils for engineering purposes, again primarily in the cooler climates. Frost action, which can move perennial plants such as alfalfa out of the ground, can also cause damage to building foundations, fence posts, sidewalks, and highways.

Soil water exerts a major influence over both soil aeration and soil temperature. It competes with soil air for the occupancy of soil pores and interferes with the diffusion of gases into and out of the soil. Soil water also resists changes in soil temperature by virtue of its high specific heat and its high energy requirement for evaporation.

STUDY QUESTIONS

1. What are the two principal gases involved with soil aeration, and how do their relative amounts change as one samples deeper into a soil profile?
2. What is aerenchyma tissue, and how does it affect plant–soil relationships?
3. If the redox potential for a soil at pH 6 is near zero, write two reactions that you would expect to take place. How would the presence of a great deal of nitrate compounds affect the occurrence of these reactions?
4. It is sometimes said that organisms in anaerobic environments will use the combined oxygen in nitrate or sulfate instead of the oxygen in O_2. Why is this statement incorrect? What actually happens when organisms reduce sulfate or nitrate?
5. If an alluvial forest soil were flooded for 10 days, and you sampled the gases evolving from the wet soil, what gases would you expect to find (other than oxygen and carbon dioxide)? In what order of appearance? Explain.
6. Explain why warm weather during periods of saturation is required in order to form a hydric soil.
7. If you were in the field trying to delineate the so-called drier end of a wetland area, what are three soil properties and three other indicators that you might look for?
8. For each of these gases, write a sentence to explain its relationship to wetland conditions: *ethylene, methane, nitrous oxide, oxygen,* and *hydrogen sulfide.*
9. What are the three major components that define a wetland?
10. Discuss four plant processes that are influenced by soil temperature.
11. Explain how a brush fire might lead to subsequent mudslides, as often occurs in California.
12. If you were to build a house below ground in order to save heating and cooling costs, would you firmly compact the soil around the house? Explain your answer.
13. If you measured a daily maximum air temperature of 28 °C at 1 P.M., what might you expect the daily maximum temperature to be at a 15-cm depth in the soil? At about what time of day would the maximum temperature occur at this depth? Explain.
14. In relation to soil temperature, explain why conservation tillage has been more popular in Missouri than in Minnesota.

REFERENCES

Arshad, A., and R. H. Azooz. 1996. "Tillage effects on soil thermal properties in a semiarid cold region," *Soil Sci. Soc. Amer. J.* **60:**561–567.

Bartlett, R. J., and B. R. James. 1993. "Redox chemistry of soils," *Advances in Agronomy* **50:**151–208.

Bartlett, R. J., and D. S. Ross. 2005. "Chemistry of redox process in soils," pp. 461–487, in A. Tabatabai and D. Sparks (eds.), *Chemical Processes in Soils.* SSSA Book Series N 8. (Madison, WI: Soil Science Society of America).

Beard, J. B., and R. L. Green. 1994. "The role of turfgrasses in environmental protection and their benefits to humans," *J. Environ. Qual.* **23:** 452–460.

CAST. 1994. *Wetland Policy Issues.* Publication No. CC1994–1. (Ames, IA: Council for Agricultural Science and Technology).

Dryness, C. T. 1976. "Effects of wildfire on soil wetability in the high cascades of Oregon," USDA Forest Service Research Paper PNW-202. (Washington, DC: USDA).

Eilperin, J. 2005. "Developers sentenced in wetlands case," p. A-14, *The Washington Post,* December 07, 2005.

Eswaren, H., P. Reich, P. Zdruli, and T. Levermann. 1996. "Global distribution of wetlands," *Amer. Soc. Agron. Abstracts* 328.

Fluker, B. J. 1958. "Soil temperature," *Soil Sci.* **86:** 35–46.

Hurt, G. W., P. M. Whited, and R. F. Pringle (eds.). 1996. *Field Indicators of Hydric Soils in the United States.* (Fort Worth, TX: USDA Natural Resources Conservation Service).

Jenkinson, B. J., and D. P. Franzmeier. 2006. "Development and evaluation of iron-coated tubes that indicate reduction in soils," *Soil Sci. Soc. Amer. J.* **70:**183–191.

Lin, X., J. E. Smerdon, A. W. England, and H. N. Pollack. 2003. "A model study of the effects of climatic precipitation changes on ground temperatures," *J. Geophys. Res.* **108**(D7):4230, doi:10.1029/2002JD002878.

Losi, M. E., C. Amrheim, and W. T. Frankenberger, Jr. 1994. "Bioremediation of chromatic contaminated groundwater by reduction and precipitation in surface soils," *J. Environ. Qual.* **23:**1141–1150.

Lowe, D. F., C. L. Oubre, and C. H. Ward (eds.). 2000. *Soil Vapor Extraction Using Radio Frequency Heating: Resource Manual and Technology Demonstration.* (New York: Lewis).

MacDonald, N. W., D. R. Zac, and K. S. Pregitzer. 1995. "Temperature effects on kinetics of microbial respiration and net nitrogen and sulfur mineralization," *Soil Sci. Soc. Amer. J.* **59:**233–240.

Melillo, J. M., P. A. Steudler, J. D. Aber, K. Newkirk, H. Lux, F. P. Bowles, C. Catricala, A. Magill, T. Ahrens, and S. Morrisseau. 2002. "Soil warming and carbon-cycle feedbacks to the climate system," *Science* **298:**2173–2176.

Patrick, W. H., Jr., and A. Jugsujinda. 1992. "Sequential reduction and oxidation of inorganic nitrogen, manganese, and iron in flooded soil," *Soil Sci. Soc. Amer. J.* **56:**1071–1073.

Richardson, J. L., and M. J. Vepraskas. 2001. *Wetland Soils—Genesis, Hydrology, Landscapes, and Classification.* (Boca Raton, FL: Lewis).

Sexstone, A. J., N. P. Revsbech, T. B. Parkin, and J. M. Tiedje. 1985. "Direct measurement of oxygen profiles and denitrification rates in soil aggregates," *Soil Sci. Soc. Amer. J.* **49:**645–651.

Tan, X., S. X. Chang, and R. Kabzems. 2005. "Effects of soil compaction and forest floor removal on soil microbial properties and N transformations in a boreal forest long-term soil productivity study," *Forest Ecology and Management* **217:**158–170.

Tindall, J. A., R. B. Beverly, and D. E. Radcliff. 1991. "Mulch effect on soil properties and tomato growth using micro-irrigation," *Agron. J.* **83:**1028–1034.

Unger, P. W. 1978. "Straw mulch effects on soil temperatures and sorghum germination and growth," *Agron. J.* **70:**858–864.

Veneman, P. L. M., D. L. Lindbo, and L. A. Spokas. 1999. "Soil moisture and redoximorphic features: A historical perspective," in M. J. Rabenhorst, J. C. Bell, and P. A. McDaniel (eds.), *Quantifying Soil Hydromorphology.* Special Publication No. 54. (Madison, WI: Soil Science Society of America).

Welsh, D., D. Smart, J. Boyer, P. Minkin, H. Smith, and T. McCandless (eds.). 1995. *Forested Wetlands: Functions, Benefits, and Use of Best Management Practices.* (Radnor, PA: USDA Forest Service).

Xu, Q., and B. Huang. 2000. "Growth and physiological responses of creeping bentgrass to changes in air and soil temperatures," *Crop Sci* **40:**1363–1368.

Yu, K., and W. H. Patrick, Jr. 2004. "Redox window with minimum global warming potential contribution from rice soils," *Soil Sci. Soc. Amer. J.* **68:**2086–2091.

Mica weathers to clay. (Serge Jolicoeur, Université de Moncton)

8
The Colloidal Fraction: Seat of Soil Chemical and Physical Activity

The landscape of the clays is like—the intricate folds of the womb—whose activity is to receive, contain, enfold, and give birth.
—WILLIAM BRYANT LOGAN

How can using sewage effluent for irrigation contribute to the safe recharge of groundwater aquifers? Why is it more difficult to restore productivity after logging a tropical rain forest on Oxisols than a temperate forest on Alfisols? Why would a nuclear power plant accident seriously contaminate food grown on some downwind soils, but not on others? The answers to these and other environmental mysteries lie in the nature of the smallest of soil particles, the clay and humus **colloids**. These particles are not just extra-small fragments of rock and organic matter. They are highly reactive materials with electrically charged surfaces. Because of their size and shape, they give the soil an enormous amount of reactive **surface area**. It is the colloids, then, that allow the soil to serve as nature's great electrostatic chemical reactor.

Each tiny colloid particle carries a swarm of positively and negatively charged ions (cations and anions) that is attracted to electrostatic charges on its surface. The ions are held tightly enough by the **soil colloids** to greatly reduce their loss in drainage waters, but loosely enough to allow plant roots access to the nutrients among them. Other modes of adsorption bind ions more tightly so that they are no longer available for plant uptake, reaction with the soil solution, or leaching loss to the environment. In addition to plant nutrient ions, soil colloids also bind with water molecules, biomolecules (e.g., DNA, antibiotics), viruses, toxic metals, pesticides, and a host of other mineral and organic substances. Hence, soil colloids greatly impact nearly all ecosystem functions.

We shall see that different soils are endowed with different types of clays that, along with humus, elicit very different types of physical and chemical behaviors. Certain clay minerals are much more reactive than others. Some are more dramatically influenced than others by the acidity of the soil and other environmental factors. Studying the soil colloids in some detail will deepen your understanding of soil architecture (Chapter 4) and soil water (Chapters 5 and 6). Knowledge of the structure, origin, and behavior of the different types of soil colloids will also help you understand soil chemical and biological processes so you can make better decisions regarding the use of soil resources.

8.1   GENERAL PROPERTIES AND TYPES OF SOIL COLLOIDS

Size

The clay and humus particles in soils are referred to collectively as the **colloidal fraction** because of their extremely small size and colloid-like behavior. Too small to be seen with an ordinary light microscope, they can be made visible only with an electron microscope. Particles behave as colloids if they are less than about 1 μm (0.000001 m) in diameter, although some soil scientists consider 2 μm to mark the upper boundary of the colloidal fraction to coincide with the definition of the clay particle size fraction.

Surface Area

As discussed in Section 4.2, the smaller the size of the particles in a given mass of soil, the greater the surface area exposed for adsorption, catalysis, precipitation, microbial colonization, and other surface phenomena. Because of their small size, all soil colloids expose a large **external surface** area per unit mass, more than 1000 times the surface area of the same mass of sand particles. Some silicate clays also possess extensive **internal surface** area between the layers of their platelike crystal units. To grasp the relative magnitude of the internal surface area, remember that these clays are structured much like this book. If you were to paint the external surfaces of this book (the covers and edges), a single brush of paint would do. However, to cover the internal surfaces (both sides of each page in the book), you might need a very large can of paint.

The total surface area of soil colloids ranges from 10 m^2/g for clays with only external surfaces, to more than 800 m^2/g for clays with extensive internal surfaces. To put this in perspective, we can calculate that the surface area exposed within 1 ha (about the size of a football field) of a 1.5-m-deep fine-textured soil (45% clay) might be as great as 8,700,000 km^2 (the land area of the entire United States).

Surface Charges

The internal and external surfaces of soil colloids carry positive and/or negative electrostatic charges. For most soil colloids, the negative charges far outnumber the positive ones. However, some mineral colloids in very acid soils have more positive charges than negative. As we shall see in Sections 8.3 to 8.7, the amount and origin of surface charge differs greatly among the different types of soil colloids and, in some cases, is influenced by changes in chemical conditions, such as soil pH. The charges on the colloid surfaces attract or repulse substances in the soil solution as well as neighboring colloid particles. These reactions, in turn, greatly influence soil chemical and physical behavior.

Adsorption of Cations and Anions

Of particular significance is the attraction of positively charged ions (**cations**) to the surfaces of negatively charged soil colloids. Each colloid particle attracts thousands of Al^{3+}, Ca^{2+}, Mg^{2+}, K^+, H^+, and Na^+ ions and lesser numbers of other cations. In moist soils the cations exist in the hydrated state (surrounded by a shell of water molecules), but for simplicity in this text, we will show just the cations (e.g., Ca^{2+} or H^+) rather than the hydrated forms (e.g., $Ca(H_2O)_6^{2+}$ or the hydronium ion, H_3O^+). These hydrated cations constantly vibrate about in a swarm near the colloid surface, held there by electrostatic attraction to the colloid's negative charges. Frequently, an individual cation will break away from the swarm and move out to the soil solution. When this happens, another cation of equal charge will simultaneously move in from the soil solution and take its place. This process of **cation exchange** will be discussed in detail (Section 8.8) because of its fundamental importance in nutrient cycling and other environmental processes. The cations swarming about near the colloidal surface are said to be **adsorbed** (loosely held) on the colloid surface. Because these cations can *exchange places* with those moving freely about in the soil solution, the term **exchangeable ions** is also used to refer to the ions in this adsorbed state.

The colloid with its adsorbed cations is sometimes described as an **ionic double layer** in which the negatively charged colloid acts as a huge anion constituting the inner ionic layer, and the swarm of adsorbed cations constitutes the outer ionic layer (Figure 8.1). Because cations from the soil solution are constantly trading places with those that are adsorbed to the colloid, the ionic composition of the soil solution reflects that of the adsorbed swarm. For example, if Ca^{2+} and Mg^{2+} dominate the exchangeable ions, they will also dominate the soil solution. Under natural conditions, the proportions of specific cations present are largely influenced by the soil parent material and the degree to which the climate has promoted the loss of cations by leaching (see Section 8.9).

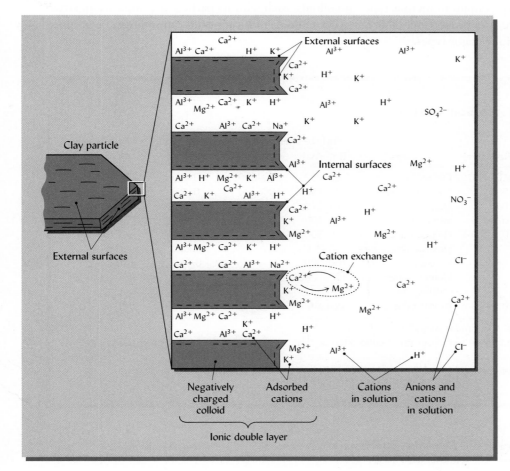

Figure 8.1

Simplified representation of a silicate clay crystal, its complement of adsorbed cations, and ions in the surrounding soil solution. The enlarged view (right) shows that the clay comprises sheetlike layers with both external and internal negatively charged surfaces. The negatively charged particle acts as a huge anion and a swarm of positively charged cations is adsorbed to it because of attraction between charges of opposite sign. Cation concentration decreases with distance from the clay. Anions (such as Cl^-, NO_3^-, and SO_4^{2-}), which are repulsed by the negative charges, can be found in the bulk soil solution farthest from the clay (far right). Some clays (not shown) also exhibit positive charges that can attract anions.

Anions such as Cl^-, NO_3^-, and SO_4^{2-} (also surrounded by water molecules, though, again, we do not show these water shells) may also be attracted to certain soil colloids that have *positive* charges on their surfaces. While adsorption of **exchangeable anions** is not as extensive as that for exchangeable cations, we shall see (Section 8.11) that it is an important mechanism for holding negatively charged constituents, especially in acid subsoils. When thinking about the colloids in soil, we should always keep in mind that they carry with them a complement of exchangeable cations and anions, along with certain other more tightly bound ions and molecules.

Adsorption of Water

In addition to adsorbing cations and anions, soil colloids attract and hold a large number of water molecules. Generally, the greater the external surface area of the soil colloids, the greater the amount of water held when the soil is air-dry. While this water may not be available for plant uptake (see Section 5.8), it does play a role in the survival of soil microorganisms, especially bacteria. The charges on the internal and external colloid surfaces attract the oppositely charged end of the polar water molecule. Some water molecules are attracted to the exchangeable cations, each of which is hydrated with a shell of water molecules. Water adsorbed between the clay layers can cause the layers to move apart, making the clay more plastic and swelling its volume. Colloids that adsorb a great deal of water may make soil unsuitable for construction purposes (see Sections 4.9 and 8.14). As a soil colloid dries, any water between the layers is removed, and the layers are brought closer together.

Types of Soil Colloids[1]

Soils contain numerous types of colloids, each with its particular composition, structure, and properties (Table 8.1). The colloids most important in soils can be grouped in four major types.

Table 8.1
MAJOR PROPERTIES OF SELECTED SOIL COLLOIDS

Colloid	Type	Size, μm	Shape	Surface area, m^2/g External	Internal	Interlayer spacing,[a] nm	Net charge,[b] $cmol_c/kg$
Smectite	2:1 silicate	0.01–1.0	Flakes	80–150	550–650	1.0–2.0	−80 to −150
Vermiculite	2:1 silicate	0.1–0.5	Plates, flakes	70–120	600–700	1.0–1.5	−100 to −200
Fine mica	2:1 silicate	0.2–2.0	Flakes	70–175	—	1.0	−10 to −40
Chlorite	2:1 silicate	0.1–2.0	Variable	70–100	—	1.41	−10 to −40
Kaolinite	1:1 silicate	0.1–5.0	Hexagonal crystals	5–30	—	0.72	−1 to −15
Gibbsite	Al-oxide	<0.1	Hexagonal crystals	80–200	—	0.48	+10 to −5
Goethite	Fe-oxide	<0.1	Variable	100–300	—	0.42	+20 to −5
Allophane & Imogolite	Noncrystalline silicates	<0.1	Hollow spheres or tubes	100–1000	—	—	+20 to −150
Humus	Organic	0.1–1.0	Amorphous	Variable[c]	—	—	−100 to −500

[a] From the top of one layer to the next similar layer, 1 nm $= 10^{-9}$ m $= 10$ Å.
[b] Centimoles of unbalanced or net charge per kilogram of colloid ($cmol_c/kg$), a measure of ion exchange capacity (see Section 8.9).
[c] It is very difficult to determine the surface area of organic matter. Different procedures give values ranging from 20 to 800 m^2/g.

---

[1] For a review of the structures and properties of the clays, see Meunier (2005), and for properties of clay and humus in soils, see Dixon and Schulze (2002).

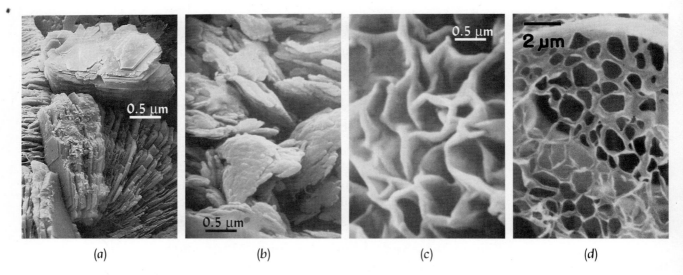

Figure 8.2

Crystals of three silicate clay minerals and a photomicrograph of humic acid found in soils, (a) kaolinite from Illinois (note hexagonal crystal at upper right), (b) a fine-grained mica from Wisconsin, (c) montmorillonite (a smectite group mineral) from Wyoming, and (d) fulvic acid (a humic acid) from Georgia. [(a)–(c) Courtesy of Dr. Bruce F. Bohor, Illinois State Geological Survey; (d) from Dr. Kim H. Tan, University of Georgia; used with permission of Soil Science Society of America]

Crystalline Silicate Clays These clays are the dominant type in most soils (except in Andisols, Oxisols, and Histosols—see Chapter 3). Their crystalline structure is layered much like pages in a book (clearly visible in Figure 8.2*a*). Each layer (page) consists of two to four sheets of closely packed and tightly bonded oxygen, silicon, and aluminum atoms. Although all are predominately negatively charged, silicate clay minerals differ widely with regard to their particle shapes (**kaolinite**, a **fine-grained mica**, and a **smectite** are shown in Figure 8.2*a–c*), intensity of charge, stickiness, plasticity, and swelling behavior.

Noncrystalline Silicate Clays These clays also consist mainly of tightly bonded silicon, aluminum, and oxygen atoms, but they do not exhibit ordered, crystalline sheets. The two principal clays of this type, **allophane** and **imogolite**, usually form from volcanic ash and are characteristic of Andisols (Section 3.7). They have high amounts of both positive and negative charge, and high water-holding capacities. Although malleable (plastic) when wet, they exhibit a very low degree of stickiness. Allophane and imogolite are also known for their extremely high capacities to strongly adsorb phosphate and other anions, especially under acid conditions.

Iron and Aluminum Oxides These are found in many soils but are especially important in the more highly weathered soils of warm, humid regions (e.g., Ultisols, Oxisols). They consist mainly of either iron or aluminum atoms coordinated with oxygen atoms (the latter are often associated with hydrogen ions to make hydroxyl groups). Some, like **gibbsite** (an Al-oxide) and **goethite** (an Fe-oxide) consist of crystalline sheets. Other oxide minerals are noncrystalline, often occurring as **amorphous** coatings on soil particles. The oxide colloids are relatively low in plasticity and stickiness. Their net charge ranges from slightly negative to moderately positive.

Organic (Humus) Organic colloids are important in nearly all soils, especially in the upper parts of the soil profile. Humus colloids are not minerals, nor are they crystalline (Figure 8.3*d*). Instead, they consist of convoluted chains and rings of carbon atoms bonded to hydrogen, oxygen, and nitrogen. Humus particles are often among the smallest of soil colloids and exhibit very high capacities to adsorb water, but almost no plasticity or stickiness. Because humus is noncohesive, soils composed mainly of humus (Histosols) have very little bearing strength and are unsuitable for making building or road foundations. Humus has high amounts of both negative and positive charge per unit mass, but the net charge is always negative and varies with soil pH.

8.2   FUNDAMENTALS OF LAYER SILICATE CLAY STRUCTURE

To see why soils rich in one silicate clay mineral, say kaolinite, behave so very differently from soils dominated by another silicate clay, say montmorillonite, it is necessary to understand the main structural features of the silicate clay minerals. We will begin by examining the main building blocks from which the layer silicates are constructed, then consider the particular arrangements that give rise to the critically important surface charges.

Silicon Tetrahedral and Aluminum-Magnesium Octahedral Sheets

Three-dimensional rotatable models of silicate clay minerals and their building blocks:
www.soils1.cses.vt.edu/MJE/VR_exports/intro.shtml

The most important silicate clays are known as **phyllosilicates** (Greek *phyllon,* leaf) because of their leaflike or planar structure. As shown in Figure 8.3, they are composed of tetrahedral and octahedral sheets. Two to four of these sheets may be stacked together in sandwich-like arrangements, with adjacent sheets strongly bound

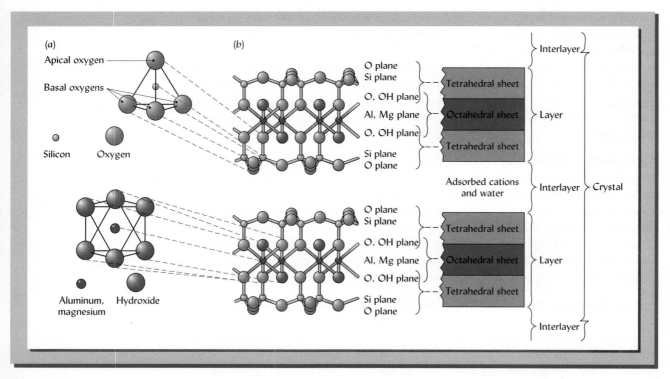

Figure 8.3

The basic molecular and structural components of silicate clays. (a) A single tetrahedron, a four-sided building block composed of a silicon ion surrounded by four oxygen atoms; and a single eight-sided octahedron, in which an aluminum (or magnesium) ion is surrounded by six hydroxy groups or oxygen atoms. (b) In clay crystals thousands of these tetrahedral and octahedral building blocks are connected to give planes of silicon and aluminum (or magnesium) ions. These planes alternate with planes of oxygen atoms and hydroxy groups. Note that apical oxygen atoms are common to adjoining tetrahedral and octahedral sheets. The silicon plane and associated oxygen–hydroxy planes make up a tetrahedral sheet. Similarly, the aluminum–magnesium plane and associated oxygen–hydroxy planes constitute the octahedral sheet. Different combinations of tetrahedral and octahedral sheets are termed layers. *In some silicate clays, these layers are separated by* interlayers *in which water and adsorbed cations are found.*

(Structural models courtesy of Darrel G. Schultze, Purdue University)

Table 8.2
IONIC RADII AND LOCATION OF ELEMENTS FOUND IN SILICATE CLAYS

Ion	Radius, nm (10^{-9} m)	Found in
Si^{4+}	0.042	
Al^{3+}	0.051	Tetrahedral sheet
Fe^{3+}	0.064	
Mg^{2+}	0.066	Octahedral sheet
Zn^{2+}	0.074	Exchange or interlayer sites
Fe^{2+}	0.076	
Na^+	0.095	
Ca^{2+}	0.099	
K^+	0.133	
O^{2-}	0.140	Both sheets
OH^-	0.155	

together by sharing some of the same oxygen atoms (see Figure 8.3). The specific nature and combination of sheets in these layers vary from one type of clay to another and largely control the physical and chemical properties exhibited. The relationship between *planes, sheets,* and *layers* shown in Figure 8.3 should be carefully studied.

Isomorphous Substitution

As clay minerals or their precursors crystallize, cations of comparable size may substitute for silicon, aluminum, and magnesium ions in the respective tetrahedral and octahedral sheets. Note from Table 8.2 that aluminum is only slightly larger than silicon. Consequently, aluminum can fit into the center of the tetrahedron in the place of the silicon without much change in the basic structure of the crystal. This process by which one element fills a position usually filled by another of similar size is called **isomorphous substitution**. This phenomenon is responsible for much of the variability in the nature of silicate clays.

Isomorphous substitution can also occur in the octahedral sheets. For example, iron and zinc ions are not much different in size from aluminum and magnesium ions (Table 8.2). Any of these ions can fit in the central position of an octahedra. In some layer silicates, isomorphous substitution occurs in both tetrahedral and octahedral sheets.

Source of Charges

Isomorphous substitution is of vital importance because it is the primary source of both negative and positive charges of silicate clays. For example, the Mg^{2+} ion is only slightly larger than the Al^{3+} ion, but it has one less positive charge. If a Mg^{2+} ion substitutes for an Al^{3+} ion in an octahedral sheet, there will be insufficient positive charges to balance the negative charges from the oxygens; hence, the lattice is left with a 1– net charge (see Figure 8.4, right). Similarly, each Al^{3+} that substitutes for a Si^{4+} in a tetrahedral sheet creates a net negative charge at that site because the negative charges from the four oxygens will be only partially balanced. In most silicate clays, the negative charges predominate (see Section 8.8). As we shall see (Sections 8.3 and 8.6), additional, more temporary charges can also develop on the edges and surfaces of tetrahedral and octahedral sheets.

Figure 8.4

Simplified diagrams of octahedral sheets in silicate minerals illustrating the effect of isomorphous substitution on the net charge. Note that for each oxygen atom, one of the two – charges is balanced by a + charge from either a H⁺ (making a hydroxyl group) or a Si atom in the tetrahedral sheet (not shown, but represented by a +). During the crystalization of the octahedral sheet shown on the right, a Mg^{2+} atom occupied one of the positions normally occupied by an Al^{3+} atom, thus leaving a –1 net charge on the sheet. Such net negative charges in the crystal can be balanced by cations from the soil solution adsorbed to the crystal surface.

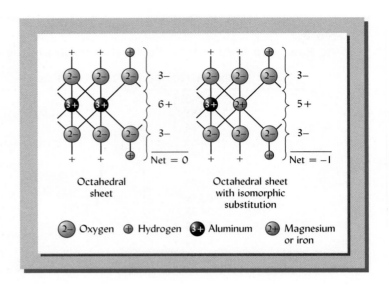

8.3 MINERALOGICAL ORGANIZATION OF SILICATE CLAYS

Based on the number and arrangement of tetrahedral (Si) and octahedral (Al, Mg, Fe) sheets contained in the crystal units or layers, crystalline clays may be classed into two main groups: **1:1 silicate clays**, in which each layer contains *one* tetrahedral and *one* octahedral sheet, and **2:1 silicate clays**, in which each layer has *one* octahedral sheet sandwiched between *two* tetrahedral sheets.

1:1-Type Silicate Clays

Interactive 3-D rotating model of kaolinite crystal: http://virtual-museum.soils .wisc.edu/kaolinite/index.html

To illustrate the properties of 1:1 silicate clays we will focus on kaolinite, which is by far the most common soil clay in this group. As implied by the term *1:1 silicate clay,* each kaolinite layer consists of one silicon tetrahedral sheet and one aluminum octahedral sheet. When the layers consisting of alternating tetrahedral and octahedral sheets are stacked on top of one another, the hydroxyls of the octahedral sheet in one layer are adjacent to the basal oxygens of the tetrahedral sheet of the next layer. These adjacent layers are bound together by **hydrogen bonding** (see Section 5.1), which prevents expansion between the layers when the clay is wetted. Therefore, cations and water generally do not enter between the structural layers of a 1:1 mineral particle. The effective surface of kaolinite is thus restricted to its outer faces or external surface area. This fact and the lack of significant isomorphous substitution in this mineral account for the relatively small capacity of kaolinite to adsorb exchangeable cations (see Table 8.1). In contrast to some 2:1 silicate clays, 1:1 clays like kaolinite exhibit less plasticity, stickiness, cohesion, shrinkage, and swelling and can also hold less water than other clays. Because of these properties, soils dominated by 1:1 clays are relatively easy to cultivate for agriculture and are well suited for use in roadbeds and building foundations (Plate 43). The nonexpanding 1:1 structure also makes kaolinite clays useful for making bricks and ceramics (see Box 8.1).

Kaolin clay used for pest control: www.nysaes.cornell.edu/ pp/resourceguide/mfs/ 07kaolin.php

Expanding 2:1-Type Silicate Clays

The four general groups of 2:1 silicate clays are characterized by *one* octahedral sheet sandwiched between *two* tetrahedral sheets. Two of these groups, **smectite** and **vermiculite**, include expanding-type minerals; the other two, **fine-grained micas (illite)** and **chlorite**, are relatively nonexpanding.

BOX 8.1
KAOLINITE CLAY—ANCIENT AND MODERN

Kaolinite, the most common of the 1:1 clay minerals, has been used for thousands of years to make pottery, roofing tiles, and bricks. The basic processes have changed little to this day. Clayey soil material is saturated with water, kneaded and molded or thrown on a potter's wheel to obtain the desired shape, and then hardened by drying or firing (Figures 8.5 and 8.6). The mass of cohering clay platelets hardens irreversibly when fired and the nonexpanding nature of kaolinite allows it to be fired without cracking from shrinkage. The heat also changes the typical gray color of the soil material to "brick red" because of the irreversible oxidation and crystallization of the iron-oxyhydroxides that often coat soil kaolinite particles. In contrast, kaolinite mined from pure deposits fires to a light, creamy color. Kaolinite is not as plastic (moldable) as some other clays, however, and so is usually mixed with more plastic types of clays for making pottery.

It was in seventh-century China that pure kaolinite deposits were first used to make objects of a translucent, lightweight, and strong ceramic called porcelain. The name *kaolinite* derives from the Chinese character *kao lin*, meaning "high ridge," as the material was first mined from a hillside in Kiangsi Province. The Chinese held a

Figure 8.6
African pottery and early 19th-century English china (inset). (Photo courtesy of R. Weil)

monopoly on porcelain-making technology (hence the term *china* for porcelain dishes) until the early 1700s. English colonists, in what is now Georgia in the United States, noted outcrops of white kaolinite clay in areas of rather unproductive soil. The colonialists soon were exporting this kaolinite as the main ingredient for making porcelain in England, where the now-famous pottery was first manufactured from the Georgia kaolinite clay.

The market for pure, white kaolinite greatly expanded when paper manufacturers started using kaolinite clay to make sizing, the coating that makes high-quality papers smoother, whiter, and more printable. Other industrial uses now include paint pigments, fillers in plastic manufacture, and ceramic materials used for electrical insulation and heat shielding (as on the belly of the space shuttle). The kaolinite in kaopectin-type medications lines the stomach walls and inactivates diarrhea-causing bacteria by adsorbing them on the clay particle surfaces. Recently, kaolinite clays have been used as a spray-on coating for plant leaves that provides nontoxic protection from insect pests and fungal diseases. Unfortunately, surface mining of kaolinite in Georgia to meet these needs has caused both environmental and social disruptions.

Figure 8.5
Kaolinitic clay soil is dug, molded, dried, stacked, and fired to make bricks. (Photo courtesy of R. Weil)

Smectite Group   The flakelike crystals of smectites (see Figure 8.2*c*) have a high amount of mostly negative charge resulting from isomorphous substitution. In contrast to kaolinite, smectites have a 2:1 structure that exposes a layer of oxygen atoms at both the top and bottom planes. Therefore, adjacent layers are only loosely bound to each other by very weak oxygen-to-oxygen and cation-to-oxygen linkages and the space between is variable (Figure 8.7). The internal surface area exposed between the

Interactive 3-D rotating model of smectite crystal: http://virtual-museum.soils.wisc.edu/soil_smectite/index.html

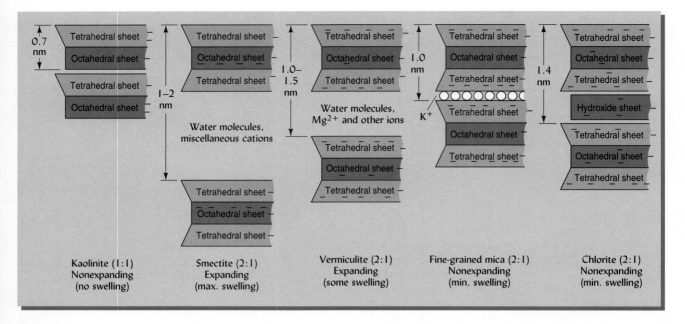

Figure 8.7

Schematic drawing illustrating the organization of tetrahedral and octahedral sheets in one 1:1-type mineral (kaolinite) and four 2:1-type minerals. The octahedral sheets in each of the 2:1-type clays can be either aluminum dominated (dioctahedral) or magnesium dominated (trioctahedral). However, in most chlorites the trioctahedral sheets are dominant while the dioctahedral sheets are generally most prominent in the other three 2:1 types. Note that kaolinite is nonexpanding, the layers being held together by hydrogen bonds. Maximum interlayer expansion is found in smectite, with somewhat less expansion in vermiculite because of the moderate binding power of numerous Mg^{2+} ions. Fine-grained mica and chlorite do not expand because K^+ ions (fine-grained mica) or an octahedral-like sheet of hydroxides of Al, Mg, Fe, and so forth (chlorite) tightly bind the 2:1 layers together. The interlayer spacings are shown in nanometers (1 nm = 10^{-9} m).

Safe storage for nuclear wastes: the Swedish approach using Bentonite clay: www.skb.se/templates/ SKBPage_8776.aspx

layers by far exceeds the external surface area of these minerals and contributes to the very high total **specific surface area** (Table 8.1). Exchangeable cations and associated water molecules are attracted to the spaces between the interlayer spaces, and the capacity to adsorb cations is very high—about 20 to 40 times that of kaolinite.

Flakelike smectite crystals tend to pile upon one another, forming wavy stacks that contain many extremely small *ultramicropores* (see Table 4.6). When soils high in smectite are wetted, adsorption of water in these ultramicropores leads to severe swelling; when they are dried, the soils shrink in volume (see Section 8.14). The expansion upon wetting contributes to the high degree of plasticity, stickiness, and cohesion that make smectitic soils very difficult to cultivate or excavate. Wide cracks commonly appear during the drying of smectite-dominated soils (such as Vertisols, Figure 3.15). The shrink–swell behavior makes smectitic soils quite undesirable for most construction activities, but they are well suited for a number of applications that require a high adsorptive capacity and the ability to form seals of very low permeability (see Section 8.14). **Montmorillonite** is the most prominent of the smectites in soils, although others are also found.

Vermiculite Group   The most common **vermiculites** are 2:1-type minerals in which the octahedral sheet is aluminum dominated. The tetrahedral sheets of most vermiculites have considerable substitution of aluminum in the silicon positions, giving rise to a cation exchange capacity that usually exceeds that of all other silicate clays, including smectites (Table 8.1).

The interlayer spaces of vermiculites usually contain strongly adsorbed water molecules, Al-hydroxy ions, and cations such as magnesium (Figure 8.7). However, these interlayer constituents act primarily as bridges to hold the units together, rather

than wedges driving them apart. For this reason, vermiculites are limited-expansion clays, expanding more than kaolinite, but much less than the smectites.

Nonexpanding 2:1 Silicate Minerals

The main nonexpanding 2:1 minerals are the **fine-grained micas** and the **chlorites**. We will discuss the fine-grained micas first.

Mica Group   Biotite and muscovite are examples of unweathered micas typically found in the sand and silt fractions. The more weathered **fine-grained micas**, such as **illite** and **glauconite**, are found in the clay fraction of soils. Their 2:1-type structures are quite similar to those of their unweathered cousins. The main source of charge in fine-grained micas is the substitution of Al^{3+} for Si^{4+} in the tetrahedral sheets. This results in a high net negative charge in the tetrahedral sheet and attracts potassium (K^+) ions that are just the right size to fit snugly into certain hexagonal "holes" between the tetrahedral oxygen groups and thereby get very close to the negatively charged sites. By their mutual attraction for the K^+ ions in between, adjacent layers in fine-grained micas are strongly bound together (Figure 8.7), making the fine-grained micas quite **nonexpansive**. Because of their nonexpansive character, the fine-grained micas are more like kaolinite than smectites with regard to their capacity to adsorb water and their degree of plasticity and stickiness.

Chlorites   In most soil **chlorites**, iron or magnesium, rather than aluminum, occupy many of the octahedral sites. Commonly, a magnesium-dominated octahedral hydroxide sheet is sandwiched in between adjacent 2:1 layers (Figure 8.7). Thus, chlorite is sometimes said to have a 2:1:1 structure. Chlorites are nonexpansive because the Mg-octahedral sheet binds the layers tightly together. The colloidal properties of the chlorites are therefore quite similar to those of the fine-grained micas (Table 8.1).

8.4   CHARACTERISTICS OF NONSILICATE COLLOIDS

Iron and Aluminum Oxides

These clays consist of modified octahedral sheets with either iron (e.g., goethite) or aluminum (e.g., gibbsite) in the cation positions. They have neither tetrahedral sheets nor silicon in their structures. Isomorphous substitution by ions of varying charge rarely occurs, so these clays do not have a large negative charge. The small amount of net charge these clays possess (positive and negative) is caused by the removal or addition of hydrogen ions at the surface oxy-hydroxyl groups (see Section 8.6). The presence of these bound oxygen and hydroxyl groups enables the surfaces of these clays to strongly adsorb and combine with anions such as phosphate or arsenate. The oxide clays are nonexpansive and generally exhibit relatively little stickiness, plasticity, and cation adsorption. They make quite stable materials for construction purposes.

In many soils, iron and aluminum oxide minerals are mixed with silicate clays. The oxides may form coatings on the external surfaces of the silicate clays, or they may occur as "islands" in the interlayer spaces of such 2:1 clays as vermiculites and smectites. In either case, the presence of iron and aluminum oxides can substantially alter the colloidal behavior of the associated silicate clays by masking charge sites, interfering with shrinkage and swelling, and providing anion-retentive surfaces.

Allophane and imogolite are clay minerals commonly associated with materials of volcanic origin, but they can also form from igneous rocks and are found in some Spodosols. Apparently, volcanic ashes release significant quantities of $Si(OH)_x$ and $Al(OH)_x$ materials that precipitate as gels in a relatively short period of time. These minerals are generally poorly crystalline in nature, imogolite being the product of a

Figure 8.8

A possible structure for humic acid, a primary constituent of colloidal humus in soils. Careful inspection will reveal the presence of many of the active OH groups illustrated in Figure 8.9, as well as certain nitrogen- and sulfur-containing groups. [From Schulten and Schnitzer (1993) with kind permission of Springer-Verlag Publishers]

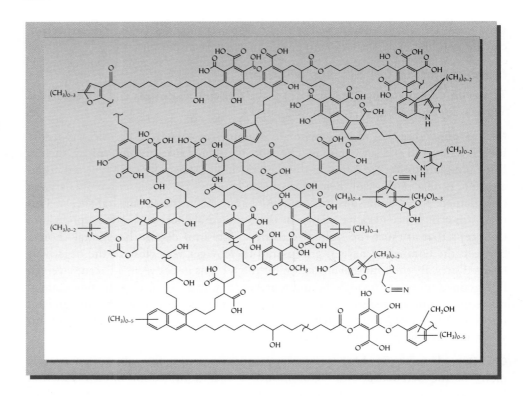

more advanced state of weathering than that which produces allophane. Both types of minerals have a pronounced capacity to strongly retain anions as well as to bind with humus, protecting it from decomposition.

Humus

All about humic substances: www.ar.wroc.pl/~weber/ humic.htm#start

Humus is a noncrystalline organic substance consisting of very large organic molecules whose chemical composition varies considerably, but generally contains 40 to 60% C, 30 to 50% O, 3 to 7% H, and 1 to 5% N. The molecular weights of humic acids, a major type of colloidal humus, range from 10,000 to 100,000 g/mol. Identification of the actual structure of humus colloids is very difficult. A proposed structure typical of humic acid is shown in Figure 8.8. Note that it contains a very complex series of carbon chains and ring structures, with numerous chemically active functional groups throughout. Figure 8.9 provides a simplified diagram to illustrate

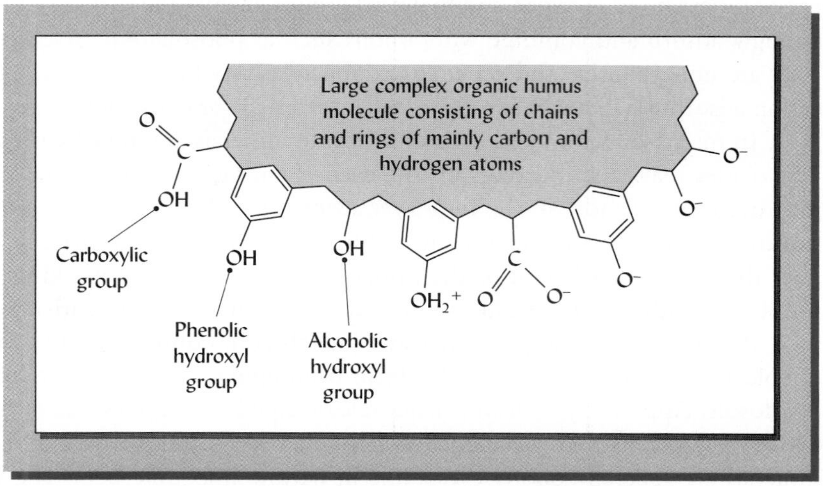

Figure 8.9

*A simplified diagram showing the principal chemical groups responsible for the high amount of negative charge on humus colloids. The three groups highlighted all include –OH that can lose its hydrogen ion by dissociation and thus become negatively charged. Note that the **carboxylic**, **phenolic**, and **alcoholic** groups on the right side of the diagram are shown in their disassociated state, while those on the left side still have their associated hydrogen ions. Note also that association with a second hydrogen ion causes a site to exhibit a net positive charge.*

(Diagram courtesy of R. Weil)

Large complex organic humus molecule consisting of chains and rings of mainly carbon and hydrogen atoms

Carboxylic group

Phenolic hydroxyl group

Alcoholic hydroxyl group

the three main types of —OH groups thought to be responsible for the high amount of charge associated with these colloids. Negative or positive charges on the humus colloid develop as H^+ ions are either lost or gained by these groups. Both cations and anions are therefore attracted to and adsorbed by the humus colloid. The negative sites always outnumber the positive ones, and a very large *net* negative charge is associated with humus (Table 8.1).

Because of its great surface area and many hydrophilic (water-loving) groups, humus can adsorb very large amounts of water per unit mass. However, humus also contains many hydrophobic sites and therefore can strongly adsorb a wide range of hydrophobic, nonpolar organic compounds (see Section 8.12). Because of its extraordinary influence on soil properties and behavior, we will delve much more deeply into the nature and function of soil humus in Chapter 11.

8.5 GENESIS AND GEOGRAPHIC DISTRIBUTION OF SOIL COLLOIDS

Genesis of Colloids

The breakdown and alteration of plant residues by microorganisms and the concurrent synthesis of new, more stable, organic compounds results in the formation of the dark-colored colloidal organic material called *humus* (see Section 11.4 for details).

The silicate clays are developed from the weathering of a wide variety of minerals by at least two distinct processes: (1) a slight physical and chemical **alteration** of certain primary minerals, and (2) a **decomposition** of primary minerals with the subsequent **recrystallization** of certain of their products into the silicate clays.

Relative Stages of Weathering   Specific conditions conducive to the formation of important clay types are shown in Figure 8.10. Note that fine-grained micas and magnesium-rich chlorites represent earlier weathering stages of the silicates, and kaolinite and (ultimately) iron and aluminum oxides the most advanced stages. The smectites (e.g., montmorillonite) represent intermediate stages. Different weathering stages may occur across climatic zones or across horizons within a single profile. As noted in Section 2.1, silicon tends to be lost as weathering progresses, leaving a lower Si:Al ratio in more highly weathered soil horizons.

Mixed and Interstratified Layers   In a given soil, it is common to find several silicate clay minerals in an intimate mixture. In fact, the properties and compositions of some mineral colloids are intermediate between those of the well-defined minerals described in Section 8.3. For example, a **mixed layer** or an **interstratified** clay mineral in which some layers are more like mica and some more like vermiculite might be called *fine-grained mica-vermiculite*.

Distribution of Clays by Geography and Soil Order

The clay of any particular soil is generally made up of a mixture of different colloidal minerals. In a given soil, the mixture may vary from horizon to horizon, because the kind of clay that develops depends not only on climatic influences and profile conditions but also on the nature of the parent material. The situation may be further complicated by the presence in the parent material itself of clays that were formed under a preceding and perhaps an entirely different type of climatic regime. Nevertheless, some broad generalizations are possible.

The well-drained and highly weathered Oxisols and Ultisols of warm humid and subhumid tropics tend to be dominated by kaolinite, along with oxides of iron and aluminum. The smectite, vermiculite, and fine-grained mica groups are more

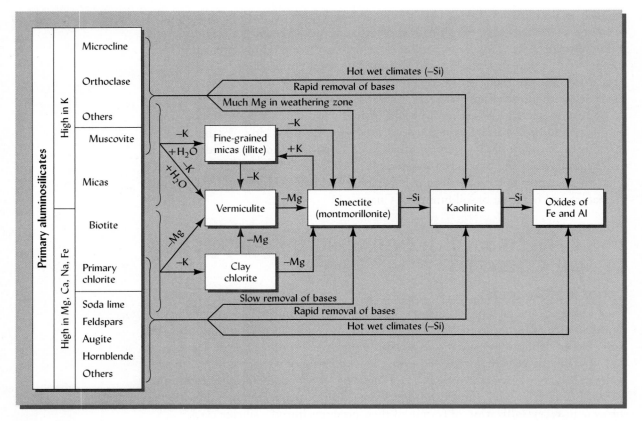

Figure 8.10

General conditions for the formation of the various layer silicate clays and oxides of iron and aluminum. Fine-grained micas, chlorite, and vermiculite are formed through rather mild weathering of primary aluminosilicate minerals, whereas kaolinite and oxides of iron and aluminum are products of much more intense weathering. Conditions of intermediate weathering intensity encourage the formation of smectite. In each case silicate clay genesis is accompanied by the removal in solution of such elements as K, Na, Ca, and Mg. Several members of this weathering series may be present in a single soil profile, with the less weathered clay in the C horizon and the more weathered clay minerals in the B or A horizons.

prominent in Alfisols, Mollisols, and Vertisols, where weathering is less intense. Where the parent material is high in micas, fine-grained micas such as illite are apt to be formed. Parent materials that are high in metallic cations (particularly magnesium) or are subject to restricted drainage, which discourages the leaching of these cations, encourage smectite formation.

8.6   SOURCES OF CHARGES ON SOIL COLLOIDS

There are two major sources of charges on soil colloids: (1) hydroxyls and other functional groups on the surfaces of the colloidal particles that by releasing or accepting H^+ ions can provide either negative or positive charges, and (2) the charge imbalance brought about by the isomorphous substitution in some clay crystal structures of one cation by another of similar size but differing in charge.

All colloids, organic and inorganic, exhibit the surface charges associated with OH^- groups, charges that are largely **pH dependent**. Most of the charges associated with humus, 1:1-type clays, the oxides of iron and aluminum, and allophane are of this type (see Figures 8.9 and 8.11 for examples in humus and kaolinite). In the case of the 2:1-type clays, however, these surface charges are complemented by a much larger number of charges emanating from the isomorphous substitution of one cation for another in the

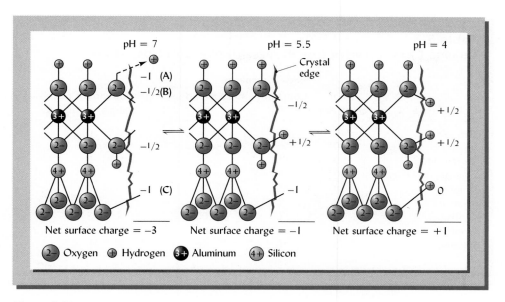

Figure 8.11

How pH-dependent charges develop at the broken edge of a kaolinite crystal. Three sources of net negative surface charge at a high pH are illustrated (left): (A) One (−1) charge from octahedral oxygen that has lost its H^+ ion by dissociation (the H broke away from the surface hydroxyl group and escaped into the soil solution). Note that such dissociation can generate negative charges all along the surface hydroxyl plane, not just at a broken edge. (B) One half (−1/2) charge from each octahedral oxygen that would normally be sharing its electrons with a second aluminum. (C) One (−1) charge from a tetrahedral oxygen atom that would normally be balanced by bonding to another silicon if it were not at the broken edge. The middle and right diagrams show the effect of acidification (lowering the pH), which increases the activity of H^+ ions in the soil solution. At the lowest pH shown (right), all of the edge oxygens have an associated H^+ ion, giving rise to a net positive charge on the crystal. These mechanisms of charge generation are similar to those illustrated for humus in Figure 8.9.

octahedral and/or tetrahedral sheets. Since these charges are not dependent on the pH, they are termed **permanent** or **constant charges**. An example of how these changes arise is given in Figure 8.4. Isomorphous substitution can also be a source of positive charges if the substituting cation has a higher charge than the ion for which it substitutes. The net charge of colloids is the balance between the negative and positive charges. In all 2:1-type silicate clays, however, the **net charge** is negative since those substitutions leading to negative charges far outweigh those producing positive charges (see Figure 8.12).

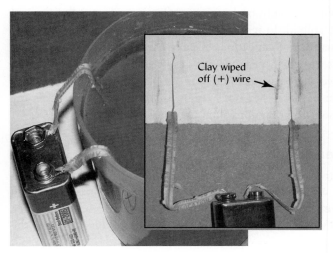

Figure 8.12

Simple demonstration of the negatively charged nature of clay. Wires connected to the (−) and (+) terminals of a 9-volt battery are dipped for a few minutes in a suspension of clayey soil in water. The wires are then wiped on a piece of paper (inset), showing that the wire on the (+) terminal has attracted the clay while the (−) wire has not. [Adapted from Weil (2009)]

Table 8.3

CHARGE CHARACTERISTICS OF REPRESENTATIVE COLLOIDS, AT pH7

Colloid type	Negative charge			pH-dependent positive charge, $cmol_c/kg$
	Total $cmol_c/kg$	Constant, %	pH dependent, %	
Organic	200	10	90	0
Smectite	100	95	5	0
Vermiculite	150	95	5	0
Fine-grained micas	30	80	20	0
Chlorite	30	80	20	0
Kaolinite	8	5	95	2
Gibbsite (Al)	4	0	100	3
Goethite (Fe)	4	0	100	3
Allophane	30	10	90	15

The charge characteristics of selected soil colloids are shown in Table 8.3. Note the high percentage of constant negative charges in some 2:1-type clays (e.g., smectites and vermiculites). Humus, kaolinite, allophane, and Fe, Al oxides have mostly variable (pH-dependent) negative charges and exhibit modest positive charges at low pH values. The negative and positive charges on soil colloids are of vital importance to the behavior of soils in nature, especially with regard to the adsorption of oppositely charged ions from the soil solution. This subject will be taken up next.

8.7 ADSORPTION OF CATIONS AND ANIONS

In soil, the negative and positive surface charges on the colloids attract and hold a complex swarm of cations and anions. Table 8.4 lists some important cations and anions. The adsorption of these ions by soil colloids greatly affects their biological availability and mobility, thereby influencing both soil fertility and environmental quality. Note that the soil solution and colloidal surfaces in most soils are dominated mainly by just a few of the cations and anions, the others being found in much smaller amounts or only in special situations such as contaminated soils. In Figure 8.1, ion adsorption was illustrated in a simplified manner, showing positive cations held on the negatively charged surfaces of a soil colloid. Actually, both cations and anions are usually attracted to different sites on the same colloid. In temperate-region soils, anions are commonly adsorbed in much smaller quantities than cations because these soils generally contain mainly 2:1-type silicate clays on which negative charges predominate. In the tropics, where soils are more highly weathered, acid, and rich in 1:1 clays and Fe, Al oxides, the amount of negative charge on the colloids is not so high, and positive charges are more abundant. Therefore, the adsorption of anions is more prominent in these soils.

Outer- and Inner-Sphere Complexes

Remembering that water molecules surround (hydrate) the cations and anions in the soil solution, we can visualize that in an **outer-sphere complex** water molecules form a bridge between the adsorbed ion and the charged colloid surface (Figure 8.13). Sometimes several layers of water molecules are involved. Thus, the

Table 8.4
SELECTED CATIONS AND ANIONS COMMONLY ADSORBED TO SOIL COLLOIDS AND IMPORTANT IN PLANT NUTRITION AND ENVIRONMENTAL QUALITY

The listed ions form inner- and/or outer-sphere complexes with soil colloids. Ions marked by an asterisk (*) are among those that predominate in most soil solutions. Many other ions may be important in certain situations.

Cation	Formula	Comments	Anion	Formula	Comments
Ammonium	NH_4^+	Plant nutrient	Arsenate	AsO_4^{3-}	Toxic to animals
Aluminum	Al^{3+} etc.[a]	Toxic to many plants	Bicarbonate	HCO_3^-	Toxic in high-pH soils
Cadmium	Cd^{2+}	Toxic pollutant	Borate	$B(OH)_4^-$	Plant nutrient, can be toxic
Calcium*	Ca^{2+}	Plant nutrient	Carbonate*	CO_3^{2-}	Forms weak acid
Cesium	Cs^+	Radioactive contaminant	Chloride*	Cl^-	Plant nutrient, toxic in large amounts
Copper	Cu^{2+}	Plant nutrient, toxic pollutant	Chromate	CrO_4^{2-}	Toxic pollutant
Hydrogen*	H^+	Causes acidity	Fluoride	Fl^-	Toxic, natural and pollutant
Iron	Fe^{2+}	Plant nutrient	Hydroxyl*	OH^-	Alkalinity factor
Lead	Pb^{2+}	Toxic to animals, plants	Molybdate	MoO_4^{2-}	Plant nutrient, can be toxic
Magnesium*	Mg^{2+}	Plant nutrient	Nitrate*	NO_3^-	Plant nutrient, pollutant in water
Manganese	Mn^{2+}	Plant nutrient	Phosphate	HPO_4^{2-}	Plant nutrient, water pollutant
Nickel	Ni^{2+}	Plant nutrient, toxic pollutant	Selenate	SeO_4^{2-}	Animal nutrient and toxic pollutant
Potassium*	K^+	Plant nutrient	Selenite	SeO_3^{2-}	Animal nutrient and toxic pollutant
Sodium*	Na^+	Used by animals, some plants, can damage soil	Silicate*	SiO_4^{4-}	Mineral weathering product, used by plants
Strontium	Sr^{2+}	Radioactive contaminant	Sulfate*	SO_4^{2-}	Plant nutrient
Zinc	Zn^{2+}	Plant nutrient, toxic pollutant	Sulfide	S^{2-}	In anaerobic soils, forms acid on oxidation

[a] Important aluminum cations include Al^{3+}, $AlOH^{2+}$, and $Al(OH)_2^+$.

ion itself never comes close enough to the colloid surface to form a bond with a specific charged site. Instead, the ion is only weakly held by electrostatic attraction, the charge on the oscillating hydrated ion balancing, in a general way, an excess charge of opposite sign on the colloid surface. Ions in an outer-sphere complex are therefore easily replaced by other similarly charged ions.

In contrast, adsorption via formation of an **inner-sphere complex** does *not* involve any intervening water molecules. Therefore, one or more direct bonds are formed between the adsorbed ion and the atoms in the colloid surface. One example already discussed is the case of the K^+ ions that fit so snugly into the spaces between silicon tetrahedra in a mica crystal. Similarly, strong inner-sphere complexes may be formed by reactions of Cu^{2+} or Ni^{2+} with the oxygen atoms in silica tetrahedra. Another important example, this time involving an anion, occurs when a $H_2PO_4^+$ ion is directly bonded by shared electrons with the octahedral aluminum in the colloid structure (Figure 8.13). Other ions cannot easily replace an ion held in an inner-sphere complex because this type of adsorption involves relatively strong bonds that are dependent on the compatible nature of specific ions and specific sites on the colloid.

Figure 8.13 illustrates only two examples of adsorption complexes on one type of colloid. In other colloids, charged silica tetrahedral surfaces form inner- and outer-sphere complexes by mechanisms similar to those shown in the figure. Permanent charges from isomorphous substitution in the interior structure of a colloid (not shown in Figure 8.13) can also cause adsorption of outer-sphere complexes.

Figure 8.13

*A diagrammatic representation of t he adsorption of ions on a colloid by the formation of outer-sphere and inner-sphere complexes. (1) Water molecules surround diffuse cations and anions (such as the Mg^{2+}, Cl^-, and HPO_4^- shown) in the soil solution. (2) In an **outer-sphere complex** (such as the adsorbed Ca^{2+} ion shown), water molecules form a bridge between the adsorbed cation and the charged colloid surface. (3) In the case of an **inner-sphere complex** (such as the adsorbed $H_2PO_4^-$ anion shown), no water molecules intervene, and the cation or anion binds directly with the metal atom (aluminum in this case) in the colloid structure.* (Diagram courtesy of R. Weil)

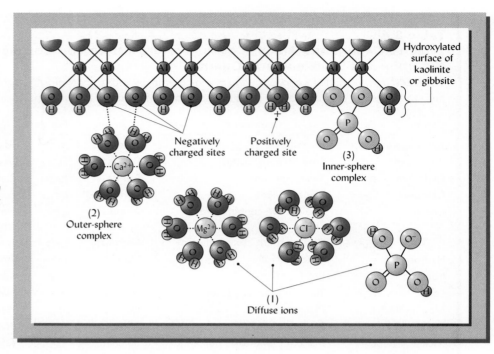

8.8   CATION EXCHANGE REACTIONS

Animation explaining cation exchange:
http://hintze-online.com/sos/1997/Articles/Art5/animat2.dcr

Let us consider the case of an outer-sphere complex between a negatively charged colloid surface and a hydrated cation (such as the Ca^{2+} ion shown in Figure 8.13). Such an outer-sphere complex is only loosely held together by electrostatic attraction, and the adsorbed ion remains in constant motion near the colloid surface. There are moments (microseconds) when the adsorbed cation is located a bit farther than average from the colloid surface. This moment provides an opportunity for another hydrated cation from the soil solution (say the Mg^{2+} ion shown in Figure 8.13) to diffuse into a position a bit closer to the negative site on the colloid. The instant this occurs, the second ion would replace the first ion, freeing the formerly adsorbed ion to diffuse out into the soil solution. In this manner, an exchange of cations between the adsorbed and diffuse state takes place. As mentioned in Section 8.1, this process is referred to as **cation exchange**. If a hydrated anion similarly replaces another hydrated anion at a positively charged colloid site, the process is called **anion exchange**. The ions held in outer-sphere complexes from which they can be replaced by exchange reactions are said to be **exchangeable** cations or anions. As a group, all the colloids in a soil, inorganic and organic, capable of holding exchangeable cations or anions are termed the cation or anion **exchangeable complex**.

Principles Governing Cation Exchange Reactions

Reversibility   We can illustrate the process of cation exchange using a simple reaction in which a hydrogen ion (perhaps generated by organic matter decomposition—see Section 9.1) displaces a sodium ion from its adsorbed state on a colloid surface:

$$\boxed{Colloid}\ Na^+ + H^+ \underset{\text{(soil solution)}}{\rightleftharpoons} \boxed{Colloid}\ H^+ + Na^+ \qquad (8.1)$$

The reaction takes place rapidly and, as shown by the double arrows, the reaction is reversible. It will go to the left if sodium is added to the system. This reversibility is a fundamental principle of cation exchange.

Charge Equivalence   Another basic principle of cation exchange reactions is that the exchange is chemically equivalent; that is, it takes place on a *charge-for-charge* basis. Therefore, although one H^+ ion exchanged with *one* Na^+ ion in the reaction just shown, it would require *two* singly charged H^+ ions to exchange with or replace *one* divalent Ca^{2+} ion. If the reaction is reversed, one Ca^{2+} ion will displace two H^+ ions. In other words, two charges from one cation species replace two charges from the other:

$$\boxed{Colloid}\ Ca^{2+} + 2\ H^+ \rightleftharpoons \boxed{Colloid}\ \begin{matrix} H^+ \\ H^+ \end{matrix} + Ca^{2+} \qquad (8.2)$$
$$\text{(soil solution)} \qquad\qquad\qquad\qquad \text{(soil solution)}$$

Note that by this principle it would require three Na^+ ions to replace a single Al^{3+} ion, and so on.

Ratio Law   Consider an exchange reaction between two similar cations, say Ca^{2+} and Mg^{2+}. If there are a large number of Ca^{2+} ions adsorbed on a colloid and some Mg^{2+} is added to the soil solution, the added Mg^{2+} ions will begin displacing the Ca^{2+} from the colloid. This will bring more Ca^{2+} into the soil solution and these Ca^{2+} ions will, in turn, displace some of the Mg^{2+} back off the colloid. The *ratio law* tells us that, at equilibrium, the ratio of Ca^{2+} to Mg^{2+} on the colloid will be the same as the ratio of Ca^{2+} to Mg^{2+} in the solution and both will be the same as the ratio in the overall system. To illustrate this concept, assume that 20 Ca^{2+} ions are initially adsorbed on a soil colloid and 5 Mg^{2+} ions are added to the system:

$$\boxed{Colloid}\ 20\ Ca^{2+} + 5\ Mg^{2+} \rightleftharpoons \boxed{Colloid}\ \begin{matrix} 16\ Ca^{2+} \\ 4\ Mg^{2+} \end{matrix} + 1\ Mg^{2+} + 4\ Ca^{2+} \qquad \text{Ratio: 4 Ca:1 Mg} \quad (8.3)$$
$$\text{(soil solution)} \qquad\qquad\qquad\qquad\qquad \text{(soil solution)}$$

If the two exchanging ions are not of the same charge (e.g., K^+ exchanging with Mg^{2+}), the reaction becomes somewhat more complicated and a modified version of the ratio law would apply.

    Up to this point, our discussion of exchange reactions has assumed that both ionic species (elements) exchanging places take part in the exchange reaction in exactly the same way. This assumption must be modified to take into account three additional factors if we are to understand how exchange reactions actually proceed in nature.

Anion Effects on Mass Action   The laws of **mass action** tell us that an exchange reaction will be more likely to proceed to the right if the released ion is prevented from reacting in the reverse direction. This may be accomplished if the released cation on the right side of the reaction either *precipitates*, *volatilizes*, or *strongly associates* with an anion. To illustrate this concept, consider the displacement of H^+ ions on an acid colloid by Ca^{2+} ions added to the soil solution as calcium carbonate:

$$\boxed{Colloid}\ \begin{matrix} H^+ \\ H^+ \end{matrix} + CaCO_3 \rightleftharpoons \boxed{Colloid}\ Ca^{2+} + H_2O + CO_2\uparrow \qquad (8.4)$$
$$\text{(added)} \qquad\qquad\qquad\qquad \text{water} \quad \text{(gas)}$$

Where $CaCO_3$ is added, when a hydrogen ion is displaced off the colloid, it combines with an oxygen atom from the $CaCO_3$ to form water. Furthermore, the CO_2 produced is a gas, which can volatilize out of the solution and leave the system. The removal of these products pulls the reaction to the right. This principle explains why $CaCO_3$ (in the form of limestone) is effective in neutralizing an acid soil, while calcium chloride is not (see Section 9.8).

Cation Selectivity   Until now, we have assumed that both cation species taking part in the exchange reaction are held with equal tenacity by the colloid and therefore have an equal chance of displacing each other. In reality, some cations are held much more

Detailed tutorial on CEC, M. J. Eick, Virginia Tech: www.soils1.cses.vt.edu/ MJE/shockwave/cec_demo/ version1.1/cec.shtml

tightly than others and so are less likely to be displaced from the colloid. In general, the higher the charge and the smaller the *hydrated radius* of the cation, the more strongly it will adsorb to the colloid. The order of strength of adsorption for commonly held cations is:

$$Al^{3+} > Sr^{2+} > Ca^{2+} > Mg^{2+} > Cs^+ > K^+ = NH_4^+ > Na^+ > Li^+$$

The less tightly held cations oscillate farther from the colloid surface and therefore are the most likely to be displaced into the soil solution and carried away by leaching. This series therefore explains why the soil colloids are dominated by Al^{3+} (and other aluminum ions) and Ca^{2+} in humid regions and by Ca^{2+} in drier regions, even though the weathering of minerals in many parent materials provides relatively larger amounts of K^+, Mg^{2+}, and Na^+ (see Section 8.10). The strength of adsorption of the H^+ ion is difficult to determine because hydrogen-dominated mineral colloids break down to form aluminum-saturated colloids.

The relative strengths of adsorption may be altered on certain colloids whose properties favor adsorption of particular cations. An important example of such colloidal "preference" for specific cations is the very high affinity for K^+ ions (and the similarly sized NH_4^+ and Cs^+ ions) exhibited by vermiculite and fine-grained micas (Section 8.3), which attract these ions to inter-tetrahedral spaces exposed at weathered crystal edges. The influence of different colloids on the adsorption of specific cations impacts the availability of cations for leaching or plant uptake. Certain metals such as copper, mercury, and lead have very high selective affinities for sites on humus and iron oxide colloids, making most soils quite efficient at removing these potential pollutants from water leaching through the profile.

Complementary Cations   In soils, colloids are always surrounded by many different adsorbed cation species. The likelihood that a given adsorbed cation will be displaced from a colloid is influenced by how strongly its neighboring cations are adsorbed to the colloid surface. For example, consider an adsorbed Mg^{2+} ion. An ion diffusing in from the soil solution is more likely to displace one of the neighboring ions rather than the Mg^{2+} ion, if the neighboring adsorbed ions (sometimes called **complementary ions**) are loosely held. If they are tightly held, then the chances are greater that the Mg^{2+} ion will be displaced. In Section 8.10, we shall discuss the influence of complementary ions on the availability of nutrient cations for plant uptake.

8.9   CATION EXCHANGE CAPACITY

Previous sections have dealt qualitatively with exchange reactions. We now turn to a consideration of the quantitative **cation exchange capacity (CEC)**. The CEC is expressed as the number of centimoles of positive charge ($cmol_c$) that can be adsorbed per unit mass. A particular soil may have a CEC of 15 $cmol_c$/kg, indicating that 1 kg of the soil can hold 15 $cmol_c$ of H^+ ions, for example, and can exchange this number of charges from H^+ ions for the same number of charges from any other cation. This means of expression emphasizes that exchange reactions take place on a charge-for-charge (not an ion-for-ion) basis. The concept of a centimole of charges and its use in CEC calculations are reviewed in Box 8.2.

Methods of Determining CEC

The CEC is an important soil chemical property that is used for classifying soils in *Soil Taxonomy* (e.g., in defining Oxic, Mollic, and Kandic diagnostic horizons, Section 3.2) and for assessing their fertility and environmental behavior.

Different methods can be used to determined CEC, and they can give different results for the same soil. Several commonly used CEC procedures call for use of a

---

BOX 8.2
CHEMICAL EXPRESSION OF CATION EXCHANGE

One mole of any atom, molecule, or charge is defined as 6.02×10^{23} (Avogadro's number) of atoms, molecules, or charges. Thus, 6.02×10^{23} negative charges associated with the soil colloidal complex would attract 1 mole of positive charge from adsorbed cations such as Ca^{2+}, Mg^{2+}, and H^+. The number of moles of the positive charge provided by the adsorbed cations in any soil gives us a measure of the cation exchange capacity (CEC) of that soil.

The CEC of soils commonly varies from 0.03 to 0.5 mole of positive charge per kilogram (3 to 50 $cmol_c$/kg).

Using the mole concept, it is easy to relate the moles of charge to the mass of ions or compounds involved in cation or anion exchange. Consider, for example, the exchange that takes place when adsorbed sodium ions in an alkaline arid-region soil are replaced by hydrogen ions:

$$\boxed{Colloid}\ Na^+ + H^+ \;\rightleftharpoons\; \boxed{Colloid}\ H^+ + Na^+$$

If 1 $cmol_c$ of adsorbed Na^+ ions per kilogram of soil were replaced by H^+ ions in this reaction, how many grams of Na^+ ions would be replaced?

Since the Na^+ ion is singly charged, the mass of Na^+ needed to provide 1 mole of charge (1 mol_c) is the gram atomic weight of sodium, or 23 g (see periodic table in Appendix B). The mass providing 1 *centimole* of charge ($cmol_c$) is 1/100 of this amount; thus, the mass of the 1 $cmol_c$ Na^+ replaced is 0.23 g Na^+/kg soil. The 0.23 g Na^+ would be replaced by only 0.01 g H, which is the mass of 1 $cmol_c$ of this much lighter element.

Another example is the replacement of H^+ ions when hydrated lime [$Ca(OH)_2$] is added to an acid soil. This time, assume that 2 $cmol_c$ H^+/kg soil is replaced by the $Ca(OH)_2$, which reacts with the acid soil as follows:

$$\boxed{Colloid}\ {}^{H^+}_{H^+} + Ca(OH)_2 \;\rightleftharpoons\; \boxed{Colloid}\ Ca^{2+} + 2H_2O$$

Since the Ca^{2+} ion in each molecule of $Ca(OH)_2$ has two positive charges, the mass of $Ca(OH)_2$ needed to replace 1 mole of charge from the H^+ ions is only one-half of the gram molecular weight of this compound, or $74/2 = 37$ g. A comparable figure for 1 *centimole* is 37/100, or 0.37 grams. Twice this amount, or 0.74 g of $Ca(OH)_2$, would be needed to replace the 2 $cmol_c$H, replaced in the reaction shown above.

In each preceding example, the number of charges provided by the replacing ion is equivalent to the number associated with the ion being replaced. Thus, 1 mole of negative charges attracts 1 mole of positive charges whether the charges come from H^+, K^+, Na^+, NH_4^+, Ca^{2+}, Mg^{2+}, Al^{3+}, or any other cation. Keep in mind, however, that only one-half the atomic weights of divalent cations, such as Ca^{2+} or Mg^{2+}, and only one-third the atomic weight of trivalent Al^{3+} are needed to provide 1 mole of charge. This *chemical equivalency* principle applies to both cation and anion exchange.

---

solution buffered to maintain a certain pH (usually either pH 7.0 using ammonium as the exchanger cation or pH 8.2 using barium as the exchanger cation). If the native soil pH is less than the pH of the buffered solution, then these methods measure not only the cation exchange sites active at the pH of the particular soil, but also any pH-dependent exchange sites (see Section 8.6) that would become negatively charged at pH 7.0 or 8.2.

Alternatively, the CEC procedure may use unbuffered solutions to allow the exchange to take place at the actual pH of the soil. The buffered methods (NH_4^+ at pH 7.0 or Ba^{2+} at pH 8.2) measure the *potential* or *maximum* cation exchange capacity of a soil. The unbuffered method measures only the **effective cation exchange capacity** (ECEC), which can hold exchangeable cations at the pH of the soil as sampled. As the different methods may yield significantly different values of CEC, it is important that the method used be known when comparing soils based on their CEC. This is especially significant if the soil pH is much below the buffer pH used.

Cation Exchange Capacities of Soils

The cation exchange capacity (CEC) of a given soil sample is determined by the relative amounts of different colloids in that soil and by the CEC of each of these colloids. Figure 8.14 illustrates the common range in CEC among different soils and

Figure 8.14

Ranges in the cation exchange capacities (at pH 7) that are typical of a variety of soils and soil materials. The high CEC of humus shows why this colloid plays such a prominent role in most soils and especially those high in kaolinite and Fe, Al oxides, and clays that have low CECs. (Diagram courtesy of R. Weil)

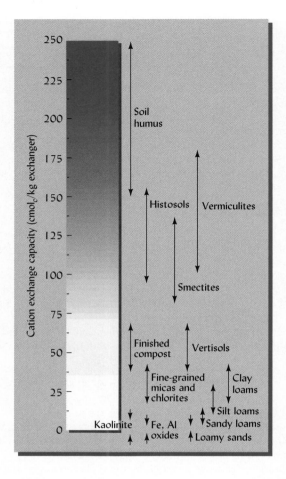

other organic and inorganic exchange materials. Note that sandy soils, which are generally low in all colloidal material, have low CECs compared to those exhibited by silt loams and clay loams. Also note the very high CECs associated with humus compared to those exhibited by the inorganic clays, especially kaolinite and Fe, Al oxides. The CEC coming from humus generally plays a prominent role, and sometimes a dominant one, in cation exchange reactions in A horizons. Using the CEC range from Figure 8.14 it is possible to estimate the CEC of a soil if the quantities of the different soil colloids in the soil are known (see Box 8.3).

pH and Cation Exchange Capacity

In previous sections, it was pointed out that the cation exchange capacity of most soils increases with pH. At very low pH values, only the permanent charges of the 2:1-type clays (see Section 8.8) and a small portion of the pH-dependent charges of organic colloids, allophane, and some 1:1-type clays hold exchangeable cations. As the pH is raised, the negative charges on some 1:1-type silicate clays, allophane, humus, and even Fe, Al oxides become more numerous, thereby increasing the cation exchange capacity (Figure 8.15).

8.10 EXCHANGEABLE CATIONS IN FIELD SOILS

The specific exchangeable cations associated with soil colloids differ from one climatic region to another—Ca^{2+}, Al^{3+}, complex aluminum hydroxy ions, and H^+ being most prominent in humid regions, and Ca^{2+}, Mg^{2+}, and Na^+ dominating in low-rainfall

---

BOX 8.3

ESTIMATING CEC FROM OTHER SOIL PROPERTIES

Data on cation exchange capacity are rather time-consuming to obtain and not always available. Fortunately, one can often estimate the CEC if other data on soil taxonomy, soil pH, clay content, and organic matter (OM) level are available.

Assume you know that a cultivated Mollisol in Iowa contains 20% clay and 4% organic matter (OM) and its pH = 7.0. The dominant clays in Mollisols are likely 2:1 types such as vermiculite and smectite. We estimate the average CEC of the clays of these types to be about 100 $cmol_c$/kg clay (Tables 8.1 and 8.3). At pH 7.0, the CEC of OM is about 200 $cmol_c$/kg (Table 8.3). Since 1 kg of this soil has 0.20 kg (20%) of clay and 0.04 kg (4%) of OM, we can calculate the CEC associated with each of these sources:

From the clays in this Mollisol:	$0.2 \text{ kg} \times 100 \text{ cmol}_c/\text{kg} = 20 \text{ cmol}_c$
From the OM in this Mollisol:	$0.04 \text{ kg} \times 200 \text{ coml}_c/\text{kg} = 8 \text{ cmol}_c$
The total CEC of this Mollisol:	$28 + 8 = 28 \text{ cmol}_c/\text{kg soil}$

---

areas (Table 8.5). The cations that dominate the exchange complex have a marked influence on soil properties.

In a given soil, the proportion of the cation exchange capacity satisfied by a particular cation is termed the **saturation percentage** for that cation. Thus, if 50% of the CEC is satisfied by Ca^{2+} ions, the exchange complex is said to have a **calcium saturation percentage** of 50.

This terminology is especially useful in identifying the relative proportions of sources of acidity and alkalinity in the soil solution. Thus, the percentage saturation with Al^{3+} and H^+ ions gives an indication of the acid conditions, while increases in the percentage **nonacid cation saturation** (sometimes referred to as the **base saturation percentage**[2]) indicate the tendency toward neutrality and alkalinity. These relationships will be discussed further in Chapter 9.

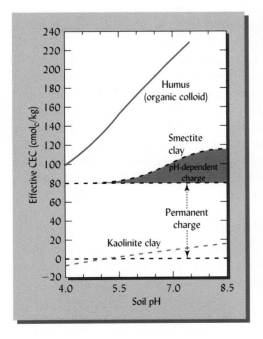

Figure 8.15

Influence of pH on representative cation exchange capacities of two clay minerals and humus. Below pH 6.0 the charge for smectite has a fairly constant charge that is due mainly to isomorphic substitution and is considered permanent. Above pH 6.0 the charge on smectite increases somewhat with pH (shaded area) because of the ionization of hydrogen from the exposed hydroxyl groups at crystal edges. In contrast, the charges on kaolinite and humus are all variable, increasing with increasing pH. Humus carries a far greater number of charges than kaolinite. At low pH kaolinite carries a net negative CEC because its positive charges outnumber the negative ones.

---

[2]Technically speaking, nonacid cations such as Ca^{2+}, Mg^{2+}, K^+, and Na^+ are not bases. When adsorbed by soil colloids in the place of H^+ ions, however, they reduce acidity and increase the soil pH. For that reason, they are traditionally referred to as *bases* and the portion of the CEC that they satisfy is often termed *base saturation percentage*.

Table 8.5

SOME EXCHANGEABLE IONS, THEIR HYDRATED SIZE, AND THEIR EXPECTED REPLACEMENT BY NH_4^+ IONS

Among ions of a given charge, the larger the hydrated radius, the more easily it is replaced.

Element	Ion	Hydrated ionic radius, (nm)[a]	Likely replacement of ion initially saturating a kaolinite clay if $cmol_c$ NH_4^+ added = CEC of the soil, (%)[b]
Lithium	Li^+	1.00	80
Sodium	Na^+	0.79	67
Ammonium	NH_4^+	0.54	50
Potassium	K^+	0.53	49
Rubidium	Rb^+	0.51	48
Cesium	Cs^+	0.50	47
Magnesium	Mg^{2+}	1.08	31
Calcium	Ca^{2+}	0.96	29
Strontium	Sr^{2+}	0.96	29
Barium	Ba^{2+}	0.88	26

[a]Not to be confused with nonhydrated radii (Table 8.2), hydrated radii are from Evangelou and Phillips (2005).

[b]Based on empirical data from various sources and assumes no special affinity by kaolinite for any of the listed ions.

Cation Saturation and Nutrient Availability

Exchangeable cations generally are available to both higher plants and microorganisms. By cation exchange, hydrogen ions from the root hairs and microorganisms replace nutrient cations from the exchange complex. The nutrient cations are forced into the soil solution, where they can be assimilated by the adsorptive surfaces of roots and soil organisms, or they may be removed by drainage water. Cation exchange reactions affecting the mobility of organic and inorganic pollutants in soils will be discussed in Section 8.12. Here we focus on the plant nutrition aspects.

The percentage saturation of essential nutrient cations such as calcium and potassium greatly influences the uptake of these elements by growing plants. For example, if the percentage calcium saturation of a soil is high, the displacement of this cation is comparatively easy and rapid. Thus, 6 $cmol_c$/kg of exchangeable calcium in a soil whose exchange capacity is 8 $cmol_c$/kg (75% calcium saturation) probably would mean ready availability, but 6 $cmol_c$/kg when the total exchange capacity of a soil is 30 $cmol_c$/kg (20% calcium saturation) would produce lower availability. This is one reason that, for calcium-loving plants such as alfalfa, the calcium saturation of at least part of the soil should approach 80 to 85%.

Influence of Complementary Cations

A second factor influencing plant uptake of a given cation is the effect of the complementary ions held on the colloids. As was discussed in Section 8.8, the strength of adsorption of common cations is in the following order for most colloids:

$$Al^{3+} > Sr^{2+} > Ca^{2+} > Mg^{2+} > Cs^+ > K^+ = NH_4^+ > Na^+ > Li^+$$

Consider, for example, the relatively loosely held ion, K^+. If the complementary ions surrounding a K^+ ion are held tightly (that is, they oscillate very close to the colloid surface), then a H^+ ion from a root is less likely to "find" a complementary ion and more likely, instead, to "bump into" and replace a K^+ ion (see Figure 8.16).

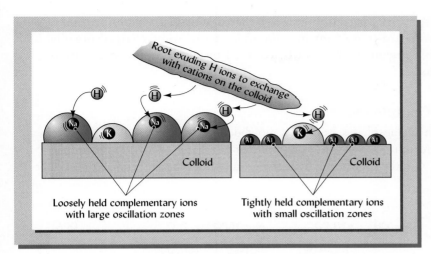

Figure 8.16

Effect of complementary ions on the availability of a particular exchangeable nutrient cation. The half spheres represent the zones in which the ion oscillates, the more loosely held ions moving within larger zones of oscillation. For simplicity, the water molecules that hydrate each ion are not shown. (Left) H$^+$ ions from the root are more likely to encounter and exchange with loosely held Na$^+$ ions rather than the more tightly held K$^+$ ion. (Right) The likelihood that H$^+$ ions from the root will encounter and exchange with a K$^+$ ion is increased by the inaccessibility of the neighboring tightly held Al^{3+} ions. The K$^+$ ion on the right colloid is comparatively more vulnerable to being replaced and sent into the soil solution and is therefore more available for plant uptake or leaching than the K$^+$ ion on the left colloid. (Diagram courtesy of R. Weil)

Plant uptake of one cation may also be reduced by excessive uptake of another cation. For instance, potassium uptake by plants is limited by high levels of calcium in some soils. Likewise, high potassium levels are known to limit the uptake of magnesium even when significant quantities of magnesium are present in the soil.

8.11   ANION EXCHANGE

Anions are held by soil colloids in two major ways. First, they are held by anion adsorption mechanisms similar to those responsible for cation adsorption. Second, they may actually react with surface oxides or hydroxides, forming more definitive **inner-sphere complexes**. We shall consider anion adsorption first.

The basic principles of **anion exchange** are similar to those of cation exchange, except that the charges on the colloids are positive and the exchange is among negatively charged anions. The positive charges associated with the surfaces of kaolinite, iron and aluminum oxides, and allophane attract anions such as SO_4^{2-} and NO_3^-. A simple example of an anion exchange reaction is as follows:

$$\boxed{\text{Colloid}} \; NO_3^- + Cl^- \rightleftharpoons \boxed{\text{Colloid}} \; Cl^- + NO_3^- \tag{8.5}$$

(positively charged   (soil solution)   (positively charged   (soil solution)
soil solid)   soil solid)

Just as in cation exchange, **equivalent** quantities of NO_3^- and Cl^- are exchanged, the reaction can be reversed, and plant nutrients so released can be absorbed by plants.

In contrast to cation exchange capacities, anion exchange capacities of soils generally *decrease* with increasing pH. In some very acid tropical soils that are high in kaolinite and iron and aluminum oxides, the anion exchange capacity may actually exceed the cation exchange capacity.

Anion exchange is very important in making anions available for plant growth while at the same time retarding the leaching of such anions from some soils. For example, anion exchange restricts the loss of sulfates from subsoils in the southern United States. Even the leaching of nitrate may be retarded by anion exchange in the subsoil of certain highly weathered soils of the humid tropics. Similarly, the downward movement into groundwater of some charged organic pollutants found in organic wastes can be retarded by such anion and/or cation exchange reactions.

Inner-Sphere Complexes

Some anions, such as phosphates, arsenates, molybdates, and sulfates, can react with particle surfaces, forming **inner-sphere complexes** (see Figure 8.13). For example, the ion may react with the protonated hydroxyl group rather than remain as an easily exchanged anion:

$$\overset{>}{\underset{\diagdown}{}}Al — OH_2^+ \;+\; H_2PO_4^- \;\longrightarrow\; \overset{>}{\underset{\diagdown}{}}Al — H_2PO_4 \;+\; H_2O \qquad\qquad (8.6)$$

(soil solid)   (soil solution)   (soil solid)   (soil solution)

This reaction actually reduces the net positive charge on the soil colloid. Also, the $H_2PO_4^-$ is held very tightly and is not readily available for plant uptake.

Anion adsorption and exchange reactions regulate the mobility and availability of many important ions. Together with cation exchange they largely determine the ability of soils to hold nutrients in a form that is accessible to plants and to retard movement of pollutants in the environment.

Weathering and CEC/AEC Levels

The range of CEC levels of different clay minerals shown in Figure 8.17 shows that clays developed under mild weathering conditions (e.g., smectites, vermiculites) have much higher CEC levels than those developed under more extreme weathering. In contrast, the AEC levels tend to be much higher in clays developed under strong weathering conditions (e.g., kaolinite) than in those found under more mild weathering. This generalized figure is helpful in obtaining a first approximation of CEC and AEC levels in soils of different climatic regions. It must be used with caution, however, since in some soils the type of clay present may reflect past rather than current weathering conditions.

Figure 8.17

The effect of weathering intensity on the charges on clay minerals and, in turn, on their cation and anion exchange capacities (CECs and AECs). Note the high CEC and very low AEC associated with mild weathering, which has encouraged the formation of 2:1-type clays such as fine-grained micas, vermiculites, and smectites. More intense weathering destroys the 2:1-type clays and leads to the formation of first kaolinite and then oxides of Fe and Al. These have much lower CECs and considerably higher AECs. Such changes in clay dominance account for the curves shown.

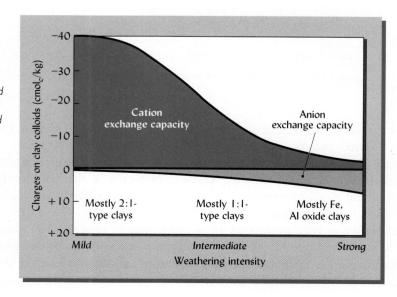

8.12   SORPTION OF ORGANIC COMPOUNDS

Soil colloids help control the movement of pesticides and other organic compounds into groundwater. The retention of these chemicals by soil colloids can prevent their downward movement through the soil or can delay that movement until the compounds are broken down by soil microbes.

By accepting or releasing protons (H^+ ions), groups such as $—OH$, $—NH_2$, and $—COOH$ in the chemical structure of some organic compounds provide positive or negative charges that stimulate anion or cation exchange reactions. Other organic compounds participate in inner-sphere complexation and adsorption reactions just as do the inorganic ions we have discussed. However, it is more common for organic colloids to be **absorbed** within the soil organic colloids by a process termed **partitioning**. The soil organic colloids tend to act as a solvent for organic chemicals, thereby partitioning their concentrations between those held on the soil colloids and those left in the soil solution.

Since we seldom know for certain the exact involvement of the adsorption, complexation, or partitioning processes, we use the general term **sorption** to describe the retention by soils of these organic compounds. Nonionic organic compounds are **hydrophobic**, meaning they are repelled by water. In moist soil, clay particles are coated with a layer of water and, therefore, cannot effectively sorb hydrophobic organic compounds. However, it is possible to replace some of the hydrated metal cations (e.g., Ca^{2+}) on such clays with large organic cations, giving rise to **organoclays** that *can* effectively retain hydrophobic organic compounds. Environmental soil scientists take advantage of this phenomenon by injecting quaternary ammonium compounds into contaminated clay soils to create organoclays that can stop the movement of organic contaminants and thus protect groundwater (see also Section 15.4).

Distribution Coefficients

The tendency of an organic contaminant to leach into the groundwater is determined by the solubility of the compound and by the ratio of the amount of chemical sorbed by the soil to that remaining in solution. This ratio is known as the **soil distribution coefficient**, K_d:

$$K_d = \frac{\text{mg chemical sorbed/kg soil}}{\text{mg chemical/L solution}} \tag{8.7}$$

The K_d therefore is typically expressed in units of L/kg. Researchers have found that the K_d for a given compound may vary widely depending on the nature of the soil in which the compound is distributed. The variation is related mainly to the amount of organic matter (organic carbon) in the soils. Therefore, most scientists prefer to use a similar ratio that focuses on sorption by organic matter. This ratio is termed the organic carbon distribution coefficient K_{oc}:

$$K_{oc} = \frac{\text{mg chemical sorbed/kg organic carbon}}{\text{mg chemical/L solution}} = \frac{K_d}{\text{g org. C/g soil}} \tag{8.8}$$

The K_{oc} can be calculated by dividing the K_d by the fraction of organic C (g/g) in the soil. Therefore, K_{oc} values are often about 100 times larger than K_d values. Higher K_d or K_{oc} values indicate that the chemical is more tightly sorbed by the soil and therefore less susceptible to leaching and movement to the groundwater. On the other hand, if the management objective is to wash the chemical out of a soil, this will be more easily accomplished for chemicals with lower coefficients. Equations 8.7 and 8.8 emphasize the importance of the sorbing power of the soil colloidal complex, and especially of humus, in the management of organic compounds added to soils.

8.13   BINDING OF BIOMOLECULES TO CLAY AND HUMUS

The enormous surface area and charged sites on the clay and humus in soils attract and bind many types of organic molecules. These molecules include such biologically active substances as DNA (genetic code material), enzymes, antibiotics, toxins, hormones, and even viruses. The data in Figure 8.18 show that adsorption of biomolecules takes place rapidly (in a matter of minutes) and the amount adsorbed is related to the type of clay mineral. The bond between the biomolecule and the colloid is often quite strong so that the biomolecule cannot be easily removed by washing or exchange reactions. In most cases, biomolecules bound to clays do not enter the interlayer spaces, but are attached to the outer planar surfaces and edges of the clay crystals.

The binding of biomolecules to soil colloids in this manner has important environmental implications for two reasons. First, such binding usually protects the biomolecules from enzymatic attack, meaning that the molecules will persist in the soil much longer than studies of unbound biomolecules might suggest. Second, it has been shown that many biomolecules retain their biological activity in the bound state. Toxins remain toxic to susceptible organisms, enzymes continue to catalyze reactions, viruses can lyse (break open) cells or transfer genetic information to host cells, and DNA strands retain the ability to transform the genetic code of living cells, even while bound to colloidal surfaces and protected from decay. The presence of such DNA is not detectable by the usual chemical tests since it is not in a living cell and therefore not expressing its genes.

When genetically modified organisms (GMOs) are introduced to the soil environment, the cryptic (hidden) genes just described may represent an undetected potential for transfer of genetic information to organisms for which it was not intended. A similar concern exists regarding plants genetically modified to produce such compounds as human drugs ("pharma crops") or insecticidal toxins. For example, millions of hectares are planted each year with corn and cotton plants given a bacterial gene that codes for production of the insecticidal toxin Bt (see Figure 8.18). The Bt toxin is released into the soil by root excretion and decomposition of plant residues containing the toxin. We know little about what effect the toxins may have on soil ecology if it accumulates in soils in the colloid-bound, but still active state.

Figure 8.18

Adsorption of a toxin called Bt, the insecticidal protein produced by the soil bacteria Bacillus thuringiensis *and used to protect some crop plants. Highly active clay minerals such as montmorillonite (a member of the smectite group of 2:1 clays) adsorb and bind much larger amounts of these biomolecules than do low-activity clays such as kaolinite (a 1:1 mineral). In both cases, the adsorption reaction was completed in 30 minutes or less. Since only 500 µg of either clay mineral was used in the experiment, it appears that the clays adsorbed an amount of the toxin equal to 30 to 80% of their mass.*
[Redrawn from Stotzky (2000)]

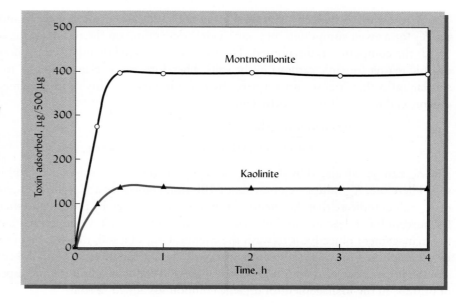

Antibiotics comprise another class of important organic compounds that sorb to soil colloids. These unique natural chemicals are irreplaceable life-saving compounds. Yet about 80% of antibiotics manufactured are *not* used to cure diseases (in animals or humans) but are used in livestock feed to stimulate faster growth of cattle, swine, and chickens. Not surprisingly, the animal manures produced on most industrial-style farms have been found to be laden with antibiotics that have passed through the animals' digestive systems. When these manures are applied to farmland, the antibiotics become sorbed on the soil colloids and may accumulate with repeated manure applications. Apparently the sorption is very strong. For example, K_d values as high as 2300 L kg^{-1} have been reported for the antibiotic tetracycline in some soils. Increasingly, research shows that even though strong sorption to soil colloids may reduce their efficacy somewhat, the soil-bound antibiotics still work against bacteria. This revelation raises concerns that the huge amounts of antibiotics exposed in the environment will select for resistant strains of "super bacteria" (including human pathogens), which would then no longer be controllable by these (once) life-saving drugs. It is clear that soil colloids and soil science have important roles to play with respect to environmental health.

8.14 PHYSICAL IMPLICATIONS OF SWELLING-TYPE CLAYS

Engineering Hazards

Soil colloids differ widely in their physical properties, including plasticity, cohesion, swelling, shrinkage, dispersion, and flocculation. These properties greatly influence the usefulness of soils for both engineering and agricultural purposes. As discussed in Sections 4.9 and 8.3, the tendency of certain clays to swell in volume when wetted is a major concern for the construction of roads and foundations. The worst clays in this regard are the smectites, which form wavy stacks or microscopic clay domains containing extremely small ultramicropores. These ultramicropores attract and hold large quantities of water (Figure 8.19), accounting for much of the swelling and plasticity of these clays. Relatively pure, mined smectite clay (often sold as "bentonite"), especially

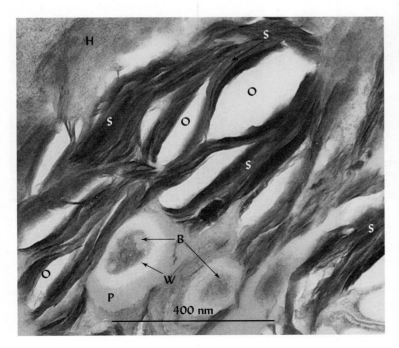

Figure 8.19

Transmission electron micrograph showing smectite clay (S) along with humus (H), bacterial cells (B), cell walls (W), and polysaccharides (P). Stacks of parallel crystals of smectite clay (S) form an open wavy clay domain structure (dark) in which ultramicropores (O) are visible as white areas. Water drawn into these ultramicropores in the clay domains accounts for most of the swelling of smectite clay upon wetting. It is less likely, as was once thought, that water causes swelling by entering the interlayers between smectite clay crystal units. Note that the entire image is about 1 micron across. (Image courtesy of M. Thompson, T. Pepper, and A. Carmo, Iowa State University)

when saturated with Na^+ ions, can have far greater potential for swelling and plasticity than the impure clays in soil.

Figure 8.20 gives an example of special steps needed to safely build homes on soils dominated by smectitic clay. The cost of building homes on smectitic soils may be double that of building on soils dominated by nonswelling clays, for which conventional foundation designs can be safely used. If preventative design measures are not taken during the construction of houses on smectite clays, homeowners will pay dearly in the future. The building foundation is likely to move with the swelling and shrinking of the soil, misaligning doors and windows and eventually cracking foundations, walls, and pipes.

The same swelling properties that make smectitic soils so problematic for construction activities make them attractive for certain environmental applications (see Box 8.4) and well suited for siting ponds and lagoons or creating wetlands. This is just one example of the critical role soil colloids play in determining the usefulness of our soils.

Figure 8.20

The different swelling tendencies of two types of clay are illustrated in the lower left. All four cylinders initially contained dry, sieved clay soil, the two on the left from the B horizon of a soil high in kaolinite, the two on the right from one of a soil high in montmorillonite. An equal amount of water was added to the two center cylinders. The kaolinitic soil settled a bit and was not able to absorb all the water. The montmorillonitic soil swelled about 25% in volume and absorbed nearly all the added water. The scenes to the right and above show a practical application of knowledge about these clay properties. Soils containing large quantities of smectite (e.g., the California Vertisol shown here) make very poor building sites. The normal-appearing homes (upper) are actually built on deep, reinforced-concrete pilings (lower right) that rest on nonexpansive substrata. Construction of the 15 to 25 such pilings needed for each home more than doubles the cost of construction. (Photos courtesy of R. Weil)

BOX 8.4
ENVIRONMENTAL USES OF SWELLING-TYPE CLAYS[a]

The physical and chemical properties of swelling-type clays make them extremely useful in certain environmental engineering applications. A common use of swelling clays—especially a mined mixture of smectite clays called bentonite—is as a sealant layer placed on the bottom and sides of ponds, waste lagoons, and landfill cells (see Section 15.10). The clay material expands when wetted and forms a highly impermeable barrier to the movement of water as well as organic and inorganic contaminants contained in the water. The contaminants are thus held in the containment structure and prevented from polluting the groundwater.

A more exotic use for swelling clays is proposed in Sweden for the final repository of that country's highly radioactive and toxic nuclear power plant wastes. The plan is to place the wastes in large (about 5 m × 1 m) copper canisters and bury them deep underground in chambers carved from solid rock. As a final defense against leakage of the highly toxic material to the groundwater, the canisters will be surrounded by a thick buffer layer of bentonite clay. The clay is packed dry around the canisters

and is expected to absorb water to saturation during the first century of storage, thus gradually swelling into a sticky, malleable mass that will fill any cavities or cracks in the rock. The clay buffer will serve three protective functions: (1) cushion the canister against small (10 cm) movements in the rock formation, (2) form a seal of extremely low permeability to keep corrosive substances in the groundwater away from the canister, and (3) act as a highly efficient electrostatic filter to adsorb and trap cationic radionuclides that might leak from the canister in some far future time.

Figure 8.21 shows how bentonite is used as a plug or sealant to prevent leakage around an environmental groundwater monitoring well. For most of the well depth, the gap between the bore hole wall and the well tube is back-filled with sand to support the tube and allow vertical movement of the groundwater to be sampled. About 30 cm below the soil surface, the space around the well casing is filled instead with air-dry granulated bentonite (white substance being poured from bucket in the photograph). As the bentonite absorbs water, it swells markedly, taking on an almost rubbery consistency and forming an impermeable seal that fits tightly against both the well casing and the soil bore hole wall. This seal prevents contaminants from the soil surface from leaking down the outside of the well casing. In the case of groundwater contaminated with volatile organics like gasoline, the bentonite also prevents vapors from escaping without being properly sampled.

Increasingly, environmental scientists are using swelling-type clays for the removal of organic chemicals from water by partitioning. For example, where there has been a spill of toxic organic chemicals, a deep trench may be dug across the slope and back-filled with a slurry of swelling clay and water to intercept a plume of polluted water. The swelling nature of the smectites prevents the rapid escape of the contaminated water while the highly reactive colloid surfaces chemically sorb the contaminants, purifying the groundwater as it slowly passes by. Chapter 15 takes a more detailed look at such "slurry walls" and other soil technologies for cleaning the environment.

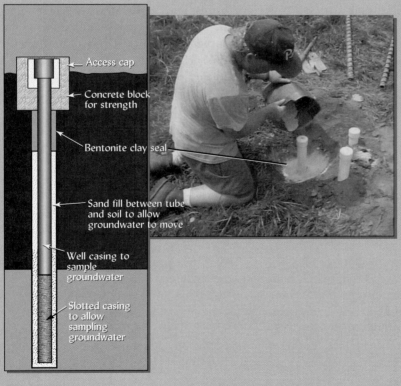

Figure 8.21
Use of swelling clay as seal for environmental monitoring well. (Diagram and photos courtesy of R. Weil)

Labels in figure:
- Access cap
- Concrete block for strength
- Bentonite clay seal
- Sand fill between tube and soil to allow groundwater to move
- Well casing to sample groundwater
- Slotted casing to allow sampling groundwater

---

[a]For more detailed information on environmental use of swelling clays, see Reid and Ulery (1998). For details of the Swedish nuclear repository use of bentonite, see Swedish Nuclear Power Inspectorate (2005) and S.K.B. (2006).

8.15   CONCLUSION

The complex structures, enormous surface area (both internal and external), and tremendous numbers of charges associated with soil colloids combine to make these tiniest of soil particles the seats of chemical and physical activity in soils. The physical activity of the colloids, their adsorption of water, swelling, shrinking, and cohesion are discussed in detail in Chapters 4, 5, and 6. Here we focused on the chemical activity of the colloids, activity that results largely from charged sites on or near colloid surfaces. These charged sites attract oppositely charged ions and molecules from the soil solution. The negative sites attract positive ions (cations) such as Ca^{2+}, Cu^{2+}, K^+, or Al^{3+} and the positive sites attract negative ion (anions) such as Cl^-, SO_4^{2-}, NO_3^-, or HPO_4^{2-}.

Although both positive and negative charges occur on colloids, in most soils the negative charges far outnumber the positive. Most elements dissolved from rocks by weathering or added to soils in lime or fertilizer will eventually end up in the oceans, but it is very fortunate for land plants and animals that attraction to soil colloids greatly slows the journey. Colloidal attraction is a major mechanism by which soils accumulate the stocks of nutrients necessary to support forests, crops, and, ultimately, civilizations. This role is especially critical for forests when nutrient storage in plant biomass is disrupted by fire or timber harvest. The colloidal attraction also enables soils to act as effective filters, sinks, and exchangers, protecting groundwater and food chains from excessive exposure to many pollutants.

When ions are attracted to a colloid, they may enter into two general types of relationships with the colloid surface. If the ion bonds directly to atoms of colloidal structure with no water molecules intervening, the relationship is termed an *inner-sphere complex*. This type of reaction is quite specific and, once created, is not easily reversed. In contrast, ions that keep their hydration shell of water molecules around them are generally attracted to colloidal surfaces with excess opposite charge. However, the attractive forces are transmitted through a chain of polar water molecules and are therefore weakened, and the interaction is less specific and quite easily reversed.

The latter type of adsorption is termed *outer-sphere complexation*. The adsorbed ion and its shell of water molecules oscillate or move about within a zone of attraction. The size of the oscillation zone depends on the strength of attraction between the particular ion and the type of charged site. Ions in such a state of dynamic adsorption are termed *exchangeable ions* because they break away from the colloid whenever another ion from the solution moves in closer and takes over, neutralizing the colloid's charges.

The replacement of one ion for another in the outer-sphere complex is termed *ion exchange*. Except in certain highly weathered, subsurface horizons, cation exchange is far greater than anion exchange. Cation and anion exchange reactions are reversible and balanced charge-for-charge (rather than ion-for-ion). The extent of the reaction is influenced by mass action, the relative charge and size of the hydrated ions, the nature of the colloid, and the nature of the other (complementary) ions already adsorbed on the colloid. Plant roots can exchange H^+ for nutrient cations or OH^- ions for nutrient anions.

The colloids in soils are both organic (humus) and mineral (clays) in nature. In most surface soils, half or more of the charges are contributed by organic matter colloids, while in most subsurface horizons, clays provide the majority of charges. The total number of negative colloid charges per unit mass is termed the *cation exchange capacity* (CEC). The CEC of different colloids varies from about 1 to over 200 $cmol_c$/kg and that of whole mineral soils commonly varies from about 1 to 50 $cmol_c$/kg. The CEC of a soil, as well as its capacity to strongly adsorb particular

Table 8.6

CATION EXCHANGE PROPERTIES TYPICAL FOR UNAMENDED CLAY LOAM SURFACE SOILS IN DIFFERENT CLIMATIC REGIONS

Note that soils with coarser textures would have less clay and organic matter and therefore lower amounts of exchangeable cations and lower CEC values.

Property	Warm, humid region (Ultisols)[a]	Cool, humid region (Alfisols)	Semiarid region (Ustolls)	Arid region (Natrargids)[b]
Exchangeable H^+ and Al^{3+}, $cmol_c$/kg (% of CEC)	7.5 (75%)	5 (28%)	0 (0%)	0 (0%)
Exchangeable Ca^{2+}, $cmol_c$/kg (% of CEC)	2.0 (20%)	9 (50%)	17 (65%)	13 (50%)
Exchangeable Mg^{2+}, $cmol_c$/kg (% of CEC)	0.4 (4%)	3 (17%)	6 (23%)	5 (19%)
Exchangeable K^+, $cmol_c$/kg (% of CEC)	0.1 (1%)	1 (5%)	2 (8%)	3 (12%)
Exchangeable Na^+, $cmol_c$/kg (% of CEC)	Tr	0.02 (0.1%)	1 (4%)	5 (19%)
Cation exchange capacity (CEC),[c] $cmol_c$/kg	10	18	26	26
Probable pH	4.5–5.0	5.0–5.5	7.0–8.0	8–10
Nonacid cations (% of CEC)[d]	25%	68%	100%	100%

[a]See Chapter 3 for explanation of soil group names.

[b]Natrargids are Aridisols with natric horizons. They are sodic soils, high in exchangeable sodium, as explained in Section 10.5.

[c]The sum of all the exchangeable cations measured at the pH of the soil. This is termed the effective CEC or ECEC (see Section 8.9).

[d]Traditionally referred to as "base" saturation.

ions (such as K^+ or HPO_4^{2-}), depends on the amount of humus in the soil and on both the amount and type of clays present. Low-activity clays (iron and aluminum oxides and 1:1-type silicate clays like kaolinite) tend to dominate highly weathered soils of warm, humid regions. High-activity clays (expanding 2:1 silicates like smectite and vermiculite, and nonexpanding 2:1 silicates like fine-grained mica and chlorite) tend to dominate soils in cooler or drier regions where weathering is less advanced. Most of the charge on humus and low-activity clays is pH-dependent (becomes more negative as pH rises), while most of the charge on high-activity clays is permanent.

The differing ability of soil colloids to adsorb ions and molecules is key to managing soils, both for plant production and to understand how CEC may regulate movement of both nutrients and toxins in the environment. Among the important properties influenced by colloids is the acidity or alkalinity of the soil, the topic of the next chapter.

STUDY QUESTIONS

1. Describe the *soil colloidal complex*, indicate its various components, and explain how it tends to serve as a "bank" for plant nutrients.

2. How do you account for the difference in surface area associated with a grain of kaolinite clay compared to that of montmorillonite, a smectite?

3. Contrast the difference in crystalline structure among *kaolinite, smectites, fine-grained micas, vermiculites*, and *chlorites*.

4. There are two basic processes by which silicate clays are formed by weathering of primary minerals. Which of these would likely be responsible for the formation of (1) fine-grained mica and (2) kaolinite from muscovite mica? Explain.

5. If you wanted to find a soil high in kaolinite, where would you go? The same for (1) smectite and (2) vermiculite?

6. Which of the silicate clay minerals would be *most* and *least* desired if one were interested in (1) a good foundation for a building, (2) a high cation exchange capacity, (3) an adequate source of potassium, and (4) a soil on which hard clods form after plowing?

7. Which of the following would you expect to be *most* and *least* sticky and plastic when wet: (1) a soil with significant sodium saturation in a semi-arid area, (2) a soil high in exchangeable calcium in a subhumid temperate area, or (3) a well-weathered acid soil in the tropics? Explain your answer.

8. A soil contains 4% humus, 10% montmorillonite, 10% vermiculite, and 10% Fe, Al oxides. What is its approximate cation exchange capacity?

9. Calculate the number of grams of Al^{3+} ions needed to replace 10 $cmol_c$ of Ca^{2+} ions from the exchange complex of 1 kg of soil.

10. A soil has been determined to contain exchangeable cations in these amounts: $Ca^{2+} = 9$ $cmol_c$, $Mg^{2+} = 3$ $cmol_c$, $K^+ = 1$ $cmol_c$, and $Al^{3+} = 3$ $cmol_c$. (a) What is the CEC of this soil? (b) What is the aluminum saturation of this soil?

11. A 100 g sample of a soil has been determined to contain the exchangeable cations in these amounts: $Ca^{2+} = 90$ mg, $Mg^{2+} = 35$ mg, $K^+ = 28$ mg, and $Al^{3+} = 60$ mg. (a) What is the CEC of this soil? (b) What is the aluminum saturation of this soil?

12. A 100 g sample of a soil was shaken with a strong solution of $BaCl_2$ buffered at pH 8.2. The soil suspension was then filtered, the filtrate was discarded, and the soil was thoroughly leached with distilled water to remove any nonexchangeable Ba^{2+}. Then the sample was shaken with a strong solution of $MgCl_2$ and again filtered. The last filtrate was found to contain 10,520 mg of Mg^{2+} and 258 mg of Ba^{2+}. What is the CEC of the soil?

13. Explain the importance of K_d and K_{oc} in assessing the potential pollution of drainage water. Which of these expressions is likely to be most consistently characteristic of the organic compounds in question regardless of the type of soil involved? Explain.

14. An accident at a nuclear power plant has contaminated soil with strontium-90 (Sr^{2+}), a dangerous radionuclide. Health officials order forages growing in the area to be cut, baled, and destroyed. However, there is concern that as the forage plants regrow, they will take up the strontium from the soil and cows eating this contaminated forage will excrete the strontium into their milk. You are the only soil scientist assigned to a risk assessment team consisting mainly of distinguished physicians and statisticians. Write a brief memo to your colleagues explaining how the properties of the soil in the area, especially those related to cation exchange, could affect the risk of contaminating the milk supply.

15. Explain why there is environmental concern about the adsorption by soil colloids of such normally beneficial substances as antibiotic drugs and natural insecticides.

REFERENCES

Dixon, J. B., and D. J. Schulze. 2002. *Soil Mineralogy with Environmental Applications.* (Madison, WI: Soil Science Society of America).

Evangelou, V. P., and R. E. Phillips. 2005. "Cation exchange in soils," pp. 343–410, in A. Tabatabai and D. Sparks (eds.), *Chemical Processes in Soils.* SSSA book series No. 8. (Madison, WI: Soil Science Society of America).

Meunier, A. 2005. *Clays.* (Berlin: Springer-Verlag).

Reid, D. A., and A. L. Ulery. 1998. "Environmental applications of smectites," in J. Dixon, D. Schultze, W. Bleam, and J. Amonette (eds.), *Environmental Soil Mineralogy.* (Madison, WI: Soil Science Society of America).

Schulten, H. R., and M. Schnitzer. 1993. "A state of the art structural concept for humic substances," *Naturwissenschaften* **80**:29–30.

S.K.B. 2006. *Final Repository—Properties of the Buffer.* Svensk Kärnbränslehantering AB (Swedish Nuclear Fuel and Waste Management Company) www.skb.se/templates/SKBPage____8762.aspx (posted 06 Feb 2006; verified 03 Jan 2009).

Stotzky, G. 2000. "Persistence and biological activity in soil of insecticidal proteins from *Bacillus thuringiensis* and of bacterial DNA bound on clays and humic acids," *J. Environ. Quality,* **29**:691–705.

Swedish Nuclear Power Inspectorate. 2005. Engineered barrier system:long-term stability of buffer and backfill. Report from a workshop in Lund, Sweden, November 15–17, 2004, synthesis and extended abstracts, Swedish Nuclear Power Inspectorate. *SKI Research Report,* 48, 120.

Weil, R. R. 2009. *Laboratory Manual for Introductory Soils,* 8th ed. (Dubuque, IA: Kendall/Hunt).

An acid soil community. (R. Weil)

9 Soil Acidity, Alkalinity, Aridity, and Salinity

The degree of soil acidity or alkalinity, expressed as soil pH, is a *master variable* that affects a wide range of soil chemical and biological properties. This chemical variable greatly influences the root uptake availability of many elements, including both nutrients and toxins. The activity of soil microorganisms is also affected. The mix of plant and even bacterial species that dominate a landscape under natural conditions often reflects the pH of the soil. For people attempting to produce crops or ornamental plants, soil pH is a major determinant of which species will grow well or even grow at all in a given site.

Soil pH affects the *mobility* of many pollutants in soil by influencing the rate of their biochemical breakdown, their solubility, and their adsorption to colloids. Thus, soil pH is a critical factor in predicting the likelihood that a given pollutant will contaminate groundwater, surface water, and food chains. Furthermore, there are certain situations in which so much acidity is generated that the acid itself becomes a significant environmental pollutant. For example, soils on certain types of disturbed land generate extremely acid drainage water that can cause massive fish kills when it reaches a lake or stream.

Acidification naturally reaches its greatest expression in regions where high rainfall promotes both the *production of H^+ ions* and the *leaching away of nonacid cations*. In addition, the solubility of the toxic element, aluminum, is inextricably tied to acidification in most soils.

In contrast, leaching in drier regions is much less extensive, producing fewer H^+ ions and allowing soils to retain the nonacid cations, Ca^{2+}, Mg^{2+}, Na^+, and K^+. Many soils of dry regions also accumulate detrimental levels of soluble salts (saline soils) or exchangeable sodium ions (sodic soils), or both. The chemical

conditions associated with alkalinity, salinity, and sodicity can lead to severe problems in the physical condition and fertility of soils in these areas.

More than 2,100 years ago the Roman armies are said to have spread salt (sodium chloride) on the lands of their vanquished enemies in the city-state of Carthage, to ensure that they would never have to fight them again. In this chapter, we will learn why sodium and salts are so damaging to soils and how they can be managed. Indeed, though only one of many problems unique to alkaline soils, we will learn that salt accumulation is perhaps the most vexing problem for long-term sustainable use of arid lands.

9.1 PROCESSES THAT CAUSE SOIL ACIDITY AND ALKALINITY

Acidity and alkalinity is all about the balance between hydrogen ions (H^+) and hydroxyl ions (OH^-) and is usually quantified using the pH scale (Box 9.1 and Figure 9.2). The two principal processes that promote soil acidification are (1) the production of H^+ ions and (2) the washing away of nonacid cations by percolating water. Since both processes are stimulated by large amounts of water entering the soil, it is not surprising that soil acidity is directly and closely related to the amount of annual precipitation.

Acidifying Processes That Produce Hydrogen Ions

Carbonic and Other Organic Acids Rainwater brings acidity to soils because as the CO_2 in air dissolves in the water, it forms carbonic acid, which subsequently disassociates

---

BOX 9.1
SOIL PH, SOIL ACIDITY, AND ALKALINITY

Whether a soil is acid, neutral, or alkaline is determined by the comparative concentrations of H^+ and OH^- ions. Pure water provides these ions in equal concentrations:

$$H_2O \rightleftharpoons H^+ + OH^-$$

The equilibrium for this reaction is far to the left; only about 1 out of every 10 million water molecules is dissociated into H^+ and OH^- ions. The product of the concentrations of the H^+ and OH^- ions is a constant (K_w), which at 25 °C is known to be 1×10^{-14}:

$$[H^+] \times [OH^-] = K_w = 10^{-14}$$

Since in pure water the concentration of H^+ ions $[H^+]$ must be equal to that of OH^- ions $[OH^-]$, this equation shows that the concentration of each is 10^{-7} ($10^{-7} \times 10^{-7} = 10^{-14}$). It also shows the inverse relationship between the concentrations of these two ions (Figure 9.1). As one increases, the other must decrease proportionately.

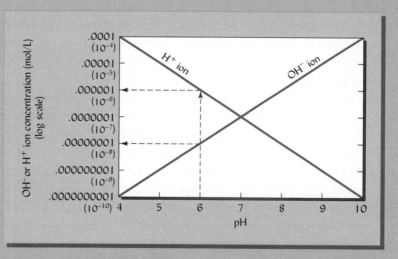

Figure 9.1
The relationship between pH, pOH, and the concentrations of hydrogen and hydroxyl ions in water solution.

Thus, if we were to increase the H^+ ion concentration $[H^+]$ by 10 times (from 10^{-7} to 10^{-6}), the $[OH^-]$ would be decreased by 10 times (from 10^{-7} to 10^{-8}) since the product of these two concentrations must equal 10^{-14}.

Scientists have simplified the means of expressing the very small concentrations of H^+ and OH^- ions by using the *negative logarithm of the H^+ ion concentration*, termed the *pH*. Thus, if the H^+ concentration in an acid medium is 10^{-5}, the *pH* is 5; if it is 10^{-9} in an alkaline medium, the *pH* is 9.

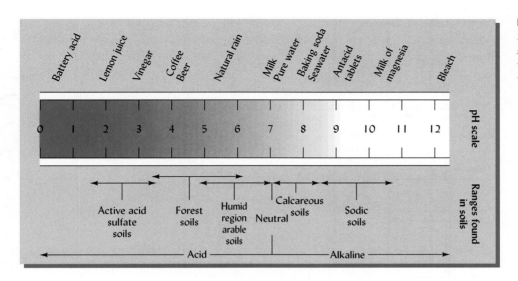

Figure 9.2
Some pH values for familiar substances (above) compared to ranges of pH typical for various types of soils (below).

to release H^+ ions. The metabolism of roots and microorganisms in the soil adds more CO_2, driving the following equation to the right, creating more acidity:

$$CO_2 + H_2O \longrightarrow H_2CO_3 \rightleftharpoons HCO_3^- + H^+ \qquad (9.1)$$

Because H_2CO_3 is a weak acid, its contribution to soil acidity is important only when the pH is greater than about 5.0. Numerous other organic acids, some weak and some quite strong, are also produced by biological activities in the soil.

Accumulation of Organic Matter   The accumulation of organic matter tends to acidify the soil for two reasons. First, organic matter forms soluble complexes with nonacid nutrient cations such as Ca^{2+} and Mg^{2+}, facilitating their loss by leaching. Second, organic matter contains numerous acid functional groups from which H^+ ions can dissociate (refer to Figure 8.9).

Oxidation of Nitrogen (Nitrification)   Oxidation reactions generally produce H^+ ions as one of their products. Reduction reactions, on the other hand, tend to consume H^+ ions and raise soil pH. Ammonium ions (NH_4^+) from organic matter or from most fertilizers are subject to oxidation that converts the N to the nitrate (NO_3^-) form. The reaction with oxygen, termed nitrification, releases two H^+ ions for each NH_4^+ ion oxidized. Because the NO_3 produced is the anion of a **strong acid** (nitric acid, HNO_3), it does not tend to recombine with the H^+ ion to make the reaction go to the left:

$$NH_4^+ + 2O_2 \rightleftharpoons H_2O + \underbrace{H^+ + H^+ + NO_3^-}_{\text{Dissociated nitric acid}} \qquad (9.2)$$

Oxidation of Sulfur   Certain plant compounds like proteins and minerals like pyrite contain chemically reduced sulfur (see Section 7.4). When such sulfur is oxidized, the reaction yields sulfuric acid (H_2SO_4). This strong acid is responsible for large amounts of acidity in certain soils which contain reduced sulfur and are exposed to increased oxygen levels because of drainage or excavation (see Section 9.5):

$$\underset{\text{Pyrite}}{FeS_2} + 3\tfrac{1}{2}O_2 + H_2O \rightleftharpoons \underset{\text{Ferrous sulfate}}{FeSO_4} + \underbrace{2H^+ + SO_4^{2-}}_{\substack{\text{Dissociated} \\ \text{sulfuric acid}}} \qquad (9.3)$$

Plant Uptake of Cations　For every positive charge taken in as a cation, a root can maintain the necessary charge balance either by taking up a negative charge as an anion or by exuding a positive charge as a different cation. When they take up far more of certain cations (e.g., K^+, NH_4^+, Ca^{2+}) than they do of anions (e.g., NO_3^-, SO_4^{2-}), plants usually exude H^+ ions into the soil solution to maintain charge balance. This exudation of H^+ acidifies the soil solution.

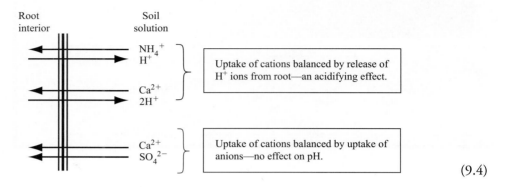

$$(9.4)$$

Alkalizing Processes That Consume Hydrogen Ions or Produce Hydroxyl Ions

The degree of acidification that actually occurs in a given soil is determined by the balance between those processes that produce H^+ ions and other processes that *consume* H^+ or produce OH^- ions (Table 9.1). In dry regions where water is scarce and organic production is low, soils become alkaline (i.e., their pH rises above 7) because more H^+ are consumed than generated (see right side of Table 9.1) and there is not enough rain to wash away the nonacid cations weathered from minerals.

Weathering of Nonacid Cations from Minerals　Mineral weathering (Section 2.1) is a long-term and very important H^+ ion–consuming process that may counteract acidification. An example is the weathering of calcium from a silicate mineral:

Table 9.1

THE MAIN PROCESSES THAT PRODUCE OR CONSUME HYDROGEN IONS (H^+) IN SOIL SYSTEMS

Production of H^+ ions increases soil acidity, while consumption of H^+ ions delays acidification and leads to alkalinity. The pH level of a soil reflects the long-term balance between these two types of processes.

Acidifying (H^+ ion–producing) processes	Alkalinizing (H^+ ion–consuming) processes
Formation of carbonic acid from CO_2	Input of bicarbonates or carbonates
Acid dissociation such as: 　$RCOOH \rightarrow RCOO^- + H^+$	Anion protonation such as: 　$RCOO^- + H^+ \rightarrow RCOOH$
Oxidation of N, S, and Fe compounds	Reduction of N, S, and Fe compounds
Atmospheric H_2SO_4 and HNO_3 deposition	Atmospheric Ca, Mg deposition
Cation uptake by plants	Anion uptake by plants
Accumulation of acidic organic matter 　(e.g., fulvic acids)	Specific (inner sphere) adsorption of anions 　(especially SO_4^{2-})
Cation precipitation such as: 　$Al^{3+} + 3H_2O \rightarrow 3H^+ + Al(OH)_3^0$ 　$SiO_2 + 2Al(OH)_3 + Ca^{2+} \rightarrow CaAl_2SiO_6 + 2H_2O + 2H^+$	Cation weathering from minerals such as: 　$3H^+ + Al(OH)_3^0 \rightarrow Al^{3+} + 3H_2O$ 　$CaAl_2SiO_6 + 2H_2O + 2H^+ \rightarrow SiO_2 + 2Al(OH)_3 + Ca^{2+}$
Deprotonation of pH-dependent charges	Protonation of pH-dependent charges

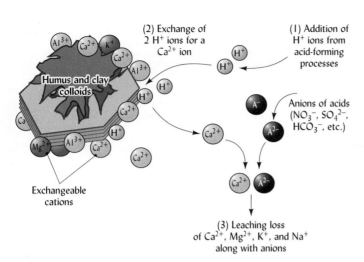

Figure 9.3

Soils become acid for two basic reasons. First, H^+ ions added to the soil solution exchange with nonacid Ca^{2+}, Mg^{2+}, K^+, and Na^+ ions held on humus and clay colloids. Second, percolating rainwater washes away the released nonacid cations in the drainage water along with accompanying anions. As a result, the exchange complex (and therefore also the soil solution) becomes increasingly dominated by acid cations (H^+ and Al^{3+}). Therefore, with greater annual precipitation, the leaching of cations is more complete, and the soils become more strongly acid. In arid regions with little or no leaching, the H^+ ions produced cause little long-term soil acidification because the Ca^{2+}, Mg^{2+}, K^+, and Na^+ are not leached, but remain in the soil where they can re-exchange with the acid cations and prevent a drop in pH level. (Diagram courtesy of R. Weil)

In the figure: (2) Exchange of 2 H^+ ions for a Ca^{2+} ion. (1) Addition of H^+ ions from acid-forming processes. Anions of acids (NO_3^-, SO_4^{2-}, HCO_3^-, etc.). Humus and clay colloids. Exchangeable cations. (3) Leaching loss of Ca^{2+}, Mg^{2+}, K^+, and Na^+ along with anions.

$$\text{Ca-Silicate} + 2H^+ \longrightarrow H_4SiO_4 + Ca^{2+} \qquad (9.5)$$

Some of the nonacid cations (Ca^{2+}, Mg^{2+}, K^+, and Na^+) released by weathering become exchangeable cations on the soil colloids. Hydrogen ions added to the soil solution from acids in rain (and other sources just discussed) may replace these cations on the exchange sites of humus and clay. The displaced nonacid cations are then subject to loss by leaching along with the anions of the added acids (Figure 9.3). The soil slowly becomes more acid if the leaching of Ca^{2+}, Mg^{2+}, K^+, and Na^+ proceeds faster than the release of these cations from weathering minerals. Thus, the formation of soil acidity is favored by higher rainfall; parent materials lower in Ca, Mg, K, and Na; and a higher degree of biological activity (favoring formation of H_2CO_3).

Accumulation of Nonacid Cations In dry regions where precipitation is less than evapotranspiration (see Section 6.3), the cations released by mineral weathering accumulate because there is not enough rain to thoroughly leach them away. The cations in solution and on the exchange complex are therefore mainly Ca^{2+}, Mg^{2+}, K^+, and Na^+. These cations are non-hydrolyzing and so do not produce acid (H^+) on reaction with water, as do the acid cations (Al^{3+} or Fe^{3+}). However, they generally do not produce OH^- ions either. Rather, their effect in water is neutral,[1] and soils dominated by them have a pH no higher than 7 unless certain *anions* are present in the soil solution.

Production of Base-Producing Anions The basic, hydroxyl (OH^-)-generating anions are principally **carbonate (CO_3^{2-})** and **bicarbonate (HCO_3^-)**. These anions originate from the dissolution of such minerals as calcite ($CaCO_3$) or from the dissociation of carbonic acid (H_2CO_3).

---

[1]The cations Ca^{2+}, Mg^{2+}, K^+, Na^+, and NH_4^+ have been traditionally called *base* or *base-forming* cations as a convenient, but inaccurate, way of distinguishing them from the acid cation, H^+, and the H^+-forming cations Al^{3+} and Fe^{3+}. It is less misleading to refer simply to acid cations (H^+, Al^{3+}, and Fe^{3+}) and nonacid cations (most other cations). Likewise, the term *nonacid saturation* should be used rather than *base saturation* to refer to the percentage of the exchange capacity satisfied by nonacid cations (usually Ca^{2+}, Mg^{2+}, K^+, and Na^+, see Section 9.3).

$$CaCO_3 \rightleftharpoons Ca^{2+} + CO_3^{2-} \qquad (9.6)$$

Calcite (Solid) — Dissolved in water — Dissolved in water

$$CO_3^{2-} + H_2O \rightleftharpoons HCO_3^- + \mathbf{OH^-} \qquad (9.7)$$

Carbonate

$$HCO_3^- + H_2O \rightleftharpoons H_2CO_3 + \mathbf{OH^-} \qquad (9.8)$$

Bicarbonate

$$H_2CO_3 \rightleftharpoons H_2O + CO_2 \uparrow \qquad (9.9)$$

Carbonic acid — (gas)

In this series of linked equilibrium reactions, carbonate and bicarbonate act as bases because they react with water to form hydroxyl ions and thus raise the pH. The importance of these reactions in **soil buffering**, or resistance to pH change, is discussed in Section 9.4.

Carbon Dioxide and Carbonates   The direction of the overall set of Reactions 9.6–9.9 determines whether OH^- ions are consumed (proceeding to the left) or produced (proceeding to the right). The reaction is controlled mainly by the precipitation or dissolution of calcite on the one end, and by the production (by respiration) or loss (by volatilization to the atmosphere) of carbon dioxide at the other end. Therefore, biological respiration in soils tends to lower the pH by driving the reaction series to the left.

Solid $CaCO_3$ precipitates out when the soil solution becomes saturated with respect to Ca^{2+} ions. Such precipitation removes Ca from the solution, again driving the reaction series to the left (lowering pH). Because of the limited solubility of $CaCO_3$, the pH of the solution cannot rise above 8.4 when the CO_2 in solution is in equilibrium with that in the atmosphere. The pH at which $CaCO_3$ precipitates in soil is typically only about 7.0 to 8.0, depending on how much the CO_2 concentration is enhanced by biological activity. This fact suggests that if *other* carbonate minerals more soluble than $CaCO_3$ (e.g., Na_2CO_3) were present, they would drive Reactions 9.6–9.9 farther to the right, producing more hydroxyl ions and thus a higher pH (see Section 9.14). Indeed **calcareous** (calcite-laden) soil horizons range in pH from 7 to 8.4 (tolerable by most plants), while **sodic** (sodium carbonate–laden) horizons may range in pH from 8.5 to as high as 10.5 (levels toxic to many plants). It is fortunate for plants that Ca^{2+} not Na^+ ions dominate the system in most soils.

Excess Anion Uptake by Roots   When plant uptake of an anion such as NO_3^- exceeds the uptake of associated cations, the roots exude the bicarbonate (HCO_3^-) anions to maintain charge balance:

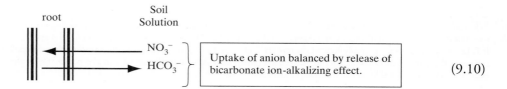

$$(9.10)$$

The resulting increased concentration of bicarbonate ions tends to *reverse* the dissociation of carbonic acid (Equation 9.1), thereby *consuming* H^+ ions and raising the pH of the soil solution. Another H^+ ion–consuming process involving nitrogen is the reduction of nitrate to nitrogen gases under anaerobic conditions (see Sections 7.3 and 12.1).

Role of Rainfall in Acidification

We have seen that soil acidification results from two basic processes that work together: (1) the *production of H^+ ions* and (2) the *removal of nonacid cations*. An abundance of rainwater plays important roles in both processes, explaining why there is such a close relationship between the amount of annual precipitation and the level of soil acidity. First, rain, snow, and fog contain a variety of acids that contribute H^+ ions to the soil receiving the precipitation. In recent decades, combustion of coal and petroleum products has significantly increased the amounts of the strong acids H_2SO_4 and HNO_3 present in precipitation (see Section 9.5). Second, greater rainfall means more water percolating through the soil profile and therefore more nonacid cations being washed away. The leaching of nonacid cations allows the incoming H^+ to dominate the soil exchange capacity and the soil to become increasingly acidic.

9.2   ROLE OF ALUMINUM IN SOIL ACIDITY

Although low pH is defined as a high concentration of H^+ ions, **aluminum** also plays a central role in soil acidity. Aluminum is a major constituent of most soil minerals (aluminosilicates and aluminum oxides), including clays. When H^+ ions are adsorbed on a clay surface, they usually do not remain as exchangeable cations for long, but instead they attack the structure of the minerals, releasing Al^{3+} ions in the process. The Al^{3+} ions then become adsorbed on the colloid's cation-exchange sites. These exchangeable Al^{3+} ions, in turn, are in equilibrium with dissolved Al^{3+} in the soil solution.

The exchangeable and soluble Al^{3+} ions play two critical roles in the soil acidity story. First, aluminum is *highly toxic* to most organisms and is responsible for much of the deleterious impact of soil acidity on plants and aquatic organisms. We will discuss this role in Section 9.7.

Second, Al^{3+} ions have a strong tendency to hydrolyze, splitting water molecules into H^+ and OH^- ions (Fe^{3+} ions do likewise at very low pH). The aluminum combines with the OH^- ions, leaving the H^+ to lower the pH of the soil solution. For this reason, Al^{3+} and H^+ together are considered **acid cations**. A single Al^{3+} ion can thus release up to three H^+ ions as the following reversible reaction series proceeds to the right in stepwise fashion:

$$Al^{3+} \rightleftharpoons AlOH^{2+} \rightleftharpoons Al(OH)_2^+ \rightleftharpoons Al(OH)_3^0 \quad \text{Gibbsite or amorphous (solid)} \tag{9.11}$$

$$pK_a = 5.0 \qquad pK_a = 5.1 \qquad pK_a = 6.7$$

Most of the hydroxy aluminum ions $[Al(OH)_x^{y+}]$ formed as the pH increases are strongly adsorbed to clay surfaces or complexed with organic matter. Often the hydroxy aluminum ions join together, forming large polymers with many positive charges. When tightly bound to the colloid's negative charge sites, these polymers are not exchangeable and so mask much of the colloid's potential cation exchange capacity.

9.3 POOLS OF SOIL ACIDITY

Principal Pools of Soil Acidity

Research suggests that three major pools of acidity are common in soils: (1) **active acidity** due to the H^+ ions in the soil solution; (2) **salt-replaceable (exchangeable) acidity**, involving the aluminum and hydrogen that are *easily exchangeable* by other cations in a simple unbuffered salt solution, such as KCl; and (3) **residual acidity**, which can be neutralized by limestone or other alkaline materials but cannot be detected by the salt-replaceable technique. These types of acidity all add up to the **total acidity** of a soil. In addition, a much less common, but sometimes very important fourth pool, namely **potential acidity**, can arise upon the oxidation of sulfur compounds in certain acid sulfate soils (see Section 9.7).

Fundamentals of acid–base chemistry: www.shodor.org/unchem/basic/ab/index.html

Active Acidity   The active acidity pool is defined by the H^+ ion activity in the soil solution. This pool is very small compared to the acidity in the exchangeable and residual pools. Even so, the active acidity is extremely important, as it determines the solubility of many substances and provides the soil solution environment to which plant roots and microbes are exposed.

Exchangeable (Salt-Replaceable) Acidity   Salt-replaceable acidity is primarily associated with exchangeable aluminum and hydrogen ions that are present in large quantities in very acid soils. These ions can be released into the soil solution by cation exchange with an unbuffered salt, such as KCl. Once released to the soil solution, the aluminum hydrolyzes to form additional H^+, as explained in Section 9.3. The chemical equivalent of salt-replaceable acidity in strongly acid soils is commonly thousands of times that of active acidity in the soil solution. Even in moderately acid soils, the limestone needed to neutralize this type of acidity is commonly more than 100 times that needed to neutralize the soil solution (active acidity). At a given pH value, exchangeable acidity is generally highest for smectites, intermediate for vermiculites, and lowest for kaolinite.

Residual Acidity   Together, exchangeable (salt-replaceable) and active acidity account for only a fraction of the total soil acidity. The remaining **residual acidity** is generally associated with hydrogen and aluminum ions (including the aluminum hydroxy ions) that are bound in nonexchangeable forms by organic matter and clays (see Figure 9.4). As the pH increases, the bound hydrogen dissociates and the bound aluminum ions are released and precipitate as amorphous $Al(OH)_3^0$. These changes free up negative cation exchange sites and increase the cation exchange capacity.

The residual acidity is commonly far greater than either the active or salt-replaceable acidity. It may be 1000 times greater than the soil solution or active acidity in a sandy soil and 50,000 or even 100,000 times greater in a clayey soil high in organic matter. The amount of ground limestone recommended to at least partly neutralize residual acidity in the upper 15 cm of soil is commonly 5 to 10 metric tons (Mg) per hectare (2.25 to 4.5 tons per acre).

Total Acidity   For most soils (not potential acid-sulfate soils), the total acidity that must be overcome to raise the soil pH to a desired value can be defined as:

$$\text{Total acidity} = \text{active acidity} + \text{salt-replaceable acidity} + \text{residual acidity} \qquad (9.12)$$

We can conclude that the pH of the soil solution is only the tip of the iceberg in determining how much lime may be needed to overcome the ill effects of soil acidity.

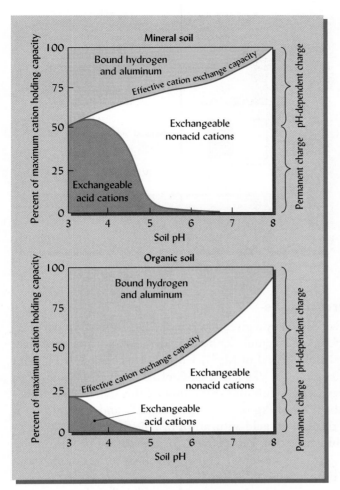

Figure 9.4

General relationship between soil pH and cations held in exchangeable form or tightly bound to colloids in two representative soils. Note that any particular soil would give somewhat different distributions. (Upper) A mineral soil with mixed mineralogy and a moderate organic matter level exhibits a moderate decrease in effective cation exchange capacity as pH is lowered, suggesting that pH-dependent charges and permanent charges (see Section 8.6 for explanation of these terms) each account for about half of the maximum CEC. At pH values above 5.5, the concentrations of exchangeable acid cations (aluminum and H^+) are too low to show in the diagram, and the effective CEC is essentially 100% saturated with exchangeable nonacid cations (Ca^{2+}, Mg^{2+}, K^+, and Na^+, the so-called base cations). As pH drops from 7.0 to about 5.5, the effective CEC is reduced because H^+ ions and $Al(OH)_x{}^{y+}$ ions (which may include $AlOH^{2+}$, $Al(OH)_2{}^+$, etc.) are tightly bound to some of the pH-dependent charge sites. As pH is further reduced from 5.5 to 4.0, aluminum ions (especially Al^{3+}), along with some H^+ ions, occupy an increasing portion of the remaining exchange sites. Exchangeable H^+ ions occupy a major portion of the exchange complex only at pH levels below 4.0. (Lower) The CEC of an organic soil is dominated by pH-dependent (variable) charges with only a small amount of permanent charge. Therefore, as pH is lowered, the effective CEC of the organic soil declines more dramatically than the effective CEC of the mineral soil. At low pH levels, exchangeable H^+ ions are more prominent and Al^{3+} less prominent on the organic soil than on the mineral soil.

(Diagram courtesy of R. Weil)

Soil pH and Cation Associations

Exchangeable and Bound Cations   Figure 9.4 illustrates the relationship between soil pH and the prevalence of two forms of hydrogen and aluminum: (1) that tightly held by the pH-dependent sites (*bound*) and (2) that associated with negative charges on the colloids (*exchangeable*). The bound forms contribute to the residual acidity pool, but only the exchangeable ions have an immediate effect on soil pH. As we shall see in Section 9.8, both forms are very much involved in determining how much lime or sulfur is needed to change soil pH.

Effective CEC and pH   Note that in both soils illustrated in Figure 9.4, the effective CEC increases as the pH level rises. This change in effective CEC results mainly from two factors: (1) the binding and release of H^+ ions on pH-dependent charge sites (as explained in Section 8.6) and (2) the hydrolysis reactions of aluminum species (as explained in Section 9.2). The change in effective CEC will be most dramatic for organic soils (Figure 9.4, *lower*) and highly weathered mineral soils dominated by iron and aluminum oxide clays. However, effective CEC changes with pH even in surface soils dominated by 2:1 clays, which carry mainly permanent charges because a substantial amount of variable charge is usually supplied by the organic matter and the weathered edges of clay minerals.

Figure 9.5

Saturation of the exchange capacity with acid and nonacid cations helps characterize the acidification of soils in the Adirondack Mountains of New York. The data represent the averages for O horizons and B horizons from more than 150 pedons in 144 watersheds. From the graph we can see that the effective cation exchange capacity (ECEC), the sum of all the exchangeable cations, was almost 30 cmol$_c$ kg^{-1} in the O horizons compared to only about 8 cmol$_c$ kg^{-1} in the B horizons. As is typical of temperate forested soils, the O horizons (which were about 90% organic) exhibited an extremely acid pH but a relatively low acid saturation, and the acid cations were mainly H$^+$. In contrast, the B horizons (which were about 90% mineral) had a more moderate pH but were 88% acid-saturated, and most of the acid cations were aluminum. [Modified from Sullivan et al. (2006)]

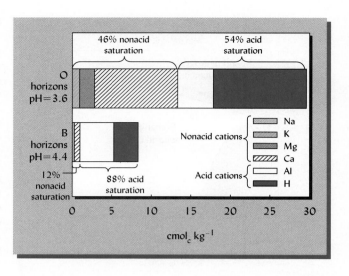

Cation Saturation Percentages

The proportion of the CEC occupied by a given ion is termed its **saturation percentage**. Consider a soil with a CEC of 20 cmol$_c$/kg holding these amounts of exchangeable cations (in cmol$_c$/kg): 10 of Ca^{2+}, 3 of Mg^{2+}, 1 of K$^+$, 1 of Na$^+$, 1 of H$^+$, and 4 of Al^{3+}. This soil, with 10 cmol$_c$ Ca^{2+}/kg and a CEC of 20 cmol$_c$/kg, is said to be 50% calcium saturated. Likewise, the aluminum saturation of this soil is 20% (4/20 = 0.20 or 20%). Together, the 4 cmol$_c$/kg of exchangeable Al^{3+} and 1 cmol$_c$/kg of exchangeable H$^+$ ions give this soil an **acid saturation** of 25% [(4 + 1)/20 = 0.25]. Similarly, the term **nonacid saturation** can be used to refer to the proportion of Ca^{2+}, Mg^{2+}, K$^+$, and Na$^+$, etc. on the CEC. Thus, the soil in our example has a nonacid saturation of 75% [(10 + 3 + 1 + 1)/20 = 0.75].

Traditionally, the nonacid cations have been referred to as **"base" cations** and their proportion on the CEC as the **percent "base" saturation**. Cations such as Ca^{2+}, Mg^{2+}, K$^+$, and Na$^+$ do not hydrolyze as Al^{3+} and Fe^{3+} do and therefore are not acid-forming cations. However, they are also *not* bases and do not necessarily form bases in the chemical sense of the word.[2] Because of this ambiguity, it is more straightforward to refer to *acid saturation* when describing the degree of acidity on the soil cation exchange complex (Figure 9.5). The relationships among these terms can be summarized as follows:

$$\text{Percent acid saturation} = \frac{\text{cmol}_c \text{ of exchangeable Al}^{3+} + \text{H}^+}{\text{cmol}_c \text{ of CEC}} \quad (9.13)$$

$$\begin{matrix}\text{Percent} \\ \text{nonacid} \\ \text{saturation}\end{matrix} = \begin{matrix}\text{Percent} \\ \text{"base"} \\ \text{saturation}\end{matrix} = \frac{\text{cmol}_c \text{ of exchangeable Ca}^{2+} + \text{Mg}^{2+} + \text{K}^+ + \text{Na}^+}{\text{cmol}_c \text{ of CEC}}$$

$$= 100 - \begin{matrix}\text{Percent acid} \\ \text{saturation}\end{matrix} \quad (9.14)$$

Acid (and Nonacid) Cation Saturation and pH

The percent saturation of a particular cation (e.g., Al^{3+}, Ca^{2+}) or class of cations (e.g., nonacid cations, acid cations) is often more closely related to the nature of the soil solution than is the absolute amount of these cations present. Generally, when the

---

[2]A base is a substance that combines with H$^+$ ions, while an acid is a substance that releases H$^+$ ions. The anions OH$^-$ and HCO$_3^-$ are strong bases because they react with H$^+$ to form the weak acids, H$_2$O and H$_2$CO$_3$, respectively.

acid cation percentage increases, the pH of the soil solution decreases. However, a number of factors can modify this relationship.

Effect of Type of Colloid   The type of clay minerals or organic matter present influences the pH of different soils at the same percent acid saturation due to differences in the ability of various colloids to furnish H^+ ions to the soil solution. For example, the dissociation of adsorbed H^+ ions from smectites is much higher than that from Fe and Al oxide clays. Consequently, the pH of soils dominated by smectites is appreciably lower than that of the oxides at the same percent acid saturation.

Effect of Method of Measuring CEC   An unfortunate ambiguity in the cation saturation percentage concept is that the actual percentage calculated depends on whether the effective CEC (which itself changes with pH) or the maximum potential CEC (which is a constant for a given soil) is used in the denominator. The different methods of measuring CEC are explained in Section 8.9.

When the concept of cation saturation was first developed, the percent nonacid saturation (then termed "base saturation") was calculated by dividing the level of these exchangeable cations by the *potential* cation exchange capacity that is measured at high pH values (7.0 or 8.2). Thus, if a representative mineral soil such as shown in Figure 9.4 has a potential CEC of 20 $cmol_c$/kg, and at pH 6 has a nonacid exchangeable cation level of 15 $cmol_c$/kg, the percent nonacid cation saturation would be calculated as 15 $cmol_c$/20 $cmol_c$ $\times$ 100 = 75%.

A second method relates the exchangeable cation levels to the *effective* CEC at the pH of the soil. As Figure 9.4 shows, the effective CEC of the representative soil at pH 6 would be only about 15 $cmol_c$/kg. At this pH level, essentially all the exchangeable sites are occupied by nonacid cations (15 $cmol_c$/kg). Using the effective CEC as our base, we find that the nonacid cation saturation is 15 $cmol_c$/15 $cmol_c$ $\times$ 100 = 100%. Thus, this soil at pH 6 is either 75% or 100% saturated with nonacid cations, depending on whether we use the potential CEC or the effective CEC in our calculations.

Uses of Cation Saturation Percentages   Which nonacid saturation percentage just described is the correct one? It depends on the purpose at hand. The first percentage (75% of the potential CEC) indicates that significant acidification has occurred and is used in soil classification (e.g., by definition Ultisols must have a nonacid, or "base," saturation of less than 35%). The second percentage (100% of the effective CEC) is more relevant to soil fertility and the availability of nutrients. It indicates what proportion of the total exchangeable cations at a given soil pH is accounted for by nonacid cations. For example, when the effective CEC of a mineral soil is less than 80% saturated with nonacid cations (i.e., more than 20% acid saturated), aluminum toxicity is likely to be a problem in many soils.

9.4   BUFFERING OF PH IN SOILS[3]

Soils tend to resist change in the pH of the soil solution when either acid or base is added. This resistance to change is called **buffering** and can be demonstrated by comparing the *titration curves* for pure water with those for various soils (Figure 9.6).

Titration Curves

A titration curve is obtained by monitoring the pH of a solution as an acid or base is added in small increments. The titration curves shown in Figure 9.6 suggest that the

---

[3]For a detailed discussion of the chemical principles behind this and related topics, see Bloom et al. (2005).

Figure 9.6

Buffering of soils against changes in pH when acid (H₂SO₄) or base (CaCO₃) is added. A well-buffered soil (C) and a moderately buffered soil (B) are compared to unbuffered water (A). Most soils are strongly buffered at low pH by the **hydrolysis and precipitation of aluminum** *compounds and at high pH by the precipitation and dissolution of* **calcium carbonate**. *Most of the buffering at intermediate pH levels (pH 4.5 to 7.5) is provided by* **cation exchange** *and* **protonation or deprotonation** *(gain or loss of H⁺ ions) of pH-dependent exchange sites on clay and humus colloids. The well-buffered soil (C) would have a higher amount of organic matter and/or highly charged clay than the moderately buffered soil (B).* [Curves based on data from Magdoff and Bartlett (1985) and Lumbanraja and Evangelou (1991)]

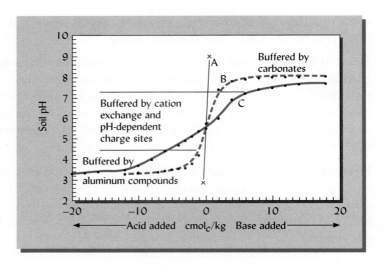

soils are most highly buffered when aluminum compounds (low pH) and carbonates (high pH) are controlling the buffer reactions. The soil is least well buffered at intermediate pH levels where H⁺ ion dissociation and cation exchange are the primary buffer mechanisms. However, considerable variability exists in the titration curves for various soils. This may be due to differences among soils with regard to the amounts and types of dominant colloids and contents of bound Al–hydroxy complexes that can absorb OH⁻ ions as the pH rises.

Mechanisms of Buffering

Compare pH changes in "water" and "buffer" by adding an acid or a base: http://michele.usc.edu/java/acidbase/acidbase.html

For soils with intermediate pH levels (5 to 7), buffering can be explained in terms of the equilibrium that exists among the three principal pools of soil acidity: active, salt-replaceable, and residual (Figure 9.7). If just enough base (e.g., lime) is applied to neutralize the H⁺ ions in the soil solution, they are largely replenished as the reactions move to the right, thereby minimizing the change in soil solution pH (Figure 9.7). Likewise, if the H⁺ ion concentration of the soil solution is increased (e.g., by organic decay or fertilizer applications), the reactions in Figure 9.7 are forced to the left, consuming H⁺ and again minimizing changes in soil solution pH. Because of the involvement of residual and exchangeable acidity, we can see that soils with higher clay and organic matter contents are likely to be better buffered in this pH range.

Throughout the entire pH range, reactions that either consume or produce H⁺ ions provide mechanisms to buffer the soil solution and prevent rapid changes in soil pH. The specific mechanisms of buffering include: (1) cation exchange reactions, (2) the hydrolysis of aluminum (Equation 9.11) at very low pH levels, (3) reactions with organic matter at moderate pH levels, (4) the dissociation of H⁺ ions from pH⁻ dependent charge sites on certain clays, and (5) the precipitation and dissolution of carbonate minerals. The latter is most important at high pH levels and is illustrated by the following reaction:

$$CaCO_3 + H_2O + H^+ \rightleftharpoons Ca^{2+} + H_2CO_3 + OH^- \qquad (9.15)$$

Importance of Soil Buffering Capacity

Soil buffering is important for two primary reasons. First, buffering tends to ensure some stability in the soil pH, preventing drastic fluctuations that might be detrimental to plants, soil microorganisms, and aquatic ecosystems. For example, well-buffered soils

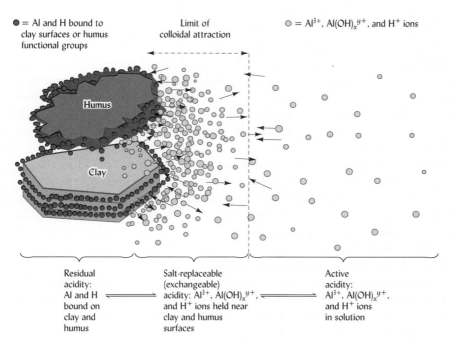

● = Al and H bound to clay surfaces or humus functional groups

Limit of colloidal attraction

○ = Al^{3+}, $Al(OH)_x^{y+}$, and H^+ ions

Residual acidity: Al and H bound on clay and humus ⇌ Salt-replaceable (exchangeable) acidity: Al^{3+}, $Al(OH)_x^{y+}$, and H^+ ions held near clay and humus surfaces ⇌ Active acidity: Al^{3+}, $Al(OH)_x^{y+}$, and H^+ ions in solution

Figure 9.7

Equilibrium relationship among residual, salt-replaceable (exchangeable), and soil solution (active) acidity in a soil with organic and mineral colloids. Note that the adsorbed (exchangeable) and residual (bound) ions are much more numerous than those in the soil solution, even when only a small portion of the ions associated with the colloids is shown. Most of the bound aluminum is in the form of $Al(OH)_x^{y+}$ ions that are held tightly on the surfaces of the clay or complexed with the humus; relatively few $Al(OH)_x^{y+}$ ions are exchangeable. Remember that the aluminum ions, by hydrolysis, also supply H^+ ions in the soil solution. It is obvious that neutralizing only the hydrogen and aluminum ions in the soil solution will be of little consequence. They will be quickly replaced by ions associated with the colloid. The soil, therefore, demonstrates high buffering capacity.

(Diagram courtesy of R. Weil)

resist the acidifying effect of acid rain, preventing the acidification of both the soil and the drainage water. Second, buffering influences the amount of amendments, such as lime or sulfur, required to bring about a desired change in soil pH.

Soils vary greatly in their buffering capacity. Other things being equal, the higher the cation exchange capacity (CEC) of a soil, the greater its buffering capacity. This relationship exists because in a soil with a high CEC, more reserve and exchangeable acidity must be neutralized or increased to affect a given change in soil pH. Thus, a clay loam soil containing 6% organic matter and 20% of a 2:1-type clay would be more highly buffered than a sandy loam with 2% organic matter and 10% kaolinite (Figure 9.8).

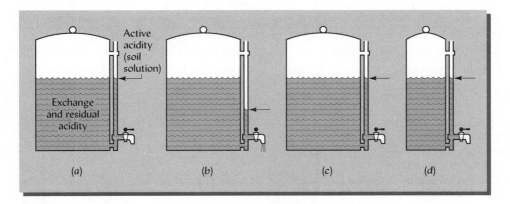

Active acidity (soil solution)

Exchange and residual acidity

(a)   (b)   (c)   (d)

Figure 9.8

The buffering capacity of soils can be described by using the analogy of a coffee dispenser. (a) The active acidity, which is represented by the coffee in the indicator tube on the outside of the urn, is small in quantity. (b) When H^+ ions are removed, this active acidity falls rapidly. (c) The active acidity is quickly restored to near the original level by movement from the exchange and residual acidity. By this process, the active acidity resists change. (d) A second soil with the same active acidity (pH) level but much less exchange and residual acidity would have a lower buffering capacity. Much less coffee would have to be added to raise the indicator level in the last dispenser. So too, much less liming material must be added to a soil with a small buffering capacity in order to achieve a given increase in the soil pH.

9.5   SOIL pH IN THE FIELD

Several methods can be used to measure soil pH: http://soils.usda.gov/technical/technotes/note8.html

More may be inferred regarding the chemical and biological conditions in a soil from the pH value than from any other single measurement. Soil pH can be easily and rapidly measured in the field or in the laboratory. Simple field kits use certain organic dyes that change color as the pH is increased or decreased. A few drops of the dye solutions are placed in contact with the soil, usually on a white spot plate (see color Plate 91), and the color of the dye is compared to a color chart that indicates the pH to within about 0.2 to 0.5 pH unit.

The most accurate method of determining soil pH is with a pH electrode. In this method, a pH-sensitive *glass* electrode and a standard reference electrode (or a probe that combines both electrodes in one) are inserted into a soil:water suspension. A special meter (called a *pH meter*) is used to measure the electrometric potential in millivolts and convert them into pH readings. Be advised that certain metallic soil probes on the market that do *not* contain a glass electrode *cannot* measure pH as claimed, and may give highly misleading readings!

Most soil-testing laboratories (see Section 13.10) in the United States measure the pH of a suspension of soil in water. This is designated the pH_{water}. Other labs (mainly in Europe and Asia) suspend the soil sample in a salt solution of either 0.02 M CaCl$_2$ (pH_{CaCl}) or 1.0 M KCl (pH_{KCl}). For normal low-salt soils, the pH_{CaCl} and pH_{KCl} readings are typically about 0.5 and 1.0 units lower than for pH_{water}. Therefore, if sub-samples of a soil were sent to three labs, the labs might report that the soil pH was 6.5, 6.0, or 5.5 (if the labs used methods for pH_{water}, pH_{CaCl}, and pH_{KCl}, respectively). All three pH values indicate the same level of acidity—and a suitable pH for most crops. Therefore, to interpret soil pH readings or compare reports from different laboratories, it is essential to know the method used. In this textbook (as in most U.S. publications), we report values for pH_{water}, unless specified otherwise.

Variability in the Field

Spatial Variation   Soil pH may vary dramatically over very small distances (millimeter or smaller). For example, plant roots may raise or lower the pH in their immediate vicinity, making the pH there quite different from that in the bulk soil just a few milimeters away (Figure 9.9 and Plate 93). Thus, the root may experience a very different chemical environment from that indicated by lab measurements of bulk soil samples.

Concentrations of fertilizers or ashes from forest fires may cause sizeable pH variations within the space of a few centimeters to a few meters. Other factors, such as

Figure 9.9

Soil pH at different distances from the roots of wheat plants receiving either ammonium (NH_4^+) or nitrate (NO_3^-) or no nitrogen fertilizer. Uptake of NH_4^+ cations causes the roots to release equivalent positive charges in the form of H^+ cations, which lower the pH (Equation 9.4). When a NO_3^- anion is taken up, the roots release a bicarbonate anion (HCO_3^-), which raises the pH (see Equation 9.10). The soil used was a calcareous sandy clay loam in the Aridisols order with pH = 8.1. In this experiment, the lowered pH near the roots using NH_4^+ markedly enhanced the plant's uptake of phosphorus by increasing the solubility of calcium phosphate minerals near the root. In more acid soils, the reduced pH might increase the toxicity of aluminum. A barrier membrane allowed soil solution to pass through, but prevented root growth into the lower soil where pH was measured. Plants were watered from the bottom by capillary rise. [Redrawn from Zhang et al. (2004) with permission of the Soil Science Society of America]

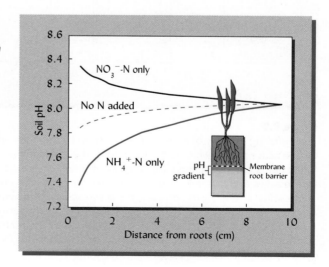

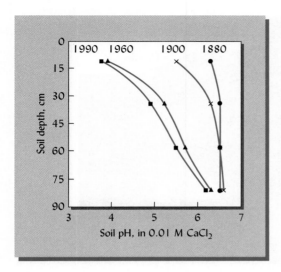

Figure 9.10

The change in soil pH (measured in 0.01 M CaCl$_2$) during a 110-year period in which a former agricultural field was allowed to revert to natural vegetation (eventually a mature oak forest). The fine-textured (clay loam to clay) Alfisol at Rothamstead in England was untilled, unfertilized, and unlimed. Note that in the first 20 years acidification was most pronounced near the soil surface. In the ensuing years acidity continued to increase most dramatically at the surface, but eventually increased throughout the profile. By 1960, the surface horizon had reached the pH range in which strong buffering by aluminum compounds probably slowed acidification.

[Drawn from data in Blake et al. (1999); used with permission of Blackwell Science, Ltd.]

erosion or drainage, may cause pH to vary considerably over larger distances (hundreds of m), often ranging over two or more pH units within a few hectares. A carefully planned sampling procedure may minimize errors due to such variability (see Section 13.10).

With Soil Depth   Different horizons, or even parts of horizons, within the same soil may exhibit substantial differences in pH. In many instances, the pH in the upper horizons is lower than in the deeper horizons (see Figure 9.5), but many patterns of variability exist. Acidifying processes usually proceed initially near the soil surface and slowly work their way down the profile.

On the other hand, human application of liming materials raises the pH mainly in the upper horizons into which the lime is incorporated. Severe acidity may occur in the subsoil beyond the depth of lime incorporation but within the reach of most plant roots. Untilled soils—including croplands managed with no-till practices, unplowed grasslands, lawns, and forest land—often show marked vertical variation in soil pH, with most of the changes in pH occurring in the upper few cm (see Figure 9.10).

For all these reasons, it is often advisable to obtain soil samples from various depth increments within the root zone and determine the pH level for each. Otherwise, serious acidity problems may be overlooked.

9.6   HUMAN-INFLUENCED SOIL ACIDIFICATION

In certain situations, the natural processes of soil acidification are greatly (and usually inadvertently) accelerated by human activities. We will consider three major types of human-influenced soil acidification: (1) nitrogen amendments, (2) acid precipitation, and (3) exposure of potential acid sulfate soils.

Nitrogen Fertilization

Chemical Fertilizers   Widely used ammonium-based fertilizers, such as ammonium sulfate [$(NH_4)_2 SO_4$] and urea [$CO(NH_2)_2$] are oxidized in the soil by microbes to produce strong inorganic acids by reactions such as the following:

$$(NH_4)_2SO_4 + 4O_2 \rightleftharpoons 2HNO_3 + H_2SO_4 + 2H_2O \tag{9.16}$$

However, since H$^+$ ions are consumed by the bicarbonate released when plants take up anions (Equation 9.10), soil acidification results largely from that portion of

Figure 9.11

Soil pH can be significantly lowered by fertilization with ammonium forms of nitrogen. Excess H^+ ions are generated during the bacterial conversion of NH_4^+ to NO_3^-. The acidification is especially severe if more NO_3^- is created than plants can take up and if most of the nonacid cations taken up by the plants are removed by harvest. As a result of the declining pH, the effective cation exchange capacity (CEC) of the soil also declines (see also Section 8.9). In the case illustrated, a Mollisol in Wisconsin was fertilized with N (urea or ammonium nitrate) for 30 years at the rates indicated. Conventional plow tillage was used to grow corn, soybean, and tobacco crops, and all the aboveground residues were removed. [Redrawn from data in Barak et al. (1997)]

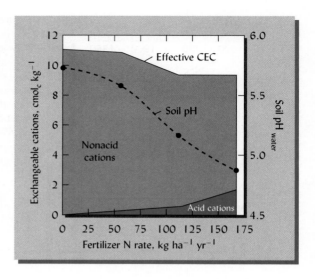

applied nitrogen that is not actually used. Excessive nitrogen fertilization rates in common use since the 1970s (Sections 12.1 and 13.2) have ensured that soil acidification from this cause is not trivial (Figure 9.11).

Acid-Forming Organic Materials   Organic materials such as leaf litter, sewage sludge, or animal manures can decrease soil pH, both by oxidation of the ammonium nitrogen released and by organic and inorganic acids formed during decomposition. Therefore, in humid regions a program of regular organic matter additions should also include regular additions of liming materials to counteract this acidification. It should be noted that some composts and plant residue contain high amounts of calcium and other nonacid cations and that certain types of sewage sludge are made with large quantities of lime to control pathogens and odors. Rather than acidify the soil, application of such organic materials may result in increased, rather than decreased, soil pH.

Acid Deposition from the Atmosphere

Animated maps showing acid deposition trends: http://nadp.sws.uiuc.edu/amaps2/

Origins of Acid Precipitation   Industrial activities such as the combustion of coal and oil in electric power generation and the combustion of fuel in vehicles emit enormous quantities of nitrogen and sulfur-containing gases into the atmosphere (Figure 9.12). The gases react in the atmosphere to form HNO_3 and H_2SO_4. These strong acids are then returned to the Earth in **acid rain** (as well as in snow, fog, and dry deposition). Normal rainwater that is in equilibrium with atmospheric carbon dioxide has a pH of about 5.5. The pH of acid rain is commonly between 4.0 and 4.5 but may be as low as 2.0.

Effects of Acid Rain   Acid rain causes expensive damage to buildings and car finishes, but the principal environmental reasons for concern about acid rain are its effects on (1) aquatic organisms and (2) forests. Since the 1970s, scientists have documented the loss of normal fish populations in thousands of lakes and streams. More recently, studies have suggested that the health of certain forest ecosystems is also suffering because of acid rain. Furthermore, scientists have learned that the health of both the lakes and the forests is not usually affected directly by the rain, but rather by the interaction of the acid rain with the soils in the watershed (Figure 9.12).

Soil Acidification   The incoming strong acids mobilize aluminum in the soil minerals, and the aluminum displaces Ca^{2+} and other nonacid cations from the exchange complex. The presence of the strong acid anions (SO_4^{2-} and NO_3^-) facilitates the leaching of the displaced Ca^{2+} ions (as explained in Figure 9.3). Soon Al^{3+} and H^+ ions,

Figure 9.12

Simplified diagram showing the formation of acid rain in urban areas and its impact on distant watersheds. Combustion of fossil fuels in electric power plants and in vehicles accounts for the largest portions of the nitrogen and sulfur emissions. About 60% of the acidity is due to sulfur gases and about 40% is due to nitrogen gases. The gases are carried hundreds of kilometers by the wind and are oxidized to form sulfuric and nitric acid in the clouds. These acids then return to Earth in precipitation and in dry deposition. The H^+ cations and NO_3^- and SO_4^{2-} anions cause acidification to occur in soils, soil aluminum to mobilize, and the loss of calcium and magnesium to accelerate. The mobilized aluminum percolates through the soil mantle, eventually reaching lakes and streams. The principal ecological effects of concern in sensitive watersheds are (1) possible decline in forest health and (2) decline or even death of aquatic ecosystems.

rather than Ca^{2+} ions, become dominant on the exchange complex, as well as in the soil solution and drainage waters. However, it is not easy to sort out how much acidification is due to natural processes internal to the soil ecosystem (see left side of Table 9.1) and how much is due to acid rain.

Effects on Forests   Some scientists are concerned that trees, which have a high requirement for calcium to synthesize wood, may eventually suffer from insufficient supplies of this and other nutrient cations in acidified soils. The leaching of calcium and the mobilization of aluminum may result in Ca/Al ratios (mol_c/mol_c) of less than 1.0, widely considered a threshold for aluminum toxicity, reduced calcium uptake, and reduced survival for forest vegetation. The scientific evidence for forest calcium deficiencies is less clear. The calcium supply in most forested soils in the humid eastern United States is being depleted as the rate of calcium loss by leaching, tree uptake, and harvest exceeds the rate of calcium deposition. However, it seems that even in very acid soils low in exchangeable Ca^{2+}, the weathering of soil minerals often releases sufficient calcium for good tree growth—at least in the short term.

Effects on Aquatic Ecosystems   The acid soil water, containing elevated levels of aluminum, and often of sulfate and nitrate, eventually drains into streams and lakes. The water in the lakes and streams becomes lower in calcium, less well-buffered, more acid, and higher in aluminum. The aluminum is directly toxic to fish, partly because it damages the gill tissues. As the lake water pH drops to about 6.0, acid-sensitive organisms in the aquatic food web die off, and reproductive performance of such fish as trout and salmon declines. With a further drop in water pH to about 5.0, virtually all fish are killed. Although the acidified water may be crystal clear (in part due to the flocculating influence of aluminum), the lake or stream is considered to be "dead" except for a few algae, mosses, and other acid-tolerant organisms.

Overview of environmental damages from acid rain: http://www.epa.gov/acidrain/effects/index.html

Ecological damage from acid rain is most likely to occur where the rain is most acid and the soils are most susceptible to acidification. The areas of the world that are most sensitive to acid rain damage are those where soils have low CEC and high acid saturation percents. Ecosystems in eastern China and Brazil are among those most likely to be damaged in the future, while those in northern Europe and northeastern North America have suffered the most in the past.

Tightening air quality standards in industrialized countries should continue to reduce acid inputs into sensitive ecosystems, eventually restoring a suitable chemical balance in the soils of these areas (and therefore in the lakes as well).

Exposure of Potential Acid Sulfate Materials[4]

Potential Acidity From Reduced Sulfur   If drainage, excavation, or other disturbance introduces oxygen into sulfur-bearing normally anaerobic soils, oxidation of the sulfur may produce large amounts of acidity. The adjective **sulfidic** is used to describe such materials with enough reduced sulfur to markedly lower the pH within two months of becoming aerated. The term **potential acidity** refers to the acidity that could be produced by such reactions.

Drainage of Certain Coastal Wetlands   Due to the microbial reduction of sulfates originally in seawater, certain coastal sediments contain significant quantities of pyrite (FeS_2), iron monosulfides (FeS), and elemental sulfur (S). Coastal wetland areas in the southeastern United States, Southeast Asia, coastal Australia, and West Africa commonly contain soils formed in such sediments. So long as waterlogged conditions prevail, the *potential* acid sulfate soils retain the sulfur and iron in their reduced forms. However, if these soils are drained for agriculture, forestry, or other development, air enters the soil pores and both the sulfur (S^0, S^-, or S^{2-}) and the iron (Fe^{2-}) are oxidized, changing the potential acid sulfate soils into *active* acid sulfate soils. Ultimately, such soils earn their name by producing prodigious quantities of sulfuric acid, resulting in soil pH values below 3.5 and in some cases as low as 2.0. The principal reactions involved are:

$$Fe^{II}S^{-I}_2 + 3\tfrac{1}{2}O_2 + H_2O \rightleftharpoons Fe^{II}S^{VI}O_4 + H_2S^{VI}O_4$$
$$\text{Pyrite} \qquad\qquad\qquad\qquad \text{Ferrous} \quad\; \text{Sulfuric}$$
$$\text{sulfate} \qquad\; \text{acid}$$

$$Fe^{II}SO_4 + \tfrac{1}{4}O_2 + 1\tfrac{1}{2}H_2O \rightleftharpoons Fe^{III}OOH + H_2SO_4 \qquad (9.17)$$
$$\text{Ferrous} \qquad\qquad\qquad\qquad\qquad \text{Iron} \qquad\quad \text{Sulfuric}$$
$$\text{sulfate} \qquad\qquad\qquad\qquad\quad \text{oxyhydroxides} \quad \text{acid}$$

$$S^0 + 1\tfrac{1}{2}O_2 + H_2O \longrightarrow H_2SO_4 \qquad (9.18)$$
$$\text{Elemental S} \qquad\qquad\qquad \text{Sulfuric}$$
$$\text{acid}$$

The iron sulfide compounds in the potential acid sulfate soils often give these soils a black color (Plates 47 and 109). The black color has sometimes led, with disastrous results, to the use of such soils by those seeking black, organic-matter-rich "topsoil" material for landscaping installations. The pH of the potential acid sulfate soils is in the neutral range (typically near 7.0) while they are still reduced, but drops precipitously within days or weeks of the soil being exposed to air. When in doubt, the pH of the soil should be monitored for several weeks while a sample is incubated in a moist, well-aerated, warm condition. See Plate 64 for an example of inappropriate, albeit inadvertent, engineering use of a potential acid sulfate clay.

---

[4]Many of these sulfide-rich soils are clayey in texture and termed cat clays. Fanning et al. (2002) describe some of the properties of these soils and discuss environmental problems that arise from their misuse. For a report on the iron and sulfur oxidizing roles of acidophile microorganisms discovered growing at pH 0.5 in acid mine drainage, see Edwards et al. (2000).

Excavation of Pyrite-Containing Materials   The sediments dredged to deepen shipping lanes in coastal harbors may also contain high concentrations of reduced sulfur compounds (see Plate 109). Furthermore, many saprolytes and sedimentary rocks (including coal-bearing shales) also contain reduced sulfur. When these once-deeply buried materials are exposed to air and water, the result, again, is the production of sulfuric acid in large quantities.

As water percolates through such oxidizing materials, it becomes an extremely acid and toxic brew known as **acid mine drainage**. Typical acid mine drainage has a pH in the range of 0.5 to 2.0, but pH values *below zero* have been measured! When this drainage water reaches a stream (as in Plate 108), iron sulfates dissolved in the drainage water continue to produce acid by oxidation and hydrolysis. The aquatic community can be devastated by the pH shock and the iron and aluminum that is mobilized. Similar problems occur when road cuts or building excavations expose buried sulfide-containing layers.

Avoidance as the Best Solution   Usually the best approach to solving this environmental challenge is to *prevent* the S oxidation in the first place. This means that sulfide-bearing wetland soils are best left undisturbed. In the case of mining or other excavation, any sulfide-bearing materials exposed must be identified and eventually deeply reburied to prevent their oxidation. If some acid drainage is unavoidable (as from abandoned, poorly designed mines), an effective treatment is to route the acid water through a wetland, either natural or constructed for the purpose (see Section 7.7). The anaerobic wetland conditions will re-reduce the iron and sulfur, causing iron sulfide to precipitate, simultaneously raising the pH of the water and reducing the iron content.

Research on acid mine drainage with pH<0:
www.pnas.org/cgi/content/full/96/7/3455

International Soil Reference and Info. Center tutorial on Acid Sulfate soils:
www.isric.org/isric/webdocs/tutorial/WHStart.htm

9.7   BIOLOGICAL EFFECTS OF SOIL pH

The pH of the soil solution is a critical environmental factor for the growth of all organisms that live in the soil, including plants, animals, and microbes.

Aluminum Toxicity[5]

Aluminum toxicity stands out as the most common and severe problem associated with strongly acid soils. Not only plants are affected; many bacteria, such as those that carry out transformations in the nitrogen cycle, are also adversely impacted by the high levels of Al^{3+} and $AlOH^{2+}$ that come into solution at low soil pH. Aluminum toxicity is rarely a problem when the soil pH is above about 5.2 (above pH_{CaCl} 4.8) because little aluminum exists in the solution or exchangeable pools above this pH level. There is an exponential increase in Al^{3+} concentration of the soil solution as pH drops from 5 to 4. Other toxic Al species, namely $AlOH^{2+}$ and $Al(OH)_2^+$, also increase in solubility below pH 5. At comparable pH levels in most organic soils (or in organic soil horizons), aluminum toxicity is much less of a problem because there is far less total aluminum in these soils—and because aluminum ions are strongly attracted and bound to the carboxylic ($R\text{-}COO^-$) and phenolic ($R\text{-}CO^-$) sites on soil organic matter, leaving much less Al^{3+} in solution.

Effects on Plants   When aluminum, which is not a plant nutrient, is taken into the root, most remains there, and little is translocated to the shoot. Therefore, analysis of leaf tissue is rarely a good diagnostic technique for aluminum toxicity. In the root, aluminum damages membranes and restricts cell wall expansion so roots cannot grow properly (Figure 9.13, *left*). Aluminum also interferes with the metabolism of phosphorus-containing compounds essential for energy transfers (ATP) and genetic coding (DNA).

The most common symptom of aluminum toxicity is a stunted root system with short, thick, stubby roots that show little branching or growth of laterals. The

---

[5]For a review of aluminum toxicity and the development of plant tolerance to aluminum, see de la Fuente-Martinez and Herrera-Estrella (2000).

Figure 9.13

Plant responses to toxicity of aluminum (left) and manganese (right) at low soil pH. (Left) As soil pH$_{water}$ drops below 5.2, exchangeable Al increases and cotton root length is severely restricted in an Ultisol. (Right) Plant shoot growth (the average of bean and cabbage) declines and Mn content of foliage increases at low pH levels in manganese-rich soils from East Africa (average data for an Andisol and an Alfisol).

[(left) From Adams and Lund (1966); (right) Redrawn from data in Weil (2000)]

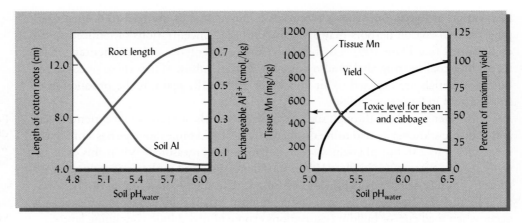

root tips and lateral roots often turn brown. In some plants, the leaves may show chlorotic (yellowish) spots. Because of the restricted root system, plants suffering from aluminum toxicity often show symptoms of drought stress and phosphorus deficiency (stunted growth, dark green foliage, and purplish stems).

Among and within plant species there exists a great deal of genetic variability in sensitivity to aluminum toxicity. Generally, plant species that originated in areas dominated by acid soils (such as most humid regions) tend to be less sensitive than species originating in areas of neutral to alkaline soils (such as the Mediterranean region). Fortunately, plant breeders have been able to find genes that confer tolerance to aluminum even in species that are typically sensitive to this toxicity.

Manganese Toxicity to Plants

Although not as widespread as aluminum toxicity, **manganese toxicity** is a serious problem for plants in acid soils derived from manganese-rich parent minerals. Unlike Al, Mn is an essential plant nutrient (see Section 12.8) that is toxic only when taken up in excessive quantities. Like aluminum, manganese becomes increasingly soluble as pH drops, but in the case of Mn, toxicity is common at pH$_{water}$ levels as high as 5.6 (about 0.5 units higher than for aluminum).

Plant species and genotypes within species vary widely with regard to their susceptibility to manganese toxicity. Symptoms of Mn toxicity may include crinkling or cupping of leaves and interveinal patches of chlorotic tissue. Unlike for Al, the leaf tissue content of Mn usually correlates with toxicity symptoms, toxicity beginning at levels that range from 200 mg/kg in sensitive plants to over 5000 mg/kg in tolerant plants. Figure 9.13 (*right*) illustrates a case in which low soil pH induced plant uptake of Mn to toxic levels.

Since the reduced form [Mn(II)] is far more soluble than the oxidized form [Mn(IV)], toxicity is greatly increased by low oxygen conditions associated with a combination of oxygen-demanding, decomposable organic matter and wetness. Manganese toxicity is also common in certain high organic matter surface horizons of volcanic soils (e.g., Melanudands). Unlike that of Al, the solubility and toxicity of Mn is commonly accentuated, rather than restricted, by higher soil organic matter (Table 9.2).

Nutrient Availability to Plants

Figure 9.14 shows in general terms the relationship between the pH of mineral soils and the availability of plant nutrients. Note that in strongly acid soils the availability of the macronutrients (Ca, Mg, K, P, N, and S) as well as the two micronutrients, Mo and B, is curtailed. In contrast, availability of the micronutrient cations (Fe, Mn, Zn, Cu, and Co) is increased by low soil pH, even to the extent of toxicity.

In slightly to moderately alkaline soils, molybdenum and all of the macronutrients (except phosphorus) are amply available, but levels of available Fe, Mn, Zn, Cu,

Table 9.2
SOIL PROPERTIES ASSOCIATED WITH ACIDITY AND SURVIVAL OF MAPLE SEEDLINGS

Al was most abundant in the low-organic-matter B horizons and Mn most abundant in the high-organic-matter O horizons. The ratio of Ca/Al was <1.0 only in the B horizons, while a ratio of Ca/Mn <30 in all horizons was associated with tree mortality.

Seedlings Surviving	Exchangeable cations, mg/kg			Ratio, mol$_c$/mol$_c$		
	Mn	Ca	Al	Ca/Al	Ca/Mn	Soil pH$_{water}$
			O Horizons			
No	188	2738	53	23.1	20	4.02
Yes	89	6371	38	74.1	98	4.45
			B Horizons			
No	15	305	279	0.5	28	4.62
Yes	8	1061	202	2.3	180	4.90

[Data from Demchik et al. (1999)]. Ca/Al and Ca/Mn ratios calculated here to give units shown. Data are averages for 18 forested sites dominated by overstory sugar maples.

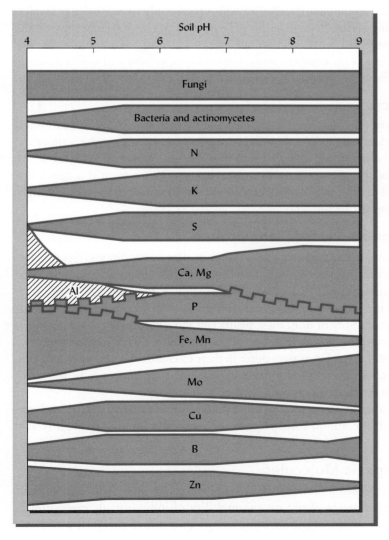

Figure 9.14

Relationships existing in mineral soils between pH and the availability of plant nutrients. The relationship with activity of certain microorganisms is also indicated. The width of the bands indicates the relative microbial activity or nutrient availability. The jagged lines between the P band and the bands for Ca, Al, and Fe represent the effect of these metals in restraining the availability of P. When the correlations are considered as a whole, a pH range of about 5.5 to perhaps 7.0 seems to be best to promote the availability of plant nutrients. In short, if the soil pH is suitably adjusted for phosphorus, the other plant nutrients, if present in adequate amounts, will be satisfactorily available in most cases.

Figure 9.15

Soil pH greatly influences the diversity of bacteria. In 98 different ecosystems in North and South America, ribosomal DNA fingerprinting was used to estimate the diversity of the bacteria communities. The index of diversity was high in soils with pH above 6 but was much reduced by more acid soil conditions. Fungal diversity was not studied here, but might be expected to show nearly the opposite trend.

[Modified from Fierer and Jackson (2006)]

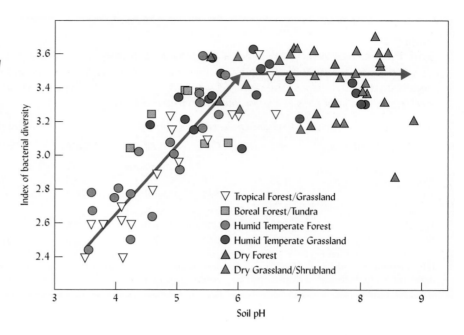

and Co are so low that plant growth is constrained. Phosphorus and boron availability is likewise reduced in moderately alkaline soil—commonly to a deficiency level.

It appears from Figure 9.14 that the pH range of 5.5 to 7.0 may provide the most satisfactory plant nutrient levels overall. However, this generalization may not be valid for all soil and plant combinations. For example, certain Mn deficiencies are common in some plants when sandy Ultisols are limed to pH values of only 6.5 to 7.0.

Microbial Effects

Fungi are particularly versatile, flourishing satisfactorily over a wide pH range. Fungal activity tends to predominate in low pH soils because bacteria are strong competitors and tend to dominate the microbial activity at intermediate and higher pH (Figure 9.15). Individual microbial species exhibit pH optima that may differ from this generality. Manipulation of soil pH can help control certain soilborne plant diseases and decomposition processes (Chapters 10 and 11).

Optimal pH Conditions for Plant Growth

Plants vary considerably in their tolerance to acid and/or alkaline conditions (Figure 9.16). Because forests exist mainly in humid regions where acid soils predominate, many forest species, such as rhododendrons, azaleas, blueberries, larch, some oaks, and most pines, are inefficient in taking up the iron they need. Since high soil pH and high calcium saturation reduce the availability of iron, these plants will show **chlorosis** (yellowing of the leaves) and other symptoms indicative of iron deficiency under these soil conditions (see Plates 86, 88, and 107). Even forest trees differ in their tolerance of soil acidity, and elm, poplar, honey locust, and the tropical legume tree *Leucaena* are known to be less tolerant to acid soils than are most other forest species.

Most cultivated crop plants (except those such as sweet potato, cassava, and others that originated in the humid tropics) grow well on soils that are just slightly acid to near neutral.

Soil pH and Organic Molecules

Soil pH influences environmental quality in many ways, but we will discuss only one example here—the influence of pH on the mobility of ionic organic molecules in soils. The molecular structure of certain ionic compounds is such that in low-pH soil,

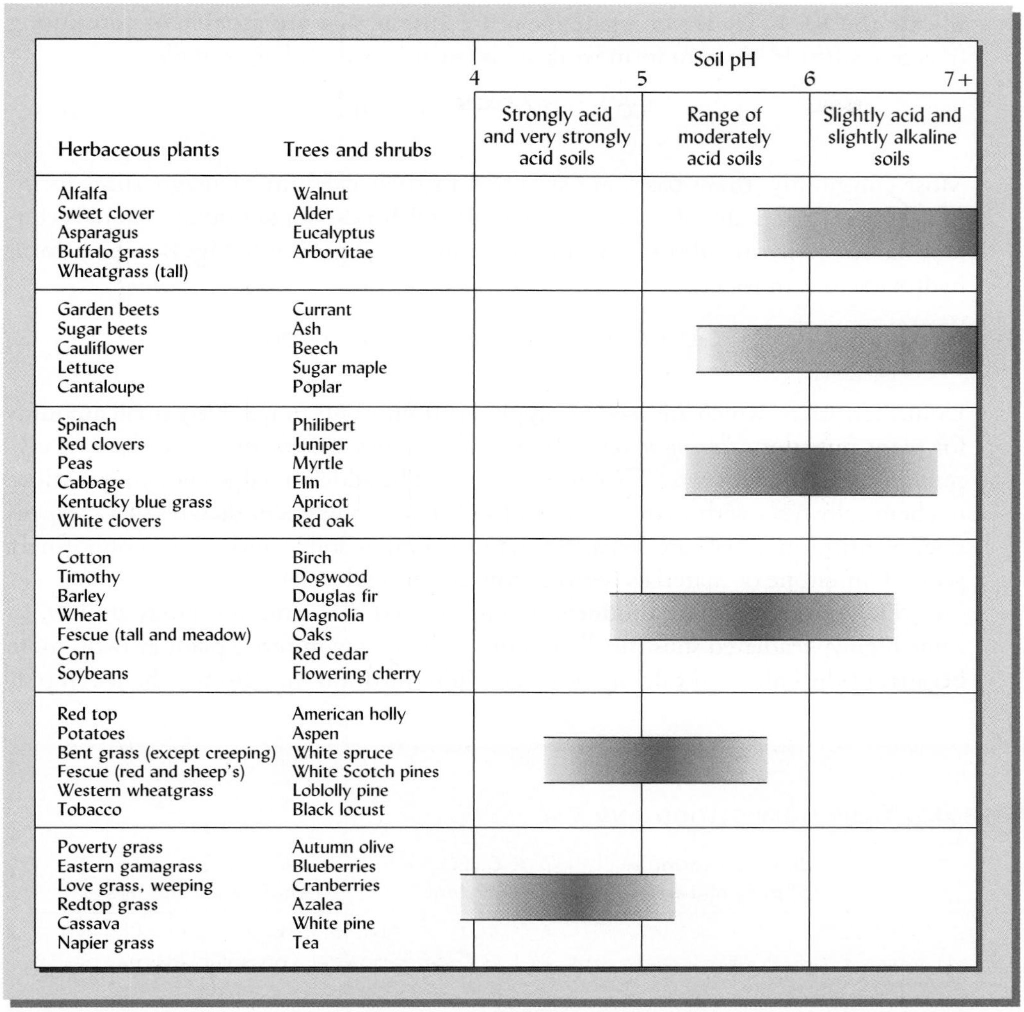

Herbaceous plants	Trees and shrubs	Soil pH		
		4 5 6 7+		
		Strongly acid and very strongly acid soils	Range of moderately acid soils	Slightly acid and slightly alkaline soils
Alfalfa Sweet clover Asparagus Buffalo grass Wheatgrass (tall)	Walnut Alder Eucalyptus Arborvitae			(bar)
Garden beets Sugar beets Cauliflower Lettuce Cantaloupe	Currant Ash Beech Sugar maple Poplar		(bar)	(bar)
Spinach Red clovers Peas Cabbage Kentucky blue grass White clovers	Philibert Juniper Myrtle Elm Apricot Red oak		(bar)	(bar)
Cotton Timothy Barley Wheat Fescue (tall and meadow) Corn Soybeans	Birch Dogwood Douglas fir Magnolia Oaks Red cedar Flowering cherry	(bar)	(bar)	(bar)
Red top Potatoes Bent grass (except creeping) Fescue (red and sheep's) Western wheatgrass Tobacco	American holly Aspen White spruce White Scotch pines Loblolly pine Black locust	(bar)	(bar)	
Poverty grass Eastern gamagrass Love grass, weeping Redtop grass Cassava Napier grass	Autumn olive Blueberries Cranberries Azalea White pine Tea	(bar)	(bar)	

Figure 9.16

Ranges of pH in mineral soils optimal for growth of selected plants.

the excess H^+ ions (protons) present in solution are attracted to and bond with these chemicals, creating positively charged sites on the molecule. This process is called **protonation**. The herbicide Atrazine is an example of a chemical whose mobility is greatly influenced by soil pH. In a low-pH environment, the positively charged molecule is adsorbed on the negatively charged soil colloids, where it is held until it can be decomposed by soil organisms. At pH values above 5.7, however, the adsorption is greatly reduced, and the tendency for the herbicide to move downward in the soil is increased. Of course, the adsorption in acidic soils also reduces the availability of Atrazine to weed roots, thus reducing its effectiveness as a weed killer.

9.8   RAISING SOIL pH BY LIMING

Agricultural Liming Materials

Liming is a very common agricultural practice in humid regions (such as the eastern United States) where the natural pH of soils is too acid for good growth of most crops. To raise the pH, the soil is usually amended with alkaline materials that provide such conjugate bases of weak acids as carbonate (CO_3^{2-}), hydroxide (OH^-), and silicate (SiO_3^{2-}). These conjugate bases are anions that are capable of consuming (reacting with) H^+ ions to form weak acids (such as water). For example:

$$CO_3^{2-} + 2H^+ \longrightarrow CO_2 + H_2O \tag{9.19}$$

Most commonly, these bases are supplied in their calcium or magnesium forms ($CaCO_3$, etc.) and are referred to as **agricultural limes**. Some liming materials contain oxides or hydroxides of alkaline earth metals (e.g., CaO, MgO), which form hydroxide ions in water:

$$CaO + H_2O \longrightarrow Ca(OH)_2 \longrightarrow Ca^{2+} + 2OH^- \tag{9.20}$$

Unlike fertilizers, which are used to supply plant nutrients in relatively small amounts for plant nutrition, *liming materials are used to change the chemical makeup of a substantial part of the root zone.* Therefore, lime must be added in large enough quantities to chemically react with a large volume of soil. This requirement dictates that inexpensive, plentiful materials are normally used for liming soils—most commonly finely ground limestone or materials derived from it (see Table 9.3).

Dolomitic limestone products should be used if magnesium levels are low. In some highly weathered soils, small amounts of lime may improve plant growth, more because of the enhanced calcium or magnesium nutrition than from a change in pH.

Table 9.3
COMMON LIMING MATERIALS: THEIR COMPOSITION AND USE

Common name of liming material	Chemical formula (of pure materials)	% $CaCO_3$ equivalent	Comments on manufacture and use
Calcitic limestone	$CaCO_3$	100	Natural rock ground to a fine powder. Low solubility; may be stored outdoors uncovered. Noncaustic, slow to react.
Dolomitic limestone	$CaMg(CO_3)_2$	95–108	Natural rock ground to a fine powder; somewhat slower reacting than calcitic limestone. Supplies Mg to plants.
Burned lime (oxide of lime)	CaO (+ MgO if made from dolomitic limestone)	178	Caustic, fast-acting, can burn foliage, expensive. Made by heating limestone. Protect from moisture.
Hydrated lime (hydroxide of lime)	$Ca(OH)_2$ (+ $Mg(OH)_2$ if made from dolomitic limestone)	134	Caustic, fast-acting, can burn foliage, expensive. Made by slaking hot CaO with water.
Basic slag	$CaSiO_3$	70	By-product of pig-iron industry. Must be finely ground. Also contains 1–7% P.
Marl	$CaCO_3$	40–70	Usually mined from shallow coastal beds, dried, and ground before use. May be mixed with soil or peat.
Wood ashes	CaO, MgO, K_2O, K(OH), etc.	40	Caustic, largely water-soluble, must be protected from water during storage.
Misc. lime-containing by-products	Usually $CaCO_3$ with various impurities	20–100	Variable composition; test for toxic impurities.

How Liming Materials React to Raise Soil pH

Chemical Reactions   Most liming materials—whether they be oxide, hydroxide, or carbonate—react with carbon dioxide and water to yield bicarbonate when applied to an acid soil. The carbon dioxide partial pressure in the soil, usually several hundred times greater than that in atmospheric air, is generally high enough to drive such reactions to the right. For example:

$$CaMg(CO_3)_2 + 2H_2O + 2CO_2 \rightleftharpoons Ca + 2HCO_3^- + Mg + 2HCO_3^- \qquad (9.21)$$

Dolomitic limestone   Bicarbonate   Bicarbonate

The Ca and Mg bicarbonates are much more soluble than are the carbonates, so the bicarbonate formed is quite reactive with the exchangeable and residual soil acidity. The Ca^{2+} and Mg^{2+} replace H^+ and Al^{3+} on the colloidal complex:

$$\boxed{\begin{array}{l}\text{Clay or}\\ \text{humus}\end{array}}\begin{array}{l}H^+\\ Al^{3+}\end{array} + 2Ca^{2+} + 4HCO_3^- \rightleftharpoons \boxed{\begin{array}{l}\text{Clay or}\\ \text{humus}\end{array}}\begin{array}{l}Ca^{2+}\\ Ca^{2+}\end{array} + Al(OH)_3 + H_2O + 4CO_2\uparrow \qquad (9.22)$$

Bicarbonate   (solid)

The insolubility of $Al(OH)_3$, the weak dissociation of water, and the release of CO_2 gas to the atmosphere all pull these reactions to the right. In addition, the adsorption of the calcium and magnesium ions lowers the percentage acid saturation of the colloidal complex, and the pH of the soil solution increases correspondingly.

    The amount of liming material required to ameliorate acid soil conditions is determined by several factors, including (1) the change required in the pH or exchangeable Al saturation, (2) the buffer capacity of the soil, (3) the amount or depth of soil to ameliorate, (4) the chemical composition of the liming materials to be used, and (5) the fineness of the liming material. The limestone requirements of soils with several different textures (and therefore likely to have different buffering capacities) are estimated in Figure 9.17. Because of the greater buffering capacity, the lime requirement for a clay loam is much higher than that of a sandy loam with the same pH value (see Section 9.4). Within the pH range of 4.5 to 7.0, the degree of change in pH brought about by additions of base to an acid soil is determined by the buffering capacity of the particular soil (see Box 9.2).

Buffer pH Methods for Lime Requirement   A rapid and inexpensive lab approach to estimating lime requirements is to equilibrate a sample of soil with a special salt solution that has a known initial pH value and is buffered to resist change in pH. The important thing to remember is that *the buffer pH indicates how much the soil acidity was able to change the buffer solution pH, but is not a measure of the soil pH itself.*

Soil acidity and liming recommendations in Ontario: www.omafra.gov/on.ca/ english/crops/pub811/ 2limeph.htm#changes

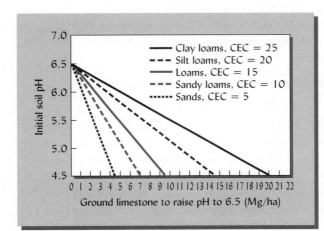

Figure 9.17

Effect of soil textural class on the amount of limestone required to raise the pH of soils from their initial level to pH 6.5. Note the very high amounts of lime needed for fine-textured soils that are strongly buffered by their high levels of clay, organic matter, and CEC. The chart is most applicable to soils in cool, humid regions where 2:1 clays predominate. In warmer regions where organic matter levels are lower and clays provide less CEC, the target pH would likely be closer to 5.8 and the amounts of lime required would be one-half to one-third of those indicated here. In any case, it is unwise to apply more than 7 to 9 Mg/ha (3 to 4 tons/acre) of liming materials in a single application. If more is needed, subsequent applications can be made at two- to three-year intervals until the desired pH is achieved.

BOX 9.2
CALCULATING LIME NEEDS BASED ON pH BUFFERING

Your client wants to grow a high-value crop of asparagus in a 2-ha field. The soil is a sandy loam with a current pH of 5.0. Asparagus is a calcium-loving crop that requires a high pH (6.8) for best production (see Figure 9.16). Since the soil texture is a sandy loam, we will assume that its buffer curve is similar to that of the moderately buffered soil B in Figure 9.6. In actual practice, a soil test laboratory using this method to calculate the lime requirement should have buffer curves for the major types of soils in its service area.

1. Extrapolating from curve B in Figure 9.6, we estimate that it will require about 2.5 $cmol_c$ of lime/kg of soil to change the soil pH from 5.0 to 6.8. (Draw a horizontal line from 5.0 on the y-axis of Figure 9.6 to curve B, then draw a vertical line from this intersection down to the x-axis. Repeat this procedure beginning at 6.8 on the y-axis. Then use the x-axis scale to compare the distance between where your two vertical lines intersect the x-axis.)
2. Each molecule of $CaCO_3$ neutralizes 2 H^+ ions: $CaCO_3 + 2H^+ \rightarrow \rightarrow Ca^{2+} + CO_2 + H_2O$
3. The mass of 2.5 $cmol_c$ of pure $CaCO_3$ can be calculated using the molecular weight of $CaCO_3$ = 100 g/mol:

$$(2.5\ cmol_c\ CaCO_3/kg\ soil) \times (100\ g/mol\ CaCO_3) \times (1\ mol\ CaCO_3/2\ mol_c) \times (0.01\ mol_c/cmol_c)$$

$$= 1.25\ g\ CaCO_3/kg\ soil$$

4. Using the conversion factor of 2×10^6 kg/ha of surface soil (see footnote 4 in chapter 4) we calculate the amount of pure $CaCO_3$ needed per hectare:

$$(1.25\ g\ CaCO_3/kg\ soil) \times (2 \times 10^6\ kg\ soil/ha) = 2,500,000\ g\ CaCO_3/ha$$

$$(2,500,000\ g\ CaCO_3/ha) \times (1\ kg\ CaCO_3/1000\ g\ CaCO_3) = 2500\ kg\ CaCO_3/ha$$

$$2500\ kg\ CaCO_3/ha = 2.5\ Mg/ha\ (or\ about\ 1.1\ tons/acre)$$

5. Since our limestone has a $CaCO_3$ equivalence of 90%, 100 kg of our limestone would be the equivalent of 90 kg of pure $CaCO_3$. Consequently, we must adjust the amount of our limestone needed by a factor of 100/90:

$$2.5\ Mg\ pure\ CaCO_3 \times 100/90 = 2.8\ Mg\ limestone/ha$$

6. Finally, because not all the $CaCO_3$ in the limestone will completely react with the soil, the amount calculated from the laboratory buffer curve is usually increased by a factor of 2:

$$(2.8\ Mg\ limestone/ha) \times 2 = 5.6\ Mg\ limestone/ha$$

(Using Appendix B, this value can be converted to about 2.5 tons/acre.)

Note that this result is very similar to the amount of lime indicated by the chart in Figure 9.17 for this degree of pH change in a sandy loam.

Exchangeable Aluminum   Liming to eliminate exchangeable aluminum, rather than to achieve a certain soil pH, has been found appropriate for highly weathered soils such as Ultisols and Oxisols. By this approach, the required amount of lime can be calculated using values for the initial CEC and the percent Al saturation. For example, if a soil has a CEC of 10 $cmol_c$/kg and is 50% Al saturated, then 5 $cmol_c$/kg of Al^{3+} ions must be displaced (and their acidity from Al hydrolysis neutralized). This would require 5 $cmol_c$/kg of $CaCO_3$:

$$5\ cmol_c/kg \times (100\ g/mol\ CaCO_3) \times (1\ mol\ CaCO_3/2\ mol_c) \times (0.01\ mol_c/cmol_c)$$
$$= 2.5\ g\ CaCO_3/kg\ soil$$

This amount is equivalent to 5000 kg/ha (2.5 g/kg $\times\ 2 \times 10^6$ kg/ha). Experience suggests that to ensure a complete reaction in the field, the amount of limestone so calculated must be multiplied by a factor of 1.5 or 2.0 to give the actual amount of lime to apply.

How Lime Is Applied

Frequency   Liming materials slowly react with soil acidity, gradually raising the pH to the desired level over a period ranging from a few weeks in the case of hydrated

Practical tutorial on agricultural liming practices: http://hubcap.clemson.edu/ ~blpprt/acidity.html

Figure 9.18
Bulk application by specially equipped trucks is the most widespread method of applying ground limestone. The scene pictured occurred on a windy day and the dispersion by wind illustrates the finely ground nature of the agricultural limestone applied. To avoid problems with heavy trucks bogging down in soft, recently tilled soils, it is often preferred to make lime applications to land that is in sod, under no-till management, or frozen hard. (Photo courtesy of R. Weil)

lime to a year or so with finely ground limestone. As Ca and Mg are removed from the soil by plants or by leaching, they are replaced by acid cations and the pH of the soil gradually drops. In humid regions the forces of acidification proceed relentlessly and application of lime to arable soils is not a one-time proposition, but must be repeated every 3 to 5 years to maintain the desired pH level.

Because of its gradual effects, lime should be spread about 6 to 12 months ahead of the crop that has the highest pH and calcium requirements. Thus, in a rotation with corn, wheat, and two years of alfalfa, the lime may be applied after the corn harvest to favorably influence the growth of the alfalfa crop that follows. However, since most lime is bulk-spread using heavy trucks (Figure 9.18), applications on the sod or hay crop rather than tilled ground will minimize soil compaction.

Liming will be most beneficial to acid-sensitive plants if as much as possible of the root environment is altered. However, in most instances it is economically and physically feasible to mix lime into only the upper 15 to 20 cm of soil. The Ca^{2+} and Mg^{2+} ions provided by limestone replace acid cations on the exchange complex and do not move readily down the profile. Therefore, for soil with a high CEC, the short-term effects of limestone are mainly limited to the soil layer into which the material was incorporated.

Overliming   A practical consideration is the danger of **overliming**—the application of so much lime that the resultant pH values are too high for optimal plant growth. Overliming is not very common on fine-textured soils with high buffer capacities, but it can occur easily on coarse-textured soils that are low in organic matter. The detrimental results of excess lime include deficiencies of iron, manganese, copper, and zinc; reduced availability of phosphate; and constraints on the plant absorption of boron from the soil solution. It is an easy matter to add a little more lime later, but quite difficult to counteract the results of applying too much. Therefore, liming materials should be added conservatively to poorly buffered soils. For some Ultisols and Oxisols, overliming may occur if the pH is raised even to 6.0.

Liming Forests   Spreading of limestone on forested watersheds is rarely practical except with very acid sandy soils, on which small applications may ameliorate the ill effects of soil acidity and provide sufficient calcium for the trees.

Untilled Soils   Some soil–plant systems, such as no-till cropland, tree farms, and turf, make it difficult to incorporate and mix limestone with the soil. Fortunately, the undisturbed residue mulches in these systems tend to encourage earthworm activity, which can help distribute lime down the profile (Plate 81).

9.9   ALTERNATIVE WAYS TO AMELIORATE THE ILL EFFECTS OF SOIL ACIDITY

Where the principal acidity problem is in the surface horizon and where sources of limestone are readily accessible, traditional liming procedures, as just described, are quite effective and economical. However, where subsoil acidity is a problem or where liming materials are not accessible or affordable, several other approaches to combating the negative effects of soil acidity may be appropriate for use with or without traditional liming. Particularly deserving of attention are the use of gypsum and organic materials to reduce aluminum toxicity and the use of plant species or genotypes that tolerate acid conditions.

Gypsum Applications

Gypsum ($CaSO_4 \cdot 2H_2O$) is a widely available material found in natural deposits or as industrial by-products. Gypsum can ameliorate aluminum toxicity despite the fact that it does not increase soil pH. In fact, gypsum has been found more effective than lime in reducing exchangeable aluminum in subsoils and thereby in improving root growth and crop yields (see Figure 9.19).

One reason for the effectiveness of gypsum is that the calcium from surface-applied gypsum moves down the soil profile more readily than that from lime. As lime dissolves, its reactions raise the pH, thus increasing the pH-dependent charges on the soil colloids, which in turn retain the released Ca^{2+}, preventing its downward leaching. In addition, the anion released by lime is either OH^- or CO_3^{2-}, both of which are largely removed by the lime reactions (either by forming water or carbon dioxide gas), thus depriving the Ca^{2+} cations of surplus anions that could accompany them in the leaching process. By contrast, gypsum as a neutral salt does not raise the soil pH and so does not increase the CEC. Furthermore, the SO_4^{2-} anion released by the dissolution of gypsum is available to accompany Ca^{2+} cations in leaching.

Using Organic Matter

Practices such as the application of organic wastes, production of cover crops (see Section 13.2), and mulching or return of crop residues all increase the organic matter in soil. In so doing, they can ameliorate the effects of soil acidity in at least three ways:

Figure 9.19

Aluminum saturation percentage in the subsoil of a fine-textured Hawaiian Ultisol after treatment of the surface soil with chicken manure, limestone, or gypsum. The soil was slowly leached with 380 mm of water. The lime raised the pH and thereby effectively reduced Al saturation, though only in the upper 10 to 20 cm. effect of gypsum extended somewhat more deeply. Gypsum does not raise the pH or increase the CEC, but is more soluble than lime and provides the SO_4^{2-} anions to accompany Ca^{2+} cations as they leach downward in solution. The greatest and deepest reduction in Al saturation resulted from the chicken manure. Another Ca-rich organic amendment, sewage sludge, gave similar results (not shown). The manure probably formed soluble organic complexes with Ca^{2+} ions, which then moved down the profile where Ca exchanged with Al to form nontoxic organic Al complexes.

[Redrawn from Hue and Licudine (1999); used with permission of the American Society of Agronomy]

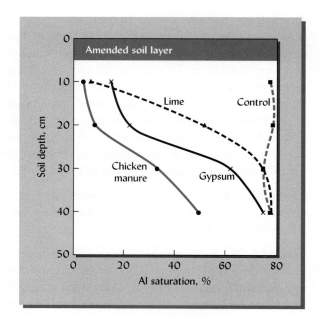

1. Humified organic matter can bind tightly with aluminum ions and prevent them from reaching toxic concentrations in the soil solution.
2. Low-molecular-weight organic acids produced by microbial decomposition or root exudation can form soluble complexes with aluminum ions that are nontoxic to plants and microbes.
3. Many organic amendments contain substantial amounts of calcium held in organic complexes that leach quite readily down the soil profile. Therefore, if such amendments as legume residues, animal manure, or sewage sludge are high in Ca, they can effectively combat aluminum toxicity and raise Ca and pH levels, not only in the surface soil where they are incorporated, but also quite deep into the subsoil (see Figure 9.19).

Enhancing these organic matter reactions may be more practical than standard liming practices for resource-poor farmers or those in areas far from limestone deposits. **Green manure** crops (vegetation grown specifically for the purpose of adding organic matter to the soil) and mulches can provide the organic matter needed to stimulate such interactions and thereby reduce the level of Al^{3+} ions in the soil solution. Aluminum-sensitive crops can then be grown following the green manure crop. One caution regarding use of organic amendments to ameliorate soil acidity is that the amounts of these materials effective for this purpose may exceed amounts suggested by nutrient management guidelines designed to prevent polluting water from excessive leaching and runoff losses of nitrogen and phosphorus (see Section 13.3).

Selecting Adapted Plants

It is often more judicious to solve soil acidity problems by changing the plant to be grown rather than by trying to change the soil pH. The choice of plant species should consider soil pH adaptation, whether revegetating a former coal mine site on acidic mining debris or landscaping a suburban front yard on alkaline desert soils.

Plant breeders and biotechnologists have developed cultivars that are quite tolerant of very acid conditions. These varieties are especially valuable in some areas of the tropics where even modest liming applications are economically impractical. These advances highlight the importance of collaboration between plant and soil scientists in enhancing plant production on acid, degraded soils.

9.10   LOWERING SOIL pH

In arid regions such as the western United States, it is often desirable to reduce the pH of highly alkaline soils. Furthermore, some acid-loving plants cannot even tolerate near-neutral pH values. For example, rhododendrons and azaleas, favorites of gardeners around the world, grow best on soils having pH values of 5.0 and below. To accommodate such plants, it is sometimes desirable to increase the acidity of even mildly acid soils. This is done by adding acid-forming organic and inorganic materials.

As organic residues decompose, organic and inorganic acids are formed. These can reduce the soil pH if the organic material is low in calcium and other nonacid cations. Leaf mold from coniferous trees, pine needles, tanbark, pine sawdust, and acid peat moss are quite satisfactory organic materials to add around ornamental plants (but see Chapter 11 for nitrogen considerations with these materials). However, some farm manures (particularly poultry manures) and leaf mold of such Ca-efficient trees as beech and maple may be alkaline and may increase the soil pH.

Inorganic Chemicals

When the addition of acid organic matter is not feasible, inorganic chemicals such as aluminum sulfate ("alum") or ferrous sulfate ($Fe^{II}SO_4$) may be used. The latter chemical provides available iron (Fe^{2+} ions) for the plant and, upon hydrolysis, enhances acidity by reactions similar to Equations 9.17 and 9.18.

Ferrous sulfate thus serves a double purpose for iron-loving plants by supplying available iron directly and by reducing the soil pH—a process that may cause a release of fixed iron present in the soil (see Plates 86 and 88). Ferrous sulfate should be worked into the soil around ornamental plants, while taking care to avoid overly disturbing the root system. Contact with ferrous sulfate (but not alum) may cause black discoloration of foliage or mulch from the formation of iron sulfides.

Another material often used to increase soil acidity is elemental sulfur (Box 9.3). As the sulfur undergoes microbial oxidation in the soil (see Section 12.2), 2 moles of acidity (as sulfuric acid) are produced for every mole of S oxidized:

$$2S + 3O_2 + 2H_2O \longrightarrow 2H_2SO_4 \qquad (9.23)$$

Under favorable conditions, sulfur is 4 or 5 times more effective, kilogram for kilogram, in developing acidity than is ferrous sulfate. Although ferrous sulfate brings about more rapid plant response, sulfur is less expensive, is easy to obtain, and is often used for other purposes. The quantities of ferrous sulfate or sulfur that should be applied will depend upon the buffering capacity of the soil and its original pH level. Figure 9.6 suggests that for each unit drop in pH desired, a well-buffered soil (e.g., a silty clay loam with 4% organic matter) will require about 4 $cmol_c$ of sulfur per kilogram of soil. This is about 1200 kg S/ha (since 1 $cmol_c$ of S = 0.32/2 = 0.16 g, the 2 mol_c/mol being based on the 2 mol of H^+ ion produced by each mole of S).

9.11 CHARACTERISTICS AND PROBLEMS OF DRY-REGION SOILS

The water-limited, high-pH, carbonate-rich nature of dry-region soils results in many characteristics and problems that are not generally found in the acid soils of more humid regions. We will begin by focusing on the nature of alkaline soils that do *not* have the excessive levels of salts or sodium covered in Sections 9.14 to 9.15.

Heterogeneity of Soils, Vegetation, and Hydrology of Noncultivated Soils

Seen from the highway, many semiarid landscapes appear to be quite densely covered with vegetation. However, if one walks out into the landscape it is immediately evident that the plants are widely spaced, with much of the surface being quite bare of vegetation (Figure 2.18). Such patchy vegetation is characteristic of environments in which water is too sparse to support a complete vegetative cover (see Section 6.4). Less obvious to the casual observer is that the soil under the plants is quite different from that in the bare spaces and that this difference is both a result of—and a cause of—the scattered plant distribution.

Islands of Fertility   Soil scientists studying arid lands have found that plants generally enhance the soil beneath them in several ways. Plants add litter, host macro- and microorganisms, and trap windblown particles. Soils under clumps of vegetation therefore become higher in organic matter, silt, and clay, as well as richer in nitrogen and other nutrients.

Increased soil organic matter from plant litter leads to greater aggregation, while generations of plant roots create a network of biopores open to the surface, and the plant canopy partially protects the surface soil structure when it rains (see Section 14.5).

BOX 9.3
COSTLY AND EMBARRASSING SOIL pH MYSTERY

Large expanding areas of ugly dead turfgrass signaled that something was terribly wrong just a week before the grand opening of an elaborate new public garden in the heart of Washington, DC. After so many rare and exotic trees, shrubs, and flowers had been planted, the horticulturalist in charge feared the worst—the soil might be toxic, need to be removed, and the whole garden started over. The garden was built partly over an underground museum. Beneath the pleasing, undulating surface topography, 1 meter of "topsoil" covered a tangle of pipes, conduits, and wires on the museum roof. If the soil needed to be removed, it would have to be done the slow and expensive way—by hand. The horticulturalist suspected some toxic factor in the soil was killing the grass and would soon start damaging other plants as well. The landscape design specifications had called for "a natural friable soil . . . with 2% organic content . . . USDA

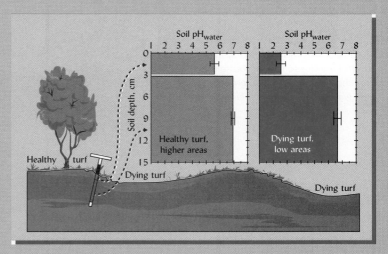

Figure 9.20
Dying turf and soil acidity in a rooftop garden. (Courtesy of R. Weil)

textural class of loam and . . . pH 5.5 to 7.0." The lowest bidder had offered to make a "topsoil" using sediments dredged from a nearby tidal river, modified with enough lime and sand to meet the pH and texture requirements. The consulting engineers had run their lab tests and determined that the material met the specifications.

In late April, as the grand opening date approached, the grass began to turn brown and die in small patches. Although the shrubs, trees, and flowers in the cultivated flower beds were still looking good, the dead patches of turfgrass grew larger with every passing day. Turf specialists were called in, but could find no diseases or pests to account for the dead patches. Now, in desperation, the horticulturalist paced nervously in the light June rain as several soil scientists worked feverishly to collect soils samples.

The soil scientists, trained as pedologists, augered deep, looking in vain for telltale signs of acid sulfate weathering (Section 9.6) they suspected might solubilize toxic aluminum and heavy metals from the river sediments. Others, noting that it was the turfgrass that seemed to suffer first, went shallow instead of deep (see Box 1.2), obtaining a separate set of samples from the topmost 3 cm. Since the turf was most damaged in the low spots, they collected pairs of soil samples from several areas of dead turf and from adjacent areas where the turf was relatively healthy. Back at the lab, they stirred each soil sample in water and measured the pH. The results were completely normal until they got to the samples of the topmost 3 cm layer. Then they couldn't believe their eyes—all these samples from dead turf areas gave readings between pH 2 and pH 3 (graph in Figure 9.20). Looking closely, they noticed small yellow flecks that smelled like sulfur. The pieces to the puzzle began to fall into place.

The previous summer, shortly after the sod had been installed, the horticulturalist had pulled "normal" 20 cm deep soil cores, which had tested at pH 7.2, considerably above the pH 6.0 to 6.5 recommended for the fine fescue turfgrass. Therefore, he had applied about 1000 kg/ha of sulfur (S) powder, as recommended, to lower the pH by about 1 unit. He pulled another set of 20 cm deep soil samples about two months later, and the pH was still about 7.0. So he repeated the S application. The lawn looked healthy during the cool, rainy winter, while the landscapers installed the valuable trees and shrubs. What the horticulturalist had failed to consider was that time and warm weather would be needed for soil microorganisms to oxidize the S and acidify the soil. The second S application had been an over-response to the normal delay and had doubled the amount of S available to oxidize. Sulfur powder is water repellent and buoyant, so rain easily washed much of it off the high areas into the low spots, thus doubling or tripling the already doubled application—giving five or six times the recommended S concentration in those areas. When warm, wet weather in spring stimulated the S-oxidizing bacteria to go into high gear, extreme acidity was produced in the thin surface layer of soil where the S was located and most of the turfgrass roots proliferated. Thankfully, the remedy would be simple and inexpensive: remove the sod along with about 5 cm of soil and lay down new sod. This they did and everyone at the opening ceremony was impressed by the beautiful lawn and garden.

The lessons learned? (1) Many soil processes are biological in nature—they respond to environmental conditions with time. (2) Taking deep soil samples may "dilute out" evidence of extreme conditions near the soil surface. Therefore, be sure to sample the upper few cm separately for untilled soils, especially if amendments have been applied.

Figure 9.21

Differential runoff and infiltration rates may give rise to "islands" of enhanced soil water availability in arid and semiarid rangelands. Slight depressions and other features initially concentrate rainwater in small areas. Once plants establish in a pocket of relatively high moisture, they tend to amplify the soil heterogeneity because their litter and roots further enhance water infiltration in their vicinity.

(Diagram courtesy of R. Weil)

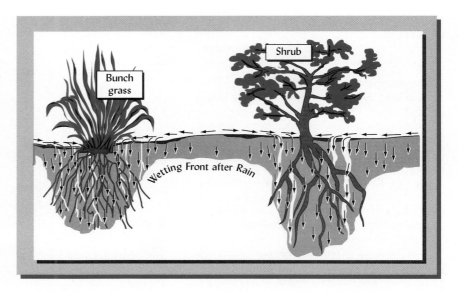

All these effects lead to a notably enhanced infiltration of water into the soils in the vegetated patches (Figure 9.21).

Over time, this *positive feedback loop* promotes a level of soil productivity in the vegetated patches distinctly more favorable than that in the unvegetated bare spaces. These so-called "islands of fertility" are a common feature of many arid land soils. Vegetation can also encourage "islands of fertility" by protecting soil from erosive desert winds.

Another special feature that influences the hydrology of arid regions' soils in their natural state is a biological crust, which often appears as a thick, dark-colored, sometimes jagged, coating on the soil surface. We shall discuss these fragile living structures in Section 10.14.

Calcium-Rich Layers

Illustrated tutorial on carbonates in soils: http://edafologia.ugr.es/carbonat/indexw.htm

Soils of low-rainfall areas commonly accumulate calcium carbonate that forms a *calcic horizon* at some depth in the soil profile (Plate 13). Calcareous soil materials (with free calcium carbonates) can be distinguished in the field by the effervescence (fizzing) that occurs if a drop of acid (10% HCl or strong vinegar) is applied. The high carbonate concentrations in these calcic horizons can inhibit root growth for some plants. In eroded spots, or in regions of very low rainfall (<25 cm/yr), carbonate concentrations may be found at or near the soil surface (Plate 99). In these cases, serious micronutrient and phosphorus deficiencies can be induced in plants that are not adapted to calcareous conditions (see following).

In other alkaline soils, one or more subsoil layers may be cemented into hard, concretelike horizons such as petrocalcic layers or duripans (Figure 3.14). Many alkaline soils also contain layers rich in calcium sulfate (gypsum), a mineral much more soluble than calcium carbonate. The depth of calcic horizons and gypsic horizons is largely determined by the age of the soil and by the amount of rainfall available to leach these minerals downward.

Colloidal Properties

Cation Exchange Capacity   The cation exchange capacities (CECs) of alkaline soils are commonly higher than those of acid soils with comparable soil textures. This is true for two reasons. First, the 2:1-type clays that are most common in alkaline soils possess high amounts of permanent charge. Second, the high pH levels of alkaline

soils stimulate high levels of pH-dependent charges on the soil colloids, especially humus (Section 8.9).

Clay Dispersion   Clays in alkaline soils are particularly subject to deflocculation or dispersion because (1) the iron and aluminum coatings that act as strong flocculating and cementing agents in acid soils are largely lacking in alkaline soils, (2) the types of clays dominant in alkaline soils are especially susceptible to dispersion, and (3) mono-valent ions (Na^+ and K^+) that are easily leached from acid soils (see Section 8.10) are still largely unleached in the soils of dry regions. Dispersion of soil clay leads to drastically reduced macroporosity, aeration, and water percolation and to the sealing of the soil surface (see Sections 9.14–9.15).

Nutrient Deficiencies

The availability of most nutrient elements is markedly influenced by soil pH, so alkaline soils can be expected to exhibit special problems regarding the solubility of plant nutrients and other elements. In addition, unlike in humid regions, the minimal weathering allows many weatherable and relatively soluble minerals to remain in the soil, in some cases contributing high levels of certain elements to the soil–plant–animal system.

The micronutrients zinc, copper, iron, and manganese are readily available in acid soils but are much less soluble at pH levels above 7. Therefore, in alkaline soils plant growth is commonly limited by deficiencies of these elements (see Plates 97–99). The low organic-matter levels of most dry-region soils further reduce the availability of these metals.

Boron deficiency is common at high pH levels in both sandy soils (because of low boron content) and clayey soils (because the boron is tightly held by the clay). In addition, plants tend to have a higher requirement for boron if calcium is abundant. For all these reasons, boron deficiencies are quite common in alkaline soils. In contrast to boron, molybdenum availability is high under alkaline conditions—so high that in some areas molybdenum toxicity is a problem. For ruminant animals (e.g., cattle and sheep) grazing on high molybdenum soils, the high solubility of molybdenum and low solubility of copper at high pH can combine to cause a disease known as **molybdenosis**.

Alkalinity: High Soil pH   There seems to be considerable confusion about the terms **alkaline** and **alkalinity**, and people often use these terms incorrectly to describe soils characterized by detrimental levels of soluble salts or sodium. Alkaline soils are simply those with a pH above 7.0. *Alkalinity* refers to the concentration of OH^- ions, much as *acidity* refers to that of H^+ ions. *Alkaline* soils should not be confused with *alkali* soils. The latter name is an obsolete term for what are now called **sodic** or **saline–sodic** soils, those with levels of sodium high enough to be detrimental to plant growth (see Section 9.14). The sources of alkalinity were discussed in Section 9.1.

9.12   DEVELOPMENT OF SALT-AFFECTED SOILS[6]

Salt-affected soils cover about 320 million ha of land throughout the world, the largest areas being found in Australia, Africa, Latin America, southwest United States, and the Near and Middle East. They typically occur in areas with precipitation-to-evaporation ratios of 0.75 or less and in low, flat areas with high water tables that may be subject to seepage from higher elevations (Plate 104). Nearly 50 million ha of cropland and pasture are currently affected by salinity, and in some regions the area of land so affected is growing by about 10% annually.

Extent and causes of salt-affected soils around the world: www.fao.org/AG/AGL/agll/spush/topic2.htm

---

[6]For a discussion of these soils, which are also referred to as *halomorphic soils*, see Abrol et al. (1988) and Szabolcs (1989).

In most cases, the soluble salts in soils originate from the weathering of primary minerals in rocks and parent materials. Salts may be transported to a developing salt-affected soil as salt-containing water moving from areas of higher to lower elevations and from soil zones that are wetter to those that are drier. The water eventually evaporates; however, the dissolved salts are left behind to accumulate in the soil. This is true in both irrigated and unirrigated landscapes.

Many salt-affected soils develop because changes in the local water balance, usually brought about by human activities, increase the input of salt-bearing water more than they increase the output of drainage water. Increased evaporation, waterlogging, and rising water tables usually result. It is worth remembering the irony that *salts usually become a problem when too much water is supplied, not too little.*

Accumulation of Salts in Nonirrigated Soils

In the United States, about one-third of the soils in arid and semiarid regions are affected by some degree of salinity. Chlorides and sulfates of calcium, magnesium, sodium, and potassium accumulate naturally in some surface soils because there is insufficient rainfall to flush them from the upper soil layers. In coastal areas, sea spray (Plate 105) and inundation with seawater can be locally important sources of salt in soils, even in humid regions.

Other localized but important sources are fossil deposits of salts laid down during geological time. These fossil salts can dissolve in underground waters and move horizontally over impervious geological layers and ultimately rise to the surface of the soil in the low-lying parts of the landscape. The low-lying areas where the saline groundwater emerges are termed **saline seeps**.

Saline seeps occur naturally in some locations, but their formation is often greatly increased when the water balance in a semiarid landscape is disturbed by bringing land under cultivation (Figure 9.22). Replacement of native, deep-rooted perennial vegetation with annual crop species greatly reduces the annual evapotranspiration, especially if the cropping system includes periods of fallow during which the soil is unvegetated. The decreased evapotranspiration allows more rainwater to percolate through the soil, thus raising the water table and increasing the flow of groundwater to lower elevations.

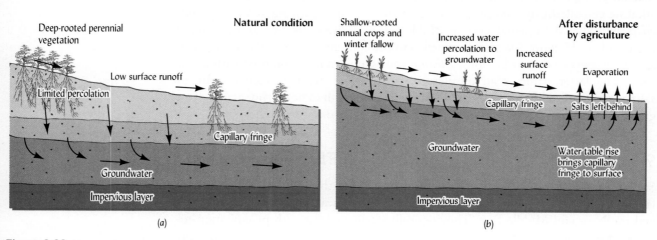

(a)   (b)

Figure 9.22

Saline seep formation in a semiarid area where the salt-rich substrata are underlain by an impermeable layer. (a) Under deep-rooted perennial vegetation, transpiration is high and the water table is kept low. (b) After conversion to agriculture, shallow-rooted annual crops take up much less water, especially if fallow is practiced, allowing more water to percolate through the salt-bearing substrata. Consequently, in lower-elevation landscape positions, the wet-season water table rises close to the soil surface. This allows the salt-laden groundwater to rise by capillary flow to the surface, from which it evaporates, leaving behind an increasing accumulation of salts. Note that the diagrams greatly exaggerate vertical distances. (Diagrams courtesy of R. Weil)

In dry regions, soils and substrata may contain substantial amounts of soluble salts that can be picked up by the percolating water. Eventually, the water table may rise to within 1 m or less of the soil surface. Capillary rise will then contribute a continuous stream of salt-laden water to replace the water lost at the surface by evaporation, leaving behind the salts, which soon will accumulate to levels that inhibit plant growth. Year by year, the evaporation zone will creep up the slope, and the barren, salinized area will become larger and more saline. Millions of hectares of land in North America, Australia, and other semiarid regions have been degraded in this fashion.

Irrigation-Induced Salinity and Alkalinity

Irrigation not only alters the water balance by bringing in more water, it also brings in more salts. Whether taken from a river or pumped from the groundwater, even the best quality freshwater contains some dissolved salts (see Section 9.16). The amount of salt brought in with the water may seem negligible, but the amounts of water applied over the course of time are huge. Again, pure water is lost by evaporation, but the salt stays and accumulates at the soil surface. The effect is accentuated in arid regions for two reasons: (1) the water available from rivers or from underground is relatively high in salts because it has flowed through dry-region soils that typically contain large amounts of easily weatherable minerals, and (2) the dry climate creates a relatively high evaporative demand, so large amounts of water are needed for irrigation. An arid-region farmer may need to apply 90 cm of water to grow an annual crop. Even if this is good-quality water relatively low in salts, it will likely add more than 6 Mg/ha (3 tons/acre) of salt on the land every year (see Section 9.18).

Australia's mirage of green pastures evaporates: www.csmonitorcom/1996/0508/050896.intl.intl.6.html

If irrigation water carries a significant proportion of Na^+ ions compared to Ca^{2+} and Mg^{2+} ions, and especially if the HCO_3^- ion is present, sodium ions may come to saturate a major part of the colloidal exchange sites, creating an unproductive **sodic** soil (Section 9.15).

During the past three decades, low-income countries in the dry regions of the world have greatly expanded the area of their land under irrigation in order to produce the food needed by their rapidly growing human populations. Initially, the expanded irrigation stimulated phenomenal increases in food-crop production. Unfortunately, many irrigation projects failed to provide for adequate drainage. As a result, the process of **salinization** has accelerated, and salts have accumulated to levels that are already adversely affecting crop production. In some areas, sodic soils have been created.

In biblical times, southeastern Iraq, an area irrigated by water from the Euphrates and Tigris, was so productive that the overall region was called the Fertile Crescent. Unfortunately, salinization set in when the societies sometimes failed to maintain the drainage ditches. Salts eventually accumulated to such a degree that crop production declined, and the area had to be abandoned.

Today, societies around the world are repeating the mistakes of the past. Some observers believe that each year, the area of previously irrigated land degraded by severe salinization is greater than the area of land newly brought under irrigation. Truly, the world needs to give serious attention to the large-scale problems associated with salt-affected soils.

9.13 MEASURING SALINITY AND SODICITY[7]

Salt-affected soils adversely affect plants because of the total concentration of salts (*salinity*) in the soil solution and because of concentrations of specific ions, especially sodium (*sodicity*).

---

[7]For an informative discussion of these methods, see Rhoades et al. (1999) or the Web site of the U.S. Salinity Laboratory at www.ars.usda.gov/main/site_main.htm?modecode=53102000

Salinity

Total Dissolved Solids   In concept, the simplest way to determine the total amount of dissolved salt in a sample of water is to heat the solution in a container until all of the water has evaporated and only a dry residue remains. The residue can then be weighed and the **total dissolved solids** (TDS) expressed as milligrams of solid residue per liter of water (mg/L). In water to be used for irrigation, TDS typically ranges from about 5 to 1000 mg/L, while in the solution extracted from a soil sample, TDS may range from about 500 to 12,000 mg/L.

Saturated paste and dilute extracts in soil testing for salt-affected turfgrass: http://gcsaa.org/gcm/2003/sept03/PDFs/09Clarify.pdf

Electrical Conductivity   Pure water is a poor conductor of electricity, but conductivity increases as more and more salt is dissolved in the water. Thus, the **electrical conductivity** (EC) of the soil solution gives us an indirect measurement of the salt content. The EC can be measured both on samples of soil (Figure 9.23) or on the bulk soil *in situ*. It is expressed in terms of deciSiemens per meter (dS/m).[8]

U.S. Soil Salinity Laboratory "News and Events": www.ars.usda.gov/pwa/?riverside/gebjsl

Mapping EC in Situ   Advances in instrumentation now allow rapid, continuous field measurement of bulk soil conductivity, which, in turn, is directly related to soil salinity. One rapid field method employs **electromagnetic induction** (EM) of electrical current in the body of the soil, the level of which is related to electrical conductivity and, in turn, to soil salinity. A small transmitter coil located in one end of the EM instrument generates a magnetic field within the soil. This magnetic field, in turn, induces small electric currents within the soil whose values are related to the soil's conductivity. These small currents generate their own secondary magnetic fields, which can be measured by a small receiving cell in the opposite end of the EM instrument. The EM instrument thus can measure ground EC to considerable depths in the soil profile without mechanically probing the soil. A handheld model of such an EM conductivity sensor is shown in Figure 9.24. The same type of instrument can be vehicle-mounted and used to rapidly map the soil salinity levels across a field.

Advances in mobile salinity sensors have made it possible to produce detailed maps of the variation in salinity within a given field. The information from these

Figure 9.23
Measuring the electrical conductivity (EC) of a soil sample in a field of wheatgrass to determine the level of salinity. A sample of the soil is stirred with pure water until a saturated paste is made. The paste is then transferred into a special conductivity cup that has a flat, circular electrode on either side (inset). This is then inserted into a stand that connects the electrodes to a conductivity meter. Note readout of 7.78 dS/m on the conductivity meter. This level of EC_p indicates a highly saline soil that would inhibit the growth of many crops. (Photos courtesy of R. Weil)

---

[8]Formerly expressed as millimhos per centimeter (mmho/cm). Since 1 S = 1 mho, 1 dS/m = 1mmho/cm.

Figure 9.24

A portable electromagnetic (EM) soil conductivity sensor used to estimate the electrical conductivity in the soil profile. When placed on the soil surface in the horizontal position (lower left), this instrument senses electrical conductivity of the soil down to about 1 m depth. When placed in the vertical position (as in the inset photo), the effective depth is about 2 m. This type of EM sensor (model EM-38, made by Geonics, Ltd., Ontario, Canada) can be mounted on a special vehicle for mobile soil salinity mapping. (Photos courtesy of R. Weil)

maps can then be used in the techniques of **precision agriculture** (see Section 13.10), which are capable of applying corrective measures tailored to match the degree of salinity in each small part of a large field.

Sodium Status

Two expressions are commonly used to characterize the sodium status of soils. The **exchangeable sodium percentage** (ESP) identifies the degree to which the exchange complex is saturated with sodium:

$$\text{ESP} = \frac{\text{Exchangeable sodium, cmol}_c/\text{kg}}{\text{Cation exchange capacity, cmol}_c/\text{kg}} \times 100 \qquad (9.24)$$

ESP levels greater than 15 are associated with severely deteriorated soil physical properties and pH values of 8.5 and above.

The **sodium adsorption ratio** (SAR) is a second, more easily measured property that is becoming even more widely used than ESP. The SAR gives information on the comparative concentrations of Na^+, Ca^{2+}, and Mg^{2+} in soil solutions. It is calculated as follows:

$$\text{SAR} = \frac{[Na^+]}{(0.5[Ca^{2+}] + 0.5[Mg^{2+}])^{1/2}} \qquad (9.25)$$

where $[Na^+]$, $[Ca^{2+}]$, and $[Mg^{2+}]$ are the concentrations (in mmol of charge per liter) of the sodium, calcium, and magnesium ions in the soil solution. An SAR value of 13 for the solution extracted from a saturated soil paste is approximately equivalent to an ESP value of 15. The SAR of a soil extract takes into consideration that the adverse effect of sodium is moderated by the presence of calcium and magnesium ions. The SAR also is used to characterize irrigation water applied to soils (see Section 9.17).

High amounts of other monovalent ions such as potassium (K^+) can also promote soil structure degradation, though less so than sodium. Therefore, some soil scientists suggest that the SAR should be modified to include the sum of (Na^+) + (K^+) in the numerator of Equation 9.25. Excessive K^+ may originate from soil minerals or irrigation water as typically is the case for Na^+, but it may also come from overapplication of potassium fertilizer or from manure generated by animals fed a high-K diet, such as the alfalfa-rich diets used by many dairy farms.

9.14   CLASSES OF SALT-AFFECTED SOILS

Using EC, ESP (or SAR), and soil pH, salt-affected soils are classified as **saline**, **saline-sodic**, and **sodic** (Figure 9.25). Soils that are not greatly salt affected are classed as **normal**.

Saline Soils

The processes that result in the accumulation of neutral soluble salts are referred to as **salinization**. The concentration of these salts sufficient to seriously interfere with plant growth (see Section 9.16) is generally defined as that which produces an electrical conductivity in the saturation extract (EC) greater than 4 dS/m. However, many sensitive plants are adversely affected when the EC is only about 2 dS/m.

Saline soils contain sufficient salinity to give EC values greater than 4 dS/m, but have an ESP less than 15 (or an SAR less than 13) in the saturation extract. Thus, the exchange complex of saline soils is dominated by calcium and magnesium, not sodium. The pH of saline soils is usually below 8.5. Because soluble salts help prevent dispersion of soil colloids, plant growth on saline soils is not generally constrained by poor infiltration, aggregate stability, or aeration. In many cases, the evaporation of water creates a white salt crust on the soil surface (see Plate 104), which accounts for the name **white alkali** that was previously used to designate saline soils.

Figure 9.25

Diagram illustrating the classification of normal, saline, saline-sodic, and sodic soils in relation to soil pH, electrical conductivity (EC), sodium adsorption ratio (SAR), and exchangeable sodium percentage (ESP). Also shown are the ranges for different degrees of sensitivity of plants to salinity.

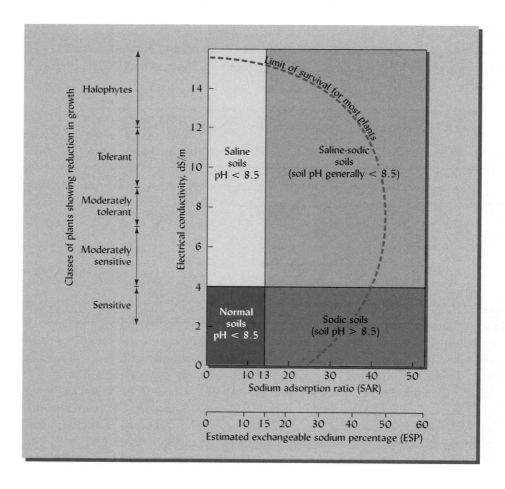

Saline-Sodic Soils

Soils that have both detrimental levels of neutral soluble salts (EC greater than 4 dS/m) *and* a high proportion of sodium ions (ESP greater than 15 or SAR greater than 13) are classified as **saline-sodic soils** (see Figure 9.25). Plant growth in these soils can be adversely affected by both excess salts and excess sodium levels.

Saline-sodic soils exhibit physical conditions intermediate between those of saline soils and those of sodic soils. The high concentration of neutral salts moderates the dispersing influence of the sodium because the excess cations move in close to the negatively charged colloidal particles, thereby reducing the tendency for particles to repel each other, or to disperse.

Unfortunately, this situation is subject to rather rapid change if the soluble salts are leached from the soil, especially if the SAR of the leaching waters is high. In such a case, salinity will drop, but the exchangeable sodium percentage will increase, and the saline-sodic soil will become a sodic soil.

Sodic Soils

Sodic soils are, perhaps, the most troublesome of the salt-affected soils (Section 9.15). While their levels of neutral soluble salts are low (EC less than 4.0 dS/m), they have relatively high levels of sodium on the exchange complex (ESP and SAR values are above 15 and 13, respectively). Some sodic soils in the order Alfisols (Natrustalfs) have a very thin A horizon overlying a clayey layer with columnar structure, a profile feature closely associated with high sodium levels (Figure 9.26). The pH values of sodic soils exceed 8.5, rising to 10 or higher in some cases.

The extremely high pH levels may cause the soil organic matter to disperse and/or dissolve. The dispersed and dissolved humus moves upward in the capillary water flow and, when the water evaporates, can give the soil surface a black color. The name **black alkali** was previously used to describe these soils. Plant growth on sodic soils is often constrained by specific toxicities of Na^+, OH^-, and HCO_3^- ions. However, the main reason for the poor plant growth—often to the point of complete barrenness—is that few plants can tolerate the extremely poor soil physical conditions and slow permeability to water and air characteristic of sodic soils.

Figure 9.26
The upper profile of a sodic soil (a Natrustalf) in a semiarid region of western Canada. Note the thin A horizon (knife handle is about 12 cm long) underlain by columnar structure in the natric (Btn) horizon. The white, rounded "caps" of the columns are composed of soil dispersed because of the high sodium saturation. The dispersed clays give the soil an almost rubbery consistency when wet. [Photo courtesy of Agriculture Canada, Canadian Soils Information System (CANSIS)]

9.15 PHYSICAL DEGRADATION OF SOIL BY SODIC CHEMICAL CONDITIONS

High sodium and low salt levels in sodic soils (and, to a lesser degree, in some "normal" soils) can cause serious degradation of soil structure and loss of macroporosity such that the movement of water and air into and through the soil is severely restricted. This structural degradation is most commonly measured in terms of the readiness of water movement—the saturated hydraulic conductivity K_{sat} of the soil (Section 5.5). The K_{sat} may be so low that the infiltration rate is reduced almost to zero, causing water to form puddles rather than soak into the soil. The soil is therefore said to be *puddled*, a condition characteristic of sodic soils. Physically, the puddled condition of a sodic soil is much like that of a rice paddy soil in which a farmer has mechanically destroyed the soil structure in order to be able to keep the paddy inundated with water.

Slaking, Swelling, and Dispersion

In most soils, the low permeability related to sodic conditions has three underlying causes. First, exchangeable sodium increases the tendency of aggregates and floccules to break up, or *slake*, upon becoming wet. The clay and silt particles released by slaking aggregates clog soil pores as they are washed down the profile. Second, when expanding-type clays (e.g., montmorillonite) become highly Na^+ saturated, their degree of swelling is increased. As these clays expand, the larger pores responsible for water drainage in the soil are squeezed shut. Third, and perhaps most importantly, sodic conditions lead to soil dispersion.

Two Causes of Soil Dispersion

Two chemical conditions promote dispersion. One is a high proportion of Na^+ ions on the exchange complex. The second is a low concentration of electrolytes (salt ions) in the soil water.

High Sodium Exchangeable Na^+ ions promote dispersion for two reasons. First, because of their single charge and large hydrated size, they are attracted only weakly to soil colloids, and so they spread out to form a relatively broad swarm of ions held in very loose outer-sphere complexes around the colloids (see Section 8.7). Second, compared to a swarm of divalent cations (which have two positive charges each), twice as many monovalent ions (with only one charge each) are needed to provide enough positive charges to counter the negative charges on a clay surface (Figure 9.27).

Low Salt Concentration A low ionic concentration in the bulk soil solution simultaneously increases the gradient causing exchangeable cations to diffuse away from the clay surface while it decreases the gradient causing anions to diffuse toward the clay. The result is a thick ionic layer or swarm of absorbed cations. Adding *any* soluble salt would increase the ionic concentration of the soil solution and encourage the opposite effects—resulting in a compressed ionic layer that allows the clay particles to come close enough together to form floccules. Thus, the damaging effects of sodium are greatest when salt concentrations are lowest.

It is worth remembering that *low salt (ion) concentrations and weakly attracted ions (e.g., sodium) encourage soil dispersion and puddling, while high salt concentrations and strongly attracted ions (e.g., calcium) promote clay flocculation and soil permeability.*

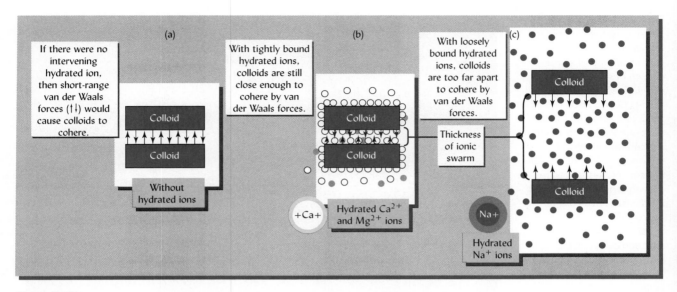

Figure 9.27

Conceptual diagrams showing how the type of cations present on the exchange complex influences clay dispersion. If colloids could approach closely (a), they would be held together (cohere) by short-range van der Waals forces. In soil, the colloids are surrounded by a swarm of hydrated exchangeable ions, which prevent the colloids from approaching so closely. If these are strongly attracted calcium and magnesium ions (b), they do not keep the colloids very far apart, so cohesive forces still have some effect. However, if they are sodium ions (c), the more spread-out ionic swarm keeps the colloids too far apart for cohesive forces to come into play. Sodium ions cause a spread-out ion swarm for two reasons: (1) their large hydration shell of water allows them to be only loosely attracted to the colloids, and (2) twice as many monovalent Na^+ ions as divalent (Ca^{2+} or Mg^{2+}) ions are attracted to a given colloid charge. When the colloid particles are separated from each other, the soil is in a dispersed condition. (Diagrams courtesy of R. Weil)

9.16 GROWTH OF PLANTS ON SALT-AFFECTED SOILS

How Salts Affect Plants

Plants respond to the various types of salt-affected soils in different ways. In addition to the nutrient-deficiency problems associated with high pH (see Sections 9.7 and 9.11), high levels of soluble salts affect plants by two primary mechanisms: **osmotic effects** and **specific ion effects**.

Soluble salts lower the osmotic potential of the soil water (see Section 5.3), making it more difficult for roots to remove water from the soil. Established plants expend more energy accumulating organic and inorganic solutes to lower the osmotic potential *inside* their cells to counteract the low osmotic potential of the soil solution outside. Plants are most susceptible to salt damage in the early stages of growth. Salinity may delay, or even prevent, the germination of seeds (Figure 9.28). Young seedlings may be killed by saline conditions that older plants of the same species could tolerate. As young plant cells encounter a soil solution high in salts, they may lose water by osmosis to the more concentrated soil solution. The cells then collapse.

The kind of salt can make a big difference in how plants respond to salinity. Certain ions, including Na^+, Cl^-, $H_3BO_4^-$, and HCO_3^-, are quite toxic to many plants. In addition to specific toxic effects, high levels of Na^+ can cause imbalances in the uptake and utilization of other cations such as K^+ or Ca^{2+}. Deterioration of physical properties in sodic soils may harm plants in at least two ways: (1) oxygen becomes deficient due to the breakdown of soil structure and the very limited air movement that results, and (2) water relations are poor due largely to the very slow infiltration and percolation rates.

Figure 9.28
Foliar symptoms on oldest leaves (inset), reduced germination, and stunted growth of soybean plants with increasing levels of soil salinity due to additions of NaCl to a sandy soil. The large numbers on the pots indicate the electrical conductivity (EC) of the soil in ds/m. Note that serious growth reductions occurred for this sensitive cultivar even at EC_e levels considered normal. The soybean cultivar used (Jackson) is more sensitive to salinity than most other cultivars of soybean. (Photos courtesy of R. Weil)

0.5   2.5   4.5

In response to excessive soil salinity, many plants become severely stunted and exhibit small dark-bluish green leaves with dull surfaces. High levels of sodium or chloride typically produce scorching or necrosis of the leaf margins and tips (see Figure 9.28, *inset*). These symptoms appear first and most severely on the oldest leaves as these have been transpiring water and accumulating salts for the longest period.

Selective Tolerance of Higher Plants to Saline and Sodic Soils

Satisfactory plant growth on salty soils depends on a number of interrelated factors, including the physiological constitution of the plant, its stage of growth, and its rooting habits. For example, old alfalfa plants are more tolerant to salt-affected soils than young ones, and deep-rooted legumes show a greater resistance to such soils than those that are shallow rooted.

Identifying a high-value crop that will grow in salt-affected soils: www.ars.usda.gov/is/AR/archive/aug04/salt0804.htm

Plant Sensitivity   While it is difficult to forecast precisely the tolerance of plant species to salty soils, numerous tests have made it possible to classify many plants into four general salt-tolerance groups (Table 9.4).

Among the plants that can be grown on salty soils are: (1) wild *halophytes* (salt-loving plants), and (2) salt-tolerant varieties developed by plant breeders. A number of wild halophytes have been found that are quite tolerant to salts and that possess qualities that could make them useful for human and/or animal consumption. They are also valuable for restoring disturbed or degraded land under saline conditions and for possible biofuel production.

Genetic Improvements   Plant breeders made a major advance with the discovery of a single gene that enables halophytes to sequester and thus tolerate high amounts of Na^+ in their cellular vacuoles (large, membrane-enclosed storage structures inside individual cells). Work is under way to use genetic engineering techniques to transfer this gene to economically important plants, with the aim of producing crops that can tolerate saline soils and the use of salty water for irrigation. However, improved plant tolerance must not be viewed as a substitute for proper salinity control, as discussed in Section 9.18.

Table 9.4
RELATIVE SALT TOLERANCE OF SELECTED PLANTS

Approximate EC_e resulting in a 10% reduction in plant growth for the most sensitive species in each column.

Tolerant, 12 dS/m	Moderately tolerant, 8 dS/m	Moderately sensitive, 4 dS/m	Sensitive, 2 dS/m
Alkali grass, Nutall	Ash (white)	Alfalfa	Alders
Alkali sacaton	Asparagus	Arborvitae	Almond
Barley (grain)	Aspen	Boxwood	Apple
Bent grass	Barley (forage)	Broad bean	Apricot
Bermuda grass	Beet (garden)	Cabbage	Azalea
Bougainvillea	Birch (black)	Cauliflower	Bean
Boxwood, Japanese	Black cherry	Celery	Beech
Canola (rapeseed)	Broccoli	Clover (alsike,	Birch
Cotton	Bromegrass	ladino, red,	Blackberry
Date	Cedar (red)	strawberry,	Burford holly
Guayule	Cowpea	and berseem)	Carrot
Hawthorn Indian	Elm	Corn	Dogwood
Jojoba	Fescue (tall)	Cucumber	Elm (American)
Kallar grass	Fig	Dallas grass	Grapefruit
Kenaf	Honeysuckle	Grape	Hemlock
Natal plum	Hydrangea	Hickory (shagbark)	Hibiscus
Oak (red and white)	Juniper	Juniper	Larch
Oleander	Kale	Lettuce	Lemon
Olive	Locust (honey)	Locust (black)	Linden
Prostrate kochia	Oak (red and white)	Maple (red)	Maple (sugar and red)
Redwort	Oats	Pea	Onion
Rescue grass	Orchard grass	Peanut	Orange
Rosemary	Pomegranate	Radish	Peach
Rugosa	Privet	Rice (paddy)	Pear
Rye (grain)	Ryegrass (perennial)	Soybean (sens. var.)	Pine (red and white)
Salt grass (desert)	Safflower	Squash	Pineapple
Sugar beet	Sorghum	Sugar cane	Plum (prune)
Tamarix	Soybean (tol. var.)	Sweet clover	Potato
Wheat grass (crested)	Squash (zucchini)	Sweet potato	Raspberry
Wheat grass (fairway)	Sudan grass	Timothy	Rose
Wheat grass (tall)	Trefoil (birdsfoot)	Tomato	Silk tree
Wild rye (altai)	Wheat	Turnip	Star jasmine
Wild rye (Russian)	Wheat grass (western)	Vetch	Strawberry
Willow	Winged bean	Viburnum	Tomato

Salt Problems Not Related to Arid Climates

Deicing Salts   In efforts to keep roads and sidewalks free of snow and ice during winter months, repeated applications of deicing salts can result in salinity levels sufficiently high as to adversely affect plants and soil organisms living alongside highways or sidewalks. In humid regions, such salt contamination is usually temporary, as the abundant rainfall leaches out the salts in a matter of weeks or months. To avoid the specific chemical and physical problems associated with sodium salts, many municipalities have switched from NaCl to KCl for deicing purposes. Sand can be used to improve traction, thereby reducing the need for deicing salt.

Containerized Plants   Salinity can also be a serious problem for indoor potted plants, particularly perennials that remain in the same pot for long periods. Greenhouse operators producing containerized plants must carefully monitor the quality of the water used for irrigation. Salts in the water, as well as those applied in fertilizers, can build up if care is not taken to flush them out occasionally with excess water. Chlorinated urban tap water used for indoor plants should be left overnight in an open container to allow some of the dissolved chlorine to escape, thus reducing the load of Cl^- ions added to the potting soil.

9.17 WATER-QUALITY CONSIDERATIONS FOR IRRIGATION[9]

Whether in a single field or in a large regional watershed, understanding the **salt balance** is a basic prerequisite for wise management of salt-affected soils. To achieve salt balance, the amount of salt coming in must be matched by the amount being removed. Meeting this condition is a fundamental challenge to the long-term sustainability of irrigated agriculture. In irrigated areas, this principally means managing the quality and amount of the irrigation water brought in and the quality and amount of soil drainage water removed.

Irrigation Water Quality

Table 9.5 provides some guidelines on water quality for irrigation. If the salt content of irrigation water is high, salt balance will be difficult to achieve. However, even very salty water can be used successfully if soils are sufficiently well drained to allow careful management of salt inputs and outputs. Where irrigation water is low in salts but has a high SAR, the formation of sodic soils is likely to accelerate. In addition, irrigation water high in carbonates or bicarbonates can reduce Ca^{2+} and Mg^{2+} concentrations

Table 9.5
SOME WATER-QUALITY GUIDELINES FOR IRRIGATION

Note that with regard to effects on physical structure of soils, higher total salinity (EC_w) in the irrigation water compensates, somewhat, for increasing sodium hazard (SAR). In addition, note that while water low in salts (low EC_w) avoids problems of restricted water availability to plants, it may worsen soil physical properties, especially if the SAR is high.

Water Property	Units	Degree of restriction on use		
		None	Slight to moderate	Severe
Salinity (affects crop water availability)				
EC_w	dS/m	<0.7	0.7–3.0	>3.0
TDS	mg/L	<450	450–2000	>2000
Physical structure and water infiltration (Evaluate using EC_w and SAR together)				
SAR = 0-3 and EC_w =	dS/m	>0.7	0.7–0.2	<0.2
SAR = 3-6 and EC_w =	dS/m	>1.2	1.2–0.3	<0.3
SAR = 6-12 and EC_w =	dS/m	>1.9	1.9–0.5	<0.5
SAR = 12-20 and EC_w=	dS/m	>2.9	2.9–1.3	<1.3
SAR = 20-40 and EC_w =	dS/m	>5.0	5.0–2.9	<2.9
Boron (B) specific ion toxicity (affects sensitive crops)				
	mg/L	<0.7	0.7–3.0	>3.0

Modified from Abrol et al. (1988) with permission of the Food and Agriculture Organization of the United Nations.

---

[9]For an overview of water-quality problems facing irrigated agriculture in California, see Letey (2000).

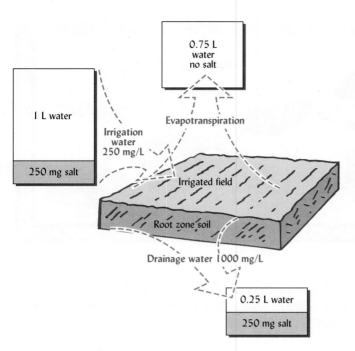

Figure 9.29

Evapotranspiration and salt balance together ensure that the drainage water from irrigated fields is much saltier than the irrigation water applied. In this example, the irrigation water contains 250 mg salts per liter. Some 75% of the applied water is lost to the atmosphere by evapotranspiration. About 25% of the water applied is used for drainage, which is necessary to maintain the salt balance (prevent the buildup of salts) in the field. The added salts are leached away with the drainage water, which then contains the same amount of salt as was added, but in only 25% of the added amount of water. The concentration of salt in the drainage water is thereby four times as great (1000 mg/L) as in the irrigation water. Disposal and/or reuse of the highly saline drainage water present challenges for any irrigation project. (Diagram courtesy of R. Weil)

in the soil solution by precipitating these ions as insoluble carbonates. This leaves a higher proportion of Na^+ in the soil solution and can increase its SAR, moving the soil toward the sodic class.

Drainage Water Salinity   Since some portion of added water must be drained away to combat salt buildup, the quality and disposition of *waste irrigation waters* must also be carefully monitored and controlled to minimize potential harm to downstream users and habitats. In any irrigation system, the drainage water leaving a field will be considerably more concentrated in salts than the irrigation water applied to the same field. (Figure 9.29 explains why.) What to do with the increasingly saline drainage water presents a major challenge to the sustainability of irrigated agriculture.

Different irrigation projects take different approaches, but rarely is the problem solved without some downstream environmental damage. Perhaps the most efficient approach would be to collect the drainage water, keep it isolated from the relatively high-quality canal water, and reuse it to irrigate a more salt-tolerant crop in a lower field. This approach provides both high-quality canal water and lower-quality drainage water for use on appropriate crops. Generally, the salt-tolerant crops are less valuable than the more salt-sensitive ones (e.g., the yield of salt-tolerant cotton from 1 ha is worth much less than the yield of salt-sensitive tomato); still, recycling drainage water saves money as well as water. Often some fresh water must be mixed with the recycled drainage water to bring its salinity level down to what can be tolerated by even the salt-tolerant crop (see Figure 9.30). After several cycles of reuse, the drainage water must be disposed of, as it will have become too saline for irrigating even the most salt-tolerant species.

Toxic Elements in Drainage Water   If either the irrigation water or the soil of the irrigated fields contains significant quantities of certain toxic trace elements, these too will become increasingly concentrated in the drainage water. The elements of concern include molybdenum (Mo), arsenic (As), boron (B), and selenium (Se). Molybdenum and selenium are necessary nutrients for animals and humans in trace amounts, but all four elements can be toxic to cattle, wildlife, or people if they become concentrated in water or food. At some locations in the western United States, such trace elements

Figure 9.30

Generalized scheme for recycling irrigation drainage water by mixing it with fresh water and irrigating salt-tolerant crops. [Redrawn from Letey et al. (2003)]

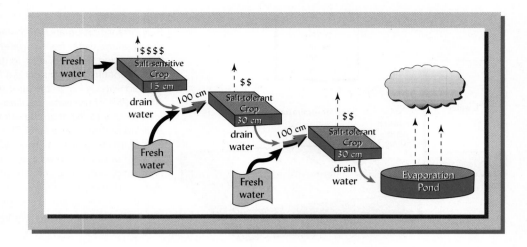

as selenium have accumulated to toxic levels in downstream wetlands or evaporative ponds. Plants growing on these affected areas can accumulate levels of these elements that are unsafe for livestock and/or wildlife.

9.18   RECLAMATION OF SALINE SOILS

Salinity stress and its mitigation: www.plantstress.com/ Articles/salinity_m/salinity_m .htm

The restoration of soil chemical and physical properties conducive to high productivity is referred to as soil **reclamation**. Reclamation of saline soils is largely dependent on the provision of effective drainage and the availability of good-quality irrigation water (see Table 9.5) so that salts can be leached from the soil. In areas where irrigation water is not available, such as in saline seeps in the Northern Plains states, the leaching of salts is not practical. In these areas, deep-rooted vegetation may be used to lower the water table and reduce the upward movement of salts.

If the natural soil drainage is inadequate to accommodate the leaching water, an artificial drainage network must be installed. Intermittent applications of excess irrigation water may be required to effectively reduce the salt content to a desired level.

Leaching Requirement

The amount of water needed to remove the excess salts from saline soils, called the **leaching requirement (*LR*)**, is determined by the characteristics of the crop to be grown, the irrigation water, and the soil. An approximation of the *LR* is given for relatively uniform salinity conditions by the ratio of the salinity of the irrigation water (expressed as its EC_{iw}) to the maximum acceptable salinity of the soil solution for the crop to be grown (expressed as EC_{dw}, the EC of the drainage water):

$$LR = \frac{EC_{iw}}{EC_{dw}} \tag{9.26}$$

The *LR* indicates water added in excess of that needed to thoroughly wet the soil and meet the crop's evapotranspiration needs. Note that if EC_{iw} is high and a salt-sensitive crop is chosen (dictating a low EC_{dw}), a very large leaching requirement *LR* will result. As mentioned in Section 9.17, disposal of the drainage water that has leached through the soil can present a major problem. It is essential that the salt-laden drainage water be removed from the soil profile, but it is also generally desirable to use

management techniques that minimize the *LR* and the amount of drainage water that requires disposal.

Management of Soil Salinity

Management of irrigated soils should aim to simultaneously minimize drainage water and protect the root zone (usually the upper meter of soil) from damaging levels of salt accumulation. These two goals are obviously in conflict. The irrigator can attempt to find the best compromise between the two and can use certain management techniques that allow plants to tolerate the presence of higher salt levels in the soil profile. One option is to plant salt-tolerant species or choose the most salt-tolerant varieties within a species.

Irrigation Timing The timing of irrigation is extremely important on saline soils, particularly early in the growing season. Germinating seeds and young seedlings are especially sensitive to salts. Therefore, irrigation should precede or immediately follow planting to move the salts downward and away from the seedling roots. The irrigator can use high-quality water to keep root-zone salinity low during the sensitive early growth stages and then switch to lower-quality water as the maturing plants become more salt tolerant.

Location of Salts in the Root Zone Tillage and planting practices can influence the location and accumulation of salts in arid-region soils. Tillage or surface-residue management practices (e.g., mulches, conservation tillage) that reduce evaporation from the soil surface should also reduce the upward transport of soluble salts. Likewise, specific techniques for applying irrigation water that direct salt concentrations away from young plant roots can allow higher levels of salt to accumulate without damage to the crop. Applying water in every other furrow and asymmetrically planting only on the wet side of the furrows can provide significant protection to young plants (Figure 9.31).

Some Limitations of the Leaching Requirement Approach

The leaching requirement approach to managing irrigated soils is only an approximation and has several inherent weaknesses. First, additional leaching may be needed, in some cases, to reduce the excess concentration of specific elements, such as boron. Second, the *LR* by itself does not take into account the rise in the water table that is likely to result from increased leaching, and so it may lead to waterlogging and, eventually, increased salinization. Third, irrigation using a simple *LR* approach usually overapplies water because an entire field is treated to avoid salt damage in its most saline spots. Fourth, the *LR* method does not consider salts that may be picked up from fossil salt deposits already in the soil and substrata. Fifth, it assumes that the EC of the drainage water is known, but in fact this may be largely unknown, since it may take years or even decades for the water applied in irrigation to reach the main drains where it can be easily sampled. In other words, the drainage water sampled today may represent the leaching conditions of several months or years ago.

An alternative approach would be to closely monitor the salinity in the soil profile by taking repeated measurements across the field, using the EM sensor methods discussed in Section 9.13. This more complex approach, combined with site-specific management techniques, seems to hold promise for future management and reclamation of salt-affected soils under irrigation.

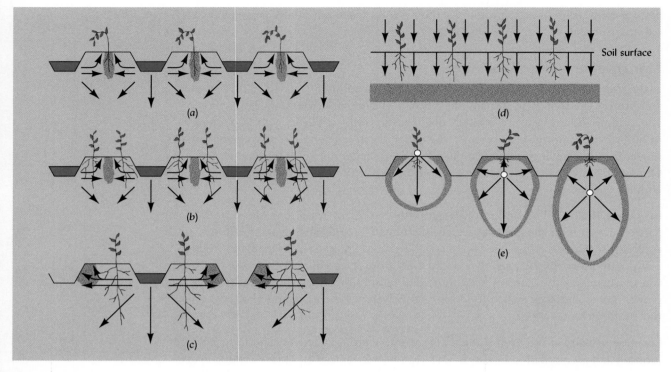

Figure 9.31

Effect of irrigation techniques on salt movement and plant growth in saline soils. (a) With irrigation water applied to furrows on both sides of the row, salts move to the center of the ridge and damage young plants. (b) Placing plants on the edges of the bed rather than in the center helps them avoid the most concentrated salts. (c) Application of water to every other furrow and placement of plants on the side of the bed nearest the water helps plants avoid the highest salt concentrations. (d) Sprinkler irrigation or uniform flooding temporarily moves salts downward out of the root zone, but the salts will return afterward as the soil surface dries out and water moves up by capillary flow. (e) Drip irrigation at low rates provides a nearly continuous flow of water, creating a low-salt soil zone with the salts concentrated at the wetting front. The placement of the drip emitters largely determines whether the salts are moved toward or away from the plant roots. (Diagram courtesy of Wesley M. Jarrell)

9.19   RECLAMATION OF SALINE-SODIC AND SODIC SOILS

Saline-sodic soils have some of the adverse properties of both saline and sodic soils. If attempts are made to leach out the soluble salts in saline-sodic soils, as was discussed for saline soils, the exchangeable Na^+ level as well as the pH would likely increase, and the soil would take on adverse characteristics of sodic soils. Consequently, for both saline-sodic and sodic soils, attention must first be given to reducing the level of exchangeable Na^+ ions and then to the problem of excess soluble salts.

Gypsum

Removing Na^+ ions from the exchange complex is most effectively accomplished by replacing them with either Ca^{2+} or H^+ ions. Providing Ca^{2+} in the form of gypsum ($CaSO_4 \cdot 2H_2O$) is the most practical way to bring about this exchange. When gypsum is added, the replaced sodium forms the soluble salt Na_2SO_4, which can be easily leached from the soil.

Several tons of gypsum per hectare are usually necessary to achieve reclamation. The soil must be kept moist to hasten the reaction, and the gypsum should be thoroughly mixed into the surface by cultivation—not simply plowed under. The treatment must be supplemented later by a thorough leaching of the soil with irrigation water to leach out most of the sodium sulfate. Gypsum is inexpensive, widely available in both natural and in industrial by-product forms, and easily handled.

Elemental sulfur and sulfuric acid can be used to reclaim sodic soils, especially where sodium bicarbonate abounds. The sulfur yields sulfuric acid, which not only changes the sodium bicarbonate to the less harmful and more leachable sodium sulfate but also decreases the pH. Sulfur and sulfuric acid have proven to be very effective in the reclamation of sodic soils, especially if large amounts of $CaCO_3$ are present. In practice, however, gypsum is much more widely used than those acid-forming materials.

Physical Condition

The effects of gypsum and sulfur on the physical condition of sodic soils is perhaps more spectacular than are the chemical effects. Sodic soils are almost impermeable to water, since the soil colloids are largely dispersed and the soil is essentially void of stable aggregates. When the exchangeable Na^+ ions are replaced by Ca^{2+} or H^+, soil aggregation and improved water infiltration results. Some research suggests that aggregate-stabilizing synthetic polymers may be helpful in at least temporarily increasing the water infiltration capacity of gypsum-treated sodic soils.

Deep-Rooted Vegetation   The reclamation effects of gypsum or sulfur are greatly accelerated by plants growing on the soil. Crops that have some degree of tolerance to saline and sodic soils, such as sugar beets, cotton, barley, sorghum, berseem clover, or rye, can be grown initially. Their roots help provide channels through which gypsum can move downward into the soil. Deep-rooted crops, such as alfalfa, are especially effective in improving the water conductivity of gypsum-treated sodic soils.

Air Injection   In addition to reduced diffusion of air because of soil dispersion, irrigated soils may provide less than optimal oxygen availability to roots because irrigation water—especially applied by drip systems—contains much less dissolved oxygen than rainwater. One approach to improving the aeration of the root zone in heavy-textured, irrigated soils is to mechanically add air. This can be easily accomplished by injecting air into drip irrigation lines and may be practical for high-value crops and ornamental landscape containers (Figure 9.32).

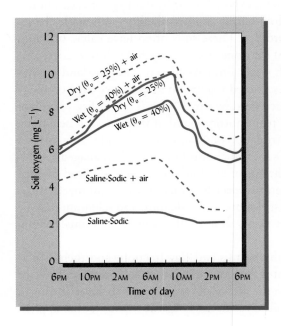

Figure 9.32

Soil oxygen content in normal and saline-sodic soils as affected by adding air to water supplied by buried drip irrigation emitters. Mature tomato plants were grown in large pots filled with heavy clay soil (a Vertisol) and either kept wet (θ_v = 40% water, near field capacity) or dry (θ_v = 25% water). In the normal soil, adding air (dotted lines) to wet soil achieved the same O_2 level as in the drier soil. Saline-sodic conditions were created by applying enough NaCl to raise the EC to 8.8 dS/m. Under these conditions, soil dispersion impeded air movement causing low O_2 levels even when air was added. Note that for all treatments, the O_2 levels declined during the day when plant roots are most actively respiring and taking up water and nutrients. Air was added into the drip lines by a venturi inlet and O_2 was measured by fiber-optic mini-sensors located at 15 cm depth from the soil surface. [Drawn from data in Bhattarai et al. (2006)]

Once salt-affected soils have been reclaimed, prudent management steps must be taken to be certain that the soils remain productive. For example, surveillance of the EC and SAR and trace element composition of the irrigation water is essential. The number and timing of irrigations should be adjusted to maintain the balance of salts entering and leaving the soil. Likewise, the maintenance of good internal drainage is essential for the removal of excess salts.

Steps should also be taken to monitor appropriate chemical characteristics of the soils, such as pH, EC, and SAR, as well as specific levels of such elements as boron, chlorine, molybdenum, and selenium that could lead to chemical toxicities. Crop and soil fertility management for satisfactory yield levels is essential to maintain the overall quality of salt-affected soils. The crop residues (roots and aboveground stalks) will help maintain organic matter levels and good physical condition of the soil.

9.20   CONCLUSION

No other soil characteristic is more important in determining the chemical environment of higher plants and soil microbes than the pH. There are few reactions involving the soil or its biological inhabitants that are not sensitive to soil pH. This sensitivity must be recognized in any soil-management system.

Acidification is a natural process in soil formation that is accentuated in humid regions where processes that produce H^+ ions outpace those that consume them. Natural acidification is largely driven by the production of organic acids (including carbonic acid) and the leaching away of the nonacid cations (Ca^{2+}, Mg^{2+}, K^+, and Na^+) that the H^+ ions from the acids displace from the exchange complex. Emissions from power plants and vehicles, as well as inputs of nitrogen into agricultural systems, are the principal means by which human activities have accelerated acidification during recent decades.

Aluminum is the other principal acid cation besides hydrogen. Its hydrolysis reactions produce H^+ ions and its toxicity comprises one of the main detrimental effects of soil acidity. The total acidity in soils is the sum of the active, exchangeable (salt replaceable), and residual pools of soil acidity. Changes in the soil solution pH (the active acidity) are buffered by the presence of the other two pools. In certain anaerobic soils and sediments, the presence of reduced sulfur provides the potential for enormous acid production if the material is exposed to air by drainage or excavation.

Arid and semiarid regions predominantly feature alkaline soils with above-neutral pH values throughout their profiles. These soils often have calcic (calcium carbonate-rich) or gypsic (gypsum-rich) horizons at some depth. Such soils cover vast areas and support important—though water-limited—desert and grassland ecosystems. Soils of dry regions interact with scattered vegetation to influence the hydrology and fertility of landscapes that are characterized by small "islands of fertility." Many of these soils are quite well endowed with plant nutrients and, if irrigated, can be among the most productive soils in the world. Irrigation of alkaline soils of arid regions almost inevitably leads to the accumulation of salts, which must be carefully managed.

The high-pH, calcium-rich conditions of alkaline soils commonly lead to deficiencies of certain essential micronutrients (especially iron and zinc) and macronutrients (especially phosphorus). In contrast, boron, molybdenum, and selenium may be so readily available as to accumulate in plants to levels that can harm animals eating the plants.

Soil pH is largely controlled by the humus and clay fractions and their associated exchangeable cations. The maintenance of satisfactory soil fertility levels in humid regions depends considerably on the judicious use of lime to balance the losses of calcium and magnesium from the soil. Gypsum and organic matter (either applied or grown) represent other tools that can be used to ameliorate soil acidity instead of, or in addition to, liming. On the other hand, it is sometimes most judicious to use acid-tolerant plants, rather than attempt to change the chemistry of the soil.

Knowing how pH is controlled, how it influences the supply and availability of essential plant nutrients as well as toxic elements, how it affects higher plants and human beings, and how it can be ameliorated, is essential for the conservation and sustainable management of soils throughout the world.

Scientists have grouped salt-affected soils into three classes based on their total salt content (indicated by electrical conductivity [EC]) and the proportion of sodium among the cations (indicated by either the sodium adsorption ration [SAR] or the exchangeable sodium percentage [ESP]). The physical conditions of saline and saline-sodic soils are satisfactory for plant growth, but the colloids in sodic soils are largely dispersed, the soil is puddled and poorly aerated, and the water infiltration rate is extremely slow.

If irrigators in arid regions apply only enough water to meet plant evapotranspiration needs, salts will build up relentlessly in the soil until the land becomes too salinized to grow crops. Therefore, just as the humid-region farmer must periodically use lime to restore a favorable balance between H^+ ion consumption and production, the irrigator must use periodic leaching to restore the balance between the import and export of salts.

Leaching and drainage are both essential components of any successful irrigation scheme. However, the leaching of salts is not a simple matter, and it inevitably leads to additional problems both in the field being leached and in sites further downstream. At best, salinity management in irrigation agriculture provides a compromise between unavoidable evils.

In order to make saline-sodic and sodic soils permeable enough for leaching to take place, the excess exchangeable sodium ions must first be removed from the exchange complex. This is accomplished by replacing the Na^+ ions with either Ca^{2+} or H^+ ions.

The world increasingly depends on irrigated agriculture in dry regions for the production of food for its growing population. Unfortunately, irrigated agriculture is in inherent disharmony with the nature of arid-region ecology. We have seen that it is therefore fraught with difficulties that require constant and careful management if human and environmental tragedies are to be avoided.

STUDY QUESTIONS

1. Soil pH gives a measure of the concentration of H^+ ions in the soil solution. What, if anything, does it tell you about the concentration of OH^- ions? Explain.

2. Describe the role of aluminum and its associated ions in soil acidity. Identify the ionic species involved and the effect of these species on the CEC of soils.

3. If you could somehow extract the soil solution from the upper 16 cm of 1 ha of moist acid soil (pH = 5), how many kg of pure $CaCO_3$ would be needed to neutralize the soil solution (bring its pH to 7.0)? Under field conditions, up to 6 Mg of

limestone may be required to bring the pH of this soil layer to a pH of 7.0. How do you explain the difference in the amounts of $CaCO_3$ involved?

4. What is meant by *buffering?* Why is it so important in soils, and what are the mechanisms by which it occurs?

5. What is acid rain, and why does it seem to have greater impact on forests than on commercial agriculture?

6. Calculate the amount of pure $CaCO_3$ that could theoretically neutralize the H^+ ions in a year's worth of acid rain if a 1-ha site received 500 mm of rain per year and the average pH of the rain was 4.0.

7. Discuss the significance of soil pH in determining specific nutrient availabilities and toxicities, as well as species composition of natural vegetation in an area.

8. How much limestone with a $CaCO_3$ equivalent of 90% would you need to apply to eliminate exchangeable aluminum in an Ultisol with CEC = 8 $cmol_c$/kg and an aluminum saturation of 60%?

9. Based on the buffer pH of your soil sample, a lab recommends that you apply 2 Mg of $CaCO_3$ equivalent to your field and plow it in 18 cm deep to achieve your target pH of 6.5. You actually plan to use the lime to prepare a large lawn and till it in only 12 cm deep. The lime you purchase has a carbonate equivalent = 85%. How much of this lime do you need for 2.5 ha?

10. A landscape contractor purchased 10 dump-truck loads of "topsoil" excavated from a black, rich-appearing soil in a coastal wetland. Before accepting the shipment, samples of the soil had been sent to a lab to be sure they met the specified properties (silt loam texture, pH 6 to 6.5) for the topsoil to be used in a landscaping job. The lab had reported that the texture and pH were in the specified range, so the topsoil was installed and an expensive landscape of beautiful plants established. Unfortunately, within a few months all the plants began to die. Replacement plants also died. The topsoil was again tested. It was still a silt loam, but

now its pH was 3.5. Explain why this pH change likely occurred and suggest appropriate management solutions.

11. A neighbor complained when his azaleas were adversely affected by a generous application of limestone to the lawn immediately surrounding the azaleas. To what do you ascribe this difficulty? How would you remedy it?

12. The ill effects of acidity in subsoils can be ameliorated by adding gypsum ($CaSO_4 \cdot 2H_2O$) to the soil surface. What are the mechanisms responsible for this effect of the gypsum?

13. What are the primary sources of alkalinity in soils? Explain.

14. Compare the availability of the following essential elements in alkaline soils with that in acid soils: (1) iron, (2) nitrogen, (3) molybdenum, and (4) phosphorus.

15. An arid-region soil, when it was first cleared for cropping, had a pH of about 8.0. After several years of irrigation, the crop yield began to decline, the soil aggregation tended to break down, and the pH had risen to 10. What is the likely explanation for this situation?

16. What physical and chemical treatments would you suggest to bring the soil described in question 15 back to its original state of productivity?

17. What are the advantages of using gypsum ($CaSO_4 \cdot 2H_2O$) in the reclamation of a sodic soil?

REFERENCES

Abrol, I. P., J. S. P. Yadov, and F. I. Massoud. 1988. "Salt-affected soils and their management," *FAO Soils Bulletin* **39**. Rome: Food and Agriculture Organization of the United Nations. Complete text available at www.fao.org/docrep/x5871e/x5871e00.htm.

Adams, F., and Z. F. Lund. 1966. "Effect of chemical activity of soil solution aluminum on cotton root penetration of acid subsoils," *Soil Sci.*, **101**: 193–198.

Barak, P., B. O. Jobe, A. R. Krueger, L. A. Peterson, and D. A. Laird. 1997. "Effects of long-term soil acidification due to nitrogen fertilizer inputs in Wisconsin," *Plant Soil*, **197**:61–69.

Bhattarai, S. P., L. Pendergast, and D. J. Midmore. 2006. "Root aeration improves yield and water use efficiency of tomato in heavy clay and saline soils," *Scientia Horticulturae*, 108: 278–288.

Blake, L., W. T. Goulding, C. J. B. Mott, and A. E. Johnston. 1999. "Changes in soil chemistry accompanying acidification over more than 100 years under woodland and grass at Rothamsted Experiment Station, UK," *European J. Soil Sci.*, **50**:401–412.

Bloom, P. R., U. L. Skyllberg, and M. E. Sumner. 2005. "Soil acidity," pp. 411–459, in A. Tabatabai and D. Sparks (eds.), *Chemical Processes in Soils*. SSSA Book Series No. 8. (Madison, Wl: Soil Science Society of America).

de la Fuente-Martinez, J. M., and L. Herrera-Estralla. 2000. "Advances in the understanding of aluminum toxicity and the development of aluminum-tolerant transgenic plants," *Advances in Agronomy*, **66**:103–120.

Demchik, M. C., W. E. Sharpe, T. Yangkey, B. R. Swistock, and S. Bubalo. 1999. "The relationship of soil Ca/Al ratio to seedling sugar maple population, root characteristics, mycorrhizal infection rate, and growth and survival," pp. 201–210, in W. E. Sharpe and J. R. Drohan (eds.), *The Effects of Acidic Deposition on Pennsylvania's Forests.* Proceedings of the Sept. 14–16, 1998, PA Acidic Deposition Conference. (University Park, Environmental Resources Research Institute, Pennsylvania State University).

Edwards, K. J., P. L. Bond, T. M. Gihring, and J. F. Banfield. 2000. "An archaeal iron-oxidizing extreme acidophile important in acid mine drainage," *Science,* **287:** 1796–1799.

Fanning, D. S., M. Rabenhorst, S. N. Burch, K. R. Islam, and S. A. Tangen. 2002. "Sulfides and sulfates," pp. 229–260, in J. B. Dixon and D. G. Schultz (eds.), *Soil Mineralogy with Environmental Applications.* SSSA Book Series No. 7. (Madison, WI: Soil Science Society of America).

Fierer, N., and R. B. Jackson. 2006. "The diversity and biogeography of soil bacterial communities," *Proceedings of the National Academy of Science* **103:**626–631.

Hue, N. V., and D. L. Licudine. 1999. "Amelioration of subsoil acidity through surface applications of organic manures," *J. Environ. Qual.,* **28:**623–632.

Letey, J. 2000. "Soil salinity poses challenges for sustainable agriculture and wildlife," *Calif. Agric.,* **54**(2):43–48.

Letey, J., D. E. Birkle, W. A. Jury, and I. Kan. 2003. "Model describes sustainable long-term recycling of saline agricultural drainage water," *Calif. Agric.,* **57:**24–27.

Likens, G. E., C. T. Driscoll, and D. C. Buso. 1996. "Long-term effects of acid rain: Response and recovery of a forest ecosystem," *Science,* **272:** 244–246.

Lumbanraja, J., and V. P. Evangelou. 1991. "Acidification and liming influence on surface charge behavior of Kentucky subsoils," *Soil Sci. Soc. Amer. J.,* **54:** 26–34.

Magdoff, F. R., and R. J. Barlett. 1985. "Soil pH buffering revisited," *Soil Sci. Soc. Amer. J.,* **49:** 145–148.

Rhoades, J. D., F. Chanduvi, and S. Lesch. 1999. *Soil Salinity Assessment Methods and Interpretation of Electrical Conductivity Measurements.* FAO Irrigation and Drainage Paper No. 57. (Rome: Food and Agriculture Organization of the United Nations).

Sullivan, T. J., I. J. Fernandez, A. T. Herlihy, C. T. Driscoll, T. C. McDonnell, N. A. Nowicki, K. U. Snyder, and J. W. Sutherland. 2006. "Acid–base characteristics of soils in the Adirondack Mountains, New York," *Soil Sci. Soc. Amer. J.,* **70:**141–152.

Szabolcs, I. 1989. *Salt-Affected Soils.* (Boca Raton, FL: CRC Press).

Weil, R. R. 2000. "Soil and plant influences on crop response to two African phosphate rocks," *Agron. J.,* **92:**1167–1175.

Zhang, F., S. Kang, J. Zhang, R. Zhang, and F. Li. 2004. "Nitrogen fertilization on uptake of soil inorganic phosphorus fractions in the wheat root zone," *Soil Sci. Soc. Amer. J.,* **68:**1890–1895.

10 ⛏

Organisms and Ecology of the Soil[1]

Under the silent, relentless chemical jaws of the fungi, the debris of the forest floor quickly disappears. . . .
—A. FORSYTH AND K. MIYATA, *TROPICAL NATURE*

Head of a bacteria-feeding nematode. (Sven Boström, Swedish Museum of Natural History)

Like a forest or an estuary, a soil is an ecosystem in which thousands of different creatures interact and contribute to the global cycles that make all life possible. This chapter will introduce some of the actors in the living drama staged largely unseen in the soil beneath our feet. This world is populated by a wild array of creatures all fiercely competing for every leaf, root, fecal pellet, and dead body that reaches the soil. Predators of all kinds lurk in the dark, some with fearsome jaws to snatch unwary victims, others whose jellylike bodies simply engulf and digest their hapless prey.

The diversity of substrates and environmental conditions found in every handful of soil spawns a diversity of adapted organisms that staggers the imagination. The collective vitality, diversity, and balance among these organisms make possible the functions of a high-quality soil. Most of the work of the soil community is carried out by creatures whose "jaws" are chemical enzymes that eat away at organic substances left in the soil by their coinhabitants.

We will learn how these organisms, both flora and fauna, interact with one another, what they eat, how they affect the soil, and how soil conditions affect them. The central theme will be how this community of organisms assimilates plant and animal materials, creating soil humus, recycling carbon and mineral nutrients, and supporting plant growth. A subtheme will be how people can manage soils to encourage a healthy, diverse soil community that efficiently makes nutrients

---

[1]For stories about how life underground affects everything on Earth, see Baskin (2005); for an introduction to the actors in the soil drama, see Nardi (2003); for soil ecology with emphasis on the meso- and microfauna, see Coleman et al. (2004); for reviews of soil microbiology, see Sylvia et al. (2005), Tate (2001), or Paul (2006).

available to higher plants, protects plant roots from pests and disease, and helps protect the global environment from some of the excesses of the human species.

10.1 THE DIVERSITY OF ORGANISMS IN THE SOIL

Soil organisms are creatures that spend all or part of their lives in the soil environment (Plate 50). Every handful of soil is likely to contain billions of organisms, with representatives of nearly every phylum of living things. A simplified, general classification of soil organisms is shown in Table 10.1. In this book we emphasize activities rather than names of organisms; consequently, we consider only very broad, simple taxonomic categories.

The term **fauna** is used in a very general way to distinguish animals (including single-celled protista) from **flora**, a term which is used in an equally general manner to refer to the true plants (including single-celled algae) as well as to all the nonanimal microorganisms. Based on similarities in genetic material, biologists classify all living organisms into three primary domains: *Eukarya* (which includes all plants, animals,

The Soil Biodiversity Program, Scotland: http://soilbio.nerc.ac.uk

Table 10.1
IMPORTANT GROUPS OF SOIL ORGANISMS BY SIZE CLASS

Generalized grouping (body width in mm)	Major taxonomic groups	Examples
Macrofauna (>2 mm)		
All heterotrophs, largely herbivores and detritivores	Vertebrates	Gophers, mice, moles
	Arthropods	Ants, beetles and their larvae, centipedes, grubs, maggots, millipedes, spiders, termites, woodlice
	Annelids	Earthworms
	Mollusks	Snails, slugs
Macroflora		
Largely autotrophs	Vascular plants	Feeder roots
	Bryophytes	Mosses
Mesofauna (0.1–2 mm)		
All heterotrophs, largely detritivores	Arthropods	Mites, collembola (springtails)
All heterotrophs, largely predators	Annelids	Enchytraeid (pot) worms
	Arthropods	Mites, protura
Microfauna (<0.1 mm)		
Detritivores, predators, fungivores, bacterivores	Nematodes	Nematodes
	Rotifera	Rotifers
	Protozoa	Amoebae, ciliates, flagellates
	Tardigrades	Water bears, *Macrobiotus sp.*
Microflora (<0.1 mm)		
Largely autotrophs	Vascular plants	Root hairs
	Algae	Greens, yellow-greens, diatoms
Largely heterotrophs	Fungi	Yeasts, mildews, molds, rusts, mushrooms
Heterotrophs and autotrophs	Bacteria	Aerobes, anaerobes
	Cyanobacteria	Blue-green algae, autotrophs
	Actinomycetes	Many kinds of actinomycetes, heterotrophs
	Archaea	Methanotrophs, *Thermoplasma sp.*, halophiles

and fungi), **Bacteria**, and **Archaea**. Organisms can also be grouped by what they "eat." Some organisms subsist on living plants (**herbivores**), others on dead plant debris (**detritivores**). Some consume animals (**predators**), some devour fungi (**fungivores**) or bacteria (**bacterivores**), and some live off of, but do not consume, other organisms (**parasites**). **Heterotrophs** rely on organic compounds for their carbon and energy needs while **autotrophs** obtain their carbon mainly from carbon dioxide and their energy from photosynthesis or oxidation of various elements.

Sizes of Organisms

The animals (*fauna*) of the soil range in size from **macrofauna** (such as moles, prairie dogs, earthworms, and millipedes) through **mesofauna** (such as tiny springtails and mites) to **microfauna** (such as nematodes and single-celled protozoans). Plants (*flora*) include the roots of higher plants, as well as microscopic algae and diatoms. Other microorganisms (too small to be seen without the aid of a microscope) such as fungi and bacteria tend to predominate in terms of numbers, mass, and metabolic capacity.

Diversity and Isolation

A typical, healthy soil might contain several species of vertebrate animals (mice, gophers, snakes, etc.), a half dozen species of earthworms, 20 to 30 species of mites, 50 to 100 species of insects (collembola, beetles, ants, etc.), dozens of species of nematodes, hundreds of species of fungi, and perhaps thousands of species of bacteria and actinomycetes. Tremendous diversity is possible because of the nearly limitless variety of foods and the wide range of habitat conditions found in soils. Within a handful of soil there may be areas of good and poor aeration, high and low acidity, cool and warm temperatures, moist and dry conditions, and localized concentrations of dissolved nutrients, organic substrates, and competing organisms. The populations of soil organisms tend to be concentrated in zones of favorable conditions, rather than evenly distributed throughout the soil. The soil aggregate (Section 4.5) can be considered a fundamental unit of habitat for meso- and microorganisms, providing a complex range of hiding places, food sources, environmental gradients, and genetic isolation on a micro-scale.

Types of Diversity

Soil scientists use the concept of biological diversity as an indicator of soil quality. A high **species diversity** indicates that the organisms present are fairly evenly distributed among a large number of species. Most ecologists believe that such complexity and species diversity are usually paralleled by a high degree of **functional diversity**—the capacity to utilize a wide variety of substrates and carry out a wide array of processes.

Ecosystem Dynamics

In most healthy soil ecosystems there are several—and in some cases many—different species capable of carrying out each of the thousands of different enzymatic or physical processes that proceed every day. This **functional redundancy**—the presence of several organisms to carry out each task—leads both to ecosystem **stability** and **resilience**. *Stability* describes the ability of soils, even in the face of wide variations in environmental conditions and inputs, to continue to perform such functions as the cycling of nutrients, assimilation of organic wastes, and maintenance of soil structure. *Resilience* describes the ability of the soil to "bounce back" to functional health after a severe disturbance has disrupted normal processes.

Given a high degree of diversity, no single organism is likely to become completely dominant. By the same token, the loss of any one species is unlikely to cripple the entire system. Nonetheless, for certain soil processes, such as ammonium oxidation

More on Soil Biodiversity from FAO: www.fao.org/ag/AGL/agll/ soilbiod/ soilbtxt.stm

Critter clips from Iowa State University: www.agron.iastate.edu/ %7Eloynachan/mov/

Arid soil biological communities: www.blm.gov/nstc/soil/ communities/index.html (see Section 12.7), methane oxidation (see Section 11.9), or the creation of aeration macropores (see Section 4.8), primary responsibility may fall to only one or two species. The activity and abundance of these **keystone species** (e.g., certain nitrifying bacteria or burrowing earthworms) merit special attention, for their populations may indicate the health of the entire soil ecosystem.

Genetic Resources

The diversity of organisms in a soil is important for reasons in addition to the safe-guarding of ecological functions; soils also make an enormous contribution to **global biodiversity**. Many scientists believe that there are more species in existence below the surface of the Earth than above it. The soil is therefore a major storehouse of the genetic innovations that nature has written into the DNA code over hundreds of millions of years. Humans have always found ways to make use of some of the genetic material in soil organisms (beer, yogurt, and antibiotics are examples). However, with recent developments in biological engineering that allow the transfer of genetic material from one type of organism to another, the soil DNA bank has taken on a much larger practical importance for human welfare. The genes from soil organisms may now be used to produce plants and animals of superior utility to the human community.

10.2 ORGANISMS IN ACTION[2]

The activities of soil flora and fauna are intimately related in what ecologists call a *food chain* or, more accurately, a *food web*. Some of these relationships are shown in Figure 10.1, which illustrates how various soil organisms participate in the degradation of residues from higher plants. As one organism eats another, nutrients and energy are said to be passed from one **trophic level** to a higher one. The first trophic level is that of the **primary producers**. The second trophic level consists of the **primary consumers** that eat those producers. The third trophic level would be those **predators** that eat the primary consumers, the fourth level would be predators that eat predators, and so on.

Virtual soil tour. Click on each number:
www.fieldmuseum
.org/undergroundadventure/
flash/VirtualTour.swf

Source of Energy and Carbon

Soil organisms can be classified as either **autotrophic** or **heterotrophic** based on where they obtain the *carbon* needed to build their cell constituents (Table 10.2). The heterotrophic soil organisms obtain their carbon from the breakdown of organic materials previously produced by other organisms. Nearly all heterotrophs also obtain their *energy* from the oxidation of the carbon in organic compounds. They are responsible for organic decay. These organisms, which include the fauna, the fungi, actinomycetes, and most other bacteria, are far more numerous in soils than the autotrophs.

The autotrophs obtain their carbon from simple carbon dioxide gas (CO_2) or carbonate minerals, rather than from carbon already fixed in organic materials. Autotrophs can be further classified based on how they obtain energy. Some use solar energy (photoautotrophs), while others use energy released by the oxidation of inorganic elements such as nitrogen, sulfur, and iron (chemoautotrophs).

As in most aboveground ecosystems, plants (and certain photosynthesizing microbes) play the principal role as **primary producers**. By combining carbon from atmospheric carbon dioxide with water, using energy from the sun, the producer

---

[2]For a succinct and well-illustrated summary of soil communities and ecological dynamics, see Tugel and Lewandowski (2000). For a colorful description of the community of soil organisms and readable, if somewhat anecdotal, advice for managing garden soils using soil organisms, see Lowenfels and Lewis (2006).

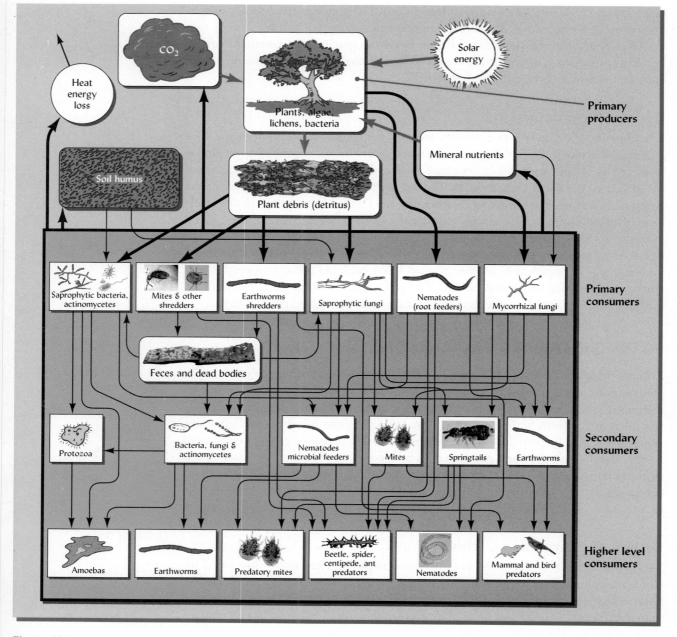

Figure 10.1

Generalized diagram of the soil food web involved in the breakdown of plant tissue, the formation of humus, and the cycling of carbon and nutrients. The large shaded compartment represents the community of soil organisms. The rectangular boxes represent various groups of organisms; the arrows represent the transfer of carbon from one group to the next as predator eats prey. The thick arrows entering the top of the shaded compartment represent primary consumption of carbon originating from the tissue of the producers—higher plants, algae, and cyanobacteria. The rounded boxes represent other inputs that support, and outputs that result from, the soil food web. Although all groups shown play important roles in the process, some 80 to 90% of the total metabolic activity in the food web can be ascribed to the fungi and the bacteria (including actinomycetes). As a result of this metabolism, soil humus is synthesized, and carbon dioxide, heat energy, and mineral nutrients are released into the soil environment. (Diagram courtesy of R. Weil)

organisms make organic molecules and living tissues. These organic materials contain both carbon and chemical energy that other organisms can utilize, either directly or indirectly, after having been passed on through intermediaries. The producers therefore form the food base for the entire food web.

Table 10.2
METABOLIC GROUPING OF SOIL ORGANISMS ACCORDING TO THEIR SOURCE OF METABOLIC ENERGY AND THEIR SOURCE OF CARBON FOR BIOCHEMICAL SYNTHESIS

	Source of energy	
Source of carbon	Biochemical oxidation	Solar radiation

Combined organic carbon	**Chemoheterotrophs:** All animals, plant roots, fungi, actinomycetes, and most other bacteria
	Examples:
	  Earthworms
	  Aspergillus sp.
	  Azotobacter sp.
	  Pseudomonas sp.
		Photoheterophs: A few algae

Carbon dioxide	**Chemoautotrophs:** Many archea and bacteria
	
	Examples:
	  Ammonia oxidizers—*Nitrosomonas sp.*
	  Sulfur oxidizers—*Thiobacillus denitrificans*
		Photoautotrophs: Plant shoots, algae, and cyanobacteria
		Examples:
		  Chorella sp.
		  Nostoc sp.

Primary Consumers

As soon as a leaf, a stalk, or a piece of bark drops to the ground, it is subject to coordinated attack by microflora and by macro- and mesofauna (see Figure 10.1). The animals, which include mites, springtail insects, woodlice, and earthworms, chew or tear holes in the tissue, opening it up to more rapid attack by the microflora. The animals and microflora that use the energy stored in the plant residues are termed *primary consumers.*

Certain soil organisms that eat live plants are called **herbivores**. Examples are parasitic nematodes and insect larvae and rodents that attack plant roots, as well as termites, ants, beetle larvae, and rodents that devour aboveground plant parts. Because they attack living plants that may be of value to humans, many of these soil herbivores are considered pests. On the other hand, certain herbivores, such as soil-dwelling immature cicadas, commonly do more good than harm for higher plants (see Plate 83).

For the vast majority of soil organisms, however, the principal source of food is the debris of dead tissues left by plants on the soil surface and within the soil pores. This debris is called **detritus**, and the animals that directly feed on it are called **detritivores**. Both herbivores and detritivores that eat the tissues of primary producers are considered primary consumers. However, many of the animals that chew up plant detritus actually get most of their nutrition from the microorganisms that live in the detritus, not from the dead plant tissues themselves. Such animals, termed *shredders*, are not really primary consumers (second trophic level), but belong to a higher trophic level as they eat detritus from other animals (fecal pellets, dead bodies) (see "Secondary Consumers," following).

On balance, most actual decomposition of dead plant and animal debris is carried out by **saprophytic** (feeding on dead tissues) microflora, both fungi and bacteria. With the assistance of the animals that physically shred and chew the plant debris (see following), the saprophytes break down all kinds of dead plant and animal compounds, from simple sugars to woody materials (Figure 10.1). Saprophytes feed on detritus, dead animals (corpses), and animal feces. They may form large colonies of microbial cells on the decaying material. These microbial colonies, in turn, soon provide nutrition to myriad other soil organisms—the secondary consumers.

Figure 10.2

A predatory mite (an Astigmatid, m) dining on its prey, a microscopic roundworm (a nematode, n). Predation of this type keeps the populations of various groups of organisms in balance and releases nutrients previously tied up in the bodies of the prey. (Photo courtesy of Marie Newman, North Carolina State University)

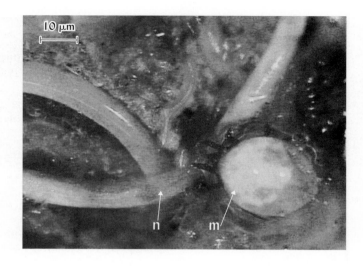

10 μm

n   m

A virtual introduction to mites:
www.sel.barc.usda.gov/acari/index.html

Secondary Consumers

Secondary consumers include microflora, such as bacteria, and fungi as well as **carnivores**, which consume other animals. Examples of carnivores include centipedes and mites that attack small insects or nematodes (see Figure 10.2), spiders, predatory nematodes, and snails. **Microbivorous feeders**, organisms that use microflora as their source of food, include certain collembola (springtail insects), mites, termites, certain nematodes (see chapter opening photo), and protozoa.

While the actions of the microflora are mostly biochemical, those of the fauna are both physical and chemical. The mesofauna and macrofauna chew the plant residues into small pieces and move them from one place to another on the soil surface and even into the soil.

The actions of these animals enhance the activity of the microflora in several ways. First, the chewing action fragments the litter, cutting through the resistant waxy coatings on many leaves to expose the more easily decomposed cell contents for microbial digestion. Second, the chewed plant tissues are thoroughly mixed with microorganisms in the animal gut, where conditions are ideal for microbial action. Third, the mobile animals carry microorganisms with them and help the latter to disperse and find new food sources to decompose.

Tertiary Consumers

In the next level of the food web, the secondary consumers are prey for still other carnivores, called *tertiary consumers*. For example, ants consume centipedes, spiders, mites, and scorpions—all of which can themselves prey on primary or secondary consumers. Many species of birds specialize in eating soil animals such as beetles and earthworms. Such mammals as moles can be effective predators of macrofauna. Whether the prey is a nematode or an earthworm, predation serves important ecological functions, including the release of nutrients tied up in the living cells.

The microflora are intimately involved in every level of decomposition. In addition to their direct attack on plant tissue (as primary consumers), microflora are active within the digestive tracts of many soil animals, helping these animals digest more resistant organic materials. The microflora also attack the finely shredded organic material in animal feces, and they decompose the bodies of dead animals. For this reason they are referred to as the ultimate decomposers.

Figure 10.3

Two dung beetles (Scarabaeidae), one on top and the other underneath, roll a ball that they have fashioned out of buffalo dung over the leaf-strewn surface of a sandy soil. The female will lay her eggs in the ball of dung and bury it in the soil (some coarse sand grains are sticking to the surface of the dung ball). Burying the dung is very important for making the nutrients therein available to the soil food web. Dung burial also prevents the reproduction of carnivorous flies and other pests of dung-producing mammals. Different dung beetles have evolved to specialize in the burial of dung from particular species of animals. (Photo courtesy of R. Weil)

Ecosystem Engineers[3]

Certain organisms make major alterations to their physical environment that influence the habitats of many other organisms in the ecosystem. These organisms are sometimes referred to as *ecosystem engineers*. For example, some of these "engineer" species are microorganisms that create an impermeable surface crust that spatially concentrates scarce water supplies in certain desert soils. Others are burrowing animals that create opportunities and challenges for other organisms by digging channels that greatly alter air and water movement in soils. Termites, ants, earthworms, and rodents are examples. Not only do burrows encourage more water and air to enter the soil, they also provide passages that plant roots can easily follow to penetrate dense subsurface soil layers.

Certain beetles of the *Scarabaeidae* family greatly enhance nutrient cycling by burying animal dung in the upper soil horizons. Many of these **dung beetles** cut round balls from large mammal feces, enabling them to roll the dung balls to a new location (Figure 10.3). The female dung beetle then lays her eggs in the ball of dung and buries it in the soil. Dispersal and burial of the dung protects the nutrients from easy loss by runoff or volatilization. Dung beetles therefore play important roles in nutrient cycling and conservation in many grazed ecosystems.

10.3 ORGANISM ABUNDANCE, BIOMASS, AND METABOLIC ACTIVITY

Soil organism numbers are influenced primarily by the amount and quality of food available. Other factors affecting their numbers include physical factors (e.g., moisture and temperature), biotic factors (e.g., predation and competition), and chemical characteristics of the soil (e.g., acidity, dissolved nutrients, and salinity). The species that inhabit the soil in a desert will certainly be different from those in a humid forest, which, in turn, will be quite different from those in a cultivated field. Acid soils are populated by species different from those in alkaline soils. Likewise, species diversification and abundance in a tropical rain forest are different from those in a cool temperate area.

Despite these variations, a few generalizations can be made. For example, forested areas usually support a more diverse soil fauna and more fungal-dominated microflora than do grasslands, although the total faunal mass per hectare and level of

---

[3]For a discussion of the ecosystem engineer concept in soils, see Jouquet et al. (2006) and Lavelle et al. (1997).

faunal activity are generally higher in grasslands. Cultivated fields are generally lower than undisturbed native lands in numbers and biomass of soil organisms, especially the fauna, partly because tillage destroys much of the soil habitat.

Total **soil biomass**, the living fraction of the soil, is generally related to the amount of organic matter present. On a dry-weight basis, the living portion is usually between 1 and 5% of the total soil organic matter. In addition, scientists commonly observe that the ratios of soil organic matter to detritus to microbial biomass to faunal biomass are approximately 1000:100:10:1.

Comparative Organism Activity

The importance of specific groups of soil organisms is commonly identified by (1) the numbers of individuals in the soil, (2) their weight (biomass) per unit volume or area of soil, and (3) their metabolic activity (often measured as the amount of carbon dioxide given off in respiration). Concentrations of microbial activity (hotspots) occur in the immediate vicinity of living plant roots and decaying bits of detritus, in the organic material lining earthworm burrows, in fecal pellets of soil fauna, and in other favored soil environments.

Mesofauna detritivores (mostly mites and collembola) translocate and partially digest organic residues and leave their excrement for microfloral degradation (see Plate 80, noting the center of the left image, where mites feeding on soft inner root tissue have left a pile of fecal pellets). By their movements in the soil, many animals rearrange soil particles to form biopores, thus favorably affecting the soil's physical condition (also visible in Plate 80). Other animals, especially certain insect larvae, feed directly on plant roots (Plate 83) and may cause considerable damage.

As might be expected, the microorganisms are the most numerous and have the highest biomass. Together with earthworms (or termites, in the case of some tropical soils), the microflora dominate the biological activity in most soils. It is estimated that about 80% of the total soil metabolism is due to the microflora, although, as previously mentioned, their activity is enhanced by the actions of soil fauna. Despite their relatively small total biomass, such microfauna as nematodes and protozoa play important roles in nutrient cycling by preying on bacteria and fungi. For these reasons, major attention will be given to the microflora and microfauna, along with earthworms, termites, and certain other fauna.

10.4   EARTHWORMS

Earthworm information:
www.sarep.ucdavis.edu/
worms/

Earthworms are probably the most important macroanimals in most soils. They (along with their much smaller mesofauna cousins, the **enchytraeid worms**) are egg-laying *hermaphrodites* (organisms without separate male and female genders) that eat detritus, soil organic matter, and microorganisms found on these materials (Figure 10.4). They do not eat living plants or their roots, and so do not act as pests to crops (and so should not be confused with root-feeding insect larvae with common names such as "army worm" or "wireworm").

The 7000 or so species of earthworms reported worldwide can be grouped according to their burrowing habits and habitat. The relatively small **epigeic** earthworms live in the litter layer or in the organic-rich soil very near the surface. Epigeic earthworms, which include the common compost worm, *Eisenia foetida*, hasten decomposition of litter but do not mix it into the mineral soil. **Endogeic** earthworms, such as the pale, pink *Allolobophora caliginosa* ("red worm"), live mainly in the upper 10 to 30 cm of mineral soil where they make shallow, largely horizontal burrows. Finally, the relatively large **anecic** earthworms make vertical, relatively permanent

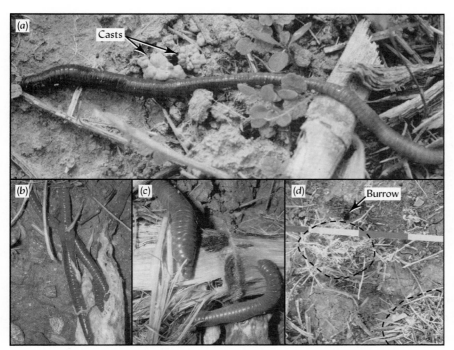

Figure 10.4

Anecic earthworm species such as these Lumbricus terrestris *(a) come to the surface to feed on litter (c), excrete soil casts (a, arrows) and reproduce (b). They incorporate large amounts of plant litter into the soil and gather plant debris into piles called* middens *to cover their burrow entrances (d, 10 cm scale markings). Earthworms are perhaps the most significant macro-organism in soils of humid temperate regions, particularly in relation to their effects on the physical conditions of soils and plant litter.* (Photos a and d courtesy of R. Weil; photos b and c courtesy of Steve Groff)

burrows as much as several meters deep. They emerge in wet weather or at night to forage on the surface for pieces of litter that they drag back into their burrows, often covering the entrance to their burrow with a **midden** of leaves (Figure 10.4*d*). The well-known "night crawler" (*Lumbricus terrestris*) was accidentally introduced to North America from Europe (perhaps in the root balls of settlers' fruit trees) and is now the most common anecic earthworm on both continents (Box 10.1).

BOX 10.1
GARDENERS' FRIEND NOT ALWAYS SO FRIENDLY[a]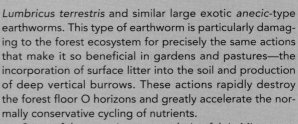

Earthworms are legendary for the many ways they enhance the productivity of gardens, pastures, and croplands. However, in other ecological settings, "nature's tillers" can be quite damaging. In particular, these active immigrants are starting to wreak havoc in some of North America's forest ecosystems. Since the Pleistocene glaciers wiped out the native earthworms some 10,000 years ago, the plant–soil communities in these boreal forests have evolved without substantial earthworm populations. Without the soil-mixing action of earthworms, these forests have developed a thick, stratified forest floor, usually consisting of several distinct O horizons. This loose, thick litter layer is essential habitat for certain native millipedes, ground-nesting birds, and salamanders. Native plants such as mayflowers, wood anemone, and trillium also depend on a thick layer of forest litter.

Recently, scientists have observed that the forest floor has all but disappeared from certain boreal forest stands. They think they know the culprit—invading populations of *Lumbricus terrestris* and similar large exotic *anecic*-type earthworms. This type of earthworm is particularly damaging to the forest ecosystem for precisely the same actions that make it so beneficial in gardens and pastures—the incorporation of surface litter into the soil and production of deep vertical burrows. These actions rapidly destroy the forest floor O horizons and greatly accelerate the normally conservative cycling of nutrients.

Some of the worst impacts are being felt in Minnesota, where European earthworms (including *Aporrectodea* sp., *Lumbricus rubellus*, and *L. terrestris*) have invaded the lake-dotted boreal forests. One approach to slowing the invasion of these uninvited soil engineers might be to surround the forest with buffer zones of habitat unsuitable for earthworms. Researchers believe the advance of these earthworms has resulted largely from their use as fishing bait. The take-home message: bring those bait worms home and dump them in your compost pile, not on the shore of your favorite fishing lake!

[a]For more on invasive earthworms, see Hendrix and Bohlen (2002) and Minnesota Worm Watch, at www.nrri.umn.edu/worms.

Figure 10.5

Influence of crop residue return to the soil on earthworm activity and soil hydraulic conductivity. The total amount of residues produced by the corn crop was about 10 Mg ha⁻¹. The lower rates of residue return represent the situations when some or all of the corn stover is collected and removed for the production of ethanol biofuel. The graph shows the effect of just 1 year of differential stover returns on numbers of earthworm middens present and the soil hydraulic conductivity (rate of water infiltration). The curves follow one another very closely, most likely because reductions in residue return negatively affect earthworm burrowing activity and fewer burrows result in less ready infiltration of rainwater. The data suggests two conclusions: (1) there exists a close positive relationship between earthworm activity and soil infiltration rate, and (2) if more than about one-third of the corn stover is removed, significant damage may be done to soil quality needed to sustain future production of crops and biofuels. Unfortunately, many models of potential biofuel production fail to take this relationship into account. The data are averages for three medium- to fine-textured soils under long-term no-till management in Ohio. [Graphed from data in Blanco-Canqui et al. (2007)]

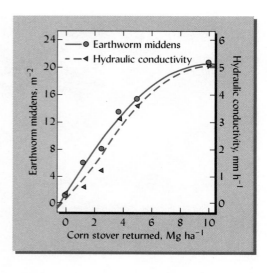

Influence on Soil Fertility, Productivity, and Environmental Quality

Burrows   Earthworms literally eat their way through the soil, creating extensive systems of burrows. Earthworm channels, whether empty or filled with casts, offer important pathways by which plant roots can penetrate dense soil layers (see Plate 81). Their extensive physical activity, which is particularly important in untilled soils (including grasslands and no-till croplands), has earned earthworms the title of "nature's tillers." Their impact on water infiltration can be dramatic (Figure 10.5).

After passing through the earthworm gut, ingested soil is expelled as globules of soil called **casts** (Figure 10.4*a*). During the passage through the earthworm's gut, organic materials are thoroughly shredded and mixed with mineral soil materials. The casting behavior of earthworms generally enhances the aggregate stability of the soil (Table 10.3). The casts are deposited within the soil profile or on the soil surface, depending on the species of earthworm, and are another sign used to assess earthworm activity in soil.

The activities of earthworms also enhance soil fertility and productivity by altering chemical conditions in the soil, especially in the upper 15 to 35 cm. Compared to the bulk soil, the casts are significantly higher in bacteria, organic matter, and available

Worm grunting: Fishing around in soil ecology: http://sciencenow .sciencemag.org/cgi/ content/full/2008/1015/1

Groff family farm no-till worm walk: www.newfarm.org/depts/ notill/features/worms.shtml

Table 10.3

CHARACTERISTICS OF EARTHWORM CASTS AND SOILS AT SIX SITES IN NIGERIA

Characteristic	Earthworm casts	Soils
Silt and clay, %	38.8	22.2
Bulk density, Mg/m³	1.11	1.28
Structural stability[a]	849	65
Cation exchange capacity, cmol/kg	13.8	3.5
Exchangeable Ca^{2+}, cmol/kg	8.9	2.0
Exchangeable K^+, cmol/kg	0.6	0.2
Soluble P, ppm	17.8	6.1
Total N, %	0.33	0.12

[a]Numbers of raindrops required to destroy structural aggregates.
From de Vleeschauwer and Lal (1981).

Table 10.4

RELATIVE NUMBERS AND BIOMASS OF FAUNA AND FLORA COMMONLY FOUND IN SURFACE SOIL HORIZONS

Microflora and earthworms dominate the life of most soils.

Organisms	Number[a]		Biomass[b]	
	Per m^2	Per gram	kg/ha	g/m^2
Microflora				
Bacteria and Archaea[c]	10^{14}–10^{15}	10^9–10^{10}	400–5000	40–500
Actinomycetes	10^{12}–10^{13}	10^7–10^8	400–5000	40–500
Fungi	10^6–10^8 m	10–10^3 m	1000–15,000	100–1500
Algae	10^9–10^{10}	10^4–10^5	10–500	1–50
Fauna				
Protozoa	10^7–10^{11}	10^2–10^6	20–300	2–30
Nematodes	10^5–10^7	1–10^2	10–300	1–30
Mites	10^3–10^6	1–10	2–500	0.2–5
Collembola	10^3–10^6	1–10	2–500	0.2–5
Earthworms	10–10^3		100–4000	10–400
Other fauna	10^2–10^4		10–100	1–10

[a]For fungi, the individual is hard to discern, so meters of hyphal length is given as a measure of abundance.
[b]Biomass values are on a liveweight basis. Dry weights are about 20 to 25% of these values.
[c]Estimated numbers of Bacteria and Archaea from Torsvik et al. (2002); others from many sources.

plant nutrients (see Table 10.3). Roots growing down earthworm burrows also find rich sources of nutrients in the casts and burrow lining material. Furthermore, when earthworms die and decay, the nutrients in their bodies are readily released into plant-available form. In addition, physical incorporation of surface residues into the soil reduces the loss of nutrients, especially of nitrogen, by erosion and volatilization.

Deleterious Effects of Earthworms

Exposure of Soil on the Surface   Not all effects of earthworms are beneficial. For example, in the process of building its middens, *Lumbricus terrestris* has been observed to leave some 60% of the soil surface bare of residues (see Figure 10.4*d*). In other situations, the middens themselves may be considered a nuisance—for example, on closely cropped golf greens. Even the burrowing action may not always be welcome—as in forests with thick litter layers (see Box 10.1). Another aspect of concern is that water rapidly percolating down vertical earthworm burrows may carry potential pollutants toward the groundwater (see Figure 6.22).

Factors Affecting Earthworm Activity

Earthworms prefer cool, moist, well-aerated soils that are well supplied with decomposable organic materials, preferably supplied as surface mulch. In temperate regions they are most active in the spring and fall, often curling into a tight ball (aestivating) to ride out hot, dry periods in summer. They do not live under anaerobic conditions, nor do they thrive in coarse sands. A few species are reasonably tolerant to low pH, but most earthworms thrive best where the soil is not too acid (pH 5.5 to 8.5) and has an abundant supply of calcium (which is an important component of their mucilage excretions). Enchytraeid worms are much more tolerant of acid conditions and are more active than earthworms in some forested Spodosols. Most earthworms are quite sensitive to excess salinity.

Other factors that depress earthworm populations include predators (moles, mice, and certain mites and millipedes), very sandy soils (partly because of the abrasive effect of sharp sand grains), direct contact with ammonia fertilizer, application of certain insecticides (especially carbamates), and tillage. The last factor is often the overriding deterrent to earthworm populations in agricultural soils.

10.5   ANTS AND TERMITES

The real kings of the jungle (video):
www.nhm.ac.uk/
nature-online/life/
insects-spiders/
insects-in-science/
ants-and-termites/index.html

How to tell ants and termites apart:
http://drdons.net/whatr.htm

Some mention has already been made (Section 10.2) of the varied food web roles of mesofauna arthropods (animals with a hard exoskeleton), namely the mites (eight-legged arachnids, which account for about 30,000 soil-dwelling species) and the springtails (wingless insects in the order *Collembola*, of which some 6500 soil species are known). We will now turn our attention to ants and termites, two groups of insects that are considered to be important ecosystem engineers.

Ants

Nearly 9000 species of soil-inhabiting ants have been identified. They are most diverse in the humid tropics, but are perhaps most functionally prominent in temperate semi-arid grasslands. Ants play important roles in forests from the tropics to the taiga. Some ant species act as detritivores, others as herbivores, and still others as predators.

Ant nest-building activity can improve soil aeration, increase water infiltration, and modify soil pH. The nests, and the microbial populations they encourage, also stimulate the cycling of soil nitrogen.

Termites

Termites are sometimes called *white ants*, although they are quite distinct from ants (one can distinguish an ant by its narrow "waist" between the abdomen and thorax). There are about 2000 species of termites, most of which use cellulose in the form of plant fiber as their primary food. Yet most termites cannot themselves digest cellulose. Instead, a termite depends on a mutualistic relationship with protozoa and bacteria that live in its gut. Metabolism under anaerobic conditions in termite guts accounts for a substantial fraction of the global production of methane (CH_4), an important greenhouse gas (see Section 11.9).

Most termite species eat decaying logs, grasses, or fallen tree leaves, but some attack the sound wood in standing trees. These groups have become quite infamous because of their habit of invading (and subsequently destroying) the houses people build of wood. The most damaging of the wood-eating species is the highly invasive Formosan subterranean termite (*Coptotermes formosanus*), which now infests wooden structures around the world in the tropical and subtropical zones. Most tree- and house-invading termites build protective tubes made of compacted soil and termite feces, which enable them to return to the soil for their daily water supply. Termite inspectors look for these earthen tunnels running up foundation walls under a house to indicate an infestation.

Termites are social animals that live in very complex labyrinths of nests, passages, and chambers that they build both below and above the soil surface. Termite mounds built from soil particles and feces cemented with saliva are characteristic features of many landscapes in Africa, Latin America, Australia, and Asia (Figure 10.6). These mounds are essentially termite "cities" with a network of underground passages and aboveground covered runways that typically spread 20 to 30 m beyond the mound. Several species, such as *Macrotermes spp.* in Africa, use plant residues to "cultivate" fungi in their mounds as a source of food.

Figure 10.6

Termites in southern Africa. (Left) A termite mound in a cultivated field constructed of soil cemented hard with termite saliva. (Right) Individual worker termites (striped) drag cut pieces of leaves into their underground nest as soldier termites (large heads and mandibles) stand guard. (Photos courtesy of R. Weil)

In building their mounds, termites transport soil from lower layers to the surface, thereby extensively mixing the soil and incorporating into it the plant residues they use as food. Scavenging a large area around each mound, these insects remove up to 4000 kg/ha of leaf and woody material annually, a substantial portion of the plant litter produced in many tropical ecosystems. They also annually move 300 to 1200 kg/ha of soil in their mound-building activities. These activities have significant impacts on soil formation, as well as on current soil fertility and productivity.

10.6 SOIL MICROANIMALS

From the viewpoint of microscopic animals, soils present many habitats that are essentially aquatic, at least intermittently so. For this reason the soil microfauna are closely related to the microfauna found in lakes and streams. The two groups exerting the greatest influence on soil processes are the nematodes and protozoa.

Nematodes

Nematodes are found in almost all soils, often in surprisingly great numbers and diversity. Some 20,000 species have been identified of the 100,000 nematode species that are thought to exist. These unsegmented roundworms are highly mobile creatures about 4 to 100 µm in cross section and up to several millimeters in length. They wriggle their way through the labyrinth of soil pores, sometimes swimming in water-filled pores (like their aquatic cousins), but more often pushing off the moist particle surfaces of partially air-filled pores. When the soil becomes too dry, nematodes survive by coiling up into a **cryptobiotic** or resting state, in which they seem to be nearly impervious to environmental conditions and use no detectable oxygen for respiration.

Most nematodes feed on fungi, bacteria, and algae or are predatory on other nematodes, protozoa, or insect larvae. The different trophic groups of nematodes can often be distinguished by the type of mouth parts present. Figure 10.7 illustrates a predator and plant parasite. Grazing by nematodes can have a marked effect on the growth and activities of fungal and bacterial populations. Since bacterial cells contain more nitrogen than the nematodes can use, nematode activity often stimulates the

Links to movies of the nematode *C. elegans* in action:
www.bio.unc.edu/faculty/goldstein/lab/movies.html

Nematode *Caenorhabditis elegans* on the move:
http://video.google.com/videoplay?docid-2019570087567872766&q=soil

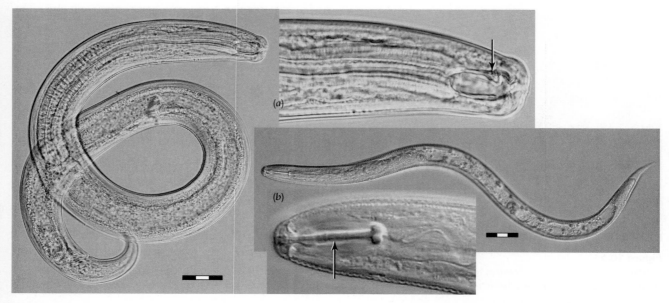

Figure 10.7

Two soil nematodes, one (a) a predator, and the other (b) a plant root parasite (soybean cyst nematode). The head and mouth parts of a nematode often reflect its trophic role, as the inset enlargements of these nematodes illustrate. Predators (such as this one in the Mononchidae family) usually have hard teeth (a, arrow) and a large mouth for capturing and swallowing prey. Nematodes that feed on plant roots or fungal hyphae are characterized by a retractable spearlike mouth part (b, arrow) that pierces the targeted cell to feed on the liquid contents. Mouth parts of a bacteria-eating nematode can be seen in the opening photo for this chapter. Scale bars marked in 10 μm units. (Photos courtesy of Lisa Stocking Gruver, University of Maryland)

cycling and release of plant-available nitrogen in the soil, accounting for 30 to 40% of the nitrogen released in some ecosystems. Certain predatory nematodes that efficiently attack insect larvae in the soil are sold for use as biological control agents, an environmentally beneficial alternative to the use of toxic pesticides. Once established, they can provide effective, long-term control of such soilborne insect pests as the corn rootworm or the grubs (e.g., Japanese beetle larvae) that destroy homeowners' lawns.

Some nematodes, especially those of the genus *Heterodera*, can infest the roots of practically all plant species by piercing the plant cells with a sharp, spearlike mouth part. Infestations beyond a certain threshold level result in serious stunting of the plant.

Until recently, the principal methods of controlling plant-parasitic nematodes were long rotations with nonhost crops (often 5 years is required for the parasitic nematode populations to sufficiently dwindle), use of genetically resistant crop varieties, and soil fumigation with highly toxic chemicals (**nematicides**). The use of soil nematicides, such as methyl bromide, has been sharply restricted because of undesirable environmental effects. New, less dangerous approaches to nematode control include the use of hardwood bark for containerized plants and interplanting or rotating susceptible crops with plants such as marigolds that produce root exudates with nematicidal properties (Figure 10.8).

Protozoa

Image gallery of protozoa:
www.pirx.com/droplet/
gallery.html

Protozoa are mobile, single-celled creatures that capture and engulf their food. With some 50,000 species in existence, they are the most varied and numerous of the soil microfauna. Most are considerably larger than bacteria (see Figure 10.9), having a diameter range of 4 to 250 μm. Soil protozoa include amoebas (which move by

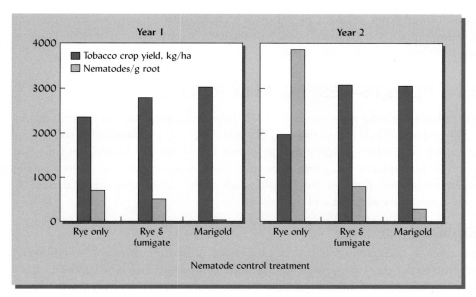

Figure 10.8

Marigolds to control plant-parasitic nematodes. A susceptible host plant (tobacco) was grown in the summers of year 1 and year 2 on a sandy soil in Ontario, Canada. Rye was grown as a cover crop each winter in the untreated (rye only) and fumigated plots (rye and fumigate). The fumigated plots were injected in year 1 and year 2 with a toxic chemical fumigant. The remaining plots (marigold) were planted to marigolds (Tagetes patula cv. "Creole") only in the summer of the year before the susceptible crop was first planted. Growing marigolds in one year controlled nematodes for the following two years. [Based on data from Reynolds et al. (2000); used with permission of the American Society of Agronomy]

extending and contracting pseudopodia), ciliates (which move by waving hairlike structures), and flagellates (which move by waving a whiplike appendage called a *flagellum*). They swim about in the water-filled pores and water films in the soil and can form resistant resting stages (called *cysts*) when the soil dries out or food becomes scarce.

Most soil-inhabiting protozoa prey upon soil bacteria, exerting a significant influence on the populations of these microflora in soils. Protozoa generally thrive best in moist, well-drained soils and are most numerous in surface horizons. Protozoa are especially active in the area immediately around plant roots. Their main influence on organic matter decay and nutrient release is through their effects on bacterial populations. Protozoa can squeeze into soil pores with openings as small as 10 μm. Still, soil aggregates often provide even smaller pores where bacterial cells can hide to escape predation.

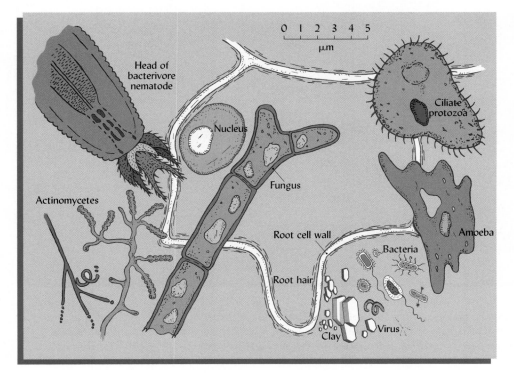

Figure 10.9

A depiction of representative groups of soil microorganisms, showing their approximate relative sizes. The large white-outlined structure in the center background is a plant root cell. (Drawing courtesy of R. Weil)

This protection from protozoa predation is a characteristic of the soil environment that helps explain the greater diversity of bacteria in soils than in aquatic habitats where such hiding places do not exist.

10.7 PLANTS—ESPECIALLY ROOTS

Higher plants store the sun's energy and are the primary producers of organic matter (see Figure 10.1). Their roots grow and die in the soil and are classified as soil organisms in this text. They typically occupy about 1% of the soil volume and may be responsible for a quarter to a third of the respiration occurring in a soil. Roots usually compete for oxygen, but they also supply much of the carbon and energy needed by the soil community of fauna and microflora.

Morphology of Roots

Depending on their size, roots may be considered to be either meso- or microorganisms. Fine feeder roots range in diameter from 100 to 400 μm, while root hairs are only 10 to 50 μm in diameter—similar in size to the strands of microscopic fungi (see Figure 10.9). Root hairs are elongated protuberances of single cells of the outer (epidermal) layer (Figure 10.10). One function of root hairs is to anchor the root as it pushes its way through the soil. Another function is to increase the amount of root surface area available to absorb water and nutrients from the soil solution.

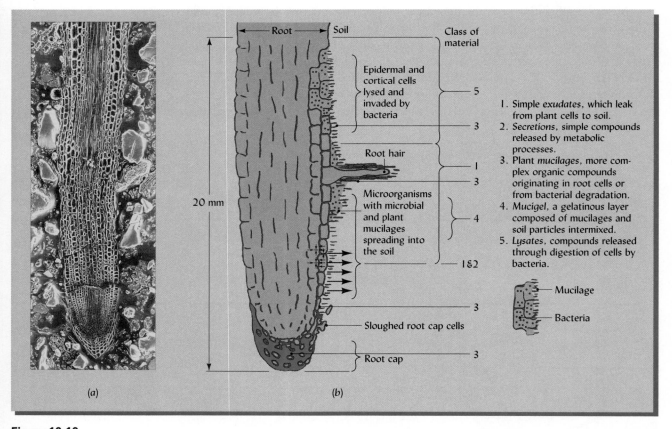

Figure 10.10

(a) Photograph of a root tip illustrating how roots penetrate soil and emphasizing the root cells through which nutrients and water move into and up the plant. (b) Diagram of a root showing the origins of organic materials in the rhizosphere. [(a) From Chino (1976), used with permission of Japanese Society of Soil Science and Plant Nutrition, Tokyo; (b) redrawn from Rovira et al. (1979), used with permission of Academic Press, London]

Roots grow by forming and expanding new cells at the growing point (meristem), which is located just behind the root tip. The root tip itself is shielded by a protective cap of expendable cells that slough off as the root pushes through the soil. Root morphology is affected by both type of plant and soil conditions. For example, fine roots may proliferate in localized areas of high nutrient concentrations. Root-hair formation is stimulated by contact with soil particles and by low nutrient supply. When soil water is scarce, plants typically put more energy into root growth than into shoot growth, decreasing the shoot-to-root ratio (thus increasing the uptake of water and minimizing its loss by transpiration). Many roots become thick and stubby in response to high soil bulk density or high aluminum concentrations in soil solution (see Chapters 4 and 9).

Effects on Soil

Living roots physically modify the soil in many ways. They follow paths of least resistance, growing between soil peds and into existing cracks and channels. Once extended into a pore, the root matures and expands, exerting lateral forces that enlarge the pore. By removing moisture from the soil, plant roots stabilize organic–mineral bonds and encourage soil shrinkage and cracking, which, in turn, increase stable soil aggregation. Roots exude many organic compounds that attract and support myriad microorganisms, which help to further stabilize soil aggregates and enrich the soil near the roots. In addition, when roots die and decompose, they provide building materials for humus, not only in the top few centimeters, but also to greater soil depths. If soil conditions permit, annual herbaceous plants (such as most crops) send their roots down 1 to 2 m. Perennial plants, especially woody species, may send some roots more than 5 m deep if restricting layers are not encountered. Generally, rooting is deepest in hot, dry climates and most shallow in boreal or wet tropical environments (see Figure 5.29).

The importance of root residues in helping to maintain soil organic matter is often overlooked. In grasslands, about 50 to 60% of the net primary production (total plant biomass) is commonly in the form of roots. In addition, prairie fires may remove most of the aboveground biomass, so that the deep, dense root systems are the main source of organic matter added to these soils. In plantation and natural forests, 40 to 70% of the total biomass production may be in the form of tree roots. In arable soils, the mass of roots remaining in the soil after crop harvest is commonly 15 to 40% that of the aboveground crop.

Rhizosphere and Rhizodeposition

The zone of soil significantly influenced by living roots is termed the **rhizosphere** and usually extends about 2 mm out from the root surface. The chemical and biological characteristics of this zone can be quite different from those of the bulk soil. Soil acidity may be 10 times higher (or lower) in the rhizosphere than in the bulk soil (e.g., Figure 9.9). Roots greatly affect the nutrient supply in this zone by withdrawing dissolved nutrients on one hand and by solubilizing nutrients from soil minerals on the other. By these and other means, roots affect the mineral nutrition of soil microbes, just as the microbes affect the nutrients available to the plant roots.

Significant quantities of organic compounds are released at the surface of young roots (see Figure 10.10). First, low-molecular-weight organic compounds are exuded by root cells, including organic acids, sugars, amino acids, and phenolic compounds. Some of these root exudates, especially the phenolics, exert growth-regulating influences on other plants and soil microorganisms in a phenomenon called **allelopathy** (see Section 11.5). Second, high-molecular-weight mucilages secreted by root-cap cells and epidermal cells near apical zones form a substance called **mucigel** when mixed with microbial cells and clay particles. This mucigel appears to have several beneficial functions: It lubricates the root's movement through the soil, it improves root–soil contact, and it provides an ideal environment for the growth of the rhizosphere microorganisms. Third, cells from the root cap and epidermis continually slough off as the root grows and enrich the rhizosphere with a wide variety of cell contents.

Taken together, these types of **rhizodeposition** typically account for 2 to 30% of total dry-matter production in young plants. The roots of common grain and vegetable plants have been observed to rhizodeposit 5 to 40% of the organic substances translocated to them from the plant shoot. Rhizodeposition decreases with plant age but increases with soil stresses, such as compaction and low nutrient supply. Because of the rhizodeposition of carbon substrates and specific growth factors (such as vitamins and amino acids), microbial numbers in the rhizosphere are typically 2 to 10 times as great as in the bulk soil. The processes just described explain why plant roots are among the most important organisms in the soil ecosystem.

Soil Algae

Directory of algal images, including those that can be found in soil: http://vis-pc.plantbio .ohiou.edu/algaeimage/ imageindex.htm

Algae, of course, do not have roots, but like vascular plants, algae consist of eukaryotic cells, those with nuclei organized inside a nuclear membrane. (Organisms formerly called *blue-green algae* are prokaryotes and therefore will be considered with the bacteria.) Also, like higher plants, algae are equipped with chlorophyll, enabling them to carry out photosynthesis. As photoautotrophs, algae need light and are therefore mostly found very near the surface of the soil. Some species can also function as heterotrophs in the dark. A few species are photoheterotrophs that use sunlight for energy but cannot synthesize all of the organic molecules they require (see Table 10.2).

Most soil algae range in size from 2 to 20 μm. Many algal species are motile and swim about in soil pore water, some by means of flagella (whiplike "tails"). Most grow best under moist to wet conditions, but some are also very important in hot or cold desert environments. Some algae (as well as certain cyanobacteria) form *lichens*, symbiotic associations with fungi. These are important in colonizing bare rock and other low-organic-matter environments. In unvegetated patches in deserts, algae commonly contribute to the formation of **microbiotic crusts** (see Section 10.13). In addition to producing a substantial amount of organic matter in some fertile soils, certain algae excrete polysaccharides that have very favorable effects on soil aggregation (see Section 4.5).

10.8   SOIL FUNGI[4]

Pictures of fungal mycelia: http://ic.ucsc.edu/~wxcheng/ wewu/soilfungi.htm

Time-lapse soil fungi: www.youtube.com/ watch?v=UvTvaxVySIE& feature=PlayList&p= 3DAF63BCB14E7C02& index=1

Scientists, analyzing DNA and fatty acids extracted from soils, estimate that there are at least 1 million fungal *species* in the soil still awaiting discovery. Because of the extensive filamentous morphology of many fungi, it is difficult to define the *numbers* of fungi in soil. For example, scientists using molecular analysis techniques determined that the fungal strands permeating the soil and tree roots of an entire 20-hectare forest stand belonged to *a single organism* that weighed more than 10,000 kg and was over 1500 years old! Instead of counting numbers, scientists use the biomass or hyphal length per m^2 as more meaningful measures of fungal presence. Total fungal biomass typically ranges from 1000 to 15,000 kg/ha in the upper 15 cm (see Table 10.4 on page 333). The fungi dominate many soils, their biomass exceeding even that of the bacteria.

---

[4]For a well-documented yet fascinating guide to fungi and their use in sustaining the productivity of forests and farms (as well as fungal roles in environmental and human health), see Stamets (2005).

Fungi are eukaryotes with a nuclear membrane and cell walls. They are aerobic heterotrophs, although some can tolerate the rather low oxygen concentrations and high levels of carbon dioxide found in wet or compacted soils. Strictly speaking, fungi are not entirely microscopic, since some of these organisms, such as mushrooms, form macroscopic structures that can easily be seen without magnification.

Molds and *mushrooms* are both considered to be filamentous fungi, because they are characterized by long, threadlike, branching chains of cells. Individual microscopic fungal filaments, called **hyphae** (Figure 10.15 on page 346), are often twisted together to form visible **mycelia** somewhat like fibers are woven into ropes. Fungal mycelia often appear as thin, white or colored strands running through decaying plant litter (Figure 10.11). Filamentous fungi reproduce by means of spores, often formed on fruiting bodies, which may be microscopic (e.g., molds) or macroscopic.

Molds often dominate the microflora in acid surface soils, where bacteria (including actinomycetes) offer only mild competition. Four of the most common genera found in soils are *Penicillium, Mucor, Fusarium,* and *Aspergillus.*

Mushroom fungi are associated with forest and grass vegetation where moisture and organic residues are ample. Although the mushrooms of many species are extremely poisonous to humans, some are edible—and a few have been domesticated. The aboveground fruiting body of most mushrooms is only a small part of the total organism. An extensive network of hyphae permeates the underlying soil or organic residue. While mushrooms are not as widely distributed as the molds, these fungi are very important, especially in the breakdown of woody tissue, and because some species form a symbiotic relationship with plant roots (see "Mycorrhizae," following).

Hunting slime molds, by Adele Conover, Smithsonian: www.smithsonianmag.com/issues/2001/march/phenom_mar01.php

Activities of Fungi

As decomposers of organic materials in soil, fungi are the most versatile and persistent of any group. Fungi play major roles in the processes of humus formation. They are quite efficient in using the organic materials they metabolize. Up to 50% of the substances decomposed by fungi may become fungal tissue, compared to about 20% for bacteria. Soil *fertility* depends in no small degree on nutrient cycling by fungi, since they continue to decompose complex organic materials under conditions that restrict many bacteria (including actinomycetes). Soil *tilth* also benefits from fungi as their

Figure 10.11

Fungal mycelia, consisting of bundles of microscopic hyphae, grow up from the soil into the leaves and woody debris of the forest floor. The ability of fungi to "reach out" from the soil in this manner helps explain why they dominate the decay of surface litter and mulch, while bacteria are more prominent in decaying organic material incorporated into the soil. Scale marked in cm. (Photo courtesy of R. Weil)

Figure 10.12

A "fairy ring" of fungal growth and the fungi's fruiting bodies (mushrooms). As the fairy ring fungi (most commonly Marasmius spp.) metabolize accumulated grass thatch and residues, they release excess nitrogen that stimulates the lush green growth of the grass. Later, bacteria decompose the aging and dead fungi, producing a second release of nitrogen. The fungus produces a chemical (thought to be hydrogen cyanide) that is toxic to itself. Therefore, each generation must grow into uncolonized soil, producing an ever-expanding circle of fungi and decay marked by an ever-larger ring of dark green grass. The grass in the center of the ring is often brown, stunted, and water-stressed, most likely because the fungi render the upper soil layers somewhat hydrophobic. (Photos courtesy of R. Weil)

hyphae stabilize soil structure (see Figure 4.9). Some of the nutrient cycling and ecological activities of soil fungi are made easily visible in the case of "fairy rings" commonly seen on lawns in early spring (Figure 10.12).

Some fungi produce compounds that kill other fungi or bacteria and provide a competitive edge over rival microorganisms in the soil. Certain species even trap nematodes (see Figure 10.13). Many fungi have proved to be highly beneficial to humankind (see Sections 10.9 and 10.13).

Unfortunately, not all the compounds produced by soil fungi benefit humans or higher plants. A few fungi produce chemicals (**mycotoxins**) that are highly toxic to

Figure 10.13

Several species of fungi prey on soil nematodes—often on those nematodes that parasitize higher plants. Some species of nematode-killing fungi attach themselves to, and slowly digest, the nematodes. Others, like this Arthrobotrys anchonia, make loops with their hyphae and wait for a nematode to swim through these lasso-like structures. The loop is then constricted, and the nematode is trapped. The nematode shown here is being crushed by two such fungal loops. Additional loops can be seen in their nonconstricted configuration. (Photo courtesy of George L. Barron, University of Guelph)

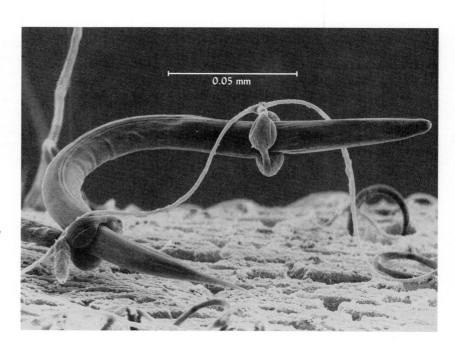

0.05 mm

plants or animals (including humans). An important example of the latter is the production of highly carcinogenic aflatoxin by the fungus *Aspergillus flavus* growing on seeds such as corn or peanuts, especially when exposed to soil and moisture. Other fungi produce compounds that allow them to invade the tissues of higher plants (see Section 10.12), causing such serious plant diseases as wilts (e.g., *Verticillium*) and root rots (e.g., *Rhizoctonia*).

On the other hand, efforts are now under way to develop the potential of certain fungi (such as *Beauveria*) as biological control agents against some insects and mites that damage higher plants. These examples merely hint at the impact of the complex array of fungal activities in the soil.

Mycorrhizae[5]

One of the most ecologically and economically important activities of soil fungi is the mutually beneficial association (**symbiosis**) between certain fungi and the roots of higher plants. This association is called **mycorrhizae**, a term meaning "fungus root." In natural ecosystems many plants are quite dependent on mycorrhizal relationships and cannot survive without them. Mycorrhizae are the rule, not the exception, for most plant species, including the majority of economically important plants.

Mycorrhizal fungi derive an enormous survival advantage from teaming up with plants. Instead of having to compete with all the other soil heterotrophs for decaying organic matter, the mycorrhizal fungi obtain sugars directly from the plant's root cells. This represents an energy cost to the plant, which may lose as much as 5 to 30% of its total photosynthate production to its mycorrhizal fungal symbiont.

In return, plants receive some extremely valuable benefits from the fungi. The fungal hyphae grow out into the soil some 5 to 15 cm from the infected root, reaching farther and into smaller pores than could the plant's own root hairs. This extension of the plant root system increases its efficiency, providing perhaps 10 times as much absorptive surface as the root system of an uninfected plant.

Mycorrhizae greatly enhance the ability of plants to take up phosphorus and other nutrients that are relatively immobile and present in low concentrations in the soil solution. Water uptake may also be improved by mycorrhizae, making plants more resistant to drought and salinity stress (e.g., Table 10.5). In soils contaminated

Overview of mycorrhizal symbioses: http://cropsoil.psu.edu/sylvia/mycorrhiza.htm

Ecology of mycorrhizae: www.anbg.gov.au/fungi/mycorrhiza.html

---

Table 10.5

EFFECT OF SEEDLING INOCULATION WITH ARBUSCULAR MYCORRHIZAE (AM) ON ROOT COLONIZATION, FRUIT YIELD, AND SHOOT NUTRIENT CONTENTS FOR TOMATOES IRRIGATED WITH NONSALINE OR SALINE WATER

Very small amounts of inoculum were added to the potting mix in seedling trays. Mycorrhizal inoculation increased all parameters, but the greatest benefits of AM accrued under saline conditions. AM-inoculated plants under saline conditions yielded 5.3 kg fruit m^{-2}, not statistically different from the 5.8 kg fruit m^{-2} yield of noninoculated plants under nonsaline conditions.

Salinity (EC) of irrigation water treatment	AM root colonization	Fruit yield	Nutrient content of plant shoot					
			Increase from AM inoculation, %					
			P	K	Na	Cu	Fe	Zn
Nonsaline (EC = 0.5)	166	29	44	33	21	93	33	51
Saline (EC = 2.4)	293	60	192	138	7	193	165	120

Data selected from Al-Karaki (2006).

---

[5]For an excellent book on mycorrhizal associations and their ecological effects, see Smith and Read (1997).

with high levels of metals, mycorrhizae protect the plants from excessive uptake of these potential toxins (see Section 15.7). There is evidence that mycorrhizae also protect plants from certain soilborne diseases and parasitic nematodes by producing antibiotics, altering the root epidermis, and competing with fungal pathogens for infection sites. For all these reasons, the use of mycorrhizae can be a powerful tool in land restoration projects as well as in some agricultural situations.

Two types of mycorrhizal associations are of considerable practical importance: **ectomycorrhiza** and **endomycorrhiza**. The ectomycorrhiza group includes hundreds of different fungal species associated primarily with temperate- or semiarid-region trees and shrubs, such as pine, birch, hemlock, beech, oak, spruce, and fir. These fungi, stimulated by root exudates, cover the surface of feeder roots with a fungal mantle. Their hyphae penetrate the roots and develop in the free space around the cells of the cortex but *do not penetrate* the cortex cell walls (hence the term *ecto*, meaning outside). Ectomycorrhizae cause the infected root system to consist primarily of visible white rootlets with a characteristic Y shape (Figure 10.14).

The most important members of the endomycorrhiza group are called **arbuscular mycorrhizae** (AM). When forming AM, fungal hyphae actually penetrate the cortical root cell walls and, once inside the plant cell, form small, highly branched structures known as **arbuscules**. These structures serve to transfer mineral nutrients from the fungi to the host plants and sugars from the plant to the fungus. Other structures, called **vesicles**, are usually also formed and serve as storage organs for the mycorrhizae (see Figure 10.14 and Plate 54).

Most native plants and agricultural crops can form AM associations and do not grow well on unfertilized soil in their absence. Two important groups of plants that do *not* form mycorrhizae are the *Brassicaceae* (mustards, cabbage, radish) and the *Chenopodiaceae* (beet and spinach). AM fungi are agriculturally most important where soils are low in nutrients, especially phosphorus (Plate 53).

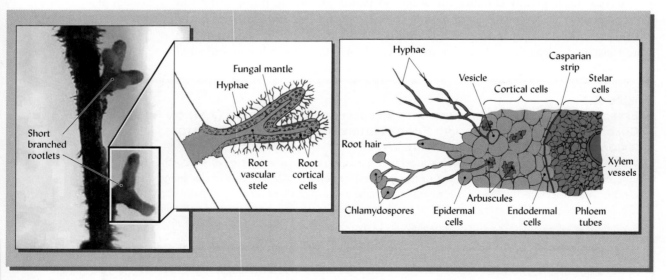

Figure 10.14

Diagram of ectomycorrhiza and arbuscular mycorrhiza (AM) associations with plant roots. (Left) The ectomycorrhiza association produces short branched rootlets that are covered with a fungal mantle, the hyphae of which extend out into the soil and between the plant cells, but do not penetrate the cells. (Right) In contrast, the AM fungi penetrate not only between cells but into certain cells as well. Within these cells, the fungi form structures known as arbuscules and vesicles. The former transfer nutrients to the plant, and the latter store these nutrients. In both types of association, the host plant provides sugars and other food for the fungi and receives in return essential mineral nutrients that the fungi absorb from the soil. [Redrawn from Menge (1981); photo courtesy of R. Weil]

The importance of mycorrhizal hyphae in stabilizing soil aggregate structure is becoming increasingly clear. Also, AM fungi have been observed to form hyphal interconnections among nearby plants and can transfer nutrients from one plant to another, sometimes resulting in complex symbiotic relationships.

Because of the near ubiquitous distribution of native mycorrhizal fungi, adding mycorrhizal inoculum rarely makes a difference in normal, biologically active soils. However, soil tillage destroys hyphal networks; therefore physical soil disruption is likely to decrease the effectiveness of native mycorrhizae. It is also best to avoid too-frequent use of nonhost species, long periods of soil bareness, or heavy fertilization with phosphorus. In addition, the buildup of effective mycorrhizae in soils is favored by growing a diversity of host plant species as continuously as possible and keeping the soil moist under a mulch.

There may be a need to inoculate soils with mycorrhizal fungi where native populations are very low or conditions for infection are unusually adverse. Examples include soils that have been subjected to broad spectrum fumigation; extreme soil heating, drying, or salinization; drastic disturbance such that subsoil layers are brought to the surface; or long periods without vegetative cover (such as surface soil stockpiled during mining or construction activities as shown in Plate 44). Successful restoration of healthy vegetation to such denuded soils often requires inoculation with effective mycorrhizal fungi.

Landscaping with mycorrhizae: www.fungi.com/mycogrow/amaranthus.html

Mycorrhizae for forestry: www.forestpests.org/nursery/mycorrhizae.html

10.9   SOIL PROKARYOTES: BACTERIA AND ARCHAEA

The organisms described in the previous sections—from mammals to molds—all belong to the Eukarya domain. The organisms in the other two domains of life, the Bacteria and the Archaea, are prokaryotes—their cells lack a nucleus surrounded by a membrane. However, despite their similar appearance under the microscope, archaea are evolutionarily quite distinct from bacteria; genetic analysis suggests that archaeans may be as closely related to plants or people as they are to bacteria! Until very recently, the archaea were thought of as rare and primitive creatures that live in only the most extreme and unusual environments on Earth—salt-saturated waters (see Plate 56), extremely acid or alkaline soils, deeply frozen ice, boiling hot water, anaerobic sediments, and the like. However, molecular identification techniques now suggest that archaeans are also common in more "normal" environments and probably represent about 10% of the microbial biomass in typical upland soils.

Into the Archaea: www.ucmp.berkeley.edu/archaea/archaea.html

Another recent insight provided by molecular identification techniques is that although we previously had no idea how little we knew, now we do! Traditionally, scientists enumerated and identified soil microbes by culturing them on agar of various kinds. We now know that the thousands of species so identified represent less than 0.1% of the species present in soils—most prokaryotes simply cannot be cultured in the lab. While the usual concept of a "species" is difficult to apply to single-celled microorganisms that reproduce asexually, "kinds" of organisms are now estimated by "genome equivalents" of DNA information. We will consider the archaea together with the bacteria in this section, calling them prokaryotes when the discussion applies to members of both domains.

Prokaryote Populations in Soils

Prokaryotes range in size from 0.5 to 5 µm, considerably smaller in diameter than most fungal hyphae (Figure 10.15). The smaller ones approach the size of the average clay particle (see Figure 10.9). Prokaryotes are found in various shapes: nearly round (coccus), rodlike (bacillus), or spiral (spirillum). In the soil, the rod-shaped prokaryotes seem to predominate. Many prokaryotes are motile, swimming about in the soil water films

Figure 10.15

Fungal hyphae associated with much smaller rod-shaped bacteria.

(Scanning electron micrograph courtesy of R. Campbell, University of Bristol, used with permission of American Phytopathological Society)

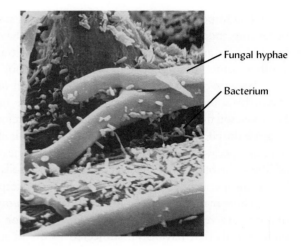

Fungal hyphae

Bacterium

by means of hairlike cilia or whiplike flagella. Others heavily colonize the nutrient-rich surface of plant roots.

The numbers of prokaryotes are extremely variable but high, ranging from a few billion to more than a trillion in each gram of soil. A biomass of a 400 to 5000 kg/ha liveweight is commonly found in the upper 15 cm of fertile soils (see Table 10.3).

Their small size and ability to form extremely resistant resting stages that survive dispersal by winds, sediments, ocean currents, and animal digestive tracts have allowed prokaryotes to spread to almost all soil environments. The prokaryote diversity in a handful of soil is said to be comparable to the diversity of insects, birds, and mammals in the Amazon Basin! Their extremely rapid reproduction (generation times of a few hours in the lab to a few days in favorable soil) enables prokaryotes to increase their populations quickly in response to favorable changes in soil environment and food availability.

Source of Energy

Microorganisms that grow under extreme conditions studied at Oak Ridge National Lab: www.ornl.gov/info/ ornlreview/rev32_3/ amazing.htm

Soil prokaryotes are either autotrophic or heterotrophic (see Section 10.2). Most soil bacteria are heterotrophic—both their energy and their carbon come from organic matter. Heterotrophic bacteria, along with fungi, account for the general breakdown of organic matter in soil. The bacteria often predominate on easily decomposed substrates, such as animal wastes and high sugar or protein plant residues. Where oxygen supplies are depleted, as in wetlands, nearly all decomposition is mediated by prokaryotes. Certain gaseous products of anaerobic metabolism, such as methane and nitrous oxide, have major effects on the global environment (see Sections 11.9 and 12.1).

Importance of Prokaryotes

Prokaryotes participate vigorously in virtually all of the organic transactions that characterize a healthy soil. Scientists are working to harness, even improve, the prokaryotes' broad range of enzymatic capabilities to help with the remediation of soils polluted by crude oil, pesticides, and various other organic toxins (see Section 15.6). The archaea are the most important group in the breakdown of hydrocarbon compounds, such as petroleum products.

Winogradsky column: perpetual life in a tube: www.biology.ed.ac.uk/ research/groups/jdeacon/ microbes/winograd.htm

Prokaryotes hold near monopolies in the oxidation or reduction of certain chemical elements in soils (see Sections 7.4 and 7.6). Some autotrophic prokaryotes obtain their energy from such inorganic oxidations, while anaerobic and facultative bacteria reduce a number of substances other than oxygen gas. Many of these biochemical oxidation and reduction reactions have significant implications for environmental quality as well as for plant nutrition. For example, through nitrogen oxidation (nitrification), selected bacteria oxidize relatively stable ammonium nitrogen to the much more mobile nitrate form of nitrogen. Likewise, certain archaeans oxidize sulfur, yielding plant-available sulfate ions, but also potentially damaging sulfuric acid (see Section 9.6). Prokaryote oxidation and reduction of inorganic ions such as iron and manganese not only influence the availability of these elements to other organisms, but also help determine soil colors (see Section 4.1). A critical process in which bacteria are prominent is nitrogen fixation—the biochemical combining of atmospheric nitrogen with hydrogen to form nitrogen compounds usable by plants (see Section 12.1).

Cyanobacteria

Previously classified as blue-green algae, **cyanobacteria** contain chlorophyll, which allows them to photosynthesize like plants. Cyanobacteria are especially numerous in rice paddies and other wetland soils and fix appreciable amounts of atmospheric nitrogen when such lands are flooded (see Section 12.1). These organisms also exhibit considerable tolerance to saline environments and are important in forming microbiotic crusts on desert soils (Section 10.13).

Soil Actinomycetes

Actinomycetes, bacteria that are filamentous and often profusely branched (see Figure 10.16), appear somewhat like tiny fungi. Generally aerobic heterotrophs, the actinomycetes live on decaying organic matter in the soil or on compounds supplied

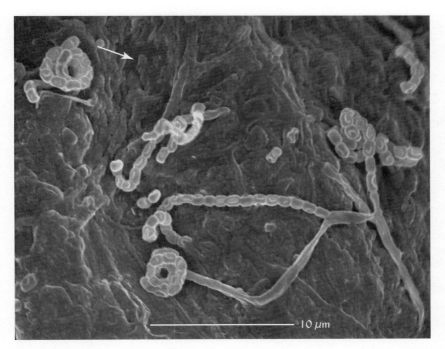

Figure 10.16
Strands of an actinomycete, a type of filamentous bacteria, growing on the surface of a soil biopore (an old root channel). The filaments, some breaking into the beadlike spores by which this organism reproduces, are about 0.8 μm in diameter. Some of the actinomycete filaments are embedded in mucilage of the soil pore (e.g., at arrow). The image is from 1.5 m deep in a clayey soil (poorly structured Alfisols) of a wheat field in eastern New South Wales, Australia. Almost all the wheat roots observed in this dense, hard subsoil were clustered into biopores made by roots of preceding alfalfa crops. The biopore surfaces are generally smooth, coated with illuvial clay and residues of old alfalfa roots. Although fungi commonly occupy old root channels, few were found in this soil, perhaps because of the antibiotic and chitinase secretions of the actinomycetes. This actinomycete image is quite unique in that the cryo scanning electron micrograph (SEM) was made directly from field material (frozen in liquid nitrogen), not from lab cultures. [SEM image courtesy of Margaret McCully, CSIRO Plant Industry, Canberria, Australia]

by plants with which certain species form parasitic or symbiotic relationships. Actinomycetes can break down even resistant compounds, such as cellulose, chitin, and phospholipids, into simpler forms. They often become dominant in the later stages of decay when the easily metabolized substrates have been used up. They are very important in the final (curing) stages of composting (see Section 11.10).

Penicillin and other antibiotics:
www.biology.ed.ac.uk/research/groups/jdeacon/microbes/penicill.htm

Special Attributes   Actinomycetes develop best in moist, warm, well-aerated soil. However, they tolerate low osmotic potential and are active in arid-region, salt-affected soils, and during periods of drought. They are generally rather sensitive to acid soil conditions, with optimum development occurring at pH values between 6.0 and 7.5. Some actinomycete species tolerate relatively high temperatures. The earthy aroma of organic-rich soils and freshly plowed land is mainly due to actinomycete-produced *geosmins*, volatile derivatives of terpene. In forest ecosystems, much of the nitrogen supply depends on actinomycetes that fix atmospheric nitrogen gas into ammonium nitrogen that is then available to plants (see Table 12.3). Many actinomycete species, especially in the genus *Streptomyces*, produce compounds that kill other microorganisms, and these "antibiotics" have become extremely important in human medicine (see Box 10.2).

BOX 10.2
A POST-ANTIBIOTIC AGE ON THE HORIZON?

Certain soil bacteria and fungi have evolved the capability of producing antibiotic compounds to provide a competitive edge in their struggle for survival in the soil. Being bathed in their own chemical warfare agents, most of these spore-forming microbes have also evolved immunities to many types of antibiotics. With the advent of antibiotic "miracle drugs" in the middle of the 20th century, these same compounds were harnessed to enable humans to all but conquer infectious bacterial diseases, which up until that time were the most common cause of human deaths. The first antibiotic compound discovered (and eventually put to use as the human medicine, penicillin) was produced by a soil fungus (*Penicillium spp.*) that contaminated some laboratory Petri dishes in 1928. In 1943, streptomycin was discovered, leading to the first of many antibiotic drugs synthesized by soil bacteria belonging to the genus *Streptomyces*. It is likely that you yourself are alive reading this book today because when you came down with a bacterial infection (perhaps pneumonia or a dirty wound), an antibiotic produced by a soil actinomycete (chloramphenicol, erythromycin, tetracycline, and vancomycin, to name a few) was available to save your life.

Unfortunately, the efficacy of these drugs is being rapidly eroded by the global spread of resistant strains of pathogenic bacteria. For example, enterococci and staphylococci that cause potentially fatal human diseases have now developed resistance to virtually every antibiotic drug available in the medical arsenal. What is causing this resistance that threatens to return humankind to the bad old days of the pre-antibiotic era? The answer is largely that gross overexposure to the various antibiotic drugs

has exerted tremendous selection pressure for resistance in the pathogen populations. Antibiotic drugs now permeate the environment—some 18 million kg are used annually in the United States alone. Part of the problem stems from overuse and misuse of human drugs by doctors, patients, hospitals, and consumers. But the principal source of antibiotics released into the environment is the enormous amount of these compounds used for nonmedical purposes. In fact, in the United States, some 87% of the antimicrobial drugs produced (nearly 15 million kg/y) is devoted to nonhuman uses, most of this as a growth-promoting additive for poultry, hog, and cattle feed in industrial-style farms.

Much of the antibiotic ingested by the livestock passes through the digestive tract unchanged and accumulates in the manure that is eventually spread on farm fields. Once in the soil, the compounds are known to retain their antibiotic activity even if adsorbed for long periods to clay surfaces. Crops growing on soils fertilized with such manure can take up small amounts of antibiotic, presenting the possibility that the antibiotics added to livestock feed may end up causing human allergic reactions and selecting for resistance in the human digestive tract. In any case, the large and continuous antibiotic presence in industrial animal facilities and in manured soils almost certainly hastens the evolution of antibiotic resistance in bacteria, including in human pathogens. While the problem has been known by scientists since the mid-1980s, policymakers in industrial countries have been slow to realize the need to eliminate such careless use of these life-saving substances.

10.10 CONDITIONS AFFECTING THE GROWTH OF SOIL MICROORGANISMS

Organic Matter Requirements

In most soils most of the time, the competetion for food among the various micro-organism is fierce. Therefore, the addition of almost any energy-rich organic sub-stance, including the compounds excreted by plant roots, is likely to stimulate an immediate increase in microbial growth and activity (see Section 11.2). In addi-tion, certain bacteria and fungi are stimulated by specific amino acids and other growth factors found in the rhizosphere or produced by other organisms. Bacteria tend to respond most rapidly to additions of simple compounds such as starch and sugars, while fungi and actinomycetes overshadow the bacteria if the added organic materials are rich in cellulose and more resistant compounds. In addition, if organic materials are left on the soil surface (as in forest litter or no-till crop residues), fungi tend dominate the microbial decomposition. Bacteria commonly play a larger role if the substrates are mixed into the soil, as by earthworms, root distribution, or tillage.

Oxygen Moisture and Temperature

Microbial activity is very sensitive to changes in soil moisture levels. The microbes are nearly quiescent in very dry soils but spring to life when water is added. The optimum balance between moisture and oxygen requirements for aerobic microbes seems to be achieved when about 60% of the soil pore space is filled with water and about 40% with air. Too high a water content will limit the oxygen supply. As can be observed in the rotting of wooden fence posts, the zone of greatest microbial activity in humid regions soils typically occurs just a few cm below the soil surface, where the oxygen supply is high but the soil is not too dry.

While most microorganisms are *aerobic* and use O_2 as the electron acceptor in their metabolism, some bacteria are *anaerobic* and use substances other than O_2 (e.g., NO_3^-, SO_4^{2-} or other electron acceptors). *Facultative* bacteria can use either aerobic or anaerobic forms of metabolism. All three of the above types of metabolism are usu-ally carried out simultaneously in different habitats within a given soil.

Microbial activity also responds markedly to soil temperature (see Section 7.8) and is generally greatest when temperatures are 20 to 40 °C. The warmer end of this range tends to favor actinomycetes. Ordinary soil temperature extremes seldom kill bacteria, and commonly only temporarily suppress their activity. However, except for certain **cryophilic** species, most microorganisms cease meta-bolic activity below about 5 °C, a temperature sometimes referred to as *biological zero* (see Section 7.9).

Exchangeable Cations and pH

Levels of exchangeable calcium and pH help determine which specific organisms thrive in a particular soil. Although in any chemical condition found in soils some bacterial species will thrive, high calcium and near-neutral pH generally result in the largest, most diverse bacterial populations (see Figure 9.15). Low pH allows fungi to become dominant. The effect of pH and calcium helps explain why fungi tend to dominate in forested soils, while bacterial biomass generally exceeds fungal biomass in most subhumid to semiarid prairie and rangeland soils. Certain metals, such as copper, which is especially toxic to bacteria and fungi, are also rendered more biologically available at low pH levels.

10.11　BENEFICIAL EFFECTS OF SOIL ORGANISMS ON PLANT COMMUNITIES

The soil fauna and flora are indispensable to plant productivity and the ecological functioning of soils. Of their many beneficial effects, only the most important can be emphasized here.

Organic Material Decomposition

Perhaps the most significant contribution of the soil fauna and flora to higher plants is the decomposition of dead leaves, roots, and other plant tissues. Soil organisms also assimilate wastes from animals (including human sewage) and other organic materials added to soils. As a by-product of their metabolism, microbes synthesize new compounds, some of which help to stabilize soil structure and others of which contribute to humus formation. The bacteria, archaeans, and fungi assimilate some of the N, P, and S in the organic materials they digest. Excess amounts of these nutrients may be excreted into the soil solution in inorganic form either by the microflora themselves or by the nematodes and protozoa that feed on them. In this manner, the soil food web converts organically bound forms of nitrogen, phosphorus, and sulfur into mineral forms that can be taken up once again by higher plants.

Breakdown of Toxic Compounds

Many organic compounds toxic to plants or animals find their way into the soil. Some of these toxins are produced by soil organisms as metabolic by-products, some are applied purposefully by humans as agri-chemicals to kill pests, and some are deposited in the soil because of unintentional environmental contamination. If these compounds accumulated unchanged, they would do enormous ecological damage. Fortunately, most biologically produced toxins do not remain long in the soil, for soil ecosystems include organisms that not only are unharmed by these compounds but can produce enzymes that allow them to use these toxins as food.

Some toxins are **xenobiotic** (artificial) compounds foreign to biological systems, and these may resist attack by commonly occurring microbial enzymes (see Section 15.5). The detoxifying activity of Prokaryotes and fungi is by far the greatest in the surface layers of soil, where microbial numbers are concentrated in response to the greater availability of organic matter and oxygen. Nontheless, some anaerobic detoxification occurs in deep soil layers and groundwater.

Inorganic Transformations

Nitrates, sulfates, and, to a lesser degree, phosphate ions are present in soils primarily due to inorganic transformations, such as the oxidation of sulfide to sulfate or ammonium to nitrate stimulated by microorganisms. Likewise, the availabilities of other essential elements, such as iron and manganese, are determined largely by microbial action. In well-drained soils, these elements are oxidized by autotrophic organisms to their higher valence states, in which forms they are quite insoluble. This keeps iron and manganese mostly in low solubility and nontoxic forms, even under fairly acid conditions. If such oxidation did not occur, plant growth would be jeopardized because of toxic quantities of these elements in solution. Microbial oxidation also controls the potential for toxicity in soil contaminated with selenium or chromium.

Table 10.6

RICE PLANTS RESPOND TO INOCULATION WITH GROWTH-PROMOTING RHIZOBACTERIA

Rhizobia or bradyrhizobia bacteria were added and colonized the rice rhizosphere, producing the plant growth hormone IAA and making the rice roots more efficient at nutrient uptake.

| Treatment | Grain yield, g/pot | Uptake of nutrients by rice plants, mg/pot | | | | IAA[a] in the rhizosphere, mg/L |
		N	P	K	Fe	
Control—no inoculation	36.7	488	111	902	18.9	1.0
Inoculated with rhizobacteria	44.3	612	134	1020	23.6	2.1
Percent change	+21	+25	+21	+13	+25	+110

[a]indol-3-acetic acid, a plant growth hormone.
Data calculated from Biswas et al. (2000).

Nitrogen Fixation

The fixation of elemental nitrogen gas, which cannot be used directly by higher plants, into compounds usable by plants is one of the most important microbial processes in soils (see Section 12.1). Actinomycetes in the genus *Frankia* fix major amounts of nitrogen in forest ecosystems; cyanobacteria are important in flooded rice paddies, wetlands, and deserts; and rhizobia bacteria are the most important group for the capture of gaseous nitrogen in agricultural soils (Table 12.2). By far the greatest amount of nitrogen fixation by these organisms occurs in root nodules or in other associations with plants.

Rhizobacteria

As pointed out in Section 10.7, the zone immediately around plant roots (the rhizosphere soil and the root surface itself, or **rhizoplane**) supports a dense population of microorganisms. Bacteria especially adapted to living in this zone are termed **rhizobacteria**, many of which are beneficial to higher plants (the so-called **plant growth-promoting rhizobacteria**). In nature, root surfaces are almost completely encrusted with bacterial cells, so little interaction between the soil and root can take place without some intervening microbial influence. The world of rhizobacteria is still largely uncharted, but research is beginning to uncover useful ways to take advantage of interactions that can benefit higher plants. In addition to those that ward off plant diseases (see Section 10.12), certain rhizobacteria promote plant growth in other ways, such as enhanced nutrient uptake or hormonal stimulation (e.g., see Table 10.6).

10.12 SOIL ORGANISMS AND DAMAGE TO HIGHER PLANTS

Although most of the activities of soil organisms are vital to a healthy soil ecosystem and economic plant production, some soil organisms affect plants in detrimental ways that cannot be overlooked. For example, soil organisms successfully compete with plants for soluble nutrients (especially for nitrogen), as well as for oxygen in poorly aerated soils. Here we focus on the soil organisms that act as herbivores, parasites, or pathogens.

Plant Pests and Parasites

The herbivorous soil fauna are by definition injurious to higher plants. Some rodents may severely damage young trees and farm crops. Snails and slugs in some climates are dreaded pests, especially of vegetables. Undoubtedly, the greatest damage to plants by

Forage brassicas for control of nematodes: www.abc.net.au/gardening/stories/s124457.htm

soil fauna is caused by the feeding of nematodes and insect larvae. To prevent or diminish such infestations, large amounts of nematicide and insecticide chemicals are used in agriculture, often with unintended ecological results (see Section 10.13).

Although bacterial blights and wilts are common, the fungi are responsible for the majority of soilborne plant diseases. Fungi of the genera *Pythium, Fusarium, Phytophthora,* and *Rhizoctonia* are especially prominent as soilborne agents of plant diseases described by such symptoms as *damping-off, root rots, leaf blights,* and *wilts.* Once a soil is infested, it is apt to remain so for a long time.

Some bacteria that live in the rhizosphere or on the rhizoplane inhibit root growth and function by various noninvasive chemical interactions. These nonparasitic **deleterious rhizobacteria** can cause stunting, wilting, foliar discoloration, nutrient deficiency, and even death of affected plants, but often the effects are subtle and difficult to detect. Their buildup may contribute to yield declines during long-term monoculture and aggravate problems in planting new trees in old orchards. On the other hand, by management that favors the deleterious rhizobacteria associated with certain weeds, scientists hope to be able to reduce weed seed germination and seedling growth and thereby reduce the use of herbicide (weed-killer) sprays on cropland and rangeland.

Plant Disease Control Through Soil Management[6]

Prevention is the best defense against soilborne diseases. Strict quarantine systems will restrict the transfer of soilborne pathogens from one area to another. Crop rotation can be very important in controlling a disease by growing nonsusceptible plants for several years between susceptible crops. Tillage may help by burying plant residues on which fungal spores might overwinter. However, disease problems are often lessened in no-tillage systems in which the soil surface remains mulched with plant residues that maintain a diverse soil community. The mulch also prevents the splashing of soil onto foliage by rain or irrigation, a major cause of plant disease spread and infection. Residues from certain green manure crops have been shown to chemically inhibit specific plant diseases. Direct management of soil physical and chemical properties can also be useful in disease control.

Soil Fertility   Regulation of soil pH is effective in controlling some diseases. For example, keeping the pH low (<5.2) can control both the actinomycete-caused *potato scab* and the fungal disease of turfgrass known as *spring dead spot.* Raising soil pH to about 7.0 can control *clubroot* disease in the cabbage family, because the spores of the fungal pathogen germinate poorly, if at all, under neutral to alkaline conditions.

Healthy, vigorous plants usually can resist or outgrow diseases better than weaker plants, so provision of an optimal level of balanced nutrition is an important step in disease management. High levels of nitrogen fertilization tend to increase plants' susceptibility to fungal diseases; high levels of ammonium (as compared to nitrate) nitrogen especially increase wilt diseases caused by *Fusarium* fungi. However, potassium fertilizers often reduce fungal disease severity, as do relatively high levels of calcium and manganese. Nutrient imbalances and micronutrient deficiencies can make plants especially susceptible to attack.

Organic Toxins   As an alternative to the use of synthetic, broad-spectrum fungicides or fumigants, certain natural organic anti-fungals can be introduced to the soil via microbial breakdown of organic amendments. An example is the rotation of cauliflower

---

[6]For a general overview of disease control by management of the soil ecosystem, see Stone et al. (2004); for a review of the scientific literature on disease and pest management using organic amendments, see Litterick and Harrier (2004).

with broccoli that has been shown, even in fields infested with the disease-causing fungus, to provide practical control of *verticillium wilt*, a serious disease of cauliflower. The broccoli leaf residues left after harvest are tilled into the soil. There they break down to release volatile compounds specifically toxic to the *Verticillium dahliae* fungus, providing a level of disease control in the following cauliflower crop equal to that achieved by synthetic fumigants.

Soil Physical Properties   Soil compaction often aggravates fungal root diseases by slowing root growth, inducing more root excretions that attract the pathogens, and by promoting wet, poorly aerated conditions. Wet, cold soils favor some seed rots and seedling diseases such as *damping-off*. Good drainage and planting on ridges can help control these diseases. High soil temperature can be used to control a number of pathogens. **Solarization**, the use of sunlight to heat soil under clear plastic sheeting, is a practical way to partially sterilize the upper few cm of soil in some field situations. Steam or chemical sterilization is a practical method of treating greenhouse potting media. It should be remembered, however, that sterilization kills beneficial microorganisms, such as mycorrhizal fungi, as well as pathogens, and so may do more harm than good!

Disease-Suppressive Soils

Research has documented the existence of **disease-suppressive soils**, in which a disease fails to develop *even though both the virulent pathogen and a susceptible host are present*. Evidence suggests that the pathogenic organisms are inhibited by **antagonism** from beneficial bacteria and fungi. Two broad types of disease suppression are recognized: general and specific.

General disease suppression is caused by high levels of overall microbial activity in a soil, especially at times critical in the development of a disease, such as when the pathogenic fungus is generating propagules or preparing to penetrate the plant cells. The presence of particular organisms is less important than the total level of activity. The mechanisms responsible for general suppression are thought to include (1) competition by beneficial microorganisms in the rhizosphere for carbon (energy) sources, (2) competition for mineral nutrients (such as nitrogen and iron), (3) colonization and decomposition of pathogen propagules (e.g., spores), (4) antibiotic production by varied actinomycete and fungal populations (see Section 10.10), and (5) lack of suitable root infection sites due to surface colonization by beneficial bacteria or previous infection by beneficial mycorrhizal fungi. In natural systems, the highly organic litter layer often provides an environment of such high microbial activity that most pathogens cannot compete. In agricultural systems, general suppression can often be encouraged by the addition of large amounts of decomposable organic matter from composts, manures, and cover crop residues, and by developing a "litter layer" through no-till and mulching techniques.

Specific suppression is attributable to the actions of a single species or a narrow group of microorganisms that inhibit or kill a particular pathogen. The effective presence of the specific suppressing organism may result from the same types of organic matter management just described or from introduction of an inoculum containing high numbers of the desired organism.

In some cases, specific disease suppressiveness has developed through long-term crop monoculture in which the buildup of the pathogen during the first few years is eventually overshadowed by a subsequent buildup of specific organisms antagonistic to the pathogen (Figure 10.17). Specific organisms known to be antagonistic to pathogens include *Trichoderma viride* fungi and certain fluorescent *Pseudomonas* bacteria, which produce antibiotics specific against pathogens or produce compounds

Rhizobacteria, underground biocontrol allies: www.ars.usda.gov/is/AR/archive/oct98/rhizo1098.htm

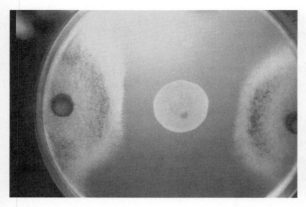

Figure 10.17

The biological basis of a disease-suppressive soil. (Left) A colony of certain Pseudomonas *bacteria (center of plate) produces an antibiotic toxic to* Gaeumannomyces graminis *(the fungal pathogen that causes take-all disease), preventing the pathogen colonies from growing to the center of the plate. (Right) Plots in eastern Washington that grew monoculture wheat for 15 years and developed high populations of organisms antagonistic to the take-all pathogen. In the 15th year of the study, the entire field was inoculated with G. graminis for experimental purposes, but the disease developed (seen as light-colored, prematurely ripened plots) only where the soil was fumigated prior to the inoculation. The fumigation killed most of the antagonistic organisms, leaving the pathogenic fungi free to infect the wheat plants.* (Photos by R. J. Cook; courtesy of the American Phytopathological Society)

that bind so tightly with iron that the pathogen spores cannot get enough of this nutrient to germinate. Despite the existence of commercial products containing beneficial microorganisms, successful suppression is usually limited by appropriate conditions in the soil rather than lack of a particular organism.

Horticulturalists have been able to control *Fusarium* diseases in containerized plants by replacing traditional potting mixes with growing media made mainly from certain well-aged **composts**. Apparently, large numbers of beneficial antagonistic organisms colonize the organic material during the final stages of composting (see Section 11.10), and the stabilized organic substrate stimulates the activity of indigenous beneficial organisms without stimulating the pathogens. Similar success in practical disease suppression has been experienced with the use of composted materials on turfgrass, especially for replacing peat (which is relatively inert and does not stimulate disease suppression) in topdressing golf course greens (see Figure 10.18).

The role of soil ecology in protecting plants from disease is not limited to belowground infections. Beneficial rhizobacteria have an intriguing mode of action

Figure 10.18

Topdressing with compost may be a practical, non-toxic means of suppressing dollar spot disease on bentgrass putting greens. The disease, caused by the fungus Sclerotinia homoeocarpa, *was controlled as well or better by the high rate of compost (4900 kg/ha topdressed every 3 weeks) as by the synthetic fungicide (Chlorthalonil, sprayed on every 2 weeks). Even the lower rate (1200 kg/ha) of compost provided some control of the disease. All the turfgrass plots were inoculated with the disease organism. The researchers suggest that topdressing with compost provided a general type of disease suppression since they found little difference among composts made from many different materials. The means from 2 years' data are shown.* [Based on data in Boulter et al. (2002)]

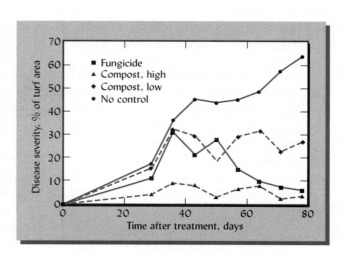

called **induced systemic resistance**, which helps plants ward off infection by diseases or insect pests both above and below ground. The process begins when a plant root system is colonized by beneficial rhizobacteria that cause the accumulation of a signaling chemical. The chemical signal is translocated up to the shoot, where it induces leaf cells to mount a chemical defense against a specific pathogen, even before the pathogen has arrived on the scene. When the pathogen (perhaps a fungal spore) does arrive on the leaf, its infection process is aborted almost before it can begin. In many cases studied so far on crop plants, the resistance-inducing organism has been a *Pseudomonas* or *Serratia* bacteria. This mechanism has been shown to effectively reduce damages by numerous fungal, bacterial, and viral pathogens and several leaf-eating insect pests.

These examples merely hint at the potential that exists for controlling plant diseases and pests through ecological management rather than applications of toxic chemicals.

10.13 ECOLOGICAL RELATIONSHIPS AMONG SOIL ORGANISMS

Mutualistic Associations

We have already mentioned a number of mutually beneficial associations between plant roots and other soil organisms (e.g., mycorrhizae and nitrogen-fixing nodules) and between several microorganisms (e.g., lichens). Other examples of such associations abound in soils. For example, photosynthetic algae reside within the cells of certain protozoans. Several types of associations, among them algal-fungal associations on or in soils and rocks, are very important cyclers of nutrients and producers of biomass in desert ecosystems. Next, we will briefly consider the nature of such associations.

Microbiotic Crusts[7]

In relatively undisturbed arid- and semiarid-region ecosystems, where the vegetation cover is quite patchy, it is common to find an irregular, usually dark-colored crust covering the soil in the areas between clumps of grasses and shrubs (see Figure 10.19). This crust is not at all like the physical-chemical crusts associated with degraded soils that have a hard, smooth surface seal (see Section 4.6). Rather, the **microbiotic crusts** of arid lands consist of mutualistic associations that usually include algae or cyanobacteria along with fungi, mosses, bacteria, and/or liverworts. An intact microbiotic crust is considered a sign of a healthy ecosystem.

Microbiotic crusts provide considerable protection against erosion by wind and water and also improve arid-region ecosystem productivity by (1) helping to conserve and cycle nutrients, (2) increasing nitrogen supplies via the nitrogen-fixing activities of the cyanobacteria, (3) enhancing water supplies in some cases by increasing infiltration and reducing evaporation, and (4) contributing to net organic matter production by crust photosynthesis, which may continue during environmental conditions that inhibit photosynthesis by higher plants in the ecosystem. The filamentous cyanobacteria make a particularly important contribution to these functions, as they not only photosynthesize but also fix from 2 to 40 kg/ha of nitrogen annually and form sticky

Soil biological crusts:
www.soilcrust.org/

Desert ecology—From biological crust to dust (audio report): www.npr.org/templates/story/story.php?storyId= 5415315

---

[7]Other names used to refer to these microbiotic crusts include *biological crusts, cryptograms, cryptobiotic crusts, microfloral crusts,* and *microphytic crusts.* All refer to the same thing. See Belnap (2003) for a brief overview and further reference citations on the ecology of microbial crusts. For a technical review of scientific knowledge and management options, see Belnap et al. (2001).

Figure 10.19

Tiny pinnacles of a microbiotic crust in Arches National Monument, Utah, seem to reflect the larger pinnacles of an arid landscape. These crusts consist of algae, cyanobacteria, fungi, and other organisms living together in a mutualistic relationship. The inset shows a scanning electron micrograph of cyanobacteria filaments that make up the backbone of many crusts. Microbiotic crusts typically cover the soil surface in the unvegetated patches between clumps of desert shrubs and grasses. The crusts improve desert productivity, but can be easily destroyed by wheels, feet, and hooves. [Large photo courtesy of Ben Waterman; inset courtesy of Jayne Belnap (U.S. Geologic Service, Moab, Utah) and John Gardner (Brigham Young University, Provo, Utah)]

polysaccharide coatings or sheaths that catch nutrient-rich dust and bind soil particles. Unfortunately, the crusts can be easily destroyed by trampling, off-road vehicles, or burial under windblown soil, and are very slow to reestablish.

Effects of Management Practices on Soil Organisms

Changes in environment affect both the number and kinds of soil organisms. Clearing forests or grasslands for cultivation drastically changes the soil environment. Monocultures or even common crop rotations greatly reduce the number of plant species and so provide a much narrower range of plant materials and rhizosphere environments than nature provides in forests or grasslands.

Case Studies and Practices for Improved Soil Biological Management: www.fao.org/landandwater/agll/soilbiod/cases.stm

While agricultural practices have different effects on different organisms, a few generalizations can be made (Table 10.7).

Table 10.7

SOIL-MANAGEMENT PRACTICES AND THE DIVERSITY AND ABUNDANCE OF SOIL ORGANISMS

Note that the practices that tend to enhance biological diversity and activity in soils are also those associated with efforts to make agricultural systems more sustainable.

Decreases biodiversity and populations	Increases biodiversity and populations
Fumigants	Balanced fertilizer use
Nematicides	Lime on acid soils
Some insecticides	Proper irrigation
Compaction	Improved drainage and aeration
Soil erosion	Animal manures and composts
Industrial wastes and heavy metals	Domestic (clean) sewage sludge
Moldboard plow–harrow tillage	Reduced or zero tillage
Monocropping	Crop rotations
Row crops	Grass–legume pastures
Bare fallows	Cover crops or mulch fallows
Residue burning or removal	Residue return to soil surface
Plastic mulches	Organic mulches

Tillage is an especially drastic disturbance of the soil ecosystem, disrupting fungal hyphae networks and earthworm burrows, as well as speeding the loss of organic matter. Reduced tillage therefore tends to increase the role of fungi at the expense of the bacteria and usually increases overall organism numbers as organic matter accumulates. Addition of animal manure or compost stimulates even higher microbial and faunal (especially earthworm) activity.

Pesticides are highly variable in their effects on soil ecology (see Section 15.4). Soil fumigants and nematicides can sharply reduce organism numbers, especially for fauna, at least on a temporary basis. On the other hand, application of a particular pesticide often stimulates the population of a specific microorganism, either because the organism can use the pesticide as food or, more likely, because the predators of that organism have been killed. A chemical that affects one group of organisms will likely affect other non-target groups as well and will eventually impact the productivity and functioning of the whole soil ecosystem. It is wise to remember that the interrelationships among soil organisms are intricate, and the effects of any perturbation of the system are difficult to predict.

Links Between Communities Above- and Belowground[8]

The communities of plants and animals we see aboveground greatly influence—and are profoundly influenced by—the communities belowground that we rarely see. The connections and interactions between them are both direct and indirect. Direct effects include the damage done to plants by pathogenic soil fungi, root-feeding nematodes, and the like. Direct effects also include the positive influences of mycorrhizae and beneficial rhizosphere bacteria, including those that cause the induced systemic resistance just described. Indirect effects include the complex feeding activities within the soil food web that eventually release nutrients that plants can use. In these and many other ways, the soil food web alters the types and productivity of plants in the aboveground world and, in so doing, affects the food supply and habitat for aboveground animals as well.

The energy that drives the soil food web comes from aboveground photosynthesis via the organic carbon in plant litter, rhizodeposition, and herbivore excretions. Therefore, the numbers and activities of the belowground organisms are highly responsive to the amount of such inputs, especially plant litter. However, it should also be noted that the *type* and *quality* of plant litter produced by the aboveground community (see also Section 11.3) has an enormous impact on the abundances of the various creatures living in the soil (Table 10.8).

Structure and function of soil biota on anthropogenic landscapes, Baltimore Ecosystem Study: www.beslter.org/ frame4-page_3a_02.html

---

Table 10.8

TYPE OF FOREST LITTER INFLUENCES THE ABUNDANCES OF SELECTED GROUPS OF FAUNA

Leaves fallen from two species of forest plants were placed in mesh litter bags (100 g dry matter per bag) and pinned to the soil surface in a temperate forest in New Zealand. After 279 days the more nutrient-rich litter supported greater numbers of most faunal groups, except predatory nematodes. The activities of these and other soil organisms will in turn influence the types and productivity of plants in the aboveground forest community.

Source of litter plant species	Nutrients in litter (%)		Numbers of organisms/litter bag			
	N	P	Microbial feeding nematodes	Predatory nematodes	Tardigrades	Coleoptera beetles
Metrosideno umbellata	0.35	0.03	6,600	210	35	4
Aristotelia serrata	2.94	0.24	12,800	73	670	364

Calculated from selected data in Wardle et al. (2006).

---

[8]For an introduction to how soil ecology influences plant invasiveness, see Wolfe and Klironomos (2005). For an authoritative treatise on the aboveground–belowground interactions, see Wardle (2002).

It is no wonder that ecologists are coming to recognize that ecosystems on land can be understood only when sufficient attention is paid to the world beneath the land's surface. For example, conservation biologists urgently need to learn what makes some exotic plants so invasive that they destroy native plant communities. It turns out that at least part of the answer may be found belowground in the community of organisms that colonize the plant rhizospheres.

10.14   CONCLUSION

The soil is a complex ecosystem with a highly diverse community of organisms that are vital to the cycle of life on Earth. Soil organisms incorporate plant and animal residues into the soil and digest them, returning carbon dioxide to the atmosphere, where it can be recycled through higher plants. Simultaneously, they create humus, the organic constituent so important to good physical and chemical soil conditions. During digestion of organic substrates, they release essential plant nutrients in inorganic forms that can be absorbed by plant roots or be leached from the soil. They also mediate the redox reactions that influence soil colors, nutrient cycling, and the production of gases that contribute to global warming.

Animals, particularly earthworms, ants, and termites, mechanically incorporate residues into the soil and leave open channels through which water and air can flow. As such, they are examples of soil ecosystem engineers that change the soil environment for all its inhabitants and create niches in which other organisms can live. Microorganisms such as fungi, archaeans, and bacteria are responsible for most organic decay, although their activity is greatly influenced by the soil fauna. Certain of the microorganisms form symbiotic associations with higher plants, playing special roles in plant nutrition and nutrient cycling. Competition for mineral nutrients among soil microbes, and between these organisms and higher plants, can result in plant nutrient deficiencies. Microbial requirements are factors in determining the success of most soil-management systems. A high level of general microbial activity fed by organic inputs can help suppress plant pathogens. Several specific fungi and bacteria produce antibiotic compounds that help them compete and also can inhibit plant pathogens, as well as form the basis for life-saving human drugs. Scientific understanding is just beginning to scratch the surface of the complex communities beneath our feet.

The soil community must have energy and nutrients if it is to function efficiently. To obtain these, soil organisms break down organic matter, aid in the production of humus, and leave behind compounds that are useful to higher plants. Organic matter, its effects on soil behavior, and its decomposition are topics of the next chapter.

STUDY QUESTIONS

1. What is *functional redundancy*, and how does it help soil ecosystems continue to function in the face of environmental shocks such as fire, clear-cutting, or tillage?

2. Give an example of an organism that plays each of these rules: *primary producer, primary consumer, secondary consumer*, and *tertiary consumer*.

3. Describe some of the ways in which mesofauna play significant roles in soil metabolism even though their biomass and respiratory activity is only a small fraction of the total in the soil.

4. What are the four main types of metabolism carried out by soil organisms relative to their sources of energy and carbon?

5. What role does O_2 play in aerobic metabolism? What elements take its place under anaerobic conditions?

6. A *mycorrhiza* is said to be a symbiotic association. What are the two parties in this symbiosis, and what are the benefits derived by each party?

7. In what ways is soil improved as a result of earthworm activity? Are there possible detrimental effects as well?

8. What is the *rhizosphere*, and in what ways does the soil in the rhizosphere differ from the rest of the soil?
9. Explain and compare the effects of tillage and manure application on the abundance and diversity of soil organisms.
10. What is *induced systemic resistance*, and how does it work?
11. What is a disease-suppressive soil? Explain the difference between general and specific forms of suppression.
12. Discuss the value and limitations of using specific inoculants for (a) mycorrhizae and (b) disease suppression. For each type of inoculation, describe a situation for which the chances would be very good for improving plant growth.

13. What are the main food web roles played by nematodes, and how can we visually (with a microscope) distinguish among nematodes that play these roles?
14. In what ways are actinomycetes like other groups of bacteria, and in what ways are they special?
15. Through appropriate extractions and counting you determine that there are 58 nematodes in a 1 g sample of soil. How many nematodes would occur in a 1.0 m^2 area of this soil? In 1 hectare? Assume the samples came from the upper 10 cm of soil with a bulk density $=1.3$ Mg/m^3.
16. Explain with two soil examples the concept of an *ecosystem engineer*.

REFERENCES

Al-Karaki, G. N. 2006. "Nursery inoculation of tomato with arbuscular mycorrhizal fungi and subsequent performance under irrigation with saline water," *Scientia Horticulturae*, **109**:1–7.

Baskin, Y. 2005. *Under Ground: How Creatures of Mud and Dirt Shape Our World* (Washington, D.C.: Island Press), 237 pp.

Belnap, J. 2003. "The world at your feet: Desert biological soil crusts," *Frontiers of Ecology and the Environment*, **1**:181–189.

Belnap, J., J. H. Kaltenecker, R. Rosentreter, J. Williams, S. Leonard, and D. Eldridge. 2001. "Biological soil crusts: Ecology and management." Technical Reference 1730-2. United States Department of the Interior, Bureau of Land Management, Denver, CO. www.soilcrust.org/ crust.pdf.

Biswas, J. C., J. K. Ladha, and F. B. Dazzo. 2000. "Rhizobia inoculation improves nutrient uptake and growth of lowland rice," *Soil Sci. Soc. Amer. J.*, **64**:1644–1650.

Blanco-Canqui, H., R. Lal, W. M. Post, R. C. Izaurralde, and M. J. Shipitalo. 2007. "Soil hydraulic properties influenced by corn stover removal from no-till corn in Ohio," *Soil Tillage Res.*, **92**:144–155.

Boulter, J. I., G. J. Boland, and J. T. Trevors. 2002. "Evaluation of composts for suppression of dollar spot (*Sclerotinia homoeocarpa*) of turfgrass," *Plant Dis.*, **86**:405–410.

Chino, M. 1976. "Electron microprobe analysis of zinc and other elements within and around rice root growth in flooded soils," *Soil Sci. and Plant Nut. J.*, **22**:449.

Coleman, D. C., D. A. J. Crossley, and P. F. Hendrix. 2004. *Fundamentals of Soil Ecology* (London: Elsevier Academic Press). 386 pp.

de Vleeschauwer, D., and R. Lal. 1981. "Properties of worm casts under secondary tropical forest regrowth," *Soil Sci.*, **132**:175–181.

Hendrix, P. F., and P. J. Bohlen. 2002. "Exotic earthworm invasions in North America: Ecological and policy implications," *BioScience*, **52**:801–811.

Jouquet, P., J. Dauber, J. Lagerlöfe, P. Lavelle, and M. Lepage. 2006. "Soil invertebrates as ecosystem engineers: Intended and accidental effects on soil and feedback loops," *Appl. Soil Ecol.*, **32**:153–164.

Lavelle, P., D. Bignell, M. Lepage, V. Wolters, P. Roger, P. Ineson, O. W. Heal, and S. Dhillion. 1997. "Soil function in a changing world: The role of invertebrate ecosystem engineers," *European Journal of Soil Biology*, **33**:159–193.

Litterick, A. M., and L. Harrier. 2004. "The role of uncomposted materials, composts, manures, and compost extracts in reducing pest and disease incidence and severity in sustainable temperate agricultural and horticultural crop production: A review," *Critical Reviews in Plant Sciences*, **23**:453–479.

Lowenfels, J., and W. Lewis. 2006. *Teaming With Microbes: A Gardener's Guide to the Soil Food Web.* (Portland, OR, Timber Press). 196 p.

Menge, J. A. 1981. "Mycorrhizae agriculture technologies," in *Background Papers for Innovative Biological Technologies for Lesser Developed Countries*, Paper No. 9., Office of Technology Assessment Workshop,

Nov. 24–25, 1980. (Washington, D.C.: U.S. Government Printing Office), pp. 383–424.

Nardi, J. B. 2003. *The World Beneath Our Feet: A Guide to Life in the Soil* (New York: Oxford University Press).

Paul, E. A. (ed.). 2006. *Soil Microbiology, Ecology and Biochemistry.* (San Diego: Academic Press).

Reynolds, L. B., J. W. Potter, and B. R. Ball-Coelho. 2000. "Crop rotation with *Tagetes* sp. is an alternative to chemical fumigation for control of root-lesion nematodes," *Agronomy J.*, **92:**957–966.

Rovira, A. D., R. C. Foster, and J. K. Martin. 1979. "Origin, nature and nomenclature of the organic materials in the rhizosphere," in J. L. Harley and R. S. Russell (eds.), *The Soil–Root Interface* (New York: Academic Press).

Smith, S. E., and D. J. Read. 1997. *Mycorrhizal Symbiosis*, 2nd ed. (San Diego: Academic Press).

Stamets, P. 2005. *Mycelium Running: How Mushrooms Can Help Save the World.* (Berkeley, California: Ten Speed Press), 339 p.

Stone, A. G., S. J. Scheuerell, and H. M. Darby. 2004. "Suppression of soil-borne fungal diseases in field agricultural systems: Organic matter management, cover cropping, and cultural practices," in F. Magdoff and R. R. Weil (eds.), *Soil Organic Matter in Sustainable Agriculture* (Boca Raton, FL: CRC Press).

Sylvia, D. M., J. J. Fuhrmann, P. G. Hartel, and D. A. Zuberer. 2005. *Principles and Applications of Soil Microbiology* (Upper Saddle River, NJ: Prentice Hall).

Tate, R. L., III. 2001. *Soil Microbiology*, 2nd ed. (New York: John Wiley).

Torsvik, V., L. Ovreas, and T. F. Thingstad. 2002. "Prokaryotic diversity—Magnitude, dynamics, and controlling factors," *Science*, **296:**1064–1066.

Tugel, A., A. Lewandowski, and D. Happe-Vonarb, (eds.) 2000. Soil biology primer. Revised Ed. Soil and Water Conservation Society, Ankeny, Iowa, [Online]soils.usda.gov/sqi/concepts/soil_biology/biology.html (verified 12 January 2009).

Wardle, D. A. 2002. *Communities and Ecosystems: Linking the Aboveground and Belowground Components* (Princeton, NJ: Princeton University Press).

Wardle, D. A., G. W. Yeates, G. M. Barker, and K. I. Bonner. 2006. "The influence of plant litter diversity on decomposer abundance and diversity," *Soil Biol. Biochem.*, **38:**1052–1062.

Wolfe, B. E., and J. N. Klironomos. 2005. "Breaking new ground: Soil communities and exotic plant invasion," *BioScience*, **55:**477–487.

Carbon cycles in a mountain meadow. (R. Weil)

11
Soil Organic Matter

I bequeath myself to the dirt to grow from the grass I love, If you want me again look for me under your boot-soles.
—WALT WHITMAN, *SONG OF MYSELF*

In most soils, the percentage of soil organic matter[1] is small, but its effects on soil function are profound. This ever-changing soil component exerts a dominant influence on many soil physical, chemical, and biological properties, especially in the surface horizons. Soil organic matter provides much of the soil's cation exchange capacity (discussed in Chapter 8) and water-holding capacity (Chapter 5). Certain components of soil organic matter are largely responsible for the formation and stabilization of soil aggregates (Chapter 4). Soil organic matter also contains large quantities of plant nutrients and acts as a slow-release nutrient storehouse, especially for nitrogen (Chapter 12). Furthermore, organic matter supplies energy and body-building constituents for most of the microorganisms whose general activities were discussed in Chapter 10. In addition to enhancing plant growth through the just-mentioned effects, certain organic compounds found in soils have direct growth-stimulating effects on plants. For all these reasons, the quantity and quality of soil organic matter are central in determining **soil quality** (Chapter 1).

Soil organic matter is a complex and varied mixture of organic substances. All organic substances, by definition, contain the element **carbon**, and, on average, carbon comprises about half of the mass of soil organic matter. Organic matter in the world's soils contains two to three times as much carbon as is found in all the world's vegetation. Soil organic matter, therefore, plays a critical

---

[1]For an explanation of the chemical nature of soil organic matter, see Clapp et al. (2005). For a broad review of the nature, function, and management of organic matter in agricultural soils, see Magdoff and Weil (2004).

role in the global carbon balance that is thought to be the major factor affecting global warming, or the **greenhouse effect**.

We will first examine the role of soil organic matter in the **global carbon cycle** and the process of **decomposition** of organic residues. Next, we will focus on inputs and losses with regard to soil carbon in specific ecosystems. Finally, we will study the processes and consequences involved in soil organic matter management.

11.1 THE GLOBAL CARBON CYCLE

The element *carbon* is the foundation of all life. From cellulose to chlorophyll, the compounds that comprise living tissues are made of carbon atoms arranged in chains or rings and associated with many other elements. The cycle of carbon on earth is the story of life on this planet. Disruption of the carbon cycle would mean disaster for all living organisms (Box 11.1).

BOX 11.1
CARBON CYCLING—UP CLOSE AND PERSONAL

Figure 11.1

The Biosphere 2 building (top) and four biospherians at work (bottom). (Photos by C. Allen Morgan, *top*, and Pascale Maslin, *bottom*, © 1995 by Decisions Investments Corp. Reprinted with permission.)

Imagine that you were one of the eight biospherians living a scientific game of survival in Biosphere 2, a giant 1.3-ha sealed glass building in the Arizona desert. Biosphere 2 contained a miniature ocean, coral reef, marsh, forest, and farm in a self-contained, self-supporting ecosystem in which the biospherians could live as part of the ecosystem they were studying. Instruments throughout the structure constantly monitored environmental parameters to study how a balanced ecosystem *really* works!

But it didn't take long for trouble to develop. First, the biospherians found it was no easy task to grow all the food they needed (Figure 11.1). Dependent on their meager harvests, they began to lose weight. They soon also began feeling short of breath. The oxygen level of the air was falling from its normal 21% to as low as 14.2%. The carbon dioxide content of the air was rising. But weren't all the green plants supposed to *use up* the carbon dioxide and *replenish* the oxygen supply? The biosphereans' blood chemistry began to resemble that of a bear in hibernation (during which both oxygen and food are limited)! Eventually, the engineers had to give up on the "fully self-contained" aspect of the project and pump in oxygen and remove carbon dioxide from the air.

What had they overlooked? It turned out that the ecosystem was thrown out of kilter by the organic-matter-rich soil hauled in for the Biosphere farm. The soil, made from a mixture of pond sediment (1.8% C), compost (22% C), and peat moss (40% C), was installed uniformly about 1 m deep. This artificial soil contained about 2.5% organic C at all depths—far more than the 0.5% C or less expected in a typical desert soil. Had the designers read this book, they would have realized that peat might be stable in a boreal wetland (cool and anaerobic) but that aerobic soil microorganisms would rapidly use up oxygen and give off carbon dioxide as they metabolized organic matter in warm, moist garden soil aerated by tillage (see Section 11.2). This tale reminds us of the importance of soils in cycling C within the real biosphere! [For more on Biosphere 2, see www.biospheres.com, Torbert and Johnson (2001), and Walford (2002)].

Pathways

The basic processes involved in the global carbon cycle are shown in Figure 11.2. Plants take in carbon dioxide from the atmosphere. Then, through the process of photosynthesis, the energy of sunlight is trapped in the carbon-to-carbon bonds of organic molecules (such as those described in Section 11.2). Some of these organic molecules are used as a source of energy (via respiration) by the plants themselves (especially by the plant roots), with the carbon being returned to the atmosphere as carbon dioxide. The remaining organic materials are stored temporarily as constituents of the standing vegetation, most of which is eventually added to the soil as plant litter (including crop residues) or root deposition (see Section 10.7). Some plant material may be eaten by animals (including humans), in which case about half of the carbon eaten is exhaled into the atmosphere as carbon dioxide. The carbon not returned to the atmosphere is eventually returned to the soil as bodily wastes or body tissues. Once deposited on or in the soil, these plant or animal tissues are metabolized (digested) by soil organisms, which gradually return this carbon to the atmosphere as carbon dioxide.

Carbon dioxide also reacts in the soil to produce carbonic acid (H_2CO_3) and carbonates and bicarbonates of calcium, potassium, magnesium, and sodium. The bicarbonates are readily soluble and may be removed in drainage. The carbonates, such as calcite ($CaCO_3$), are much less soluble and tend to accumulate in soils under alkaline conditions. Although this chapter focuses on the organic C in soils, the inorganic C content of soils (mainly as carbonates) may be substantial, especially in arid regions (Table 11.1). Eventually, as with the C in soil organic matter, most of the bicarbonate C and some of the carbonate C in soils is returned to the atmosphere as CO_2.

Microbial metabolism and interaction with clay in the soil produces some organic materials of such stability that decades or even centuries may pass before the carbon in them is returned to the atmosphere as carbon dioxide. Such resistance to decay allows organic matter to accumulate in soils.

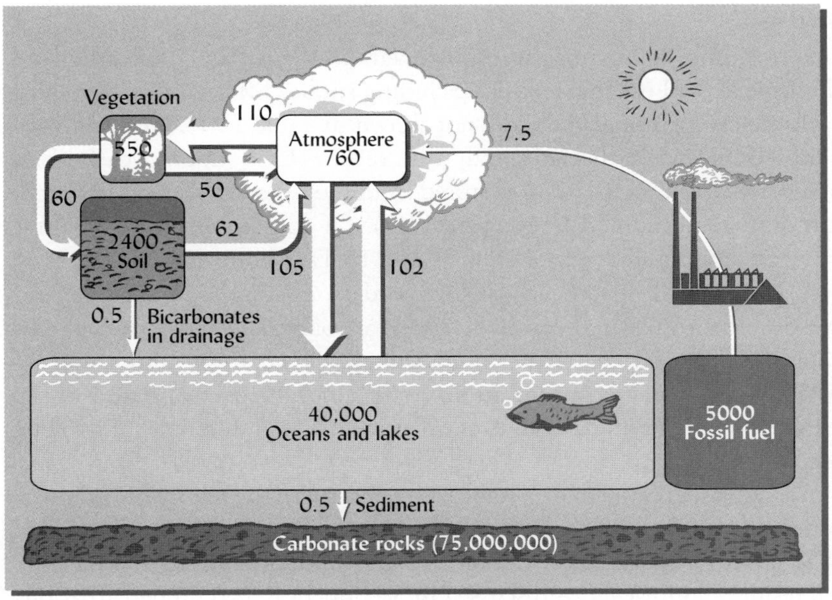

Satellite imagery showing C moving from the atmosphere to the biosphere: www.gsfc.nasa.gov/topstory/20010327colors_of_life.html

Figure 11.2
A simplified representation of the global carbon cycle emphasizing those pools of carbon which interact with the atmosphere. The numbers in the boxes indicate the petagrams (Pg = 10^{15} g) of carbon stored in the major pools. The numbers by the arrows show the amount of carbon annually flowing (Pg/yr) by various processes between the pools. Note that the soil contains almost twice as much carbon as the vegetation and the atmosphere combined. Imbalances caused by human activities can be seen in the flow of carbon to the atmosphere from fossil fuel burning (7.5) and in the fact that more carbon is leaving (62 + 0.5) than entering (60) the soil. These imbalances are only partially offset by increased absorption of carbon by the oceans. The end result is that a total of 221.5 Pg/yr enters the atmosphere while only 215 Pg/yr of carbon is removed. It is easy to see why carbon dioxide levels in the atmosphere are rising. [Data from IPCC (2007); soil carbon estimate from Batjes (1996)]

Table 11.1
MASS OF ORGANIC AND INORGANIC CARBON IN THE WORLD'S SOILS

Values for the upper 1 m represent 75 to 90% of the carbon in most soil profiles. Inorganic carbon is present mainly as calcium carbonates in soils of dry regions. Wetland soils as a group contain 468 Pg of organic C, some 30.3% of the total organic C in global soils.

Soil order	Global area, 10^3 km^2	Global carbon[a] in upper 100 cm Organic	Inorganic	Total	Total %
		Pg			
Entisols	21,137	90	263	353	14.2
Inceptisols	12,863	190	34	224	9.0
Histosols	1526	179	0	180	7.2
Andisols	912	20	0	20	0.8
Gelisols	11,260	316	7	323	12.9
Vertisols	3160	42	21	64	2.6
Aridisols	15,699	59	456	515	20.6
Mollisols	9005	121	116	237	9.5
Spodosols	3353	64	0	64	2.6
Alfisols	12,620	158	43	201	8.0
Ultisols	11,052	137	0	137	5.5
Oxisols	9810	126	0	126	5.1
Misc. land	18,398	24	0	24	1.0
Total	130,795	1526	940	2468	100.0

[a]Organic matter can be roughly estimated as 2.0 times this value, although the multiplier traditionally used is 1.72. Organic nitrogen may also be estimated from organic carbon values by dividing by 12 for most soils, but see Section 11.3.
Pg = petagram = 10^{15} g.
Data selected from Eswaran et al. (2000).

Carbon Sources

Globally, at any one time, approximately 2400 petagrams (Pg or 10^{15}g) of carbon are stored in soil profiles as soil organic matter (excluding surface litter), about one-third of that at depths below 1 m (and therefore not shown in Table 11.1). An additional 940 Pg are stored as soil carbonates that can release CO_2 upon weathering. Altogether, about twice as much carbon is stored in the soil than in the world's vegetation and atmosphere combined (see Figure 11.2). Of course, this carbon is not equally distributed among all types of soils (Table 11.1). About 45% of the total organic carbon is contained in soils of just three orders, Histosols, Inceptisols, and Gelisols. Histosols (and Histels in the order Gelisols) are of limited extent but contain very large amounts of organic matter per unit land area. Inceptisols (and nonhistic Gelisols) contain only moderate concentrations of carbon, but cover vast areas of the globe. The reasons for the varying amounts of organic carbon in different soils will be detailed in Section 11.8.

In a mature natural ecosystem or a stable agroecosystem, the release of carbon as carbon dioxide by oxidation of soil organic matter (mostly by microbial respiration) is balanced by the input of carbon into the soil as plant residues (and, to a far smaller degree, animal residues). However, as discussed in Section 11.8, certain perturbations of the system, such as deforestation, some types of fires, tillage, and artificial drainage, result in a net loss of carbon from the soil system.

Figure 11.2 shows that, globally, the release of carbon from soils into the atmosphere is about 62 Pg/yr, while only about 60 Pg/yr enter the soils from the atmosphere via plant residues. This imbalance of about 2 Pg/yr, along with about 7.5 Pg/yr of carbon released by the burning of fossil fuels (in which carbon was sequestered from the atmosphere millions of years ago) is only partially offset by increased absorption of atmospheric carbon dioxide by the ocean. Fossil fuel burning and degrading land-use practices have increased the concentration of carbon dioxide in the atmosphere at an accelerating rate, from 290 to 390 ppm during the past century alone. The implications of carbon dioxide imbalances and of other gaseous emissions on the greenhouse effect will be discussed in Section 11.9, after we consider the processes involved in the carbon cycle.

C emissions data, forecasts, and analyses: www.eia.doe.gov/environment.html

11.2   THE PROCESS OF DECOMPOSITION IN SOILS

Composition of Plant Residues

Plant residues are the principal material undergoing decomposition in soils and, hence, are the primary source of soil organic matter. Green plant tissues contain from 60 to 90% water by weight (Figure 11.3). If plant tissues are dried to remove all water, the *dry matter* remaining consists mostly (at least 90 to 95%) of carbon, oxygen, and hydrogen. During photosynthesis, plants obtain these elements from carbon dioxide and water. If plant dry matter is burned (oxidized), these elements become carbon dioxide and water once more. Of course, some ash and smoke will also be formed upon burning, accounting for the remaining 5 to 10% of the dry matter. In the ash and smoke can be found the many nutrient elements originally taken up by the plants from the soil. The essential nutrient elements in the ash will be given detailed consideration in Chapter 12.

The organic compounds in plant tissue can be grouped into broad classes (Figure 11.3). Carbohydrates, which range in complexity from simple sugars and starches to cellulose, are usually the most plentiful of plant organic compounds. **Lignins** and **polyphenols** are notoriously resistant to decomposition. Certain plant parts, especially seed and leaf coatings, contain significant amounts of fats, waxes, and

Genetic diversity associated with chitin degradation in soil: http://soilbio.nerc.ac.uk/Download/newsletter5.PDF

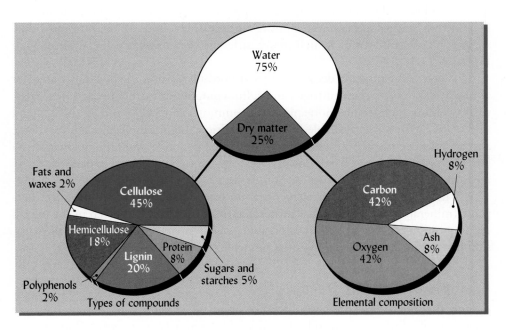

Figure 11.3
Typical composition of representative green-plant materials. The major types of organic compounds are indicated at left and the elemental composition at right. The ash is considered to include all the constituent elements other than carbon, oxygen, and hydrogen (nitrogen, sulfur, calcium, etc.).

oils, which are more complex than carbohydrates but less so than lignins. **Proteins** contain about 16% nitrogen and decompose easily.

Organic compounds can be listed in terms of ease of decomposition as follows:

1. Sugars, starches, and simple proteins   Rapid decomposition
2. Crude proteins
3. Hemicellulose
4. Cellulose
5. Fats and waxes
6. Lignins and phenolic compounds   Very slow decomposition

Decomposition of Organic Compounds in Aerobic Soils

Decomposition involves the breakdown of large organic molecules into smaller, simpler components. When organic tissue is added to an aerobic soil, three general microbiological reactions take place: (1) Enzymatic oxidation of carbon compounds to produce carbon dioxide, water, energy, and decomposer biomass; (2) Release and/or immobilization of the essential nutrient elements, such as nitrogen, phosphorus, and sulfur, by a series of specific reactions that are relatively unique for each element; (3) Formation of compounds very resistant to microbial action, either through modification of compounds in the original tissue or by microbial synthesis.

Decomposition: An Oxidation Process   In a well-aerated soil, all of the organic compounds found in plant residues are subject to oxidation:

$$R - (C, 4H) + 2O_2 \xrightarrow[\text{oxidation}]{\text{Enzymatic}} CO_2\uparrow + 2H_2O + \text{energy (478 kJ mol}^{-1}\text{ C)} \qquad (11.1)$$

Carbon and hydrogen-containing compounds

Many intermediate steps are involved in this overall reaction, and it is accompanied by important side reactions that involve elements other than carbon and hydrogen. Even so, this basic reaction accounts for most of the organic matter decomposition in the soil, as well as for the oxygen consumption and CO_2 release.

Cellulose and starch are long chains (polymers) of sugar molecules that are broken down by rather specialized organisms into short chains and then into individual sugar (glucose) molecules, which many different organisms can metabolize, as in Equation 11.1.

When plant proteins decay, they yield not only carbon dioxide and water but amino acids. In turn, these nitrogen and sulfur compounds further break down, eventually yielding such simple inorganic ions as ammonium (NH_4^+), nitrate (NO_3^-), and sulfate (SO_4^{2-}), forms available for plant nutrition. Lignin molecules are very large and complex, consisting of hundreds of interlinked phenolic ring subunits that only a few microorganisms can break down. Once the lignin subunits are separated, many types of microorganisms participate in their decay.

The process that releases elements from organic compounds to produce inorganic (mineral) forms is known as **mineralization**, usually the last step in the overall decomposition process. The decay of organic tissues is an important source of nitrogen, sulfur, phosphorus, and other essential elements for plants.

Example of Organic Decay

Assume that the soil has not been disturbed or amended with plant residues for some time. Initially, little or no readily decomposable materials are present. Competition for food is severe and microbial activity is relatively low, as reflected in the low **soil respiration** rate or level of CO_2 emission from the soil. The supply of soil carbon is steadily being depleted. Small populations of **k-strategist** microorganisms survive by slowly digesting the very resistant, stable soil organic matter. These organisms are so named because they produce enzymes with high affinity constants (k) for specific resistant compounds.

Now suppose that deciduous trees in a forest begin to lose their leaves in fall or a farmer plows in residues of a harvested crop. The appearance of easily decomposable and often water-soluble compounds, such as sugars, starches, and amino acids, stimulates an almost immediate increase in metabolic activity among the soil microbes. Soon the slower-acting k-strategists are overtaken by rapidly multiplying populations of *opportunist* or *colonizing* organisms that have been awakened from their dormant state by the presence of new food supplies. These organisms are known as **r-strategists**, so named for their rapid rate (r) of growth and reproduction that allows them to take advantage of a sudden influx of food. Microbial numbers and carbon dioxide evolution from microbial respiration both increase exponentially in response to the new food resource (upper panel in Figure 11.4). Soon microbial activity is at peak intensity, energy is being rapidly liberated, and carbon dioxide is being formed in large quantities. As organisms multiply, they increase the **microbial biomass** and also synthesize new exocellular organic compounds. The microbial biomass at this point may account for as much as one-sixth of the organic matter in the soil. The intense microbial activity may even stimulate the breakdown of some resistant soil organic matter, a phenomenon known as the **priming effect**.

With all this frenetic microbial activity, the easily decomposed compounds are soon exhausted. While the specialized k-strategists continue their slow work, degrading

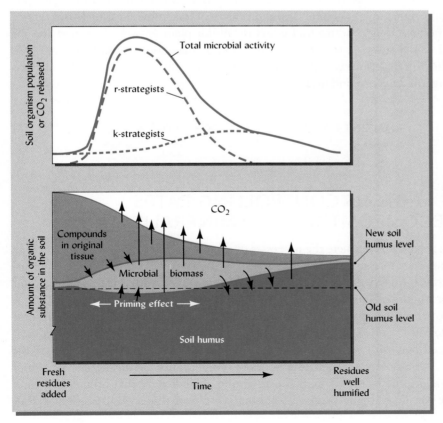

Figure 11.4

Schematic of the general changes occurring when fresh plant residues are added to a soil. The arrows indicate transfers of carbon among compartments. The upper panel shows the relative growth or activity of r-strategist (opportunist), k-strategist (more specialized) microorganisms, and the sum of these two groups. The time required for the process will depend on the nature of the residues and the soil. Most of the carbon released during the initial rapid breakdown of the residues is converted to carbon dioxide, but smaller amounts of carbon are converted into microbial biomass (and synthesis products) and, eventually, into soil humus. The peak level of microbial activity appears to accelerate the decay of the original humus, a phenomenon known as the priming effect. However, the humus level is increased by the end of the process. Where vegetation, environment, and management remain stable for a long time, the soil humus content will reach an equilibrium level at which the carbon added to the humus pool through the decomposition of plant residues is balanced by carbon lost through the decomposition of existing soil humus.

cellulose and lignin, r-strategists begin to die of starvation. As microbial populations plummet, the dead cells provide a readily digestible food source for the survivors, which continue to evolve carbon dioxide and water. The decomposition of the dead microbial cells is also associated with the **mineralization** or release of simple inorganic products, such as nitrates and sulfates. As food supplies are further reduced, microbial activity continues to decline, and the general-purpose r-strategists again sink back into comparative quiescence. A little of the original residue material persists, mainly as tiny particles that have become **physically protected** from decay by lodging inside soil pores too tight to allow access by most organisms. Some of the remaining carbon has also been **chemically protected** by conversion into **soil humus**, which is a dark-colored, heterogeneous, mostly colloidal mixture of modified lignin and newly synthesized organic compounds. Humus is highly resistant to attack and may be further protected by binding strongly to clay particles. Thus, a small percentage of the carbon in the added residues has been retained, increasing slightly the pool of stable soil organic matter. In a mature ecosystem, this increase will likely be offset during each annual cycle by slow, steady k-strategy decomposition, resulting in little net change in the level of soil organic matter from year to year.

Decomposition in Anaerobic Soils

Without sufficient oxygen present, aerobic organisms cannot function, so anaerobic or facultative organisms become dominant. Under low-oxygen or anaerobic conditions, decomposition takes place much more slowly than when oxygen is plentiful. Hence, wet, anaerobic soils tend to accumulate large amounts of organic matter in a partially decomposed condition.

The products of anaerobic decomposition include a wide variety of partially oxidized organic compounds. Anaerobic decomposition releases relatively little energy for the organisms involved; therefore, the end products (alcohols and methane gas) still contain much energy. Some of the products of anaerobic decomposition are of concern because they produce foul odors or inhibit plant growth. The methane gas produced in wet soils is a major contributor to the greenhouse effect (Section 11.9). The following reaction is typical of those carried out in wet soils by various **methanogenic bacteria** and **archaea**:

$$4C_2H_5COOH + 2H_2O \xrightarrow{\text{Bacteria}} 4CH_3COOH + CO_2 \uparrow + 3CH_4 \uparrow \qquad (11.2)$$

Propionate   Acetate   Carbon   Methane
                                 dioxide

11.3   FACTORS CONTROLLING RATES OF DECOMPOSITION AND MINERALIZATION

The time needed to complete the processes of decomposition and mineralization may range from days to years, depending mainly on two broad factors: (1) the environmental conditions in the soil, and (2) the quality of the added residues as a food source for soil organisms.

The environmental conditions conducive to rapid decomposition and mineralization include sufficient soil moisture, good aeration (about 60% of the soil pore space filled with water), and warm temperatures (25 to 35 °C, Section 7.8). Ironically, periodic stresses such as episodes of severe drying actually accelerate overall mineralization due to the dramatic burst of microbial activity that occurs each time the soil re-wets (e.g., Figure 12.5). These conditions were discussed in Section 10.9 in relation to microbial activity and will be considered again in Section 11.8 as they affect the levels of organic matter accumulating in soils. Here we will focus on factors that determine the quality of the residues as a food resource for microbes, including the physical condition of the residues, their C/N ratio, and their content of lignins and polyphenols.

Physical Factors Influencing Residue Quality

The location of residues in or on the soil is a physical factor that has a critical impact on decomposition rates. Surface placement of plant residues, as in forest litter or conservation tillage mulch, usually results in slower, more variable rates of decomposition than where similar residues are incorporated into the soil by root deposition, faunal action, or tillage. Surface residues are subject to drying, as well as extremes of temperature. Surface residues are physically out of reach for most soil organisms, save the larger fauna such as earthworms and fungal mycelia (see Figure 10.11). Compared to surface residue, incorporated residues are in intimate contact with soil moisture and soil organisms, decompose more quickly, and may lose nutrients more easily by leaching.

Residue particle size is another important physical factor—the smaller the particles, the more rapid the decomposition. Diminution of residues into smaller particles physically exposes more surface area and cell contents to decomposition. In addition, some organic materials exhibit hydrophobicity (water repellency), making them slow to wet and difficult to attack by water-soluble microbial enzymes.

Carbon/Nitrogen Ratio of Organic Materials and Soils

The carbon content of typical plant dry matter is about 42% (see Figure 11.3). The nitrogen content of plant residues is much lower and varies widely (from <1 to >6%). The C/N ratio in plant residues ranges from between 10:1 to 30:1 in legumes and young green leaves to as high as 600:1 in some kinds of sawdust (Table 11.2).

Table 11.2
TYPICAL CARBON AND NITROGEN CONTENTS AND C/N RATIOS OF SOME ORGANIC MATERIALS

Organic material	% C	% N	C/N
Spruce sawdust	50	0.05	600
Newspaper	39	0.3	120
Wheat straw	38	0.5	80
Corn stover	40	0.7	57
Maple leaf litter	48	1.4	34
Rotted barnyard manure	41	2.1	20
Bluegrass from fertilized lawn	42	2.2	20
Broccoli residues	35	1.9	18
Young alfalfa hay	40	3.0	13
Hairy vetch cover crop	40	3.5	11
Digested municipal sewage sludge	31	4.5	7
Soil microorganisms			
Bacteria	50	10.0	5
Fungi	50	5.0	10
Soil organic matter			
Average forest O horizons	50	1.3	45
Average forest A horizons	50	2.8	20
Mollisol Ap horizon	56	4.9	11
Average B horizon	46	5.1	9

Generally, as plants mature, the proportion of protein in their tissues declines, while the proportion of lignin and cellulose, and the C/N ratio, increase. These differences in composition have pronounced effects on the rate of decay when plant residues are added to the soil. Among microorganisms, bacteria are generally richer in protein than fungi and, consequently, have a lower C/N ratio (5:1 vs. 10:1).

The C/N ratio in the organic matter of arable (cultivated) surface (Ap) horizons commonly ranges from 8:1 to 15:1, the median being near 12:1. The ratio is generally lower for subsoils than for surface layers in a soil profile. In a given climatic region, little variation occurs in the C/N ratio for similarly managed soils. For instance, in calcium-rich soils of semiarid grasslands (e.g., Mollisols, tropical Alfisols), the C/N ratio is relatively narrow. In more severely leached and acidic A horizons in humid regions, the C/N is relatively wide; C/N ratios as high as 30:1 are not uncommon. Forest O horizons commonly have C/N ratios of 30:1 to 50:1. When such soils are brought under cultivation and limed to increase their pH and calcium content, the enhanced decomposition tends to lower the C/N ratio to near 12:1.

Soil microbes, like other organisms, require a balance of nutrients from which to build their cells and extract energy. Soil organisms need carbon for building essential organic compounds and to obtain energy. However, organisms must also obtain sufficient nitrogen to synthesize nitrogen-containing cellular components, such as amino acids, enzymes, and DNA. On the average, soil microbes must incorporate into their cells about eight parts of carbon for every one part of nitrogen. Because only about one-third of the carbon metabolized by microbes is incorporated into their cells (the remainder is respired and lost as CO_2), the microbes need to find about 1 g of N for every 24 g of C in their "food."

This requirement results in two extremely important practical consequences. First, if the C/N ratio of organic material added to soil exceeds about 25:1, the soil microbes will have to scavenge the soil solution to obtain enough nitrogen. Thus, the incorporation of high C/N residues will deplete the soil's supply of soluble nitrogen, causing plants to suffer from nitrogen deficiency. Second, the decay of organic materials can be delayed if sufficient nitrogen to support microbial growth is neither present in the material undergoing decomposition nor available in the soil solution.

Examples of Inorganic Nitrogen Release During Decay

The practical significance of the C/N ratio becomes apparent if we compare the changes that take place in the soil when residues of either high or low C/N ratio are added (Figure 11.5). Consider a soil with a moderate level of soluble nitrogen (mostly nitrates). Microbial activity in this soil is low, as evidenced by low CO_2 production. If no nitrogen were lost or taken up by plants, the level of nitrates would very slowly increase as the native soil organic matter decays.

Now consider what happens when a large quantity of readily decomposable organic material is added to this soil. If this material has a C/N ratio greater than 25, changes will occur according to the pattern shown in Figure 11.5a. In the example shown, the initial C/N ratio of the residues is about 60, typical for many kinds of leaf litter. As soon as the residues contact the soil, the microbial community responds to the new food supply (see Section 11.2). Heterotrophic r-strategists become active, multiply rapidly, and yield CO_2 in large quantities. Because of the microbial demand for nitrogen, little or no mineral nitrogen (NH_4^+ or NO_3^-) is available to higher plants during this **nitrate depression period**. As the microbes use up the supply of easily oxidizable carbon, their numbers decrease, carbon dioxide formation drops off, and nitrogen competition becomes less acute. As decay proceeds, the C/N ratio of the

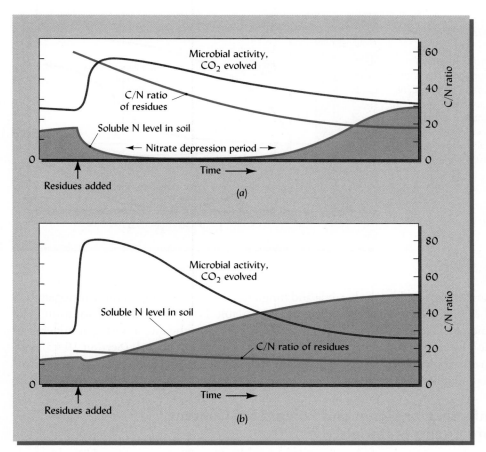

(a)

(b)

Figure 11.5

Changes in microbial activity, in soluble nitrogen level, and in residual C/N ratio following the addition of either high (a) or low (b) C/N ratio organic materials. Where the C/N ratio of added residues is above 25, microbes digesting the residues must supplement the nitrogen contained in the residues with soluble nitrogen from the soil. During the resulting nitrate depression period, competition between higher plants and microbes would be severe enough to cause nitrogen deficiency in the plants. Note that in both cases soluble N in the soil ultimately increases from its original level once the decomposition process has run its course. The trends shown are for soils without growing plants, which, if present, would continually remove a portion of the soluble nitrogen as soon as it is released.

remaining plant material decreases because carbon is being lost (by respiration) and nitrogen is being conserved (by incorporation into microbial cells). Generally, when the C/N ratio drops below about 20, nitrates appear again in quantity, and the original conditions will return, except that the soil is somewhat richer in both nitrogen and humus.

The nitrate depression period may last for a few days, a few weeks, or even several months. To avoid producing seedlings that are stunted, chlorotic, and nitrogen-starved, planting should be delayed until after the nitrate depression period or additional sources of nitrogen can be applied to satisfy the nutritional requirements of both the microbes and the plants.

With organic materials of low C/N ratio (Figure 11.5*b*), more than enough nitrogen is present to meet the needs of the decomposing organisms. Therefore, soon after decomposition begins, some of the nitrogen from organic compounds is released into the soil solution, augmenting the level of soluble nitrogen available for plant uptake. Generally, nitrogen-rich materials decompose quite rapidly, resulting in a period of intense microbial growth and activity, but no nitrate depression period.

Influence of Soil Ecology

In nature, the process of nitrogen mineralization involves the entire food web (see Section 10.2), not just the saprophytic bacteria and fungi. For example, when bacteria and fungi grow rapidly on a food source, the large biomass of bacterial and fungal cells contains much of the nitrogen originally in the residues. This nitrogen is immobilized and not available to plants. However, when certain nematodes, protozoa, and

Figure 11.6

Temporal patterns of nitrogen mineralization or immobilization with organic residues differing in quality based on their C/N ratios and contents of lignin and polyphenols. Lignin contents greater than 20%, polyphenol contents greater than 3%, and C/N ratios greater than 30 would all be considered high in the context of this diagram, the combination of these properties characterizing litter of poor quality—that is, litter that has a limited potential for microbial decomposition and mineralization of plant nutrients. (Diagram courtesy of R. Weil)

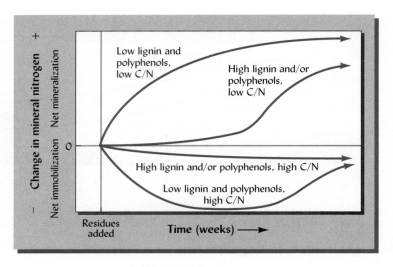

earthworms feed on the bacteria and fungi, the animals soon ingest more nitrogen than they can use. They then excrete plant-available NH_4^+ into the soil solution. The microbial feeding activity of soil animals may increase the rate of nitrogen mineralization by 100%. Soil management that favors a complex food web (Section 10.13) with many trophic levels can be expected to enhance the cycling and efficient use of nitrogen (and of other nutrients).

Influence of Lignin and Polyphenol Content of Organic Materials

The lignin contents of plant litter range from less than 2% to more than 50%. Materials with high lignin content decompose very slowly. Polyphenol compounds found in plant litter at up to 5 to 10% of the dry weight may also inhibit decomposition.

Because they support only low levels of microbial activity and biomass, residues high in phenols and/or lignin are considered to be *poor quality resources* for the soil organisms that cycle carbon and nutrients. The production of such slow-to-decompose residues by certain forest plants may help explain the accumulation of extremely high levels of humified nitrogen and carbon in the soils of mature boreal forests.

The lignin and phenol contents also influence the decomposition and release of nitrogen from **green manures**—plant residues used to enrich agricultural soils. Figure 11.6 illustrates the combined effects of C/N ratio and lignin or phenol content on the balance between immobilization and mineralization of nitrogen during plant residue decomposition.

Complexities of decay and nutrient release from logs in Northwest forests: http://oregonstate.edu/ dept/ncs/newsarch/2005/ Aug05/decay.htm

11.4   GENESIS AND NATURE OF SOIL ORGANIC MATTER AND HUMUS

Everything about humus—a view from Poland: www.ar.wroc.pl/~weber/ humic.htm

We use the general term **soil organic matter** (**SOM**) to encompass all the organic components of a soil: (1) living **biomass** (intact plant and animal tissues and microorganisms); (2) dead roots and other recognizable plant residues or litter (in practice, particles larger than 2-mm sieve openings are often excluded); and (3) a largely amorphous and colloidal mixture of complex organic substances no longer identifiable as tissues. Only the third category of organic material is properly referred to as **soil humus** (Figure 11.7). Since the element carbon (C) plays a prominent role in the chemical structure of all organic substances, it is not surprising that the term **soil**

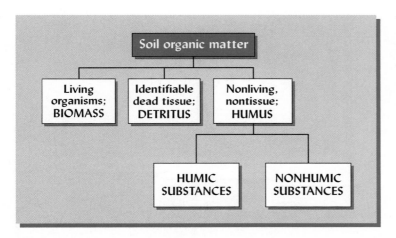

Figure 11.7
Classification of soil organic matter components separable by chemical and physical criteria. Although surface residues (litter) are not universally considered to be part of the soil organic matter, we include them because they are the principal components of the O horizons in soil profiles.

organic carbon (**SOC**) is often used to refer to the C component of soil organic matter. Since soil organic matter commonly contains about half carbon by weight (50% C), it is usually appropriate to estimate soil organic matter as two times the organic C (SOM = 2 × SOC).

Microbial Transformations

As decomposition of plant residues proceeds, microbes polymerize (link together) some of the simpler new compounds with each other and with the complex residual products into long, complex chains that resist further decomposition. These high-molecular-weight compounds interact with nitrogen-containing amino compounds, giving rise to a significant component of resistant humus. The presence of colloidal clays stimulates the complex polymerization. These ill-defined, complex, resistant, polymeric compounds are called **humic substances**. The term **nonhumic substances** refers to the group of identifiable biomolecules that are mainly produced by microbial action and are generally less resistant to breakdown.

One year after plant residues are added to the soil, most of the carbon has returned to the atmosphere as CO_2, but one-fifth to one-third is likely to remain in the soil either as live biomass (~5%) or as the humic (~20%) and nonhumic (~5%) fractions of soil humus (Figure 11.8). The proportion remaining from root residues tends to be somewhat higher than that remaining from incorporated leaf litter.

Humic Substances

Humic substances comprise 60 to 80% of the soil organic matter. They are composed of huge molecules with variable structures characterized by aromatic rings. Humic substances generally are dark-colored, amorphous substances with molecular weights varying from 2000 to 300,000 g/mol. Because of their complexity, they are the organic materials most resistant to microbial attack. Humus colloids exhibit very high capacities to hold both water and cations. The highly complex molecules absorb nearly all wavelengths of visible light, giving the substance its characteristic black color (see the humus-rich A horizons in Plates 2 and 8). Depending on the environment, the half-life (the time required to destroy half the amount of a substance) of humic substances may range from decades to centuries.

Nonhumic Substances

About 20 to 30% of the humus in soils consists of nonhumic substances. These substances are generally less complex and less resistant to microbial attack than those of

Figure 11.8

Disposition of 100 g of organic carbon in residues one year after they were incorporated into the soil. More than two-thirds of the carbon has been oxidized to CO_2, and less than one-third remains in the soil—some in the cells of soil organisms, but a larger component as soil humus. The amount converted to CO_2 is generally greater for aboveground residues than for belowground (root) residues.

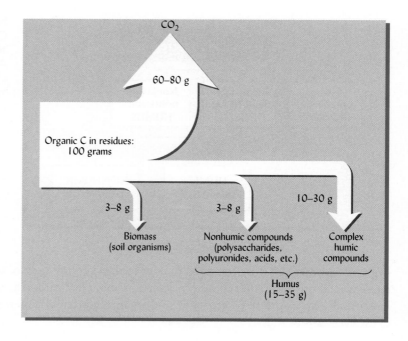

3-D models of biomolecules—check out the DNA-protein complex and water: www.umass.edu/microbio/chime

the humic group. Unlike humic substances, they are comprised of specific biomolecules with definite physical and chemical properties. Some of these nonhumic substances are microbially modified plant compounds, while others are compounds synthesized by soil microbes. An important example is called *glomalin*, a group of sugar–protein compounds produced by mycorrhizal fungi (see Section 10.8).

Some simpler compounds (such as low-molecular-weight organic acids and some proteinlike materials) are part of the nonhumic group. Although none of these simpler materials are present in large quantities, they may influence the availability of plant nutrients, such as nitrogen and iron, and may also directly affect plant growth.

Stability of Humus

Studies using radioactive isotopes have shown that some organic carbon incorporated into humus thousands of years ago is still present in soils, evidence that humic materials can be extremely resistant to microbial attack. This resistance to oxidation is important in maintaining soil organic matter levels and in protecting associated nitrogen against rapid mineralization and loss. Nonetheless, without annual additions of sufficient plant residues, microbial oxidation will inevitably reduce soil organic matter levels.

Clay–Humus Combinations   Interaction with clay minerals provides another means of stabilizing soil organic matter and the nitrogen it contains. Organic matter that is entrapped in the ultra-micropores (<1 μm) formed by clay particles is physically inaccessible to decomposing organisms (see Figure 8.18). Although the extent and mechanisms are not yet fully understood, clay–humus interactions undoubtedly contribute to the high organic matter content of clay soils (see Section 11.8).

11.5   INFLUENCES OF ORGANIC MATTER ON PLANT GROWTH AND SOILS

Long ago, the observation that plants generally grow better on organic-matter-rich soils led people to think that plants derive much of their nutrition by absorbing humus from the soil. We now know that higher plants derive their carbon from carbon dioxide and that most of their nutrients come from inorganic ions dissolved in the soil solution. In fact, plants can complete their life cycles growing totally without humus, or even without soil (as in soilless, or **hydroponic**, production systems using only aerated nutrient solutions). This is not to say that soil organic matter is less important to plants than was once supposed but rather that the benefits accrue to plants indirectly through the many influences of organic matter on soil properties. These will be discussed later in this section, after we consider two types of direct organic matter effects on plants.

Direct Influence of Humus on Plants

It is well established that some organic compounds are absorbed by higher plants. For example, plants can absorb a varying proportion of their nitrogen and phosphorus needs as soluble organic compounds. In addition various growth-promoting compounds such as vitamins, amino acids, auxins, and gibberellins, are formed as organic matter decays. These substances may at times stimulate growth in both higher plants and microorganisms.

Small quantities of both humic substances in the soil solution are known to enhance certain aspects of plant growth. Some scientists have suggested that they may act as hormone-like regulators of specific plant-growth functions. However, little evidence is available that supports this mechanism. Other scientists suggest that the humic substances stimulate plant growth by improving the availability of micronutrients, especially iron and zinc. Commercial humate products have been marketed with claims that small amounts enhance plant growth, but scientific tests of many of these products have failed to show any benefit from their use, perhaps because effective levels of humic substances are naturally present in most soils.

Allelochemical Effects[2]

Allelopathy is the process by which one plant infuses the soil with a chemical that affects the growth of other plants. The plant may do this by directly exuding **allelochemicals**, or the compounds may be leached out of the plant foliage by throughfall rainwater. In other cases, microbial metabolism of dead plant tissues (residues) forms the allelochemicals (Figure 11.9).

Allelochemicals present in the soil are apparently responsible for many of the effects observed when various plants grow in association with one another. Because they produce such chemicals, certain weeds (e.g., Johnsongrass and giant foxtail) damage crops far out of proportion to the size and number of weeds present. Crop

A review of allelopathy: www.colostate.edu/Depts/ Entomology/courses/ en570/papers_2002/ mccollum.htm

Figure 11.9
Positive and negative allelopathic effects of winged beans on grain amaranth plants. In the pot on the left (T4) amaranth is growing in fresh soil (no association with winged beans). In the center pot (T20) amaranth is growing in soil previously used to grow winged beans (positive effect). In the pot on the right (T28) the amaranth is growing in fresh soil, but the plant was watered three times with a water extract of winged bean tissue (negative effect). The average dry weight of the amaranth plants for each treatment is shown. All pots were watered with a complete nutrient solution. [From Weil and Belmont (1987)]

[2]For a comprehensive review of allelopathy, see Inderjit et al. (1999). For the role of allelopathy in plant species invasiveness, see Hierro and Callaway (2003).

residues left on the soil surface may inhibit the germination and growth of the next crop planted (e.g., wheat residues often inhibit sorghum plants).

Other allelopathic interactions influence the succession of species in natural ecosystems. Allelopathy may be partially responsible for the invasiveness of certain exotic plant species that rapidly dominate a new ecosystem to which they have been recently introduced. The invaders' allelochemicals may be more effective in the new ecosystem than they were in their territory of origin where neighboring plant species had time to evolve tolerance.

Allelopathic interactions are usually very specific, involving only certain species, or even varieties, on both the producing and receiving ends. While they vary in chemical composition, most allelochemicals are relatively simple phenolic or organic acid compounds that could be included among the nonhumic substances found in soils. Because most of these compounds can be rapidly destroyed by soil microorganisms or easily leached out of the root zone, effects are usually relatively short-lived once the source is removed.

The walnut tree—allelopathic effects and tolerant plants: www.ext.vt.edu/pubs/nursery/430-021/430-021.html

Influence of Organic Matter on Soil Properties and Indirectly on Plants

Even small amounts of soil organic matter profoundly affect so many soil properties and processes that they are mentioned in every chapter in this book! Figure 11.10 summarizes some of the more important effects of organic matter on soil properties and on soil–environment interactions. Often one effect leads to another, so that a complex chain of multiple benefits results from the addition of organic matter to soils. For example (beginning at the upper left in Figure 11.10), adding organic mulch to the soil surface encourages earthworm activity, which in turn leads to the production of burrows and other biopores, which in turn increases the infiltration of water and decreases its loss as runoff, a result that finally leads to less pollution of streams and lakes. Study Figure 11.10 carefully and try following some of the other pathways of influence on the soil and environment.

Influence on Soil Physical Properties   Humus tends to give surface horizons dark brown to black colors. Granulation and aggregate stability are encouraged, especially by the nonhumic substances produced by bacteria and fungi (see Section 4.5). The humic fractions help reduce the plasticity, cohesion, and stickiness of clayey soils, making these soils easier to manipulate. Soil water retention is also improved, since organic matter increases both infiltration rate and water-holding capacity. Organic matter has an especially pronounced effect on the water-holding capacity of very sandy soils, which can often be improved by adding stable organic amendments (Figure 11.11).

Influence on Soil Chemical Properties   Humus generally accounts for 50 to 90% of the cation-adsorbing power of mineral surface soils. Like clays, humus colloids hold nutrient cations (potassium, calcium, magnesium, etc.) in easily exchangeable form, wherein they can be used by plants but are not too readily leached out of the profile by percolating waters. Through its cation exchange capacity and acid and base functional groups, organic matter also provides much of the pH buffering capacity in soils (see Section 9.4). In addition, nitrogen, phosphorus, sulfur, and micronutrients are stored as constituents of soil organic matter, from which they are slowly released by mineralization.

Humic acids also attack soil minerals and accelerate their decomposition, thereby releasing essential nutrients as exchangeable cations. Organic acids, polysaccharides, and fulvic acids all can attract such cations as Fe^{3+}, Cu^{2+}, Zn^{2+}, and Mn^{2+} from the edges of mineral structures and **chelate** or bind them in stable organomineral

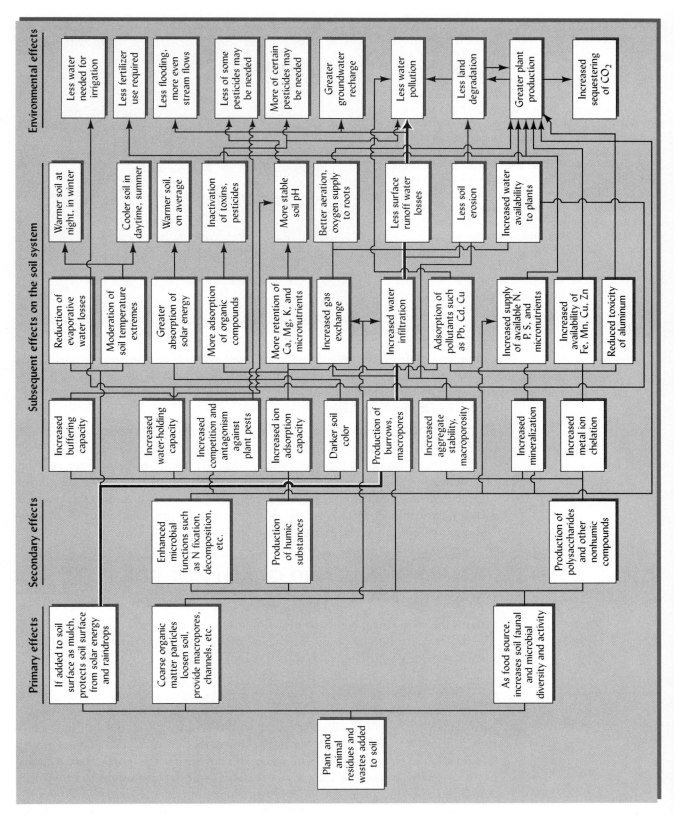

Figure 11.10

Some of the ways in which soil organic matter influences soil properties, plant productivity, and environmental quality. Many of the effects are indirect, the arrows indicating the cause-and-effect relationships. It can readily be seen that the influences of soil organic matter are far out of proportion to the relatively small amounts present in most soils. Many of these influences are discussed in this and other chapters in this book. The thicker line shows the sequence of effects referred to in the text in this section. (Diagram courtesy of R. Weil)

Figure 11.11

Effect of organic amendment on soil water retention. Columns 30 cm tall were filled with medium-sized sand of the type used for the root zone in golf course greens. The sand in some columns was mixed with peat (20% by volume). The columns were saturated with water and allowed to drain freely for 24 hours before the volumetric water was measured. While both columns were near 100% saturation at the lowest depth (where the water potential was near zero, see Section 5.3), the peat more than doubled the water held in the upper part of the profile. Improved water retention is a major reason why amending and topdressing sand-based golf greens are popular practices. However, the use of large amounts of peat mined from sphagnum wetlands cannot be considered environmentally sustainable. Compost made from various organic wastes (Sections 10.12 and 11.10) is a practical and environmentally beneficial substitute for peat in this kind of application. [Redrawn from data in Bigelow et al. (2004)]

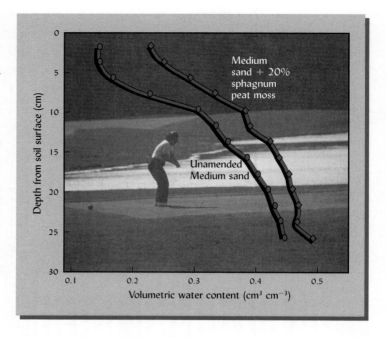

complexes. Some of these metals are made more available to plants as micronutrients because they are kept in soluble, chelated form (see Chapter 12). In very acid soils, organic matter alleviates aluminum toxicity by binding the aluminum ions in nontoxic complexes (see Sections 9.2 and 9.9).

Biological Effects Soil organic matter—especially the detritus fraction—provides most of the food for the community of heterotrophic soil organisms described in Chapter 10. In Section 11.3 it was shown that the quality of plant litter and soil organic matter markedly affects decomposition rates and, therefore, the amount of organic matter accumulating in soils. The type and diversity of organic residues added to a soil can influence the type and diversity of organisms that make up the soil community, thus indirectly affecting plants and animals.

11.6 AMOUNTS AND QUALITY OF SOIL ORGANIC MATTER

Perhaps the most useful approach to defining soil organic matter quality is to recognize different portions or **pools** of organic carbon that vary in their susceptibility to microbial metabolism (Figure 11.12).

Active Organic Matter

The **active pool** of soil organic matter consists of labile (easily decomposed) materials with half-lives (the time it takes for half of a mass of material to decay) of only a few days to a few years. Organic matter in the active pool includes such organic matter fractions as the living biomass, tiny pieces of detritus (termed **particulate organic matter**, [**POM**]), most of the polysaccharides, and other nonhumic substances described in Section 11.4. This pool provides most of the readily accessible food for soil organisms and most of the readily mineralizable nitrogen. It is responsible for most of the beneficial effects on structural stability that lead to enhanced infiltration of water, resistance to erosion, and ease of tillage. The active pool can be readily increased by the addition of fresh plant and animal residues, but it is also very readily lost when such additions are reduced or tillage is intensified. This pool rarely comprises more than 10 to 20% of the total soil organic matter.

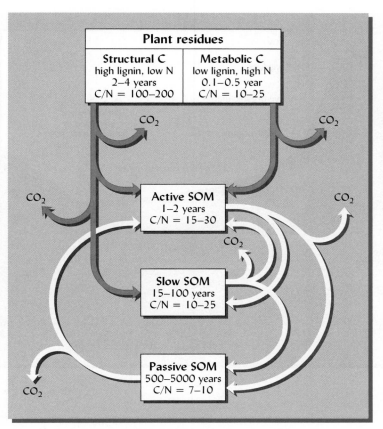

Figure 11.12

A conceptual model that recognizes various pools of soil organic matter (SOM) differing by their susceptibility to microbial metabolism. Models that incorporate active, slow, and passive pools of soil organic matter have proven very useful in explaining and predicting real changes in soil organic matter levels and in associated soil properties. Note that microbial action can transfer organic carbon from one pool to another. For example, when the nonhumic substances and other components of the active fraction are rapidly broken down, some resistant, complex by-products may be formed, adding to the slow and passive pools. Note that all these metabolic changes result in some loss of carbon from the soil as CO_2. [Adapted from Paustian et al. (1992)]

Passive and Slow Organic Matter

The **passive pool** of soil organic matter consists of very stable materials remaining in the soil for hundreds or even thousands of years. This pool includes most of the material physically protected in clay–humus complexes. The passive pool accounts for 60 to 90% of the organic matter in most soils, and its quantity is increased or diminished only slowly. The passive pool is most closely associated with the colloidal properties of soil humus, and it is responsible for most of the cation- and water-holding capacities contributed to the soil by organic matter.

Intermediate in properties between the active and passive pools is the **slow pool** of soil organic matter. This pool probably includes the finest fractions of particulate organic matter that are high in lignin and other slowly decomposable and chemically resistant components. The half-lives of these materials are typically measured in decades. The slow pool is an important source of mineralizable nitrogen and other plant nutrients, and it provides much of the underlying food source for the steady metabolism of the K-strategist soil microbes (see Section 11.2).

Changes in Active and Passive Pools with Soil Management

Soil scientists have consistently observed that productive soils managed with conservation-oriented practices contain relatively high amounts of organic matter fractions associated with the active pool, including microbial biomass, particulate organic matter, and oxidizable sugars.

Soil-management practices that cause only very small changes in total soil organic matter often cause rather pronounced alterations in aggregate stability, nitrogen mineralization rate, or other soil properties attributed to organic matter. This occurs because the relatively small pool of active organic matter may undergo a large percentage increase or decrease without changing a large percentage of the much larger pool

Figure 11.13

Changes in various pools of organic matter in the upper 25 cm of a representative soil after bringing virgin land under cultivation. Initially, under natural vegetation, this soil contained about 91 Mg/ha of total organic matter. The resistant passive pool accounted for about 44 Mg/ha, or about half of the total soil organic matter. The rapidly decomposing active pool accounted for about 14 Mg, or about 16% of the total soil organic matter. After about 40 years of cultivation, the passive pool had declined by about 11% to about 39 Mg/ha, while the active pool had lost 90% of its mass, declining to only 1.4 Mg/ha. Note that much of the organic loss due to the change in land management came at the expense of the active pool. This was also the pool that most quickly increased when improved organic matter management was adopted after the 100th year. The susceptibility of the active pool to rapid change explains why even relatively small changes in total soil organic matter can produce dramatic changes in important soil properties, such as aggregate stability and nitrogen mineralization. (Diagram courtesy of R. Weil)

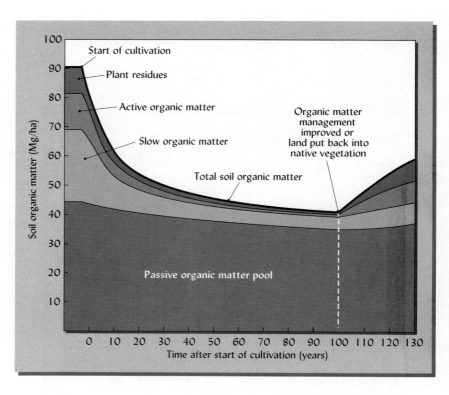

of total organic matter (Figure 11.13). As a result, the soil organic matter remaining after some years of active organic matter losses is far less effective in promoting structural stability and nutrient cycling than the original organic matter in the virgin soil. If a favorable change in environmental conditions or management regime occurs, the plant litter and active pools of soil organic matter are also the first to positively respond (see right-hand portion of Figure 11.13). Therefore, monitoring the active C pool can serve as an early warning of soil quality changes.

11.7 CARBON BALANCE IN THE SOIL–PLANT–ATMOSPHERE SYSTEM

Whether the goal is to reduce greenhouse gas emissions or to enhance soil quality and plant production, proper management of soil organic matter requires an understanding of the cycling and balance of carbon. Although each type of ecosystem, whether a deciduous forest, a wetland, or a wheat field, will emphasize particular compartments and pathways in the carbon cycle, consideration of a specific example, such as that described in Box 11.2 will be helpful.

The rate at which soil organic matter either increases or decreases is determined by the balance between *gains* and *losses* of carbon. The gains come primarily from plant residues grown in place and from applied organic materials. The losses are due mainly to respiration (CO_2 losses), plant removals, and erosion (Table 11.3).

Agroecosystems

In order to halt or reverse the net carbon loss shown in Figure 11.14, management practices would have to be implemented that would either *increase the additions* of carbon to the soil or *decrease the losses* of carbon from the soil. Since all crop residues

BOX 11.2
CARBON BALANCE—AN AGROECOSYSTEM EXAMPLE

The principal carbon pools and annual flows in a terrestrial ecosystem are illustrated in Figure 11.14 using a hypothetical cornfield in a warm temperate region. During a growing season the corn plants produce (by photosynthesis) 17,500 kg/ha of dry matter containing 7500 kg/ha of carbon (C). This C is equally distributed (2500 kg/ha each) among the roots, grain, and unharvested aboveground residues. In this example, the harvested grain is fed to cattle, which oxidize and release as CO_2 about 50% of this C (1250 kg/ha), assimilate a small portion as weight gain, and void the remainder (1100 kg/ha) as manure. The corn stover and roots are left in the field and, along with the manure from the cattle, are incorporated into the soil by tillage or by earthworms.

The soil microbes decompose the crop residues (including the roots) and manure, releasing as CO_2 some 75% of the manure C, 67% of the root C, and 85% of the C in the surface residues. The remaining C in these pools is assimilated into the soil as humus. Thus, during the course of one year, some 1475 kg/ha of C enters the humus pool (825 kg from roots, plus 375 from stover, plus 275 from manure). These values are in general agreement with Figure 11.8, but they will vary widely among different soil conditions and ecosystems.

At the beginning of the year, the upper 30 cm of soil in our example contained 65,000 kg/ha organic C in humus. Such a soil cultivated for row crops would typically lose about 2.5% of its organic C by soil respiration each year. In our example this loss amounts to some 1625 kg/ha of C. Smaller losses of soil organic C occur by soil erosion (160 kg/ha), leaching (10 kg/ha), and formation of carbonates and bicarbonates (10 kg/ha).

Comparing total losses (1805 kg/ha) with the total gains (1475 kg/ha) for the pool of soil humus, we see that the soil in our example suffered a net annual loss of 330 kg/ha of C, or 0.5% of the total C stored in the soil humus. If this rate of loss were to continue, degradation of soil quality and productivity would surely result.

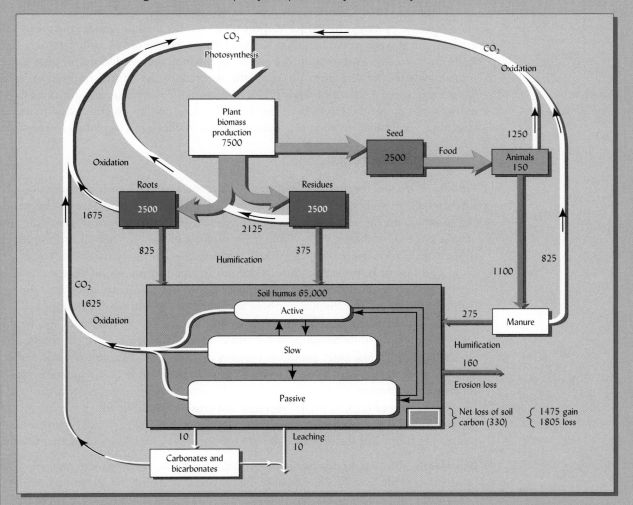

Figure 11.14
Carbon cycling in an agroecosystem.

Table 11.3

FACTORS AFFECTING THE BALANCE BETWEEN GAINS AND LOSSES OF ORGANIC MATTER IN SOILS

Factors promoting gains	Factors promoting losses
Green manures or cover crops	Erosion
Conservation tillage	Intensive tillage
Return of plant residues	Whole plant removal
Low temperatures and shading	High temperatures and exposure to sun
Controlled grazing	Overgrazing
High soil moisture	Low soil moisture
Surface mulches	Fire
Application of compost and manures	Application of only inorganic materials
Appropriate nitrogen levels	Excessive mineral nitrogen
High plant productivity	Low plant productivity
High plant root:shoot ratio	Low plant root:shoot ratio

and animal manures in the example are already being returned to the soil, additional carbon inputs could most practically be achieved by growing more plant material (i.e., increasing crop production or growing cover crops during the winter).

Specific practices to reduce carbon losses would include better control of soil erosion and the use of conservation tillage. Using a no-till production system would leave crop residues as mulch on the soil surface where they would decompose much more slowly. Refraining from tillage might also reduce the annual respiration losses from the original 2.5% to perhaps 1.5%. A combination of these changes in management would convert the system in our example from one in which organic matter is degrading (declining) in the upper horizons to one in which it is aggrading (increasing).

Natural Ecosystems

Forests If the soil fertility were not too low, the total annual biomass production of a forest would probably be similar to that of the cornfield in Figure 11.14. The standing biomass, on the other hand, would be much greater in the forest since the tree crop is not removed each year. While some litter would fall to the soil surface, much of the annual biomass production would remain stored in the trees.

The rate of humus oxidation in the undisturbed forest would be considerably lower than in the tilled field because the litter would not be incorporated into the soil through tillage, and the absence of physical disturbance would result in slower soil respiration. The litter from certain tree species may also be rich in phenolics and lignin, factors that greatly slow decomposition and C losses (see Figure 11.6). In forest soils, decomposition of leaf litter produces copious quantities of dissolved organic carbon (DOC) compounds and 5 to 40% of the total C losses may occur by leaching—a much greater proportion than from all but the most heavily manured cropland soils. However, losses of organic matter through soil erosion would be much smaller on the forested site. Taken together, these factors allow annual net gains in soil organic matter in a young forest and maintenance of high soil organic matter levels in mature forests.

Grasslands Similar trends occur in natural grasslands, although the total biomass production is likely to be considerably less, depending mainly on the annual rainfall. Among the principles illustrated in Box 11.2, and applicable to most ecosystems, is the dominant role that plant root biomass plays in maintaining soil organic matter levels. In a grassland, the contribution from the plant roots is relatively more important than in a forest. Therefore, a greater proportion of the total biomass produced tends to accumulate as soil organic matter, and this soil organic C is distributed more uniformly with depth. Fires that burn dead aboveground plant material are usually thought to reduce soil organic matter inputs to the soil, but in perennial grasslands studies have shown that increased root growth stimulated by the fires may contribute at least as much carbon to the soil as that lost in the blaze itself.

Wetlands   Wetlands, whether dominated by woody or herbaceous vegetation, exhibit among the highest levels of primary productivity of any ecosystems. However, microbial decomposition is severely retarded by lack of oxygen in the strongly anaerobic wetland soils. In addition, certain products of anaerobic decay (alcohols, organic acids, etc.) actually act as preservatives and inhibit even anaerobic organisms. As a result, organic carbon accumulates rapidly (300 to 3000 kg/ha annually) and may continue to do so for thousands of years in some cases (see Sections 7.7 and 11.9). This prodigious level of carbon sequestration can be reduced or even reversed by practices, such as ditching, drainage, or peat mining, that increase the soil's oxygen content. Converting wetlands to agricultural use, except perhaps for some types of flooded rice production, dramatically increases losses and decreases gains of carbon. The effect of prescribed burning on marsh soil carbon is still under study.

11.8 FACTORS AND PRACTICES INFLUENCING SOIL ORGANIC LEVELS

As indicated in Section 11.7, the level to which organic matter accumulates in soils is determined by the balance of gains and losses of organic carbon.

Influence of Climate

Temperature   The processes of organic matter production (plant growth) and destruction (microbial decomposition) respond differently to increases in temperature. At low temperatures, plant growth outstrips decomposition and organic matter accumulates. The opposite is true where mean annual temperature exceeds approximately 25 to 35 °C. Within zones of comparable moisture and vegetation, the average soil organic carbon and nitrogen contents increase from two to three times for each 10 °C decline in mean annual temperature. This temperature effect can be readily observed by noting the greater organic carbon content of well-drained surface soils as one travels from south (Louisiana) to north (Minnesota) in the humid grasslands of the North American Great Plains region. Similar changes in soil organic carbon are evident as one climbs from warm lowlands to cooler highlands in mountainous regions.

Moisture   Under comparable conditions, soil organic carbon and nitrogen increase as the effective moisture becomes greater. These relationships are evidenced by the darker and thicker A horizons encountered as one travels across the North American Great Plains region from the drier zones in the West (Colorado) to the more humid East (Illinois). The explanation lies mainly in the sparser vegetation of the drier regions. The lowest levels of soil organic matter and the greatest difficulty in maintaining those levels are found where annual mean temperature is high and rainfall is low. These relationships are extremely important to the relative difficulty of sustainable natural resource management. Organic matter levels are influenced not only by temperature and precipitation, but also by vegetation, drainage, aspect, texture, and soil management.

Influence of Natural Vegetation

The greater plant productivity engendered by a well-watered environment leads to greater additions to the pool of soil organic matter. Grasslands generally dominate the subhumid and semiarid areas, while trees are dominant in humid regions. In climatic

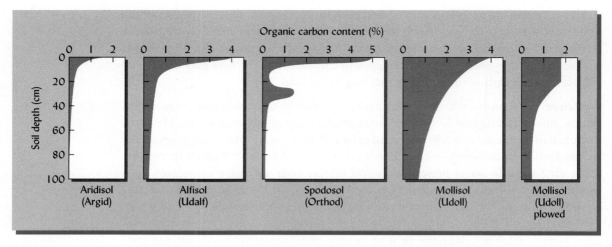

Figure 11.15

Vertical distribution of organic carbon in well-drained soils of four soil orders. Note the higher content and deeper distribution of organic carbon in the soils formed under grassland (Mollisols) compared to the Alfisol and Spodosol, which formed under forests. Also note the bulge of organic carbon in the Spodosol subsoil due to illuvial humus in the spodic horizon (see Chapter 3). The Aridisol has very little organic carbon in the profile, as is typical of dry-region soils.

zones where the natural vegetation includes both forests and grasslands, the total organic matter is higher in soils developed under grasslands than under forests (see Figure 11.15). With grassland vegetation, a relatively high proportion of the plant residues consist of root matter, which decomposes more slowly and contributes more efficiently to soil humus formation than does forest leaf litter.

Effects of Texture and Drainage

Soil texture and drainage are often responsible for marked differences in soil organic matter within a local landscape. Under aerobic conditions, soils high in clay and silt are generally richer in organic matter than are nearby sandy soils (Figure 11.16). The finer-textured soils accumulate more organic matter for several reasons: (1) they produce more plant biomass, (2) they lose less organic matter because they are less well aerated, and (3) more of the organic material is protected from decomposition by

Figure 11.16

Soils high in silt and clay tend to contain high levels of organic carbon. The data shown are for surface soils in 279 tilled maize fields in subhumid regions of Malawi (▼) and Honduras (•). All soils were moderately well-to well-drained and tilled. Variability (scatter of data points) among soils with the same silt + clay content is probably due to differences in (1) the type of clay minerals present (2:1 silicates tend to stabilize more organic carbon), (2) site elevation (cooler, high elevation locations being conducive to greater organic carbon accumulation), and (3) years since cultivation began (longer history of cultivation leading to lower organic carbon levels). (Data courtesy of R. Weil and M. A. Stine, University of Maryland, and S. K. Mughogho, University of Malawi)

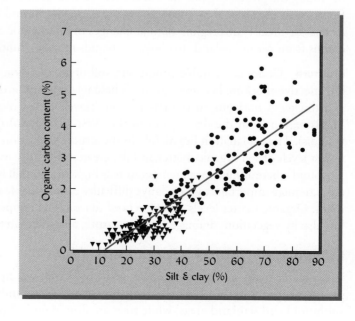

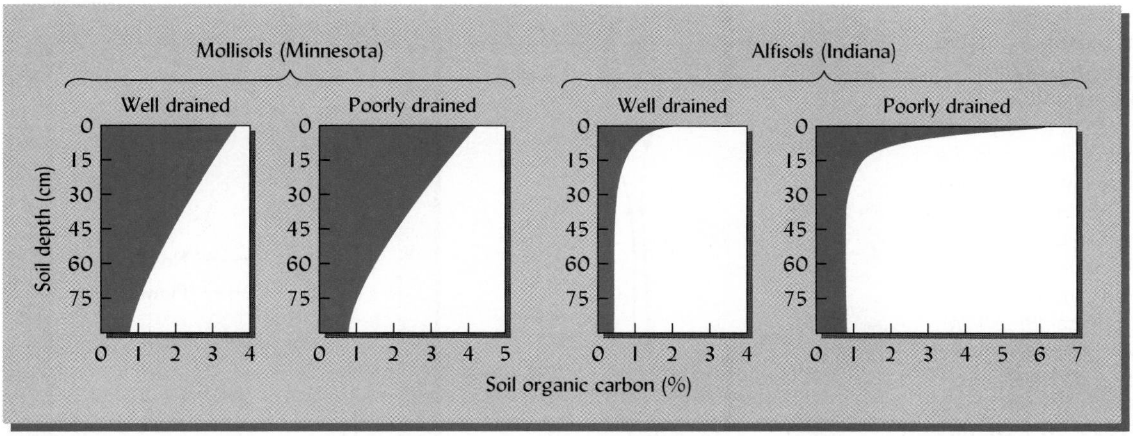

Figure 11.17
Distribution of organic carbon in four soil profiles, two well drained and two poorly drained. Poor drainage results in higher organic carbon content, particularly in the surface horizon.

being bound in clay–humus complexes (see Section 11.4) or sequestered inside soil aggregates. A given amount and type of clay can be expected to have a finite capacity to stabilize organic matter in organomineral complexes. Once this capacity is saturated, further additions of organic matter are likely to add little to humus accumulation, as they will remain readily accessible to microbial decomposition.

Drainage Effects   In poorly drained soils, the high moisture supply promotes plant dry-matter production and relatively poor aeration inhibits organic-matter decomposition. Poorly drained soils therefore generally accumulate much higher levels of organic matter and nitrogen than similar but better-aerated soils (Figure 11.17).

Influence of Agricultural Management and Tillage

Except where barren deserts are brought under irrigation, cultivated land contains much lower levels of organic matter than do comparable areas under natural vegetation. This is not surprising; under natural conditions all the organic matter produced by the vegetation is returned to the soil, and the soil is not disturbed by tillage. By contrast, in cultivated areas much of the plant material is removed for human or animal food and relatively less finds its way back to the land. Also, soil tillage aerates the soil and breaks up the organic residues, making them more accessible to microbial decomposition. The rapid decline in organic matter upon conversion of natural vegetation to cropland is illustrated in Figure 11.13. Conversion from tillage to no-till management can result in rapid carbon increases near the soil surface (Figure 11.18).

No-till farming—carbon as a new cash crop: www.washingtonpost.com/ac2/wp-dyn?pagename= article&node=contentId= A55389-2002Aug23

Many long-term experimental plots demonstrate that cultivated soils kept highly productive by supplemental applications of nutrients, lime, and manure and by the choice of high-yielding cropping systems are likely to have more organic matter than comparable, less productive soils. High productivity is sustainable if it involves not only greater economic harvests, but also larger amounts of roots and shoots returned to the soil.

The Conundrum of Soil Organic Matter Management

Although contradictory, the goals of using and conserving must be pursued simultaneously. The *decomposition* of SOM is necessary for its use as a source of nutrients for plant growth and organic compounds that promote biological diversity, disease suppression, aggregate stability, and metal chelation. In contrast, the *accumulation* of

Figure 11.18

Less tillage means more soil organic carbon. In each case, the no-till system had been used on the experimental plots for 8 to 10 years when the data were collected. In the plowed plots, the soil was disturbed annually by tillage to about 20 cm deep. The soils in Maryland (a) and Brazil (b) were well-drained Ultisols and the climate was temperate (Maryland) to subtropical (Brazil). In Maryland, corn was grown every year with a rye cover crop. In Brazil, oats were rotated with corn using legume cover crops in between. In both cases, no-till encouraged the accumulation of organic C, but only in the upper 5 to 10 cm of the soil. [Data from Weil et al. (1988) and Bayer et al. (2000)]

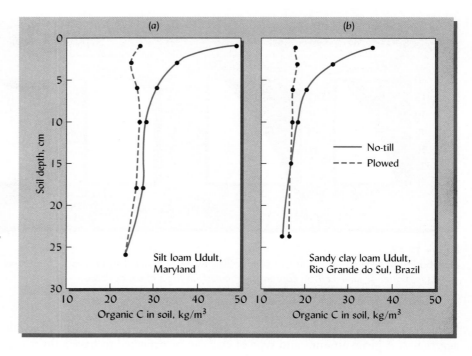

SOM is necessary for these functions in the long term, as well as for the sequestering of C, the enhancement of soil water-holding, the adsorption of exchangeable cations, the immobilization of pesticides, and the detoxification of metals.

General Guidelines for Managing Soil Organic Matter

Can farmers make money with C sequestration? (Texas A & M. University): www.oznet.ksu.edu/ctec/ CASMGSnewsletter/ Jan04–3.htm

A continuous supply of plant residues (roots and tops), animal manures, composts, and other materials must be added to the soil to maintain an appropriate level of soil organic matter, especially in the active pool. It is almost always preferable to keep the soil vegetated than to keep it in bare fallow. Even if some plant parts are removed in harvest, vigorously growing plants provide below- and aboveground residues as major sources of organic matter for the soil. Moderate applications of lime and nutrients may be needed to help free plant growth from the constraints imposed by chemical toxicities and nutrient deficiencies. Where climate permits, cover crops often present a tremendous opportunity to provide protective cover and additional organic material for the soil.

There is no "ideal" amount of soil organic matter. It is generally not practical to try to maintain higher soil organic matter levels than the soil–plant–climate control mechanisms dictate. For example, 1.5% organic matter might be an excellent level for a sandy soil in a warm climate, but would be indicative of a very poor condition for a finer-textured soil in a cool climate. It would be foolhardy to try to achieve as high a level of organic matter in a well-drained Texas silt loam soil as might be desirable for a similar soil found in Canada.

Adequate nitrogen is requisite for adequate organic matter because of the relationship between nitrogen and carbon in humus and because of the positive effect of nitrogen on plant productivity. Accordingly, the inclusion of leguminous plants and the judicious use of nitrogen-containing fertilizers to enhance high plant productivity are two desirable practices. At the same time, steps must be taken to minimize the loss of nitrogen by leaching, erosion, or volatilization (see Chapter 12).

Tillage should be eliminated or limited to that needed to control weeds and to maintain adequate soil aeration. The more tillage that is performed, the faster organic matter is lost from the surface horizons.

Perennial vegetation, especially natural ecosystems, should be encouraged and maintained wherever feasible. Improved agricultural production on existing farmlands should be pursued to allow land currently supporting natural ecosystems to be left relatively undisturbed. In addition, there should be no hesitation about taking land out of cultivation and encouraging its return to natural vegetation where such a move is appropriate. The fact is that large areas of land under cultivation today in every continent never should have been cleared.

11.9 THE GREENHOUSE EFFECT: SOILS AND CLIMATE CHANGE[3]

Soil is a major component of the Earth's system of self-regulation that has created (and, we hope, will continue to maintain) the environmental conditions necessary for life on this planet. Biological processes occurring in soils have major long-term effects on the composition of the Earth's atmosphere, which in turn influences all living things, including those in the soil.

Global Climate Change

Of particular concern today are increases in the levels of certain gases in the Earth's atmosphere. Known as **greenhouse gases**, they cause the Earth to be much warmer than it would otherwise be. Like the glass panes of a greenhouse, these gases allow short-wavelength solar radiation in but trap much of the outgoing long-wavelength radiation. This heat-trapping **greenhouse effect** of the atmosphere is a major determinant of global temperature and, hence, global climates. Gases produced by biological processes, such as those occurring in the soil, account for approximately half of the rising greenhouse effect. Of the five primary greenhouse gases (carbon dioxide, methane, nitrous oxide, ozone, and CFCs), only chlorofluorocarbons (CFCs) are exclusively of industrial origin.

Predicting changes in global temperature is complicated by numerous factors, such as cloud cover and volcanic dust, which can counteract the heat-trapping effects of the greenhouse gases. However, the average global temperature has increased by 0.5 to 1.0 °C during the past century, and is likely to increase by another 1 to 2 °C in this century. Major changes in the Earth's climate are sure to result, including changes in rainfall distribution and growing season length, increases in sea level, and greater frequency and severity of storms. The rise in sea level alone, as predicted by some climate models, would threaten the homes of hundreds of millions of people living in coastal areas, mainly in Asia and North America. Through national programs and international treaties, much effort and expense are currently being directed at reducing the anthropogenic (human-caused) contributions to climate change. Soil science has the potential to contribute greatly to our ability to deal with global warming and the increasing levels of greenhouse gases.

Carbon Dioxide

In 2009, the atmosphere contained about 395 ppm CO_2, as compared to about 280 ppm before the Industrial Revolution. Levels are increasing at about 0.5% per year. Although the burning of fossil fuels is a major contributor, much of the increase in atmospheric CO_2 levels comes from a net loss of organic matter from the world's soils. Research (Figure 11.19) indicates that the feedback between the soil and atmosphere works both ways—changes in the levels of gases beneficial or harmful to plants influence the rate at which carbon accumulates in soil organic matter.

Storing carbon in soil: Why and how?
www.agiweb.org/geotimes/jan02/feature_carbon.html

Free-air CO_2 enrichment (FACE) research:
http://public.ornl.gov/face/index.html

Intergovernmental Panel on Climate Change:
www.ipcc.ch

Climate change will affect agriculture—the view from Australia:
www.greenhouse.gov.au/impacts/agriculture.html

Photo documentation of global climate change. See links under "References":
www.worldviewof globalwarming.org/

---

[3]For a review of the potential for soil management to mitigate climate changes, see Paustian and Babcock (2004).

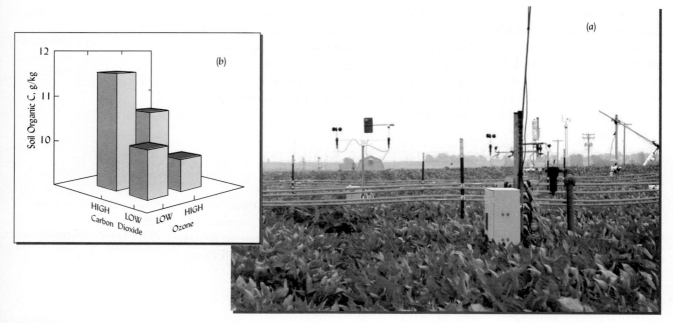

Figure 11.19

Changing atmospheric composition affects the C cycle. (a) Within a circular FACE (free air carbon enrichment) experimental plot on Mollisols at the University of Illinois, soybeans experience air altered to simulate the atmospheric composition expected in 50 years. Horizontal tubing and vertical poles supply computer-controlled concentrations of carbon dioxide and/or ozone, which are continuously monitored by various sensors. Increasing atmospheric CO_2 from low (350 mg/L, the ambient level) to high (500 mg/L, the level expected by the year 2050) enhanced photosynthesis and plant growth, thus increasing the amount of fixed carbon available for translocation to the roots and eventually to the soil. (b) Data from a different carbon-ozone enrichment experiment on Ultisols in Maryland show measurably increased soil organic carbon after 5 years of soybean and wheat crops grown in high CO_2 air. Increased root:shoot ratio, rhizodeposition of carbon compounds, and overall greater litter inputs all played a role. Ozone, a pollutant at ground level, injures plants, reduces photosynthesis, and therefore impacts the soil in a manner opposite to that of CO_2. The data suggest the full effect of CO_2 is seen only when ozone remains at low levels. [Data from Weil et al. (2000); photo courtesy of R. Weil]

Global climate change and agricultural production:
www.fao.org/docrep/W5183E/W5183E00.htm

We have already discussed many ways that land managers can increase levels of soil organic matter (Section 11.8) by changing the balance between gains and losses. The opportunities for *sequestering* carbon are greatest for degraded soils that currently contain only a small portion of the organic matter levels they contained originally under natural conditions. Reforestation of denuded areas is one such opportunity. Others include switching cropland from conventional tillage to no-tillage, which commonly sequesters 0.2 to 0.5 Mg ha^{-1} y^{-1} of carbon during the first decade. Conversion of cultivated land to perennial vegetation and restoration of wetlands can sequester carbon at much higher rates.

Calculate your greenhouse gases:
www.b-e-f.org/offsets/calculator

By slowly increasing soil organic matter to near precultivation levels, such management changes could significantly enhance society's efforts to stem the rise in atmospheric CO_2 and at the same time improve soil quality and plant productivity. Some estimates suggest that during a 50-year period, improved management of agricultural lands could provide about 15% of the CO_2 emission reductions that the United States will need to make. It should be noted, however, that in accordance with the factors discussed in Section 11.8, soils have only a finite capacity to assimilate carbon into stable soil organic matter. Therefore carbon sequestration in soils can only buy time before other kinds of actions (shifts to renewable energy sources, increased fuel efficiency, etc.) are fully implemented to reduce carbon emissions to levels that will not threaten climate stability.

Biofuels Production   The production of *biofuels* as substitutes for gasoline and diesel fuel is attracting much attention and money. One such fuel, biodiesel, consists of fuel-grade oil generally extracted from the same crops (e.g., soybeans, rapeseed, sunflower) used to provide cooking oil. Even more attention is turning toward the production of ethanol, not from grain as has been done for decades (and is known to be an energy loser!) but from cellulosic residues such as corn stover (the stalks and leaves) and switchgrass, which can be grown much more inexpensively and abundantly than grain. Much controversy still surrounds the question as to whether such fuels will actually produce more energy than is consumed in the fossil fuel used to grow and process the crop. However, one aspect that should *not* be controversial is the critical importance of leaving enough crop residues on the soil to maintain the levels of soil organic matter consistent with high quality and continued productivity of the soil resource. Research suggests that soil organic matter levels, aggregate stability, and other soil properties critical to sustaining productivity may rapidly suffer if all, or even half, of the corn stover is removed to make biofuels (Table 11.4). Unfortunately, the need to "share" the plant residues with the soil has been overlooked by many energy engineers and planners advising policymakers on biofuel production strategies. Some have advised not only that all aboveground residues be harvested to make the biofuels, but that the crops be genetically altered to reduce the ratio of roots to shoots—just the opposite of what sustainable soil management requires!

Wetland Soils   Although they cover only about 2% of the world's land area, Histosols (and Histels—permafrost soils with organic surface horizons) are important in the global carbon cycle because they hold about 20% of global soil carbon. Drainage of these wetland soils (as for high-value horticultural production) speeds the oxidation of organic matter, which over time destroys the soil itself. Draining Histosols or mining them for peat can make major contributions to the rise in atmospheric CO_2. On the other hand, Histosol formation may buffer climate change because increased global temperatures will cause the sea level to rise as seawater expands and ice caps melt. The rising sea level will flood more coastal land area, creating more Histosols in tidal marshes. The organic matter accumulation in these new

Soil-related concerns in biofuel production: www.culturechange.org/cms/index.php?option=com_content&task=view&id=107&Itemid=1

Environmentalist assessment of climate change impacts in U.S. Gulf Coast: www.ucsusa.org/gulf/

Table 11.4

DETERIORATION OF SOIL ORGANIC CARBON AND STRUCTURAL PROPERTIES IN THE UPPER 5 TO 10 CM OF THREE OHIO SOILS AFTER ONE YEAR OF CORN STOVER REMOVAL FOR BIOFUEL PRODUCTION[a]

Soil series, Great Group	Cropping history	Stover returned, Mg/ha[b]	Soil organic C, g/kg	Soil bulk density, Mg/m^3	Tensile strength of soil aggregates, kPa	Water stable aggregates <4.75 mm diameter, %
Rayne silt loam (Hapludults)	33-yr. continuous no-till corn	5	29	1.46	140	18
		0	19	1.50	50	13
Hoytville clay loam (Epiaqualfs)	8-yr. corn/soybean minimum till	5	26	1.31	380	20
		0	22	1.49	120	8
Celina silt loam (Hapludalfs)	15-yr. corn/soybean no-till	5	28	1.25	225	36
		0	21	1.42	80	13

[a] Data selected from Blanco-Canqui et al. (2006).
[b] 5 Mg/ha corn stover represented the normal practice of returning all aboveground residues after corn grain harvest. Return of 0 Mg/ha represented removal of all stover for biofuel production.

Histosols would represent a significant sequestering of CO_2 that might help to buffer the global warming trend (as would the increased absorption of CO_2 by the greater volume of ocean waters). However, such predictions are complicated by the involvement of another greenhouse gas, methane, which is emitted by some Histosols and other wetland soils.

Methane   Methane (CH_4) occurs in the atmosphere in far smaller amounts than CO_2. However, methane's contribution to the greenhouse effect is nearly half as great as that from CO_2 because each molecule of CH_4 is about 25 times as effective as CO_2 in trapping outgoing radiation. The level of CH_4 is rising at about 0.6% per year. Soils both add CH_4 and remove it from the atmosphere.

Biological soil processes account for much of the methane emitted into the atmosphere. When soils are strongly anaerobic, as in most wetlands and rice paddies, bacteria produce CH_4, rather than CO_2, as they decompose organic matter (see Sections 7.4 and 11.2). Among the factors influencing the amount of CH_4 released to the atmosphere from wet soils are (1) the maintenance of a redox potential (E_h) near 0 mV, (2) the availability of easily oxidizable carbon, and (3) the nature and management of the plants growing on these soils (70 to 80% of the CH_4 released from flooded soils escapes to the atmosphere through the hollow stems of wetland plants). Periodically draining rice paddies prevents the development of extremely anaerobic conditions and therefore can substantially decrease CH_4 emissions. Such management practices should be given serious consideration, as rice paddies are thought to be responsible for up to 25% of global CH_4 production.

Salt marshes (such as those along the coast of Louisiana) tend to emit much less methane than do freshwater swamps; therefore, the balance of greenhouse gases entering and leaving coastal marshes is likely to be more favorable for mitigating global warming. Sulfate reducers derive much more energy by using sulfate as their alternative electron acceptor than methanogens can derive by reducing carbon dioxide to methane. Therefore, the high levels of sulfate in seawater allow sulfate-reducing prokaryotes to outcompete the methanogens for available carbon substrates in salt marshes. The large amount of sulfate reduction also tends to *poise* the redox potential, preventing it from falling quite low enough to favor methane production (see Section 7.3).

Wetland soils are not the only ones that contribute to atmospheric CH_4. Significant quantities of methane are also produced by the anaerobic decomposition of cellulose in the guts of termites living in well-aerated soils (see Section 10.5), and of garbage buried deep in landfills (see Section 15.10).

In well-aerated soils, certain **methanotrophic bacteria** produce the enzyme *methane monooxygenase*, which allows them to obtain their energy by oxidizing methane to methanol. This reaction, which is largely carried out in soils, reduces the global greenhouse gas burden by about 1 billion Mg of methane annually. Unfortunately, the long-term inputs of inorganic nitrogen on cropland, pastures, and forests can greatly reduce the capacity of the soil to oxidize methane (Figure 11.20). The evidence suggests that the rapid availability of ammonium from fertilizer stimulates ammonium-oxidizing bacteria at the expense of the methane-oxidizing bacteria. Long-term experiments indicate that supplying nitrogen in organic form (as manure) actually enhances the soil's capacity for methane oxidation.

Nitrous oxide (N_2O) is another greenhouse gas produced by microorganisms in poorly aerated soils, but since it is not directly involved in the carbon cycle, it will be discussed in the next chapter, when the nitrogen cycle is taken up.

Because the soil can act as a major source or sink for carbon dioxide, methane, and nitrous oxide, it is clear that, together with steps to modify industrial outputs, soil management has a major role to play in controlling the atmospheric levels of greenhouse gases.

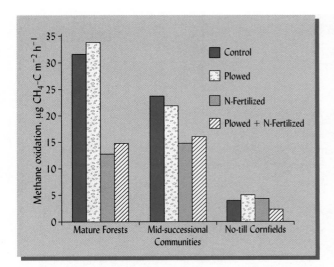

Figure 11.20

Mineral nitrogen reduced the capacity of soils to oxidize methane and thereby remove this potent greenhouse gas from the atmosphere. Forested soils exhibited the highest rates of methane oxidization and the greatest impairment due to the addition of N. A one-time physical disturbance (plowing) had little impact. Nitrogen was applied as a solution of ammonium nitrate (100 kg N ha⁻¹). The study was on sandy loam soils (Typic Hapludalfs) in southern Michigan. [Modified from Suwanwaree and Robertson (2005)]

11.10 COMPOSTS AND COMPOSTING

Composting is the practice of creating humuslike organic materials by aerobic decomposition outside the soil. The finished product, **compost**, is popular as a mulching material, as an ingredient for potting mixes, as an organic soil conditioner, and as a slow-release fertilizer. High-quality compost can be made at ambient temperatures by a slow decomposition process, or more quickly by a process called **vermicomposting**, in which certain litter-dwelling (epigeic) earthworms are added to help transform the material. We will focus, however, on the most commonly practiced type of composting, in which intense decomposition activity occurs within large, well-aerated piles. This approach is called **thermophilic composting** because the large mass of rapidly decomposing material, combined with the insulating properties of the pile, results in a considerable buildup of heat.

Composting Process   Thermophilic compost typically undergoes a three-stage process of decomposition. (1) During a brief, initial **mesophilic** stage, sugars and readily available microbial food sources are rapidly metabolized, causing the temperature in the compost pile to gradually rise from ambient levels to over 40 °C. (2) A *thermophilic* stage occurs during the next few weeks or months, during which temperatures rise to 50 to 75 °C while oxygen-using thermophilic organisms decompose cellulose and other more resistant materials. Frequent mixing during this stage is essential to maintain oxygen supplies and assure even heating of all the material. The easily decomposed compounds are used up, and humuslike compounds are formed during this stage. (3) A second mesophilic or *curing stage* follows for several weeks to months, during which the temperature falls back to near ambient, and the material is recolonized by mesophilic organisms, including certain beneficial microorganisms that produce plant-growth-stimulating compounds or are antagonistic to plant pathogenic fungi (see Section 10.13).

Nature of the Compost Produced   As raw organic materials are humified in a compost pile, the content of nonhumic substances declines, and the content of humic acid increases markedly. During the composting process, the C/N ratio of organic materials in the pile decreases until a fairly stable ratio, in the range of 10:1 to 20:1, is achieved.

Although 50 to 75% of the carbon in the initial material is typically lost during composting, mineral nutrients are mostly conserved. The mineral content (referred to as *ash* content) therefore increases over time, making finished compost more concentrated

Making compost, step-by-step:
www.klickitatcounty.org/SolidWaste/default.asp?fCategoryIDSelected=965105457

Controlled microbial composting—the Luebke method:
www.ibiblio.org/steved/Luebke/Luebke-compost2.html

BOX 11.3
MANAGEMENT OF A COMPOST PILE

Materials to Use   Materials good for home composting include tree leaves (best if shredded), grass clippings, weeds (best before going to seed), kitchen scraps, wood shavings, gutter cleanings, pine needles, spoiled hay, straw, and even vacuum cleaner dust. Large-scale commercial compost is often made from materials such as municipal garbage, sewage sludge, wood chips, animal manures, municipal leaves, and food-processing wastes.

Materials to Avoid   Some materials to avoid include meat scraps (odors, rodents), cat droppings (carry microbes harmful to infants and pregnant women), sawdust from pressure-treated lumber and plywood (heavy metals and arsenic), and plastics and glass (nonbiodegradable; nuisance or dangerous in final product).

Balancing Nutrients   Although highly carbonaceous materials can be made to compost satisfactorily if they are turned frequently and kept moist, best results are obtained if high-C/N-ratio materials (e.g., brown leaves, straw, or paper) are mixed with low-C/N-ratio materials (e.g., green grass clippings, legume hay, blood meal, sewage sludge, or livestock manure) so as to achieve an overall C/N ratio between 20 and 30. Nitrogen fertilizer can also be added to lower the C/N ratio.

Other materials commonly added to improve nutrient balance and content include mixed fertilizers (N, P, and K), wood ashes (K, Ca, and Mg), bone meal or phosphate rock powder (P and Ca), and seaweed (K, Mg, Ca, and micronutrients). Some of these materials contain enough soluble salts to necessitate leaching of the finished compost before using it on salt-sensitive plants.

Composting Methods   Provide good aeration throughout the pile, but build the pile large enough to provide mass sufficient to generate heat and prevent excessive drying. Backyard compost piles should be at least 1 m square and 1 m high. Compost bins are available to make turning the compost easier. Large-scale composting is usually carried out in windrows, about 2 to 3 m wide, 1 to 2 m tall, and many meters long (Figure 11.21). In dry climates, compost may be made in pits dug about 1 m deep to protect the material

Figure 11.21

(a) An efficient and easily managed method of composting suitable for homeowners is the three-bin method in which materials are turned with a pitchfork from one bin to the next. The perforated white plastic pipes enhance aeration. The leftmost bin contains relatively fresh materials, while the one at the right contains finished compost. (b) A special machine turns large-scale compost windrows (direction of travel is away from the reader) to mix the material and maintain well-aerated conditions at a facility in North Carolina where university dining hall food scraps are processed into compost for campus landscaping. (Photos courtesy of R. Weil)

from drying. The various materials to be composted can be mixed together or applied in thin layers. Often, a small amount of garden soil or finished compost is added to ensure that plenty of decomposer organisms will be immediately available. Compost activators containing microbial inoculum or herbal extracts are available, but while some may speed the initial heating of the pile, scientific tests rarely show any other advantage to using these preparations.

Oxygen and Moisture Control   Low oxygen levels, usually due to inadequate turning combined with excessive moisture, can produce putrid odors as anaerobic decomposition takes over. Monitoring temperature and oxygen levels in the pile can help avoid this situation. To promote good aeration it is best to mix in a bulking agent such as wood chips, avoid excessive packing, and either turn the pile or pull a stream of air through it (Figure 11.21). If turning is used, this should be done during the thermophilic stage whenever the temperature begins to drop. The compost water content should be maintained at 50 to 70%. Properly moist compost will feel damp—but not dripping wet—when squeezed. Turning the pile during dry weather can help reduce excess moisture, while turning it during rain can help moisten a too-dry pile.

in nutrients than the initial combination of raw materials used. Properly prepared finished compost should be free of viable weed seeds and pathogenic organisms, as these are generally destroyed during the thermophilic phase. However, inorganic contaminants such as heavy metals are *not* destroyed by composting. Proper management of compost is essential if the finished product is to be desirable for use as a potting media or soil amendment (see Box 11.3).

Effects of Composting   Although making compost may involve more work and expense than applying uncomposted organic materials directly to the soil, the process offers several distinct advantages. (1) Composting provides a means of safely storing organic materials with a minimum of odor release until it is convenient to apply them to soils. (2) Compost is easier to handle than the raw materials as a result of the 30 to 60% smaller volume and greater uniformity of the resulting material. (3) For residues with a high initial C/N ratio, proper composting ensures that any nitrate depression period will occur in the compost pile, not in the soil, thereby avoiding induced plant nitrogen deficiency. (4) **Cocomposting** low-C/N-ratio materials (such as livestock manure and sewage sludge) with high-C/N-ratio materials (such as sawdust, wood chips, senescent tree leaves, or municipal solid waste) provides sufficient carbon for microbes to immobilize the excess nitrogen and minimize any nitrate leaching hazard from the low-C/N materials. (5) High temperatures during the thermophilic stage in well-managed compost piles kill most weed seeds and pathogenic organisms in a matter of a few days. Under less ideal conditions, temperatures in parts of the pile may not exceed 40 to 50 °C, so weeks or months may be required to achieve the same results. (6) Most toxic compounds that may be in organic wastes (pesticides, natural phytotoxic chemicals, etc.) are destroyed by the time the compost is mature and ready to use. Composting is used as a method of biological treatment of polluted soils and wastes (see Section 15.5). (7) Some composts can effectively suppress soilborne plant diseases by encouraging microbial antagonisms (see Section 10.12). Most success in disease suppression has occurred when well-cured compost is used as a main component of potting mixes for greenhouse-grown plants. (8) Because compost is made from organic waste materials that recently used up CO_2 in the process of their production (plant photosynthesis), compost is considered carbon-neutral, making it a much more environmentally sustainable choice than peat for potting mixes.

Disadvantages of Compost   Compost often has low nutrient contents and very low availability of the nutrients present. It also usually has a relatively high P to N ratio in comparison to plant needs. Because of this ratio, attempts to use compost as the principal source of plant nutrients can easily result in the application of potentially polluting levels of phosphorus (see Chapter 13). In addition, because the more labile organic substances are decomposed during the composting process, compost usually provides less benefit to soil aggregation than would the fresh residues from which the compost was made.

11.11   CONCLUSION

Organic matter is a complex and dynamic soil component that exerts major and generally beneficial influences on soil behavior, properties, and functions in the ecosystem. Because of the enormous amount of carbon stored in soil organic matter and the dynamic nature of this soil component, soil management may be an important tool for moderating the global greenhouse effect.

Organic residue decay, nutrient release, and humus formation are controlled by environmental factors and by the quality of the organic materials. High contents of lignin and polyphenols, along with high C/N ratios, markedly slow the decomposition

To compost or not to compost?
www.organicaginfo.org/ upload/Compost .MarkMeasures.pdf

The Don't Bag It composting plan to save landfill space: http://aggie-horticulture .tamu.edu/earthknd/ compost/compost.html

process, causing organic matter to accumulate while reducing the availability of nutrients.

Soil organic matter comprises three major pools of organic compounds. The *active pool* consists of relatively easily decomposed compounds and plays a major role in nutrient cycling, micronutrient chelation, maintenance of structural stability and soil tilth, and as a food source underpinning biological diversity and activity in soils. Most of the organic matter is in the *passive pool*, which contains very stable materials that resist microbial attack and may persist in the soil for centuries. This pool provides cation exchange and water-holding capacities, but is relatively inert biologically. When soil is cleared of natural vegetation and brought under cultivation, the initial decline in soil organic matter is principally at the expense of the active pool. The passive pool is depleted very slowly and over very long periods of time.

The level of soil organic matter is influenced by climate (being higher in cool, moist regions), drainage (being higher in poorly drained soils), and by vegetation type (being generally higher where root biomass is greatest, as under grasses).

The maintenance of soil organic matter, especially the active pools, in mineral soils is one of the great challenges in natural resource management around the world. By encouraging vigorous growth of crops or other vegetation, abundant residues (which contain both carbon and nitrogen) can be returned to the soil directly or through feed-consuming animals. Also, the rate of destruction of soil organic matter can be minimized by restricting soil tillage, controlling erosion, and keeping most of the plant residues at or near the soil surface.

For some purposes, it is advantageous to manage the decomposition of organic matter outside of the soil in a process known as *composting*. Composting transforms various organic waste materials into a humuslike product that can be used as a soil amendment or a component of potting mixes.

The decay and mineralization of soil organic matter is one of the main processes governing the cycling of nutrients in soils—the subject of the next chapter.

STUDY QUESTIONS

1. Compare the amounts of carbon in Earth's standing vegetation, soils, and atmosphere.

2. If you wanted to apply an organic material that would make a long-lasting mulch on the soil surface, you would choose an organic material with what chemical and physical characteristics?

3. Describe how the addition of certain types of organic materials to soil can cause a nitrate depression period. What are the ramifications of this phenomenon for plant growth?

4. In addition to humic substances, what other categories of organic materials are found in soils?

5. Some scientists include plant litter (surface residues) in their definition of soil organic matter, while others do not. Write two brief paragraphs, one justifying the inclusion of litter as soil organic matter and one justifying its exclusion.

6. What soil properties are mainly influenced by the active and passive pools, respectively, of organic matter?

7. In this book and elsewhere, the terms *soil organic carbon* and *soil organic matter* are used to mean almost the same thing. How are these terms related, conceptually and quantitatively? Why is the term *organic carbon* generally more appropriate for quantitative scientific discussions?

8. Explain, in terms of the balance between gains and losses, why cultivated soils generally contain much lower levels of organic carbon than similar soils under natural vegetation.

9. In what ways are soils involved in the greenhouse effect that is warming up the Earth? What are some common soil-management practices that could be changed to reduce the negative effects and increase the beneficial effects of soils on the greenhouse effect?

10. Explain why compost is more environmentally sustainable than peat for use in potting media and as an amendment for golf course greens.

REFERENCES

Batjes, N. H. 1996. "Total carbon and nitrogen in the soils of the world," *European J. Soil Sci.,* **47**:151–163.

Bayer, C., J. Mielniczuk, T. Amado, L. Martin-Neto, and S. Fernandes. 2000. "Organic matter storage in a sandy clay loam Acrisol affected by tillage and cropping system in southern Brazil," *Soil and Tillage Research,* **54**:101–109.

Bigelow, C. A., D. C. Bowman, and D. K. Cassel. 2004. "Physical properties of sand amended with inorganic materials or sphagnum peat moss." *USGA Turfgrass and Environmental Research Online,* **3**(6):1–14. http://usgatero.msu.edu/v03/n06.pdf (posted 15 March 2004; verified 20 July 2006).

Blanco-Canqui, H., R. Lal, W. M. Post, R. C. Izaurralde, and L. B. Owens. 2006. "Rapid changes in soil carbon and structural properties due to stover removal from no-till corn plots," *Soil Sci.,* **171**: 468–482.

Clapp, C. E., M. H. B. Hayes, A. J. Simpson, and W. L. Kingery. 2005. "Chemistry of soil organic matter," pp. 1–150, in M. A. Tabatabai and D. L. Sparks (eds.), *Chemical Processes in Soils.* SSSA Book Series No. 8 (Madison, WI: Soil Science Society of America).

Eswaran, H., P. F. Reich, J. Kimble, F. H. Beinroth, E. Padmanabhan, and P. Moncharoen. 2000. "Global carbon stocks," pp. 15–26, in R. Lal et al. (eds), *Global Climate Change and Pedogenic Carbonates* (Boca Raton, FL: Lewis Publishers).

Hierro, J. L., and R. M. Callaway. 2003. "Allelopathy and exotic plant invasion," *Plant Soil,* **256**:29–39.

Inderjit, K. M., M. Dakshini, and C. L. Foy (eds.). 1999. *Principles and Practices in Plant Ecology: Allelochemical Interactions* (Boca Raton, FL: CRC Press).

IPCC. 2007. Climate change 2007: The physical science basis. Summary for policymakers. [Online]. Available from the Intergovernmental Panel on Climate Change, United Nations. www.ipcc.ch/SPM2feb07.pdf (posted 2 February 2007; verified 3 February 2007).

Magdoff, F., and R. R. Weil (eds.). 2004. *Soil Organic Matter in Sustainable Agriculture* (Boca Raton, FL: CRC Press).

Paustian, K., and B. Babcock. 2004. *Climate Change and Greenhouse Gas Mitigation: Challenges and Opportunities,* Task Force Report 141. (Ames, IA: Council on Agricultural Science and Technology).

Paustian, K., W. J. Parton, and J. Persson. 1992. "Modeling soil organic matter-amended and nitrogen-fertilized long-term plots," *Soil Sci. Soc. Amer. J.,* **56**:476–488.

Suwanwaree, P., and G. P. Robertson. 2005. "Methane oxidation in forest, successional, and no-till agricultural ecosystems: Effects of nitrogen and soil disturbance," *Soil Sci. Soc. Am. J.,* **69**:1722–1729.

Torbert, H., and H. Johnson. 2001. "Soil of the intensive agriculture biome of Biosphere 2. *J. Soil Water Conservation,* **56**:4–11.

Walford, R. L. 2002. "Biosphere 2 as voyage of discovery: The serendipity from inside," *BioScience,* **52**:259–263.

Weil, R. R., and G. S. Belmont. 1987. "Interactions between winged bean and grain amaranth," *Amaranth Newsletter,* **3**(1):3–6.

Weil, R. R., P. W. Benedetto, L. J. Sikora, and V. A. Bandel. 1988. "Influence of tillage practices on phosphorus distribution and forms in three Ultisols," *Agron. J.,* **80**:503–509.

Weil, R. R., K. R. Islam, and C. L. Mulchi. 2000. "Impact of elevated CO_2 and ozone on C cycling processes in soil," p. 47, in *Agronomy Abstracts* (Madison, WI: American Society of Agronomy).

12

Nutrient Cycles and Soil Fertility

No sólo son raíces bajo las piedras teñidas de sangre, no sólo sus pobres huesos derribados definitivamente trabajan en la tierra . . .

(They are not only roots beneath the bloodstained stones, not only do their poor demolished bones definitely till the soil . . .)

—PABLO NERUDA, CHILEAN POET, *ESPAÑA EN CORAZON (SPAIN IN OUR HEARTS)*

Skeleton of Chincoteague pony returns phosphorus to soil. (R. Weil)

Soils are at the very hub of the biogeochemical cycles that transform, transport, and renew the supplies of the mineral nutrients so essential for the growth of terrestrial plants. As each nutrient cycles through the soil, a given atom may appear in many different chemical forms, each with its own properties, behaviors, and consequences for soil development and for the ecosystem. For some elements, such as nitrogen and sulfur, the cycles are exceedingly complex, involving many different biologically mediated transformations and movement into and out of the soil as solid particles, in solution, and as gases. The cycling of phosphorus also involves a fascinating set of complex interactions among chemical and biological processes. For calcium, magnesium, and potassium, the weathering of minerals and cation exchange reactions dominate the cycles. For the micronutrients, mobility and bioavailability are controlled mainly by soil pH, redox potential, and reactions with soluble organic compounds.

These cycles impact not only soil fertility but also the health of aquatic ecosystems and the health and survival of humans on Earth. This cycling of nutrients explains why vegetation (and, indirectly, animals) can continue to remove nutrients from a soil for millennia without depleting the soil of its supply of essential elements. The biosphere does not quickly run out of nitrogen or magnesium because it uses the same supply over and over again. When human activities short-circuit or break these cycles, soils do become impoverished, as do the people who depend upon them. It is also worth noting that the impacts of mismanaging these nutrient cycles are not confined to terrestrial systems. In fact, undue leakage of N and P from the soil phase of their cycles is responsible for some of the most devastating water pollution problems on the planet.

Scarcity of N is the most widely occurring nutritional limit on the productivity of terrestrial ecosystems. Nitrogen is also the element most widely overapplied to agro-ecosystems and the most widely responsible for deterioration of water quality. Phosphorus is the second most widely limiting nutrient and is generally even more scarce than N. It, too, has a history of overapplication in modern agriculture and there-fore has become responsible for widespread pollution of aquatic systems.

In this chapter we discuss the soil processes and principles governing the nutrient cycles. Then in Chapter 13, we will apply this knowledge to the practical manage-ment of soil fertility and environmental quality.

12.1   NITROGEN IN THE SOIL SYSTEM[1]

Nitrogen and Plant Growth and Development

Nitrogen (N) is major part of all proteins—including the enzymes, which in turn control virtually all biological processes. Other critical nitrogenous plant components include the nucleic acids and chlorophyll. Nitrogen is also essential for carbohydrate use in plants.

Plants deficient in N tend to exhibit **chlorosis** (yellowish or pale green leaf col-ors), a stunted appearance, and thin, spindly stems (see Plates 87 and 94). The older leaves are the first to turn yellowish. Nitrogen-deficient plants often have a low shoot-to-root ratio, and they mature more quickly than healthy plants.

When too much N is supplied, excessive vegetative growth occurs, the top-heavy plants are prone to falling over (lodging), plant maturity may be delayed, and the plants may become more susceptible to diseases and pests. These problems are especially noticeable if other nutrients, such as potassium, are in relatively low supply.

Forms of Nitrogen Taken Up by Plants   Plant roots take up N from the soil princi-pally as dissolved nitrate (NO_3^-) and ammonium (NH_4^+) ions, a relatively equal mix-ture of the two ions giving the best results with most plants. As explained in Sections 9.1 and 9.5, uptake of ammonium markedly lowers the pH of the rhizosphere soil, while uptake of nitrate tends to raise it. These pH changes, in turn, influence the uptake of other ions such as phosphates and micronutrients (Sections 12.3 and 12.6). The direct uptake of soluble organic N compounds also occurs and is of particular significance in natural grasslands and forests.

Distribution and Cycling in Soils

The atmosphere, which is 78% gaseous nitrogen (N_2), appears to be a virtually limitless reservoir of this element. However, the very strong triple bond between the two nitrogen atoms ($N\equiv N$) makes this gas extremely inert and not directly usable by plants or ani-mals. Little nitrogen would be found in soils and little vegetation would grow in terrestrial ecosystems around the world were it not for certain natural processes (principally micro-bial nitrogen fixation and lightning) that can break this triple bond and form **reactive nitrogen**, which includes any form of nitrogen that is readily available to living organisms.

The nitrogen cycle has long been the subject of intense scientific investigation, since understanding the translocations and transformations of this element is funda-mental to solving many environmental, agricultural, and natural-resource-related problems. The principal pools and forms of nitrogen, and the processes by which

Rate nitrogen sufficiency for corn:
www.ipm.iastate.edu/ipm/icm/2006/9-18/ntool.html

Global nitrogen: Cycling out of control:
www.ehponline.org/members/2004/112-10/focus.html

Nitrogen sources and transformations:
www.ext.colostate.edu/pubs/crops/00550.html

Nitrogen species found in soil:
www.sws.uiuc.edu/nitro/nspecies.asp

---

[1]For a review of nitrogen in agriculture and the environment, see Schepers and Raun (2008). For a broad assessment of natural and anthropogenic reactive N in global ecosystems, see Galloway et al. (2003).

they interact in the cycle, are illustrated in Figure 12.1. This figure deserves careful study; we will refer to it frequently as we discuss each of the major divisions of the nitrogen cycle.

Most of the nitrogen in terrestrial systems is found in the soil organic matter, which typically contains about 5% N. Therefore, the distribution of nitrogen in soil profiles closely parallels that of soil organic matter (see Sections 11.3 and 11.10). Except where large amounts of chemical fertilizers have been applied, inorganic (i.e., mineral) nitrogen seldom accounts for more than 1 to 2% of the N in the soil.

Ammonium and nitrate are two critical forms of inorganic nitrogen in the cycle (Figure 12.1). In addition to its possible loss by erosion and runoff, the N in ammonium is subject to five major fates: (1) *immobilization* by microorganisms; (2) removal

Nitrogen cycle links:
http://users.rcn.com/jkimball
.ma.ultranet/BiologyPages/
N/NitrogenCycle.html

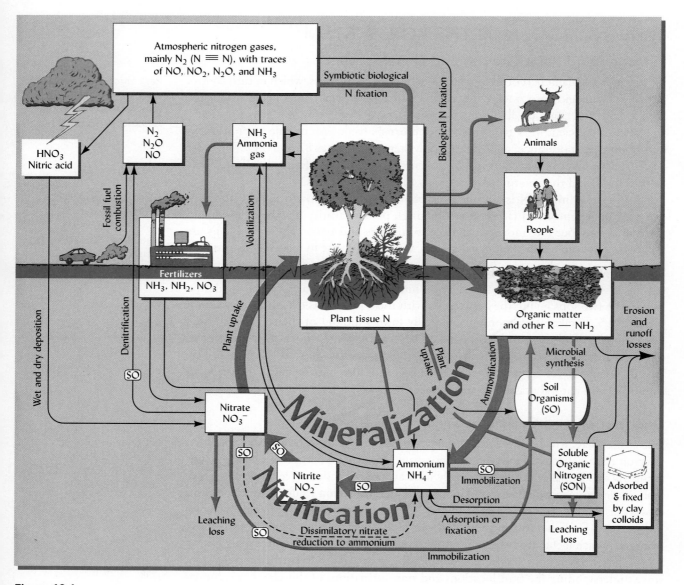

Figure 12.1

The nitrogen cycle, emphasizing the primary cycle (thickest arrows) in which organic nitrogen is mineralized, plants take up the mineral nitrogen, and eventually organic nitrogen is returned to the soil as plant residues. Note the processes by which soil nitrogen is lost and replenished. The boxes represent various forms of nitrogen; the arrows represent processes by which one form is transformed into another. Soil organisms, whose enzymes drive most of the reactions in the cycle, are represented as rounded boxes labeled "SO." (Diagram courtesy of R. Weil)

by *plant uptake*; (3) *fixation* in the interlayers of certain 2:1 clay minerals; (4) *volatilization* after being transformed into ammonia gas; and (5) oxidation to nitrite and subsequently to nitrate by a microbial process called *nitrification*. Similarly, the N in nitrate is subject to five major fates: (1) *immobilization* by microorganisms; (2) removal by *plant uptake*; (3) loss to groundwater by *leaching* in drainage water; (4) *volatilization* to the atmosphere as several nitrogen-containing gases formed by *denitrification*; or (5) dissimilatory (nitrate) reduction to ammonium (DNRA) by anaerobic organisms.

Immobilization and Mineralization

Nitrogen in organic compounds is protected from loss but is largely unavailable to higher plants. Much of this nitrogen is present as amine groups ($-C-NH_2$), largely in proteins or as part of humic compounds. The decomposition process involves the breakdown of large, insoluble N-containing organic molecules into smaller and smaller units with the eventual release of the nitrogen as NH_4^+. This cycling of N is highly visible in Figure 10.12, which shows a dark green circle of grass plants that have responded to the fungal release of N from the residues of previous grass growth. The enzymes that bring about this process are produced mainly by microorganisms. These enzymes may carry out the reactions inside microbial cells, but most often they are excreted by the microbes and work extracellularly in the soil solution or while adsorbed to colloidal surfaces. This enzymatic process termed **mineralization** (Figure 12.1) may be indicated as follows, using an amino compound ($R-NH_2$) as an example of the organic nitrogen source:

Effects of field conditions: http://muextension.missouri .edu/explore/envqual/ wq0260.htm

$$R-NH_2 \underset{-2H_2O}{\overset{+2H_2O}{\rightleftharpoons}} OH^- + R-OH + NH_4^+ + \underset{-O_2}{\overset{+O_2}{\rightleftharpoons}} 4H^+ + energy + NO_2^- \underset{-\frac{1}{2}O_2}{\overset{+\frac{1}{2}O_2}{\rightleftharpoons}} energy + NO_3^- \quad (12.1)$$

$\xrightarrow{\qquad\qquad\text{Mineralization}\qquad\qquad}$ (above)

$\xleftarrow{\qquad\qquad\text{Immobilization}\qquad\qquad}$ (below)

Typically, only about 1.5 to 3.5% of the organic nitrogen of a soil mineralizes annually (Box 12.1). Even so, this rate of mineralization provides sufficient mineral nitrogen for normal growth of natural vegetation in most soils excepting those with low organic matter, such as the soils of deserts and sandy areas. Furthermore, *isotope tracer studies* of farm soils that have been amended with synthetic nitrogen fertilizers show that mineralization of soil nitrogen provides a major part of the nitrogen taken up by crops.

The opposite of mineralization is **immobilization**, the conversion of inorganic nitrogen ions (NO_3^- and NH_4^+) into organic forms (see Equation 12.1 and Figure 12.1). Immobilization can take place by both biological and nonbiological (abiotic) processes. The latter probably involves rapid chemical reactions with high C/N ratio soil organic matter and can be quite important in forested soils. Biological immobilization occurs when microorganisms decomposing organic residues require more N than they can obtain from the residues they are metabolizing. The microorganisms then scavenge NO_3^- and NH_4^+ ions from the soil solution to incorporate into such cellular components as proteins, leaving the soil solution essentially devoid of mineral N (see also Section 11.3). When the organisms die, some of the organic nitrogen in their cells may be converted into forms that make up the humus complex, and some may undergo mineralization to NO_3^- and NH_4^+ ions. Mineralization and immobilization occur simultaneously in the soil; whether the *net* effect is an increase or a decrease in the mineral N available depends primarily on the C/N ratio in the organic residues undergoing decomposition (see Section 11.3).

BOX 12.1
CALCULATION OF NITROGEN MINERALIZATION

If the climate, organic matter content, management practices, and texture of a soil are known, it is possible to make a rough estimate of the amount of N likely to be mineralized each year. To do so, we make assumptions such as the following:

- The soil organic matter (SOM) concentration (percentage) may range from close to zero to over 75% (in a Histosol) (see Section 3.9). Values between 0.5 and 5% are most common. ☛ Assume a value of 2.5% (2.5 kg SOM/100 kg soil) for our example.

- Most N used by plants is likely to come from the upper horizon. If this is 15 cm deep, 2×10^6 kg/ha is a reasonable estimate of its weight per hectare. See Section 4.7 to calculate the weight of this horizon if bulk density of a soil is known. ☛ Assume 2×10^6 kg soil/ha 15 cm deep in our example.

- Assume the concentration of nitrogen in the SOM (see Section 11.3) to be about 5 kg N/100 kg SOM.

- The percentage of SOM likely to be mineralized in one year for a given soil depends upon the soil texture, climate, and management practices. Values of around 2% are typical for a fine-textured soil, while values of around 3.5% are typical for coarse-textured soils. Slightly higher values occur in warm climates; slightly lower values occur in cool climates. ☛ Assume a value of 2.5 kg SOM mineralized/100 kg SOM for our example.

The amount of nitrogen likely to be released by mineralization during a typical growing season may be calculated using the sample values assumed above:

$$\frac{\text{kg N mineralized}}{\text{ha 15 cm deep}} = \left(\frac{2.5 \text{ kg SOM}}{100 \text{ kg soil}}\right)\left(\frac{2 \times 10^6 \text{ kg soil}}{\text{ha 15 cm deep}}\right)\left(\frac{5 \text{ kg N}}{100 \text{ kg SOM}}\right)\left(\frac{2.5 \text{ kg SOM mineralized}}{100 \text{ kg SOM}}\right) \quad (12.2)$$

$$\frac{\text{kg N mineralized}}{\text{ha}} = \left(\frac{2.5}{100}\right)\left(\frac{2 \times 10^6}{1}\right)\left(\frac{5}{100}\right)\left(\frac{2.5}{100}\right) = 62.5 \text{ kg N/ha}$$

Most N mineralization occurs during the growing season when the soil is relatively moist and warm. Contributions from the deeper layers of this soil might be expected to bring total N mineralized in the root zone of this soil during a growing season to over 120 kg N/ha.

These calculations estimate the N mineralized annually from the native soil organic matter. Animal manures, legume residues, or other nitrogen-rich organic soil amendments would mineralize much more rapidly than the native soil organic matter and thus would substantially increase the amount of mineral N available in the soil.

Soluble Organic Nitrogen (SON)[2]

Soluble organic nitrogen (SON) compounds are subject to plant uptake and leaching losses in both natural and agriculturalecosystems. They account for about 0.3 to 1.5% of the total organic nitrogen in soils, a pool size similar to that of mineral nitrogen (NH_4^+ and NO_3^-). In fact, where organic manures have been applied to arable soils or where permanent grassland has been grown for many years, SON contents are often considerably higher than those of mineral nitrogen. Microbial uptake of both SON and mineral N take place concurrently in soils, giving rise in some soils to direct competition between plants and microbes for both forms of nitrogen.

Soluble organic nitrogen is also a significant component of the N lost by leaching. For example, SON may comprise nearly all the N leached from some pristine forests and typically 30 to 60% of that leached from dairy farms and beef feedlots. In fact, SON comprises about 25% of the N carried by the Mississippi River into the Gulf of Mexico. Thus, the SON likely contributes to the environmental problems downstream and should be studied along with nitrate N to understand and solve nitrogen-pollution problems.

---

[2]For reviews of the nature, significance, and analyses of soluble organic nitrogen, see van Kessel et al. (2009). For SON in natural ecosystems, see Neff et al. (2003).

The chemical constituents of SON have not been fully identified. However, we know that some of the compounds are hydrophilic and that others are hydrophobic. This suggests some may be able to interact with inorganic colloids, but others would react primarily with the soil organic matter.

Plant Absorption of SON   In nitrogen-limited ecosystems, such as those in strongly acidic and infertile soils (including some organic soils), SON may be the primary source of N for plants. This helps explain the fact that plant growth, particularly of some forest species, is considerably greater than one would expect based on the limited supply of inorganic nitrogen available at any one time. SON may be taken up directly by plant roots, or it may be assimilated through mycorrhizal associations.

Ammonium Fixation by Clay Minerals

Like other positively charged ions, ammonium ions are attracted to the negatively charged surfaces of clay and humus, where they are held in exchangeable form, available for plant uptake, but partially protected from leaching. However, because of the particular size of the ammonium ion (and potassium also), it can become entrapped within cavities in the crystal structure of several 2:1-type clay minerals, especially vermiculites (see Figures 8.7 and 12.32). Ammonium and potassium ions *fixed* in the rigid part of a crystal structure are held in a nonexchangeable form, from which they are released only slowly.

In soils with considerable 2:1 clay content, interlayer-fixed NH_4^+ typically accounts for 5 to 10% of the total nitrogen in the surface soil and up to 20 to 40% of the nitrogen in the subsoil. In highly weathered soils, on the other hand, ammonium fixation is minor because little 2:1 clay is present. In some forest soils, about half the nitrogen in the O and A horizons is immobilized by either ammonium fixation or chemical reactions with humus. While ammonium fixation may be considered an advantage because it provides a means of conserving nitrogen, the rate of release of the fixed ammonium is often too slow to be of much practical value in fulfilling the needs of fast-growing annual plants.

Ammonia Volatilization

Ammonia gas (NH_3) can be produced from the breakdown of organic materials and from such fertilizers as anhydrous ammonia and urea. The ammonia gas is in equilibrium with dissolved ammonium ions according to the following reversible reaction:

$$NH_4^+ + OH^- \rightleftharpoons H_2O + NH_3 \uparrow \qquad (12.3)$$

$$\underset{\text{Dissolved ions}}{} \qquad \underset{\text{Gas}}{}$$

From Reaction 12.3, we can draw two conclusions. First, ammonia volatilization will be more pronounced at high pH levels (i.e., OH^- ions drive the reaction to the right); second, ammonia-gas-producing amendments will drive the reaction to the left, raising the pH of the solution in which they are dissolved.

Soil colloids, both clay and humus, adsorb ammonia gas, so ammonia losses are greatest in soils with little of these colloids or where the ammonia is not in close contact with the soil. For these reasons, ammonia losses can be quite large from sandy soils and from alkaline or calcareous soils, especially when the ammonia-producing materials are left at or near the soil surface and when the soil is drying out. High temperatures, as often occur on the surface of the soil, also favor the volatilization of ammonia (Figure 12.2).

Figure 12.2

Ammonia volatilization is markedly affected by temperature and pH. Here, urea fertilizer (NH₂–CO–NH₂) was applied to a silt loam soil surface. Urea absorbs moisture from the air or soil and then hydrolyzes to form ammonia. The loss of ammonia gas is especially rapid when pH exceeds 7 and temperature exceeds 16 °C. Ammonia loss can be even faster from animal feces (manure) than from urea. Ammonia-forming amendments should not be left on the surface of warm, high pH soil for more than a day. [Redrawn from Glibert et al. (2006) using data in Franzen (2004)]

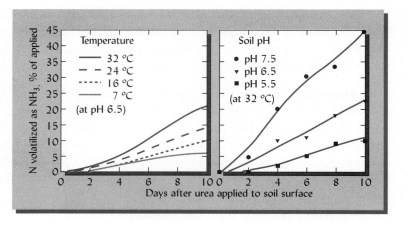

Incorporation of manure and fertilizers into the top few centimeters of soil can reduce ammonia losses by 25 to 75% from those that occur when the materials are left on the soil surface. In natural grasslands and pastures, incorporation of animal wastes by earthworms and dung beetles is critical in maintaining a favorable nitrogen balance and a high animal-carrying capacity in these ecosystems (see Section 10.2).

Wetlands (including rice paddies) can lose much ammonia, especially on warm days when photosynthesizing algae use up all the dissolved CO_2, eliminating carbonic acid from the surface water and thus greatly elevating the pH level.

By the reverse of the ammonium loss mechanism just described, both soils and plants can absorb ammonia from the atmosphere. Thus the soil–plant system can help cleanse ammonia from the air, while deriving usable nitrogen for plants and soil microbes.

Nitrification

Ammonium ions in the soil may be enzymatically oxidized by certain soil bacteria, yielding first nitrites and then nitrates. These bacteria are classed as **autotrophs** because they obtain their energy from oxidizing the ammonium ions rather than organic matter. The process termed **nitrification** consists of two main sequential steps. The first step results in the conversion of ammonium to nitrite by a specific group of autotrophic bacteria (***Nitrosomonas***). The nitrite so formed is then immediately acted upon by a second group of autotrophs, ***Nitrobacter***. Therefore, when NH_4^+ is released into the soil, it is usually converted rapidly into NO_3^- (see Figure 12.3). The enzymatic oxidation releases energy for these bacteria as follows:

$$NH_4^+ + 1\tfrac{1}{2}O_2 \xrightarrow[\text{bacteria}]{\text{Nitrosomonas}} NO_2^- + 2H^+ + H_2O + 275 \text{ kJ energy} \qquad (12.4)$$
Ammonium   Nitrite

$$NO_2^- + \tfrac{1}{2}O_2 \xrightarrow[\text{bacteria}]{\text{Nitrobacter}} NO_3^- + 76 \text{ kJ energy} \qquad (12.5)$$
Nitrite   Nitrate

So long as conditions are favorable for both reactions, the second follows the first closely enough to prevent accumulation of much nitrite. This is fortunate, because even at concentrations of just a few mg/kg, nitrite is quite toxic to most plants. When oxygen supplies are marginal, the nitrifying bacteria may also produce some NO and N_2O, which are potent greenhouse gases.

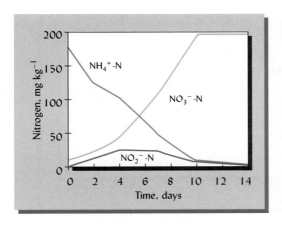

Figure 12.3

Transformation of ammonium into nitrite and nitrate by nitrification. On day zero, the silt loam soil was amended with enough $(NH_4)_2SO_4$ to supply 170 mg of N/kg soil. It then underwent a warm, well-aerated incubation for 14 days. Every second day, soil samples were extracted and analyzed for various forms of nitrogen. Note that the increase in nitrate-N (NO_3^--N) almost mirrored the decline in ammonium-N (NH_4^+-N), except for the small amount of nitrite-N (NO_2^--N) that accumulated temporarily between days 2 and 10. This pattern is consistent with the two-step process depicted by Equations 12.4 and 12.5. No plants were grown during the study. [Data selected from Khalil et al. (2004)]

Regardless of the source of ammonium, nitrification will significantly increase soil acidity by producing H$^+$ ions, as shown in Reaction 12.4. See also Section 9.6.

Nitrification can be "reversed" by several bacterial processes, the best known of which is **denitrification**, an anaerobic process by which heterotrophic bacteria reduce nitrate to such gases as NO, N_2O, and N_2 (see below). **Dissimilatory nitrate reduction to ammonium (DNRA)** is another anaerobic microbial process that in effect reverses nitrification; it reduces NO_3^- to NO_2^- and then to NH_4^+.

Soil Conditions Affecting Nitrification   The nitrifying bacteria are much more sensitive to environmental conditions than are the broad groups of heterotrophic organisms responsible for the release of ammonium from organic nitrogen compounds (**ammonification**). Nitrification requires a supply of ammonium ions and oxygen to make NO_2^- and NO_3^- ions and is therefore favored in well-drained soils. The optimum moisture for nitrification is about 60% of the pore space filled with water (Figure 12.4). Since nitrifiers are autotrophs, their carbon sources are bicarbonates and CO_2. They perform best between 20 and 30 °C and perform very slowly if the soil is cold (below 5 °C).

Chemicals have been found that inhibit the nitrification process. Such compounds, as well as others that retard the dissolution of urea, can be blended with fertilizers to retard the formation of nitrate and thereby reduce its loss by leaching or denitrification.

Irrigation of an initially dry soil, the first rains after a long dry season, the thawing and rapid warming of frozen soils in spring, and sudden aeration by tillage are

Nitrification inhibitors:
www.extension.iastate.edu/
Publications/NCH55.pdf

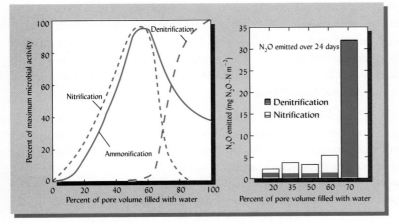

Figure 12.4

(Left) Rates of nitrification, ammonification, and denitrification are closely related to the availability of oxygen and water as depicted by percentage of water-filled pore space. Both nitrification and ammonification proceed at their maximal rates near 55 to 60% water-filled pore space; however, ammonification proceeds in soils too waterlogged for active nitrification. Only a small overlap exists in the conditions suitable for nitrification and denitrification. (Right) The greenhouse gas, nitrous oxide (N_2O), is mainly produced by denitrification, but it is also a minor by-product of nitrification. An experiment that used ^{15}N tracers shows the abrupt shift from one process to the other at water-filled porosities between 60 and 70%. [Bar graph from Bateman and Baggs (2005)]

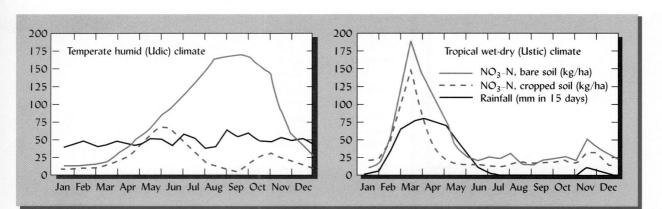

Figure 12.5

Seasonal patterns of nitrate-N concentration in representative surface soils with and without growing plants. (Left) In a representative humid temperate region with cool winters and rainfall rather uniformly distributed throughout the year, NO_3^--N accumulates as the soil warms up in May and June. The nitrates are lost by leaching in the fall. (Right) In a representative tropical region with four rainy months followed by eight months of dry, hot weather, a large flush of NO_3^--N appears when the rains first moisten the dry soil. This nitrate flush is caused by the rapid decomposition and mineralization of the dead cells of microorganisms previously killed by the dry, hot conditions. Note that soil nitrate is lower in both climates when plants are grown, because much of the nitrate formed is removed by plant uptake. (Diagrams courtesy of R. Weil)

examples of environmental fluctuations that typically cause a flush of mineralization and nitrification (Figure 12.5). The growth patterns of natural vegetation and the optimum planting dates for crops are greatly influenced by resulting seasonal fluctuations in nitrate levels.

The Nitrate Leaching Problem

Nitrates and water quality:
www.soil.ncsu.edu/
publications/Soilfacts/
AG-439-02/

In contrast to ammonium cations, nitrate anions are not adsorbed by the negatively charged colloids that dominate most soils. Therefore, nitrate ions move downward freely with drainage water and are readily leached from the soil. The loss of nitrogen in this manner involves three basic concerns: (1) waste that impoverishes the ecosystem; (2) acidification of the soils and the co-leaching of such cations as Ca^{2+}, Mg^{2+}, and K^+ (as described in Section 9.6); and (3) the movement of nitrate to groundwater causes several serious water-quality problems downstream. We will examine the nature of these environmental impacts in Section 13.2.

Nitrate pollution threatens amphibians:
www.on.ec.gc.ca/wildlife/
factsheets/nitrate-e.html

Much of the nitrate mineralized in certain highly weathered, tropical Oxisols and Ultisols leaches below the root zone before annual crops such as corn can take it up. Soil scientists in Africa recently discovered that some of this leached nitrate is not lost to groundwater. Instead, the highly weathered and acid clays deep in the subsoil adsorb the nitrate on their anion exchange sites (see Section 8.11). Deep-rooted trees such as *Sesbania* are capable of taking up this subsoil nitrate. If grown in rotation with annual food crops, the trees subsequently enrich the surface soil when they shed their leaves, making this pool of once-leached nitrogen available again for food production. Such **agroforestry** practices as this have the potential to improve both crop production and environmental quality in the humid tropics.

Water-Quality Impacts   Water-quality problems caused by nitrogen are mainly associated with the movement of nitrate with drainage waters to the groundwater. The nitrate may contaminate drinking water causing health hazards (see Section 13.2) for people (especially babies) as well as livestock. The nitrates may also eventually flow underground to surface waters, such as streams, lakes, and estuaries. The key factor for health hazards is *concentration* of nitrate in the drinking water and the level of exposure (amount of water ingested, especially over long periods). Even more widespread are the damages to water quality and to the health of aquatic ecosystems, especially those with salty or brackish water. The key factor for this kind of damage is often the *total load* (mass flux) of nitrogen delivered to the sensitive ecosystem. Most of the total N load is usually delivered by nitrate and dissolved organic N in drainage waters.

Gaseous Losses by Denitrification

Nitrogen may be lost to the atmosphere when nitrate ions are converted to gaseous forms of nitrogen by a series of widely occurring biochemical reduction reactions termed **denitrification.**[3] The organisms that carry out this process are commonly present in large numbers and are mostly facultative anaerobic bacterial *heterotrophs*, which obtain their energy and carbon from the oxidation of organic compounds. Other denitrifying bacteria are *autotrophs*, such as *Thiobacillus denitrificans*, which obtain their energy from the oxidation of sulfide. The general series of reductions can be shown as:

$$2NO_3^- \xrightarrow{-2O} 2NO_2^- \xrightarrow{-2O} 2NO\uparrow \xrightarrow{-O} N_2O\uparrow \xrightarrow{-O} N_2\uparrow \qquad (12.6)$$

| Nitrate ions | Nitrite ions | Nitric oxide gas | Nitrous oxide gas | Dinitrogen gas | |
| (+5) | (+3) | (+2) | (+1) | (0) | ← Valence state of nitrogen |

For these reactions to take place, sources of organic residues or sulfides should be available to provide the energy the denitrifiers need. The soil air in the microsites where denitrification occurs should contain no more than 10% oxygen, and lower levels of oxygen are preferred. Optimum temperatures for denitrification are from 25 to 35 °C, but the process will occur between 2 and 50 °C. Very strong acidity (pH < 5.0) inhibits rapid denitrification and favors the formation of N_2O.

Generally, when oxygen levels are very low, the end product released from the overall denitrification process is dinitrogen gas (N_2). It should be noted, however, that NO and N_2O are commonly also released during denitrification under the fluctuating aeration conditions that often occur in the field. The proportion of the three main gaseous products seems to be dependent on the prevalent pH, temperature, degree of oxygen depletion, and concentration of nitrate and nitrite ions available.

Atmospheric Pollution   The question of how much of each nitrogen-containing gas is produced is not merely of academic interest. Dinitrogen gas is quite inert and environmentally harmless, but the oxides of nitrogen are very reactive gases and have the potential to do serious environmental damage in at least four ways. First, NO and N_2O released into the atmosphere by denitrification can contribute to the formation of nitric acid, one of the principal components of acid rain. Second, the nitrogen oxide gases can react with volatile organic pollutants to form ground-level ozone, a major air pollutant in the photochemical smog that plagues many urban areas. Third, when NO rises into the upper atmosphere, it contributes to the greenhouse effect (as much as 300 times that of an equal amount of CO_2) by absorbing infrared radiation that would otherwise escape into space (see Section 11.9). Finally, as N_2O moves up into the stratosphere, it participates in reactions that result in the destruction of ozone (O_3), a gas that helps shield the Earth from cancer-causing ultraviolet solar radiation. As this protective layer is further degraded, thousands of additional cases of skin cancer are likely to occur annually. While there are other important sources of N_2O, such as automobile exhaust fumes, a major contribution to the problem is being made by denitrification in soils, especially in rice paddies, wetlands, and heavily fertilized or manured agricultural soils.

N and greenhouse gases from row crops: www.oznet.ksu.edu/ctec/ CASMGSnewsletter/ Jan04-1.htm

---

[3]Nitrate can also be reduced to nitrite and to nitrous oxide gas by nonbiological chemical reactions and under some circumstances by bacteria carrying out nitrification, but these reactions are quite minor in comparison with biological denitrification. A recently discovered bacterial process, the anaerobic oxidation of ammonium (**anammox**), converts ammonium and nitrate to N_2 gas. It is widespread in ocean sediments and may be important in hydric soils as well.

Quantity of Nitrogen Lost Through Denitrification   During periods of adequate soil moisture, denitrification results in a slow but relatively steady loss of N from undisturbed forest systems. In contrast, losses from agricultural soils are highly variable in both time and space. Most of the annual N loss often occurs during just a few days in summer, when heavy rain has temporarily caused the warm soils to become poorly aerated. Low-lying, organic-rich areas and other hot spots may lose nitrogen 10 times as fast as the average rate for a typical field. Well-drained, humid-region soils rarely lose more than 5 to 15 kg N/ha annually by denitrification, but where drainage is restricted and where large amounts of nitrogen fertilizer or N-rich organic materials are applied, losses of 30 to 60 kg N/ha/yr of nitrogen have been observed (Plate 94).

Denitrification in Flooded Soils   In flooded soils, such as those found in natural wetlands or rice paddies, losses by denitrification may be very high, especially where soils are subject to alternate periods of wetting and drying. Nitrates that are produced by nitrification during the dry periods are subject to denitrification when the soils are submerged. Even in permanently submerged soils, both reactions take place at once— nitrification at the soil–water interface where some oxygen is present and denitrification at lower soil depths (see Figure 12.6). However, nitrogen losses can be dramatically reduced by keeping the soil flooded and by deep placement of the fertilizer into the reduced zone of the soil. In this zone, because there is insufficient oxygen to allow nitrification to proceed, nitrogen remains in the ammonium form and is not susceptible to loss by denitrification.

Tidal wetlands, which become alternately anaerobic and aerated as the water level rises and falls, have particularly high potentials for converting nitrogen to gaseous forms. Often the resulting rapid loss of nitrogen is considered to be a beneficial function of wetlands, in that the process protects estuaries and lakes from the eutrophying effects of too much nitrogen. In fact, wastewater high in organic carbon and nitrogen can be cleaned up quite efficiently by allowing it to flow slowly over a

Figure 12.6

Nitrification–denitrification reactions and kinetics of the related processes controlling nitrogen loss from the aerobic–anaerobic layers of a flooded soil system. Nitrates, which form in the thin aerobic soil layer just below the soil–water interface, diffuse into the anaerobic (reduced) soil layer below and are denitrified to the N_2 and N_2O gaseous forms, which are lost to the atmosphere. Placing the urea or ammonium-containing fertilizers deep in the anaerobic layer prevents oxidation of ammonium-N to nitrate-N, thereby greatly reducing N loss. [Modified from Patrick (1982)]

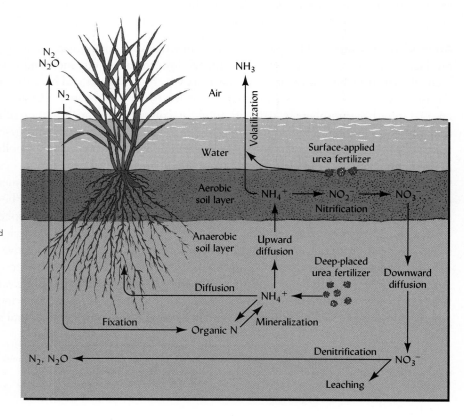

specially designed water-saturated soil system in a process known as *overland flow wastewater treatment*.

Denitrification in Groundwater   Many studies have documented the significance of denitrification taking place in the poorly drained soils under **riparian** vegetation (mainly woodlands adjacent to streams). Contaminated groundwater may lose most of its nitrate load as it flows through the riparian zone on its way to the stream. Most of nitrate is believed to be lost by denitrification, stimulated by organic compounds leached from the decomposing forest litter and by the anaerobic conditions that prevail in the wet riparian zone soils.

Constructed wetlands can be used to reduce the nitrate content of surface waters moving toward streams. When coupled with buffer strips, such wetlands can remove half or more of the nitrates of the surface water before it enters the stream channel.

We have just discussed a number of biological processes that lead to losses of nitrogen from the soil system. We turn next to the principal biological process by which soil nitrogen is replenished.

Biological Nitrogen Fixation

Biological nitrogen fixation converts the inert dinitrogen gas of the atmosphere (N_2) to reactive nitrogen that becomes available to all forms of life through the nitrogen cycle. The process is carried out by a limited number of bacteria, including several species of *Rhizobium*, actinomycetes, and cyanobacteria (formerly termed blue-green algae). Globally, enormous amounts of nitrogen are fixed biologically each year. Terrestrial systems alone fix an estimated 139 million Mg. However, the amount that is fixed in the manufacture of fertilizers is now nearly as great (see Figure 12.7).

Regardless of the organisms involved, the key to biological nitrogen fixation is the enzyme *nitrogenase*, which catalyzes the reduction of dinitrogen gas to ammonia:

$$N_2 + 8H^+ + 6e^- \xrightarrow[\text{(Fe,Mo)}]{\text{(Nitrogenase)}} 2NH_3 + H_2 \qquad (12.7)$$

The ammonia, in turn, is combined with organic acids to form amino acids and, ultimately, proteins.

Breaking the $N \equiv N$ triple bond in N_2 gas requires a great deal of energy. Therefore, the process is greatly enhanced by association with higher plants, which can supply this energy from photosynthesis. Nitrogen-fixing organisms have a relatively high requirement for molybdenum, iron, phosphorus, and sulfur, because these

Nitrogen fixation links: http://academic.reed.edu/biology/Nitrogen/Nfix1.html

Lesson on N fixation: www.soils.umn.edu/academics/classes/soil2125/doc/s9chap2.htm

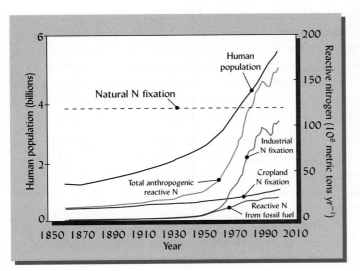

Figure 12.7

Changes in the human population and its contribution to global reactive nitrogen. Note that by the early 1980s human-caused N fixation (industrial fertilizer production, agricultural crop legumes, and combustion of fossil fuels) had surpassed natural N fixation (by legumes, cyanobacteria, and actinomycetes in natural terrestrial ecosystems as well as by lightning). [Modified from Lambert and Driscoll (2003), based on data in Galloway and Cowling (2002), with permission from Hubbard Brook Research Foundation]

Table 12.1
INFORMATION ON DIFFERENT SYSTEMS OF BIOLOGICAL NITROGEN FIXATION

N-fixing systems	Organisms involved	Plants involved	Site of fixation
Symbiotic			
Obligatory			
Legumes	Bacteria *Rhizobia* and *Bradyrhizobia*	Legumes	Root nodules
Nonlegumes (angiosperms)	Actinomycetes (*Frankia*)	Nonlegumes (angiosperms)	Root nodules
Associative			
Morphological involvement, internal	Cyanobacteria, bacteria	Various higher plants and microorganisms	Leaf and root nodules, lichens
Nonmorphological involvement, external	Cyanobacteria, bacteria	Various higher plants and microorganisms	Rhizosphere (root environment) Phyllosphere (leaf environment)
Nonmorphological involvement, internal	Endophytic diazotrophic bacteria	Tropical grasses, etc.	Plant root cells
Nonsymbiotic	Cyanobacteria, bacteria	Not involved with plants	Soil, water independent of plants

nutrients are either part of the nitrogenase molecule or are needed for its synthesis and use. Nitrogenase is destroyed by free O_2, so organisms that fix nitrogen must protect the enzyme from exposure to oxygen. When nitrogen fixation takes place in root nodules (see below), one means of protecting the enzyme from free oxygen is the formation of *leghemoglobin*. This compound, which gives active legume nodules a red interior color (Plate 103), is virtually the same molecule as the hemoglobin that gives human blood its red color when oxygenated. Leghemoglobin binds oxygen in such a way as to protect the nitrogenase while making oxygen available for respiration in other parts of the nodule tissue.

Biological nitrogen fixation occurs through a number of microbial systems that may or may not be directly or indirectly associated with higher plants (Table 12.1). Although the legume–bacteria symbiotic systems have received the most attention, recent findings suggest that the other systems involve many more families of plants worldwide and may even rival the legume-associated systems as suppliers of biological nitrogen to the soil. Each major system will be discussed briefly.

Symbiotic Fixation with Legumes

The process with legumes: http://academic.reed.edu/biology/Nitrogen/Nfix1(legumes).html

The **symbiosis** (mutually beneficial relationship) of plants in the legume family and bacteria of the genera *Rhizobium* and *Bradyrhizobium* provide the major biological source of fixed nitrogen in agricultural soils. These organisms infect the root hairs and the cortical cells, ultimately inducing the formation of **root nodules** that serve as the site of nitrogen fixation (Figure 12.8 and Plate 103). In this mutually beneficial association, the host plant supplies the bacteria with carbohydrates for energy, and the bacteria reciprocate by supplying the plant with reactive-nitrogen compounds (Plate 51).

Organisms Involved A given *Rhizobium* or *Bradyrhizobium* species will infect some legumes but not others. Crop legumes that can be inoculated by a given *Rhizobium* species are included in the same cross-inoculation group (see Table 12.2). In areas where a given legume has been grown for several years, the appropriate species of *Rhizobium* is probably present in the soil. However, if the natural *Rhizobium* population in the soil is too low, special mixtures of the appropriate *Rhizobium* and *Bradyrhizobium* inoculant may be applied, either by coating the legume seeds or by applying the inoculant directly to the soil. Effective and competitive strains of

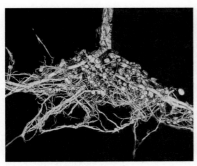  

Figure 12.8

Photos illustrating soybean nodules. (Left) the nodules are seen on the roots of the soybean plant, and a closeup (center) shows a few of the nodules associated with the roots. A scanning electron micrograph (right) shows a single plant cell within the nodule stuffed with the bacterium Bradyrhizobium japonicum. *(Courtesy of W. J. Brill, University of Wisconsin)*

Rhizobium, which are available commercially, often give significant yield increases, but only if used on the proper crops.

The legume–Rhizobium associations generally function best on soils that are not too acid (although *Bradyrhizobium* associations generally can tolerate considerable acidity) and that are well supplied with essential nutrients. However, high levels of available nitrogen, whether from the soil or added in fertilizers, tend to depress biological nitrogen fixation, as plants make the heavy energy investment required for symbiotic nitrogen fixation only when short supplies of mineral nitrogen make it necessary.

The amount of N fixed varies widely with soil conditions and among bacterial strains and legume species (Table 12.3). Such legumes as alfalfa and soybean can fix more than 150 kg N per ha, usually all the N needed for maximum plant production. Other legumes such as phaseolus beans normally fix only 30 to 50 kg N per ha and may need supplemental N applications to achieve optimal yields. The capacity to fix

Table 12.2
RHIZOBIA BACTERIA AND ASSOCIATED LEGUME CROSS-INOCULATION GROUPS

Genus	Bacteria Species/subgroup	Host legume
Rhizobium	R. leguminosarum	
	bv. *viceae*	*Vicia* (vetch), *Pisum* (peas), *Lens* (lentils), *Lathyrus* (sweet pea)
	bv. *trifolii*	*Trifolium* spp. (most clovers)
	bv. *phaseoli*	*Phaseolus* spp. (dry bean, runner bean, etc.)
	R. Meliloti	*Melilotus* (sweet clover, etc.), *Medicago* (alfalfa), *Trigonella* (fenugreek)
	R. loti	*Lotus* (trefoils), *Lupinus* (lupins), *Cicer* (chickpea), *Anthyllis*, *Leucaena*, and many other tropical trees
	R. Fredii	*Glycine* spp. (e.g., soybean)
Bradyrhizobium	B. japonicum	*Glycine* spp. (e.g., soybean)
	B. sp.	*Vigna* (cowpeas), *Arachis* (peanut), *Cajanus* (pigeon pea), *Pueraria* (kudzu), *Crotolaria* (crotolaria), and many other tropical legumes
Azhorhizobium	A. sp	*Sesbania* trees

Table 12.3
TYPICAL LEVELS OF NITROGEN FIXATION FROM DIFFERENT SYSTEMS

Crop or plant	Associated organism	Typical fixation, kg N/ha/yr
Symbiotic		
Legumes (nodulated)		
Ipil-ipil tree (*Leucaena leucocephala*)	Bacteria (*Rhizobium*)	100–500
Locust tree (*Robina* spp.)		75–200
Alfalfa (*Medicago sativa*)		150–250
Vetch (*Vicia vileosa*)		50–150
Bean (*Phaseolus vulgaris*)		30–50
Cowpea (*Vigna unguiculata*)	Bacteria (*Bradyrhizobium*)	50–100
Soybean (*Glycine max L.*)		50–200
Pigeon pea (*Cajunus*)		150–280
Kudzu (*Pueraria*)		100–140
Nonlegumes (nodulated)		
Alders (Alnus)	Actinomycetes (*Frankia*)	50–150
Species of Gunnera	Cyanobacteria (*Nostoc*)	10–20
Nonlegumes (nonnodulated)		
Bahia grass (*Paspalum notatum*)	Bacteria (*Azobacter*)	5–30
Azolla	Cyanobacteria (*Anabena*)	150–300
Sugarcane	Bacteria (endophytic)	30–80
Nonsymbiotic	Bacteria (*Azobacter, Clostridium*)	5–20
	Cyanobacteria (various)	10–50

N from the air gives legumes a competitive advantage over nonleguminous plants in N poor soils. The presence of N fixing legume species in crop rotations, meadow plant communities, or forests generally enriches the soil with N, especially where the soil was initially low in N and the legumes are strong N fixers. Legumes can thus improve the growth of other, non-N-fixing plant species growing together or in sequence with the legumes. Such effects can be seen in pine forests with a locust tree understory or in hayfields where grasses are mixed with clovers. The N fixed by the legumes is generally made available to other plants when the legume roots and their nodules die and these N-rich tissues decay. However, some more direct transfers to nonlegumes can also take place through mycorrhizal connections (see Section 10.8). The presence of legumes, therefore, hastens the natural reforestation of degraded landscapes (see Plate 101) and reduces the requirement for N fertilizers in crop rotations. Soil enrichment in N can be maximized by incorporating the entire plant biomass of a leguminous **green manure**, a crop grown for the expressed purpose of improving the soil with its root and shoot residues. On the other hand, if most of the N fixed is removed from a site by harvesting the entire shoot (e.g., hay) or the protein-rich seed (e.g., soybean, peanut), little N may remain to enrich the soil.

Symbiotic Fixation with Nodule-Forming Nonlegumes

Nearly 200 species from more than a dozen genera of nonlegume plants are known to develop nodules and to accommodate symbiotic nitrogen fixation. Included are several important groups of angiosperms. These plants, which are present in certain forested areas and wetlands, form distinctive nodules when their root hairs are invaded by soil actinomycetes of the genus *Frankia*.

The rates of nitrogen fixation per hectare compare favorably with those of the legume–*Rhizobium* associations (see Table 12.3). On a worldwide basis, the total nitrogen fixed in this way may even exceed that fixed by agricultural legumes. Because of their nitrogen-fixing ability, certain of the tree–actinomycete associations are able to colonize infertile soils and newly forming soils on disturbed lands, which may have extremely low fertility as well as other conditions that limit plant growth.

Symbiotic Nitrogen Fixation Without Nodules

Among the most significant nonnodule nitrogen-fixing systems are those involving cyanobacteria, such as the *Azolla–Anabaena* complex, which flourishes in certain tropical rice paddies. The *Anabaena* cyanobacteria inhabit cavities in the leaves of the floating fern *Azolla* and fix quantities of nitrogen comparable to those of the *Rhizobium*–legume complexes (see Table 12.3).

While nitrogen fixation by the legume–*rhizobium* association has been studied and managed for more than a century, scientists are only beginning to learn about non-nodule-forming nitrogen fixing bacteria that live inside certain, mainly tropical, plants. Two important tropical grasses (family Poacaea), sugarcane and rice, are known to benefit greatly from N fixed by such **endophytic** (*inside plants*) **diazotrophic bacteria**. The bacteria live inside or in between root cells, forming associations that are inhibited by high levels of N fertilizer use. This may be one of the reasons they are abundant in some soils but not in others. Their activity is also apparently controlled by plant genetics even within a particular plant species. This source of N is thought to account for the ability of sugarcane in Brazil to produce high yields with very little N fertilizer as compared to that grown in many other countries.

A more widespread but less intense nitrogen-fixing phenomenon is that which occurs in the *rhizosphere* of certain grasses and other nonlegume plants. The organisms responsible are bacteria, especially those of the *Spirillum* and *Azotobacter* genera (see Table 12.3). Plant root exudates supply these microorganisms with energy for their nitrogen-fixing activities.

Nonsymbiotic Nitrogen Fixation[4]

Certain free-living microorganisms present in soils and water are able to fix nitrogen without being directly associated with higher plants. The transformation is referred to as *nonsymbiotic* or *free-living*.

Fixation by Heterotrophs   Several different groups of bacteria and cyanobacteria are able to fix nitrogen nonsymbiotically. In upland mineral soils, the major fixation is brought about by species of several genera of heterotrophic aerobic bacteria, *Azotobacter* and *Azospirillum* (in temperate zones) and *Beijerinckia* (in tropical soils). Certain anaerobic bacteria of the genus *Clostridium* are also active in fixing nitrogen.

The amount of N fixed by such heterotrophs is typically only 5 to 20 kg per ha annually, enough to significantly affect natural forests and grassland but not enough to be very important in agricultural systems.

Fixation by Autotrophs   In the presence of light, certain photosynthetic bacteria and cyanobacteria are able to fix carbon dioxide and nitrogen simultaneously. The contribution of cyanobacteria is thought to be of some significance, especially in wetlands (including in rice paddies). In some cases, cyanobacteria contribute a major part of the nitrogen needs of rice, but nonsymbiotic species rarely fix more than 20 to 30 kg N/ha/yr. Nitrogen fixation by cyanobacteria in upland soils also occurs, but at much lower levels than found under wetland conditions.

Video and more on *Azotobacter*: www.microbiologybytes.com/video/Azotobacter.html

---

[4]For insights into the exploitation of nonsymbiotic N fixation for agriculture, see Kennedy et al. (2004).

Figure 12.9

Global distribution of reactive nitrogen deposition to land from the atmosphere. Excessive N deposition can damage forests and other natural ecosystems. Highest deposition (darkest shading) occurs in high-rainfall regions downwind of intensive livestock feeding, rice production, and/or industrialized population centers. Ammonia from manure and nitrogen oxides from rice paddies and fossil fuel combustion are principal sources of the N that falls with rain, snow, and dust. Although variable, there is usually more ammonium than nitrate deposited. [From Eickhout et al. (2006)]

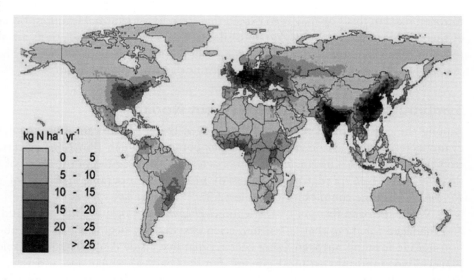

kg N ha^{-1} yr^{-1}

	0 - 5
	5 - 10
	10 - 15
	15 - 20
	20 - 25
	> 25

Nitrogen Deposition from the Atmosphere

The atmosphere contains small quantities of ammonia and nitrogen oxide gases released from soils, plants, and fossil fuel combustion (especially in vehicle engines—see Section 9.6), as well as nitrates formed by lightning strikes. The term *nitrogen deposition* refers to the addition of these atmosphere-borne nitrogen compounds to soil and water (usually after transformation to the NH_4^+ or NO_3^- forms) through rain, snow, dust, and gaseous absorption.

The quantity of N deposited (see Figure 12.9) is greatest in high-rainfall areas downwind from cities and concentrated animal farming areas. As the Earth's growing human population puts ever more reactive nitrogen into circulation, its unwanted deposition is becoming an increasingly serious global environmental problem.

Although the N may stimulate greater plant growth in agricultural systems, the effects on forests, grasslands, and aquatic ecosystems are quite damaging. The nitrates in particular are associated with acidification of rain (Section 9.6), but since the ammonium soon nitrifies, both forms lead to soil acidification. Nitrogen added by deposition also impacts methane oxidation (see Figure 11.20). Methane is an important greenhouse gas affecting climate change, and its removal from the atmosphere by soil oxidation helps maintain its global balance (see also Section 11.9). Forested soils have particularly high rates of methane oxidation, but also may be hardest hit by additions of mineral nitrogen. The methane oxidation capacities of grasslands and croplands are also significantly reduced by input of mineral nitrogen.

12.2   SULFUR AND THE SOIL SYSTEM

Sulfur in Plants and Animals

Role of sulfur:
www.soil.ncsu.edu/
publications/Soilfacts/
AG-439-15-Archived/

Sulfur is a constituent of certain essential amino acids, vitamins, enzymes, and aromatic oils. Among plants, the legume, cabbage, and onion families require especially large amounts of sulfur. Healthy plant foliage generally contains 0.15 to 0.45% sulfur, or approximately one-tenth as much sulfur as nitrogen. Plants deficient in sulfur tend to become spindly and light green or yellow in appearance (Plate 92). However, unlike nitrogen, sulfur is relatively immobile in the plant, so the chlorosis develops first on the youngest rather than oldest leaves as sulfur supplies are depleted. Also, unlike nitrogen-deficient plants, sulfur-deficient plants tend to have low sugar but high nitrate contents in their sap.

Sulfur deficiencies in agricultural plants have become increasingly common because of three simultaneous trends: the cleaning up of sulfur dioxide air pollution, the elimination of most S "impurities" from N-P-K fertilizers, and the greatly increased removals of S by higher-yielding crops.

Sulfur deficiencies are most prevalent where soil parent materials are low in S, where extreme weathering and leaching have removed this element, or where there is little replenishment of sulfur from polluted air. In many tropical countries, one or more of these conditions prevail, and sulfur-deficient areas are common.

Soils of dry savannas are particularly deficient in sulfur as a result of the annual burning that converts most of the S in the plant residues to sulfur dioxide. This S gas is subsequently carried by the wind hundreds of kilometers away to areas covered by rain forest, where some of the sulfur dioxide is absorbed by moist soils and foliage and some is deposited with rainfall.

Sources of Sulfur in Soils

The three major natural sources of sulfur that can become available for plant uptake are (1) *organic matter*, (2) *soil minerals*, and (3) *sulfur gases in the atmosphere*. In natural ecosystems where most of the sulfur taken up by plants is eventually returned to the same soil, these three sources combined are usually sufficient to supply the 5 to 20 Kg S per ha needed by growing plants. These three sources of sulfur will be considered in order.

Organic Matter[5]   Three principal groups of organic sulfur compounds exist in soil organic matter: (1) highly reduced S bonded to carbon in compounds such as sulfides, and thiols including amino acids such as cysteine, cystine, and methionine; (2) sulfoxides and sulfonates in which the sulfur is bonded to carbon but also to oxygen (C—S—O); and (3) highly oxidized S in ester sulfates (C—O—S).

Over time, soil microorganisms break down these organic sulfur compounds into soluble forms analogous to the release of ammonium and nitrate from organic matter, discussed in Section 12.1. As with nitrogen, most A horizon soil sulfur is organic, with only 2 to 10% in the mineralized (sulfate) form, even in the sandy Spodosols.

In dry regions, gypsum $(CaSO_4 \cdot 2H_2O)$, which supplies inorganic sulfur, is often present in the subsurface horizons. Therefore, the proportion of organic sulfur is not likely to be as high in arid- and semiarid-region soils as it is in humid-region soils. This is especially true in the subsoils, where organic sulfur may be scarce and where gypsum is prominent.

Soil Minerals   The inorganic forms of sulfur include the soluble compounds on which plants and microbes depend. The two most common inorganic sulfur compounds are sulfates and sulfides. The sulfate minerals are most easily solubilized, and the sulfate ion (SO_4^{2-}) is easily assimilated by plants. Sulfate minerals are most common in regions of low rainfall, where they accumulate in the lower horizons of some Mollisols and Aridisols (see Figure 12.10). They may also accumulate as neutral salts in the surface horizons of saline soils in arid and semiarid regions.

Sulfides that are found in some humid-region soils with restricted drainage must be oxidized to the sulfate form before the sulfur can be assimilated by plants. When these soils are drained, oxidation can occur, and ample available sulfur is released. In some cases, so much sulfur is oxidized that problems of extreme acidity result (see following).

Sulfur in nature, see sulfur section in:
www.iitap.iastate.edu/gcp/chem/nitro/nitro_lecture.html

---

[5]For a study of organic sulfur forms in a range of soils, see Zhao et al. (2006).

Figure 12.10

The distribution of organic and inorganic sulfur in representative soil profiles of the soil orders Mollisols, Spodosols, and Oxisols. In each, soil organic forms dominate the surface horizon. Considerable inorganic sulfur, both as adsorbed sulfate and calcium sulfate minerals, exists in the lower horizons of Mollisols. Relatively little inorganic sulfur exists in Spodosols. However, the bulk of the profile sulfur in the humid tropics (Oxisols) is present as sulfate adsorbed to colloidal surfaces in the subsoil. (Diagram courtesy of R. Weil)

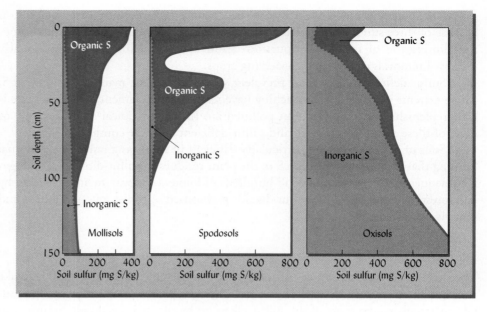

Another mineral source of sulfur is the clay fraction of some soils high in Fe, Al oxides and kaolinite. These clays are able to strongly adsorb sulfate from soil solution and subsequently release it slowly by anion exchange, especially at low pH. Oxisols and other highly weathered soils of the humid tropics and subtropics may contain large stores of sulfate, especially in their subsoil horizons (Figure 12.10). Considerable sulfate may also be bound by the metal oxides in the spodic horizons under certain temperate forests.

Atmospheric Sulfur[6]   The atmosphere contains varying quantities of carbonyl sulfide (COS), hydrogen sulfide (H_2S), sulfur dioxide (SO_2), and other sulfur gases, as well as sulfur-containing dust particles. These atmospheric forms of sulfur arise from volcanic eruptions, ocean spray, biomass fires, and industrial plants (such as electricity generation stations fired by high-sulfur coal, and metal smelters). In the past half century, the contribution from industrial sources has dominated sulfur deposition in certain locations.

In the atmosphere, most of the sulfur materials are eventually oxidized to sulfates, forming sulfuric acid. The "acid rain" problem caused by this atmospheric sulfur (as well as nitrogen) was discussed in Section 9.6 and is highlighted in Figure 12.11.

After watching forests and lakes become seriously damaged by acid rain in the 1970s and 1980s, governments in North America and parts of Europe established regulatory programs that have reduced sulfur emissions by almost half since the late 1980s (although nitrogen oxide emissions have not been equally addressed). Today in eastern North America, annual S deposition is more commonly less than 8 to 10 kg S/ha (40 to 50 kg sulfate ha^{-1}). On the other hand, sulfur emissions are on the rise in China, India, and other newly industrializing regions, where burning of coal and oil cause as much as 50 to 75 kg S/ha to come down in a year. In other areas little affected by industrial emissions, deposition is generally as little as 1 to 5 kg S/ha/yr.

The deposition high in H_2SO_4 is mostly absorbed by soils, with some also absorbed through plant foliage. In some cases, 25 to 35% of plant sulfur comes from

---

[6]For a discussion of global sulfur deposition trends, see Stern (2006).

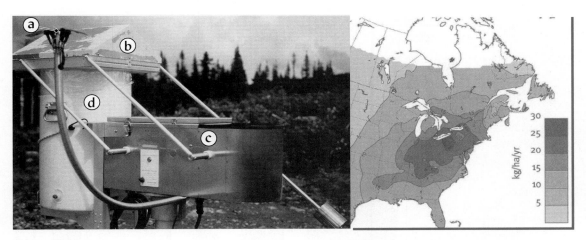

Figure 12.11

This apparatus collects both wet and dry sulfur deposition. A sensor (a) triggers the small roof (b) to move over and cover the dry deposition collection chamber (c) at the first sign of precipitation. The wet deposition chamber (d) is then exposed to collect precipitation. When the precipitation ceases, the sensor triggers the roof to move back over the wet deposition collection chamber so that dry deposition can again be collected. The map shows the average geographic distribution of sulfate deposition from the atmosphere in eastern North America. [Photo courtesy of R. Weil; map modified from International Joint Commission (2002)]

this source, even if available soil sulfate is adequate. In sulfur-deficient soils, about half of the plant needs can come from the atmosphere.

Cycling of Sulfur Compounds in Soils

The major transformations that sulfur undergoes in soils are shown in Figure 12.12. The inner circle shows the relationships among the four major forms of this element: (1) *sulfides*, (2) *sulfates*, (3) *organic sulfur*, and (4) *elemental sulfur*. The outer portions show the most important sources of sulfur and how this element is lost from the system.

Reactions in the sulfur cycle: http://filebox.vt.edu/users/chagedor/biol_4684/Cycles/Scycle.html

Considerable similarity to the nitrogen cycle is evident (compare Figures 12.1 and 12.12). In each case, the atmosphere is an important source of the element in question. Both elements are held largely in the soil organic matter, both are subject to microbial oxidation and reduction, both can enter and leave the soil in gaseous forms, and both are subject to some degree of leaching in the anionic form. Microbial activities are responsible for many of the transformations that determine the fates of both N and S.

Mineralization   Sulfur behaves much like nitrogen as it is absorbed by plants and microorganisms and moves through the sulfur cycle. The organic forms of S must be mineralized by soil organisms if the S is to be used by plants. The rate at which this occurs depends on the same environmental factors that affect N mineralization, including moisture, aeration, temperature, and pH. When conditions are favorable for general microbial activity, S mineralization occurs. Some of the more easily decomposed organic compounds in the soil are sulfate esters, from which microorganisms release sulfate ions directly. However, in much of the soil organic matter, sulfur in the reduced state is bonded to carbon atoms in protein and amino acid compounds. In the latter case the mineralization reaction might be expressed as follows:

$$\underset{\substack{\text{Proteins and}\\\text{other organic}\\\text{combinations}}}{\text{Organic sulfur}} \longrightarrow \underset{\substack{\text{H}_2\text{S and other}\\\text{sulfides are}\\\text{simple examples}}}{\text{decay products}} \xrightarrow{\text{O}_2} \underset{\text{Sulfates}}{\text{SO}_4^{2-} + 2\text{H}^+} \qquad (12.8)$$

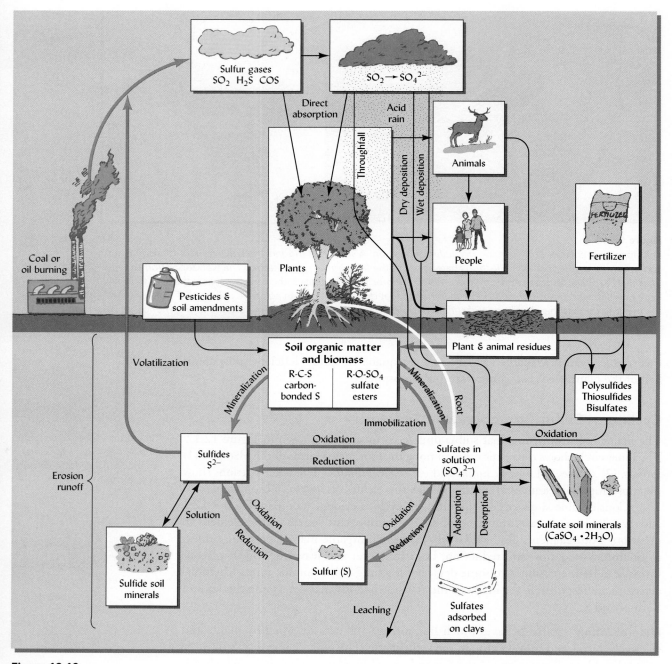

Figure 12.12

The sulfur cycle, showing some of the transformations that occur as this element is cycled through the soil–plant–animal–atmosphere system. In the surface horizons of all but a few types of arid-region soils, the great bulk of sulfur is in organic forms. However, in deeper horizons or in excavated soil materials, various inorganic forms may dominate. The oxidation and reduction reactions that transform sulfur from one form to another are mainly mediated by soil microorganisms. (Diagram courtesy of R. Weil)

Because this release of available sulfate is mainly dependent on microbial processes, the supply of available sulfate in soils fluctuates with seasonal, and sometimes daily, changes in environmental conditions (Figure 12.13).

Immobilization Microbial immobilization of inorganic forms of sulfur occurs when low-sulfur, carbon-rich organic materials are added to soils. A C/S ratio greater than 400:1 generally leads to such immobilization of sulfur. When the microbial activity

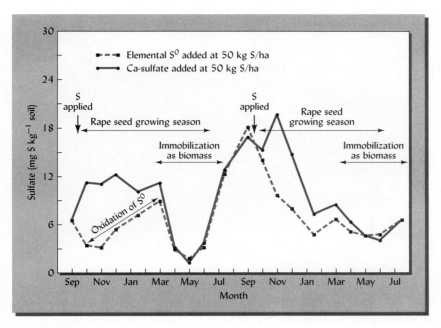

Figure 12.13

Seasonal changes in the sulfate form of sulfur available in the surface horizon of a soil (Argixeroll) in Oregon used to grow the oilseed crop, rape. This crop is sown in the fall, grows slowly during the winter, and then grows rapidly during the cool spring months. Data are shown for plots that were fertilized with either elemental S or calcium sulfate. The vertical arrows indicate dates on which these amendments were applied. Note that sulfate concentration was greater in the calcium sulfate-fertilized soils for the first few months after each application while the elemental S was slowly converted to sulfate by microbial oxidation. A distinct depression in sulfate concentration occurred each spring as the soil warmed up, stimulating both immobilization of sulfate into microbial biomass and uptake of sulfate into the rapeseed crop. Sulfate concentrations reached a peak in late summer and early fall, when crop uptake ceased after harvest and microbial mineralization was rapidly occurring. Movement of dissolved sulfate from the lower horizons up into the surface soil may also have occurred during hot, dry weather.

[Modified from Castellano and Dick (1991)]

subsides, the inorganic sulfate reappears in the soil solution. Stable soil organic matter generally contains C, N, and S, in a ratio approximately 100:8:1.

During the microbial breakdown of organic materials in wetland soils, several sulfur-containing gases are formed. Hydrogen sulfide (H_2S) is commonly produced by reduction of sulfates by anaerobic bacteria. Although these gases can be adsorbed by soil colloids, some escape to the atmosphere.

Sulfur Oxidation and Reduction

The Oxidation Process   The oxidation of iron sulfides to form sulfuric acid was illustrated by Equations 9.16 and 9.17. The oxidation of hydrogen sulfide and elemental sulfur may be illustrated as follows:

$$H_2S + 2O_2 \longrightarrow H_2SO_4 \longrightarrow 2H^+ + SO_4^{2-} \tag{12.9}$$

$$2S + 3O_2 + 2H_2O \longrightarrow 2H_2SO_4 \longrightarrow 4H^+ + SO_4^{2-} \tag{12.10}$$

Oxidation rates of commercial sulfur products: http://soil.scijournals.org/cgi/content/full/65/1/239

Most sulfur oxidation in soils is *bio-chemical* in nature, carried out by a number of autotrophic bacteria, which are active over a wide range of soil conditions. For example, sulfur oxidation may occur at pH values ranging from <2 to >9. This flexibility

Figure 12.14
Construction of this highway cut through several layers of sedimentary rock. One of these layers contained reduced sulfide materials. Now exposed to the air and water, this layer is producing copious quantities of sulfuric acid as the sulfide materials are oxidized. Note the failure of vegetation to grow below the zone from which the acid is draining. (Photo courtesy of R. Weil)

is in contrast to the comparable nitrogen oxidation process, nitrification, which requires a rather narrow pH range closer to neutral.

Extreme acid may be produced when reduced S compounds are exposed to oxidation because of excavation or drainage of anaerobic soils (see Section 9.6 and Figure 12.14).

If allowed to proceed unchecked, the acids may wash into nearby streams. Thousands of kilometers of streams have been seriously polluted in this manner, the water and rocks in such streams often exhibiting orange colors from the iron compounds in the acid drainage (see Plates 108 and 109).

The Reduction Process   Like nitrate ions, sulfate ions tend to be unstable in anaerobic environments. They are reduced to sulfide ions by a number of bacteria of two genera, *Desulfovibrio* and *Desulfotomaculum*. A representative reaction showing the reduction of sulfur coupled with organic matter oxidation follows:

$$2R—CH_2OH + SO_4^{2-} \longrightarrow 2R—COOH + 2H_2O + S^{2-} \qquad (12.11)$$

Organic alcohol   Sulfate   Organic acid   Sulfide

In anaerobic soils, the sulfide ion reacts immediately with reduced soluble iron or manganese to form insoluble sulfides. Sulfide ions will also undergo hydrolysis to form gaseous hydrogen sulfide, which causes the rotten-egg smell of swampy or marshy areas.

Sulfur Retention and Exchange

Since many sulfate compounds are quite soluble, the sulfate would be readily leached from the soil, especially in humid regions, were it not for its adsorption by the soil colloids. As was pointed out in Chapter 8, most soils have some anion exchange capacity that is associated with positive charges on iron and aluminum oxides and, to a limited extent, 1:1-type silicate clays. Sulfate ions also react directly with hydroxy groups exposed on the surfaces of these clays. Figure 12.15 illustrates how sulfate adsorption on the surface of some Fe, Al oxides and 1:1-type clays increases at lower pH values.

Much sulfate may be thusly held by such clays that tend to accumulate in the subsoil horizons of Ultisols and Oxisols. For example, symptoms of sulfur deficiency commonly occur early in the growing season on sandy Ultisols in the southeastern United States. However, the symptoms may disappear as the roots reach the deeper horizons where sulfate is retained.

Figure 12.15

Effect of decreasing soil pH on the adsorption of sulfates by 1:1-type silicate clays and oxides of Fe and Al (reaction with a surface-layer Al is illustrated). At high pH levels (a), the particles are negatively charged, the cation exchange capacity is high, and non-acid cations are adsorbed. Sulfates are repelled by the negative charges. As acidity is increased (b), more H^+ ions are attracted to the particle surface and the negative charge is satisfied, but the SO_4^{2-} ions are still not attracted. At still lower pH values (c), additional H^+ ions are attracted to the particle surface, resulting in a positive charge that attracts the SO_4^{2-} ion. This is easily exchanged with other anions. At still lower pH levels, the SO_4^{2-} reacts directly with Al and becomes a part of the crystal structure. Such sulfate is tightly bound, and it is removed very slowly, if at all.

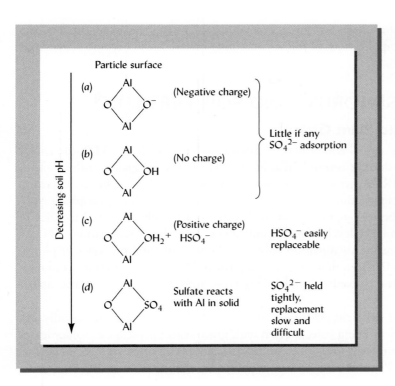

Sulfate Adsorption and Leaching of Non-Acid Cations   When sulfate ions leach from a soil, they are usually accompanied by equivalent quantities of cations, including Ca and Mg and other nonacid cations. In soils with high sulfate adsorption capacities, sulfate leaching is low and the loss of companion cations is also low. This is of considerable importance in soils of forested areas that receive acid precipitation.

Sulfur and Soil Fertility Maintenance

Figure 12.16 depicts the major gains and losses of sulfur from soils. Maintaining adequate quantities of sulfur for mineral nutrition of plants is becoming increasingly difficult. In some parts of the world (especially in certain semiarid grasslands), sulfur is already the next most limiting nutrient after nitrogen. Crop residues and farmyard manures can help replenish the S removed in harvests, but these sources generally can help to recycle only those S supplies that already exist within a farm. In regions with

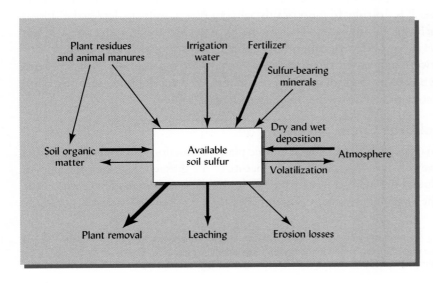

Figure 12.16

Major gains and losses of available soil sulfur. The thickness of the arrows indicates the relative amounts of sulfur involved in each process under average conditions. Considerable variation occurs in the field.

low-sulfur soils and clean air, greater dependence must be placed on fertilizer additions. Regular applications of sulfur-containing materials will become increasingly necessary.

12.3   PHOSPHORUS[7] AND SOIL FERTILITY[8]

Phosphorus and Plant Growth

Phosphorus is an essential component of **adenosine triphosphate** (**ATP**, the *energy currency* of cells), **deoxyribonucleic acid** (**DNA**, the seat of genetic inheritance), **ribonucleic acid** (**RNA**, which directs protein synthesis) and phospholipids (critical in cellular membranes). Bones and teeth are made of the calcium-phosphate compound apatite. In healthy plants, leaf tissue P content is usually about 0.2 to 0.4% of the dry matter, similar to levels of S, but only about 1/10 the concentration of N.

Adequate phosphorus nutrition enhances the fundamental processes of photosynthesis, nitrogen fixation, flowering, fruiting (including seed production), and maturation. Phosphorus is needed in especially large amounts in meristematic tissues such as root tips.

A phosphorus-deficient plant is usually stunted, thin-stemmed, and spindly, but its foliage, rather than being pale, is often dark green with purple areas. Older leaves show deficiency symptoms first. Phosphorus-deficient plants often seem quite normal in appearance (see Plates 90, 96, and 102), except for their size.

The Phosphorus Problem in Soil Fertility

Phosphorus has long been a problem for soil fertility for three reasons. *First*, the total P content of soils is relatively low, ranging from 200 to 2000 kg in the upper 15 cm of 1 ha of soil. *Second*, the phosphorus compounds common in soils are mostly unavailable for plant uptake, often because they are highly insoluble. *Third*, when soluble sources of P, such as those in fertilizers and manures, are added to soils, they are fixed (changed to unavailable, highly insoluble compounds).

Early research showed that fixation reactions with soil minerals allowed only 10 to 15% of the P applied in fertilizers and manures to be taken up by plants in the year of application. Consequently, farmers who could afford to do so typically applied two to four times as much phosphorus as was removed in the crop harvest. Repeated over many years, such practices can saturate the P-fixation capacity and build up the level of available soil P to the point where it can be a source of environmental pollution.

Where poor farmers cannot afford to buy fertilizer, underuse rather than overuse of fertilizer P is the rule. In most of sub-Saharan Africa, soils are depleted of P year after year, such that in some areas, the decline in per-capita food production will not likely be reversed until the critical phosphorus deficiency problems are solved.

Thus two major environmental problems related to soil phosphorus are **land degradation** caused by too little available phosphorus and **accelerated eutrophication** of lakes and streams caused by too much (see Chapter 13).

The Phosphorus Cycle

In order to manage phosphorus for economic plant production and for environmental protection, we will have to understand the nature of the different forms of phosphorus found in soils and the manner in which these forms of phosphorus interact within the soil and in the larger environment (Figure 12.17).

---

[7]For a fascinating historical account about all aspects of this element, see Emsley (2002). For in-depth technical reviews of environmental, biogeochemical, and agricultural aspects of phosphorus, see Tiessen (1995).
[8]For a review of the plant availability of this element, see Sharpley (2000).

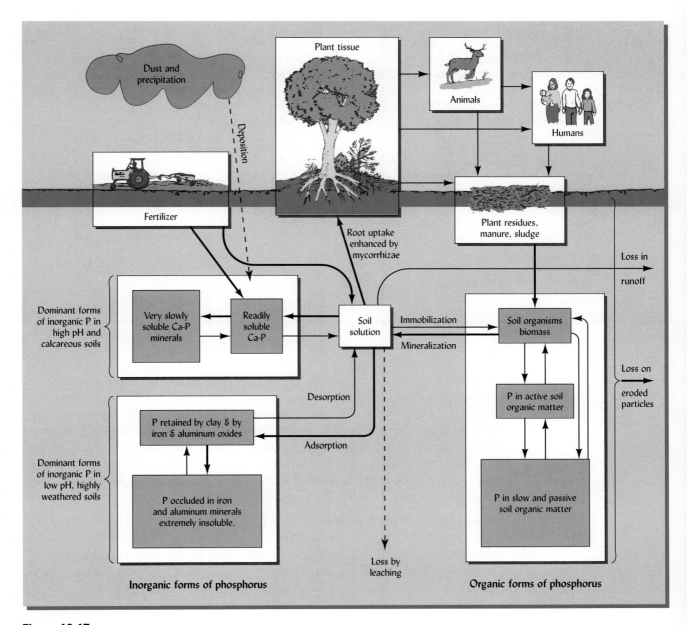

Figure 12.17

The phosphorus cycle in soils. The boxes represent pools of the various forms of phosphorus in the cycle, while the arrows represent translocations and transformations among these pools. The three largest white boxes indicate the principal groups of phosphorus-containing compounds found in soils. Within each of these groups, the less soluble, less available forms tend to dominate. Thick arrows represent the principal pathways. (Diagram courtesy of R. Weil)

Phosphorus in Soil Solution   Compared to other macronutrients, the concentration of phosphorus in the soil solution is very low, generally ranging from 0.001 mg/L in very infertile soils to about 1 mg/L in rich, well-fertilized soils. Plant roots absorb phosphorus dissolved in the soil solution, mainly as phosphate ions (HPO_4^{2-} in alkaline soils and $H_2PO_4^-$ in acid soils). Some soluble organic phosphorus compounds are also taken up.

Uptake by Roots and Mycorrhizae   Plant uptake of phosphate ions from the soil solution is curtailed by their slow movement to root surfaces. This is overcome in part by root proliferation into zones where the ions are held. Also, phosphate ions move to the roots of many plants through symbiosis with mycorrhizal fungi. The microscopic,

Table 12.4

EFFECT OF PREVIOUS LAND USE (FALLOW VS. CROP) ON EARLY SEASON MYCORRHIZAL COLONIZATION, SEEDLING GROWTH, AND GRAIN YIELD OF CORN ON A SOIL TESTING HIGH IN AVAILABLE PHOSPHORUS

In each year, the lack of a continuous host for mycorrhizal fungi due to previous fallow resulted in reduced corn root colonization at the three-leaf stage, which in turn depressed growth and P uptake by the six-leaf stage, leading to lower final grain yields. Practices that encourage mycorrhizae may help plants get off to a quick start with a minimum of starter fertilizer.

	Previous land use	Mycorrhizal root colonization on three-leaf stage corn, %	Shoot dry wt. at six-leaf stage, kg/ha	P concentration at six-leaf stage, %	P uptake at six-leaf stage, g/ha	Grain yield, kg/ha
Year 1	Crop	20.2	193	0.284	563	2903
	Fallow	11.0	142	0.228	337	2378
Year 2	Crop	46.9	103	0.262	273	7176
	Fallow	12.8	81	0.178	148	6677
Year 3	Crop	17.2	261	0.336	882	5495
	Fallow	8.0	158	0.293	469	4980

Data selected from Bittman et al. (2006).

threadlike mycorrhizal hyphae extend out into the soil several centimeters from the root surfaces (Plate 54 and Section 10.9) and absorb phosphorus ions as the ions enter the soil solution. The hyphae then transport the phosphate inside the hyphal cells, where soil-retention mechanisms cannot interfere. Generally, this mycorrhizal association is best developed where host plants are growing undisturbed in soils with low phosphorus availability. However, the fungi can benefit early-season growth of annual plants even in soils testing high in available P if the soil is kept vegetated with plants that serve as suitable hosts for the fungi (e.g., Table 12.4).

Decomposition of Plant Residues   Once in the plant, a portion of the phosphorus is translocated to the plant shoots, where it becomes part of the plant tissues. As the plants shed leaves and their roots die, or when they are eaten by people or animals, phosphorus returns to the soil in the form of plant residues, leaf litter, and wastes from animals and people. Microorganisms that decompose the residues temporarily tie up at least part of the P in their cells (microbial biomass-P), but eventually release a portion of it through mineralization (see Section 11.2). Some of it becomes associated with the active and passive fractions of the soil organic matter (see Section 11.6), where it is subject to storage and future release. These organic forms also slowly mineralize to release phosphate ions that plant roots can absorb, thereby repeating the cycle.

Chemical Forms in Soils   In most soils, the amount of phosphorus available to plants from the soil solution at any one time is very low, seldom exceeding about 0.01% of the total phosphorus in the soil. The bulk of the soil phosphorus exists in three general groups of compounds—namely, *organic phosphorus, calcium-bound inorganic phosphorus,* and *iron- or aluminum-bound inorganic phosphorus* (see Figure 12.17). Of the inorganic phosphorus, the calcium compounds predominate in most alkaline soils, while the iron and aluminum forms are most important in acidic soils. All three groups of compounds slowly contribute phosphorus to the soil solution, but most of the phosphorus in each group is of very low solubility and not readily available for plant uptake.

Unlike nitrogen and sulfur, phosphorus is not generally lost from the soil in gaseous form. Because soluble inorganic forms of phosphorus are strongly adsorbed by mineral surfaces, leaching losses of inorganic phosphorus are generally very small,

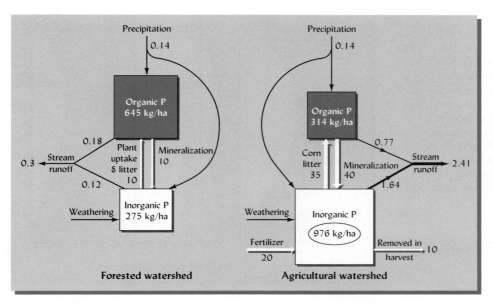

Figure 12.18

Phosphorus balance in surface soils (Ultisols) of adjacent forested and agricultural watersheds. The forest consisted primarily of mature hardwoods that had remained relatively undisturbed for 45 or more years. The agricultural land was producing row crops for more than 100 years. It appears that in the agricultural soil, about half of the organic P has been converted into inorganic forms or lost from the system since cultivation began. At the same time, substantial amounts of inorganic P accumulated from fertilizer inputs. Compared to the forested soil, mineralization of organic phosphorus was about four times as great in the agricultural soil, and the amount of P lost to the stream was eight times as great. Flows of P, represented by arrows, are given as kg/ha/yr. Although not shown in the diagram, it is interesting to note that nearly all (95%) of the P lost from the agricultural soil was in particulate form, while losses from the forest soil were 33% dissolved and 67% particulate. [Data from Vaithiyanathan and Correll (1992)]

but in heavily manured soils, leaching of P may be sufficient to stimulate eutrophication in downstream waters.

Gains and Losses   The principal pathways by which phosphorus is lost from the soil system are plant removal (5 to 50 kg ha^{-1} yr^{-1} in harvested biomass), erosion of phosphorus-carrying soil particles (0.1 to 10 kg ha^{-1} yr^{-1}), phosphorus dissolved in surface runoff water (0.01 to 3.0 kg ha^{-1} yr^{-1}), and leaching to groundwater (0.0001 to 0.4 kg ha^{-1} yr^{-1}). For each pathway, the higher figures cited for annual phosphorus loss would most likely apply to cultivated soils (see Figure 12.18).

The amount of phosphorus that enters the soil from the atmosphere (sorbed on dust particles) is quite small (0.05 to 0.5 kg ha^{-1} yr^{-1}), but may nearly balance the losses from the soil in undisturbed forest and grassland ecosystems. As already discussed, in an agroecosystem, optimal crop production may initially require the input from fertilizer to exceed the removal in crop harvest, but only until enough phosphorus accumulates to reduce the P-fixing capacity of the soil. The level of soil fertility and severity of environmental pollution are largely determined by the balance—or lack of balance—between inputs from fertilizer and feed and outputs as plant and animal products (see Section 13.1).

Organic Phosphorus in Soils

Both inorganic and organic forms of phosphorus occur in soils, and both are important to plants as sources of this element. The organic fraction generally constitutes 20 to 80% of the total phosphorus in surface soil horizons (Figure 12.19). The deeper

Figure 12.19

Phosphorus contents of representative soil profiles from three soil orders. All three soils contain a high proportion of organic phosphorus in their surface horizons. The Aridisol has a high inorganic phosphorus content throughout the profile because rainfall during soil formation was insufficient to leach much of the inorganic phosphorus compounds from the soil. The increased phosphorus in the subsoil of the Ultisol is due to adsorption of inorganic phosphorus by iron and aluminum oxides in the B horizon. In both the Mollisol and Aridisol, most of the subsoil phosphorus is in the form of inorganic calcium–phosphate compounds.

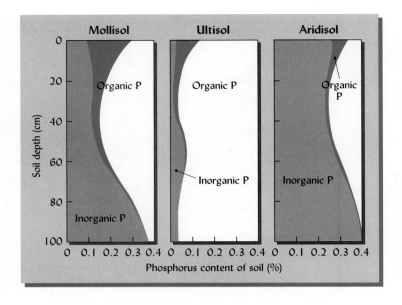

horizons may hold large amounts of inorganic Ca-phosphates, especially in soils from arid and semiarid regions.

Three broad groups of organic phosphorus compounds are known to exist in soils: (1) inositol phosphates or phosphate esters of a sugarlike compound, inositol $[C_6H_6(OH)_6]$; (2) nucleic acids; and (3) phospholipids. While other organic phosphorus compounds are present in soils, the identity and amounts present are less well understood.

Most of the soil organic phosphorus occurs in compounds not yet identified, but our ignorance of the specific compounds involved does not detract from their importance as suppliers of phosphorus through microbial breakdown.

Much of the phosphorus in the soil solution and in leachates from soils that have received large quantities of animal wastes is present as **dissolved organic phosphorus (DOP)**. DOP is generally more mobile than soluble inorganic phosphates, probably because it is not so readily adsorbed by iron, aluminum, clays, and $CaCO_3$ in the soil. In the lower horizons of such soils, the DOP commonly makes up more than 50% of the total soil solution phosphorus. In heavily manured sandy soils with high water tables, the DOP can move to the groundwater and then to streams and lakes, thereby contributing significantly to eutrophication.

Mineralization of Organic P   Phosphorus held in organic forms can be mineralized and immobilized by the same general processes that release nitrogen and sulfur from soil organic matter. Net immobilization of soluble inorganic phosphorus is most likely to occur if residues added to the soil have a C/P ratio greater than 300:1, while net mineralization is likely if the ratio is below 200:1. Mineralization of organic phosphorus in soils is subject to many of the same influences that control the general decomposition of soil organic matter—such as temperature, moisture, and tillage (see Section 11.3). In temperate regions, mineralization of organic phosphorus in soils typically releases 5 to 20 kg P/ha/yr, most of which is readily absorbed by growing plants. When forested soils are first brought under cultivation in tropical climates, the amount of phosphorus released by mineralization may exceed 50 kg/ha/yr, but unless phosphorus is added from outside sources these high rates of mineralization will soon decline due to the depletion of readily decomposable soil organic matter. Florida Histosols drained for agricultural use release about 80 kg P/ha/yr. These organic soils possess little capacity to retain dissolved phosphorus, so water draining from them contains up to 1.5 mg P/L and contributes to the degradation of the Everglades wetland system.

Contribution of Organic Phosphorus to Plant Needs   Recent evidence indicates that the readily decomposable or easily soluble fractions of soil *organic phosphorus* are often the most important factor in supplying phosphorus to plants in *highly weathered soils* (e.g., Ultisols, Oxisols), even though the total organic matter content of these soils may not be especially high. In contrast, the more soluble *inorganic forms* of phosphorus play the biggest role in P fertility of *less weathered soils* (e.g., Mollisols, Vertisols), even though these generally contain relatively high amounts of soil organic matter. Green manure crops and cover crop residues left on the soil surface as a mulch can improve phosphorus availability in both groups of soils. In temperate regions, freezing and thawing can lyse plant cells in growing plants and fresh residues, rapidly releasing soluble P that may be lost in winter runoff water before the spring flush of plant growth can take it up.

Inorganic Phosphorus in Soils

Two phenomena tend to control the concentration of inorganic P in the soil solution and the movement of phosphorus in soils: (1) the solubility of phosphorus-containing minerals and (2) the fixation or adsorption of phosphate ions on the surface of soil particles.

Fixation and Retention   Dissolved phosphate ions in mineral soils are subject to many types of reactions that tend to remove the ions from the soil solution and produce phosphorus-containing compounds of very low solubility. These reactions are sometimes collectively referred to by the general terms *phosphorus fixation* and *phosphorus retention*. *Phosphorus retention* is a somewhat more general term that includes both precipitation and fixation reactions. Phosphorus fixation may be viewed as troublesome if it prevents plants from using all but a small fraction of fertilizer phosphorus applied. On the other hand, phosphorus fixation can be viewed as a benefit if it causes most of the dissolved phosphorus to be removed from phosphorus-rich wastewater applied to a soil (Box 12.2).

Inorganic Phosphorus Compounds   As a group, the calcium phosphate compounds become more soluble as soil pH decreases; hence, they tend to dissolve and disappear from acid soils. On the other hand, the calcium phosphates are quite stable and very insoluble at higher pH and so become the dominant forms of inorganic phosphorus present in neutral to alkaline soils.

Of the common calcium compounds containing phosphorus, the **apatite** minerals are the least soluble and are therefore the least available source of phosphorus. The simpler mono- and dicalcium phosphates are readily available for plant uptake. Except on recently fertilized soils, however, these compounds are present in only extremely small quantities because they easily revert to the more insoluble forms.

In contrast to calcium phosphates, the iron and aluminum hydroxy phosphate minerals, **strengite** ($FePO_4 \cdot 2H_2O$) and **variscite** ($AlPO_4 \cdot 2H_2O$), have very low solubilities in strongly acid soils and become more soluble as soil pH rises. These minerals would therefore be quite unstable in alkaline soils, but are prominent in acid soils, in which they are quite insoluble and stable.

Effect of Aging on Inorganic Phosphate Availability   Usually, when soluble phosphorus is added to a soil, a rapid reaction removes the phosphorus from solution (*fixes* the phosphorus) in the first few hours. The freshly fixed phosphorus may be slightly soluble and of some value to plants. With time, the solubility of the fixed phosphorus tends to decrease to extremely low levels.

BOX 12.2
PHOSPHORUS REMOVAL FROM WASTEWATER

Removal of phosphorus from municipal wastes takes advantage of some of the same reactions that bind phosphorus in soils. After primary and secondary sewage treatment that removes solids and oxidizes most of the organic matter, tertiary treatment in huge, specially designed tanks (Figure 12.20) causes phosphorus to precipitate through reactions with iron and aluminum compounds:

$$Al_2(SO_4)_3 \cdot 14H_2O + 2PO_4^{3-} \rightarrow 2AlPO_4 + 3SO_4^{2-} + 14H_2O \qquad (12.12)$$

Alum   Soluble   Insoluble AlP
            phosphate

$$FeCl_3 + PO_4^{3-} \rightarrow FePO_4 + 3Cl^- \qquad (12.13)$$

Ferric   Soluble   Insoluble FeP
chloride phosphate

The insoluble aluminum and iron phosphates settle out of solution and are later mixed with other solids from the wastewater to form sewage sludge. The low-phosphorus water (effluent), after minor processing, is returned to the river. Other less-expensive tertiary treatment approaches involve the spraying of the wastewater on vegetated soils. As the wastewater flows over the soil surface or percolates down through the soil, the combination of plant uptake and soil sorption reactions very cost-effectively remove P from the wastewater.

Figure 12.20
Modern sewage plants include tertiary treatment for phosphorus removal. The chemical reactions in the sewage treatment process are similar to those affecting phosphorus availability in soils.

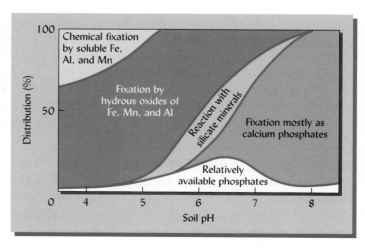

Solubility of Inorganic Phosphorus

The particular types of reactions that fix phosphorus in relatively unavailable forms differ from soil to soil and are closely related to soil pH (Figure 12.21). In acid soils these reactions involve mostly Al, Fe, or Mn. In alkaline and calcareous soils, the reactions primarily involve precipitation as various calcium phosphate minerals or adsorption to the iron impurities on the surfaces of carbonates and clays. At moderate pH values, adsorption on the edges of kaolinite or on the iron oxide coating on kaolinite clays plays an important role.

Figure 12.21
Inorganic fixation of added phosphates at various soil pH values. Average conditions are postulated, and it is not to be inferred that any particular soil would have exactly this distribution. The actual proportion of the phosphorus remaining in an available form will depend upon contact with the soil, time for reaction, and other factors. It should be kept in mind that some of the added phosphorus may be changed to an organic form in which it may be temporarily unavailable.

Precipitation by Iron, Aluminum, and Manganese Ions

Probably the easiest type of phosphorus-fixation reaction to visualize is the simple reaction of $H_2PO_4^-$ ions with dissolved Fe^{3+}, Al^{3+}, and Mn^{3+} ions to form insoluble hydroxy phosphate precipitates (Figure 12.22a). In strongly acid soils, enough soluble Al, Fe, or Mn is usually present to cause the chemical precipitation of nearly all dissolved $H_2PO_4^-$ ions by reactions such as the following (using the aluminum cation as an example):

$$Al^{3+} + H_2PO_4^- + 2H_2O \rightleftharpoons 2H^+ + Al(OH)_2H_2PO_4 \qquad (12.14)$$

$$\text{(soluble)} \qquad\qquad\qquad\qquad \text{(insoluble)}$$

Freshly precipitated hydroxy phosphates are temporarily soluble enough to supply some P to roots because they have a great deal of surface area exposed to the soil solution.

Reaction with Hydrous Oxides and Silicate Clays

Most of the phosphorus fixation in acid soils probably occurs when $H_2PO_4^-$ ions react with, or become adsorbed to, the surfaces of insoluble oxides of iron, aluminum, and manganese [such as gibbsite ($Al_2O_3 \cdot 3H_2O$) and goethite ($Fe_2O_3 \cdot 3H_2O$)] and 1:1 type silicate clays. These hydrous oxides occur as crystalline and noncrystalline particles and as coatings on the interlayer and external surfaces of clay particles. Fixation of phosphorus by clays probably takes place over a relatively wide pH range. The large quantities of Fe, Al oxides and 1:1 clays present in many soils make possible the fixation of extremely large amounts of phosphorus by these reactions.

Although all the exact mechanisms have not been identified, $H_2PO_4^-$ ions are known to react with iron and aluminum mineral surfaces in several different ways, resulting in different degrees of phosphorus fixation. Some of these reactions are shown diagrammatically in Figure 12.22.

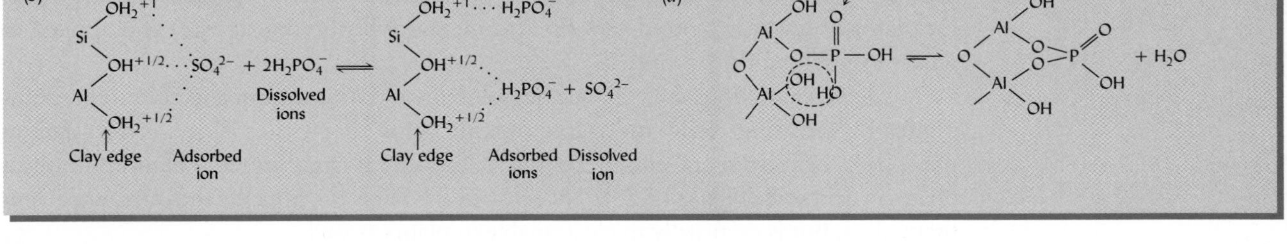

Figure 12.22

How phosphate ions are removed from the soil solution and fixed by reaction with iron and aluminum in various hydrous oxides. Freshly precipitated Al, Fe, and Mn phosphates (a) are relatively soluble, though over time they become increasingly insoluble. In (b) the phosphate is reversibly adsorbed by anion exchange. In (c) a phosphate ion replaces an —OH_2or an —OH group in the surface structure of Al or Fe hydrous oxide minerals. In (d) the phosphate further penetrates the mineral surface by forming a stable binuclear bridge. The adsorption reactions (b, c, d) are shown in order—from those that bind phosphate with the least tenacity (relatively reversible and somewhat more plant-available) to those that bind phosphate most tightly (almost irreversible and least plant-available). It is probable that, over time, phosphate ions added to a soil may undergo an entire sequence of these reactions, becoming increasingly unavailable. Note that (b) illustrates an outer-sphere complex, while (c) and (d) are examples of inner-sphere complexes (see Figure 8.13).

Finally, as more time passes, the precipitation of additional iron or aluminum hydrous oxide may bury the phosphate deep inside the oxide particle. Such phosphate is termed *occluded* and is the least available form of phosphorus in most acid soils.

Precipitation reactions similar to those just described are responsible for the rapid reduction in availability of phosphorus added to soil as soluble $Ca(H_2PO_4)_2 \cdot H_2O$ in fertilizers. This type of reaction can also be used to control the solubility of phosphorus in wastewater (see Box 12.2).

Effect of Iron Reduction Under Wet Conditions

Phosphorus bound to iron oxides by the mechanisms just discussed is very insoluble under well-aerated conditions. However, prolonged anaerobic conditions can reduce the iron in these complexes from Fe^{3+} to Fe^{2+}, making the iron–phosphate complex much more soluble and causing it to release phosphorus into solution. While these reactions may improve the phosphorus availability for wetland plants and paddy rice, they are of special relevance to water quality. Phosphorus bound to soil particles may accumulate in river- and lake-bottom sediments, along with organic matter and other debris. As the sediments become anoxic, the reducing environment may cause the gradual release of phosphorus held by hydrous iron oxides. Thus, the phosphorus eroded from soils today may aggravate the problem of eutrophication for years to come, even after the erosion and loss of phosphorus from the land has been brought under control.

Inorganic Phosphorus Availability at High pH Values

The availability of phosphorus in alkaline soils is determined principally by the solubility of the various calcium phosphate compounds present. In alkaline soils (e.g., pH = 8), soluble $H_2PO_4^-$ quickly reacts with calcium to form a sequence of products of decreasing solubility. For instance, highly soluble monocalcium phosphate $[Ca(H_2PO_4)_2 \cdot H_2O]$ added as fertilizer rapidly reacts with calcium carbonate in the soil to form first slightly soluble dicalcium phosphate ($CaHPO_4 \cdot 2H_2O$) and then very low solubility tricalcium phosphate $[Ca_3(PO_4)_2]$.

Although tricalcium phosphate has very low solubility, it may undergo further reactions to form compounds even more insoluble, such as the hydroxy-, oxy-, carbonate-, and fluorapatite compounds (apatites). These compounds are thousands of times less soluble than freshly formed tricalcium phosphates. The extreme insolubility of apatites in neutral or alkaline soils generally makes powdered phosphate rock (which consists mainly of apatite minerals) quite ineffective as a source of phosphorus for plants unless it is ground very fine (to increase weathering surface) and applied to relatively acidic soils.

Bacteria and fungi can enhance the solubility of both calcium and aluminum phosphates by releasing citric and other organic acids that either dissolve the calcium phosphates or form metal complexes that release the P from iron and aluminum phosphates in acid soils (Figure 12.23). The released P is likely used first by the microorganisms themselves, but is eventually made available to plants as well.

Phosphorus-Fixation Capacity of Soils

The phosphorus-fixation capacity of a soil may be conceptualized as the total number of sites on soil particle surfaces capable of reacting with phosphate ions. Phosphorus fixation may also be due to reactive soluble iron, aluminum, or manganese. The different types of fixation mechanisms are illustrated schematically in Figure 12.24.

One way of determining the phosphorus-fixing capacity of a particular soil is to shake a known quantity of the soil in a phosphorus solution of known concentration.

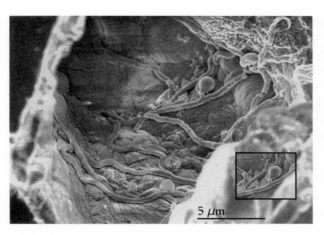 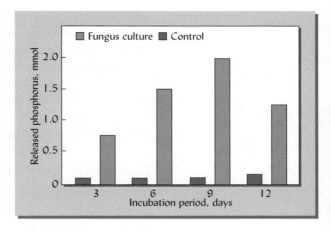

Figure 12.23

Certain soil microorganisms can increase the availability of phosphorus in minerals such as rock phosphate and aluminum phosphates that normally hold the phosphorus in very insoluble forms. (Left) A micrograph of a fungus growing on the surface of an aluminum phosphate found in soils. The fungus is thought to produce organic acids that help release some soluble phosphorus. (Right) In another experiment, phosphorus is released from phosphate rock by a culture of a fungus (Aspergillus niger) that had been isolated from a tropical soil. [Micrograph courtesy of Dr. Anne Taunton, University of Wisconsin; graph drawn from data in Goenadi et al. (2000)]

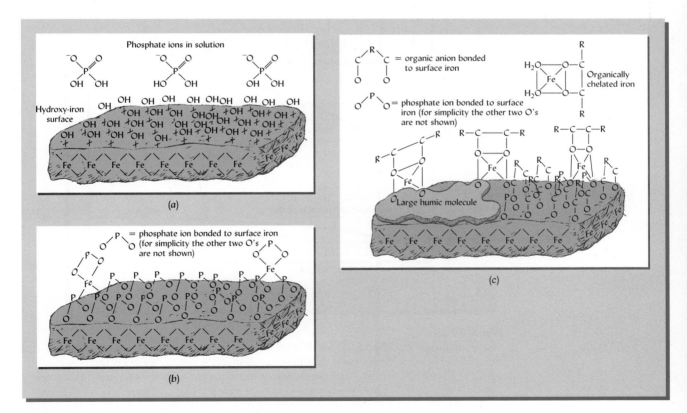

Figure 12.24

Schematic illustrations of phosphorus-fixation sites on a soil particle surface showing hydrous iron oxide as the primary fixing agent. In part (a) the sites are shown as + symbols, indicating positive charges or hydrous metal oxide sites, each capable of fixing a phosphate ion. In part (b) the fixation sites are all occupied by phosphate ions (the soil's fixation capacity is satisfied). Part (c) illustrates how organic anions, larger organic molecules, and certain strongly fixed inorganic anions can reduce the sites available for fixing phosphorus. Such mechanisms partially account for the reduced phosphorus fixation and greater phosphorus availability brought about when mulches and other organic materials are added to a soil. (Diagram courtesy of R. Weil)

After about 24 hours an equilibrium will be approached, and the concentration of phosphorus remaining in the solution (the **equilibrium phosphorus concentration** [**EPC**]) can be determined. The difference between the initial and final (*equilibrium*) solution phosphorus concentrations represents the amount of phosphorus fixed by the soil. If this procedure is repeated using a series of solutions with different initial phosphorus concentrations, the results can be plotted as a phosphorus-fixation curve (Figure 12.25), and the maximum phosphorus-fixation capacity can be extrapolated from the value at which the curve levels off.

Desorption from Soil to Water

If a portion of the fixed phosphorus is present in relatively soluble forms and most of the fixation sites are already occupied phosphate ions, some release of phosphorus to solution is likely to occur when the soil is exposed to water with a very low phosphorus concentration. This release (often called *desorption*) of phosphorus is indicated in Figure 12.25 where the curve for soil A crosses the zero fixation line and becomes negative (negative fixation = release). The solution concentration (*x*-axis) at which zero fixation occurs (phosphorus is neither released nor retained) is called the EPC_0. The EPC_0 is an important parameter for both soil fertility and environmental assessment because it indicates (1) the capacity to replenish the soil solution as it is depleted of P by plant roots, and (2) the rate at which the soil will release phosphorus into runoff and leaching waters.

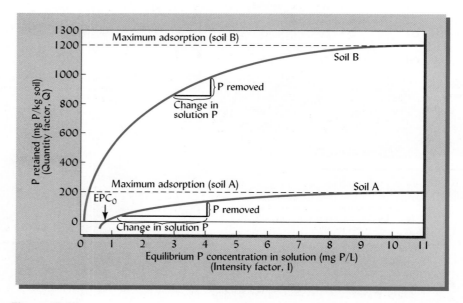

Figure 12.25

The relationship between P fixed and P in solution when two different soils (A and B) are equilibrated with solutions of various initial P concentrations. Initially, each soil removes nearly all of the P from solution, and as more and more concentrated solutions are used, the soil fixes greater amounts of P. However, eventually solutions are used that contain so much P that most of the P fixation sites are satisfied, and much of the dissolved P remains in solution. The amount fixed by the soil levels off as the maximum P-fixing capacity of the soil is reached (see horizontal dashed lines: for soil A, 200 mg P/kg soil; for soil B, 1200 mg P/kg soil). If the initial P concentration of a solution is equal to the equilibrium P concentration (EPC) for a particular soil, that soil will neither remove P from nor release P to the solution (i.e., phosphorus fixation = 0 and EPC = EPC_0). If the solution P concentration is less than the EPC_0, the soil will release some P (i.e., the fixation will be negative). In this example, soil B has a much higher P-fixing capacity and a much lower EPC than does soil A. It can also be said that soil B is highly buffered.

Factors Affecting the Extent of Phosphorus Fixation in Soils   Soils that remove more than 350 mg P/kg of soil (i.e., a phosphorus-fixing capacity of about 700 kg P/ha) from solution are generally considered to be high phosphorus-fixing soils. High phosphorus-fixing soils tend to maintain low phosphorus concentrations in the soil solution and in runoff water.

If soils with similar pH values and mineralogy are compared, phosphorus fixation tends to be more pronounced, and ease of phosphorus release tends to be lowest, in those soils with higher clay contents. Generally, those clays that possess greater anion exchange capacity (due to positive surface charges) have a greater affinity for phosphate ions. Thus, the soil components responsible for phosphorus-fixing capacity are, in order of increasing extent and degree of fixation:

$$2{:}1 \text{ clays} \ll 1{:}1 \text{ clays} < \text{carbonate crystals} < \text{crystalline Al, Fe, Mn oxides}$$
$$< \text{amorphous Al, Fe, and Mn oxides, allophone}$$

As a general rule, in mineral soils, phosphate fixation is at its lowest (and plant availability is highest) when soil pH is maintained in the 6.0 to 7.0 range (see Figure 12.21).

The low recovery by plants of phosphates added to field mineral soils in a given season is partially due to this fixation. A much higher recovery would be expected in organic soils and in many potting mixes, where calcium, iron, and aluminum concentrations are not as high as in mineral soils.

Effect of Organic Matter   Organic matter has little capacity to strongly fix phosphate ions. To the contrary, amending soil with organic matter, especially decomposable material, is likely to reduce phosphorus fixation by several mechanisms (see Figure 12.24). As a result, adding organic material reduces the slope of sorption curve and increases EPC_0 (level of phosphorus in solution). The principles of soil phosphorus behavior just discussed suggest a number of approaches to ameliorate soil deficiencies and prevent excessive losses of this critical element.

Enhancing Phosphorus Availability on Low P Soils

Where phosphorus-fixing capacity is grossly unsaturated, optimum plant growth will likely require additions that considerably exceed uptake. However, as the excess phosphorus begins to saturate fixation sites, rates of application should be lowered to supply no more than what plants take up. P fertilizer placed in a localized zone is less likely to undergo fixation reactions than if it were mixed into the bulk soil. Therefore, less fertilizer is required if it is placed in narrow bands or in small holes. In untilled systems, broadcasting on the surface effectively creates a horizontal "band."

When ammonium and phosphorus fertilizers are mixed in a band, the acidity produced by oxidation of ammonium ions (see Section 12.1) and by uptake of excess cations as ammonium (see Section 9.1) keeps the P in more soluble compounds and enhances plant P uptake. Phosphorus availability can also be optimized in most soils by proper liming or acidification to a pH level between 6 and 7.

During the microbial breakdown of organic materials, phosphorus is released slowly and can be taken up by plants or mycorrhizae before it can be fixed by the soil. In addition, organic compounds can reduce soil P fixation capacity (Figure 12.24). Residues or prunings from certain phosphorus-efficient plants (e.g., the African shrub *Tithonia diversifolia*) can be harvested from a donor site and transferred to a low-fertility receiving site where phosphorus will be supplied as the transferred plant material decays.

Practices that enhance mycorrhizal symbiosis usually improve the utilization of soil phosphorus. Such practices range from including good host plants, to reducing tillage, to inoculation with appropriate fungi (see Section 10.9).

Differences Among Plant Species   Different species enhance phosphorus uptake by at least four strategies. (1) Monocots exhibit extensive fibrous root systems and mycorrhizal associations. (2) N-fixing legumes use little nitrate and take up an excess of cations over anions, leading to rhizosphere acidification and subsequent release of P from low-solubility Ca-phosphates. (3) Certain species excrete specific compounds (such as *piscidic acid* produced by pigeon pea) that complex with Fe to greatly increase the availability of iron-bound soil phosphorus. (4) Plants in the *Brassicaceae* family (mustard, radish, etc.) compensate for their very poor mycorrhizal properties by excreting citric and malic acids, forming extensive fine root hairs, and taking up high amounts of Ca^{2+}. Knowledge of these plant characteristics can aid in choosing plants for restoration ecology, as well as for low-income farmers who can afford little fertilizer.

Reducing Phosphorus Losses to Water

Avoiding **excess accumulation** requires carefully keeping the sum of all P inputs (deposition, fertilizers, organic amendments, plant residues, and animal feed) from consistently exceeding plant removals or accumulating beyond the levels needed to support near-optimum plant growth (see Section 13.1). Use of conservation tillage practices can minimize runoff and erosion, *especially from land already high in phosphorus.* Cover crops and plant residues can increase infiltration and reduce runoff (see Section 6.2). Application of manure or fertilizer to frozen soils should be avoided.

Natural or constructed wetlands (Section 7.7) and riparian (shoreline) buffer strips (Section 13.2) can tie up some phosphorus before runoff enters sensitive lakes or streams. These measures remove some dissolved P, but mainly P bound to sediment.

Various iron-, aluminum-, or calcium-containing materials react with dissolved P to form highly insoluble compounds, similar to those described in Box 12.2. Treating poultry litter in the chicken house with aluminum sulfate (alum) or spreading such inorganic amendments on the surface of grass sod are practices that greatly reduce the loss of phosphorus dissolved from P-rich organic fertilizers such as manure or compost (Figure 12.26).

Figure 12.26

Effect of surface application of iron sulfate, gypsum, or lime on the concentration of phosphorus dissolved in runoff water. Bermuda grass sod was fertilized with composted dairy manure and subjected to two 30-minute rain events. The iron sulfate quickly reacted with phosphate ions as they were released from the compost, forming highly insoluble iron phosphate compounds. Gypsum and lime, which formed some calcium-phosphates, were far less effective in reducing phosphorus in the runoff water.

[Drawn from data in Torbert et al. (2005)]

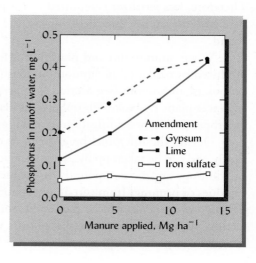

12.4   POTASSIUM IN SOILS AND PLANTS[9]

The potassium story differs in many ways from that of phosphorus, sulfur, and nitrogen. Unlike these other macronutrients, potassium is present in the soil solution only as a positively charged cation, K^+. Like phosphorus, potassium does not form gases that could be lost to the atmosphere. Its behavior in the soil is influenced primarily by soil cation exchange properties (see Chapter 8) and mineral weathering (Chapter 2), rather than by microbiological processes. Unlike nitrogen and phosphorus, potassium causes no off-site environmental problems when it leaves the soil system. It is not toxic and does not cause eutrophication in aquatic systems.

Potassium in Plant and Animal Nutrition

Potassium is not actually incorporated into the structures of organic compounds. Instead, potassium remains in the dissolved ionic form (K^+) in the cell or acts as an activator for enzymes. Potassium is known to activate over 80 different enzymes responsible for such plant and animal processes as energy metabolism, starch synthesis, nitrate reduction, photosynthesis, and sugar degradation.

As a component of the plant cytoplasmic solution, potassium plays a critical role in lowering cellular osmotic water potentials (see Section 5.3), thereby reducing the loss of water from leaf stomata and increasing the ability of root cells to take up water from the soil. The K content of healthy leaf tissue can be expected to be in the range of 1 to 4% in most plants, similar to that of N but an order of magnitude greater than that of P or S.

Potassium is especially important in helping plants adapt to environmental stresses. Good potassium nutrition is linked to improved drought tolerance, improved winter hardiness, better resistance to certain fungal diseases, and greater tolerance to insect pests (Figure 12.27). Potassium also enhances the quality of flowers, fruits, and vegetables by improving flavor and color and strengthening stems (thereby reducing lodging). In many of these respects, potassium seems to counteract some of the detrimental effects of excess nitrogen.

In animals, including humans, potassium plays critical roles in regulating the nervous system and in the maintenance of healthy blood vessels. Human diets that include such high-potassium foods as bananas, potatoes, orange juice, and leafy green vegetables have been shown to lower risk of stroke and heart disease. Maintaining a balance between potassium and sodium is especially important in human diets.

Deficiency Symptoms in Plants   Compared to deficiencies of P, a deficiency of potassium is relatively easy to recognize in most plants. Specific foliar symptoms associated with potassium deficiency occur earliest and most severely on the oldest leaves.

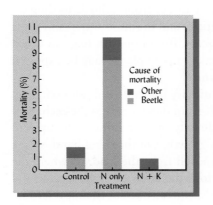

Figure 12.27

Influence of potassium and nitrogen fertilizer treatments on the percentage of Ponderosa pine trees dying from beetle damage and other causes in the first four years after planting in western Montana. Nitrogen used alone (224 kg N/ha) stimulated a large increase in tree mortality, but adding potassium (224 kg K/ha) completely counteracted this effect. Both fertilization treatments stimulated the growth of the surviving trees. [From Mandzak and Moore (1994)]

---

[9]For further information on this topic, see Mengel and Kirkby (2001).

Figure 12.28
Potassium deficiency often produces easily recognized foliar symptoms, mainly on older leaves: (left) chlorotic leaf margins on soybean; (right) small, white necrotic spots on hairy vetch leaflets. (Photos courtesy of Potash and Phosphate Institute, *left*, and R. Weil, *right*)

In general, the tips and edges of the oldest leaves begin to yellow (chlorosis) and then die (necrosis), so that the leaves appear to have been burned on the edges (Figure 12.28 and Plate 87). In several important legume species, white necrotic spots form a unique pattern along the leaflet margins—a symptom that people often mistake for insect damage (see Figure 12.28, *right*).

The Potassium Cycle

Figure 12.29 shows the major forms in which potassium is held in soils and the changes it undergoes as it is cycled through the soil–plant system. The original sources of potassium are the primary minerals, such as micas (biotite and muscovite) and potassium feldspar (orthoclase and microcline). As these minerals weather, their rigid lattice structures become more pliable. For example, potassium held between the 2:1-type crystal layers of mica is in time made more available, first as nonexchangeable but slowly available forms near the weathered edges of minerals and, eventually, as the readily exchangeable forms and the soil solution forms from which it is absorbed by plant roots.

Potassium is taken up by plants in large quantities. Depending on the type of ecosystem under consideration, a portion of this potassium is leached from plant foliage by rainwater (throughfall) and returned to the soil, and a portion is returned to the soil with the plant residues. In natural ecosystems, most of the K taken up by plants is returned in these ways or as wastes (mainly urine) from animals feeding on the vegetation. Some K is lost with eroded soil particles and in runoff water, and some is lost to groundwater by leaching. In agroecosystems, from one-fifth (e.g., in cereal grains) to nearly all (e.g., in hay crops) of the K taken up by plants may be exported to distant markets, from which it is unlikely to return.

At any one time, most soil potassium is in primary minerals and nonexchangeable forms. In relatively fertile soils, the release of K from these forms to the exchangeable and soil solution forms that plants can use directly may be sufficiently rapid to keep plants supplied with enough K for optimum growth. On the other hand, where high yields of agricultural crops or timber are removed from the land, or where the content of weatherable K-containing minerals is low, the levels of exchangeable and solution K may have to be supplemented by outside sources, such as chemical fertilizers, poultry manure, or wood ashes. Without these additions, the supply of available K will likely be depleted over a period of years, and the productivity of the soil will

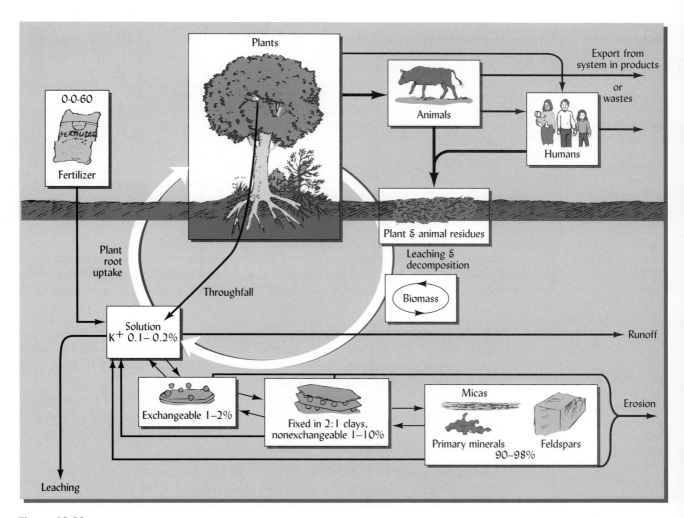

Figure 12.29

Major components of the potassium cycle in soils. The large circular arrow emphasizes the biological cycling of potassium from the soil solution to plants and back to the soil via plant residues or animal wastes. Primary and secondary minerals are the original sources of the element. Exchangeable potassium may include those ions held and released by both clay and humus colloids, but potassium is not a structural component of soil humus. The interactions among solution, exchangeable, nonexchangeable, and structural potassium in primary minerals are shown. The bulk of soil potassium occurs in the primary and secondary minerals and is released very slowly by weathering processes. (Diagram courtesy of R. Weil)

likewise decline. An example of depletion and restoration of available soil K is given in Figure 12.30.

Deep-rooted perennials often act as "nutrient pumps," taking potassium (or other nutrients) from deep subsoil horizons into their root systems, translocating it to their leaves, and then recycling it back to the surface of the soil via leaf fall and leaching. In most mature natural ecosystems, the small (1 to 5 kg/ha) annual losses of potassium by leaching and erosion are more than balanced by weathering of potassium from primary minerals and nonexchangeable forms in the soil profile, followed by vegetative translocation to the surface of the soil. In agricultural systems, leaching is high because high exchangeable K levels are maintained, and crop roots are active for only part of the year.

The Potassium Problem in Soil Fertility

Availability of Potassium   In contrast to phosphorus, potassium is found in comparatively high levels in most mineral soils, except those consisting mostly of quartz sand. In fact, the total quantity of this element is generally greater than that of any other

Figure 12.30

The general pattern of depletion of A-horizon exchangeable potassium by decades of exploitative farming, followed by its restoration under forest vegetation. Farming without fertilizers for over a century depleted the exchangeable potassium in a sandy soil in New York. After the supply of available potassium was exhausted, the land was abandoned in the 1920s. In the 1920s and 1930s, a red pine forest was planted. The trees in some plots were fertilized with potassium, causing the expected rapid recovery of exchangeable potassium to its preagricultural level. Even where the trees were not fertilized, the level of exchangeable potassium in the surface soil was replenished, but slowly, over a period of about 80 years. The trees were planted on a Plainfield loamy sand (Udipsamment), a soil with a very low cation exchange capacity and low levels of exchangeable K^+. [From Nowak et al. (1991)]

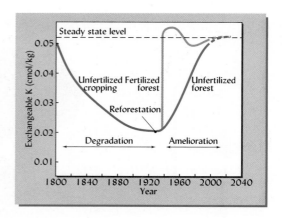

major nutrient element (see Table 1.2). Yet the quantity of K held in an easily exchangeable condition at any one time may be very small. Most K is held rigidly as part of the primary minerals or is fixed in forms that are, at best, only moderately available to plants.

Leaching Losses   Potassium is much more susceptible to leaching than is P, but it is less so than N. Humid-region soils growing annual crops and receiving only moderate rates of fertilizer commonly lose by leaching about 25 to 50 kg/ha of K, the greater values being typical of acid, sandy soils.

Losses would undoubtedly be much larger were the leaching of potassium not slowed by the attraction of the K^+ ions to the negatively charged cation exchange sites on clay and humus surfaces. Liming an acid soil to raise its pH can reduce the leaching losses of K because of the *complementary ion effect* (see Figure 8.16).

Luxury Consumption   Many plants take up about the same amount of K as they do N. If most or all of the aboveground plant parts are removed in harvest, the drain on the soil supply of potassium can be very large. This situation is made even more critical by the tendency of plants to take up soluble K far in excess of their needs if sufficiently large quantities are present. This tendency is termed *luxury consumption* because the excess K absorbed does not increase plant growth. The principles involved in luxury consumption are shown in Figure 12.31. Luxury consumption of K may depress Ca and Mg uptake and cause nutritional imbalances both in plants and in animals that consume them.

Figure 12.31

The general relationship between available potassium level in soil, plant growth, and plant uptake of potassium. If available soil potassium is raised above the level needed for maximum plant growth, many plants will continue to increase their uptake of potassium without any corresponding increase in growth. The potassium taken up in excess of that needed for optimum growth is termed luxury consumption. Such luxury consumption may be wasteful, especially if the plants are completely removed from the soil. It may also cause dietary imbalance in grazing animals. (Diagram courtesy of R. Weil)

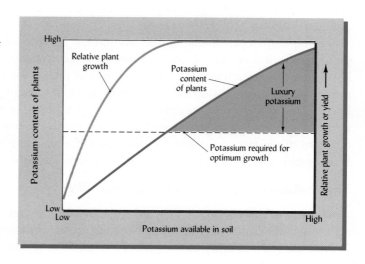

Table 12.5

THE INFLUENCE OF DOMINANT CLAY MINERALS ON THE AMOUNTS OF WATER-SOLUBLE, EXCHANGEABLE, FIXED (NONEXCHANGEABLE), AND TOTAL POTASSIUM IN SOILS

Potassium pool	Dominant clay mineralogy of soils, mg K/kg soil		
	Kaolinitic (26 soils)	Mixed (53 soils)	Smectitic (23 soils)
Total potassium	3340	8920	15,780
Exchangeable potassium	45	224	183
Water-soluble potassium	2	5	4

Data from Sharpley (1990).

In summary, then, the problem of potassium is at least threefold: (1) a very large proportion is relatively unavailable to higher plants; (2) it is subject to leaching losses; and (3) the removal of potassium by plants is high, especially when luxury quantities of this element are supplied. With these ideas as a background, the various forms and availabilities of potassium in soils will now be considered.

Forms and Availability of Potassium in Soils

The total amount of K in a soil and its distribution among the four major pools shown in Figure 12.29 is largely a function of the kinds of clay minerals present in a soil. Generally, soils dominated by 2:1 clays contain the most potassium; those dominated by kaolinite contain the least (Table 12.5). In terms of availability for plant uptake, K in primary mineral structures is *unavailable*, nonexchangeable K in secondary minerals is *slowly available*, and exchangeable K on soil colloids and K soluble in water are *readily available*.

In the presence of vermiculite, smectite, and other 2:1-type minerals, the K^+ ions (as well as the similarly sized NH_4^+ ions) in the soil solution (or added as fertilizers) not only become adsorbed but also may become definitely fixed by the soil colloids (Figure 12.32). Potassium (and ammonium) ions fit in between layers in the

Potassium in soils: www.soils.wisc.edu/~barak/soilscience326/potassium.htm

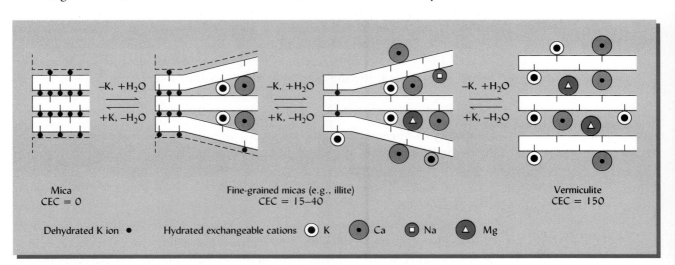

Figure 12.32

Diagrammatic illustration of the release and fixation of potassium between primary micas, fine-grained mica (illite clay), and vermiculite. In the diagram, the release of potassium proceeds to the right, while the fixation process proceeds to the left. Note that the dehydrated potassium ion is much smaller than the hydrated ions of Na^+, Ca^{2+}, Mg^{2+}, etc. Thus, when potassium is added to a soil containing 2:1-type minerals such as vermiculite, the reaction may go to the left and potassium ions will be tightly held (fixed) in between layers within the crystal, producing a fine-grained mica structure. Ammonium ions (NH_4^+, not shown) are of a similar size and charge to potassium ions and may be fixed by similar reactions. [Modified from McLean (1978)]

crystals of these normally expanding clays and become an integral part of the crystal. These ions cannot be replaced by ordinary exchange processes and consequently are referred to as *nonexchangeable ions*. As such, these ions are not readily available to most higher plants. This form is in equilibrium, however, with the more available forms and consequently acts as an extremely important reservoir of slowly available nutrients.

As a result of this equilibrium, very sandy soils with low CEC are poorly buffered with respect to K. That is, such soils have little capacity to maintain the K concentration as plants remove dissolved K from the soil solution. In soils with a greater CEC and buffering capacity, the initial solution concentration of K may be somewhat lower, but the soil is capable of maintaining a fairly constant supply of solution K ions throughout the growing season.

All plants can easily utilize the readily available forms, but the ability to obtain potassium held in the slowly available and unavailable forms differs greatly among plant species.

Factors Affecting Potassium Fixation in Soils

Kaolinite and other 1:1-type clays fix little potassium. On the other hand, clays of the 2:1 type, such as vermiculite, fine-grained mica (illite), and smectite, fix potassium very readily and in large quantities. Even silt-sized fractions of some micaceous minerals fix and subsequently release potassium.

Alternating wetting/drying or freezing/thawing enhances both the fixation of K in nonexchangeable forms and the release of previously fixed K to the soil solution.

Influence of pH   Applications of lime sometimes increase potassium fixation capacity of soils. As the pH increases, H^+ and hydroxyl aluminum ions associated with colloidal surfaces are removed or neutralized, making it is easier for K ions to move closer to the colloidal surfaces, where they are more susceptible to fixation in 2:1 clays. Furthermore, high Ca and Mg levels in the soil solution may reduce potassium uptake by the plant because cations tend to compete against one another for uptake by roots.

Practical Aspects of Potassium Management

Except in very sandy soils, the problem of potassium fertility is rarely one of total supply, but rather one of adequate *rate* of transformation from nonavailable to available forms. Where crops are removed, especially if little plant residue is returned, then the plant–soil cycle must be supplemented by release of potassium from less available mineral forms and, to some degree, by fertilization. If high yields of forage legumes are to be produced, the soil may have to be capable of supplying K for very high uptake rates during certain periods, resulting in the need for high levels of fertilization even on soils well supplied with weatherable minerals.

Frequency of Application   Although a heavy dressing applied every few years may be most convenient, more frequent light applications of K may offer the advantages of reduced luxury consumption of K by some plants, reduced losses of this element by leaching, and reduced opportunity for fixation in unavailable forms before plants have had a chance to use the K applied.

Potassium-Supplying Power of Soils   The idea that each kilogram of potassium removed by plants or leaching must be returned in fertilizers may not always be correct. In many soils, the large quantities of moderately available forms already present can be utilized so that only a part of the total amount removed by harvest need be replaced by fertilizer.

However, continued crop removal can deplete the available K pools even in these soils. The problem of maintaining soil potassium is diagrammed in Figure 12.33. In less developed agricultural regions of the world, potassium fertilizer use will have to increase for many years to come if yields are to be increased or even maintained.

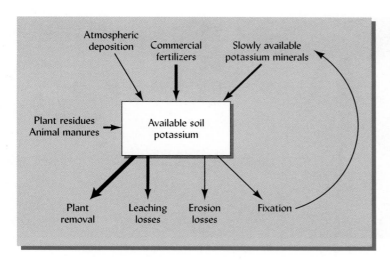

Figure 12.33

Gains and losses of available soil potassium under average field conditions. The approximate magnitude of the changes is represented by the width of the arrows. For any specific case, the actual amounts of potassium added or lost undoubtedly may vary considerably from this representation. As was the case with nitrogen and phosphorus, commercial fertilizers can be important in meeting plant demands where intensive agriculture is practiced.

12.5 CALCIUM AS ESSENTIAL NUTRIENT

Calcium (Ca) is an essential macronutrient for all plants, and its status in soils has a major influence on the species composition and productivity of terrestrial ecosystems. For animals, the Ca content of plants they eat is important because it is a major component of bones and teeth and plays important roles in many physiological processes. It has even been suggested that the relatively higher calcium status of soils in Africa may account for the occurrence of such large herbivores as elephants, zebras, and giraffes in semiarid savannas in Africa but not in South America.

Calcium in Plants

Most monocots are considered to be *calcifuge* (calcium avoiding) plants and grow well with 0.15 to 0.5% Ca in leaf tissues. Many dicots are considered to be *calcicoles* (Latin for "chalk dwelling") and need 1 to 3% Ca in their leaves for optimal growth. Trees store a great deal of calcium in their woody tissues; the net calcium uptake by many trees is close to that of nitrogen.

Physiological Roles   Calcium is a major component of cell walls. H is also intimately involved with cell elongation and division, membrane permeability, and the activation of several critical enzymes. Calcium is taken up almost exclusively by young roots and moves within the plant mainly with the transpiration water in the xylem rather than in the phloem.

Deficiencies of calcium are quite rare for most plants, except in very acid soils. When calcium deficiency does occur, it is usually associated with growing points (meristems) such as buds, unfolding leaves, fruits, and root tips (Figure 12.34). When calcium is deficient, normally harmless levels of other metals can become toxic to the plant. These effects can cause sensitive forest trees to lose branches and eventually die. In very acid soils, Ca deficiency is often accompanied by Al or Mn toxicity, and the related effects on plants are difficult to tell apart.

Some Ca deficiencies are related to the transport of Ca within the plant. Blossom-end rot on melons and tomatoes is caused by inadequate Ca associated with unevenness in the water supply that interrupts the flow of Ca.

Soil Forms and Processes

Calcium in the soil is found mainly in three pools that resupply the soil solution: (1) calcium-containing minerals (such as calcite or plagioclase), (2) calcium complexed with soil humus, and (3) calcium held by cation exchange on the clay and humus

Figure 12.34

Root growth was almost completely inhibited by low calcium in the nutrient solution (left) compared to healthy roots in the same nutrient solution but with calcium added (right). If the ratio of calcium to all other cations in solution drops below 5:1, the integrity of root membranes is lost, causing many other elements to become toxic to the plants. (Photo courtesy of R. Weil)

colloids. The cycling of calcium among these and other soil pools, and the gains and losses of calcium by such mechanisms as plant uptake, atmospheric deposition as dust and soot, liming, and leaching, comprise the calcium cycle, as illustrated in Figure 12.35.

In arid and semiarid regions, the high pH, high carbonate nature of the soil solution greatly diminishes the solubility of calcium-containing minerals. Deposition of dust from coal burning and wind erosion of calcareous desert soils can make substantial contributions to soil calcium thousands of kilometers downwind. Such calcium deposition can partially offset acidification caused by nitrogen and sulfur deposition. Central and southeastern China provide some of the most extreme examples of Ca deposition as environmental policies struggle to keep pace with rapid industrialization.

Figure 12.35

Simplified diagram of the cycling of calcium and magnesium in soils. The rectangular compartments represent pools of these elements in various forms, while the arrows represent processes by which the elements are transformed or transported from one pool to another. (Diagram courtesy of R. Weil)

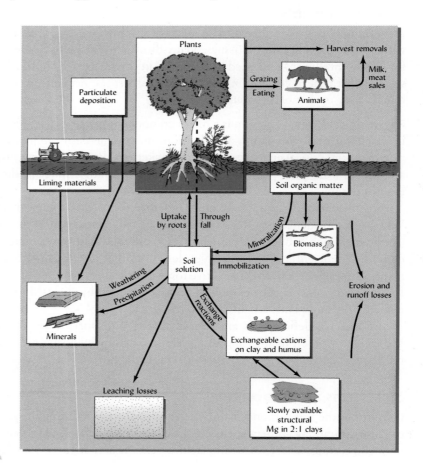

Table 12.6
CASCADE OF CALCIUM-RELATED EFFECTS FROM ACID DEPOSITION ON FORESTS

Biogeochemical response to acid deposition	Physiological response	Effect on forest function
Leaching of Ca from leaf membrane	Reduced cold tolerance of needles in red spruce	Loss of current-year needles in red spruce
Reduction of the ratios of Ca to Al and Ca to Mn in soil and in soil solutions	Dysfunction in fine roots of red spruce, blocking uptake of Ca	Decreased growth and increased susceptibility to stress in red spruce
Reduction of the ratios of Ca to Al and Ca to Mn in soil and in soil solutions	Greater energy use to acquire Ca in soils with low Ca:Al ratios	Decreased growth and increased photosynthate allocation to roots
Reduction of the availability of nutrient cations in marginal soils	Sugar maples on drought-prone or nutrient-poor soils are less able to withstand stresses	Episodic dieback and growth impairment in sugar maple

Modified from Fenn et al. (2006).

Leaching Losses

The need for repeated applications of limestone in humid regions (Section 9.8) suggests significant losses of calcium and magnesium from the soil. Leaching, crop removal, and soil erosion in humid-region agricultural soils combine to cause Ca losses of about 1 Mg/ha of $CaCo_3$ equivalents.

Similarly, timber harvest methods that leave the soil open to erosion and nutrient leaching lead to rapid losses of calcium (and magnesium) from forest ecosystems. Scientists are concerned that for some soils in the humid regions, the release of Ca from mineral weathering will not be able to keep up with the losses and that acid rain combined with intensive timber harvesting may be depleting the Ca reserves in the more poorly buffered watersheds. The acidification affects a number of biogeochemical processes involving calcium. These processes in turn influence tree physiology, which ultimately impacts the ecological functioning of forests (Table 12.6). Nonetheless, research in this area is still inconclusive and up to now forests have rarely shown a positive growth response to applied calcium.

Fungi may help trees mine Ca: http://news.bbc.co.uk/1/hi/sci/tech/2040623.stm

12.6   MAGNESIUM AS A PLANT NUTRIENT

Magnesium in Plants

Plants generally take up Mg in similar or somewhat smaller amounts (0.15 to 0.75% of dry matter) than Ca. About one-fifth of the magnesium in plant tissue is found in the chlorophyll molecule and so is intimately involved with photosynthesis. Magnesium also plays critical roles in the synthesis of oils and proteins and in the activation of enzymes involved in energy metabolism.

The most common symptom of Mg deficiency is interveinal chlorosis on the older leaves, which appears as a mottled green and yellow coloring in dicots (Plate 89) and a striping in monocots. Common on very sandy soils with low CEC, these symptoms are sometimes termed *sand drown* as they appear somewhat like those caused by oxygen-starvation in a waterlogged soil. Spruce and fir trees growing on soils low in exchangeable Mg have exhibited reduced Mg in needle tissue, stunted growth, and needles that turn yellow, especially on the tips.

Forages with low contents of Mg compared to Ca and K can cause grazing animals to suffer from a sometimes-fatal Mg deficiency known as *grass tetany*. High levels of K can aggravate this problem by reducing Mg uptake.

Magnesium in Soil

The main source of plant-available Mg in most soils is the pool of exchangeable Mg on the clay-humus complex (Figure 12.35). As plants and leaching remove this Mg, the easily exchangeable pool is replenished by Mg weathered from minerals (e.g., dolomite, biotite, hornblende, serpentine). In some soils, replenishment also takes place from a pool of slowly available Mg in the structure of certain 2:1 clays (see Section 8.3). Variable amounts of Mg are made available by the breakdown of plant residues and soil organic matter. In unpolluted forests, research suggests that atmospheric deposition, rather than rock weathering, may supply much of the magnesium used by trees.

Ratio of Calcium to Magnesium[10]

Being less tightly held (more easily leached) than Ca, exchangeable Mg commonly saturates only 5 to 20% of the effective CEC, as compared to the 60 to 90% typical for Ca in neutral to moderately acid soils (see Figure 9.5). Some agriculturists believe that optimum plant growth and soil tilth require a ratio of exchangeable Ca:Mg very near 6:1 (65% Ca and 10% Mg saturation of the CEC). This belief can lead to the wasteful use of soil amendments in an effort to achieve this so-called ideal Ca:Mg ratio. Numerous research studies have shown that, in fact, plants grow very well and meet their Ca and Mg needs in soils with Ca:Mg ratios anywhere from 1:1 to 15:1. Soil aggregation and biological activity are also largely unaffected by a similarly wide range in this ratio. However, even though plant and soil health are not likely to be affected by soil Ca:Mg ratios, the ratio of Ca to Mg content in plant tissue may be altered enough to influence the nutrition of grazing animals, and Ca or Mg mineral supplements may be needed in animal diets.

Soils formed on serpentine rock, which is rich in Mg but contains little or no Ca, offer an unusual but dramatic exception to the above statements. Having a Ca:Mg ratio much smaller than 1.0, serpentine-derived soils may exhibit severe deficiency of Ca and toxicity of Mg for all but the few plant species that evolved to grow in this unique soil environment.

12.7 MICRONUTRIENTS IN THE SOIL–PLANT SYSTEM

Deficiency Versus Toxicity

When a nutrient is present at very low levels, plant growth may be restricted by insufficient supplies of the nutrient (*deficiency range*). As the level of nutrient is increased, plants respond by taking up more of the nutrient and increasing their growth. If a level of nutrient availability has been reached that is sufficient to meet the plants' needs (*sufficiency range*), raising the level further will have little effect on plant growth, although the concentration of the nutrient may continue to increase in the plant tissue. At some level of availability, the plant will take up too much of the nutrient for its own good (*toxicity range*), causing adverse physiological reactions to take place (Figure 12.36).

For macronutrients, the sufficiency range is very broad, and toxicity seldom occurs. However, for micronutrients the difference between deficient and toxic levels may be narrow, making the possibility of toxicity quite real. For example, in the cases of boron and molybdenum, severe toxicity may result from applying as little as 3 to 4 kg/ha of available nutrient to a soil initially deficient in these elements. While the sufficiency range for other micronutrients is much wider and toxicities are not as likely

Deficiency and toxicity symptoms:
http://hort.ufl.edu/teach/orh3254/DefSymptoms.htm

---

[10]For an objective review of how Ca/Mg ratios and cation balancing in general became widely, but wrongly, promoted for fertility management, see Kopittke and Menzies (2007).

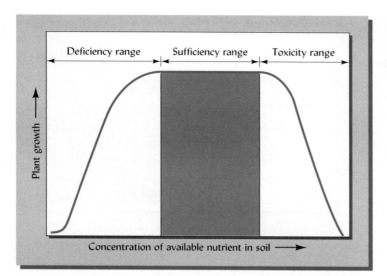

Figure 12.36

The relationship between the amount of a micronutrient available for plant uptake and the growth of the plant. Within the deficiency range, as nutrient availability increases so does plant growth (and uptake, which is not shown here). Within the sufficiency range, plants can get all of the nutrient they need, and so their growth is little affected by changes within this range. At higher levels of availability a threshold is crossed into the toxicity range, in which the amount of nutrient present is excessive and causes adverse physiological reactions that lead to reduced growth and even death of the plant. (Diagram courtesy of R. Weil)

from overfertilization, toxicities of copper, zinc, and nickel have been observed on soils contaminated by industrial sludges, manure from concentrated hog facilities, metal smelter wastes, and long-term application of copper sulfate fungicide. Toxicity of manganese is quite common in association with low soil pH (see Section 9.7). In certain forests it appears that the high ratios of manganese to magnesium or calcium are associated with the decline and death of sensitive tree species.

High levels of molybdenum may occur naturally in certain poorly drained alkaline soils. In some cases, enough of this element may be taken up by plants to cause toxicity, not only to susceptible plants but also to livestock grazing forages on these soils. Boron, too, may occur naturally in alkaline soils at levels high enough to cause plant toxicity. Although somewhat larger amounts of most other micronutrients are required and can be tolerated by plants, great care needs to be exercised in applying micronutrients, especially for maintaining nutrient balance.

In addition, irrigation water in dry regions may contain enough dissolved boron, molybdenum, or selenium to damage sensitive crops, even if the original levels of these trace elements in the soil were not very high. Therefore, it is prudent to monitor the content of these elements in water used for irrigation. Selenium does not appear essential for plants, but it is required by animals and can be toxic to both (see Chapter 15).

Micronutrient Roles in Plants[11]

Micronutrients are required in plant tissue levels at one or more orders of magnitude lower than for the macronutrients. The ranges of plant tissue concentrations considered deficient, adequate, and toxic for several micronutrients are illustrated in Figure 12.37.

Intensive plant production practices have increased crop yields, resulting in greater removal of micronutrients from soils. The trend toward more concentrated, high analysis fertilizers has reduced the use of impure salts and organic manures, which formerly supplied significant amounts of micronutrients. Increased knowledge of plant nutrition and improved methods of analysis in the laboratory are helping in the diagnosis of micronutrient deficiencies that might formerly have gone unnoticed. Increasing evidence indicates that food grown on soils with low levels of trace elements

---

[11]For an excellent overview of micronutrients in global agriculture, see Fageria et al. (2002). For a detailed review of their physiological roles in plants, see Welch (1995). For a classic reference for soil and plant aspects of micronutrients in agriculture, see Mortvedt et al. (1991).

Figure 12.37

Deficiency, normal, and toxicity levels in plants for seven micronutrients. Note that the range is shown on a logarithmic scale and that the upper limit for manganese is about 10,000 times the lower range for molybdenum and nickel. In using this figure, keep in mind the remarkable differences in the ability of different plant species and cultivars to accumulate and tolerate different levels of micronutrients. (Based on data from many sources)

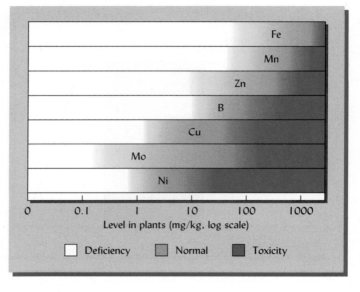

may provide insufficient human dietary levels of certain elements, even though the crop plants themselves show no signs of deficiency. For all these reasons, increasing attention is directed toward micronutrients in soils.

Physiological Roles in Plants

Micronutrients play many complex roles in plant nutrition. While most of the micronutrients participate in the functioning of a number of enzyme systems (Table 12.7), there is considerable variation in the specific functions of the various micronutrients

Table 12.7
FUNCTIONS OF SEVERAL MICRONUTRIENTS IN HIGHER PLANTS

Micronutrient	Functions in higher plants
Zinc	Present in several dehydrogenase, proteinase, and peptidase enzymes; promotes growth hormones and starch formation; promotes seed maturation and production.
Iron	Present in several peroxidase, catalase, and cytochrome oxidase enzymes; found in ferredoxin, which participates in oxidation-reduction reactions (e.g., NO_3^- and SO_4^{2-} reduction and N fixation); important in chlorophyll formation.
Copper	Present in laccase and several other oxidase enzymes; important in photosynthesis, protein and carbohydrate metabolism, and probably nitrogen fixation.
Manganese	Activates decarboxylase, dehydrogenase, and oxidase enzymes; important in photosynthesis, nitrogen metabolism, and nitrogen assimilation.
Nickel	Essential for urease, hydrogenases, and methyl reductase; needed for grain filling, seed viability, iron absorption, and urea and ureide metabolism (to avoid toxic levels of these nitrogen-fixation products in legumes).
Boron	Activates certain dehydrogenase enzymes; facilitates sugar translocation and synthesis of nucleic acids and plant hormones; essential for cell division and development.
Molybdenum	Present in nitrogenase (nitrogen fixation) and nitrate reductase enzymes; essential for nitrogen fixation and nitrogen assimilation.
Cobalt	Essential for nitrogen fixation by associative bacteria; found in vitamin B_{12}.
Chloride	Essential for photosynthesis and enzyme activation. Plays role in regulation of water uptake on salt-affected soils.

Table 12.8

EFFECT OF MANGANESE LEVELS ON THE INCIDENCE OF ROOT ROT AND ROOT PEROXIDASE ACTIVITY IN COWPEA[a]

Manganese catalyses peroxidase, an exo-enzyme that aids the synthesis of both lignin and monophenols. The lignin acts as a mechanical and chemical barrier against fungal invasion, while monophenols are fungal toxins that may also act in defense of the plant. Mn fertilizer controlled root rot almost as well as the chemical fungicide Carbendazim.

| Treatment | Cowpea tissue conc. | | Disease incidence 10 days after sowing (%) | Peroxidase activity index |
	Mn (mg/kg)	N (%)		
Mn, 0 mg/kg soil	45	3.15	75	16
Mn, 5 mg/kg soil	78	2.54	47	24
Mn, 10 mg/kg soil	92	2.85	44	27
Carbendazim, 0.2% on seeds	45	3.43	31	31

[a]Grown in low-fertility, alkaline (pH 8.6) loamy sand inoculated with *Rhizoctonia bataticola*. Data selected from Kalim et al. (2003).

in plant and microbial growth processes. For example, through its catalysis of certain enzymes, manganese plays an important role in the mechanisms by which plants defend themselves from pathogen attack (Table 12.8).

Deficiency Symptoms

Insufficient supply of a nutrient is commonly expressed by visible plant symptoms that can be used to diagnose the micronutrient deficiencies. Most of the micronutrients are relatively immobile in the plant (Plate 88). That is, the plant cannot efficiently transfer the nutrient from older leaves to newer ones. Therefore, the concentration of the nutrient tends to be lowest, and the symptoms of deficiency most pronounced, in the younger leaves that develop after the supply of the nutrient has run low. This pattern is in contrast to most of the macronutrients (except sulfur), which are more easily translocated by the plants and so become most deficient in the older leaves. The iron-deficient sorghum in Plate 99 and the zinc-deficient peach leaves in Plate 97 illustrate the pattern of pronounced deficiency symptoms on the younger leaves.

In the case of zinc deficiency in corn (Plate 98), broad white bands on both sides of the midrib are typical in young corn plants, but the symptoms may disappear as the soil warms up and the maturing plant's root system expands into a larger volume of soil.

Photos of mineral deficiencies in plants: www.hbci.com/~wenonah/min-def/list.htm

Sources of Micronutrients

Inorganic Forms Deficiencies and toxicities of micronutrients may be related to the total contents of these elements in the soil. More often, however, these problems result from the chemical forms of the elements in the soil and, particularly, their solubility and availability to plants.

The mineral forms of micronutrients are altered as mineral decomposition and soil formation occur. Oxides and, in some cases, sulfides of elements such as iron, manganese, and zinc are formed. Secondary silicates, including the clay minerals, may contain considerable quantities of iron and manganese and smaller quantities of zinc and cobalt. Ultramafic rocks, especially serpentinite, are high in nickel. The micronutrient cations released as weathering occurs are subject to colloidal adsorption, just as are the calcium or aluminum ions (see Section 8.7). Anions such as borate and molybdate in soils may undergo adsorption or reactions similar to those of the phosphates. The chemical forms commonly taken up by plant roots are shown in Table 1.1.

Organic Forms   Organic matter is an important secondary source of some of the trace elements. Several of them tend to be held as complex combinations by organic (humus) colloids. Copper is especially tightly held by organic matter—so much so that its availability can be very low in organic soils (Histosols). In uncultivated profiles, there is a somewhat greater concentration of micronutrients in the surface soil, much of it presumably in the organic fraction. Correlations between soil organic matter and contents of copper, molybdenum, and zinc have been noted. Although the elements thus held are not always readily available to plants, their release through decomposition is undoubtedly an important fertility factor. Animal droppings are a good source of micronutrients, much of it present in organic forms.

The cycling of micronutrients through the soil–plant–animal system is illustrated in a generalized way by Figure 12.38. Although not every micronutrient will participate in every pathway shown in this figure, it can be seen that organic chelates

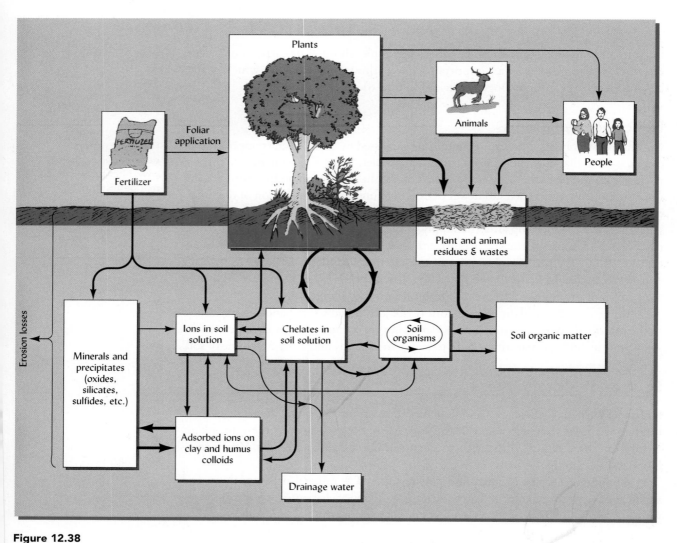

Figure 12.38

Cycling and transformations of micronutrients in the soil–plant–animal system. Although all micronutrients may not follow each of the pathways shown, most are involved in the major components of the cycle. The formation of organic chelates, which keep most of these elements in soluble forms, is a unique feature of this cycle. (Diagram courtesy of R. Weil)

(see below), soil colloids, soil organic matter, and soil minerals all contribute micronutrients to the soil solution and, in turn, to growing plants.

General Conditions Conducive to Deficiency/Toxicity

Strongly leached, acid, sandy soils are low in most micronutrients for the same reasons they are deficient in most of the macronutrients—their parent materials were initially low in the elements, and acid leaching has removed much of what was originally present. In the case of molybdenum, acid soil conditions markedly decrease availability. The micronutrient contents of *organic soils* depend on the extent of the accumulation of these elements as the soils were formed. The ability of organic soils to bind copper accentuates deficiencies of this element. *Intensive cropping*, in which large amounts of plant nutrients are removed in the harvest, accelerates the depletion of micronutrient reserves. *Extremes of soil pH* can markedly affect micronutrient availability (see Figure 9.14).

Soil erosion influences micronutrient availability by removing topsoil in which much of the potentially available micronutrient is held. Erosion also exposes subsoil horizons that are often higher in pH than the topsoil, a condition that leads to deficiencies of iron and zinc. Eroded ridges or hillsides are common sites of micronutrient deficiencies in some areas.

Micronutrient deficiencies and toxicities are often related to the level of these elements in the *parent materials* from which the soils form or to the minerals in the watersheds that might be transported to the soil in question. Trace elements (including micronutrients) are often applied to soils in industrial and domestic wastes. Small quantities of trace elements applied in these wastes can help alleviate nutrient deficiencies. However, repeated land applications of large quantities of wastes have increased the soil levels of some trace elements to their toxicity ranges. These levels are adversely affecting not only the plants, but the animals that consume them as well (see Chapter 15).

12.8   FACTORS INFLUENCING THE AVAILABILITY OF MICRONUTRIENT CATIONS

Iron, manganese, zinc, copper, and nickel are each influenced in a characteristic way by the soil environment. However, certain soil factors have the same general effects on the availability of all of them.

Soil pH

At low pH values, the solubility of micronutrient cations is high, and as the pH is raised, their solubility and availability to plants decrease. As the pH is increased, the ionic forms of the micronutrient cations are changed first to the hydroxy ions and, finally, to the insoluble hydroxides or oxides of the elements. Overliming of an acid soil often leads to deficiencies of iron, manganese, zinc, copper, and sometimes boron. Such deficiencies associated with high pH occur naturally in many of the calcareous soils of arid regions.

The general desirability of a slightly acid soil (with a pH between 6 and 7) largely stems from the fact that for most plants, this pH condition allows micronutrient cations to be soluble enough to satisfy plant needs without becoming so soluble as to be toxic (see Figure 9.14). Section 9.7 gives more specific information on the pH preferences of various plants.

Oxidation State and pH

The trace element cations iron, manganese, nickel, and copper occur in soils in more than one valence state. In the lower valence states, the elements are considered *reduced*; in the higher valence states, they are *oxidized*. Oxidation and reduction reactions in

relation to soil drainage are discussed in Sections 7.4 and 7.5. The changes from one valence state to another are, in most cases, brought about by microorganisms and organic matter. At pH values common in soils, the oxidized states of iron, manganese, and copper are generally much less soluble than are the reduced states.

The interaction of soil acidity and aeration in determining micronutrient availability is of great practical importance. Iron, manganese, and copper are generally more available under conditions of restricted drainage or in flooded soils. Very acid soils that are poorly drained may supply toxic quantities of iron and manganese.

At the high end of the soil pH range, good drainage and aeration often have the opposite effect. Well-oxidized calcareous soils are sometimes deficient in *available* iron, zinc, or manganese even though large *total* quantities of these trace elements are present. The hydroxides of the high-valence forms of these elements are too insoluble to supply the ions needed for plant growth.

There are marked differences in the sensitivity of different plant varieties to iron deficiency in soils with high pH. This is apparently caused by differences in their ability to solubilize iron immediately around the roots. Efficient varieties responds to iron stress by acidifying the immediate vicinity of the roots and by excreting compounds capable of reducing the iron to a more soluble form, with a resultant increase in its availability.

Organic Matter

Diagnosing copper deficiency: www.1.agric.gov.ab.ca/ $department/deptdocs.nsf/ all/agdex3476? opendocument

Organic matter, organic residues, and manure can supply organic compounds that react with micronutrient cations to form water-insoluble complexes that protect the nutrients from interactions with mineral particles that could bind them in even more insoluble forms. Other complexes provide slowly available nutrients as they undergo microbial breakdown. Organic complexes that enhance micronutrient availability are considered below. Reactions with organic matter and amorphous iron are the major controls on Cu sorption, whereas Zn is held mainly by the iron and probably by cation exchange reactions.

Examples of zinc and copper: www.extension.umn.edu/ distribution/cropsystems/ DC0720.html

The microbial decomposition of organic plant residues and animal manures can result in the release of micronutrients by the same mechanisms that stimulate the release of macronutrient ions. As was the case for macronutrients such as nitrogen, however, temporary deficiencies of the trace elements may occur when the residues are added due to the assimilation of micronutrients in the bodies of the active microorganisms.

Micronutrient-enriched organic products have been used as a nutrient source on soils deficient in available trace elements. For example, composts of iron-enriched organic materials, such as forest by-products, peat, animal manures, and plant residues, have been found to be effective on iron-deficient soils.

Role of Mycorrhizae

The nature of the mycorrhizal symbiosis and its importance in phosphorus nutrition were described in Sections 10.9 and 12.3, but it is worth mentioning here that mycorrhizae have also been shown to increase plant uptake of micronutrients (Table 10.5). Crop rotations and other practices that encourage a diversity of mycorrhizal fungi may thereby improve micronutrient nutrition of plants.

Surprisingly, mycorrhizae also appear to protect plants from excessive uptake of micronutrients and other trace elements where these elements are present in potentially toxic concentrations. Seedlings of such trees as birch, pine, and spruce are able to grow well on sites contaminated with high levels of zinc, copper, nickel, and aluminum only if their roots are sheathed by ectomycorrhizae. The mycorrhizae apparently help exclude these metallic cations from the root stele and prevent long-distance transport of metal cations within the plant.

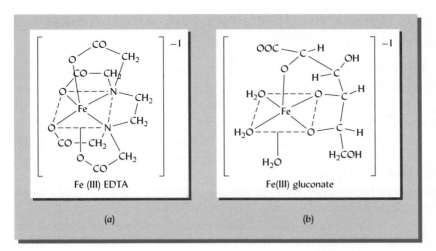

Figure 12.39
Structural formula for two common iron chelates, ferric ethylenediaminetetra-acetate (Fe-EDTA) (a) and ferric gluconate (b). In both chelates, the iron is protected and yet can be used by plants. [Diagrams from Clemens et al. (1990); reprinted by permission of Kluwer Academic Publishers]

Organic Compounds as Chelates

The cationic micronutrients react with certain organic molecules to form organometallic complexes called **chelates**. If these complexes are soluble, they increase the availability of the micronutrient and protect it from precipitation reactions. Conversely, formation of an insoluble complex will decrease the availability of the micronutrient.

A chelate (from the Greek *chele*, "claw") is an organic compound in which two or more atoms are capable of bonding to the same metal atom, thus forming a ring. These organic molecules may be synthesized by plant roots and released to the surrounding soil, may be present in the soil humus, or may be synthetic compounds added to the soil to enhance micronutrient availability. In complexed form, the cations are protected from reaction with inorganic soil constituents that would make them unavailable for uptake by plants. Iron, zinc, copper, and manganese are among the cations that form chelate complexes. Two examples of an iron chelate ring structure are shown in Figure 12.39.

Chelating agents:
http://scifun.chem.wise.edu/chemweek/Chelates/Chelates.html

Stability of Chelates

Chelates vary in their stability and therefore in their suitability as sources of micronutrients. The stability of a chelate is measured by its stability constant, which is related to the tenacity with which a metal ion is bound in the chelate.

The stability constant is useful in predicting which chelate is best for supplying which micronutrient. An added metal chelate must be reasonably stable within the soil if it is to have lasting advantage. For example, the stability constant for EDDHA-Fe^{3+} is 33.9, but that for EDDHA-Zn^{2+} is only 16.8. We can therefore predict that if EDDHA-Zn were added to a soil, the Zn in the chelate would be rapidly and almost completely replaced by Fe^{3+} from the soil, leaving the Zn in the unchelated form and subject to precipitation.

It should not be inferred that only iron chelates are effective. The chelates of other micronutrients, including zinc, manganese, and copper, have been used successfully to supply these nutrients. Synthetic chelates, because of their high cost, are mainly used to ameliorate micronutrient deficiencies of fruit trees and ornamentals. Some of the chelators used in micronutrient fertilizers, such as gluconate, are naturally occurring and can supply certain micronutrients much more economically than can the more expensive aminopolycarboxylate compounds (e.g., EDDHA).

In addition, selection of plant varieties whose roots produce their own chelating agents and practices that encourage the production of natural chelating agents from decomposing organic matter may take increasing advantage of chelation phenomena to improve micronutrient fertility of soils.

Stability, chelation, and the chelate effect:
www.chem.uwimona.edu.jm:1104/courses/chelate.html

12.9   FACTORS INFLUENCING THE AVAILABILITY OF THE MICRONUTRIENT ANIONS

Chlorine, molybdenum, and boron are quite different chemically, so few generalizations can be made about their reactions in soils.

Chlorine

Chlorine is taken up by plants in larger quantities than any of the micronutrients except iron. Most of the chlorine in soils is in the form of the chloride ion, which leaches rather freely from humid-region soils. In semiarid and arid regions, a higher concentration might be expected, with the amount reaching the point of salt toxicity in some of the poorly drained saline soils.

There are no known natural soil conditions that reduce the availability and use of this element. Accretions of chlorine from the atmosphere, along with those from fertilizer salts such as potassium chloride, are sufficient to meet most crop needs. However, beneficial effects of chlorine on plant growth occur. Tropical palms, adapted to growth in coastal soils where ocean spray contributes much chlorine, sometimes show chlorine deficiency if they are grown on inland soils with relatively low chlorine levels. Chlorine helps to control several fungal diseases in plants and also has an indirect effect on plant nutrition since it tends to suppress nitrification. This leads to a higher NH_4^+ to NO_3^- ratio in the soil solution, and, as the NH_4^+ ion is taken up by the plants, to decrease in rhizosphere pH. This greater acidity increases the availability and uptake of manganese that may, in turn, suppress the "take all" disease.

Plants such as tomato and potato that require high levels of potassium can suffer from Cl^- toxicity if their potassium is supplied in its most common fertilizer form, potassium chloride (KCl). In such cases, at least part of the needed K should be supplied as potassium sulfate, thereby reducing the inadvertent chloride exposure.

Boron

Boron is one of the most commonly deficient of all the micronutrients. The availability of boron is related to the soil pH, being greatest in acid soils. While it is most available at low pH, boron is also rather easily leached from acid, sandy soils. Therefore, deficiency of boron on acid, sandy soils usually occurs because of the low supply of total boron rather than because of low availability of the boron present. Dicot plants require much more B than do monocots.

Soluble boron is present in soils mostly as boric acid $[B(OH)_3]$ or as $B(OH)_4^-$. These compounds can exchange with the OH groups on the edges and surfaces of low-activity clays and organic matter. The boron so adsorbed is quite tightly bound, especially between pH 7 and 9, the range of lowest availability of this element. This probably accounts for the lime-induced boron deficiency noted on some soils as the pH is raised to pH 7 and above.

Boron is also a component of soil organic matter that is released by microbial mineralization. Consequently, organic matter serves as a major reservoir for boron in many soils and exerts considerable control over the availability of this nutrient.

Boron availability is generally impaired by dry conditions, and B deficiencies are common in calcareous Aridisols.

Molybdenum

Soil pH is the most important factor influencing the availability and plant uptake of molybdenum. At low pH values, molybdenum is adsorbed by silicate clays and, more especially, by oxides of iron and aluminum through *ligand exchange* with hydroxide

Boron:
www.extension.umn.edu/distribution/cropsystems/DC0723.html

Boron management by plant type:
www.borax.com/agriculture/boron1.html

ions on the surface of the colloidal particles. The liming of acid soils will usually increase the availability of molybdenum. The effect is so striking that in some cases, especially in Australia and New Zealand, a primary reason for liming very acid soils is to supply molybdenum. In some instances, 30 g of molybdenum added to acid soils has given about the same increase in legume growth, as has the application of several megagrams of lime.

At high soil pH values, molybdenum availability may be excessive. **Molybdenosis** is a potentially fatal disorder caused by excessive molybdenum in the diet of livestock grazing plants grown on certain very high pH soils.

12.10 CONCLUSION

Although the cycling of each nutrient element involves specific and unique transformations, several general processes contribute to the larger picture of soil fertility. We have seen the important roles that soil biology plays in the release of plant-available mineral nutrients from organic compounds in soil organic matter and plant litter. Soil biology makes additional contributions by the microbial transformations so important in the cycling of nitrogen and sulfur and by the formation of organic chelating agents so important in making micronutrient cations available for plant uptake. For some nutrients (e.g., Ca, Mg, K), the chemical processes of cation exchange and mineral weathering largely control the availability for plant uptake and leaching. For others (e.g., N, S, Fe, Mn), (bio)chemical oxidation and reduction reactions play critical roles. Finally, soil pH affects the solubility of nearly all the nutrients but is often the dominant influence on the micronutrients Fe, Mn, Zn, and Mo.

Ultimately, optimal plant and human health requires not only an adequate supply of each essential element but also a proper balance among them. Imbalances and deficiencies of mineral nutrients not only hamper plant growth but also adversely affect the health of the animals and people for whom the plants serve as food. Micronutrient deficiencies are becoming increasingly common in agriculture as a result of higher levels of removal by ever-more-productive crops combined with breeding for higher yields, at the expense of micronutrient acquisition efficiency. In addition, inadvertent applications of micronutrients have actually decreased with declining use of micronutrient-containing organic amendments, low-analysis fertilizers, and inorganic amendments. With sufficient political and economic attention, innovations in soil management and plant breeding have the potential to bring help to the millions of people worldwide who suffer from serious dietary deficiencies of micronutrients, particularly iron and zinc.

The biogeochemical cycles of plant nutrients reveal many complex connections between soils and plants and between the soil–plant system and the larger environment. Among the plant nutrient elements, N, P, S, and Cu probably pose the greatest problems for the larger environment. Anaerobic microorganisms as well as fire can convert nitrogen and sulfur into gaseous forms. Once in the atmosphere, some of these gases contribute to ecosystem acidification, and some (e.g., NO_2) contribute to global warming. Excessive buildup of soil N and P, particularly in agricultural soils associated with concentrated animal production, are major sources of the nutrients that cause eutrophication of aquatic systems. Copper, if mismanaged, can poison both land (where it can accumulate to levels toxic to many plants) and water (where even small amounts are highly toxic to fish).

For any one of the nutrients, but especially for the macronutrients, there exists a critical need to balance exports from with imports to the soil system, lest soils become depleted and ecosystems go into decline. Nitrogen is unique in this regard, in that certain microorganisms can fix N_2, so plentiful in air, into $-NH_2$ forms that are biologically available. Industrial fertilizer manufacture can do the same, but at great cost in fossil fuels. Unfortunately, there is no comparable process for replenishing the supply of other nutrients, which must therefore be considered non-renewable resources to be carefully husbanded, conserved, and recycled. Phosphorus may be the nutrient in most critical need of conservation as current use rates are predicted to nearly exhaust its known reserves within this century.

The principles learned from this chapter will be applied in the next, which explores the practical management of nutrients to achieve optimal plant productivity and soil quality while minimizing economic and environmental costs.

STUDY QUESTIONS

1. A sandy loam soil under a golf course fairway has an organic matter content of 3% by weight. Calculate the approximate amount of nitrogen (in kg N/ha) you would expect this soil to provide for plant uptake during a typical year. Show your work and state what assumptions or estimates you made to do this calculation.

2. The grass in the fairway referred to in question 1 is mowed weekly from May through October and produces an average of 200 kg/ha dry matter in clippings each time it is mowed. The clippings contain 2.5% N on average (dry weight basis). How much N from fertilizer would need to be applied to maintain this growth pattern? Show your calculations and state what assumptions or estimates you made.

3. Significant amounts of both sulfur and nitrogen are added to soils by atmospheric deposition. In what situations is this phenomenon beneficial and under what circumstances is it detrimental?

4. Tests showed a soil to contain 25 kg nitrate-N per ha. About 2000 kg of wheat straw was applied to 1 ha of this land. The straw contained 0.4% N. How much N was applied in the straw? Explain why two weeks after the straw was applied, new tests showed no detectable nitrate N. Show your work and state what assumptions or estimates you made to do this calculation.

5. Chemical fertilizers and manures with high N contents are commonly added to agricultural soils. Yet these soils are often lower in total N than are nearby soils under natural forest or grassland vegetation. Explain why this is the case.

6. Why are S deficiencies in agricultural crops more widespread today than 20 years ago?

7. How can riparian forests or wetlands help reduce nitrate contamination of streams and rivers?

8. In some tropical regions, agroforestry systems that involve mixed cropping of deep rooted trees and shallow-rooted food crops are used. What advantages in nitrogen management do such systems have over monocropping systems that do not involve trees?

9. You have learned that nitrogen, potassium, and phosphorus are all "fixed" in the soil. Compare the processes of these fixations and the benefits and constraints they each provide.

10. Assume that you add a soluble phosphate fertilizer to an Oxisol and to an Aridisol. In each case, within a few months most of the phosphorus has been changed to insoluble forms. Indicate what these forms are and the respective compounds in each soil responsible for their formation.

11. How does the phosphorus content of cultivated soils in the eastern United States or northern Europe compare with that of nearby forest soils that have never been cleared? What is the reason for this difference?

12. Which is likely to have the higher buffering capacity for phosphorus and potassium, a sandy loam or a clay? Explain.

13. In the spring, a certain surface soil showed the following soil test: soil solution K = 20 kg/ha; exchangeable K = 200 kg/ha. After two crops of alfalfa hay that contained 250 kg/ha of potassium were harvested and removed, a second soil test showed soil solution K = 15 kg/ha and exchangeable K = 150 kg/ha. Explain why there was not a greater reduction in soil solution and exchangeable K levels.

14. What is the effect of soil pH on the availability of phosphorus, molybdenum, and iron, and what are the unavailable forms at the different pH levels?

15. What is *luxury consumption* of plant nutrients, and what are its advantages and disadvantages?

16. How does phosphorus that forms relatively insoluble inorganic compounds in soils find its way as a pollutant into streams and other waterways?

17. Compare the organic P levels in the upper horizons of a forested soil with those of a nearby soil that has been cultivated for 25 years. Explain the difference.

18. What portion of the plant would you look at to find symptoms of Mn and Mg deficiencies, respectively?

19. During a year's time, some 250 kg nitrogen and only 30 g molybdenum have been taken up by the trees growing on a hectare of land. Would you therefore conclude that the nitrogen was more essential for tree growth? Explain.

20. Since only small quantities of micronutrients are needed annually for normal plant growth, would it be wise to add large quantities of these elements now to satisfy future plant needs? Explain.

21. Iron deficiency is common for peaches and other fruits grown on highly alkaline irrigated soils of arid regions, even though these soils are quite high in iron. How do you account for this situation, and what would you do to alleviate the difficulty?

22. How do Fe and Al oxides effect the availability of Mo and B in soils? Explain.

23. Give an example of a fungal-caused plant disease that can be effectively reduced by fertilizing with a micronutrient. Name the micronutrient and explain why it helps control the disease.

24. What are *chelates*, how do they function, and what are their sources?

25. Two Aridisols, both at pH 8, were developed from the same parent material, one having restricted drainage, the other being well drained. Plants growing on the well-drained soils showed iron deficiency symptoms while those on the less-well-drained soil did not. What is the likely explanation for this?

26. Since boron is required for the production of good-quality table beets, some companies purchase only beets that have been fertilized with specified amounts of this element. Unfortunately, an oat crop following the beets does very poorly compared to oats following unfertilized beets. Give possible explanations for this situation.

REFERENCES

Bateman, E. J., and E. M. Baggs. 2005. "Contributions of nitrification and denitrification to N_2O emissions from soils at different water-filled pore space," *Biol. Fertil. Soils,* **41**:379–388.

Bittman, S., C. G. Kowalenko, D. E. Hunt, T. A. Forge, and X. Wu. 2006. "Starter phosphorus and broadcast nutrients on corn with contrasting colonization by mycorrhizae," *Agron. J.,* **98**:394–401.

Castellano, S. D., and R. P. Dick. 1991. "Cropping and sulfur fertilization influence on sulfur transformations in soil," *Soil Sci. Soc. Amer. J.,* **54**:114–121.

Clemens, D. F., B. M. Whitehurst, and G. B. Whitehurst. 1990. "Chelates in agriculture," *Fertilizer Research,* **25**:127–131.

Eickhout, B., A. F. Bouwman, and H. Van Zeijts. 2006. "The role of nitrogen in world food production and environmental sustainability," *Agric. Ecosyst. Environ.,* **116**:4–14.

Emsley, J. 2002. *The 13th Element: The Sordid Tale of Murder, Fire, and Phosphorus,* (New York: John Wiley & Sons).

Fageria, N. K., V. C. Baligar, and R. B. Clark. 2002. "Micronutrients in crop production," *Adv. Agron,* **77**:185–268.

Fenn, M. E., T. G. Huntington, S. B. Mclaughlin, C. Eagar, and R. B. Cook. 2006. "Status of soil acidification in North America," *Journal of Forest Science,* **52**:3–13.

Franzen, D. W. 2004. *Volatilization of Urea Affected by Temperatures and Soil pH.* North Dakota State University Extension Service. www.ag.ndsu.edu/procrop/fer/ureavo05.htm.

Galloway, J. N., and E. B. Cowling. 2002. "Reactive nitrogen and the world: 200 years of change," *Ambio,* **31**:64–71.

Glibert, P., J. Harrison, C. Heil, and S. Seitzinger. 2006. "Escalating worldwide use of urea: A global change contributing to coastal eutrophication," *Biogeochemistry,* **77**:441–463.

Goenadi, D. H., Siswanto, and Y. Sugiarto. 2000. "Bioactivation of poorly soluble phosphate rocks with a phosphorus-solubilizing fungus," *Soil Sci. Soc. Amer. J.,* **64**:927–932.

Guillard, K., and K. L. Kopp. 2004. "Nitrogen fertilizer form and associated nitrate leaching from cool-season lawn turf," *J. Environ. Qual.,* **33**:1822–1827.

International Joint Commission. 2002. *The Canada–United States Air Quality Agreement 2002 Progress Report.* International Joint Commission.

Jasinski, S. M. 2006. *Phosphate Rock,* (Reston, VA: U.S. Department of the Interior, U.S. Geological Survey), pp. 124–125. http://minerals.usgs.gov/minerals/pubs/commodity/phosphate_rock/phospmcs06.pdf.

Kalim, S., Y. P. Luthra, and S. K. Gandhi. 2003. "Cowpea root rot severity and metabolic changes in relation to manganese application," *Journal of Phytopathology*, **151**:92–97.

Kennedy, I. R., A. T. M. A. Choudhury, and M. L. Kecskes. 2004. "Non-symbiotic bacterial diazotrophs in crop-farming systems: Can their potential for plant growth promotion be better exploited?" *Soil Biol. Biochem.*, **36**:1229–1244.

Khalil, K., B. Mary, and P. Renault. 2004. "Nitrous oxide production by nitrification and denitrification in soil aggregates as affected by O_2 concentration," *Soil Biol. Biochem.*, **36**:687–699.

Kopittke, P. M., and N. W. Menzies. 2007. "A review of the use of the basic cation saturation ratio and the 'ideal' soil," *Soil Sci. Soc. Amer. J.*, **71**:259–265.

Lambert, K. F., and C. Driscoll. 2003. "Nitrogen pollution: From the sources to the sea," *Science Links Publication*, Vol. 1, No. 2 (Hanover, NH: Hubbard Brook Research Foundation).

Mandzak, J. M., and J. A. Moore. 1994. "The role of nutrition in the health of inland Western forests," *J. Sustainable Forestry*, **2**:191–210.

Marschner, H. 1995. *Mineral Nutrition of Higher Plants*, 2nd ed. (New York: Academic Press).

McLean, E. O. 1978. "Influence of clay content and clay composition on potassium availability, pp. 1–19, in G.S. Sekhon (ed.), *Potassium in Soils and Crops*. (New Delhi, India. Potash Research Institute of India).

Mengel, K., and E. A. Kirkby. 2001. *Principles of Plant Nutrition*, 5th ed. (Dordrecht, The Netherlands: Kluwer Academic Publishers).

Mortvedt, J. J., F. R. Cox, L. M. Shuman, and R. M. Welch (eds.). 1991. *Micronutrients in Agriculture*. SSSA Book Series, No. 4 (Madison, WI: Soil Science Society of America).

Munson, R. D. (ed.). 1985. *Potassium in Agriculture* (Madison, WI: American Society of Agronomy).

Neff, J. C., F. S. Chapin, and P. M. Vitousek. 2003. "Breaks in the cycle: Dissolved organic nitrogen in terrestrial ecosystems," *Frontiers of Ecology and the Environment*, **1**:205–211.

Nowak, C. A., R. B. Downard, Jr., and E. H. White. 1991. "Potassium trends in red pine plantations at Pack Forest, New York," *Soil Sci. Soc. Amer. J.*, **55**:847–850.

Patrick, W. H., Jr. 1982. "Nitrogen transformations in submerged soils," in F. J. Stevenson (ed.), *Nitrogen in Agricultural Soils*. Agronomy Series No. 27 (Madison, WI: Amer. Soc. Agron., Crop Sci. Soc. Amer., Soil Sci. Soc. Amer.).

Santamaria, P. 2006. "Nitrate in vegetables: Toxicity, content, intake and EC regulation," *Journal of the Science of Food and Agriculture*, **86**:10–17.

Schepers, J. S., and W. R. Raun (eds.). 2008. *Nitrogen in Agricultural Systems*. Agronomy Monograph 49 (Madison. WI: American Society of Agronomy).

Sharpley, A. N. 1990. "Reaction of fertilizer potassium in soils of differing mineralogy," *Soli Sci.*, **49**:44–51.

Sharpley, A. 2000. "Phosphorus availability," pp. D-18–D-38, in M. E. Summer (ed.), *Handbook of Soil Science* (New York: CRC Press).

Spencer, J. E. 2000. "Arsenic in groundwater," *Arizona Geology*, **30**(3).

Stern, D. I. 2006. "Reversal of the trend in global anthropogenic sulfur emissions," *Global Environmental Change*, **16**:207–220.

Stevenson, F. J. 1986. *Cycles of Soil Carbon, Nitrogen, Phosphorus, Sulfur, and Micronutrients* (New York: Wiley).

Tabatabai, S. J. 1986. *Sulfur in Agriculture*. Agronomy Series No. 27 (Madison, WI: Amer. Soc. Agron., Crop Sci. Soc. Amer., Soil Sci. Soc. Amer.).

Tiessen, H. (ed.). 1995. *Phosphorus in the Global Environment—Transfers, Cycles and Management* (New York: John Wiley and Sons). www.icsu-scope.org/downloadpubs/scope54/TOC.htm.

Torbert, H. A., K. W. King, and R. D. Harmel. 2005. "Impact of soil amendments on reducing phosphorus losses from runoff in sod," *J. Environ Qual.*, **34**:1415–1421.

Vaithiyanathan, P., and D. L. Correll. 1992. "The Rhode River watershed: Phosphorus distribution and export in forest and agricultural soils," *J. Environ. Qual.*, **21**:280–288.

van Kessel, C., T. Clough, and J.W. van Groenigen. 2009. "Dissolved Organic Nitrogen: An Overlooked Pathway of Nitrogen Loss from Agricultural Systems?" *J. Environ Qual.*, **38**:393–401.

Viets, F. J., Jr. 1965. "The plants' need for and use of nitrogen," in *Soil Nitrogen* (*Agronomy*, No. 10) (Madison, WI: American Society of Agronomy).

Weil, R. R., P. W. Benedetto, L. J. Sikora, and V. A. Bandell. 1988. "Influence of tillage practices on phosphorus distribution and forms in three Ultisols," *Agron. J.*, **80**:503–509.

Welch, R. M. 1995. "Micronutrient nutrition of plants," *Critical Reviews in Plant Science*, **14**(1):49–82.

Wood, B. W., C. C. Reilly, and A. P. Nyczepir. 2004. "Mouse-ear of pecan: A nickel deficiency," *Hortscience*, **39**:1238–1242.

Zhao, F. J., J. Lehmann, D. Solomon, M. A. Fox, and S. P. McGrath. 2006. "Sulphur speciation a*nd turnover in soils: Evidence from sulphur K-edge* xanes spectroscopy and isotope dilution studies," *Soil Biol. Biochem.*, **38**:1000–1007.

Pastoral scene belies unmanaged nutrients. (R. Weil)

13
Practical Nutrient Management

For every atom lost to the sea, the prairie pulls another out of the decaying rocks. The only certain truth is that its creatures must suck hard, live fast, and die often, lest its losses exceed its gains.
—ALDO LEOPOLD, *A SAND COUNTY ALMANAC* (1949)

As stewards of the land, soil managers must keep nutrient cycles in balance. By doing so they maintain the soil's capacity to supply the nutritional needs of plants and, indirectly, of us all. While undisturbed ecosystems may need no intervention, few ecosystems are so undisturbed. More often, human hands have directed the output of the ecosystem for human ends. Forests, farms, fairways, and flower gardens are ecosystems modified to provide us with lumber, food, recreational opportunities, and aesthetic satisfaction. By their very nature, managed ecosystems need management.

In managed ecosystems, nutrient cycles can become unbalanced through increased removals (e.g., harvest of timber and crops), through increased system leakage (e.g., leaching and runoff), through simplification (e.g., monoculture, be it of pine tree or cotton), through increased demands for rapid plant growth (whether the soil is naturally fertile or not), and through increased animal density (especially if imported feed brings in nutrients from outside the ecosystem). Some of the greatest impacts of land management are felt not on the land but in the water, where excess nutrients play havoc with aquatic ecosystems. The land manager is, of necessity, also a nutrient and environmental manager.

Nutrient management is one aspect of a holistic approach to managing soils in the larger environment. It aims to achieve four broad, interrelated goals: (1) cost-effective production of high-quality plants and animals, (2) efficient use and conservation of nutrient resources, (3) maintenance or enhancement of soil quality, and (4) protection of the environment beyond the soil.

In this chapter we will discuss methods of enhancing nutrient recycling, as well as sources of additional nutrients that can be applied

to soils systems. We will learn how to diagnose nutritional disorders of plants and correct soil fertility problems. Building on the principles set out earlier, this chapter contains much practical information on profitable production of abundant, high-quality plant products and on maintaining the quality of both the soil and the rest of the environment.

13.1 GOALS OF NUTRIENT MANAGEMENT[1]

Nutrient Budgets

150-year study of soil fertility impacts on plant ecology: http://news.bbc.co.uk/2/hi/sci/tech/4766081.stm

Managing farm nutrient balance: www.ew.govt.nz/Environmental-information/Land-and-soil/Managing-Land-and-Soil/Managing-farm-nutrients/

A useful step in planning nutrient management is to conceptualize the nutrient flows for the particular system under consideration. Such a flowchart should attempt to account for all the major inputs and outputs of nutrients. Simplified examples of such nutrient budgets are shown in Figure 13.1.

Addressing nutrient imbalances, shortages, and surpluses may call for analysis of nutrient flows, not just on a single farm or enterprise, but on a watershed, regional, or even national scale. For example, many countries in Africa are net exporters of nutrients. That is, exports of agricultural and forest products carry with them more nutrients than are imported into the country as fertilizers, food, or animal feed. During the past 30 years, an average of 22 kg/ha N, 2.5 kg/ha P, and 15 kg/ha K have been lost *annually* from about 200 million ha of cultivated land in sub-Saharan Africa (excluding South Africa). This net *negative* nutrient balance appears to be a contributing factor in the impoverishment of African soils, the reduction of agricultural productivity, and the stagnation or decline of national economies.

By contrast, in temperate regions, the cultivated land on average receives nutrients in excess of those removed in crops, runoff, and erosion. During the past 30 years, the nearly 300 million ha of cultivated temperate-region soils had a net *positive* nutrient balance of at least 60 kg/ha N, 20 kg/ha P, and 30 kg/ha K. While nutrient

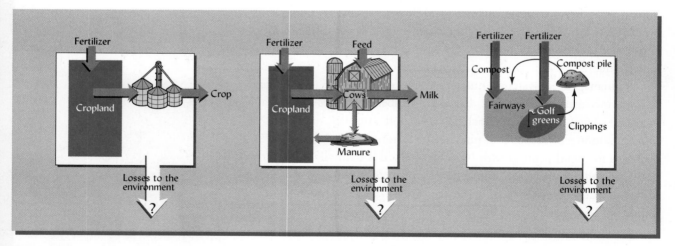

Figure 13.1

Conceptual nutrient flowcharts for a cash-grain farm, a dairy farm, and a golf course. Only the managed inputs, key recycling flows, and outputs are shown. Unmanaged inputs, such as nutrient deposition in rainfall, are not shown but must be taken into consideration in developing a complete nutrient management plan. Outputs that are difficult to manage, such as leaching and runoff losses to the environment, are shown as being variable. Such flowcharts are a starting point in identifying imbalances between inputs and outputs that could lead to wasted resources, reduced profitability, and environmental damage.

[1]For an overview of issues and advances in nutrient management for agriculture and environmental quality, see Magdoff et al. (1997). For a standard textbook on management of agricultural soil fertility and fertilizers, see Havlin et al. (2005).

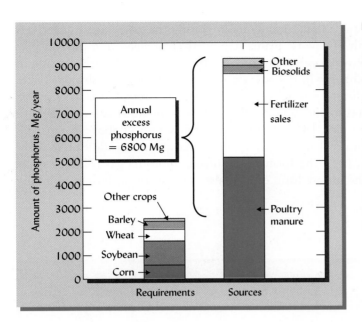

Figure 13.2

Phosphorus balance for the State of Delaware. Crop requirements estimated from recommended fertilizer rates based on soil tests and expected crop responses (see Section 13.10) totaled about 2600 Mg P. The manure generated by the state's poultry industry alone contained about twice as much P as required by all the crops in the state. The total amount of P applied in fertilizers and manures (plus some other P-containing wastes) totaled more than 9400 Mg P, leaving a surplus of more than 6800 Mg P. If the total P sources were spread on all 217,900 ha of cropland in the state, there would be 31 kg/ha of excess P each year. Most of this excess P came from poultry feed, and has resulted in P buildup in soils and increased P losses in runoff. [Data from Beegle et al. (2002)]

deficiencies still occur in some fields, the great majority of soils have experienced a nutrient buildup. Some of the excess nutrients move into streams, lakes, or the atmosphere, where they can contribute to environmental damage (Section 13.2).

Changes in the structure of agricultural production in some countries have led to *regional* nutrient imbalances and concomitant serious water-pollution problems. The concentration of livestock production facilities in a region that must import feed from other areas is a case in point. The animal manure produced in these areas contains nutrients far in excess of the amounts that can be used in an efficient and environmentally safe manner by crops in nearby fields. The concentrated poultry industry in Delaware is one example for which the imbalance between phosphorus sources and utilization potential has been documented (Figure 13.2).

Soil Quality and Productivity

The concept of using nutrient management to enhance soil quality goes far beyond simply supplying nutrients for the current year's plant growth. Rather, it includes the long-term nutrient-supplying and nutrient-cycling capacity of the soil, improvement of soil physical properties or tilth, maintenance of above- and belowground biological functions and diversity, and the avoidance of chemical toxicities. Likewise, the management tools employed go far beyond the application of various fertilizers (although this may be an important component of nutrient management). Nutrient management requires the integrated management of physical, chemical, and biological processes. The effects of tillage on organic matter accumulation (Chapter 11), the increase in nutrient availability brought about by earthworm activity (Chapter 10), the role of mycorrhizal fungi in phosphorus uptake by plants (Chapter 12), and the impact of fire on soil nutrient and water supplies (Chapter 7) are all examples of components of integrated nutrient management.

13.2   ENVIRONMENTAL QUALITY

Nutrient management impacts the environment most directly with water-quality problems caused by N and P. Together, these two nutrients are the most widespread cause of water-quality impairment in lakes and estuaries, and are second only to

Nutrients in the nation's waters—too much of a good thing? http://pubs.usgs.gov/circ/circ1136

sediment (see Chapter 14) among pollutants impairing the water quality of rivers and streams. The growth of aquatic plants (e.g., seaweeds, algae, and phytoplankton) often explodes when concentrations of N or P exceed critical levels, leading to numerous undesirable changes in the aquatic ecosystem. In most freshwater (lakes and streams), P is the limiting nutrient that can set off eutrophication (see Box 13.1). In saltier waters (estuaries and coastal areas), N is the nutrient most likely to cause eutrophication (see Box 13.2). Levels of total dissolved N above 2 mg/L are often considered above normal and damaging to the ecosystem. In addition to stimulating eutrophication, N in the form of dissolved ammonia gas (NH_3) can be directly toxic to fish. The nitrate form of N is also of concern for human drinking water (see Box 13.3).

BOX 13.1
PHOSPHORUS AND EUTROPHICATION

In unpolluted lakes and streams, the water is commonly clear, free of excess growth of algae and other aquatic plants, and is inhabited by highly diverse communities of organisms. When phosphorus is added to a phosphorus-limited lake, it stimulates a burst of algal growth (referred to as an *algal bloom*) and, often, a shift in the dominant algal species. The overfertilization is termed **eutrophication** (from the Greek *eutruphos*, meaning "well-nourished" or "nourishing"). **Natural eutrophication**—the slow accumulation of nutrients over centuries—causes lakes to slowly fill in with dead plants, eventually forming Histosols (see Figure 2.15). Excessive input of nutrients under human influence, called **cultural eutrophication**, tremendously speeds this process. Critical levels of phosphorus in water, above which eutrophication is likely to be triggered, are approximately 0.03 mg/L of dissolved phosphorus and 0.1 mg/L of total phosphorus.

During eutrophication, phosphorus-stimulated algae and plants may suddenly cover the surface of the water with what resembles a mat of algal scum and floating plants (Plate 95). When these aquatic weeds and algal mats die, they sink to the bottom, where their decomposition by microorganisms uses up the oxygen dissolved in the water. The process is accelerated by warm water temperatures. The decrease in oxygen (anoxic conditions) severely limits the growth of many aquatic organisms, especially fish. Such eutrophic lakes often become turbid, limiting growth of beneficial submerged aquatic vegetation and benthic (bottom-feeding) organisms that serve as food for much of the fish community. In extreme cases eutrophication can lead to massive fish kills (Figure 13.3).

Eutrophic conditions favor the growth of *Cyanobacter* (blue-green algae) at the expense of zooplankton, a major food source for fish. These *Cyanobacter* produce toxins and bad-tasting and bad-smelling compounds that make the water unsuitable for human or animal consumption. Filamentous algae can clog water treatment intake filters and thereby increase the cost of water remediation. Dense growth of both algae and aquatic weeds may make the water useless for boating and swimming. Furthermore, eutrophic waters generally have a reduced level of biological diversity (fewer species) and fewer fish of desirable species. (See also Box 13.2 on eutrophication and nitrogen.) Thus, eutrophication can transform clear, oxygen-rich, good-tasting water into cloudy, oxygen-poor, foul-smelling, bad-tasting, and possibly toxic water in which a healthy aquatic community cannot survive.

Figure 13.3
Massive fish kills can occur in sensitive waters. (Photo courtesy of R. Weil)

BOX 13.2
NITROGEN POLLUTION: DEAD ZONE IN THE GULF OF MEXICO

Under natural conditions, low levels of nitrogen limit aquatic algae growth—especially in salty and brackish water, which inhibits N-fixing algae. Increased human input of nitrogen can remove this constraint. The resulting degradation of estuaries and coastal waters is the most widespread water-quality problem induced by nitrogen pollution. The mouth of the Mississippi River in the Gulf of Mexico provides a major example. Off the coast of Louisiana, an enormous "dead zone" of water some 4 to 60 m deep reaches from the mouth of the Mississippi River westward nearly 500 km to Texas. Nutrient-rich freshwater carried by the river glides over the cooler, saltier (and therefore heavier) Gulf water. The nutrients (mainly N, but also P and Si) stimulate explosive growth of algae, which sink to the bottom when they die (Figure 13.4). In decomposing this dead tissue, microorganisms deplete the

oxygen dissolved in the water to levels unable to sustain animal life. Fish, shrimp, and other aquatic species either migrate out of the zone or die. This state of low oxygen in the water (less than 2 to 3 mg O_2/L) is known as **hypoxia**, and the process that brings it about is called **eutrophication** (see also Box 13.1).

Concentrations of N in the Mississippi River have tripled in the past 30 years, mainly due to human activities, especially those in agriculture. Critical assessments suggest that only about 11% of the N delivered by the river comes from sewage treatment plants and other point sources; nearly 50% comes from fertilizers and the rest mainly from farmland runoff and manure. Major efforts are required to help farmers and others improve their N use efficiency and reduce the transformation of this valuable nutrient into a pollutant.

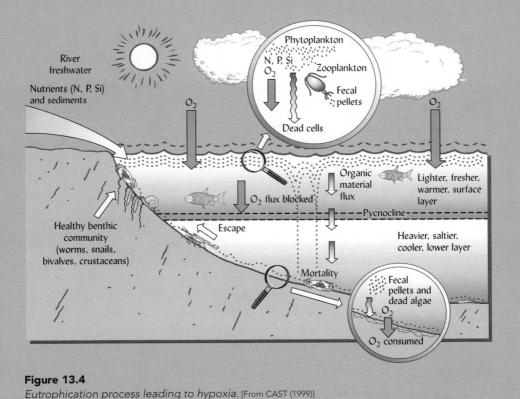

Figure 13.4
Eutrophication process leading to hypoxia. [From CAST (1999)]

The health hazard is determined by *concentration* of nitrate in the drinking water and the level of exposure (amount of water ingested, especially over long periods). The key factor for eutrophication damage to water quality is the *total load* (mass flux) of N delivered to the sensitive ecosystem. The total nitrogen load may be comprised partly of organic and ammonium forms of N transferred from the land in surface runoff or on eroded soil material, but N leached through the soil profile as nitrate (along with soluble organic N) is often the main contributor.

Most industrialized countries have made great strides in reducing nutrient pollution from factory and municipal sewage outfalls (called **point sources** because these

Nitrate pollution threatens amphibians: www.on.ec.gc.ca/wildlife/factsheets/nitrate-e.html

BOX 13.3
SOIL, NITRATE, AND YOUR HEALTH[a]

Mismanagement of soil nitrogen can result in levels of nitrates in drinking water (usually in groundwater) and food (mainly leafy vegetables) that may seriously threaten human health. While nitrate itself is not directly toxic, once ingested, a portion of the nitrate is reduced by bacterial enzymes to nitrite, which *is* considered toxic.

The most widely known (though actually quite rare) malady caused by nitrite is **methemoglobinemia**, in which the nitrites decrease the ability of hemoglobin in the blood to carry oxygen to the body cells. Inadequately oxygenated blood is blue rather than red. This, and the fact that infants under three months of age are much more susceptible to this illness than older individuals, accounts for the condition being commonly referred to as "blue baby syndrome." Most deaths from this disease have been caused by infant formula made with high-nitrate water. With the aim of protecting infants from methemoglobinemia, governments have set standards to limit the nitrate concentrations allowed in drinking water (Figure 13.5). In the United States this limit is 10 mg/L NO_3^-–N (= 45 mg/L nitrate), and in the European Union it is 50 mg/L nitrate (= 11 mg/L NO_3^-–N).

Of greater potential concern is the tendency of nitrate to form N-nitroso compounds in the stomach by binding with such organic precursors as amines derived from proteins. Certain N-nitroso compounds are known to be highly toxic, causing cancer in some 40 species of test animals, so the threat to humans must be given serious consideration.

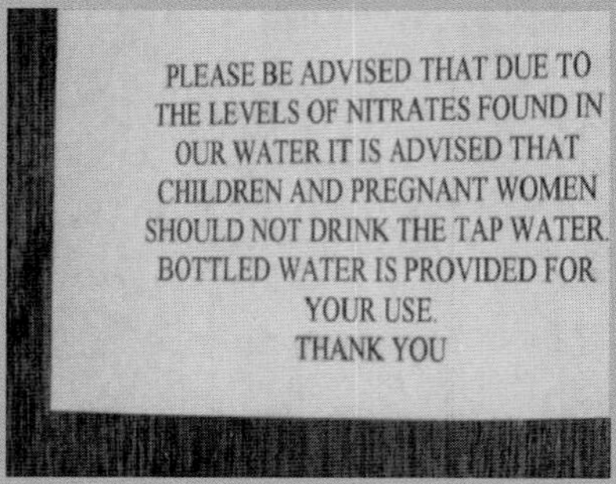

PLEASE BE ADVISED THAT DUE TO THE LEVELS OF NITRATES FOUND IN OUR WATER IT IS ADVISED THAT CHILDREN AND PREGNANT WOMEN SHOULD NOT DRINK THE TAP WATER. BOTTLED WATER IS PROVIDED FOR YOUR USE. THANK YOU

Figure 13.5
Hotel warning sign.

Nitrates (or nitrites formed therefrom) have also been reported to promote certain types of diabetes, stomach cancers, interference with iodine uptake by the thyroid gland, and certain birth defects. While documenting cause and effect in chronic diseases is always uncertain, many of these effects appear to be associated with nitrate concentrations much lower than the drinking water limits just mentioned.

On the other hand, several research studies suggest that ingestion of nitrate does no harm and may actually provide protection against bacterial infections and some forms of cardiovascular diseases and stomach cancers. Therefore, while a precautionary approach is probably wise, we must conclude that the "jury is still out" on the health risks of nitrates in drinking water and vegetables.

[a]For differing perspectives on nitrate effects on health, see Santamaria (2006), L'hirondel and L'hirondel (2002), and Addiscott (2006).

sources are clearly localized). However, much less has been accomplished with regard to controlling nutrients in the runoff water coming from the landscape (called **nonpoint sources** because these sources are diffuse and not easily identified). For example, agricultural activities account for some 40% of the N and P loads in the 170,000 km^2 Chesapeake Bay watershed (on the U.S. East Coast), while point sources account for about 20%. Streams draining forestland and rangeland are generally far lower in nutrients than those draining watersheds dominated by agricultural and urban land uses.

Nutrient Management Plans

One tool for reducing nonpoint source N and P pollution is a nutrient management plan—a document that records an integrated strategy and specific practices for how nutrients will be used in plant production (Table 13.1). In some cases these plans are legal documents used to implement government regulations. A nutrient management plan is prepared by a specially trained soil scientist who consults closely with the landowner to meet both environmental goals and practical needs. The plan attempts to balance the inputs of N and P (and other nutrients) with their desirable outputs (i.e., removal in harvested products). A goal is to prevent undesirable outputs (runoff, leaching) that exceed **maximum allowable daily loadings (MDL)**, the largest amount of nutrient runoff and leaching (in g ha^{-1} day^{-1}) permitted from an area of

Table 13.1
TYPICAL COMPONENTS OF A NUTRIENT MANAGEMENT PLAN

This chapter explains many of the tools used to plan for nutrient application to land in a manner that maximizes nutrient-use efficiency and minimizes water pollution risks.

- Aerial site photographs or maps and a soil map
- Current and/or planned plant production sequences or crop rotations
- Soil test results and recommended nutrient application rates
- Plant tissue analysis results
- Nutrient analysis of manure or other soil amendments

- Realistic yield goals and a description of how they were determined
- A complete nutrient budget for N, P, and K in the production system
- An accounting of all nutrient inputs such as fertilizers, animal manure, sewage sludge, irrigation water, compost, and atmospheric deposition

- Planned rates, methods, and timing of nutrient applications
- Location of environmentally sensitive areas or resources, if present
- The potential risk of N and/or P water pollution as assessed by a *N Leaching Index*, a *P Site Index*, or other acceptable assessment tools

land. An important component of many nutrient management plans, the *P site index*, is explained in Section 13.11. In the United States, practices officially sanctioned to implement these strategies are known as **best management practices (BMPs)**. Four general types of practices will now be briefly considered: (1) buffer strips, (2) cover crops, (3) conservation tillage, and (4) forest stand management.

Riparian Buffer Strips   Buffer strips of dense vegetation situated along the bank (the **riparian zone**) of a stream or lake are a simple and generally cost-effective method of protecting water from the polluting effects of a nutrient-generating land use. Fertilized cropland, poultry or livestock operations, farmland that has been amended with organic wastes, forest harvest operations, and urban development are examples of land uses that have the potential to generate nutrient or sediment loadings. The vegetation in the buffer strips may consist of natural or planted species, including grasses, shrubs, trees (Plate 60), or a combination of these vegetation types (Figure 13.6).

Water running off the surface of the nutrient-rich land passes through the riparian buffer strip before it reaches the stream. Trees (and their litter layer) or grass plants

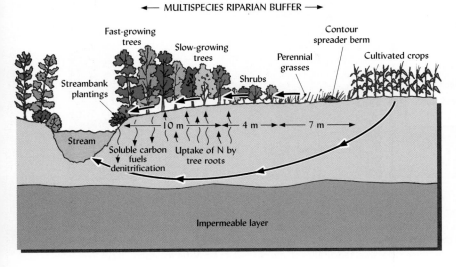

Figure 13.6

A general plan for a multispecies riparian buffer strip designed to protect the stream from nutrients and sediment in cropland runoff while also providing wildlife habitat benefits. A grass-covered level berm spreads runoff water evenly to avoid gullies. The perennial grasses filter out sediments and take up dissolved nutrients. Deep tree roots remove some nutrients from shallow groundwater. Soluble carbon from tree litter percolates downward to provide an energy source for anaerobic denitrifying bacteria that can remove additional nitrogen from shallow groundwater. The woody vegetation also provides wildlife habitat, shades the stream, and provides in-stream woody debris for fish habitat. A total buffer width of 10 to 20 m is usually sufficient to obtain most of the potential environmental benefits. (Diagram courtesy of R. Weil)

(and their thatch layer) reduce water velocity so most of the sediment and attached nutrients settle out. In addition, dissolved nutrients are adsorbed by the soil, immobilized by microorganisms, or are taken up by the buffer strip plants. Occasional mowing of grass or thinning of tree stands may help maintain high rates of nutrient uptake. The decreased flow velocity also increases the retention time—the length of time during which microbial action can work to break down chemicals such as pesticides before they reach the stream. Under some circumstances, buffer strips along streams encourage denitrification that removes nitrate from the groundwater flowing under them (although emission of nitrous oxide to the atmosphere may be an undesirable consequence of such nitrate reduction; see Figure 12.1).

Cover Crops

Dynamic cover crop bibliography: www.nal.usda.gov/wqic/Bibliographies/dynamic.html#cover

Instead of being harvested, a *cover crop* is grown to provide vegetative cover for the soil and then is killed and either left on the surface as a mulch or tilled into the soil as a *green manure*. Numerous benefits can be achieved in comparison to leaving the soil unvegetated for the off-season. The plants may provide habitat for wildlife and for beneficial insects; they protect the soil from the erosive forces of wind and rain; they add to the soil organic matter; they increase the rate of water infiltration; and, if leguminous, they may increase the nitrogen available in the soil. In addition, cover crops can reduce the loss of nutrients and sediment in surface runoff.

Cover Crops Reduce Leaching Losses   Cover crops can also serve as important nutrient management tools to reduce the leaching losses of nutrients, principally nitrogen (Figure 13.7). In many temperate humid regions, the greatest potential for leaching of nitrate from cropland occurs during the fall, winter, and early spring, after harvest and before planting of the main crop. During this time of vulnerability, an actively growing cover crop will reduce percolation of water and remove much of the nitrogen from the water that does percolate, incorporating this nutrient into plant tissue. For this purpose, an ideal cover crop should rapidly produce an extensive root system once the main crop has ceased growth. Largely because of their more rapid root growth in fall, winter annual cereals (rye, wheat, oats) and Brassicas (rape, forage radish, mustards) have proven to be more efficient than legumes (vetch, clover, etc.) at mopping up leftover soluble nitrogen.

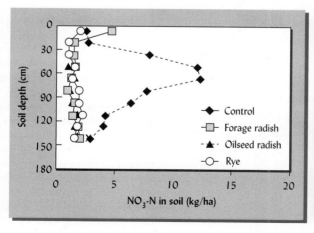

Figure 13.7

Cover crops can capture soluble nitrogen (N) left in the soil profile after the main cropping season. They thereby substantially reduce N leaching to groundwater during the winter. Forage radish and rye are among the temperate region cover crops (photo) capable of capturing more than 100 kg/ha of residual N in fall, cleaning the soil profile of soluble N to considerable depths. The graph shows nitrate-N in November in a sandy Ultisol, expressed as kg N/ha for each 15 cm depth increment. For comparison, the control plots had some weeds, but no cover crop. [From Dean and Weil (2009)]

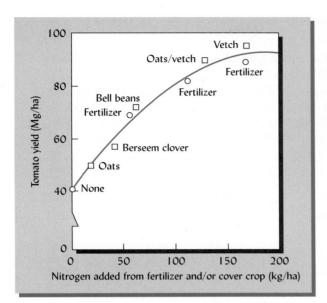

Figure 13.8

Effects of winter cover crops and inorganic fertilizer on the yield of the following main crop of processing tomatoes grown on a Xeralf soil in California. The amount of nitrogen shown on the x-axis was added either as inorganic fertilizer or as aboveground residues of the cover crops. Note that nitrogen from either source seemed to be equally effective. For the tomato crop, which requires nitrogen over a long period of time, vetch alone or vetch mixed with oats produced enough nitrogen for near-optimal yields. [Data abstracted from Stivers et al. (1993). © Lewis Publishers, an imprint of CRC Press, Boca Raton, Florida]

Legume Cover Crops Reduce the Need for Nitrogen Fertilizer

A legume cover crop system may be able to replace part or all of the nitrogen fertilizer normally used to grow the main crop (Figure 13.8). The cover crop is usually killed mechanically or by a herbicide spray about a week before the main crop is planted. The rate of N release from cover crop residues depends on whether they are left on the surface as a mulch in a no-till system (see Section 14.6) or plowed under in a conventional tillage system. When evaluating the feasibility of using cover crops to nitrogen supply, consideration should be given to the many other benefits as well as the costs and risks of growing the cover crop.

Nitrogen is supplied to the following main crop over a period of a month or more as the cover crop residues decay, rather than all at once like a soluble fertilizer application. In most cases the low C/N ratio legume residues decay rapidly enough to keep up with the nitrogen demands of the main crop. Legume cover crop systems have been adapted for vineyards, rice paddies, grain crops, vegetable fields, and home gardens wherever there is sufficient water to support both the cover crop and the main (summer) crop. Cover crops are increasingly important in organic farming systems because environmental restrictions on excessive phosphorus application (Section 13.12) are forcing organic farmers to reduce their traditional dependence for nitrogen on purchased manure and compost (which contain high levels of phosphorus).

In temperate regions, winter annual legumes, such as vetch, clovers, and peas, can be sown in fall after the main crop harvest or, if the growing season is short, they can be seeded by airplane while the main crop is still in the field. After surviving the coldest part of the winter in a dormant state, the cover crop will resume growth in spring and associated microorganisms will fix as much as 3 kg/ha of nitrogen daily during the warmer spring weather. The amount of nitrogen provided is therefore partially determined by how long the cover crop is allowed to grow.

Calculating N credits for legumes:
www.extension.umn.edu/distribution/cropsystems/DC3769.html

Conservation Tillage

The term **conservation tillage** applies to agricultural practices that keep at least 30% of the soil surface covered by plant residues. The effects of conservation tillage on soil properties and on the prevention of soil erosion are discussed elsewhere in this textbook (see Sections 6.4 and 14.6). Here, we emphasize the effects on nutrient losses.

Table 13.2

INFLUENCE OF WHEAT PRODUCTION AND TILLAGE ON ANNUAL LOSSES OF PHOSPHORUS IN RUNOFF WATER AND ERODED SEDIMENTS COMING FROM SOILS IN THE SOUTHERN GREAT PLAINS

The total phosphorus lost includes the phosphorus dissolved in the runoff water and the phosphorus adsorbed to the eroded particles.[a] Although cattle grazing on the natural grasslands probably increased losses of phosphorus from these watersheds, the losses from the agricultural watersheds were about 10 times as great. The no-till wheat fields lost much less particulate phosphorus, but more dissolved phosphorus, than the conventionally tilled wheat fields.

Location and soil	Management	Dissolved P	Particulate P	Total P
		kg P/ha/yr		
El Reno, Okla. Paleustolls, 3% slope	Wheat with conventional plow and disk	0.21	3.51	3.72
	Wheat with no-till	1.04	0.43	1.42
	Native grass, heavily grazed	0.14	0.10	0.24
Woodward, Okla. Ustochrepts, 8% slope	Wheat with conventional sweep plow and disk	0.23	5.44	5.67
	Wheat with no-till	0.49	0.70	1.19
	Native grass, moderately grazed	0.02	0.07	0.09

[a]Wheat was fertilized with up to 23 kg/ha of P each fall.
Data from Smith et al. (1991).

Compared to plowed fields with little residue cover, conservation tillage usually reduces the total amount of water running off the land surface, and reduces even more the load of nutrients and sediment carried by that runoff (Table 13.2). When combined with a cover crop, the reductions are greater still. The relatively small amounts of nutrients lost from untilled land (no-till cropland, pastures, and forests) tend to be mostly dissolved in the water rather than attached to sediment particles, while the reverse is true for tilled land. Because of large, sediment-associated nutrient losses, the total nutrient loss in surface runoff from conventionally tilled land generally is far greater than that from land where no-till or conservation tillage methods are used. On the other hand, the loss of nitrate by leaching may be somewhat greater with conservation tillage than conventional tillage. Also, animal manure applied to the surface of no-till cropland may lose much dissolved phosphorus if a heavy rain causes surface runoff to occur shortly after the application.

Nutrient Losses Associated with Forest Management

Undisturbed forests lose nutrients primarily by (1) leaching and runoff of dissolved ions and organic compounds, (2) erosion of nutrient-containing organic litter and mineral particles, and (3) volatilization of certain nutrients, especially during fires. However, weathering from soil and rock minerals, combined with atmospheric deposition, usually can maintain the plant-available supply of most elements. Nitrogen losses from forested ecosystems commonly range from about 1 to 5 kg/ha each year, while atmospheric inputs of nitrogen (not including biological nitrogen fixation) are usually two to three times as great. Annual losses of phosphorus from forest soils are typically very low (<0.1 kg/ha), and are closely balanced by inputs of this element from the atmosphere and by slow release from mineral weathering.

Management of forests to produce marketable wood products tends to increase losses by all three pathways just mentioned, plus it adds a fourth very significant pathway, namely, removal of nutrients in forest products (whole trees, logs, or pine straw).

Managed Forests for Healthy Ecosystems: www.utextension.utk.edu/ publications/pbfiles/pb1574 .pdf

Forest management practices that physically disturb the soil (and therefore tend to increase nutrient losses mainly by erosion) include building roads for timber harvesting, dragging logs on the ground (skidding), and after-harvest site preparation for planting new trees (see Section 14.9).

Disturbance   Disturbance of forest soil not only leaves it more vulnerable to erosion, but also can alter the nutrient balance in several other ways. Carefully planned tree harvesting and regeneration using methods that minimize soil disturbance and hasten revegetation can keep nutrient losses to low levels. Two practices should generally be avoided, as they can be particularly damaging to the forest nutrient cycle: First, extended suppression of unwanted vegetation with repeated use of herbicides is inadvisable because it delays the repopulation of the soil with active roots. Second, windrowing of stumps and slash (tree branches and tops left after harvesting logs) can be detrimental even if it clears land for easy replanting. The slash in most forests has a very high C/N ratio (see Section 12.3), so its slow decay promotes immobilization and reduces the loss of nitrogen mineralized in the forest floor. Wildfires may increase the loss of phosphorus, primarily via eroded sediment (Figure 13.9). Erosion tends to transport predominantly the clay and organic matter fractions of the soil (which are relatively rich in phosphorus), leaving behind the coarser, lower-phosphorus fractions.

Nutrient Losses   Because of increased litter decomposition and reduced plant uptake, streams draining many clear-cut watersheds periodically carry elevated levels of nitrogen and other nutrients for many years after the timber harvest has occurred (Figure 13.10). The concentrations of nitrogen are usually (but not always) too small (<2 mg N/L) to immediately threaten stream water quality. However, this loss of nutrients, combined with nutrient removal in the harvested trees, raises concerns about soil depletion and site productivity, especially on sites that are nutrient-poor to begin with.

Fertilization   Fertilizer application to forests is becoming an increasingly common practice. As might be expected, increases in peak nutrient exports can often be detected when forested watersheds undergo fertilization. Although the effects of forest fertilization on water quality do not yet approach those associated with fertilizer use in agriculture, foresters would be well served to study the lessons learned from fertilizer use on farms and so avoid making the nutrient management mistakes that have plagued agriculture.

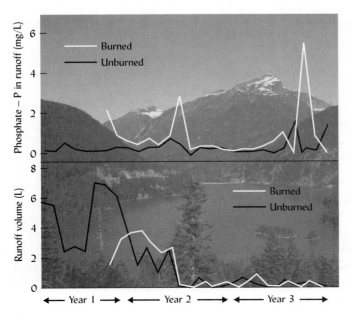

Figure 13.9

Effect of wildfires on nutrient runoff in the Sierra Nevada Mountains surrounding Lake Tahoe. The renowned clarity of Lake Tahoe's waters has suffered in recent years, and research suggests that runoff of phosphorus (as well as N) from the mountain slopes is partly responsible for the eutrophication of the lake. Note that although wildfire had little effect on the runoff volumes generated, runoff from burned areas sporadically contained much more reactive phosphorus (PO_4–P), even several years after the fire. [Redrawn from Miller et al. (2006); Photo courtesy of R. Weil]

Figure 13.10

Mean monthly nitrate-N concentrations in stream water from two commercial forested watersheds in North Carolina. Data are shown for a calibration period when both watersheds were managed identically (1971–1976), a treatment period (1976–1977) when one watershed was clear-cut and replanted, and a post-harvest period (1978–1996) when both watersheds were left undisturbed. Nitrate-N concentrations increased in the logged watershed and continued to be higher even after 20 years of regrowth, suggesting that nutrient-recycling processes were disrupted in some fundamental way. However, when concentrations were multiplied by streamflow volume (not shown) to calculate nitrate-N mass losses, the differences between watersheds were relatively small, nitrate-N losses in the logged watershed ranging from 0.25 to 1.27 kg N ha^{-1} yr^{-1} more than the control watershed during the first 5 years after logging. These differences can be compared to 4.5 kg N ha^{-1} yr^{-1} received in atmospheric deposition. [Redrawn from Swank et al. (2001)]

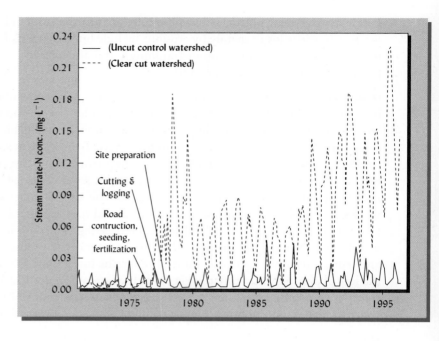

13.3   RECYCLING NUTRIENTS THROUGH ANIMAL MANURES

Huge quantities of farm manure are available each year for the recycling of essential elements to the land. For each kilogram of liveweight, farm animals produce about 4 kg dry weight of manure per year. In the United States, the farm animal population voids some 350 million Mg of manure solids per year, about 10 times as much as does the human population.

For centuries, the use of farm manure has been synonymous with a successful and stable agriculture. In this context, manure supplies organic matter and plant nutrients to the soil and is associated with the production of soil-conserving forage crops used to feed animals. About half of the solar energy captured by plants grown for animal feed ultimately is embodied in animal manure, which if returned to the soil can be a major driver of soil quality.

Where animal and crop production are integrated on a farm, manure handling is not too much of a problem. The use of carefully managed pasture for cattle can be maximized so that the animals themselves spread much of the manure while grazing. The total amount of nutrients in the manure produced on the farm is likely to be somewhat less than that needed to grow the crops; thus, modest amounts of inorganic fertilizers may be needed to make up the difference.

Concentrated Animal-Feeding Operations (CAFOs)

Unfortunately, in most industrialized countries, the advent of huge, concentrated animal-feeding operations (CAFOs) has changed the perception of animal manure from an *opportunity for recycling nutrients* as efficiently as possible to an *obligation for disposing of wastes* with as little cost and environmental damage as possible (Box 13.4, Figure 13.11). Due to inadequate manure-holding and disposal practices, the groundwater under such CAFOs as cattle feedlots is often polluted with nitrates and pathogens, and water from nearby wells may be unfit to drink (Table 13.3).

BOX 13.4
MANURE HAPPENS: *DIS*-INTEGRATING ANIMALS IN AGRICULTURE

To visualize the enormity of the manure disposal problem, consider a 100,000-head beef feedlot. We can estimate that the feedlot produces 200,000 Mg of manure (dry matter) per year:

$$\frac{4 \text{ Mg manure}}{\text{Mg liveweight}} \times \frac{0.5 \text{ Mg liveweight}}{\text{animal}} \times 100{,}000 \text{ animals} \simeq 200{,}000 \text{ Mg manure}$$

If this manure contains 2% N and the corn silage (whole plants) grown to feed the cattle removes some 240 kg N/ha, then we can estimate that the manure should be applied at 12 Mg/ha (~6 tons/acre):

$$\frac{1 \text{ Mg manure}}{0.02 \text{ Mg N}} \times \frac{240 \text{ kg N}}{\text{ha}} \times \frac{1 \text{ mg}}{1000 \text{ kg}} = \frac{12 \text{ Mg manure}}{\text{ha}}$$

In the unlikely event that the land had not been previously manured, some N fertilizer might be needed for the corn silage in the first year or two to supplement the N released from this amount of manure, but soon N released from previous years' applications would make supplementary fertilizer unnecessary. Utilization of the manure in this manner would require some 17,000 ha of land:

$$\frac{1 \text{ ha}}{12 \text{ Mg manure}} \times 200{,}000 \text{ Mg manure} \simeq 17{,}000 \text{ ha}$$

If the corn were grown in rotation with soybean (a legume crop that does not need applied N), then the total amount of cropland required for manure utilization would double to 34,000 ha or 340 km². To find this much cropland, some of the manure would have to be hauled 20 km (12 miles) or more from the feedlot! Finally, if soil phosphorus is already at (or above) optimal levels from previous manuring, as is usually the case near CAFOs, manure should be applied at a much lower rate tailored to meet the P (not N) needs of the crops, thus requiring an even larger land base.

In practice, to save transportation costs and time, manure is often applied to nearby fields at higher-than-needed rates. However, applications at these higher rates will likely result in the pollution of surface and groundwater by nitrogen and phosphorus and may cause salinity damage to crops and soils.

Figure 13.11
Feedlot in Colorado where 100,000 cattle are fed grain from distant farms. (Photo courtesy of R. Weil)

Table 13.3
SOIL AND SITE CHARACTERISTICS AFFECT NITRATES IN WELLS IN FIVE MIDWESTERN STATES
Sites with sandy soils, near cropland, near barnyards, and having shallow wells had the highest nitrate-N levels.

Characteristics	Texture of soils		Proximity to cropland		Proximity to feedlot or barnyard		Well depth		Shallow wells near barnyard or feedlot
	Sandy	Clayey	<6 m	Out of sight	<6 m	Out of sight	Deep >30 m	Shallow <15 m	
No. of wells	2412	6415	1684	3098	704	7520	5106	3467	158
Percent with nitrate-N levels >10 mg/L	7.2	3.1	6.4	1.8	12.2	2.8	1.1	9.7	25.3

Data from Richards et al. (1996).

Poultry and Swine Manure   Even more concentration of nutrients exists in the poultry and swine industries. Nearly all chickens and most hogs produced in the United States are grown in large "factory farms" that are concentrated near meat-processing plants. Not only do such CAFOs import nutrients in feed grains, but they also import calcium-P mineral feed supplements to compensate for the inability of their non-ruminant animals to digest *phytic acid*, the form of P found in most seeds. The manure produced therefore contains more nitrogen and far more phosphorus than the local cropland base can properly utilize. As a result, watersheds with large CAFOs tend to have very high levels of N and P in both the soil and in the water draining from the land.

Some Stop-Gap Measures   Although probably not long-term solutions to an unbalanced agricultural system, the public welfare may be served by such approaches as the following: (1) Discourage further manure applications to fields already saturated with nutrients; instead, facilitate transportation of manure to areas with low P soils. (2) Encourage the use of new corn varieties that contain less phytic acid P and more inorganic P, allowing better assimilation by nonruminant animals and making it less necessary to purchase P feed supplements, thereby reducing the amount of P excreted. (3) Promote composting of manure to reduce the volume of material and the solubility of the nutrients in it. (4) Eliminate the overfeeding of P supplements to all types of livestock, in order to reduce the concentration of P in the manure. (5) Mix iron or aluminum compounds with the manure to reduce the solubility of its phosphorus (see Section 12.3).

Maryland's manure-matching service helps manure do the most good: www.mda.state.md.us/resource_conservation/financial_assistance/manure_management/index.php

Nutrient Composition of Animal Manures

Generally, about 75% of the N, 80% of the P, and 90% of the K ingested by animals passes through the digestive system and appears in the excreta. For this reason, animal manures are valuable sources of both macro- and micronutrients. For a particular type of animal, the actual water and nutrient content of a load of manure will depend on the nutritional quality of the animals' feed, how the manure was handled, and the conditions under which it has been stored. The variability in nutrient content from one type of animal manure to another (e.g., poultry manure compared to horse manure) is even greater. Therefore, one has to be cautious with general statements about the value and use of manure.

Both the urine (except for poultry, which produce solid uric acid instead of urine) and feces are valuable components of animal manure. On the average, about *one-half of the N, one-fifth of the P, and three-fifths of the K* are found in the urine. Effective nutrient conservation requires that manure handling and storage minimize the loss of this valuable liquid portion.

The data in Table 13.4 show that on a dry-weight basis, animal manures contain from 2 to 5% N, 0.5 to 2% P, and 1 to 3% K. These values are one-half to one-tenth as great as are typical for commercial fertilizers.

Furthermore, as it comes from the animal, the water content is 30 to 50% for poultry to 70 or 85% for cattle (see Table 13.4). If the fresh manure is handled as a solid and spread directly on the land (Figure 13.12, *right*), the high water content is a nuisance that adds to the expense of hauling. If the manure is handled and digested in a liquid form or slurry and applied to the land as such, even more water is involved (Figure 13.12, *left*). All this water dilutes the nutrient content of manure, as normally spread in the field, to values much lower than those cited for dry manure in Table 13.4. The high content of water and low content of nutrients makes it difficult to economically justify transporting manure to distant fields where it might do the most good. However, the value of the micronutrients in manure (Table 13.4) and the nonnutrient

Table 13.4
COMMONLY USED ORGANIC NUTRIENT SOURCES: THEIR APPROXIMATE NUTRIENT CONTENTS AND OTHER CHARACTERISTICS

Along with nitrogen-fixing legumes grown in rotation and as cover crops, materials such as these (except sewage sludge and municipal solid wastes) provide the mainstay of nutrient supply in organic farming. The nutrient contents shown for animal manures are typical of well-fed livestock in confinement production systems. Manure from free-range animals not given feed supplements may be considerably lower in both nitrogen and phosphorus.

Material	Water,[a] %	Percent of dry weight						g/Mg of dry weight						Other Characteristics
		Total N	P	K	Ca	Mg	S	Fe	Mn	Zn	Cu	B	Mo	
Activated sewage sludge	<10	6	1.5	0.5	—	—	—	—	—	450	—	—	—	Most common form is Milorganite, N available over 2 to 6 months.
Coffee grounds[d]	60	1.6	0.01	0.04	0.08	0.01	0.11	330	50	15	40	—	—	May acidify soil.
Cottonseed meal	<15	7	1.5	1.5	—	—	—	—	—	—	—	—	—	Acidifies the soil. Commonly used as livestock feed.
Dairy cow manure[b]	75	2.4	0.7	2.1	1.4	0.8	0.3	1,800	165	165	30	20	—	May contain high-C bedding.
Dried blood	<10	13	1	1	—	—	—	—	—	—	—	—	—	Slaughterhouse by-product, N is available quickly.
Dried fish meal	<15	10	3	3	—	—	—	—	—	—	—	—	—	Incorporate or compost because of bad odors. Can feed to livestock.
Feedlot cattle manure[c]	80	1.9	0.7	2.0	1.3	0.7	0.5	5,000	40	8	2	14	1	May contain soil and soluble salts.
Hardwood tree leaves[f]	20	1.0	0.1	0.4	1.6	0.2	0.1	1,500	550	80	10	38	—	High Pb for some street trees.
Horse manure[c]	63	1.4	0.4	1.0	1.6	0.6	0.3	—	200	125	25	—	—	May contain high-C bedding.
Municipal solid waste compost[e]	40	1.2	0.3	0.4	3.1	0.3	0.2	14,000	500	650	280	60	7	May have high C/N and contain heavy metals, plastic, and glass.
Poultry (broiler) manure[b]	35	4.4	2.1	2.6	2.3	1.0	0.6	1,000	413	480	172	40	0.7	May contain high-C bedding, high soluble salts, arsenic, and ammonia.
Sewage sludge	80	4.5	2.0	0.3	1.5[g]	0.2	0.2	16,000[g]	200	700	500	100	15	May contain high soluble salts, toxic heavy metals.
Sheep manure[c]	68	3.5	0.6	1.0	0.5	0.2	0.2	—	150	175	30	30	—	May contain weed seeds.
Spoiled legume hay	40	2.5	0.2	1.8	0.2	0.2	0.2	100	100	50	10	1,500	3	May contain elevated Cu levels.
Swine manure[c]	72	2.1	0.8	1.2	1.6	0.3	0.3	1,100	182	390	150	75	0.6	
Wood wastes	—	—	0.2	0.2	0.2	1.1	0.2	2,000	8,000	500	50	30	—	Very high C/N ratio; must be supplemented by other N.
Young rye green manure	85	2.5	0.2	2.1	0.1	0.05	0.04	100	50	40	5	5	.05	Nutrient content decreases with advanced growth stage.

[a]Water content given for fresh materials. Processing and storage methods may alter water content to less than 5% (heat-dried) or to more than 93% (slurry).
[b]Broiler and dairy manure composition estimated from means of approximately 800 and 400 samples analyzed by the University of Maryland manure analysis program 1985–1990.
[c]Composition of swine, sheep, and horse manure calculated from North Carolina Cooperative Extension Service Soil Fact Sheets prepared by Zublena et al. (1993).
[d]Coffee grinds data from Krogmann et al. (2003).
[e]Composition of municipal solid waste compost based on mean values for the products of 10 composting facilities in the United States, as reported by He et al. (1995). Sulfur as sulfate-S.
[f]Hardwood leaf data from Heckman and Kluchinski (1996).
[g]Sludge contents of Ca and Fe may vary 10-fold depending on the wastewater treatment processes used.
Data derived from many sources.

Figure 13.12

Spreading dairy manure as liquid slurry after lagoon storage (left) and as a solid after storage in a pile (right). Such methods of manure spreading are effective means of recycling nutrients but are labor- and time-consuming. Many loads of manure will be hauled to fertilize each field. The manure should not be spread when the soil is frozen and should be incorporated as soon as possible after spreading. Calibration of the spreaders is important to prevent unintentional overapplication of nutrients.
(Photos courtesy of R. Weil)

Manure woes on Michigan dairy farm with cracking soils:
www.pmac.net/AM/big_stink.html

Hog manure handling:
www.epa.gov/agriculture/ag101/porkmanure.html

benefits of its organic matter (see Section 11.5) may be even greater than that of its N-P-K content, and should be included in any economic evaluation of manure transport.

In the future, improved manure handling technologies such as anaerobic digestion may provide energy from manure. This, along with more integrated animal farming systems, may help to redress the serious nutrient imbalances that have developed with regard to manure production in industrialized agriculture.

13.4   INDUSTRIAL AND MUNICIPAL BY-PRODUCTS

In addition to farm manures, four major types of organic wastes are of significance in land application: (1) municipal garbage, (2) sewage effluents and sludges (see Sections 16.5 and 15.7), (3) food-processing wastes, and (4) wastes of the lumber industry. Because of their uncertain content of toxic chemicals, these and other industrial wastes may or may not be acceptable for land application.

Society's concern for environmental quality has forced waste generators to seek nonpolluting, but still affordable, ways of disposing of these materials. Although once seen as mere waste products to be flushed into rivers and out to sea, these materials are increasingly seen as sources of nutrients and organic matter that can be used beneficially to promote soil productivity in agriculture, forestry, landscaping, and disturbed-land reclamation.

Garbage and Food-Processing Wastes

Municipal garbage has been used for centuries to enhance soil fertility in Asian countries. Most municipal solid waste (MSW) in industrial countries is incinerated or landfilled (see Section 15.10), but growing concerns about air quality and scarcity of landfill space are raising the level of interest in using soil application as a means of disposal. About 50 to 60% of MSW consists of decomposable materials (paper, food scraps, yard waste, street tree leaves, etc.). Once the inorganic glass, metals, and so forth are removed, MSW can be **composted** (Section 11.10), sometimes in conjunction with such nutrient-rich materials as sewage sludge or animal manure, to produce MSW compost that is then applied to the land. Because of its very low nutrient content (see Table 13.4), MSW compost is an expensive way to distribute nutrients. However, alternative disposal options are often even more expensive. Potentially, the entire annual production of organic-material MSW, some 80 million cubic meters in the United States, could be recycled as a soil amendment using less than 10% of the agricultural land in the country. Increasing numbers of compost operations of all sizes are being run by communities, small businesses, or even individuals.

Land application of food-processing wastes is being practiced in selected locations, but the practice is focused almost entirely on pollution abatement and not on soil enhancement. Liquid wastes are commonly applied through sprinkle irrigation to permanently grassed fields.

Wood Wastes

Sawdust, wood chips, and shredded bark from the lumber industry have long been sources of soil amendments and mulches, especially for home gardeners and landscapers. Because of their high C/N ratios and high lignin contents, these materials decompose very slowly. They make good mulching material, but do not readily supply plant nutrients. In fact, sawdust incorporated into soils to improve soil physical properties may cause plants to become nitrogen deficient unless an additional source of nitrogen is applied (see Section 11.3).

Wastewater Treatment By-Products

Sewage treatment has evolved over the past century to help society avoid polluting rivers and oceans with pathogens, oxygen-demanding organic debris, and eutrophying nutrients. The amount of material *removed* from the wastewater during the treatment process has increased tremendously. This solid material, known as sewage **sludge**, must also be disposed of safely. Advanced wastewater treatment is designed to remove nutrients (mainly P, but increasingly also N) from the sewage **effluent** (the treated water that is returned to the stream). There is a growing interest in using soils to assist with the sewage problem in two ways: (1) as a system of assimilating, recycling, or disposing of the solid sludge; and (2) as a means of carrying out the final removal of nutrients and organics from the liquid effluent.

Some cities operate sewage farms on which they produce crops, usually animal feeds and forages, that offset part of the expense of effluent disposal. Forest irrigation is a cost-effective method of final effluent cleanup and produces enhanced tree growth as a bonus (Figure 13.13). The rate of wood production is greatly increased as a result of both the additional water and the additional nutrients supplied therewith. This method of advanced wastewater treatment is used by a number of cities around the world.

In a carefully planned and managed effluent irrigation system, the combination of (1) nutrient uptake by the plants, (2) adsorption of inorganic and organic constituents by soil colloids, and (3) degradation of organic compounds by soil microorganisms results in the purification of the wastewater. Percolation of the purified water eventually replenishes the groundwater supply.

Sewage sludge has been spread on the land for decades, and its use will likely increase in the future. If sewage sludge has been treated to meet certain land-application standards (low pathogen and contaminant levels), the term **biosolids** may be applied. The product Milorganite, a dried, activated (oxygenated) sludge sold by the Milwaukee Sewerage Commission, has been widely used as a slow-release fertilizer in North America since 1927, especially on turfgrass. Numerous other cities market composted sludge products to landscaping and other specialty

Figure 13.13

Final treatment of sewage effluent and recharge of groundwater are being accomplished by natural soil and plant processes in this effluent-irrigated forest on Ultisols near Atlanta, Georgia. Nutrient flows, groundwater quality, and tree growth are carefully monitored. Some of the greatly increased production of wood is used as an energy source to run the sewage treatment plant. (Photo courtesy of R. Weil)

Biosolids to fertilize farmland (University of Nebraska): http://lancaster.unl.edu/enviro/biosolids/overview.htm

users. However, the great bulk of sewage sludge used on land is applied as liquid slurry or as partially dried cake.

Composition of Sewage Sludge   The composition of sludge varies from one sewage treatment plant to another, depending on the nature of treatment the sewage receives, especially the degree to which the organic material is allowed to digest. Representative values for plant nutrients are given in Table 13.4. Levels of plant micronutrient metals (zinc, copper, iron, manganese, and nickel) as well as other heavy metals (cadmium, chromium, lead, etc.) are determined largely by the degree to which industrial wastes have been mixed in with domestic wastes. In the United States, the levels of metals in sewage are far lower than they were in the past, because of source-reduction programs that require industrial facilities to remove pollutants *before* sending their sewage to municipal treatment plants. Nonetheless, vigilance must be maintained to avoid sludges too contaminated for safe land application.

In comparison with inorganic fertilizers, sludges are generally low in nutrients, especially potassium (which is soluble and found mainly in the effluent). Representative levels of N, P, and K are 4, 2, and 0.3%, respectively (see Table 13.4). If the sewage treatment precipitates phosphorus by reactions with iron or aluminum compounds, the phosphorus in the sludge will likely have a very low availability to plants.

Integrated Recycling of Wastes

In most industrialized countries, widespread recycling of organic wastes other than animal manures is a relatively recent phenomenon. In heavily populated areas of Asia, however, and particularly in China and Japan, such recycling has long been practiced (Figure 13.14). There, organic "wastes" have traditionally been used for biogas production, as food for fish, and as a source of heat from compost piles. The plant nutrients and organic matter are recycled and returned to the soil. Despite China's increasing use of chemical fertilizers, the traditional respect for what others might see as wastes supports the complex recycling systems that continue to supply about 50% of the nutrients used in China's agriculture. As they look to achieving a more sustainable future, Western countries have much to learn from traditional Chinese attitudes and practices.

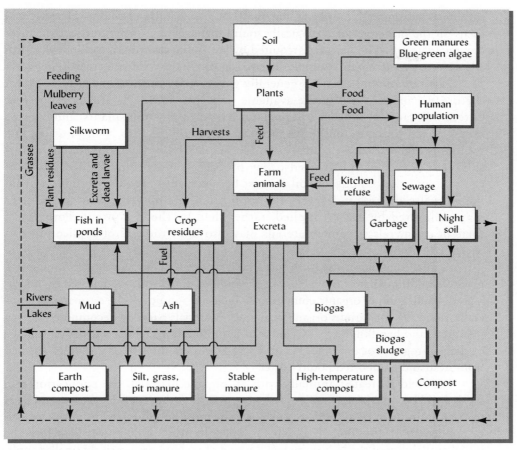

Figure 13.14
Traditional recycling of organic wastes and nutrient elements in China. Note the degree to which the soil is involved in the recycling processes. [Concepts from FAO (1977) and Yang (2006)]

13.5   PRACTICAL UTILIZATION OF ORGANIC NUTRIENT SOURCES[2]

Organic Farming   In this section we use the term "organic" to describe soil amendments that are organic in the chemical sense; any compound containing carbon is considered to be organic. When used to describe a type of farming system, the term "organic" refers to how the diverse components of a farm are integrated organically into a whole in the manner of a living organism. Principles and practices used in **organic farming** (see web link in margin) have been codified by certifying programs, including one administered by the United States Department of Agriculture. Not all soil amendments used in organic farming are chemically organic and not all chemically organic materials are approved for use in organic farming. For instance inorganic phosphate rock is approved for use, sewage sludge (a largely organic material) is not allowed at all, and raw (uncomposted) animal manures are restricted in their use. In general, organic farming allows only "naturally occuring" soil amendments and is more restrictive in the use of organic materials than is conventional farming. Some of the practices discussed in this section may not be approved for organic farming.

Principles of organic farming: www.ifoam.org/about_ ifoam/principles/index.html

Nutrient-Based Application Rates   In Section 11.5 we discussed the many beneficial effects on soil physical and chemical properties that can result from amendment of

---

[2]For practical information on managing soil fertility specifically for organic farming, see Heckman et al. (2009).

soils with decomposable organic materials, such as manure or sludge. Here we will focus on the principles of ecologically sound management of the nutrients in these organic materials. The first step (required by law in the case of sewage sludge) is usually to have a representative sample analyzed so you know what you are working with.

The rate of application is generally governed by the amount of N or P that the organic material will make available to plants. Nitrogen usually is the first criterion because N is needed in the largest quantity by most plants, and because excess N can present a pollution problem. The ratio of P to N in most organic sources is higher than in plant tissue. Consequently, if organic materials supply sufficient N to meet plant needs, they probably supply more P than the plant can use, and the buildup of soil P must be taken into account in the long run. For soils already high in P levels, the application rate for an organic amendment may be limited by the P supplied, rather than by the N content. Potentially, toxic heavy metals in some materials may also limit the rate of application (see Section 15.7).

A small fraction of the N in manure or sludge may be soluble (ammonium or nitrate) and immediately available, but the bulk of the N must be released by microbial mineralization of organic compounds. Table 13.5 indicates the estimated N mineralization rates for various organic materials. Materials partially decomposed during treatment and handling (e.g., by composting or digestion) release a lower percentage of their nitrogen. For example, Figure 13.15 compares the rate of nitrate-N released from fresh and composted poultry litter.

If a field is treated annually with an organic material, the application rate needed will become progressively smaller because, after the first year, the amount of N released from material applied in previous years must be subtracted from the total to be applied afresh (see Box 13.5). This is especially true for composts for which the initial availability of the N is quite low. Instead of making progressively smaller applications, another practical strategy is to use a moderate application every year, but supplement with N from other sources in the first few years until N release from previous and current applications can supply the entire requirement.

Table 13.5
RELEASE OF MINERAL NITROGEN FROM VARIOUS ORGANIC MATERIALS APPLIED TO SOILS, AS A PERCENTAGE OF THE ORGANIC NITROGEN ORIGINALLY PRESENT[a]

For example, if 10 Mg of poultry floor litter initially contains 300 kg N in organic forms, 50%, or 150 kg, of N would be mineralized in year 1. Another 15% (0.15 × 300), or 45 kg, of N would be released in year 2.

Organic nitrogen source	Year 1	Year 2	Year 3	Year 4
Poultry floor litter	50	15	8	3
Dairy manure (fresh solid)	25	18	9	4
Swine manure lagoon liquid	45	12	6	2
Feedlot cattle manure	35	15	6	2
Composted feedlot manure	20	8	4	1
Lime-stabilized, aerobically digested sewage sludge	40	12	5	2
Anaerobically digested sewage sludge	20	8	4	1
Composted sewage sludge	10	5	3	2
Activated, unstabilized sewage sludge	45	15	4	2

[a]These values are approximate and may need to be increased for warm climates or sandy soils and decreased for cold or dry climates or heavy clay soils.
Sources of data: Eghball et al. (2002) and Brady and Weil (1996).

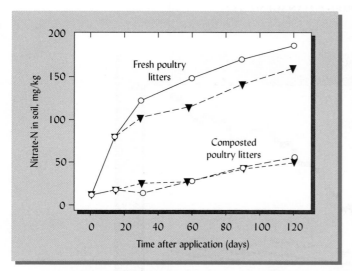

Figure 13.15

Nitrate nitrogen accumulation in a silt loam soil (Hapludalfs) incubated with composted or fresh poultry litter from two sources. Each of the litters was mixed with soil in amounts sufficient to provide 0.23 g of total N per kg of dry soil (approximately equivalent to 300 kg N/ha). Soil-litter mixtures were incubated at 25 °C for 120 days in controlled environment chambers. [Modified from Preusch et al. (2002)]

BOX 13.5
CALCULATION OF AMOUNT OF ORGANIC NUTRIENT SOURCE NEEDED TO SUPPLY NITROGEN FOR A CROP

If the soil phosphorus level is not already above the optimal range, the rate of release of available nitrogen usually determines the proper amount of manure, sludge, or other organic nutrient source to apply. The amount of nitrogen made available in any year should meet, but not exceed, the amount of nitrogen that plants can use for optimum growth. Our example here is a field producing corn two years in a row. The goal is to produce 7000 kg/ha of grain each year. This yield normally requires the application of 120 kg/ha of available nitrogen (about 58 kg of grain per kg N applied). We expect to obtain this N from a lime-stabilized sewage sludge containing 4.5% total N and 0.2% mineral N (ammonium and nitrate).

Year 1
Calculation of amount of sludge to apply per hectare:

% organic N in sludge $\quad$ = total N $-$ mineral N = 4.5% $-$ 0.2% = 4.3%.

Organic N in 1 Mg of sludge = 0.043 $\times$ 1000 kg = 43 kg N.

Mineral N in 1 Mg of sludge = 0.002 $\times$ 1000 kg = 2 kg N.

Mineralization rate for lime-stabilized sludge in first year (Table 13.5) = 40% of organic N.

Available N mineralized from 1 Mg sludge in first year = 0.40 $\times$ 43 kg N = 17.2 kg N.

Total available N from 1 Mg sludge = mineral N + mineralized N = 2.0 + 17.2 = 19.2 kg N.

Amount of (dry) sludge needed = 120 kg N/(19.2 kg available N/Mg dry sludge) = 6.25 Mg dry sludge.

Adjust for moisture content of sludge (e.g., assume sludge has 25% solids and 75% water).

Amount of wet sludge to apply: 6.25 Mg dry sludge/(0.25 Mg dry sludge/Mg wet sludge)
$\qquad$ = 6.25/0.25 = 25 Mg wet sludge.

Year 2
Calculate amount of N mineralized in year 2 from sludge applied in year 1:

Second year mineralization rate (Table 13.5) = 12% of original organic N.

N mineralized from sludge in year 2 = 0.12 $\times$ 43 Kg N/Mg $\times$ 6.25 Mg dry sludge
$\qquad$ = 2.25 kg N from sludge in year 2.

Calculate amount of sludge to apply in year 2:

N needed from sludge applied in year 2 = N needed by corn $-$ N released from sludge applied in year 1
$\qquad$ = 120 kg $-$ 32.25 kg = 87.75 kg N needed/ha.

Amount of (dry) sludge needed per ha = 87.75 kg N/(19.2 kg available N/Mg dry sludge) = 87.75/19.2
$\qquad$ = 4.57 Mg dry sludge/ha.

Adjust for moisture content of sludge (e.g., assume sludge has 25% solids and 75% water).

Mg of wet sludge to apply: 4.57 Mg dry sludge/(0.25 Mg dry sludge/Mg wet sludge) = 4.57/0.25
$\qquad$ = 18.3 Mg wet sludge.

Note that the 10.82 Mg dry sludge (6.25 + 4.57) also provided plenty of P: 216 kg P/ha (assuming 2% P; see Table 13.4), an amount that greatly exceeds the crop requirement and will soon lead to excessive buildup of P.

Figure 13.16

World fertilizer use since 1960 by major nutrient element. The use of nitrogen (N) has increased much faster than that of phosphorus (P) and potassium (K). The dip in world fertilizer use in the early 1990s was due mainly to drastic reductions in Russia and Ukraine after the collapse of the Soviet Union. In most industrialized countries fertilizer use has leveled off or declined, but it continues to increase in China, India, and much of the developing world. In the latter countries, the removal of nutrients in crop harvests may still exceed the amounts returned to the soil. [Data selected from FAO (2006)]

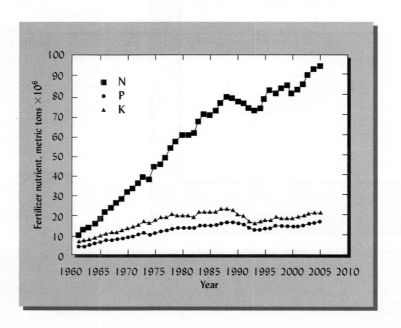

13.6   INORGANIC COMMERCIAL FERTILIZERS

The worldwide use of fertilizers on farms increased dramatically since the middle of the 20th century (Figure 13.16), accounting for a significant part of the equally dramatic increases in food production during the same period. The need to supplement forest soil fertility is also increasing as the demands for forest products increase the removal of nutrients and competing uses of land leave forestry with more infertile, marginal sites. Fertilization of new forest plantings and existing stands is increasingly practiced in Japan, northwestern and southeastern United States, the Scandinavian countries, and Australia, with the fertilizer spread by helicopter.

Regional Use of Fertilizers

Fertilizer use statistics within a region can suggest whether fertilizers are contributing to soil quality enhancement or to environmental degradation. For example, in Europe and East Asia where moisture is abundant and intensified cropping is common, fertilizer nutrient application rates are nearly triple the world average. In the Netherlands, nitrogen additions from fertilizers and manures are more than four times the removal of this element in harvested crops. In contrast, soils in sub-Saharan Africa are literally being mined—far less nutrients are being added from all sources than are being removed in the crops. The fertilizer nutrient rate in this region is only about 10% of the world average. We must, therefore, be as site specific as possible in evaluating the role of fertilizers in meeting humanitarian and environmental goals.

Properties and Use of Inorganic Fertilizers

Most fertilizers are inorganic salts containing readily available plant nutrient elements. Some are manufactured, but others, such as phosphorus and potassium, are found in natural geological deposits. Under very high temperatures and pressures, nitrogen gas in the atmosphere is fixed with hydrogen from natural gas to produce ammonia gas, NH_3. This is the starting point for the manufacture of most nitrogen fertilizers, including urea, ammonium nitrate, ammonium sulfate, sodium nitrate, and liquid mixtures known as "nitrogen solutions." The industrial process of nitrogen fixation has greatly altered the recycling of nitrogen on Earth (see Section 12.1).

In most cases, fertilizers are used to supply plants with the macronutrients nitrogen, phosphorus, and/or potassium—sometimes called the *primary fertilizer elements.* Fertilizers that supply sulfur, magnesium, and the micronutrients are also manufactured. The composition of inorganic commercial fertilizers (Table 13.6) is much more precisely defined than is the case for the organic material discussed above.

It can be seen from the data in Table 13.6 that a particular nutrient (say, nitrogen) can be supplied by many different *carriers,* or fertilizer compounds. Decisions as to which fertilizers to use must take into account not only the nutrients they contain, but also a number of other characteristics of the individual carriers. Table 13.6 provides

Table 13.6
COMMONLY USED INORGANIC FERTILIZER MATERIALS: THEIR NUTRIENT CONTENTS AND OTHER CHARACTERISTICS

| Fertilizer | Percent by weight | | | | Salt hazard | Acid formation[b] | Other nutrients and comments |
	N	P	K	S			
Primarily sources of nitrogen							
Anhydrous ammonia (NH_3)	82				Low	−148	Pressurized equipment needed; toxic gas; must be injected into soil.
Urea [$CO(NH_2)_2$]	45				Moderate	−84	Soluble; hydrolyses to ammonium forms. Volatilizes if left on soil surface.
Ammonium nitrate (NH_4NO_3)	33				High	−59	Absorbs moisture from air; can be left on soil surface. Can explode if mixed with organic dust or S.
Sulfur-coated urea	30–40			13–16	Low	−110	Variable slow rate of release.
UF (urea formaldehyde)	30–40				Very low	−68	Slowly soluble; faster with warm temperatures.
UAN solution	30				Moderate	−52	Most commonly used liquid N.
IBDU (isobutylidene diurea)	30				Very low	—	Slowly soluble.
Ammonium sulfate [$(NH_4)_2SO_4$]	21			24	High	−110	Rapidly lowers soil pH; very easy to handle.
Sodium nitrate ($NaNO_3$)	16				Very high	+29	Hardens, disperses soil structure.
Potassium nitrate (KNO_3)	13		36	0.2	Very high	+26	Very rapid plant response.
Primarily sources of phosphorus							
Monoammonium phosphate ($NH_4H_2PO_4$)	11	21–23		1–2	Low	−65	Best as starter.
Diammonium phosphate [$(NH_4)_2HPO_4$]	18–21	20–23		0–1	Moderate	−70	Best as starter.
Triple superphosphate		19–22		1–3	Low	0	15% Ca.
Phosphate rock [$Ca_3(PO_4)_2 \cdot CaX$]		8–18[a]			Very low	Variable	Low to extremely low availability. Best as fine powder on acid soils. 30% Ca. Contains some Cd, F, etc.
Single superphosphate		7–9		11	Low	0	Nonburning, can place with seed. 20% Ca.
Bonemeal	1–3[a]	10[a]	0.4		Very low	—	Slow availability of N, P as for phosphate rock. 20% Ca.
Colloidal phosphate		8[a]			Very low	—	P availability as for phosphate rock. 20% Ca.

(continued)

Table 13.6

COMMONLY USED INORGANIC FERTILIZER MATERIALS: THEIR NUTRIENT CONTENTS AND OTHER CHARACTERISTICS (CONTINUED)

Fertilizer	N	P	K	S	Salt hazard	Acid formation[b]	Other nutrients & comments
	Percent by weight						
	N	P	K	S			
Primarily sources of potassium							
Potassium chloride (KCl)			50		High	0	47% Cl—may reduce some diseases.
Potassium sulfate (K_2SO_4)			42	17	Moderate	0	Use where Cl not desirable.
Wood ashes	0.5–1	1–4			Moderate to high	+40	About ½ the liming value of limestone; caustic. 10–20% Ca, 2–5% Mg, 0.2% Fe, 0.8% Mn.
Greensand		0.6	6		Very low	0	Very low availability.
Granite dust			4		Very low	0	Very slow availability.
Primarily sources of other nutrients							
Basic slag		1–7			Low	+70	10% Fe, 2% Mn, slow availability; best on acid soils. 3–30% Ca, 3% Mg.
Gypsum ($CaSO_4 \cdot 2H_2O$)				19	Low	0	Stabilizes soil structure; no effect on pH; Ca and S readily available. 23% Ca.
Calcitic limestone ($CaCO_3$)					Very low	+95	Slow availability; raises pH. 36% Ca.
Dolomitic limestone [$CaMg(CO_3)_2$]					Very low	+95	Very slow availability; raises pH. ~24% Ca, ~12% Mg.
Epsom salts ($MgSO_4 \cdot 7H_2O$)				13	Moderate	0	No effect on pH; water soluble. 2% Ca, 10% Mg.
Sulfur, flowers (S)				95	—	−300	Irritates eyes; very acidifying; slow acting; requires microbial oxidation.
Solubor					Moderate	—	Very soluble; compatible with foliar sprays. 20.5% B.
Borax ($Na_2B_4O_7 \cdot 10H_2O$)					Moderate	—	Very soluble. 11% B, 9% Na.
EDTA chelates					—	—	See label. Usually 13% Cu or 10% Fe or 12% Mn or 12% Zn.
Cu, Fe, Mn, or Zn sulfates				13–20			25% Cu, 19% Fe, 27% Mn, or 35% Zn; very soluble.

[a]Highly variable contents.

[b]A negative number indicates that acidity is produced; a positive number indicates that alkalinity is produced; kg $CaCO_3$/100 kg material needed to neutralize acidity.

information about some of these characteristics, such as the salt hazard, acid-forming tendency, tendency to volatilize, ease of solubility, and content of nutrients other than the principal one. Of the nitrogen carriers, anhydrous ammonia, nitrogen solutions, and ureas are the most widely used. Diammonium phosphate and potassium chloride supply the bulk of the phosphorus and potassium used.

Fertilizer Grade

Every fertilizer label states the **grade** as a three-number code, such as 10-5-10 or 6-24-24. These numbers stand for percentages indicating the *total* nitrogen (N) content, the *available* phosphate (P_2O_5) content, and the *soluble* potash (K_2O) content. However,

BOX 13.6
HOW MUCH NITROGEN, PHOSPHORUS, AND POTASSIUM ARE IN A BAG OF 6-24-24?

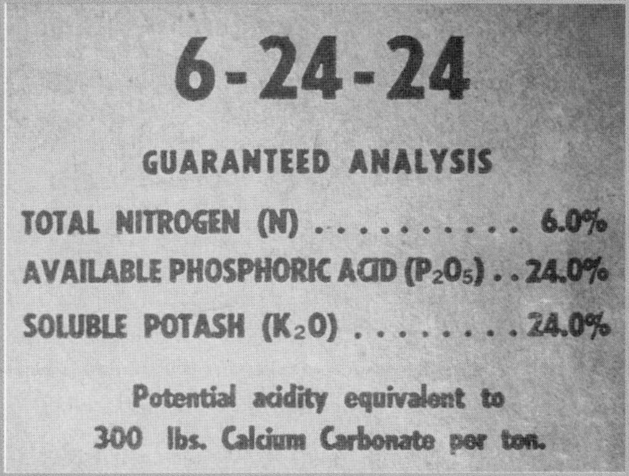

6-24-24

GUARANTEED ANALYSIS

TOTAL NITROGEN (N) 6.0%

AVAILABLE PHOSPHORIC ACID (P_2O_5) . . 24.0%

SOLUBLE POTASH (K_2O) 24.0%

Potential acidity equivalent to
300 lbs. Calcium Carbonate per ton.

Figure 13.17
A typical commercial fertilizer label. Note that a calculation must be performed to determine the percentage of the nutrient elements P and K in the fertilizer since the contents are expressed as if the nutrients were in the forms of P_2O_5 and K_2O. Also note that after interacting with the plant and soil, this material would cause an increase in soil acidity that could be neutralized by 300 units of $CaCO_3$ per 2000 units (1 ton = 2000 lbs) of fertilizer material.

Conventional labeling of fertilizer products reports percentage total N, citrate soluble P_2O_5, and water soluble K_2O. Thus, a fertilizer package (Figure 13.17) labeled as 6-24-24 (6% nitrogen, 24% P_2O_5, 24% K_2O) actually contains 6% total N, 10.5% citrate soluble P, and 19.9% water soluble K (see calculations below).

To determine the amount of fertilizer needed to supply the recommended amount of a given nutrient, first convert percent P_2O_5 and percent K_2O to percent P and K, by calculating the proportion of P_2O_5 that is P and the proportion of K_2O that is K. The following calculations may be used:

Given that the molecular weights of P, K, and O are 31, 39, and 16 g/mol, respectively:

Molecular weight of P_2O_5 = 2(31) + 5(16) = 142 g/mol

$$\text{Proportion P in } P_2O_5 = \frac{2P}{P_2O_5} = \frac{2(31)}{2(31) + 5(16)} = 0.44$$

To convert $P_2O_5 \rightarrow P$, multiply percent P_2O_5 by 0.44:
Molecular weight of K_2O = 2(39) + 16 = 94

$$\text{Proportion K in } K_2O = \frac{2K}{K_2O} = \frac{2(39)}{2(39) + (16)} = 0.83$$

To convert $K_2O \rightarrow K$, multiply percent K_2O by 0.83:
Thus, if the bag in Figure 13.17 contains 25 kg of the 6-24-24 fertilizer, it will supply 1.5 kg N (0.06 × 25); 2.6 kg P(0.24 × 0.44 × 25); and 5 kg K (0.24 × 0.83 × 25).

---

plants do not take up phosphorus and potassium in these chemical forms, nor do any fertilizers actually contain P_2O_5 or K_2O.

The oxide expressions for phosphorus (P_2O_5) and potassium (K_2O) are relics of the days when geochemists reported the contents of rocks and minerals in terms of the oxides formed upon heating. Unfortunately, these expressions found their ways into laws governing the sale of fertilizers, and there is considerable resistance to changing them, although some progress is being made. In scientific work and in this textbook, the simple elemental contents (P and K) are used wherever possible. Box 13.6 explains how to convert between the elemental and oxide forms of expression. Economic comparisons among different equally suitable fertilizers should be based on the price per kilogram of nutrient, not the price per kilogram of fertilizer.

Fate of Fertilizer Nutrients

A common misconception about fertilizers suggests that inorganic fertilizers applied to soil directly feed the plant, and that therefore the biological cycling of nutrients, such as described by Figures 12.1 for nitrogen, are of little consequence where inorganic fertilizers are used. The reality is that even when the application of fertilizer greatly increases both plant growth and nutrient uptake, the fertilizer stimulates increased cycling of the nutrients, and the nutrient ions taken up by the plant come largely from various pools in the soil and not directly from the fertilizer. For example, some of the added N may go to satisfy the needs of microorganisms, preventing them from competing with plants for other pools of N. This knowledge has been obtained

Table 13.7

SOURCE OF NITROGEN IN CORN PLANTS GROWN IN NORTH CAROLINA ON AN ENON SANDY LOAM SOIL (ULTIC HAPLUDALF) FERTILIZED WITH THREE RATES OF NITROGEN AS AMMONIUM NITRATE

The source of the nitrogen in the corn plant was determined by using fertilizer tagged with the isotope ^{15}N. Moderate fertilizer use increased the uptake of N already in the soil system as well as that derived from the fertilizer.

Fertilizer nitrogen applied, kg/ha	Corn grain yield, Mg/ha	Total N in corn plant, kg/ha	Fertilizer-derived N in corn, kg/ha	Soil-derived N in corn, kg/ha	Fertilizer-derived N in corn as percent of total N in corn	Fertilizer-derived N in corn as percent of N applied
50	3.9	85	28	60	33	56
100	4.6	146	55	91	38	55
200	5.5	157	86	71	55	43

Calculated from Reddy and Reddy (1993).

by careful analysis of dozens of nutrient studies that used fertilizer with isotopically tagged nutrients. Results from such a study are summarized in Table 13.7, which shows somewhat more N uptake from fertilizer than is typically reported. Generally, as fertilizer rates are increased, the efficiency of fertilizer nutrient use decreases, leaving behind in the soil an increasing proportion of the added nutrient.

The Concept of the Limiting Factor

Two German chemists (Justus von Liebig and Carl Sprengel) are credited with first publishing, in the mid-1800s, "the law of the minimum," which holds that *plant production can be no greater than that level allowed by the growth factor present in the lowest amount relative to the optimum amount for that factor.* This growth factor, be it temperature, nitrogen, or water supply, will limit the amount of growth that can occur and is therefore called the **limiting factor** (Figure 13.18).

If a factor is not the limiting one, increasing it will do little or nothing to enhance plant growth. In fact, increasing the amount of a nonlimiting factor may actually reduce plant growth by throwing the system further out of balance. For

Figure 13.18

An illustration of the law of the minimum and the concept of the limiting factor. Plant growth is constrained by the essential element (or other factor) that is most limiting. The level of water in the barrel represents the level of plant production. (Left) Phosphorus is represented as being the factor that is most limiting. Even though the other elements are present in more than adequate amounts, plant growth can be no greater than that allowed by the level of phosphorus available. (Right) When phosphorus is added, the level of plant production is raised until another factor becomes most limiting—in this case, nitrogen.

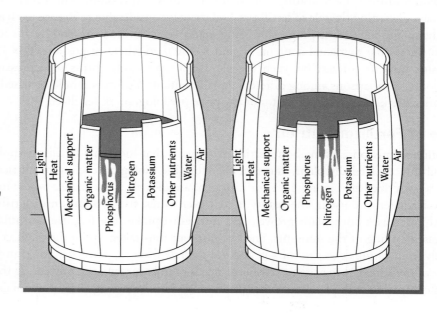

example, if a plant is limited by lack of phosphorus, adding more nitrogen may only aggravate the phosphorus deficiency.

Looked at another way, applying available phosphorus (the first limiting nutrient in this example) may allow the plant to respond positively to a subsequent addition of nitrogen. Thus, the increased growth obtained by applying two nutrients together often is much greater than the sum of the growth increases obtained by applying each of the two nutrients individually, suggesting an *interaction* or a *synergy* between the two.

13.7 FERTILIZER APPLICATION METHODS

There are three general approaches to applying fertilizers (Figure 13.19): (1) *broadcast application*, (2) *localized placement*, and (3) *foliar application*. Each method has some advantages and disadvantages and may be particularly suitable for different situations. Often some combination of the three methods is used.

In many instances, fertilizer is spread evenly over the entire field or area to be fertilized. This method is called **broadcasting**. Often the broadcast fertilizer is mixed into the plow layer by means of tillage, but in some situations it is left on the soil surface and allowed to be carried into the root zone by percolating rain or irrigation

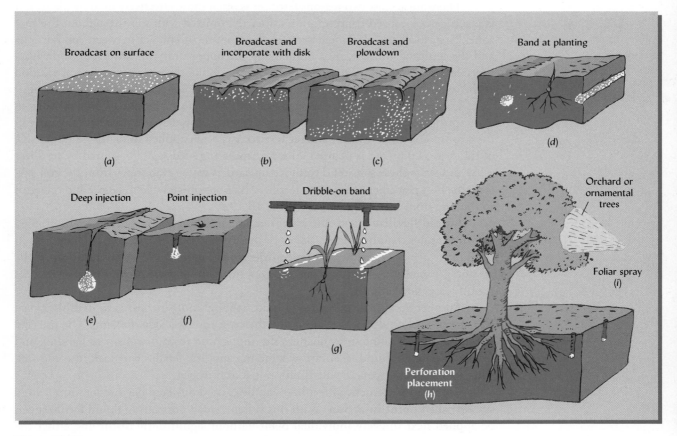

Figure 13.19

Fertilizers may be applied by many different methods, depending on the situation. Methods (a) to (c) represent broadcast fertilizer, with or without incorporation. Methods (d) to (h) are variations of localized placement. Method (i) is foliar application and has special advantages, but also limitations. Commonly, two or three of these methods may be used in sequence. For example, a field may be prepared with (c) before planting; (d) may be used during the planting operation; (g) may be used as a **side dressing** *early in the growing season; and, finally, (i) may used to correct a micronutrient deficiency that shows up in the middle of the season.*

water. The broadcast method is most appropriate when a large amount of fertilizer is being applied with the aim of raising the fertility level of the soil over a long period of time. Broadcasting is the most economical way to spread large amounts of fertilizer over wide areas.

For phosphorus, zinc, manganese, and other nutrients that tend to be strongly retained by the soil, broadcast applications are usually much less efficient than localized placement. Often 2 to 3 kg of fertilizer must be broadcast to achieve the same response as from 1 kg that is placed in a localized area. A heavy one-time application of phosphorus and potassium fertilizer, broadcast and worked into the soil, is a good preparation for establishing perennial plants such as lawns, pastures, and orchards. Where sprinkler irrigation is practiced, liquid fertilizers can be broadcast in the irrigation water, a practice sometimes called **fertigation**.

Although it is commonly thought that nutrients must be thoroughly mixed throughout the root zone, research has clearly shown that a plant can easily obtain its entire supply of a nutrient from a concentrated localized source in contact with only a small fraction of its root system. In fact, a small portion of a plant's root system can grow and proliferate in a band of fertilizer even though the salinity level caused by the fertilizer would be fatal to a germinating seed or to a mature plant were a large part of its root system exposed. This finding allowed the development of techniques for localized fertilizer placement.

There are at least two reasons fertilizer is often more effectively used by plants if it is placed in a localized concentration rather than mixed with soil throughout the root zone. First, localized placement reduces the amount of contact between soil particles and the fertilizer nutrient, thus minimizing the opportunity for adverse fixation reactions. Second, in the fertilized zone the concentration of the nutrient in the soil solution at the root surface will be very high, resulting in greatly enhanced uptake by the roots.

Localized placement is especially effective for young seedlings, in cool soils in early spring, and for plants that grow rapidly with a big demand for nutrients early in the season. For these reasons, **starter fertilizer** is often applied in bands on either side of the seed as the crop is planted. Since germinating seeds can be injured by fertilizer salts, and since these salts tend to move upward as water evaporates from the soil surface, the best placement for starter fertilizer is approximately 5 cm below and 5 cm off to the side from the seed row (see Figure 13.19*d*).

Liquid fertilizers and slurries of manure and sewage sludge can also be applied in bands rather than broadcast. Bands of these liquids are placed 10 to 30 cm deep in the soil by a process known as **knife injection** (Figure 13.19*e*). In addition to the advantages mentioned for banding fertilizer, injection of these organic slurries reduces runoff losses and odor problems.

Another approach to banding liquids (though not slurries) is to **dribble** a narrow stream of liquid fertilizer alongside the crop row as a side dressing. The use of a stream instead of a fine spray changes the application from broadcast to banding and results in enough liquid in a narrow zone to cause the fertilizer to soak into the soil. This action greatly reduces volatilization loss of nitrogen.

Localized placement of fertilizer can be carried one step beyond banding with a system called **point injection**. With this system, small portions of liquid fertilizer can be applied next to every individual plant without significantly disturbing either the plant root or the surface residue cover left by conservation tillage. The point injection system is a modern version of the age-old dibble stick with which peasant farmers plant seeds and later apply a portion of fertilizer in the soil next to each plant, all with a minimum of disturbance of the surface mulch.

The use of **drip irrigation** systems (see Section 6.9) has greatly facilitated the localized application of nutrients in irrigation water. Because drip fertigation is applied at frequent intervals, the plants are essentially spoon-fed, and the efficiency of nutrient use is quite high.

Trees in orchards and ornamental plantings are best treated individually, the fertilizer being applied around each tree within the spread of the branches but beginning approximately 1 m from the trunk (see Figure 13.19*h*). The fertilizer is best applied by what is called the *perforation* method. Numerous small holes are dug around each tree within the outer half of the branch-spread zone and extending down into the upper subsoil where the fertilizer is placed. Special large fertilizer pellets are available for this purpose. This method of application places the nutrients within the tree root zone and avoids an undesirable stimulation of the grass or cover that may be growing around the trees. If the cover crop or lawn around the trees needs fertilization, it is treated separately, the fertilizer being drilled in at the time of seeding or broadcast later.

Plants are capable of absorbing nutrients through their leaves in limited quantities. Under certain circumstances, the best way to apply a nutrient is *foliar application*—spraying a dilute nutrient solution directly onto the plant leaves (Figure 13.19*i*). Diluted NPK fertilizers, micronutrients, or small quantities of urea can be used as foliar sprays, although care must be taken to avoid significant concentrations of Cl^- or NO_3^-, which can be toxic to some plants. Foliar fertilization may conveniently fit in with other field operations for horticultural crops, because the fertilizer is often applied simultaneously with pesticide sprays.

The amount of nutrients that can be sprayed on leaves in a single application is quite limited. Therefore, while a few spray applications may deliver the entire season's requirement for a micronutrient, only a small portion of the macronutrient needs can be supplied in this manner.

13.8   TIMING OF FERTILIZER APPLICATION

Availability When the Plants Need It

For mobile nutrients such as nitrogen (and to some degree potassium), the general rule is to make applications as close as possible to the period of rapid plant nutrient uptake. For rapid-growing summer annuals, such as corn, this means making only a small starter application at planting time and applying most of the needed nitrogen as a side dressing just before the plants enter the rapid nutrient accumulation phase, usually about four to six weeks after planting. For cool-season plants, such as winter wheat or certain turf grasses, most of the nitrogen should be applied about the time of spring "green-up," when the plants resume a rapid growth rate. For trees, the best time is when new leaves are forming. With slow-release organic sources, time should be allowed for mineralization to take place prior to the plants' period of maximum uptake.

Environmentally Sensitive Periods

In temperate (**Udic** and **Xeric**) climates, most leaching takes place in the winter and early spring when precipitation is high and evapotranspiration is low. Nitrates left over after plant uptake has ceased have the potential for leaching during this period. In this regard it should be noted that, for grain crops, the rate of nutrient uptake begins to decline during grain-filling stages and has virtually ceased long before the crop is ready for harvest. With inorganic nitrogen fertilizers, avoiding leftover nitrates is largely a matter of limiting the amount applied to what the plants are expected to take up. However, for slow-release organic sources applied in late spring or early summer, mineralization is likely to continue to release nitrates after the crop has matured and ceased taking them up. To the extent that this timing of nitrate release is unavoidable, nonleguminous cover crops should be planted in the fall to absorb the excess nitrate being released (Figure 13.20).

Figure 13.20

Two nitrogen fertilizer systems for corn production. The system most commonly used in the past (solid lines) involves very high N applications and continuous corn culture. At least 150 kg N/ha is applied—part as starter fertilizer at planting time, the remainder as a side dressing just before the most rapid growth stage. Unfortunately, when the crop matures and is harvested, much soluble nitrogen remains in the soil, probably in the form of nitrates. The excess N is subject to leaching during the fall and winter months. More environmentally sound systems involve crop rotation that include legumes, or if corn is grown continuously, the rate of nitrogen fertilizer is greatly reduced (broken lines). A presidedress nitrate soil test (PSNT) is used to determine the amounts of nitrogen to apply. At the end of the season, a cornstalk nitrate test (CSNT) is used to assess plant nitrogen status and a winter cover crop is planted to capture leftover nitrate. Economic returns are about the same, but the nitrogen remaining in the soil at crop harvest is low and nitrate contamination is minimized.

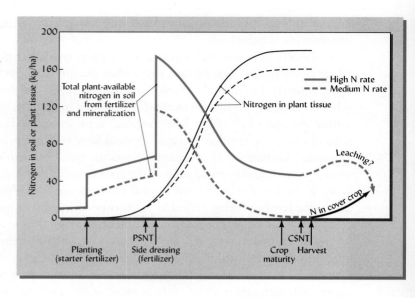

In high-rainfall conditions and on permeable soils, dividing a large dose of fertilizer into two or more split applications may avoid leaching losses prior to the crop's establishment of a deep root system. In cold climates, another environmentally sensitive period occurs during early spring, when snowmelt over frozen or saturated soils results in torrents of runoff water. This runoff may pollute rivers and streams with soluble nutrients from manure or fertilizer that is at or near the soil surface.

Application of urea fertilizers to mature forests is usually carried out when rains can be expected to wash the nutrients into the soil and minimize volatilization losses. Furthermore, nitrogen fertilization of forests should occur just prior to the onset of the growing season (early spring, or in warm climates, winter) so that the tree roots will have the entire growing season to utilize the nitrogen. The pulse of nitrate leaving the watershed can be reduced if a 10- to 15-m unfertilized buffer is maintained along all streams (see Section 13.2).

Sometimes it is simply not possible to apply fertilizers at the ideal time of the year. For example, although a crop may respond to a late-season side dressing, such an application will be difficult if the plants are too tall to drive over without damaging them. Using an airplane may allow more flexibility in fertilizer timing. Early spring applications may be limited by the need to avoid compacting wet soils. Economic costs or the time demands of other activities may also require that compromises be made in the timing of nutrient application.

13.9   DIAGNOSTIC TOOLS AND METHODS

Three basic tools are available for diagnosing soil fertility problems: (1) *field observations*, (2) *plant tissue analysis*, and (3) *soil analysis* (soil testing). To effectively guide the application of nutrients, as well as to diagnose problems as they arise in the field, all three approaches should be integrated. There is no substitute for careful observation and *recording* of circumstantial evidence and symptoms in the field. Effective observation and interpretation requires skill and experience, as well as an open mind. It is not uncommon for a supposed soil fertility problem to actually be caused by soil compaction, weather conditions, pest damage, or human error. The task of the diagnostician is to use all the tools available in order to identify the factor that is limiting plant growth, and then devise a course of action to alleviate the limitation.

Plant Symptoms and Field Observations

The detectivelike work involved can be one of the more exciting and challenging aspects of nutrient management. To be an effective soil fertility diagnostician, several general guidelines are helpful:

Field diagnosis guide for tropical rice: www.knowledgebank.irri .org/ricedoctor/

1. Develop an organized way to *record* your observations. The information you collect may be needed to properly interpret soil and plant analytical results obtained at a later date.

2. Talk to the person who owns or manages the land. Ask when the problem was first observed and if any recent changes have taken place. Obtain records on plant growth or crop yield from previous years, and ascertain the history of management of the site for as many years as possible. It is often useful to sketch a map of the site showing features you have observed and the distribution of symptoms.

3. Look for *spatial patterns*—how the problem seems to be distributed in the landscape and in individual plants. Linear patterns across a field may indicate a problem related to tillage, drain tiles, or the incorrect spreading of lime or fertilizer. Poor growth concentrated in low-lying areas may relate to the effects of soil aeration. Poor growth on the high spots in a field may reflect the effects of erosion and possibly exposure of subsoil material with an unfavorable pH.

4. Closely examine individual plant leaves to characterize any foliar symptoms. Nutrient deficiencies can produce characteristic symptoms on leaves and other plant parts. Examples of such symptoms are shown in several figures in Chapter 12 and in Plates 87 to 107. Determine whether the symptoms are most pronounced on the younger leaves (as is the case for most of the micronutrient cations) or on the older leaves (as is the case for nitrogen, potassium, and magnesium). Some nutrient deficiencies are quite reliably identified from foliar symptoms, while others produce symptoms that may be confused with herbicide damage, insect damage, or damage from poor aeration.

5. Observe and *measure* differences in plant growth and crop yield that may reflect different levels of soil fertility, even though no leaf symptoms are apparent. Check *both* aboveground and belowground growth. Are mycorrhizae associated with tree roots? Are legumes well nodulated? Is root growth restricted in any way?

Plant Tissue Analysis

Nutrient Concentrations   The concentration of essential elements in plant tissue is related to plant growth or crop yield, as shown in Figure 13.21. The range of tissue concentrations at which the supply of a nutrient is sufficient for optimal plant growth is termed the **sufficiency range**. At the upper end of the sufficiency range, plants may be participating in luxury consumption, as the additional nutrient uptake has not produced additional plant growth (see Figure 12.31). At concentrations

Figure 13.21

The relationship between plant growth or yield and the concentration of an essential element in the plant tissue. For most nutrients there is a relatively wide range of values associated with normal, healthy plants (the sufficiency range). Beyond this range, plant growth suffers from either too little or too much of the nutrient. The critical range (CR) is commonly used for the diagnosis of nutrient deficiency. Nutrient concentrations below the CR are likely to reduce plant growth even if no deficiency symptoms are visible. This moderate level of deficiency is sometimes called hidden hunger. *The odd hook at the lower left of the curve is the result of the so-called* dilution effect *that is often observed when extremely stunted, deficient plants are given a small dose of the limiting nutrient. The growth response may be so great that even though somewhat more of the element is taken up, it is diluted in a much greater plant mass.*

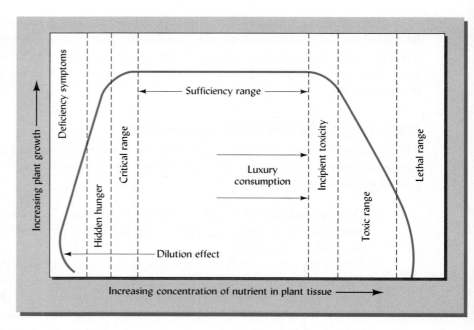

Plant analysis handbook for agronomic and horticultural crops: (University of Georgia): http://aesl.ces.uga.edu/publications/plant/plant.html

Plant tissue testing in Asian agriculture: www.fftc.agnet.org/library/ac/1994f/html#eb536t4

above the sufficiency range, plant growth may decline as nutrient elements reach concentrations that are toxic to plant cells or interfere with the use of other nutrients. If tissue concentrations are in the **critical range**, the supply is just marginal and growth is expected to decline if the nutrient becomes any less available, even though visible foliar symptoms may not be exhibited ("hidden hunger"). Plants with tissue concentrations below the sufficiency range for a nutrient are likely to respond to additions of that nutrient if no other factor is more limiting. The sufficiency range and critical range have been well characterized for many plants, especially for agronomic and major horticultural crops. Less is known about forest trees and ornamentals.[3] Sufficiency ranges for 11 essential elements in a variety of plants are listed in Table 13.8.

Tissue Analysis   Tissue analysis can be a powerful tool for identifying plant nutrient problems if several simple precautions are taken. First, it is critical that the correct plant part be sampled. Second, the plant part must be sampled at the specified stage of growth, because the concentrations of most nutrients decrease considerably as the plant matures. Third, it must be recognized that the concentration of one nutrient may be affected by that of another nutrient, and that sometimes the ratio of one nutrient to another (e.g., Mg/K, N/S, or Fe/Mn) may be the most reliable guide to plant nutritional status (Figure 13.22). In fact, several elaborate mathematical systems for assessing the ratios or balance among nutrients have proven useful for certain plant species.[4] Because of the uncertainties and complexities in interpreting tissue concentration data, it is wise to sample plants from the best and worst areas in a field or stand. The differences between samples may provide valuable clues concerning the nature of the nutrient problem.

---

[3]Detailed information on tissue analysis for a large number of plant species can be found in Reuter and Robinson (1986).

[4]The best developed of the multinutrient ratio systems is known as the Diagnostic Recommendation Integrated System (DRIS). For details, see Walworth and Sumner (1987).

Table 13.8
A GUIDE TO SUFFICIENCY RANGES FOR TISSUE ANALYSIS OF SELECTED PLANT SPECIES

Values apply only to the indicated plant parts and stage of growth. Normally, 6 to 20 plants should be sampled. Leaves should be washed briefly in distilled water to remove any soil or dust and then dried before submitting for analysis.

Plant species and Part to sample	Content, %						Content, µg/g				
	N	P	K	Ca	Mg	S	Fe	Mn	Zn	B	Cu
Pine trees (*Pinus* spp.) Current-year needles near terminal	1.2–1.4	0.10–0.18	0.3–0.5	0.13–0.16	0.05–0.09	0.08–0.12	20–100	50–600	20–50	3–9	2–6
Turfgrasses, cool season Clippings	3.0–5.0	0.3–0.4	2–4	0.3–0.8	0.2–0.4	0.25–0.8	40–500	20–100	20–50	5–20	6–30
Corn (*Zea mays*) Ear-leaf at tasseling	2.5–3.5	0.20–0.50	1.5–3.0	0.2–1.0	0.16–0.40	0.16–0.50	25–300	20–200	20–70	6–40	6–40
Soybean (*Glycine max*) Youngest mature leaf at flowering	4.0–5.0	0.31–0.50	2.0–3.0	0.45–2.0	0.25–0.55	0.25–0.55	50–250	30–200	25–50	25–60	8–20
Apple (*Malus* spp.) Leaf at base of nonfruiting shoots	1.8–2.4	0.15–0.30	1.2–2.0	1.0–1.5	0.25–0.50	0.13–0.30	50–250	35–100	20–50	20–50	5–20
Rice (*Oryza sativa*) Youngest mature leaf at tillering	2.8–3.6	0.14–0.27	1.5–3.0	0.16–0.40	0.12–0.22	0.17–0.25	90–200	40–800	20–160	5–25	6–25
Alfalfa (*Medicago sativa*) Upper third of plant at first flower	3.0–4.5	0.25–0.50	2.5–3.8	1.0–2.5	0.3–0.8	0.3–0.5	50–250	25–100	25–70	6–20	30–80

Data derived from many sources.

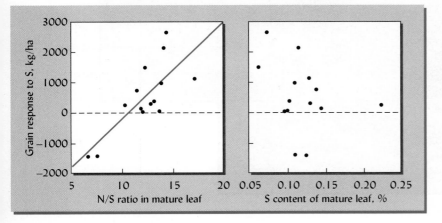

Figure 13.22
Relationship between sulfur content in leaves (right) or the N/S ratio in leaves (left) and the corn grain increases produced by application of sulfur. The ratio of two interacting elements is often a better guide to plant nutrient status than the tissue contents of either element alone. Nitrogen and sulfur are both needed to synthesize plant proteins. As the N/S ratio of the unfertilized corn increased, so did the positive response to application of sulfur (as gypsum). There was no clear relationship between response to sulfur application and the leaf S content by itself, even though most of the corn had S levels below the sufficiency range indicated in Table 13.8. The data are from experiments on 14 small farms in Malawi. Sulfur deficiencies are widespread in central African soils.
[Data from Weil and Mughogho (2000)]

13.10   SOIL ANALYSIS

Since the *total* amount of an element present tells us very little about the ability of a soil to supply that element to plants (Section 1.15), more meaningful *partial soil analyses* have been developed. *Soil testing* is the routine partial analysis of soils for the purpose of guiding nutrient management.

Sampling the Soil

Sampling turf for soil testing:
http://mulch.cropsoil.uga.edu/turf/Sampling/sampling.html

Soil sampling is widely acknowledged to be one of the weakest links in the soil testing process. About a teaspoonful of soil is eventually used to represent millions of kilograms of soil in the field. Since soils are highly variable, both horizontally and vertically, it is essential to carefully follow the sampling instructions from the soil testing laboratory.

Because of the variability in nutrient levels from spot to spot, it is always advisable to divide a given field or parcel of land into as many distinct areas as practical, taking soil samples from each to determine nutrient needs. For example, suppose a 20-ha field has a 2-ha low spot in the middle, and 5 ha at one end that used to be a permanent pasture. These two areas should be sampled, and later managed, separately from the remainder of the field. Similarly, a homeowner should sample flower beds separately from lawn areas, low spots separately from sloping areas, and so on. On the other hand, known areas of unusual soil that are too small or irregular to be *managed separately* should be avoided and *not* included in the composite sample from the whole field.

Composite Sample   Usually, a soil probe is used to remove a thin cylindrical core of soil from at least 12 to 15 randomly scattered places within the land area to be represented (Figure 13.23). The 12 to 15 subsamples are thoroughly mixed in a plastic

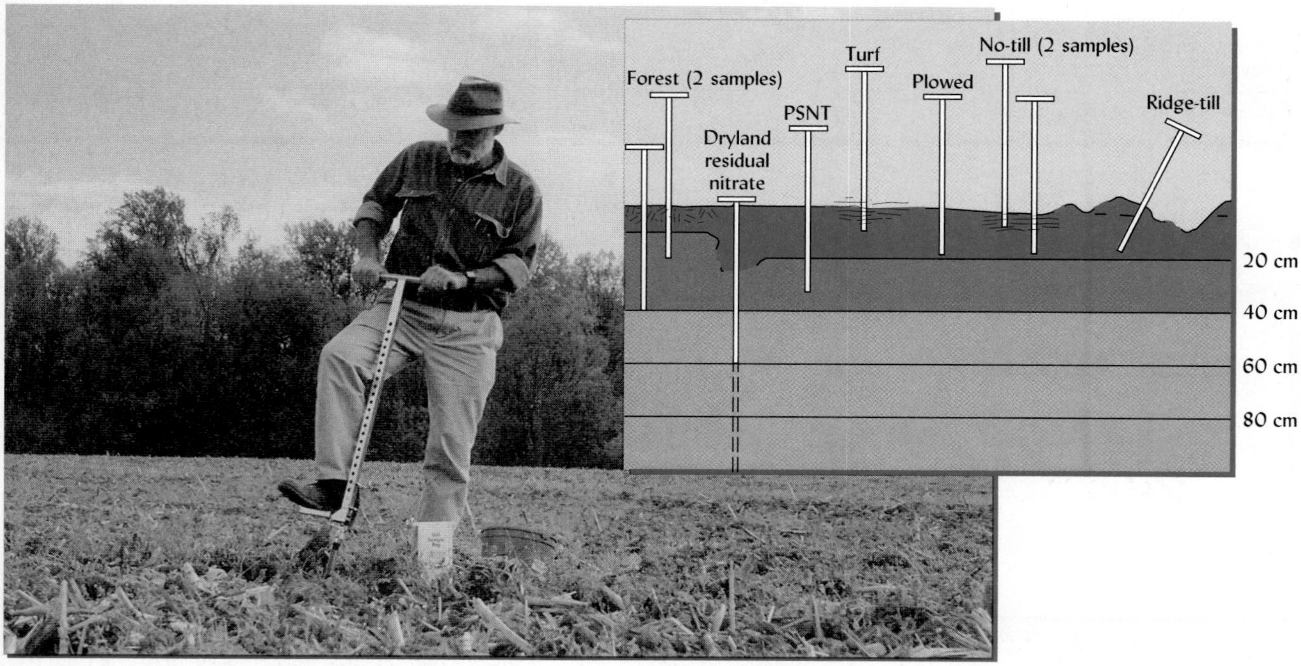

Figure 13.23

Taking the soil sample in the field is often the most error-prone step in the soil testing process because soil properties vary greatly both with depth and from place to place in even a uniform-appearing field. Areas that are obviously unusual (wet spots, places where manure was piled, eroded spots, etc.) should be avoided. (diagram) The proper depth to sample depends on the purpose of the soil test and the nature of the soil. Some suggested depths for different situations are shown. (Photo courtesy of R. Weil)

bucket, and about 0.5 L of the soil is placed in a labeled container and sent to the lab. If the soil is moist, it should be air-dried without sun or heat prior to packaging for routine soil tests. Heating the sample might cause falsely high results for certain nutrients.

Depth to Sample The standard depth of sampling for a plowed soil is the depth of the plowed layer, about 15 to 20 cm, but various other depths are also used (see Figure 13.23). Because in many unplowed soils nutrients are stratified in contrasting layers, the depth of sampling can greatly alter the results obtained.

Time of Year Seasonal changes are often observed in soil test results for a given area. For example, in temperate regions, the potassium level is usually highest in early spring, after freezing and thawing has released some fixed K ions from clay interlayers, and lowest in late summer, after plants have removed much of the readily available supply. The time of sampling is especially important if year-to-year comparisons are to be made. A good practice is to sample each area every year or two (always at the same time of year), so that the soil test levels can be tracked over the years to determine whether nutrient levels are being maintained, increased, or depleted.

Timing for Special Nitrogen Tests Timing is especially critical to determine the amount of mineralized nitrogen in the root zone. In relatively dry, cold regions (e.g., the Great Plains), the *residual nitrate* test is done on 60-cm-deep samples obtained sometime between fall and before planting in spring. In humid regions, where nitrate leaching is more pronounced, a special test has been developed to determine whether a soil will mineralize enough nitrogen for corn (and some other crops). The samples for this *presidedress nitrate test* (PSNT) are taken from the upper 30 cm when the corn is about 30 cm tall, just in time to determine how much nitrogen to apply as the crop enters its period of most rapid nitrogen uptake. In this case, the soil must be sampled during the narrow window of time when spring mineralization has peaked, but plant uptake has not yet begun to deplete the nitrate produced (see Figure 13.20).

Site-Specific Management Zones The just-described standard procedure of compositing many small soil samples into a single mixed sample to represent the "average" soil in a large field must be recognized for the compromise that it is. Fertilizer recommendations based on the *average* soil condition in the field will likely be either too high or too low for almost any particular spot in that field. Using *geographic information systems* (GIS) computer technology, fertilizer rates can be much more precisely tailored to account for soil variations within a field. However, the benefits of doing this are not always worth the costs. The costs include field labor and fees for collecting and processing a large number of soil samples collected in a grid pattern within the field. Since each sample is geo-referenced as to its specific location (using a satellite-based *global positioning systems* [GPS]), computer software can generate maps showing management zones with defined soil properties and fertilizer needs. Special computer-controlled fertilizer-spreading equipment can be automatically adjusted "on the go" to spread more or less fertilizer as called for by the map of management zones (Figure 13.24).

For example, areas mapped as testing low in phosphorus might be slated to receive higher-than-average application rates of phosphorus fertilizer, while areas mapped as being high in phosphorus might receive no phosphorus fertilizer at all. Similarly, reduced rates of nitrogen fertilizer application might be mapped for areas of sandy soils with high leaching potential but low yield potential. The application-rate maps are then programmed into a computer onboard the machine that spreads the fertilizer.

Site-specific management systems can also control insects and weeds and modify plant seeding rates and depths, allowing precision farming to manage land on a site-specific basis, rather than on a whole field basis.

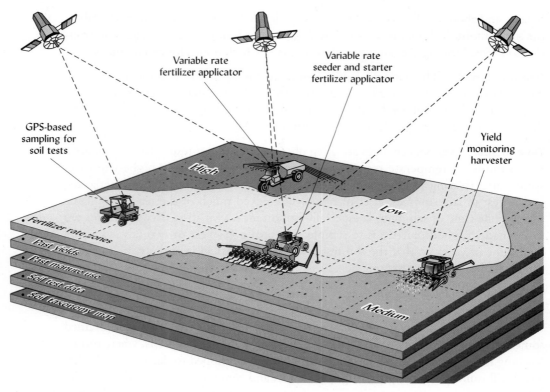

Figure 13.24

High technology used to facilitate site-specific nutrient-management systems. The global positioning system (GPS) can plot the location of many soil sampling and plant production sites on a grid basis within a field. A soil sample (composite of 20 cores) is taken from each cell (about 1 ha) and is analyzed. With soil analysis and other data from these cells, computers can help create maps such as the one shown here for a 18-ha field. The top map combines data from the other layers to define "fertilizer rate zones." Satellite/ computer systems can then be used to control variable-rate fertilizer applicators that apply only the amounts of nutrients that the soil tests and past soil management suggest are needed. At harvest time, similar satellite/computer connections make possible the monitoring of crop yields on the same grid basis when the harvest machine traverses the field. The yield data are used to create yield maps, which can then be used to further refine the nutrient-management system. [Modified from PPI (1996)]

Chemical Analysis of the Sample

Sampling for site-specific management (Manitoba): www.gov.mb.ca/agriculture/ soilwater/soilfert/fbd01s02 .html

The most common and reliable soil tests are those for soil pH, potassium, phosphorus, and magnesium. Micronutrients are sometimes extracted using chelating agents, especially for calcareous soils in the more arid regions. While the nitrate and sulfate present in the soil at the time of sampling can be measured, predicting the availability of nitrogen and sulfur is considerably more difficult because of the many biological factors involved in their cycles (see Sections 12.1 and 12.2).

Because the methods used by different labs may be appropriate for different types of soils, it is advisable to send a soil sample to a lab in the same region from which the soil originated. Such a lab is likely to use procedures appropriate for the soils of the region and should have access to data correlating the analytical results to plant responses on soils of a similar nature.

Soil tests designed for use on soils or soil-based potting media generally do not give meaningful results when used on artificial peat-based soilless potting media. Special extraction procedures must be used for the latter, and the results must then be correlated to nutrient uptake and growth of plants grown in similar media. It is important to provide the soil testing lab with complete information concerning the nature of your soil, its management history, and your plans for its future use.

Interpreting the Results to Make a Recommendation

The soil test values themselves are merely *indices* of nutrient-supplying power. They do *not* indicate the *actual amount* of nutrient that will be supplied. For this reason, it is best to think of soil test reports as more indicative than quantitative.

Many years of field experimentation at many sites are needed to determine the soil test levels that indicate a low, medium, or high capacity to supply the nutrient tested. Such categories are used to predict the likelihood of obtaining a profitable response from an application of the nutrient tested (Figure 13.25). Since the actual units of measurement (ppm, mg/L, or lb/acre) have little actual meaning to the end user, soil test results are increasingly reported as index values on a relative scale (often with 75 to 100 considered optimal). The report shown in Figure 13.26 uses both interpretative categories and a relative index scale.

Merits of Soil Testing

When the precautions already described are observed, soil testing is an invaluable tool in making fertilizer recommendations. These tests are most useful when they are correlated with the results of field fertilization experiments. Adding amendments to achieve some "ideal" balance of nutrients in the soil is often wasteful of money and resources. Rather, soil tests used correctly in conjunction with calibration experiments can indicate what level of amendment needs to be added, if any, to allow the soil to supply sufficient nutrients for optimal plant growth.

Dependability in predicting how plants will respond to soil amendments varies with the particular soil test, the tests for some parameters being much more reliable than others because of the consistency and breadth of field correlation data. In general, the tests for pH (need for liming or acidification), P, K, Mg, B, and Zn are quite reliable. Some soil test labs offer tests for additional soil properties such as other micronutrients (Cu, Mn, Fe, Mo, etc.), humus fractions (fulvic, humic, etc.), nonacid cation ("base") saturation, and even various microbial populations (fungi, bacteria,

Problems with "cation balancing" (University of Wisconsin–Madison): www.soils.wisc.edu/ extension/FAPM/ approvedppt2004/ Kelling1.pdf

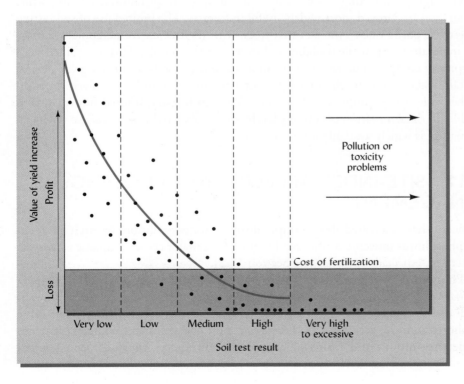

Figure 13.25

The relationship between soil test results for a nutrient and the extra yield obtained by fertilizing with that nutrient. Each data point represents the difference in plant yield between the fertilized and the unfertilized soil. Because many factors affect yield and because soil tests can only approximately predict nutrient availability, the relationship is not precise, but the data points are scattered about the trend line. If the point falls above the fertilizer cost line, the extra yield was worth more than the cost of the fertilizer and a profit would be made. For a soil testing in the very low and low categories, a profitable response to fertilizer is very likely. For a soil testing medium, a profitable response is a 50:50 proposition. For soils testing in the high category, a profitable response is unlikely.

Figure 13.26

A typical soil test report with index values spanning such categories as low, medium, and optimum or high. The report includes the nature and history of the field, the levels of certain nutrients and soil properties that were measured in the lab, and recommendations for the amounts and types of soil amendments to apply for a specific crop. For the particular field sampled here, the following inter-pretations can be made. The pH is low and should be raised to near 6.0. Some dolomitic limestone should be included to add Mg, which is in low to medium supply. Moderate amounts of K are likely to give an economic response. The P level is highly excessive, so P additions of any kind should be avoided, and steps should be taken to manage transport of P to waterways. As Delaware is in a humid region where soluble N may leach over winter, N application is recommended based on yield goal and soil properties that affect mineralization, but not on an analytical test for N. In an arid or semiarid region, a soil test lab would likely also measure and report nitrate and electrical conductivity in the soil. Sulfur and B are not part of the standard analysis package in Delaware, but might be included if requested. (Report courtesy of K. L. Gartley, University of Delaware)

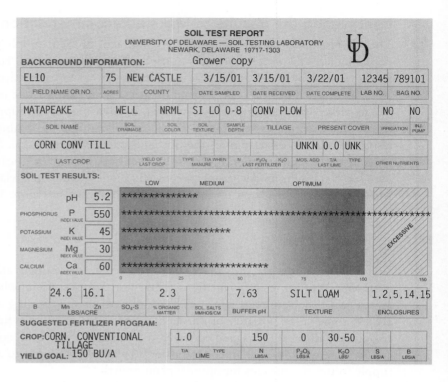

nonparasitic nematodes, etc.). However, the basis for making valid and practical rec-ommendations for managing soil fertility based on such soil test parameters is cur-rently very scant, at best.

Generally, soil testing has been most relied upon in agricultural systems, while foliar analysis has proved more widely useful in forestry. The limited use of soil testing in forestry is partly due to the fact that trees, with their extensive perennial root sys-tems, integrate nutrient bioavailability throughout the profile. This, combined with the typically complex nutrient stratification in forested soils, creates a great deal of uncertainty about how to obtain a representative sample of soil for analysis. In addi-tion, because of the comparably long time frame in forestry, limited information is available on the correlation of soil test levels with timber yields in the sense that such information is widely available for agronomic crops.

13.11 SITE-INDEX APPROACH TO PHOSPHORUS MANAGEMENT

As concerns have increased about nonpoint source water pollution, research has iden-tified phosphorus movement from land to water as a major cause of aquatic ecosystem degradation, especially for lakes (see Section 13.2).

Phosphorus movement from land to water is determined by *phosphorus-transport*, *phosphorus-source*, and *phosphorus-management* parameters. Transport parameters include the mechanisms that govern how rain, snow-melt, or irrigation water cause phosphorus to move across the landscape in runoff and sediment. Source parameters include the amount and forms of phosphorus on and in the soil. Management parameters include the method of application, timing, and placement of phosphorus and such disturbances as tillage.

Overenrichment of Soils

In Section 12.3 we discussed how phosphorus, so scarce in nature, has accumulated over a period of decades to excessive levels in many agricultural soils as a result of two historical trends: the long-term overapplication of phosphorus fertilizer and the concentration of livestock and the subsequent heavy applications of animal manure to soils nearby (see Section 13.2). Because of these trends, some farm fields have become so high in phosphorus that rain or irrigation water interacting with the soil carries away enough phosphorus to impair the ecology of receiving waters. It is imperative that sites with such a high potential to cause phosphorus pollution be identified and appropriately managed to reduce their environmental impact. It may take decades to lower soil test P levels back to the optimum range (Figure 13.27).

Transport of Phosphorus from Land to Water

A review of the phosphorus cycle (Section 12.3) reminds us that phosphorus can move from land to water by three principal pathways: (1) Attached to eroded soil particles (the main P-loss pathway from tilled soils), (2) dissolved in water running off the surface of the land (a major pathway for pastures, woodland, and no-till cropland), and (3) dissolved or attached to suspended colloidal particles in water percolating down the profile and through the groundwater aquifers that feed streams or lakes from below (a significant pathway in very sandy or poorly drained soils with shallow water tables and possibly in heavily manured soils with many continuous macropores).

Phosphorus Soil Test Level as Indicator of Potential Losses

As described in Section 13.10, soil testing was designed to determine how well a soil can supply nutrients to plant roots. However, routinely used phosphorus soil tests have been shown to also provide a useful (though not perfect) indication of how readily phosphorus will desorb from a soil and dissolve in runoff water. The relationship

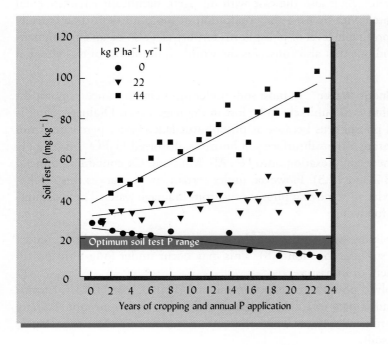

Figure 13.27

Changes in soil test P as a result of annual applications of 0, 22, or 44 kg P ha^{-1} to a corn–soybean rotation on an Iowa Mollisol. Data from this and other sites in the study suggest that application of 12 to 20 kg P ha^{-1} yr^{-1} is sufficient to balance P removed in crop harvests and maintain soil test P in the optimal range for many cropping systems. The soil tests were conducted using the Bray-1 method for which the optimal levels are 16 to 22 mg P extracted per kg soil. [Data from Dodd and Mallarino (2005) with permission of the Soil Science Society of America]

Figure 13.28

The generalized relationship between levels of plant-available phosphorus in soils (soil test P level) and environmental losses of P dissolved in surface runoff and subsurface drainage waters. The generalized relationship between traditional plant nutrient supply interpretation categories and environmental interpretation categories is also indicated. The diagram suggests that, fortunately, soil P levels can be achieved that are both conducive to optimum plant growth and protective of the environment. If P losses by soil erosion are controlled (although not shown, this is a big if), significant quantities of dissolved P would be lost only when soils contain P levels in excess of those needed for optimum plant growth. Losses of P by leaching in drainage water would be significant only at very high soil P levels, as this pathway rapidly increases only after the P-sorption capacity of the soil is substantially saturated. Note the threshold levels (vertical arrows) for P losses. The vertical axes are not to scale. [Figure based on data and concepts discussed in Sharpley (1997), Beegle et al. (2000), and Higgs et al. (2000)]

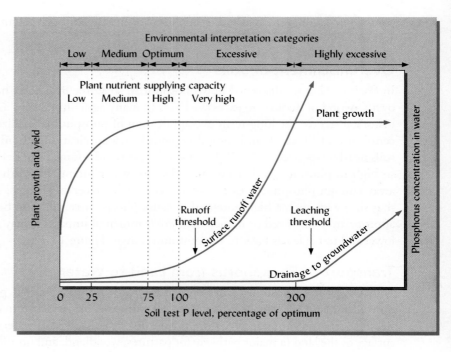

between phosphorus soil test level in the upper centimeter or two of soil and phosphorus in runoff water is not linear (Figure 13.28). Rather, the relationship exhibits a threshold effect such that little phosphorus is lost to runoff if soils are near or below the optimum range for phosphorus, but the amount released from the soil to runoff water increases exponentially as phosphorus soil test levels rise much higher than this. Therefore, as suggested by Figure 13.28, it appears that soils can be maintained at phosphorus levels sufficient for optimal plant growth without undue losses of phosphorus in runoff water. As is also the case with nitrogen, significant environmental damages are associated with excess—the application of more nutrient than is really needed. The most profitable level of phosphorus fertility can also be an environmentally sound level, and one that also conserves the world's finite stocks of this essential element.

Phosphorus in Drainage Water   In most soils under most circumstances, significant quantities of phosphorus are unlikely to be lost in drainage water. Drainage water is usually very low in phosphorus because as phosphorus-laden water percolates from the surface down through the subsurface soil horizons, dissolved $H_2PO_4^-$ is strongly sorbed by inner-sphere complexation onto Fe-, Al-, Mn-, and Ca-containing mineral surfaces (Sections 8.7 and 12.3). However, under certain circumstances this mechanism is not so effective in removing phosphorus from drainage water. These circumstances include the following:

1. The percolating water is flowing through macropores and has little contact with soil surfaces (Section 6.8). This may occur under long-term no-till management or in pastures.
2. The dissolved phosphorus is not in mineral ($H_2PO_4^-$ or HPO_4^{2-}) forms but instead is part of soluble organic molecules, which are not strongly sorbed by mineral surfaces. This situation is common under forests or heavily manured soils.

3. The phosphorus is sorbed onto the surface of dispersed colloidal particles that are so small that they remain suspended in the percolating water.

4. The phosphorus-fixation capacity of the soil is so small (as in sands, Histosols, and some waterlogged soils with reduced iron) or already so saturated with phosphorus (as in soils overloaded with phosphorus after many years of excessive manure and fertilizer applications) that little more phosphorus can be sorbed (see Figure 13.28).

Site Characteristics Influencing Transport of Phosphorus

To cause environmental damage, the phosphorus-rich sites must also have characteristics conducive to the transport of phosphorus to sensitive bodies of water. Such characteristics might include close proximity to the water body, the absence of a protective vegetated buffer, erodible soils, or high runoff rates caused by low permeability and steep slopes. Control of soil erosion by conservation tillage, mulches, and other practices discussed in Chapter 14 can greatly reduce the transport of sediment-associated phosphorus, but will have little effect on, or may even increase, the loss of phosphorus dissolved in runoff or drainage water. In fact, when no-till fields are fertilized, the phosphorus tends to accumulate in the upper 1 to 2 cm of soil, where it is most susceptible to desorption into runoff water. The beneficial influence of buffers in removing both dissolved and sediment-associated phosphorus from runoff was discussed in Section 13.2.

Phosphorus Site Index[5]

Usually, most of the phosphorus entering a river or lake originates from only a small fraction of the watershed land area. Effective environmental management requires identification of the high-risk sites. As just discussed, these sites can be expected to be those with high-risk phosphorus source characteristics (large amounts of phosphorus-containing materials are applied and/or phosphorus soil test levels are excessively high) and high phosphorus transport characteristics. Researchers have attempted to integrate phosphorus source, transport, and management characteristics of a site into an index of phosphorus pollution risk, commonly referred to as the *phosphorus site index*. The phosphorus index is designed to identify sites where the risk of phosphorus movement may be relatively higher than that of other sites.

13.12   CONCLUSION

The continuous availability of plant nutrients is critical for the sustainability of most ecosystems. The challenge of nutrient management is threefold: (1) to provide adequate nutrients for plants in the system; (2) to simultaneously ensure that inputs are in balance with plant utilization of nutrients, thereby conserving nutrient resources; and (3) to prevent contamination of the environment with unutilized nutrients.

The recycling of plant nutrients must receive primary attention in any ecologically sound management system. This can be accomplished in part by returning plant residues to the soil. These residues can be supplemented by judicious application of the organic wastes that are produced in abundance by municipal, industrial, and agricultural operations worldwide. The use of cover crops grown specifically to be returned to the soil is an additional organic means of recycling nutrients.

For sites from which crops or forest products are removed, nutrient losses commonly exceed the inputs from recycling. Inorganic fertilizers will continue to supplement natural and managed recycling to replace these losses and to increase the

---

[5]See, for example, USDA (2007).

level of soil fertility so as to enable humankind to not only survive, but to flourish on this planet. In extensive areas of the world, fertilizer use will have to be increased above current levels to avoid soil and ecosystem degradation, to remediate degraded soils, and to enable profitable production of food and fiber.

The use of fertilizers, both inorganic and organic, should not be done in a simply habitual manner or for so-called insurance purposes. Rather, soil testing and other diagnostic tools should be used to determine the true need for added nutrients. In managing nitrogen and phosphorus, increasing attention will have to be paid to the potential for transport of these nutrients from soils where they are applied to waterways where they can become pollutants. If soils are low in available nutrients, fertilizers often return several dollars' worth of improved yield for every dollar invested. However, where the nutrient-supplying power of the soil is already sufficient, adding fertilizers is likely to be damaging both to the bottom line and to the environment.

STUDY QUESTIONS

1. The groundwater under a heavily manured field is high in nitrates, but by the time it reaches a stream bordering the field, the nitrate concentration has declined to acceptable levels. What are likely explanations for the reduction in nitrate?

2. You want to plant a cover crop in fall to minimize nitrate leaching after the harvest of your corn crop. What characteristics would you look for in choosing a cover crop to ameliorate this situation?

3. What management practices on forested sites can lead to significant nitrogen losses, and how can the losses be prevented?

4. What effect do forest fires have on nutrient availabilities and losses to streams?

5. Compare the resource-conservation and environmental-quality issues related to each of the three so-called fertilizer elements, N, P, and K.

6. A park manager wants to fertilize an area of turfgrass with nitrogen and phosphorus at the rates of 60 kg/ha of N and 20 kg/ha of P. He has stocks of two types of fertilizers: urea (labeled 45-0-0) and diammonium phosphate (labeled 18-46-0). How much of each should he blend together to fertilize a 10-ha area of turfgrass?

7. How much phosphorus (P) is there in a 25-kg bag of fertilizer labeled "20-20-10"?

8. Compare the relative advantages and disadvantages of organic and inorganic nutrient sources.

9. A certified organic grower plans to grow a crop that requires the application of 120 kg of plant-available N/ha and 20 kg/ha of P. She has a source of compost that contains 1.5% total N (with 10% of this available in the first year) and 1.1% total P (with 80% of this available in the first year). (a) Assuming her soil has a low P soil test level, how much compost should she apply to provide the needed N and P? (b) If her soil is already optimal in P, how can she provide the needed amounts of both N and P *without* causing further P buildup?

10. Discuss the concept of the *limiting factor* and indicate its importance in enhancing or constraining plant growth.

11. Why are nutrient-cycling problems in agricultural systems more prominent than those in forested areas?

12. Discuss how GIS-based, site-specific nutrient-application technology might improve profitability and reduce environmental degradation.

13. Discuss the value and limitations of soil tests as indicators of plant nutrient needs and water pollution risks.

14. When might the use of plant tissue analyses have advantages over soil testing for correcting nutrient imbalances?

REFERENCES

Aber, J., et al. 2000. "Applying ecological principles to management of the U.S. National Forests," *Issues in Ecology*, **6**. http://esa.org/science_resources/issues/FileEnglish/Issue6.pdf (confirmed 12 September 2006) (Washington, DC: Ecological Society of America).

Addiscott, T. M. 2006. "Is it nitrate that threatens life or the scare about nitrate?" *Journal of the Science of Food and Agriculture,* **86**:2005–2009.

Beegle, D. B., O. T. Carton, and J. S. Bailey. 2000. "Nutrient management planning: Justification, theory, practice," *J. Environ. Quality,* **29**:72–79.

Beegle, D. B., L. E. Lanyon, and J. T. Sims. 2002. "Nutrient balances," pp. 171–193, in P. M. Haygarth and S. C. Jarvis (eds.), *Agriculture, Hydrology and Water Quality.* (Wallingford, UK: CAB International).

Brady, N. C., and R. R. Weil. 1996. *The Nature and Properties of Soils*, 11th ed. (Upper Saddle River, NJ: Prentice Hall).

CAST. 1999. *Gulf of Mexico hypoxia: Land and sea interactions.* Task Force Report 134 (Ames, IA: Council for Agricultural Science and Technology).

Dean, J. E., and R. R. Weil. 2009. "Brassica cover crops for nitrogen retention in the Mid-Atlantic coastal plain," *J. Environ. Qual.* **38**:520–528.

Dodd, J. R., and A. P. Mallarino. 2005. "Soil-test phosphorus and crop grain yield responses to long-term phosphorus fertilization for corn-soybean rotations," *Soil Sci. Soc. Am. J.*, **69**:1118–1128.

Eghball, B., B. J. Wienbold, J. E. Gilley, and R. A. Eigenberg. 2002. "Mineralization of manure nutrients," *J. Soil Water Conserv.*, **57**:470–473.

FAO. 2006. *FAOSTAT.* Food and Agriculture Organization of the United Nations. http://faostat.fao.org (verified 20 August 2006).

FAO. 1977. *China: Recycling of Organic Wastes in Agriculture.* F. A. O. Soils Bulletin **40** (Rome: U.N. Food and Agriculture Organization).

Havlin, J. L., J. D. Beaton, S. L. Tisdale, and W. L. Nelson. 2005. *Soil Fertility and Fertilizers—An Introduction to Nutrient Management*, 7th ed. (Upper Saddle River, NJ: Prentice Hall).

He, Xin-Tao, T. Logan, and S. Traina. 1995. "Physical and chemical characteristics of selected U.S. municipal solid waste composts," *J. Environ. Qual.*, **24**:543–552.

Heckman, J. R., and D. Kluchinski. 1996. "Chemical composition of municipal leaf waste and hand-collected urban leaf litter," *J. Environ. Qual.*, **25**:355–362.

Heckman, J. R., R. R. Weil, and F. Magdoff. 2009. "Practical steps to soil fertility for organic agriculture," Chapter 7, in C. A. Francis, ed., *Ecology in Organic Farming Systems* (Madison, WI: American Society of Agronomy/Crop Science Society of America, and Soil Science Society of America).

Higgs, B., A. E. Johnston, J. L. Salter, and C. J. Dawson. 2000. "Some aspects of achieving sustainable phosphorus use in agriculture," *J. Environ. Qual.*, **29**:80–87.

Krogmann, U., B. F. Rogers, L. S. Boyles, W. J. Bamka, and J. R. Heckman. 2003. "Guidelines for land application of non-traditional organic wastes (food processing by-products and municipal yard wastes) on farmlands in New Jersey," *Bulletin e281*. Rutgers Cooperative Extension, New Jersey Agricultural Experiment Station, Rutgers, The State University of New Jersey. (posted June 2003; verified 08 February 2009).

L'hirondel, J., and J.-L. L'hirondel. 2002. *Nitrate and Man: Toxic, Harmless or Beneficial?* (Wallingford, UK: CABI).

Magdoff, F., L. Lanyon, and B. Liebhardt. 1997. "Nutrient cycling, transformations, and flows: Implications for a more sustainable agriculture," *Advances in Agronomy*, **60**:2–73.

Miller, W. W., D. W. Johnson, T. M. Loupe, J. S. Sedinger, E. M. Carroll, J. D. Murphy, R. F. Walker, and D. Glass. 2006. "Nutrients flow from runoff at burned forest site in Lake Tahoe basin," *Calif. Agric.*, **60**:65–71.

Potash and Phosphate Institute. 1996. "Site-specific nutrient management systems for the 1990's." Pamphlet. (Norcross, GA.: Potash and Phosphate Institute and Foundation for Agronomic Research).

Preusch, P. L., P. R. Adler, L. J. Sikora, and T. J. Tworkoski. 2002. "Nitrogen and phosphorus availability in composted and uncomposted poultry litter," *J. Environ. Qual.*, **31**:2051–2057.

Reddy, G. B., and K. R. Reddy. 1993. "Fate of nitrogen-15 enriched ammonium nitrate applied to corn," *Soil Sci. Soc. Amer. J.*, **57**:111–115.

Reuter, D. J., and J. B. Robinson. 1986. *Plant Analysis: An Interpretation Manual* (Melbourne, Australia: Inkata Press).

Richards, R. P. et al. 1996. "Well water, well vulnerability, and agricultural contamination in the midwestern United States," *J. Environ. Qual.*, **25**:389–402.

Santamaria, P. 2006. "Nitrate in vegetables: Toxicity, content, intake and EC regulation," *Journal of the Science of Food and Agriculture*, **86**:10–17.

Sharpley, A. N. 1997. "Rainfall frequency and nitrogen and phosphorus runoff from soil amended with poultry litter," *J. Environ. Qual.*, **26**:1127–1132.

Smith, S. J., A. N. Sharpley, J. W. Naney, W. A. Berg, and O. R. Jones. 1991. "Water quality impacts associated with wheat culture in the Southern Plains," *J. Environ. Qual.*, **20**:244–249.

Stivers, L. J., C. Shennen, E. Jackson, K. Groody, and C. J. Griffin. 1993. "Winter cover cropping in vegetable production systems in California," in M. G. Paoletti, et al. (eds.), *Soil Biota, Nutrient Cycling and Farming Systems* (Boca Raton, FL: Lewis Press).

Swank, W. T., J. M. Vose, and K. J. Elliot. 2001. "Long-term hydrologic and water quality responses

following commercial clear cutting of mixed hard-woods on a southern Appalachian catchment," *Forest Ecology and Management,* **143**:163–178.

USDA Natural Resources Conservation Service. 2006. *The P Index: A Phosphorus Assessment Tool.* www.nrcs.usda.gov/TECHNICAL/ECS/nutrient/pindex.html (confirmed July 2007).

Walworth, J. L., and M. E. Sumner. 1987. "The diagnosis and recommendation integrated system (DRIS)," *Advances in Soil Science,* **6**:149–187.

Weil, R. R., and S. K. Mughogho. 2000. "Sulfur nutrition of maize in four regions of Malawi," *Agron. J,* **92**:649–656.

Yang, H. S. 2006. "Resource management, soil fertility and sustainable crop production: Experiences of China," *Agric. Ecosyst. Environ.,* **116**:27–33.

Zublena, J. P., J. C. Barker, and T. A. Carter. 1993. "Poultry manure as a fertilizer source," *Soil Facts* (Raleigh, NC: North Carolina Cooperative Extension Service, North Carolina State University).

Wind erosion degrades desert soil and vegetation. (R. Weil)

14
Soil Erosion and Its Control

No soil phenomenon is more destructive worldwide than the erosion caused by wind and water. Since prehistoric times people have brought the scourge of soil erosion upon themselves, suffering impoverishment and hunger in its wake. Past civilizations have disintegrated as their soils, once deep and productive, washed away, leaving only thin, rocky relics of the past. It is hard to imagine that agricultural communities once flourished in the now nearly barren hills in parts of India, Greece, Lebanon, or Syria.

Since 1960, farmers have had to more than double world food output to feed the unprecedented numbers of people on Earth. As the ratio of people to land steadily rises, poor people see little choice but to clear and burn steep, forested slopes or plow up natural grasslands to plant their crops. Population pressures have also led to overgrazing of rangelands and overexploitation of timber resources. All these activities lead to a downward spiral of ecological deterioration, land degradation, and deepening poverty. The impoverished crops and rangelands leave little if any residues to protect the soil, leading to further erosion, driving ever-more-desperate people to clear and cultivate—and degrade—still more land. Add to this the intense, concentrated erosion on sites disturbed by construction or mining activity, as well as globalization pressures to expand cropping and logging to marginal lands, and it is clear that the current threat of soil erosion is more ominous than at any other time in history.

The degraded productivity of farm, forest, range, and urban lands tells only part of the sad erosion story. Soil particles washed or blown from the eroding areas are subsequently deposited elsewhere—in nearby low-lying sites within the landscape or far away—even on other continents. Far downstream or downwind,

the sediment and dust cause major water and air pollution and bring enormous economic and social costs to society.

Combating soil erosion is everybody's business. Fortunately, much has been learned about the mechanisms of erosion and techniques have been developed that can effectively and economically control soil loss in most situations. This chapter will equip you with some of the concepts and tools you will need to do your part in solving this pressing world problem.

14.1   SIGNIFICANCE OF SOIL EROSION AND LAND DEGRADATION[1]

Land Degradation

Erosion and Runoff—A video from Pennsylvania: www.greentreks.org/watershedstv/smil/ws_wk180.ram

During the past half century, human land use and associated activities have degraded some 5 billion ha (about 43%) of the Earth's land. Degraded lands may suffer from destruction of native vegetation communities, reduced agricultural yields, lowered animal production, and simplification of once-diverse natural ecosystems with or without accompanying degradation of the soil resource. On about 2 billion ha of the degraded lands in the world, soil degradation is a major part of the problem. In some cases the soil degradation occurs mainly as deterioration of physical properties by compaction or surface crusting (see Sections 4.7 and 4.6), or as deterioration of chemical properties by acidification (see Section 9.6) or salt accumulation (see Section 9.12). However, most (~85%) soil degradation stems from erosion—the destructive action of wind and water.

The two main components of land degradation—damage to plant communities and deterioration of soil—interact to cause a downward spiral of accelerating ecosystem damage and human poverty (Figure 14.1). Improvements in both soil and vegetation

Figure 14.1

The downward spiral of land degradation resulting from the feedback loop between soil and vegetation. As the natural vegetation is disturbed, soil becomes exposed to raindrops and wind leading to erosion and loss of soil, including organic matter and nutrients. The now impoverished soil can support only stunted crops or other vegetation, which leaves the soil with even less protective cover and root mass than before. Soil loss becomes severe, such that the soil depth and the capacity to hold water are greatly reduced and vegetation can barely survive, leaving extremely degraded soil. Incapable of providing nutrients and water needed to support healthy growth of natural vegetation or crops, the site continues to erode, polluting rivers with sediment and impoverishing the people who attempt to grow their food on the land.
(Diagram courtesy of R. Weil)

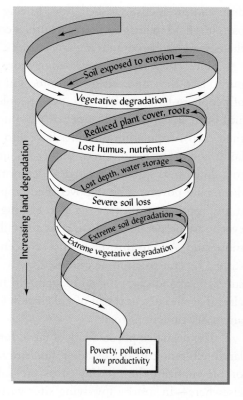

---

[1]For readable accounts of historical degradation of land and water resources, see Hillel (1991) and Montgomery (2007).

management must go hand-in-hand if the productive potential of the land is to be protected—or even be restored by moving *up* rather than *down* the spiral—a prospect that is attainable.

Geological Versus Accelerated Erosion

Geological Erosion   Erosion is a process that transforms soil into **sediment**. Soil erosion that takes place naturally, without the influence of human activities, is termed **geological erosion**. It is a natural leveling process. It inexorably wears down hills and mountains, and through subsequent deposition of the eroded sediments, it fills in valleys, lakes, and bays. In most settings, new soil forms from the underlying rock or regolith a bit faster than the old soil is lost. The very existence of soil profiles bears witness to the net accumulation of soil and the effectiveness of undisturbed natural vegetation in protecting the land surface from erosion.

Geological erosion by water tends to be greatest in semiarid regions where rainfall is enough to be damaging, but not enough to support dense, protective vegetation. Areas blanketed by deep deposits of silts may have exceptionally high erosion rates under such conditions.

Sediment Loads   Rainfall, geology, and other factors (including human activities) influence the sediment loads carried by the world's great rivers. Although rivers like the Mississippi and Yangtze were muddy before humans disturbed their watersheds, current sediment loads are far greater than before. To gain some perspective on the enormous amount of soil transported to the sea by these rivers, consider the Mississippi's sediment load (only a fifth as great as that of the Yangtze or the Ganges). If this 300 million Mg of sediment were carried to the Gulf of Mexico by dump trucks, it would take a continuous, year-round caravan of more than 80,000 large trucks, stretching all the way from Wisconsin to New Orleans (1600 km) and back, with a 20-Mg load being dumped into the Gulf about every 2 seconds.

Human-Accelerated Erosion   We stand in awe at the edge of the Grand Canyon, which was formed over millennia by geologic erosion—yet few realize that humankind has become the preeminent force on the landscape, now moving nearly twice as much soil per year as global geologic processes, and two-thirds of that *unintentionally* through erosion, mainly associated with agricultural activities (Figure 14.2).

Soil erosion in U.S. agriculture: www.epa.gov/agriculture/ag101/cropsoil.html#envconcerns

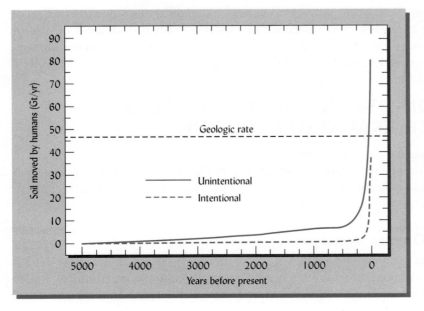

Figure 14.2

Estimates of the total amount of soil material moved annually by humans as a function of time. Intentional soil movement refers mainly to construction and excavation activities. Unintentional soil movement refers mainly to soil loss due to agricultural activities such as land clearing, tillage, overgrazing, and long periods without vegetative cover. Humans now move more soil material than all natural processes combined. This may not mean that sediment loads of major rivers have increased this dramatically, as movement of soil within a landscape does not necessarily lead to sediment in rivers. [For comparisons of agricultural to geologic soil movement, see Wilkinson and McElroy (2007). Graph redrawn from Hooke. (2000)]

Figure 14.3

Sediment deposition rates into Lake Pepin (Minnesota–Wisconsin border) from watersheds of the Minnesota river (92% cultivated) and the upper Mississippi and St. Croix rivers (42% cultivated). Neither watershed deposited much sediment before about 1830 because little of the land had been cleared for farming. But as agriculture expanded, deposition rates increased, especially from the Minnesota watershed where intensive row-crop agriculture became dominant. [Redrawn from Kelley and Nater (2000)]

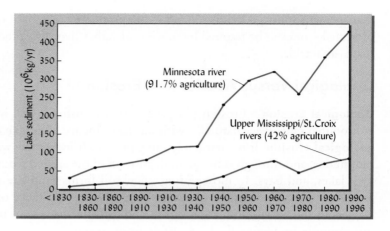

Accelerated erosion occurs when people disturb the soil or the natural vegetation by overgrazing livestock, cutting forests for agricultural use (Figure 14.3), plowing hillsides, or tearing up land for construction of roads and buildings. Accelerated erosion is often 10 to 1000 times as destructive as geological erosion, especially on sloping lands in regions of high rainfall. In the United States, the average erosion rate on cropland is about 12 Mg/ha—7 Mg by water and 5 Mg by wind. Cultivated soils in Africa and Asia are eroding at 10 times these average rates. In comparison, erosion on undisturbed humid-region grasslands and forests generally occurs at rates considerably below 0.1 Mg/ha.

About 4 billion Mg of soil is moved annually by soil erosion in the United States. More than half of the movement by water and about 60% of the movement by wind take place on croplands that produce most of the country's food. Much of the remainder comes from semiarid rangelands, from logging roads and timber harvest on forest lands, and from soils disturbed for highway and building construction. Although progress has been made in reducing erosion, current high losses are simply not sustainable.

Under the influence of accelerated erosion, soil is commonly washed or blown away faster than new soil can form by weathering or deposition. As a result, the soil depth suitable for plant roots is often reduced. In severe cases, gently rolling terrain may become scarred by deep gullies, and once-forested hillsides may be stripped down to bare rock. Accelerated erosion often makes the soils in a landscape more heterogeneous. An example can be seen in the striking differences in surface soil color that develop as ridgetop soils are truncated, exposing material from the B or C horizon at the land surface, while soils lower in the landscape are buried under organic-matter-enriched sediment (Figure 14.4 and Plates 36 and 65).

14.2   ON-SITE AND OFF-SITE EFFECTS OF ACCELERATED SOIL EROSION

Erosion damages the site on which it occurs and also has undesirable effects off-site in the larger environment. While the costs associated with these types of damages may not be immediately apparent, they are real and grow with time. Landowners and society as a whole must eventually foot the bill.

Types of On-Site Damages

Loss of soil is the most obvious damage from erosion. In reality, the damage done is greater than the amount of soil lost would suggest because the soil material eroded away is almost always more valuable than that left behind. Surface horizons erode,

Eroded ridge

Figure 14.4

Erosion and deposition occur simultaneously across a landscape. (Left) The soil on this ridgetop was worn down by erosion during nearly 300 years of cultivation. The surface soil exposed on the ridgetop consists mainly of light-colored C horizon material. (Right) Erosion on the sloping wheat field in the background has deposited a thick layer of sediment in the foreground, burying the plants at the foot of the hill. (Photos courtesy of R. Weil, *left*, and USDA Natural Resources Conservation Service, *right*)

while less fertile subsurface horizons remain untouched. Remaining topsoil is also impaired as erosion selectively removes organic matter and fine mineral particles, while leaving behind mainly relatively less active, coarser fractions.

The deterioration of soil structure often leaves a dense crust on the soil surface (see Section 4.6), which, in turn, greatly reduces water infiltration and increases water runoff. Newly planted seeds and seedlings may be washed downhill, trees may be uprooted, and small plants may be buried in sediment. In the case of wind erosion, fruits and foliage may be damaged by the sandblasting effect of blowing soil particles. Finally, gullies carve up badly eroded land and may undercut pavements and building foundations, causing unsafe conditions and expensive repairs.

Types of Off-Site Damages

Erosion moves sediment and nutrients off the land, creating the two most widespread water pollution problems in our rivers and lakes. The nutrients impact water quality largely through the process of eutrophication caused by excessive nitrogen and phosphorus, as was discussed in Section 13.2. In addition to nutrients, sediment and runoff water may also carry toxic metals and organic compounds, such as pesticides. The sediment itself is a major water pollutant, causing a wide range of environmental damages.

Ecosystem threats to Big Darby Creek watershed: www.nature.org/wherework/ northamerica/states/ ohio/bigdarby/habitat/

Damages from Sediment   Sediment deposited on the land may smother crops and other low-growing vegetation (Figure 14.4). It fills in roadside drainage ditches and creates hazardous driving conditions where mud covers the roadway.

Sediment that washes into streams makes the water cloudy or turbid (Plate 110). High **turbidity** prevents sunlight from penetrating the water and thus reduces photosynthesis and survival of the *submerged aquatic vegetation* (SAV). The demise of the SAV, in turn, degrades the fish habitat and upsets the aquatic food chain. The muddy water also fouls the gills of some fish. Sediment deposited on the stream bottom can have a disastrous effect on many freshwater fish by burying the pebbles and rocks among which they normally spawn. The buildup of bottom sediments can actually raise the level of the river, so that flooding becomes more frequent and more severe. For example, to counter the rising river bottom, flood-control levees along the Mississippi must be constantly enlarged.

It is estimated that 1.5 billion Mg of sediment are deposited each year in U.S. reservoirs, alone. Similarly, harbors and shipping channels fill in and become impassible. The loss of function and the costs of dredging, excavation, filtering, and construction activities necessary to remedy these situations run into the billions of dollars every year.

Windblown Sand and Dust   Wind erosion also has its off-site effects. Blowing sands may bury roads and fill in drainage ditches, necessitating expensive maintenance. The sandblasting effect of wind-borne soil particles may damage the fruits and foliage of crops in neighboring fields, as well as the paint on vehicles and buildings many kilometers downwind from the eroding site. Finer wind-blown dust with clay-size particles causes the most expensive and far-reaching damages. Much of this dust—especially particulate matter (PM) with diameters between 2.5 and 10 microns (PM_{10})—arises from wind erosion on cropland, rangelands, and construction sites (as well as from traffic on unpaved roads). Even more damaging are particles smaller than 2.5 microns ($PM_{2.5}$), which arise mainly from vehicle exhaust and smoke from fires and industrial plants. The off-site damages from these dust particles include aesthetics-related costs and major health hazards.

Health Hazards from PM_{10} and $PM_{2.5}$ While silt-sized particles are generally fil-tered by nose hairs or trapped in the mucous of the windpipe and bronchial tubes, smaller clay-sized particles often pass through these defenses and lodge in the alveoli (air sacs) of the lungs. The particles themselves cause inflammation of the lungs, and they may also often carry toxic substances that cause further lung damage. For exam-ple, airborne clay particles adsorb water vapor and may become coated with sulfuric or nitric acids found in the atmosphere. Human pathogens may also adhere to dust particles and travel with them to spread disease. Wind-blown dust is a global problem such that wind erosion in the Sahara desert in Africa and Gobi desert in China has been implicated in respiratory diseases in North America.

Maintenance of Soil Productivity   Although extreme soil erosion can reduce soil productivity to almost zero, in most cases the effect is too subtle to notice between one year and the next. Where farmers can afford to do so, they compensate for the loss of nutrients by increasing the use of fertilizer. The losses of organic matter and water-holding capacity are much more difficult to overcome. Over the long term, accelerated soil erosion that exceeds the rate of soil formation leads to declining pro-ductivity on most soils. In the United States, crop yields on severely eroded soils are often 20 to 40% lower than on similar soils with only slight erosion.

Ultimately, the rate of decline of soil productivity, or the cost of maintaining con-stant crop-yield levels, is determined by such soil properties as *depth to a root-restricting layer* and *permeability of the subsoil*. A deep, well-drained, and well-managed soil may not decline much in productivity even though it suffers some erosion. In contrast, ero-sion on a shallow, low-permeability soil may bring about a rapid productivity decline.

14.3   MECHANICS OF WATER EROSION

Movie of raindrop impact:
www.public.asu.edu/
~mschmeec/rainsplash
.html

Soil erosion by water is fundamentally a three-step process (Figure 14.5): (1) *detachment* of soil particles from the soil mass; (2) *transportation* of the detached particles down-hill by floating, rolling, dragging, and splashing; and (3) *deposition* of the transported particles at some place lower in elevation.

On comparatively smooth soil surfaces, the beating action of raindrops causes most of the detachment. Where water is concentrated into channels, the cutting

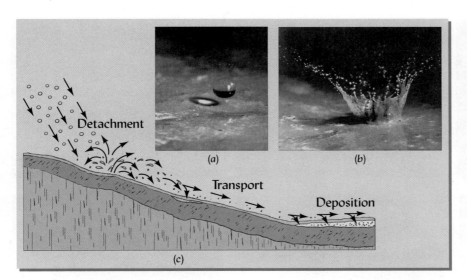

Figure 14.5
The three-step process of soil erosion by water begins with the impact of raindrops on wet soil. (a) A raindrop speeding toward the ground. (b) The splash that results when the drop strikes a wet, bare soil. Such raindrop impact destroys soil aggregates, encouraging sheet and interrill erosion. Also, considerable soil may be moved by the splashing process itself. The raindrop detaches soil particles, which are then transported and eventually deposited in locations downhill (c).

action of turbulent, flowing water can also detach soil particles. In some situations, freezing-thawing action contributes to soil detachment as well.

Influence of Raindrops

History may someday record that one of the truly significant scientific advances of the 20th century was the realization that most erosion is initiated by the impact of raindrops, rather than the flow of running water. For centuries prior to this realization, soil conservation efforts aimed at controlling the more visible flow of water across the land, rather than protecting the soil surface from the impact of raindrops.

A raindrop accelerates as it falls until it reaches *terminal velocity*—the speed at which the friction between the drop and the air balances the force of gravity. Larger raindrops fall faster. Speeding raindrops impact the soil with explosive force (see Figure 14.5).

Raindrop impact detaches soil, destroys granulation, and causes an appreciable transportation of soil. So great is the force exerted by raindrops that they not only loosen and detach soil granules but may even beat the granules to pieces.

Transportation of Soil

Raindrop Splash Effects   When raindrops strike a wet soil surface, they detach soil particles and send them flying in all directions (see Figure 14.5). If the land is sloping or if the wind is blowing, this splashing may be greater in one direction, leading to considerable net horizontal movement of soil.

Role of Running Water   If the rate of rainfall exceeds the soil's infiltration capacity, water will pond on the surface and begin running downslope. The soil particles sent flying by raindrop impact will then land in flowing water, which will carry them down the slope. So long as the water is flowing smoothly in a thin layer (sheet flow), it has little power to detach soil. However, in most cases the water is soon channeled by irregularities in the soil surface which cause it to increase in both velocity and turbulence. The channelized flow then not only carries along soil splashed by raindrops, but also begins to detach particles as it cuts into the soil mass. This is an accelerating process, for as a channel is cut deeper, it fills with greater and greater volumes of turbulent water. So familiar is the power of runoff water to cut and carry that the public generally ascribes to it all the damage done by heavy rainfall.

(a) Sheet erosion   (b) Rill erosion   (c) Gully erosion

Figure 14.6

Three major types of soil erosion. Sheet erosion is relatively uniform erosion from the entire soil surface. Note that the perched stones and pebbles have protected the soil underneath from rain drop impact. The pencil gives a sense of scale. Rill erosion is initiated when the water concentrates in small channels (rills) as it runs off the soil. Subsequent cultivation may erase rills, but it does not replace the lost soil. Gully erosion creates deep channels that cannot be erased by cultivation. Although gully erosion looks the most catastrophic of the three, far more total soil is lost by the less obvious sheet and rill erosion. [Drawings from FAO (1987); photos courtesy of USDA Natural Resources Conservation Service]

Types of Water Erosion

Three types of water erosion are generally recognized: (1) *sheet,* (2) *rill,* and (3) *gully* (Figure 14.6). In **sheet erosion**, splashed soil is removed more or less uniformly, except that tiny columns of soil often remain where pebbles intercept the raindrops (see Figure 14.6*a*). However, as the sheet flow is concentrated into tiny channels (termed **rills**), **rill erosion** becomes dominant. Rills are especially common on bare land, whether newly planted or in fallow (see Figure 14.6*b*). Rills are channels small enough to be smoothed by normal tillage, but the damage is already done—the soil is lost. When sheet erosion takes place primarily between irregularly spaced rills, it is called **interrill erosion**.

Where the volume of runoff is further concentrated, the rushing water cuts deeper into the soil, deepening and coalescing the rills into larger channels termed **gullies** (see Figure 14.6*c*). This is **gully erosion**. Gullies on cropland are obstacles for tractors and cannot be removed by ordinary tillage practices. All three types may be serious, but sheet and rill erosion, although less noticeable than gully erosion, are responsible for most of the soil moved.

Deposition of Eroded Soil

Erosion may send soil particles on a journey of a thousand kilometers or more—off the hills, into creeks, and down great muddy rivers to the ocean. On the other hand, eroded soil may travel only a meter or two before coming to rest in a slight depression on a hillside or at the foot of a slope (as was shown in Figure 14.4). The amount of soil delivered to a stream, divided by the amount initially eroded, is termed the **delivery ratio**. As much as 60% of eroded soil may reach a stream (delivery ratio = 0.60) in certain watersheds where valley slopes are very steep. As little as 1% may reach the streams draining a gently sloping coastal plain. Typically, the delivery ratio is larger for small watersheds than for large ones, because the latter provide many more opportunities for deposition before a major stream is reached. It is estimated that about 5 to 10% of all eroded soil in North America is washed out to sea. The remainder is deposited in reservoirs, river beds, on flood plains, or on relatively level land farther up the watershed.

Soil from an eroding PA streambank travels to the Chesapeake Bay: www.bayjournal.com/ article.cfm?article=699

14.4 MODELS TO PREDICT THE EXTENT OF WATER-INDUCED EROSION[2]

Land managers and policymakers need to predict the extent of soil erosion so that they can plan for the best management of soil resources, evaluate the consequences of alternative tillage practices, determine compliance with environmental regulations, develop sediment-control plans for construction projects, and estimate the years it will take to silt-in dams and channels.

The WEPP model explained and available for download: http://topsoil.nserl.purdue .edu/nserlweb/weppmain/ wepp.html

The detachment, transport, and deposition processes of soil erosion can be predicted mathematically by soil erosion *models*. These are equations—or sets of linked equations—that interrelate information about the rainfall, soil, topography, vegetation, and management of a site with the amount of soil likely to be lost by erosion. The most ambitious and sophisticated of the erosion models developed so far is a complex, process-based computer program called the Water Erosion Prediction Project (WEPP). It is based on an understanding of the fundamental mechanisms involved with each process leading to soil erosion.

The Universal Soil-Loss Equation (USLE)

In contrast to the process-based operation of WEPP, most predictions of soil erosion continue to rely on much simpler models that statistically relate soil erosion to a number of easily observed factors. Scientists can make such *empirical* models if they know that certain conditions are associated with soil erosion, even if they do not understand the details of *why* this is so. At the heart of these models is the realization that water-induced erosion results from the interaction of rain and soil. Decades of erosion research have clearly identified the major factors affecting this interaction. These factors are quantified in the **universal soil-loss equation (USLE)**:

$$A = R \times K \times LS \times C \times P \qquad (14.1)$$

A, the predicted annual soil loss, is the product of

R = rainfall erosivity } Rain-related factor

K = soil erodibility
L = slope length } Soil-related factors
S = slope gradient or steepness

C = cover and management } Land-management factors
P = erosion-control practices

[2]For discussion of the original USLE, see Wischmeier and Smith (1978), and for the RUSLE, see Renard et al. (1997). In this textbook, we use the scientifically acceptable SI units for the *R* and *K* factors in our discussion of these erosion equations. However, since these soil-loss equations were published in the United States for use by landowners and the general public, most maps, tables, and computer programs available supply values for the *R* and *K* factors in customary English units, rather than in SI units. When using English units for the *R* and *K* factors, the soil loss *A* is expressed in tons (2000 lb) per acre, which can be easily converted to Mg/ha by multiplying by 2.24. For details on converting the customary English units to SI units, see Foster et al. (1981).

Working together, these factors determine how much water enters the soil, how much runs off, how much soil is transported, and when and where it is redeposited. Note that because the factors are multiplied together, *if any one factor could be reduced to zero, the resulting amount of erosion (A) would also be reduced to zero.*

Unlike the WEPP program, the USLE was designed to predict only the amount of soil loss by sheet and rill erosion in an average year for a given location. It cannot predict erosion from a specific year or storm, nor can it predict the extent of gully erosion and sediment delivery to streams. It can, however, show how varying any combination of the soil- and land-management-related factors might be expected to influence soil erosion, and therefore can be used as a decision-making aid in choosing the most effective strategies to conserve soil.

The USLE has been used widely since the 1970s. More recently, the basic USLE was updated and computerized to create an erosion-prediction tool called the **revised universal soil-loss equation (RUSLE)**. The RUSLE uses the same basic factors of the USLE just shown, although some are better defined and interrelationships among them improve the accuracy of soil-loss prediction. The RUSLE is a computer software package that is constantly being improved and modified as experience is gained from its use around the world.

RUSLE-2 download and training: http://fargo.nserl.purdue .edu/rusle2_dataweb/ RUSLE2_Index.htm

14.5 FACTORS AFFECTING INTERRILL AND RILL EROSION

Rainfall Erosivity Factor, *R*

The rainfall **erosivity** factor, *R*, represents the driving force for sheet and rill erosion. It takes into consideration the total rainfall and, more important, the intensity and seasonal distribution of the rain. Rainfall index values for locations in the United States are shown in Figure 14.7. Conservation practices based on the predictions of the USLE or RUSLE using long-term average *R* factors may not be sufficient to limit erosion damages from relatively rare, but extremely damaging, storms.

Soil Erodibility Factor, *K*

The soil **erodibility** factor, *K* (Table 14.1), indicates a soil's inherent susceptibility to erosion. The *K* value assigned to a particular type of soil indicates the amount of soil lost per unit of erosive energy in the rainfall, assuming a standard research plot (22 m long, 9% slope) on which the soil is kept continuously bare by tillage.

The two most significant and closely related soil characteristics influencing erodibility are (1) *infiltration capacity* and (2) *structural stability*. High infiltration means that less water will be available for runoff, and the surface is less likely to be ponded (which would make it more susceptible to splashing). Stable soil aggregates resist the beating action of rain, and thereby save soil even though runoff may occur. Certain tropical clay soils high in hydrous oxides of iron and aluminum are known for their highly stable aggregates that resist the action of torrential rains. Downpours of a similar magnitude on swelling-type clays would be disastrous.

Topographic Factor, *LS*

The topographic factor, *LS*, reflects the influence of length and steepness of slope on soil erosion. It is expressed as a unitless ratio with soil loss from the area in question in the numerator, and that from a standard plot (9% slope, 22 m long) in the denominator. The longer the slope, the greater the opportunity for concentration of the runoff water.

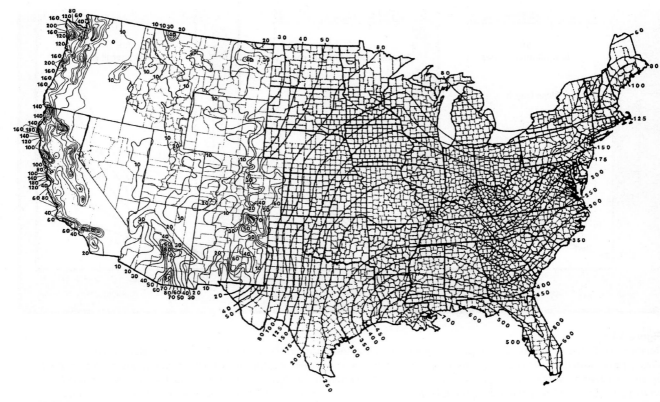

Figure 14.7

The geographic distribution of R values for rainfall erosivity in the continental United States. Note the very high values in the humid, subtropical Southeast, where annual rainfall is high and intense storms are common. Similar amounts of annual rainfall along the coast of Oregon and Washington in the Northwest result in much lower R values because there the rain mostly falls gently over long periods. The complex patterns in the West are mainly due to the effects of mountain ranges. Values on map are in units of 100 (ft·ton·in.)/(acre·yr). To convert to SI units of (MJ·mm)/(ha·h·yr), multiply by 17.02. [Redrawn from USDA (1995)]

Figure 14.8 illustrates the increases in LS factors that occur as slope length and steepness increase for sites with low, moderate, and high ratios of rill to interrill (sheet) erosion. Most sites cultivated to row crops have moderate rill to interrill erosion ratios. On sites where this ratio is low, such as rangelands, more of the soil movement occurs by interrill

Table 14.1
COMPUTED *K* VALUES (IN SI UNITS) FOR SOILS AT DIFFERENT LOCATIONS

Soil	Location	Compounds[a] *K*
Udalf (Dunkirk silt loam)	Geneva, NY	0.091
Udalf (Keene silt loam)	Zanesville, OH	0.063
Udoll (Marshall silt loam)	Clarinda, IA	0.044
Aqualf (Mexico silt loam)	McCredie, MO	0.034
Udult (Cecil sandy loam)	Watkinsville, GA	0.030
Alfisols	Indonesia	0.018
Oxisols	Ivory Coast	0.013
Ultisols	Hawaii	0.012
Ultisols	Nigeria	0.005
Oxisols	Puerto Rico	0.001

[a]To convert *K* values from (Mg · ha · h)/(ha · MJ · mm) to English units of (ton · acre · h)/(100 acres · ft-ton · in.), multiply values in this table by 7.6. From Wischmeier and Smith (1978) and Cassel and Lal (1992).

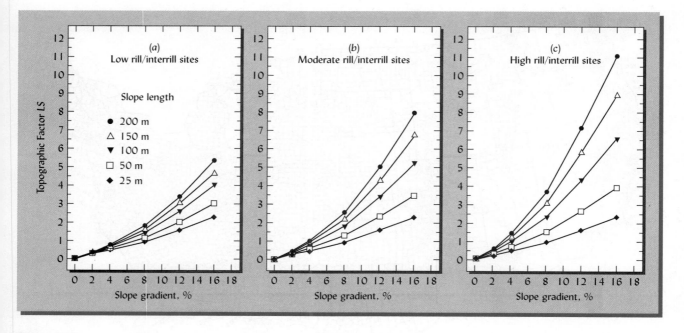

Figure 14.8

Relationship between values of the topographic factor LS and the slope gradient for several lengths of slope on three types of sites: (a) sites with low ratios of rill to interrill erosion, such as many rangelands; (b) sites with moderate ratios of rill to interrill erosion, such as most tilled row-crop land; and (c) sites with high ratios of rill to interrill erosion, such as freshly disturbed construction sites and new seedbeds. The LS values extrapolated from these graphs can be used in the Universal Soil-Loss Equation. [Graphs based on data in Renard et al. (1997)]

erosion. On these sites, slope steepness (%) has a relatively greater influence on erosion, while the slope length has a relatively smaller influence. The opposite is true for freshly excavated construction areas and other highly disturbed sites, which have high rill to interrill erosion ratios. Here, where rill erosion predominates, slope length has a greater influence.

Cover and Management Factor, *C*

Erosion and runoff are markedly affected by different types of vegetative cover and cropping systems. Undisturbed forests and dense grass provide the best soil protection and are about equal in their effectiveness. Perennial forage crops (both legumes and grasses) are next in effectiveness because of their relatively dense year-round cover. Annual crops, such as corn, soybeans, cotton, or potatoes, offer relatively little living cover during the early growth stages and thereby leave the soil susceptible to erosion unless residues from previous crops cover the soil surface.

Cover crops consist of plants that are similar to the forage crops just mentioned. They can provide soil protection during the time of year between the growing seasons for annual crops. For widely spaced perennial plantings such as orchards and vineyards, cover crops can permanently protect the soil between rows of trees or vines. A mulch of plant residue or applied materials is also effective in protecting soils. Even small increases in surface cover result in large reductions in soil erosion, particularly interrill erosion (Figure 14.9).

Regulation of grazing to maintain a dense vegetative cover on range- and pastureland and the inclusion of close-growing hay crops in rotation with row crops on arable land will help control both erosion and runoff. Likewise, the use of conservation tillage systems, which leave most of the plant residues on the surface, greatly decreases erosion hazards.

The *C* factor in the USLE or RUSLE is the ratio of soil loss under the conditions in question to that which would occur under continuously bare soil. This ratio

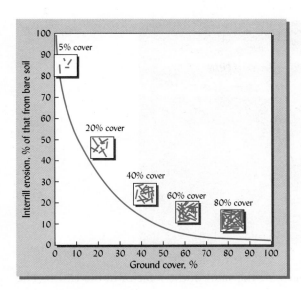

Figure 14.9

Reduction in interrill erosion achieved by increasing ground cover percentage. The diagrams above the graph illustrate 5, 20, 40, 60, and 80% ground cover. Note that even a light covering of mulch has a major effect on soil erosion. The graph applies to interrill erosion. On steep slopes, some rill erosion may occur even if the soil is well covered. (Generalized relationship based on results from many studies)

C will approach 1.0 where there is little soil cover (e.g., a bare seedbed in the spring or freshly graded bare soil on a construction site). It will be low (e.g., <0.10) where large amounts of plant residues are left on the land or in areas of dense perennial vegetation. Examples of C values are given in Table 14.2.

Support Practice Factor, P[3]

On some sites with long and/or steep slopes, erosion control achieved by management of vegetative cover, residues, and tillage must be augmented by the construction of physical structures or other steps aimed at guiding and slowing the flow of runoff

---

Table 14.2
EXAMPLES OF *C* VALUES FOR THE COVER AND VEGETATION MANAGEMENT FACTOR
The C values indicate the ratio of soil eroded from a particular vegetation system to that expected if the soil were kept completely bare.

Vegetation	Management/condition	C value
Range grasses and low (<1 m) shrubs	75% canopy cover, no surface litter	0.17
	75% canopy cover, 60% cover with decaying litter	0.032
Scrub brush about 2 m tall	25% canopy cover, no litter	0.40
	75% canopy cover, no litter	0.28
Trees with no understory, about 4 m drop fall	75% canopy cover, no litter	0.36
	75% canopy cover, 40% leaf litter cover	0.09
Woodland with understory	90% canopy cover, 100% litter cover	0.001
Permanent pasture	Dense stand of grass sod	0.003
Corn–soybean rotation	Fall plowing, conventional tillage, residues removed	0.53
	Spring chisel plow–plant conservation tillage, 2500 kg/ha surface residues after planting	0.22
	No-till planting, 5000 kg/ha surface residues after planting	0.06
Corn–oats–hay–hay rotation	Spring conventional plowing before planting	0.05
	No-till planting	0.03

Values typical of midwestern United States. Based on Wischmeier and Smith (1978) and Schwab et al. (1996).

---

[3]Many of the erosion-control practices or management techniques discussed with regard to the C and P factors and in later sections of this chapter are considered to be **best management practices (BMPs)** (see also Section 13.2) under provisions of the Clean Water Act in the United States. The act defines BMPs as "optimal operating methods and practices for reducing or eliminating water pollution" from land-use activities.

Figure 14.10

Contour ridges must be carefully laid out with sufficient height to hold back water from even heavy rainfall. Here, surface retention is fast becoming surface runoff.

(Photo courtesy of R. Weil)

water. These **support practices** determine the value of the *P* factor in the USLE. The *P* factor is the ratio of soil loss with a given support practice to the corresponding loss without that practice. If there are no support practices, the *P* factor is 1.0. Support practices include tillage on the contour, contour strip-cropping, terrace systems, and grassed waterways, all of which will tend to reduce the *P* factor.

Contour Cultivation Rows of plants slow the flow of runoff water if they follow the contours across the slope gradient (but the rows *encourage* channelization and gullies if they run up and down the slope). Even more effective is planting on ridges built up of soil along the contours. However, ridges must be designed to carry heavy runoff safely from the field (Figure 14.10).

On long slopes subject to sheet and rill erosion, the fields may be laid out in narrow strips across the incline, alternating the tilled crops, such as corn and potatoes, with hay and small grains. Water cannot achieve an undue velocity on the narrow strips of tilled land, and the hay and grain crops check the rate of runoff. Such a layout is called **strip-cropping** and is the basis for erosion control in many hilly agricultural areas (Plate 72). This arrangement can be thought of as shortening the effective slope length.

When the cross strips are laid out rather definitely on the contours, the system is called **contour strip-cropping**. The width of the strips will depend primarily on the degree of slope, the permeability of the land, and the soil erodibility. Widths of 30 to 125 m are common. Contour strip-cropping is often augmented by diversion ditches and waterways between fields. Permanent sod established in the swales produces **grassed waterways** that can safely carry water off the land without the formation of gullies (Plate 72).

Terraces Construction of various types of terraces reduces the effective length and gradient of a slope. **Bench terraces** are used where nearly complete control of the water runoff must be achieved, such as in rice paddies. Where farmers use large machinery and need to farm all the land in a field, **broad-based terraces** are more common. Broad-based terraces waste little or no land and are quite effective if properly maintained. Water collected behind each terrace flows gently across (rather than down) the field in a terrace channel, which has a drop of only about 50 cm in 100 m (0.5%). The terrace channel usually guides the runoff water to a grassed waterway, through which the water moves downhill to a nearby canal, stream, or river.

Table 14.3

P FACTORS FOR CONTOUR AND STRIP-CROPPING AT DIFFERENT SLOPES AND THE TERRACE SUBFACTOR AT DIFFERENT TERRACE INTERVALS

The product of the contour or strip-cropping factors and the terrace subfactor gives the P value for terraced fields.

Slope, %	Contour P factor	Strip-cropping P factor	Terrace interval, m	Terrace subfactor Closed outlets	Terrace subfactor Open outlets
1–2	0.60	0.30	33	0.5	0.7
3–8	0.50	0.25	33–44	0.6	0.8
9–12	0.60	0.30	43–54	0.7	0.8
13–16	0.70	0.35	55–68	0.8	0.9
17–20	0.80	0.40	69–90	0.9	0.9
21–25	0.90	0.45	90	1.0	1.0

Contour and strip-cropping factors from Wischmeier and Smith (1978); terrace subfactor from Foster and Highfill (1983).

Examples of *P* values for contour tillage and strip-cropping at different slope gradients are shown in Table 14.3. The five factors of the USLE (*R, K, LS, C,* and *P*) can suggest many approaches to the practical control of soil erosion. A sample calculation is shown in Box 14.1 to illustrate how the USLE can help evaluate erosion-control options.

14.6 CONSERVATION TILLAGE

For centuries, conventional agricultural practice around the world encouraged extensive soil tillage that leaves the soil bare and unprotected from the ravages of erosion. During the last half century, two technological developments have allowed many farmers to avoid this problem by managing their soils with greatly reduced tillage—or no tillage at all. First came the development of herbicides that could kill weeds chemically rather than mechanically. Second, farmers and equipment manufacturers developed machinery that could plant crop seeds even if the soil was covered by plant

Conservation tillage pros, cons, and methods: www.ncsu.edu/sustainable/ tillage/tillage.html

---

BOX 14.1

CALCULATION OF EXPECTED SOIL LOSS USING USLE

The principles involved in both USLE and RUSLE can be verified by making calculations using USLE and its associated factors. Note that the factors in the USLE are related to each other in a multiplicative fashion. Therefore, if any one factor can be made to be near zero, the amount of soil loss A will be near zero.

Assume, for example, a location in Iowa on a Marshall silt loam with an average slope of 6% and an average slope length of 100 m. Assume further that the land is clean-tilled and fallowed.

Figure 14.7 shows that the R factor for this location is about 150 in English units or (150 × 17) 2550 in SI units.

The K factor for a Marshall silt loam in central Iowa is 0.044 (Table 14.2) and the topographic factor LS from Figure 14.8 is 1.7 (high rill to interrill ratio on soil kept bare). The C factor is 1.0, since there is no cover or other management practice to discourage erosion. If we assume the tillage is up and down the hill, the P value is also 1.0. Thus, the anticipated soil loss can be calculated by the USLE (A = RKLSCP):

$$A = (2550)(0.044)(1.7)(1.0)(1.0)$$
$$= 191 \text{ Mg/ha or } 85.2 \text{ tons/acre}$$

residues. These developments obviated two of the main reasons that farmers tilled their soils. Farmer interest in reduced tillage heightened as it was shown that these systems produced equal or even higher crop yields in many regions while saving time, fuel, money—and soil. The latter attribute earned these systems the name of **conservation tillage**.

Conservation Tillage Systems

Pursuing conservation tillage in organic farming systems:
http://attra.ncat.org/attra-pub/organicmatters/conservationtillage.html

While there are numerous conservation tillage systems in use today, all have in common that they leave significant amounts of organic residues on the soil surface after planting. Keep in mind that conventional tillage involves first moldboard plowing (Figure 14.11, *left*) to completely bury weeds and residues, followed by one to three passes with a harrow to break up large clods, then planting the crop, and subsequently several cultivations between crop rows to kill weeds. Every pass with a tillage implement bares the soil anew and also weakens the structure that helps soil resist water erosion.

Conservation tillage systems range from those that merely reduce excess tillage to the no-till system, which uses no tillage beyond the slight soil disturbance that occurs as the planter cuts a planting slit through the residues to a depth of several cm into the soil (Figure 14.12, *inset*). The conventional moldboard plow was designed to leave the field "clean"; that is, free of surface residues. In contrast, conservation tillage systems, such as **chisel plowing** (Figure 14.11, *right*), stir the soil but only partially incorporate surface residues, leaving more than 30% of the soil covered. **Stubble mulching**, whose water-conserving attributes were highlighted in Section 6.4, is another example. **Ridge tillage** is a conservation system in which crops are planted on top of permanent 15- to 20-cm-high ridges. About 30% soil coverage is maintained, even though the ridges are scraped off a bit for planting and then built up again by shallow tillage to control weeds.

With **no-till** systems, we can expect 50 to 100% of the surface to remain covered. Well-managed continuous no-till systems in humid regions include cover crops during the winter and high-residue-producing crops in the rotation. Such systems keep the soil completely covered at all times and build up organic surface layers somewhat like those found in forested soils.

Figure 14.11
Conventional inversion tillage and conservation tillage in action. (Left) In conventional tillage, a moldboard plow inverts the upper soil horizon, burying all plant residues and producing a bare soil surface. (Right) A chisel plow, one type of conservation tillage implement, stirs the soil but leaves a good deal of the crop residues on the soil surface. (Photos courtesy of R. Weil)

Figure 14.12

In no-till systems, one crop is planted directly into the residue of a cover crop or of a previous cash crop, with only a narrow band of soil disturbed. No-till systems leave virtually all of the residue on the soil surface, providing up to 100% cover and nearly eliminating erosion losses. Here corn was planted into a cover crop killed with a herbicide (weed-killing chemical) to form a surface mulch. The inset shows a closeup of the no-till planter in action (direction of travel is to the right). The rolling furrow openers cut a slot through the residue and soil into which the seed is placed at a depth set by the depth wheel. Snug seed-to-soil contact is ensured by the press wheel that closes the slot. (Photos courtesy of R. Weil)

Conservation tillage systems generally provide yields equal to or greater than those from conventional tillage, provided the soil is not poorly drained and in a cool region. However, during the transition from conventional tillage to no-tillage, crop yields may decline slightly for several years for reasons associated with some of the effects outlined in the following subsections.

No-till systems, especially, have spread to nearly all regions of the USA and are now used in some form on almost half of all the conservation tillage hectares. The no-till system has been used continuously on some farms in the eastern United States since about 1970 (more than 40 years without any tillage). One of the most significant examples of no-till expansion has been in Argentina and southern Brazil. Thousands of small-scale soybean and corn farmers there have successfully adapted cover-crop-based no-till systems using animal traction or small tractors.

Conservation Tillage Effects on Soil

Since conservation tillage systems were initiated, hundreds of field trials have demonstrated that these tillage systems allow much less soil erosion than do conventional tillage methods. Surface runoff is also decreased, although the differences are not as pronounced as with soil erosion (Figure 14.13). These differences are reflected in the much lower *C* factor values assigned to conservation tillage systems (see Table 14.2).

The erosion-control value of an undisturbed surface residue mulch was discussed in the previous section. Conservation tillage also significantly reduces the loss of nutrients dissolved in runoff water or attached to sediment.

When soil management is converted from plow tillage to conservation tillage (especially no-till), numerous soil properties are affected, mostly in favorable ways. The changes are most pronounced in the upper few centimeters of soil. Generally, the changes are greatest for systems that use cover crops and produce large amounts of plant residue (especially corn and small grains in humid regions), retain residue coverage, and cause little to no soil disturbance. Many of these changes are illustrated in other chapters of this textbook.

The abundance, activity, and diversity of soil organisms tend to be greatest in conservation tillage systems characterized by high levels of surface residue and little

No-till without herbicides. Rodale Research Institute: www.newfarm.org/depts/ NFfield_trials/1103/ notillroller.shtml

Video on no-till for sustainable rural development: http://info.worldbank.org/ etools/bspan/PresentationView .asp?PID=665&EID=339

Conservation tillage in Zambia: www.fao.org/ag/ags/agse/ agse_s/3ero/namibia1/til_ nam.htm

Figure 14.13

Effect of tillage systems on soil erosion and runoff from corn plots in Illinois following corn and following soybeans. Soil loss by erosion was dramatically reduced by the conservation tillage practices. The period of runoff was reduced most by the disk chisel system where corn was grown after corn. The soil was a Typic Argiudoll (Catlin silt loam), 5% slope, planted up- and downslope, tested in early spring. [Data from Oschwald and Siemens (1976)]

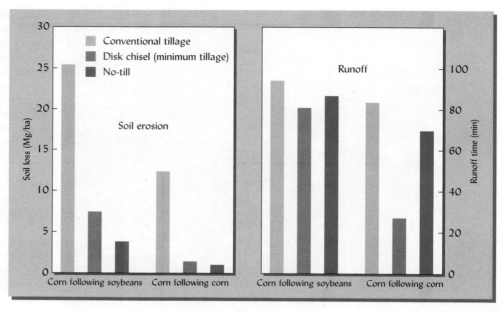

physical soil disturbance (see Section 10.14). Earthworms and fungi, both important for soil structure, are especially favored. However, organic residues left on the surface in no-till are actually more slowly decomposed than those incorporated by conventional tillage. No-till residues are in less intimate contact with the soil particles so their breakdown is delayed, and they remain as a protective surface barrier for a longer period of time.

14.7 VEGETATIVE BARRIERS

Narrow rows of permanent vegetation (usually grasses or shrubs) planted on the contour can be used to slow down runoff, trap sediment, and eventually build up "natural" or "living" terraces (Figure 14.14). In some situations, tropical grasses (e.g., a deep-rooted,

Figure 14.14

The use of vegetative barriers to create natural terraces. (Photo) A tropical grass (vetiver) has been planted vegetatively on the contour in a cassava field by pushing root cuttings into the soil. The grass is planted in a hedge running perpendicular to the slope direction. In a year or so the grass will be well established and its dense root and shoot growth will serve as a barrier to hold soil particles while permitting some water to pass on through. Note the buildup of soil above the grass, basically forming a terrace wall (diagram). (Photo courtesy of Centro Internacional de Agricultura Tropical in Cali, Colombia)

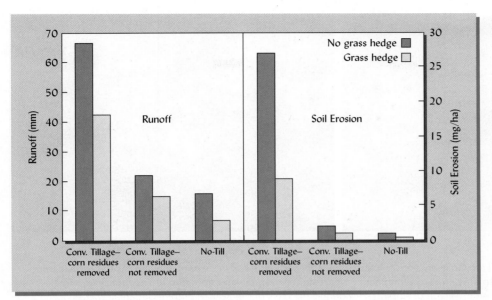

Figure 14.15

Narrow grass hedges can be effective tools to reduce losses of runoff water and soil. Switchgrass hedges 0.72 m wide were established across the slope every 16 m in a field of continuous corn that then received a total of 120 mm of simulated rainfall. The soil was a Typic Hapludoll in Iowa with an average slope of 12%. Corn residues left on the soil surface as well as no-till systems also drastically reduced losses, especially of the soil. [Data estimated from a figure in Gilley et al. (2000)]

drought-tolerant species called *vetiver grass*) have shown considerable promise as an affordable alternative to the construction of terraces.

The deep-rooted grass plants with dense, stiff stems tend to filter out soil particles from the muddy runoff. This sediment accumulates on the upslope side of the grass barrier and, in time, actually creates a terrace that may be more than 1 m above the soil surface on the downslope side of the plants. Narrow grass hedges can be effective in reducing runoff and erosion from soils in many regions, including the U.S. Midwest (Figure 14.15).

14.8 CONTROL OF GULLY EROSION AND MASS WASTING

Gullies rarely form in soils protected by healthy, dense, forest or sod vegetation, but are common on deserts, rangeland, and open woodland in which the soil is only partially covered. Gullies also readily form in soils exposed by tillage or grading if small rills are allowed to coalesce so that running water eats into the land (Figure 14.16, *left*). Water concentrated by poorly designed roads and trails may cause gullies to form even in dense forests. In many cases, neglected gullies will continue to grow and after a few years, devastate the landscape (see Figure 14.16, *right*). On the other hand, in some stony soils coarse fragments left behind in the channel bottom may protect it from further cutting action.

Remedial Treatment of Gullies

If small enough, gullies can be filled in, shaped for smooth water flow, sown to grass, and thereafter left undisturbed to serve as grassed waterways. When the gully erosion is too active to be checked in this manner, more extensive treatment may be required. If the gully is still small, a series of check dams about 0.5 m high may be constructed at intervals of 4 to 9 m, depending on the slope. These small dams may be constructed from materials available on site, such as large rocks, rotted hay bales, brush, or logs. Wire netting may be used to stabilize these structures. Check dams, whether large or small, should be constructed with the general features

Figure 14.16

The devastation of gully erosion. (Left) Gully erosion in action on a highly erodible soil in western Tennessee. The roots of the small wheat plants are powerless to prevent the cutting action of the concentrated water flow. (Right) The legacy of neglect of human-induced accelerated erosion. Tillage of sloping soils during the days of the Roman Empire began a process of accelerated erosion that eventually turned swales into jagged gullies that continue to cut into this Italian landscape with each heavy rain. For a sense of scale, note the olive trees and houses on the grassy, gentle slopes of the relatively uneroded hilltops.
(Photos courtesy of the USDA Natural Resources Conservation Service, *left*, and R. Weil, *right*)

illustrated in Figure 14.17. After a time, enough sediment may collect behind the dams to form a series of bench terraces and the ditch may be filled in and put into permanent sod.

With very large gullies, it may be necessary to divert the runoff away from the head of the channel and install more permanent dams of earth, concrete, or stone in the channel itself. Again, sediment deposited above the dams will slowly fill in the gully. Semipermanent check dams, flumes, and riprap-lined channels are also used on construction sites, but are generally too expensive for extensive use on agricultural land.

Mass Wasting on Unstable Slopes

Liquefaction of saturated soils:
www.ce.washington.edu/
~liquefaction/html/what/
what1.html

The downhill movement of large masses of unstable soil (**mass wasting**) is quite different from the erosion of the soil surface, which is the main topic of this chapter. Mass wasting can be a problem on very steep slopes (usually greater than 60% slope). While this type of soil loss sometimes occurs on steep pastures, it is most common on

Figure 14.17

A schematic drawing of a check dam used to arrest gully erosion. Whether made from rock, brush, concrete, or other materials, a check dam should have the general features shown. The structure should be dug into the walls of the gully to prevent water from going around it. The center of the dam should be lower so that water will spill over there and not wash out the soil of the gully walls. An erosion-resistant apron made from densely bundled brush, concrete, large rocks, or similar material should be installed beneath the center of the dam to prevent the overflow from undercutting the structure. In contrast to the gully-healing effect of a well-designed check dam, the haphazard dumping of rocks, brush, or junked cars into a gully will make matters worse, not better. (Diagram courtesy of R. Weil)

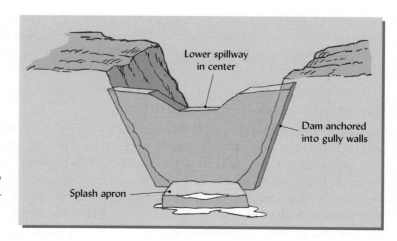

Lower spillway
in center

Dam anchored
into gully walls

Splash apron

Plate 61 *An urban turf profile with a thick thatch layer.*

Plate 62 *Water-stressed soybeans show extent of oak tree root competition for water.*

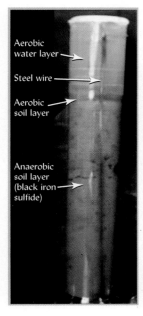

Aerobic water layer

Steel wire

Aerobic soil layer

Anaerobic soil layer (black iron sulfide)

Plate 63 *Oxidized (orange iron oxide) and reduced (black iron sulfide) zones along steel wire embedded in Winogradsky column.*

Topsoil

Ap1

Ap2

25

'2C

Sand layer

50

'2Bs

Sulfuric horizon

'3Bw

75

Sulfidic clay layer

'3Cgd

Plate 64 *Profile of 6-year-old landfill cap with layers of topsoil, sand, and (sulfidic) clay.*

Ap

Bt

C

Plate 65 *Ultisol profile in Maryland truncated by 300 years of erosion under tillage.*

Plate 66 *Grass turned green by septage, a sign of septic drain field malfunction.*

Plate 67 *Dark green grass growing over septic drain lines. Texas.*

Plate 68 *Soil quality kit tests for soil respiration, bulk density, and infiltration.*

Plate 69 *Finger flow—the uneven water infiltration and movement due to hydrophobic organic coatings on sand particles.*

Plate 70 *Forest floor or O horizons in Vermont.*

Plate 71 *The darker surface soil was brushed aside to expose a hydrophobic layer caused by burning the chaparral vegetation. Water beads up rather than soaking into this layer.*

Plate 72 *Contour strip fields and grassed waterway (arrow) in New York.*

Plate 73 *Sod includes about 1 cm of soil removed from the sod farm.*

Plate 74 *Efficient drip irrigation of a young apple orchard in Mexico.*

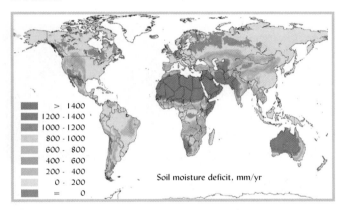

Plate 76 *Annual soil moisture deficit (in mm).*

Plate 78 *Constructed Entisols (world reference group, Technosols) in urban plaza, including layers of geotextile and sand.*

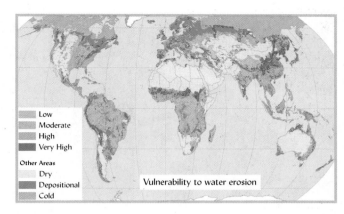

Plate 75 *Global soil vulnerability to erosion by water.*

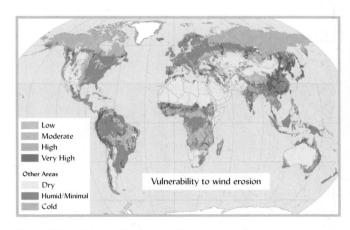

Plate 77 *Global soil vulnerability to erosion by wind.*

Plate 79 *House in Africa made of several kinds of soil.*

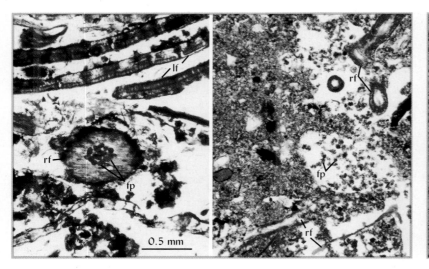

Plate 80 Influence of soil fauna on microstructure in O (left) and A (right) horizons of a forested Ultisol in Tennessee. Particulate organic matter (POM) includes leaf fragments (lf), fecal pellets (fp), and root fragments (rf).

Plate 81 Roots from sweet pepper plants follow organic matter–lined earthworm burrows through the compacted B subsoil of a Pennsylvania Inceptisol. Earthworm activity was encouraged by 15 years of no-till practices and cover crops.

Plate 83 This cicada nymph nestled 60 cm deep in the B horizon of a forested Ultisol will feed by sucking sap from oak tree roots for several years before emerging as an adult. Free water in the macropore bathes the cicada whose burrowing promotes drainage and enhances root growth.

Plate 82 Plant roots grow along the gleyed coating of a fragipan prism whose reddish interior is too dense for roots to grow. The roots are squeezing flat between prisms.

Plate 84 Extension of tree roots far beyond the drip line can be seen by competition for water between trees and grass in the surface soil. Contrast this pattern of drought stress to that shown in Plate 85.

Plate 85 Drought-stressed grass shows the importance of soil depth. The rectangular area of brown grass is underlain by a shallow (25 cm) layer of soil atop the roof of an underground library. The trees and green grass at right grow in deeper soil.

Plate 86 *The soil on the right of this hydrangea was limed; that on the left was acidified (with FeSO₄). After a year, blue flowers formed on the low-pH side, pink on the high-pH side.*

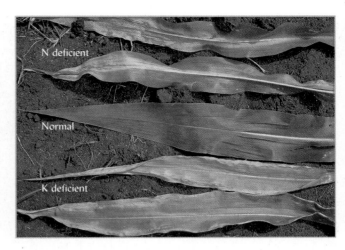

Plate 87 *Leaves from near the bottom of N-deficient (yellow tip and mid-rib), K-deficient (necrotic leaf edges), and normal corn plants. All the leaves came from the same field.*

Plate 88 *This iron-deficient azalea was sprayed with FeSO₄ on one side three days before being photographed. Soil pH higher than 5.5 can induce such Fe deficiency.*

Plate 89 *Magnesium deficiency causes interveinal chlorosis on the older leaves. Poinsettia.*

Plate 90 *Phosphorus deficiency causes severe stunting and purpling of older leaves. Tomato.*

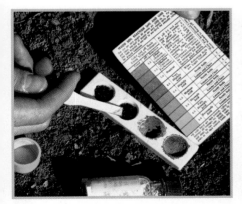

Plate 91 *After wetting soil with pH-sensitive dye, the color is compared to a chart (yellow at pH 4.0 to purple at pH 8.5) in order to estimate soil pH in the field. This soil has a pH of about 7.*

Plate 92 *Sulfur deficient typically sorghum plant on the left.*

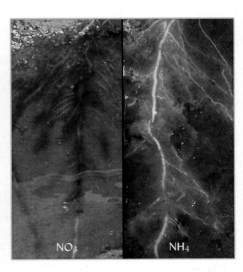

Plate 93 *Dye shows grass plant rhizosphere pH as affected by nitrogen form used. Color key as in Plate 91.*

Plate 94 *Nitrogen-deficient corn on Udolls in central Illinois. Ponded water after heavy rains resulted in nitrogen loss by denitrification and leaching.*

Plate 95 *Slow-moving coastal plain stream choked with algal bloom caused by nitrogen and phosphorus from upstream farmland.*

Plate 96 *Normal (left) and phosphorus-deficient (right) corn plants. Note stunting and purple color.*

Plate 97 *Zinc deficiency on peach tree. Note whorl of small, misshaped leaves.*

Plate 98 *Zinc deficiency on sweet corn. Note broad, whitish bands.*

Plate 99 *Eroded calcareous soil (Ustolls) with iron-deficient sorghum.*

Plate 100 *Boron deficiency on alfalfa. Note reddish foliage.*

Plate 101 *Pink blooms belong to pioneering redbud (Cercis canadensis L.) trees, a nitrogen-fixing legume that enriches the soil for the other species during succession.*

Plate 102 *Phosphorus-deficient grape leaves.*

Plate 103 *The large soybean root nodule was cut open to show its red interior, indicative of active nitrogen fixation. The red, iron-coordinated compound is very similar to the hemoglobin that makes human blood red.*

Plate 104 *Looking like snow, the salt crust covering this soil formed when salt-laden groundwater in this salt marsh rose by capillarity and evaporated from the soil surface, leaving the dissolved salt behind. Near the Great Salt Lake in Utah.*

Plate 105 *Sea spray has caused salt injury (brown leaves) despite the high salt tolerance of this Bermuda grass at Pebble Beach Golf Course in California.*

Plate 106 *It's a good thing this homeowner readjusted his spreader before he finished fertilizing the lawn. Salt "burn" from too much N fertilizer.*

Plate 107 *Iron deficiency causes yellowing with sharply contrasting green veins on the younger leaves. Rose growing in soil with pH 6.8.*

Plate 108 *Stream polluted by acid drainage caused by sulfuricization of soils forming in coal mine spoil. $FeSO_4$ in the acid drainage oxides in the stream causes the orange color.*

Plate 109 *Early stages of soil formation in material dredged from Baltimore Harbor. Sulfidic materials (black), acid drainage (orange liquid), salt accumulations (whitish crust), and initiation of prismatic structure (cracks) are all evident.*

Plate 110 *Landsat Thematic Mapper image of Washington, DC (upper left) and the sediment-laden Potomac River (center). Composite image with natural colors.*

Plate 111 *Landsat Thematic Mapper image of Palo Verde Valley, California, irrigation scheme. Composite image using bands 2, 3, and 4. Lush vegetation appears bright red.*

Plates 1–4, 7, 10, 11, 13, 15–23, 25–36, 38–53, 55–63, 65–68, 70–74, 78–79, 81–92, 94–101, and 103–109 courtesy of R. Weil; plates 8 and 12 courtesy of R. W. Simonson; plates 5 and 14 courtesy of Chien-Lu Ping, Agriculture and Forestry Experiment Station, University of Alaska–Fairbanks; plates 6 and 9 courtesy of Soil Science Society of America; plate 24 courtesy of Carlos F. Dorronsoro Fernandez, Univ. of Granada, Spain; plates 53, 54, and 102 courtesy of P. R. Schreiner, Oregon State Univ.; plate 64 courtesy of Chris Smith, USDA-NRCS; plate 69 courtesy of Stefan Doerr, Univ. of Swansea Wales; plates 75 and 77 courtesy of USDA/NRCS; plate 76 from Tao et al. (2003) courtesy of Swedish Academy of Science; plate 80 courtesy of Debra Phillips, Oak Ridge National Laboratory, Tennessee; plate 93 courtesy of Joseph Heckman, Rutgers Univ.; plate 110 courtesy Space Imaging, Inc.; plate 111 courtesy of Earth Satellite Corp, Rockville, Maryland.

non-agricultural land. Mass wasting can take several forms. **Soil creep** (Plate 46) is the slow deformation (without shear failure) of the soil profile as the upper layers move imperceptibly downhill. **Landslides** occur with the sudden shear failure and downhill movement of a mass of soil, usually under very wet conditions (Plates 41 and 42). **Mud flows** involve the partial liquefaction and flow of saturated soil due to loss of cohesion between particles.

Mass wasting is sometimes triggered by human activities that undermine natural stabilizing forces or cause the soil to become water saturated as a result of concentrated water flow. The rotting of large soil-anchoring tree roots several years after clear-cutting a forest or the construction of a road cut at the toe of a steep slope are all-too-common examples.

14.9 CONTROL OF ACCELERATED EROSION ON RANGE- AND FORESTLAND

Rangeland Problems

Many semiarid rangelands lose large amounts of soil under natural conditions, but accelerated erosion can lead to even greater losses if human influences are not carefully managed. Overgrazing, which leads to the deterioration of the vegetative cover on rangelands, is a prime example. Grass cover generally protects the soil better than the scattered shrubs that usually replace it under the influence of poorly managed livestock grazing. In addition, cattle congregating around poorly distributed water sources and salt licks may completely denude the soil. Cattle trails, as well as ruts from off-road vehicles, can channelize runoff water and spawn gullies that eat into the landscape. Because of the prevalence of dry conditions, wind erosion (to be discussed in Sections 14.11 and 14.12), also plays a major role in the deterioration of rangeland soils.

Erosion on Forestlands

In contrast to deserts and rangelands, land under healthy, undisturbed forests lose very small amounts of soil. However, accelerated erosion can be a serious problem on forested land, both because the rates of soil loss may be quite high and because the amount of land involved is often enormous. The main cause of accelerated erosion in forested watersheds is usually the construction of logging roads, timber-harvest operations, and the trampling of trails and off-trail areas by large numbers of recreational users (or cattle, in some areas).

To understand and correct these problems, it is necessary to realize that the secret of low natural erosion from forested land is the undisturbed forest floor, the O horizons that protect the soil from the impact of raindrops and allow such high infiltration rates that surface runoff is very small or absent. Contrary to the common perception, it is the forest floor, rather than the tree canopy or roots, that protects the soil from erosion (Figure 14.18, *left*). In fact, rainwater dripping from the leaves of tall trees often forms very large drops that reach terminal velocity and impact the ground with more energy than direct rain from even the most intense of storms. If the forest floor has been disturbed and mineral soil exposed, serious splash erosion can result (see Figure 14.18, *right*). Gully erosion can also occur under the forest canopy if water is concentrated, as by poorly designed roads.

The main sources of eroded soil from timber production are *logging roads* (that are built to provide access to the area by trucks), *skid trails* (the paths along which logs are dragged), and *yarding areas* (places where collected logs are sized and loaded onto trucks). Relatively little erosion results directly from the mere

Forest Service WEPP Interfaces for roads, forests, etc.:
http://forest.moscowfsl.wsu.edu/fswepp/

Figure 14.18

The leaf mulch on the forest floor, rather than the tree roots or canopy, provides most of the protection against erosion in a wooded ecosystem. (Left) An undisturbed temperate deciduous forest floor (as seen through a rotten stump). The leafless canopy will do little to intercept rain during winter months. During the summer, rainwater dripping from the foliage of tall trees may impact the forest floor with as much energy as unimpeded rain. (Right) Severe erosion has taken place under the tree canopy in a wooded area where the protective forest floor has been destroyed by foot traffic. The exposed tree roots indicate that nearly 25 cm of the soil profile has washed away. (Photos courtesy of R. Weil)

felling of the trees (except where large tree roots are needed to anchor the soil against mass wasting). Strategies to control erosion should include consideration of (1) intensity of timber harvest, (2) methods used to remove logs, (3) scheduling of timber harvests, and (4) design and management of roads and trails. Soil disturbance in preparation for tree regeneration (such as tillage to eliminate weed competition or provide better seed-to-soil contact) must also be limited to sites with low susceptibility to erosion.

The least expensive and most commonly used method of tree removal is by wheeled tractors called *skidders* (see Chapter 4, Figure 4.19). This method generally disrupts the forest floor, exposing the mineral soil on perhaps 30 to 50% of the harvested area. In contrast, more expensive methods using cables to lift one end of the log off the ground are likely to expose mineral soil on only 15 to 25% of the area. Occasionally, for very sensitive sites, logs are lifted to yarding areas by balloon or helicopter, practices which are very expensive but result in as little as 4 to 8% bare mineral soil.

Design and Management of Roads   Poorly built logging roads may lose as much as 100 Mg/ha of soil by erosion of the road surface, the drainage ditch walls, or the soil exposed by road cuts into the hillside. Roads also collect and channelize large volumes of water, which can cause severe gullying. Roads should be so aligned as to avoid these problems. Although expensive, placing gravel on the road surface, lining the ditches with rocks, and planting perennial vegetation on exposed road cuts can eliminate up to 99% of the soil loss. A much less expensive measure is to provide cross channels (shallow ditches or **water bars**, as shown in Figure 14.19, *right*) every 25 to 100 m to prevent excessive accumulation of water and safely spread it out onto areas protected by natural vegetation. After timber harvest is complete, the roads in an area should be grassed over and closed to traffic.

Design of Skidding Trails   Skidding trails that lead runoff water downhill toward a yarding area invite the formation of gullies. Repeated trips dragging logs over the same secondary trails also greatly increase the amount of mineral soil exposed to erosive forces. Both practices should be avoided, and yarding areas should be

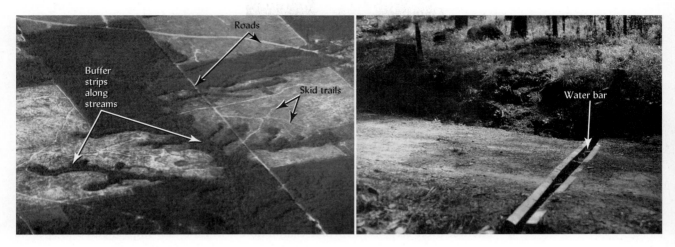

Figure 14.19
Two important forestry practices designed to minimize damage from erosion due to timber harvests. (Left) An aerial view of clear-cut and unharvested blocks of pine forest in Alabama. Narrow skid trails can be seen that lead up to a staging area located on high ground and spread apart going downhill toward the streams. This is in contrast to the common and easier practice of dragging logs downhill so that skid trails converge at a low point, inviting runoff water to concentrate into gully-cutting torrents. Also visible are several dark-colored buffer strips where the trees were left undisturbed along streams to provide protection for water quality. (Right) An open-top culvert (also called a water bar) in a well-designed logging road in Montana. (Photos courtesy of R. Weil)

located on the highest-elevation, most level, and most well-drained areas available. (Figure 14.19, *left*).

Buffer Strips Along Stream Channels   When forests are harvested, buffer strips as wide as 1.5 times the height of the tallest trees should generally be left untouched along all streams (Figure 14.19, *left*). As discussed in Section 16.2, buffer strips of dense vegetation have a high capacity to remove sediment and nutrients from runoff water. Forested buffers also protect the stream from excessive logging debris. In addition, streamside trees shade the water, protecting it from the undesirable heating that would result from exposure to direct sunlight.

14.10   EROSION AND SEDIMENT CONTROL ON CONSTRUCTION SITES

Although active construction sites cover relatively little land in most watersheds, they may still be a major source of eroded sediment because the potential erosion per hectare on drastically disturbed land is commonly 100 times that on agricultural land. Heavy sediment loads are characteristic of rivers draining watersheds in which land use is changing from farm and forest to built-up land (Plate 110). Historically, once urbanization of a watershed is complete (all land being either paved over or covered by well-tended lawns), sedimentation rates return to levels as low (or lower) than before the development took place.

To prevent serious sediment pollution from construction sites, governments in the United States (e.g., through state laws and the Federal Clean Water Act of 1992) and in many other countries require that contractors develop detailed erosion- or sediment-control plans before initiating construction projects. The goals of erosion control on construction sites are (1) to avoid on-site damage, such as undercutting of foundations or finished grades and loss of topsoil needed for eventual landscaping; and (2) to retain eroded sediment on-site so as to avoid all the environmental damages (and liabilities) that would result from deposition of sediment on neighboring land and roads, and in ditches, reservoirs, and streams.

Erosion and sediment
control planning for
construction sites:
www.civil.ryerson.ca/
stormwater/menu_5/
index.htm

Principles of Erosion Control on Construction Sites

Five basic steps are useful in developing plans to meet the aforementioned goals:

1. When possible, schedule the main excavation activities for low-rainfall periods of the year.
2. Divide the project into as many phases as possible, so that only a few small areas must be cleared of vegetation and graded at any one time.
3. Cover disturbed soils as completely as possible, using vegetation or other materials.
4. Control the flow of runoff to move the water safely off the site without destructive gully formation.
5. Trap the sediment before releasing the runoff water off-site.

The last three steps bear further elaboration. They are best implemented as specific practices integrated into an overall erosion-control plan for the site.

Keeping the Disturbed Soil Covered

Once a section of the site is graded, any sloping areas not directly involved in the construction should be sodded or sown to fast-growing grass species adapted to the soil and climatic conditions. (Plate 43 shows erosion on an unprotected road bank.)

Seeded areas should be covered with **mulch** or specially manufactured **erosion blankets** (Figure 14.20). Erosion blankets, made of various biodegradable or non-biodegradable materials, provide instant soil cover and protect the seed from being washed away.

A technology commonly used for protecting steep slopes and areas difficult to access, such as road cuts, is a **hydroseeder**, which sprays out a mixture of seed, fertilizer, lime (if needed), mulching material, and sticky polymers. Good construction-site management includes removal and stockpiling of the A-horizon material before an area is graded (see Plate 44). This soil material is often quite high in fertility and is a potential source of sediment and nutrient pollution. The stockpile should therefore be hydroseeded to provide a grass cover which will protect it from erosion until the topsoil is used for landscaping around the finished structures.

Figure 14.20

Total sediment generated by five summer storms on construction site soil left bare or protected by straw mulch or various types of commercial erosion blankets. The clayey soil met the State Department of Transportation specifications for topsoil. It was raked, fertilized, and seeded to grass mixture. The experimental plots had a 35% slope gradient and were 9.75 m long. Immediately after seeding, the soils were covered by the various erosion-control materials. The best grass vegetative cover and biomass was achieved on straw mulch plots in the first year. The best short-term sediment control was achieved by the commercial blankets, such as the wood-fiber blanket shown in the background photo. [Drawn from data in Benik et al. (2003); photo courtesy of R. Weil]

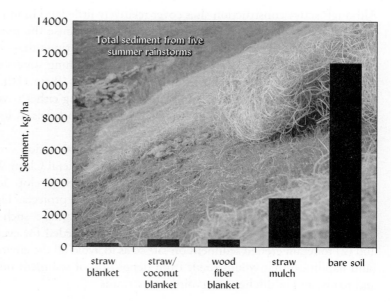

Controlling the Runoff

Freshly exposed and disturbed subsoil material is highly susceptible to the cutting action of flowing water. The gullies so formed may ruin a grading job, undercut pavements and foundations, and produce enormous sediment loads. The flow of runoff water must be controlled by carefully planned grading, terracing, and channel construction. Most construction sites require a perimeter waterway to catch runoff before it leaves the site and to channel it to a retention basin.

The sides and bottom of such channels must be covered with "armor" to withstand the cutting force of flowing water. Where high water velocities are expected, the soil must be protected with **hard armor** such as **riprap** (large, angular rocks, such as are shown in Figure 14.21), **gabions** (rectangular wire-mesh containers filled with fist-size stone), or interlocking concrete blocks. The soil is first covered with a **geotextile** filter cloth (a tough nonwoven material) to prevent mixing of the soil into the rock or stone.

In smaller channels, and on more gentle slopes where relatively low water velocities will be encountered, **soft armor**, such as grass sod or erosion blankets, can be used. Generally, soft armor is cheaper and more aesthetically appealing than hard armor. Newer approaches to erosion control often involve reinforced vegetation (e.g., trees or grasses planted in openings between concrete blocks or in tough erosion mats).

The term **bioengineering** describes techniques that use vegetation (locally native, noninvasive species are preferred) and natural biodegradable materials to protect channels subject to rather high water velocities. Examples include the use of **brush mattresses** to stabilize steep slopes. In this technique, live tree branches are tightly bundled together, staked down flat using long wooden pegs, and partially covered with soil. The so-called **live stake** technique (Figure 14.22) is another example of a bioengineering approach commonly used to stabilize soil along channels subject to high velocity water. In both cases, the soil is provided some immediate physical protection from scouring water, and eventually the dormant cuttings take root to provide permanent, deep-rooted vegetative protection.

grass sod

riprap

Figure 14.21
The flow of runoff from large areas of bare soil must be carefully controlled if off-site pollution is to be avoided. Here, a carefully designed channel with a grass sod bottom and sides lined with large rocks (riprap) prevents gully erosion, reduces soil loss, and guides runoff around the perimeter of a construction site. (Photo courtesy of R. Weil)

Figure 14.22
An example of bioengineering along a stream bed that had been disturbed during the construction of a commercial airport in Illinois. Living willow branches are pounded into the soft erodible stream bank to anchor the soil and reduce the scouring power of the water during high flows (right). Eventually the willow branches will take root and sprout (left), providing trees that will permanently stabilize the stream bank and improve wildlife habitat. (Photos courtesy of R. Weil)

Trapping the Sediment

For small areas of disturbed soils, several forms of sediment barriers can be used to filter the runoff before it is released. The most commonly used types of silt barriers are straw bales and woven fabric silt fences. If installed properly, both can effectively slow the water flow so that most of the sediment is deposited on the uphill side of the barrier, while relatively clear water passes through.

On large construction sites, a system of protected slopes and channels leads storm runoff water to one or more retention and sedimentation ponds located at the lowest elevation of the site. As the flowing water meets the still water in the pond it drops most of its sediment load, allowing the relatively clear water to be skimmed off the top and released to the next pond or off the site. Wetlands (Section 7.7) are often constructed to help purify the overflow from sedimentation ponds before the water is released into the natural stream or river.

14.11   WIND EROSION: IMPORTANCE AND FACTORS AFFECTING IT

Losing Soil—Earth Policy Institute: www.earth-policy.org/ Books/Seg/PB2ch05_ss3 .htm

Erosion of soil by wind is nearly as great a problem as erosion by water. Wind erosion is greatest in arid and semiarid regions, but also occurs on certain humid region soils. When strong wind blows across a dry soil surface, it detaches and carries soil particles along for varying distances. While the larger sands roll and bounce across the land surface, the finer silts and clays may be picked up by the wind and carried to great heights and for great distances—the windborne dust may even cross oceans from one continent to another.

Wind erosion causes widespread damage, not only to the vegetation and soils of the eroding site, but also to anything that can be damaged by the abrasiveness of soil-laden wind, and finally to the offsite area where the eroded soil settles back to Earth (Figure 14.23). In the semiarid Great Plains states, annual wind erosion exceeds water

Figure 14.23

Wind erosion in action. (Left) Much wind erosion comes from dust storms such as this one moving across the High Plains of Texas. The swirling black cloud consists of fine particles eroded from the soil by high winds sweeping across the flat farmland and range. Apparently, much of the land was not as well covered as the wheat field in the foreground. (Right) Soil eroded by wind during a single dust storm has piled up to a depth of nearly 1 m along a fencerow in Idaho. The existing soil and plants are covered by deposits that are quite unproductive because the soil structure has been destroyed. Also, these deposits are subject to further movement when the wind picks up or changes direction. (Photos courtesy of Dr. Chen Weinan, USDA Agricultural Research Service, Warm Springs, TX, left, and R. Weil, right)

erosion on cropland, averaging from 4 Mg/ha in Nebraska to 29 Mg/ha in New Mexico, where mismanagement of plowed lands and overgrazing of range grasses have greatly increased the susceptibility of the soils to wind action. In dry years, the results have been most deplorable.

Even in humid regions, certain soils suffer significant wind erosion when their surface layer dries out and wind velocity is high. The movement of sand dunes along the Atlantic coast and on the eastern shore of Lake Michigan are examples. Wind erosion also damages cultivated sandy or peaty soils when conditions are dry. The on-site and off-site damages caused by wind erosion were discussed in Section 14.2.

The erosive force of the wind tends to be much greater in some regions and seasons than in others. For example, the semiarid Great Plains region of the United States is subject to winds with 5 to 10 times the erosive force of the winds common in the humid East. In the Great Plains, the winds are most powerful in the winter season. In other regions, high winds occur most commonly during the hot summers.

Mechanics of Wind Erosion

Like water erosion, wind erosion involves three processes: (1) *detachment*, (2) *transportation*, and (3) *deposition*. The moving air, itself, results in some detachment of tiny soil grains from the granules or clods of which they are a part. However, once the moving air is laden with soil particles, its abrasive power is greatly increased. The impact of these rapidly moving grains dislodges other particles from soil clods and aggregates. These dislodged particles are now ready for one of the three modes of wind-induced transportation, depending mostly on their size.

Animation of saltation in action: http://plantandsoil.unl.edu/croptechnology2005/soil_sci/animationOut.cgi?anim_name=saltation-modd.swf

The first and most important mode of particle transportation is that of **saltation**, or the movement of soil by a series of short bounces along the ground surface (Figure 14.24). The particles remain fairly close to the ground as they bounce, seldom rising more than 30 cm or so. Depending on conditions, this process may account for 50 to 90% of the total movement of soil.

Saltation also encourages **soil creep**, or the rolling and sliding of the larger particles along the surface. The bouncing particles carried by saltation strike soil aggregates

Figure 14.24

How particles move during wind erosion. As indicated by the straight arrows, the wind is blowing from left to right and is slowed somewhat by friction and obstructions near the soil surface. Fine particles are picked up from the soil surface and carried into the atmosphere, where they remain suspended until the wind velocity is reduced. Medium-sized particles or aggregates, being too large to be carried up in suspension, are bounced along the soil surface. When they strike larger soil aggregates, they break up to release particles of various sizes. The finer particles move in suspension up into the air above, and the medium-sized particles continue to bounce along the soil surface. This process of particle movement stimulated by medium-sized particles skipping along the surface is termed saltation. (Diagram courtesy of R. Weil)

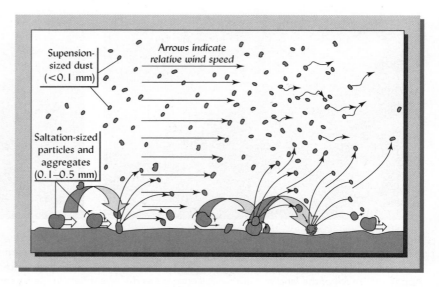

and speed up their movement along the surface. Soil creep accounts for the movement of particles up to about 1.0 mm in diameter, which may amount to 5 to 25% of the total movement.

The most spectacular method of transporting soil particles is by movement in **suspension**. Here, dust particles of a fine-sand size and smaller are moved parallel to the ground surface and upward. Although some of them are carried at a height no greater than a few meters, the turbulent action of the wind results in fine particles being carried kilometers upward into the atmosphere and many hundreds of kilometers horizontally. These particles return to the earth only when the wind subsides and/or when precipitation washes them down.

Factors Affecting Wind Erosion

Wet soils do not blow because of the adhesion between water and soil particles. Dry winds generally lower the moisture content to below the wilting point before wind erosion takes place. Other factors that influence wind erosion are (1) wind velocity and turbulence, (2) soil surface conditions, (3) soil characteristics, and (4) the nature and orientation of the vegetation.

Wind Velocity   The rate of wind movement, especially gusts having greater than average velocity, will influence erosion. The *threshold velocity*—the wind speed required to initiate soil movement—is usually about 25 km/h (7 m/s). At higher wind speeds, soil movement is proportional to the cube of the wind velocity. Thus, the quantity of soil carried by wind increases dramatically as speeds above 30 km/h are reached. Although the wind itself has some direct influence in picking up fine soil, the impact of wind-carried particles as they strike the soil is probably more important.

Surface Roughness   Wind erosion is less severe where the soil surface is rough. This roughness can be obtained by proper tillage methods, which create large clods or ridges. Leaving a stubble mulch (see Section 6.4) is an even more effective way of reducing windborne soil losses.

Soil Properties   In addition to moisture content, wind erosion is also influenced by (1) mechanical stability of soil clods, aggregates, and crusts; (2) bulk density; and (3) size of erodible soil fractions. Some clods resist the abrasive action of wind-carried particles. If a natural biological soil crust or a physical crust resulting from a previous rain is present,

Video of dust devils on Mars: http://cc.jpl.nasa.gov/mer/050505-DustDevil.qtl

it, too, may be able to withstand the wind's erosive power. The presence of clay, organic matter, and other cementing agents is also important in helping clods and aggregates resist abrasion. This is one reason why sandy soils, which are low in such agents, are so easily eroded by wind. Because they participate in saltation, soil particles or aggregates about 0.1 mm in diameter are more erodible than those much larger or smaller in size.

Vegetation   Vegetation or stubble mulch will reduce wind erosion hazards, especially if in rows that run perpendicular to the prevailing wind direction. This effectively slows wind movement near the soil surface. In addition, plant roots help bind the soil and make it less susceptible to wind damage.

14.12 PREDICTING AND CONTROLLING WIND EROSION

A wind erosion equation (WEQ) has been in use since the late 1960s to predict amounts of erosion (E):

$$E = f(I \times C \times K \times L \times V) \tag{14.2}$$

The WEQ considers how these factors interact with each other. Consequently, it is not as simple as is the USLE for water erosion. The **soil erodibility factor**, I, relates to the properties of the soil and to the degree of slope of the site in question. The **soil-ridge-roughness factor**, K, takes into consideration the cloddiness of the soil surface, vegetative cover, V, and ridges on the soil surface. The **climatic factor**, C, involves wind velocity, soil temperature, and precipitation (which helps control soil moisture). The **width of field factor**, L, is the width of a field in the downwind direction. Except for a round field, the width changes as the direction of the wind changes, so the prevailing wind direction is generally used. The **vegetative cover**, V, relates not only to the degree of soil surface covered with residues, but to the nature of the cover—whether it is living or dead, still standing, or flat on the ground.

A revised, more complex, and more accurate computer-based prediction model has been developed, and is known as the **revised wind erosion equation (RWEQ)**. It is still an empirical model based on many years of research to characterize the relationship between observable conditions and resulting wind erosion severity. The RWEQ (like RUSLE) calculates the erosion hazard during 15-day intervals throughout the year. For each interval of time, the RWEQ makes adjustments in the residue, soil erodibility, and soil roughness parameters, based on the input information about management operations and weather conditions. For example, it assumes that residues decompose over time, that tillage operations flatten standing residues, and that rainfall shakes clods to reduce soil roughness.

Scientists and engineers around the world are also cooperating in the development of a much more complex process-based model known as the **Wind Erosion Prediction System (WEPS)**. Like its water erosion sister, WEPP, this computer program simulates all the basic processes of wind interaction with soil. Scientists are continually improving the model and testing its predictions against data observed in the real world.

Wind erosion prediction model: www.weru.ksu.edu/weps/wepshome.html

Control of Wind Erosion

Soil Moisture   The factors of the wind erosion equation give clues to methods of reducing wind erosion. For example, since soil moisture increases cohesiveness, the wind speed required to detach soil particles increases dramatically as soil moisture increases. Therefore, where irrigation water is available, it is common to moisten the soil surface when high winds are predicted (Figure 14.25). Unfortunately, most wind

Figure 14.25

Wind erosion control in an area of productive Histosols (Saprists) in central Michigan. Prior to being cleared and drained, this area was a partially forested bog. When their surface layer is dry, cultivated Histosols are very light and fluffy and susceptible to wind erosion. The rows of trees (mainly willows) were planted perpendicular to the prevailing winds to slow the wind velocity and protect these valuable organic soils from erosion. Wetting the soil surface is another effective means of reducing wind erosion, as seen by the darker-colored field in the background (where the water table was raised) and the darker circles in the left foreground where sprinkler irrigation was used. Note that the photo was taken in early spring before most crops were planted and before the trees had fully leafed out. (Photo courtesy of USDA Natural Resources Conservation Service)

erosion occurs in dry regions without available irrigation. A vegetative cover also discourages soil blowing, especially if the plant roots are well established. In dry-farming areas where summer fallow is practiced on some of the land, hot winds dry the bare soil surface of fallowed fields, making them especially susceptible to wind erosion.

Tillage   Certain conservation tillage practices described in Section 14.6 were used for wind erosion control long before they became popular as water erosion control practices. Keeping the soil surface rough and maintaining some vegetative cover is accomplished by using appropriate tillage practices. However, the vegetation should be well anchored into the soil to prevent it from blowing away. Stubble mulch has proven to be effective for this purpose (see Section 6.4).

Tillage can greatly reduce wind erosion if it is done while there is sufficient soil water to cause large clods to form. Tillage on a dry soil may produce a fine, dusty surface that aggravates the erosion problem. Tillage, strip-cropping, and alternate strips of cropped and fallowed land should be perpendicular to the wind.

Barriers   Barriers such as shelter belts (Figure 14.25) are effective in reducing wind velocities for short distances and for trapping drifting soil. Various devices are used to control blowing of sands, sandy loams, and cultivated peat soils (even in humid regions). Windbreaks and tenacious grasses and shrubs are especially effective. Picket fences and woven screens, though less efficient as windbreaks than such trees as willows, are often preferred because they can be moved from place to place as crops and cropping practices are varied. Rye, planted in narrow strips across the field, is sometimes used on peat lands and on sandy soils. Narrow rows of such perennial grasses as tall wheatgrass are being evaluated for a combination of wind erosion control and capturing of winter snows in cold, semiarid areas.

14.13   PROGRESS IN SOIL CONSERVATION

Land Capability Classification

A land capability classification system developed by the USDA half a century ago is still helpful in identifying land uses and management practices that can minimize soil erosion, especially that induced by rainfall. The eight **land capability classes** indicate the *degree* of limitation imposed on land uses (Figure 14.26), with Class I the least

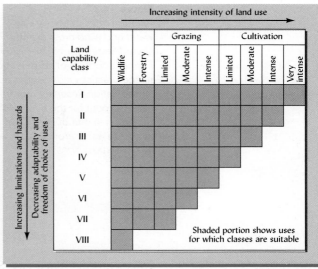

Figure 14.26

(Left) A landscape in San Mateo County, California, illustrating six of the eight land capability classes. (Right) A chart indicating the land use intensity appropriate for each capability class without incurring undue problems or soil degradation. Note the increasing limitations on safe land use as one moves from Class I to Class VIII land. (Photo from USDA/NRSC; chart modified from Hockensmith and Steel, 1949)

limited and Class VIII the most limited. Each land use class may have four subclasses that indicate the *type* of limitation encountered: risks of erosion (e); wetness, drainage, or flooding (w); root-zone limitations, such as acidity, density, and shallowness (s); and climate limitations, such as a short growing season (c). The erosion (e) subclasses are the most common, and they will be the focus of our attention here. For example, Class IIe land is slightly susceptible to erosion, while Class VIIIe is extremely susceptible. Figure 14.26 (*right*) shows the intensity of land use allowable for each capability class if erosion losses (or problems associated with the other subclasses) are to be avoided. Land in classes I and II is considered to be *prime agricultural land.*

Soil Erosion Control

Soil erosion in the United States accelerated when the first European settlers chopped down trees and began to farm the sloping lands of the humid eastern part of the country. Soil erosion was a factor in the declining productivity of these lands that, in time, led to their abandonment and the westward migration of people in search of new farmlands.

It wasn't until the worldwide depression and widespread droughts of the early 1930s accentuated rural poverty and displaced millions of people that governments began to pay attention to the rapid deterioration of soils. In 1930, Dr. H. H. Bennett and associates recognized the damage being done and obtained U.S. government support for erosion-control efforts. Since then, considerable reductions in erosion have been achieved in the United States and elsewhere.

During the 1940s and 1950s, such physical practices as contour strips, terraces, and windbreaks were installed with much persuasion and assistance from government agencies. Some of this progress was reversed as the terraces and windbreaks appeared to stand in the way of the "fence row to fence row" all-out crop production policies of the 1970s. But since 1982, rather remarkable progress has been achieved in reducing soil erosion (Figure 14.27), largely as a result of two factors: (1) the spread of

W. C. Lowdermilk's 1939 travelogue about soil erosion in Europe, North Africa, and the Middle East: www.soilandhealth.org/ 01aglibrary/010119lowdermilk .usda/cls.html

Figure 14.27

Average rates of soil loss by water and wind erosion in the United States from 1982 to 2003. Farmer adoption of conservation tillage practices along with the Conservation Reserve Program likely account for the nearly 40% reduction in rate of combined wind and water erosion. [Calculated from data in USDA/NRCS (2006)]

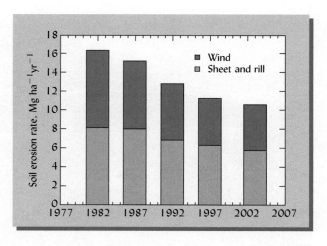

conservation tillage (see Section 14.6), and (2) the implementation of land-use changes as part of the conservation reserve program. Progress needs to continue on both fronts.

However, about one-third of the cultivated cropland in the United States is still losing more than 11 Mg/ha/yr, the maximum loss that can be sustained without serious loss of productivity on most soils. After some 80 years of soil conservation efforts, soil erosion is still a major problem on about half the cropland of the United States, and in much of the world the problem has actually worsened.

The Conservation Reserve Program

A major part (about 60%) of the reduction in soil erosion experienced in the United States since 1982 is due to government programs that have paid farmers to shift some land from crops to grasses and forests. Establishing grass or trees on these former croplands reduced the sheet and rill erosion from an average of 19.3 to 1.3 Mg/ha and the wind erosion from 24 to 2.9 Mg/ha. Between 1982 and 2006, 14 million ha of cropland were diverted to such noncultivated uses through the **Conservation Reserve Program (CRP)**. The CRP is reoriented to target **highly erodible land (HEL)** and other environmentally sensitive areas.

The CRP is basically an arrangement by which the U.S. taxpayers pay rent to farmers to forego cropping part of their farmland and instead plant grass or trees on it (more rent is paid for trees). The rental agreements (leases) run for 10 to 15 years, during which time the land is undisturbed. The benefits to the nation include greatly reduced soil losses and sediment pollution as well as a dramatic upswing in wild bird and animal populations in the newly restored habitat. Where strips of land along streams (riparian zone buffers) have been incorporated into the CRP, water-quality benefits have also been amplified.

Conservation Management to Enhance Soil Quality

In a broad sense, conservation management practices are those that improve soil quality in more ways than just by protecting the soil from erosion. Properties that indicate the level of soil quality, especially those associated with soil organic matter, can be enhanced by such conservation measures as minimizing tillage, maximizing residue cover of the soil surface, providing for diversity of plant types, keeping soil under grass sod vegetation for at least part of the time, adding organic amendments where practical, and maintaining balanced soil fertility. Improved soil quality, in turn, enhances the soil's capacity to support plants, resist erosion, prevent environmental contamination,

and conserve water. Conversation management therefore can lead to the upward spiral of soil and environmental improvement referred to in Section 14.1.

Adapting Soil Conservation to the Needs of Resource-Poor Farmers

As satisfying as the progress of the past decades has been, soil losses by erosion are still much too high. Continued efforts must be made to protect the soil and to hold it in place. In the United States, some 30 million ha of highly erodible cropland continue to lose an average of more than 15 Mg/ha of soil each year from water erosion, and an equal amount from wind erosion. In spite of remarkable progress, conservation tillage systems have *not* been adopted for more than half of the nation's cropland. And no one knows what will happen to the CRP lands when the rental leases expire. With the recent demand for biofuel crops and the rise in crop prices, farmers will be tempted to return their CRP lands to production and risk losing all these benefits. The battle to bring erosion under control has just begun, not only in the United States but throughout the world.

In much of the world, so little land is available that many farmers must use *all* land capable of some food production (regardless of land capability class) simply to stave off starvation and impoverishment. These farmers often realize that farming erodible land jeopardizes their future livelihood and that of their children, but they see no choice. It is imperative to either find nonagricultural employment for these people or to find farming systems that are sustainable on these erodible lands.

Fortunately, some farmers and scientists have developed, through long traditions of adaptation or through innovation and research, farming systems that *can* produce food and profits while conserving such erodible soil resources. Examples include the no-till cover cropping systems discussed in Section 14.6. The traditional Kandy Home Gardens of Sri Lanka's humid mountains provide another example in which a rain forest–like mixed stand of tall fruit and nut trees is combined with an understory of pepper vines, coffee bushes, and spice plants to provide valuable harvests while keeping the soil under perennial vegetative protection. In Central America, farmers have learned to plant thick stands of velvet bean (*Mucuna*) or other viny legumes that can be chopped down by machete to leave a soil-protecting, water-conserving, weed-inhibiting mulch on steep farmlands. In Asia, steep lands have been carefully terraced in ways that allow production of food, even paddy rice, on very steep land without causing significant erosion.

When governments cajole, pay, or force farmers to install soil conservation measures on their land, the results are unlikely to be long-lasting. Usually, farmers will abandon the unwanted practices as soon as the pressure is off. On the other hand, if scientists and conservationists work *with* farmers to help them develop and adapt conservation systems that the farmers feel are of benefit to them and their land, then effective and lasting progress can be made. Experience with conservation tillage systems in the United States, mulch farming systems in Central America, and vegetative contour barriers in Asia have shown that farmers can help develop practices that are good for their land and for their profits: a win–win situation.

Interactive map depicting barriers posed by soil and climate conditions, including erosion risk:
www.sciencemag.org/cgi/content/full/304/5677/1616/DC1#map

14.14   CONCLUSION

Accelerated soil erosion is one of the most critical environmental and social problems facing humanity today. Erosion degrades soils, making them less capable of producing the plants on which animals and people depend. Equally important, erosion causes great damage downstream in reservoirs, lakes, waterways, harbors, and municipal water supplies. Wind erosion also causes fugitive dust that may be very harmful to human health.

Nearly 4 billion Mg of soil is eroded each year on land in the United States alone. Half of this erosion occurs on croplands, and the remainder on harvested timber areas, rangelands, and construction sites. Some one-third of U.S. cropland still suffers from erosion that exceeds levels thought to be tolerable.

Water carries away most of the sediment in humid areas by sheet and rill erosion. Gullies created by infrequent, but violent, storms account for much of the erosion in drier areas. Wind is the primary erosion agent in many drier areas, especially where the soil is bare and low in moisture during the season when strong winds blow.

Protecting soil from the ravages of wind or water is by far the most effective way to constrain erosion. In croplands and forests, such protection is due mainly to the cover of plants and their residues. Conservation tillage practices maintain vegetative cover on at least 30% of the soil surface, and the widening adoption of these practices has contributed to the significant reductions in soil erosion achieved over the past two decades. Crop rotations that include perennial sod and close-growing crops, coupled with such practices as contour tillage, strip cropping, and terracing, also help combat erosion on farmland.

In forested areas, most erosion is associated with timber-harvesting practices and forest road construction. For the sake of future forest productivity and current water quality, foresters must become more selective in their harvest practices and invest more in proper road construction.

Construction sites for roads, buildings, and other engineering projects lay bare many scattered areas of soils that add up to a serious erosion problem. Control of sediment from construction sites requires carefully phased land clearing, along with vegetative and artificial soil covers, and installation of various barriers and sediment-holding ponds. These measures may be expensive to implement, but the costs to society that result when sediment is not controlled are too high to ignore. Once construction is completed, erosion rates on urban areas are commonly as low as those on areas under undisturbed native vegetation.

Erosion-control systems must be developed in collaboration with those who use the land, and especially the poor, for whom immediate needs overshadow concerns for the future. As put succinctly in *The River*, a classic 1930s documentary film produced during the rebirth of American soil erosion consciousness, "Poor land makes poor people, and poor people make poor land."

Video archive, *The River*:
http://archive.org/details/
RiverThe1937

STUDY QUESTIONS

1. Explain the distinction between *geologic erosion* and *accelerated erosion*. Is the difference between the two greater in humid or arid regions?

2. When erosion takes place by wind or water, what are three important types of damages that result on the land whose soils are eroding? What are five important types of damages that erosion causes in locations away from the eroding site?

3. What is a common T value, and what is meant by this term? Explain why certain soils have been assigned a higher T value than other soils.

4. Describe the three main steps in the water erosion process.

5. Many people assume that the amount of soil eroded on the land in a watershed (A in the universal soil-loss equation) is the same as the amount of sediment carried away by the stream draining that watershed. What factor is missing that makes this assumption incorrect? Do you think that this means the USLE should be renamed?

6. Why is the total annual rainfall in an area *not* a very good guide to the amount of erosion that will take place on a particular type of bare soil?

7. Contrast the properties you would expect in a soil with either a very high K value or a very low K value.

8. How much soil is likely to be eroded from a Keene silt loam in central Ohio, on a 12% slope, 100 m long, if it is in dense permanent pasture and has no support practices applied to the land? Use the information available in this chapter to calculate an answer.

9. What type of conservation tillage leaves the greatest amount of soil cover by crop residues? What are the advantages and disadvantages of this system?

10. Why are narrow strips of grass planted on the contour sometimes called a "living terrace"?

11. In most forests, which component of the ecosystem provides the primary protection against soil erosion by water, the *tree canopy*, *tree roots*, or *leaf litter*?

12. Certain soil properties generally make land susceptible to erosion by wind or erosion by water. List four properties that characterize soils highly susceptible to wind erosion. Indicate which two of these properties should also characterize soils highly susceptible to water erosion, and which two should not.

13. Which two factors in the wind erosion prediction equation (WEQ) can be affected by tillage? Explain.

14. Describe a soil in land capability Class IIw in comparison with one in Class IVe.

15. Why is it important that there be a close relationship between land in the CRP and that considered to be HEL?

REFERENCES

Benik, S. R., B. N. Wilson, D. D. Biesboer, B. Hansen, and D. Stenlund. 2003. "Evaluation of erosion control products using natural rainfall events," *J. Soil Water Conserv.*, **58**:98–104.

Cassel, D. K., and R. Lal. 1992. "Soil physical properties of the tropics: Common beliefs and management constraints," in R. Lal and P. A. Sanchez (eds.), *Myths and Science of Soils of the Tropics.* SSA Special Publication no. 29 (Madison, WI: Soil Science Society of Amer.), pp. 61–89.

Daily, G. C., T. Söderqvist, S. Aniyar, K. Arrow, P. Dasgupta, P. R. Ehrlich, C. Folke, A. Jansson, B-O. Jansson, N. Kautsky, S. Levin, J. Lubchenco, K-G. Mäler, D. Simpson, D. Starrett, D. Tilman, and B. Walker. 2000. "The value of nature and the nature of value," *Science,* **289**:395–396.

FAO. 1987. *Protect and Produce* (Rome: U.N. Food and Agriculture Organization).

FAO. 2001. "The economics of conservation agriculture," FAO Y2781/E. Food and Agriculture Organization of the United Nations, Rome 73 pp. www.fao.org/docrep/004/Y2781E/y2781e00 .htm#toc.

Foster, G. R., D. K. McCool, K. G. Renard, and W. C. Moldenhauer. 1981. "Conversion of the universal soil loss equation to SI metric units," *J. Soil Water Cons.,* **36**:355–359.

Foster, G. R., and R. E. Highfill. 1983. "Effect of terraces on soil loss: USLEP factor values for terraces," *J. Soil Water Cons.,* **38**:48–51.

Gilley, J. E., B. Eghball, L. A. Kramer, and T. B. Moorman. 2000. "Narrow grass hedge effects on runoff and soil loss," *J. Soil Water Cons.,* **55**:190–196.

Hillel, D. 1991. *Out of the Earth: Civilization and the Life of the Soil* (New York: The Free Press).

Hockensmith, R. D., and J. G. Steele. 1949. "Recent trends in the use of the land-capability classification," *Soil Sci. Soc. Amer. Proc.,* **14**:383–388.

Hooke, R. L. 2000. "On the history of humans as geomorphic agents," *Geology,* **28**:843–846.

Hudson, N. 1995. *Soil Conservation,* 3rd ed. (Ames, IA: Iowa State University Press).

Kelley, D. W., and E. A. Nater. 2000. "Historical sediment flux from three watersheds into Lake Pepin, Minnesota, USA," *J. Environ. Qual.,* **29**:561–568.

Montgomery, D. R. 2007. *Dirt: The Erosion of Civilizations.* (Berkeley, CA: University of California Press).

Oschwald, W. R., and J. C. Siemens. 1976. "Conservation tillage: A perspective," *Agronomy Facts SM-30* (Urbana, IL: University of Illinois).

Renard, K. G., G. Foster, D. Yoder, and D. McCool. 1994. "RUSLE revisited: Status, questions, answers and the future," *J. Soil Water Cons.,* **49**:213–220.

Renard K. G., G. R. Foster, G. A. Weesies, D. K. McCool, and D. C. Yoder. 1997. *Predicting Soil Erosion by Water: A Guide to Conservation Planning with the Revised Universal Soil Loss Equation (RUSLE).* Agricultural Handbook no. 703 (Washington, DC: USDA).

Schwab, G. O., D. D. Fangmeirer, and W. J. Elliot. 1996. *Soil and Water Management Systems,* 4th ed. (New York: Wiley).

Stout, J. E., and J. A. Lee. 2003. "Indirect evidence of wind erosion trends on the southern high plains of North America," *Journal of Arid Environments,* **55**:43–61.

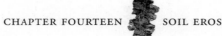

USDA. 1995. Agricultural Handbook no. 703 (Washington, DC: U.S. Department of Agriculture).

U.S. Department of Agriculture, Natural Resources Conservation Service. 2007. *Annual national resources inventory for 2003: Soil erosion* www.nrcs.usda.gov/technical/NRI/2003/nri03eros-mrb.html (posted February 2007; verified 13 February 2009).

Wilkinson, B. H., and B. J. McElroy. 2007. "The impact of humans on continental erosion and sedimentation," *Geological Society of America Bulletin,* **119**:140–150.

Wischmeier, W. J., and D. D. Smith. 1978. *Predicting Rainfall Erosion Loss—A Guide to Conservation Planning.* Agricultural Handbook no. 537 (Washington, DC: USDA).

Polluted soil awaits remediation near abandoned oil refinery. (R. Weil)

Soils and Chemical Pollution

Black and portentous
this humor prove,
unless good counsel
may the cause remove . . .
—W. SHAKESPEARE, *ROMEO AND JULIET*

The soil is a primary recipient by design or accident of a myriad of wastes, chemicals and products used in modern society, many of which we have conveniently "thrown away." Every year, millions of tons of industrial, domestic, and agricultural products find their way into the world's soils. Once there, they become part of biological cycles that affect all forms of life.

In previous chapters we highlighted the enormous capacity of soils to accommodate added organic and inorganic chemicals. Tons of organic residues are broken down by soil microbes each year (Chapter 11), and large quantities of inorganic chemicals are fixed or bound tightly by soil minerals (Chapter 12). But we also learned of the limits of the soil's capacity to sorb these chemicals, and how environmental quality suffers when these limits are exceeded (Chapters 8 and 13).

We have seen how soil processes affect the production and sequestering of greenhouse gases, such as nitrous oxide, methane, and carbon dioxide (see Chapters 11 and 12). Other nitrogen- and sulfur-containing gases come to earth in acid rain (see, e.g., Section 9.6). Mismanaged irrigation projects on arid-region soils result in the accumulation of salts (see Chapter 9).

We have also seen how fertilizer and manure applications that leave excess quantities of nutrients in the soil can result in the contamination of ground and surface waters with nitrates (Section 12.1) and phosphates (Section 12.3). The eutrophication of lakes, estuaries, and slow-moving rivers is evidence of these nutrient buildups. Huge "animal factories" for meat and poultry production produce mountains of manure that must be disposed of without loading the environment with unwanted chemicals and with pathogens that are harmful to humans and other animals (Section 13.4).

In this chapter we will focus on chemicals that contaminate and degrade soils, including some whose damage extends to water, air, and living things. The brief review of soil pollution is intended as an introduction to the nature of the major pollutants, their reactions in soils, and alternative means of managing, destroying, or inactivating them.

15.1   TOXIC ORGANIC CHEMICALS

Comprehensive information on cleaning contaminated soils at abandoned industrial sites: www.epa.gov/brownfields/

Industrialized societies have synthesized thousands of organic (carbon-containing) compounds for many uses. Included are plastics and plasticizers, lubricants and refrigerants, fuels and solvents, pesticides and preservatives. Some are extremely toxic to humans and other life. Through accidental leakage and spills or through planned spraying or other treatments, synthetic organic chemicals can be found in virtually every corner of our environment—in the soil, in the groundwater, in the plants, and in our own bodies.

Environmental Damage from Organic Chemicals

Reducing the environmental impact of your household hazardous wastes: www.klickitatcounty.org/ SolidWaste/default.asp? fCategoryIDSelected= -1671944469

These artificially synthesized compounds are termed **xenobiotics** because they are unfamiliar to the living world (Greek *xeno*, "strange"). Being nonnatural, many xenobiotics are both toxic to living organisms and resistant to biological decay. Some xenobiotic compounds are relatively inert and harmless, but others are biologically damaging even in very small concentrations. Those that find their way into soils may inhibit or kill soil organisms, thereby undermining the balance of the soil community (see Chapter 10). Other chemicals may be transported from the soil to the air, water, or vegetation, where they may be contacted, inhaled, or ingested by any number of organisms, including people. It is imperative, therefore, that we control the release of organic chemicals and that we learn of their fate and effects once they enter the soil.

Organic chemicals may enter the soil as contaminants in industrial and municipal organic wastes applied to or spilled on soils, as components of discarded machinery, in large or small lubricant and fuel leaks, as military explosives, or as sprays applied to control pests in terrestrial ecosystems. Pesticides are probably the most widespread organic pollutants associated with soils. In the United States, pesticides are used on some 150 million ha of land, three-fourths of which is agricultural land. Soil contamination by other organic chemicals is usually much more localized.

The Nature of the Pesticide Problem

Pest management practices in U.S. agriculture, 1990–1997: www.ers.usda.gov/ publications/sb969/ sb969d.pdf

Pesticides are chemicals that are designed to kill pests (that is, any organism that the pesticide user perceives to be unwelcome). Some 600 chemicals in about 50,000 formulations are used to control pests. They are used extensively in all parts of the world. About 600,000 Mg of organic pesticide chemicals are used annually in the United States, with more than three times that amount used in the rest of the world. Although the total amount of pesticides used has remained relatively constant or even dropped since the 1980s, formulations in use today are generally more potent, so that smaller quantities are applied per hectare to achieve toxicity to the pest.

Benefits of Pesticides   Pesticides have provided many benefits to society. They have helped control mosquitoes and other vectors of such human diseases as yellow fever and malaria. They have protected crops and livestock against insects and diseases. Without the control of weeds by chemicals called *herbicides,* conservation tillage (especially no-till) would be much more difficult to adopt; much of the progress made in controlling soil erosion probably would not have come about without herbicides.

Problems With Pesticides   While the benefits to society from pesticides are great, so are the known costs (Table 15.1). Unknown, long-term costs, especially to human health, may be even greater. Widespread and heavy use of pesticides on agricultural soils and suburban and urban landscapes has led to contamination of both surface and groundwater. Therefore, when pesticides are used, they should be chosen for low toxicity to humans and wildlife, low mobility in soils, and low persistence (see Section 15.3). Even then, the use of pesticides often has wide-ranging detrimental effects on the microbial and faunal communities. In fact, the harm done, though not always obvious, may outweigh the benefits. Examples include insecticides that kill natural enemies of pest species as well as the target pest (sometimes creating new major pests from species formerly controlled by natural enemies) and fungicides that kill both disease-causing and beneficial mycorrhizal fungi (see Section 10.9). Given these facts, it should not come as a surprise that despite the widespread use of pesticides, about the same proportion of crop production is lost to insects, diseases, and weeds in the United States as before synthetic organic pesticides were in use.

Alternatives to Pesticides   Pesticides should not be seen as a panacea, or even as indispensable. Some farmers, most notably the increasing number who practice **organic farming**, produce profitable, high-quality yields without the use of synthetic pesticides. The term *organic farming* has little to do with the chemical definition of *organic*, which simply indicates that a compound contains carbon. Rather, it refers to a system and philosophy of farming that eschews the use of synthetic chemicals while it emphasizes soil organic matter and biological interactions to manage agroecosystems. In managing the effects of pests in any type of plant community (agricultural, ornamental, or forest), chemical pesticides should be used as a *last* resort, rather than as a *first* resort. Before resorting to the use of an insecticide or herbicide, every effort should be made to minimize the detrimental effects of insects and weeds by means of crop diversification, establishment of habitat for beneficial insects, application of organic soil amendments, implementation of cultural practices to reduce weed competition, and selection of pest-resistant plant cultivars. Too often, because pesticides are available as a convenient crutch, these more sophisticated approaches to plant management are not thoroughly explored.

Integrated pest management (University of California): www.ipm.ucdavis.edu

---

Table 15.1
TOTAL ESTIMATED ENVIRONMENTAL AND SOCIAL COSTS FROM PESTICIDE USE IN THE UNITED STATES

Type of impact[a]	Cost, $ million/yr
Public health impacts	1400
Domestic animal deaths and contamination	55
Loss of natural enemies	1000
Cost of pesticide resistance	2500
Honeybee and pollination losses	550
Crop losses	1700
Fishery losses	42
Groundwater contamination and cleanup costs	3200
Cost of government regulations to prevent damage	350
Total	10800

[a]The death of an estimated 60 million wild birds may represent an additional substantial cost in lost revenues from hunters and bird watchers.
Data adjusted for inflation to 2009 dollars, from Pimental et al. (1992). © American Institute of Biological Sciences.

Farmscaping: An alternative to pesticides (ATTRA): www.attra.org/attra-pub/farmscape.html

Nontarget Damages   Although some pesticides are intentionally applied to soils, most reach the soil because they miss the insect or plant leaf that is the application target. When pesticides are sprayed in the field, most of the chemical misses the target organism. For pesticides aerially applied to forests, about 25% reaches the tree foliage, and far less than 1% reaches a target insect. About 30% may reach the soil, while about half of the chemical applied is likely to be lost into the atmosphere or in runoff water.

Designed to kill living things, many of these chemicals are potentially toxic to organisms other than the pests for which they are intended. Some are detrimental to nontarget organisms, such as beneficial insects and certain soil organisms. Those chemicals that do not quickly break down may be biologically magnified as they move up the food chain. For example, as earthworms ingest contaminated soil, the chemicals tend to concentrate in the earthworm bodies. When birds, rodents, or fish eat the earthworms, the pesticides can build up further to lethal levels. The near extinction of certain birds of prey (including the American bald eagle) during the 1960s and 1970s called public attention to the sometimes-devastating environmental consequences of pesticide use. More recently, evidence is mounting to suggest that human endocrine (hormone) balance may be disrupted by the minute traces of some pesticides found in soil, water, air, and food.

15.2   KINDS OF ORGANIC CONTAMINANTS

Industrial Organics

Industrial organics that often end up contaminating soils by accident or neglect include petroleum products used for fuel [gasoline components such as benzene, and more complex polycyclic aromatic hydrocarbons, (PAHs)], solvents used in manufacturing processes [such as trichloroethylene (TCE)], and military explosives such as trinitrotoluene (TNT). Polychlorinated biphenyls (PCBs) constitute a particularly troublesome class of widely dispersed compounds. These compounds can disrupt reproduction in birds and cause cancer and hormone effects in humans and other animals. Several hundred varieties of liquid or resinous PCBs were produced from 1930 to 1980 and used as specialized lubricants, hydraulic fluids, and electrical transformer insulators, as well as in certain epoxy paints and many other industrial and commercial applications. Because of their extreme resistance to natural decay and their ability to enter food chains, even today soil and water all over the globe contain at least traces of PCBs.

Leaking underground storage tanks—the threat to public health and the environment: www.sierraclub.org/toxics/Leaking_USTs/index.asp

The sites most intensely contaminated with organic pollutants are usually located near chemical manufacturing plants or oil storage facilities, but railway, shipping, and highway accidents also produce hot spots of contamination. Thousands of neighborhood gas stations represent potential or actual sites of soil and groundwater contamination as gasoline leaks from old, rusting underground storage tanks (Figure 15.1). However, as already mentioned, by far the most widely dispersed xenobiotics are those designed to kill unwanted organisms (i.e., pests).

Pesticides

Pesticides are commonly classified according to the group of pest organisms targeted: (1) *insecticides,* (2) *fungicides,* (3) *herbicides* (weed killers), (4) *rodenticides,* and (5) *nematocides.* In practice, all find their way into soils. Since the first three are used in the largest quantities and are therefore more likely to contaminate soils, they will be given primary consideration. Most pesticides contain aromatic rings of some kind, but there is great variability in pesticide chemical structures.

Insecticides   Most of these chemicals are included in three general groups. The *chlorinated hydrocarbons,* such as DDT, were the most extensively used until the early

Figure 15.1
Leaking underground storage tank (LUST) replacement at a gas station in California. The old rusting steel tanks have been removed and replaced by more corrosion-resistant fiberglass tanks, which are set in the ground and covered with pea gravel. The soil and groundwater aquifer beneath the tanks were cleaned up using special techniques to stimulate soil microorganisms and to pump out volatile organics such as benzene vapors. Remediation and replacement typically costs $700,000 for a single gas station. (Photo courtesy of R. Weil)

1970s, when their use was banned or severely restricted in many countries due to their low biodegradability and persistence, as well as their toxicity to birds and fish. The *organophosphate* pesticides are generally biodegradable, and thus less likely to build up in soils and water. However, they are extremely toxic to humans, so great care must be used in handling and applying them. The *carbamates* are considered least dangerous because of their ready biodegradability and relatively low mammalian toxicity. However, they are highly toxic to honeybees and other beneficial insects and to earthworms.

Fungicides   Fungicides are used mainly to control diseases of fruit and vegetable crops and as seed coatings to protect against seed rots. Some are also used to protect harvested fruits and vegetables from decay, to prevent wood decay, and to protect clothing from mildew. Organic materials such as the thiocarbamates and triazoles are currently in use.

Herbicides   The quantity of herbicides used in the United States exceeds that of the other types of pesticides combined. Starting with 2,4-D (a chlorinated phenoxyalkanoic acid), dozens of chemicals in literally hundreds of formulations have been placed on the market. These include the *triazines*, used mainly for weed control in corn; *substituted ureas*; some *carbamates*; the relatively new *sulfonylureas*, which are potent at very low rates; *dinitroanilines*; and *acetanilides*, which have proved to be quite mobile in the environment. One of the most widely used herbicides, *glyphosate* (e.g., Roundup®), does not belong to any of the aforementioned chemical groups. Unlike most herbicides, it is nonselective, meaning that it will kill almost any plant, including crops. However, a gene that confers resistance to its effects has been discovered and engineered into several major crops. These genetically engineered crops can then be grown with a very simple, convenient method of weed control that usually consists of one or two sprayings of glyphosate that will kill all plants other than the resistant crop.

As one might expect, this wide variation in chemical makeup provides an equally wide variation in properties. Most herbicides are biodegradable, and most (but not all) of them are relatively low in mammalian toxicity. However, some are quite toxic to fish, soil fauna, and perhaps to other wildlife. They can also have deleterious effects on beneficial aquatic vegetation that provides food and habitat for fish and shellfish.

Nematocides   Although nematocides are not as widely used as herbicides and insecticides, some of them are known to contaminate soils and the water draining from

The future role of pesticides in U.S. agriculture: http://books.nap.edu/books/0309065267/html/index.html

treated soils. For example, some carbamate nematocides dissolve readily in water, are not adsorbed onto soil surfaces, and consequently easily leach downward and into the groundwater. Other nematicidal chemicals are volatile soil fumigants that kill virtually all life in the soil, both the helpful and the harmful. Happily, the search for nontoxic substitutes has led to the development of many nonchemical means to manage the pests once controlled by this highly toxic chemical (e.g., see Figure 10.8).

15.3   BEHAVIOR OF ORGANIC CHEMICALS IN SOIL

Once they reach the soil, organic chemicals, such as pesticides or hydrocarbons, follow one or more of seven pathways (Figure 15.2): (1) they may vaporize into the atmosphere without chemical change; (2) they may be absorbed by soils; (3) they may move downward through the soil in liquid or solution form and be lost from the soil by leaching; (4) they may undergo chemical reactions within or on the surface of the soil; (5) they may be broken down by soil microorganisms; (6) they may wash into streams and rivers in surface runoff; and (7) they may be taken up by plants or soil animals and move up the food chain. The specific fate of these chemicals will be determined at least in part by their chemical structures, which are highly variable.

Organic chemicals vary greatly in their volatility and subsequent susceptibility to atmospheric loss. Some soil fumigants, such as methyl bromide (now banned from most uses), were selected because of their very high vapor pressure, which permits

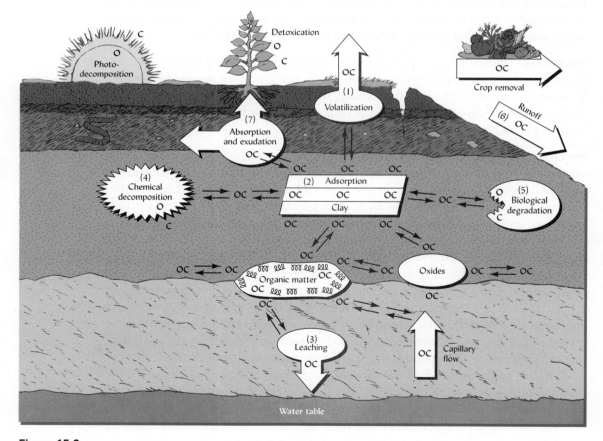

Figure 15.2

Processes affecting the dissipation of organic chemicals (OC) in soils. Note that the OC symbol is split up by decomposition (both by light and chemical reaction) and degradation by microorganisms, indicating that these processes alter or destroy the organic chemical. In transfer processes, the OC remains intact. [From Weber and Miller (1989)]

them to penetrate soil pores to contact the target organisms. This same characteristic encourages rapid loss to the atmosphere after treatment, unless the soil is covered or sealed. A few herbicides (e.g., trifluralin) and fungicides (e.g., PCNB) are sufficiently volatile to make vaporization a primary means of their loss from soil. The lighter fractions of crude oil (e.g., gasoline and diesel) and many solvents vaporize to a large degree when spilled on the soil.

The assumption that disappearance of pesticides from soils is evidence of their breakdown is questionable. Some chemicals lost to the atmosphere are known to return to the soil or to surface waters with the rain.

The adsorption of organic chemicals by soil is determined largely by the characteristics of the compound and of the soils to which they are added. Soil organic matter and high-surface-area clays tend to be the strongest adsorbents for some compounds, while oxide coatings on soil particles strongly adsorb others. Everything else being equal, larger organic molecules with many charged sites are more strongly adsorbed.

Some organic chemicals with positively charged groups, such as the herbicides diquat and paraquat, are strongly adsorbed by silicate clays. Adsorption by clays of some pesticides tends to increase at low pH, which encourages adding H^+ ions to functional groups (e.g.,—NH_2) to yield a positive charge on the herbicide.

The tendency of organic chemicals to leach from soils is closely related to their solubility in water and their potential for adsorption. Some compounds, such as chloroform and phenoxyacetic acid, are a million times more water soluble than others, such as DDT and PCBs, which are quite soluble in oil but not in water. High water-solubility favors leaching losses.

Strongly adsorbed molecules are not likely to move down the profile (Table 15.2). Likewise, conditions that encourage such adsorption will discourage leaching. Leaching is apt to be favored by water movement, the greatest leaching hazard occurring in highly permeable, sandy soils that are also low in organic matter. Periods of high rainfall around the time of application of the chemical promote both leaching and runoff losses (Table 15.3). With some notable exceptions, herbicides seem to be somewhat more mobile than most fungicides or insecticides, and therefore are more likely to find their way to groundwater supplies and streams (Figure 15.3).

Pesticide leaching potential by watershed for 13 crops grown in the U.S.: www.unl.edu/nac/atlas/ Map_Html/Clean_Water/ National/NRI_%20Pesticide_ Leaching_1992/Pesticide_ leaching.htm

Contamination and Persistence

Experts once maintained that contamination of groundwater by pesticides occurred only from accidents such as spills, but it is now known that many pesticides reach the groundwater from normal agricultural use. Since many people (e.g., 40% of Americans) depend on groundwater for their drinking supply, leaching

---

Table 15.2
THE DEGREE OF ADSORPTION TO SOIL OF SELECTED HERBICIDES

Common name	Example trade name	Adsorptivity to soil colloids
Dalapon	Dowpon	None
Chloramben	Amiben	Weak
2,4-D	Several	Moderate
Propachlor	Ramrod	Moderate
Atrazine	AAtrex	Strong
Alachlor	Lasso	Strong
Glyphosate	Roundup	Very strong
Paraquat	Paraquat	Very strong

Table 15.3
SURFACE RUNOFF AND LEACHING LOSSES (THROUGH DRAIN TILES) OF THE HERBICIDE ATRAZINE FROM A CLAY LOAM LACUSTRINE SOIL (ALFISOLS) IN ONTARIO, CANADA

Year of study	Surface runoff loss	Drainage water loss	Total dissolved loss	Percent of total applied, %	Rainfall, May–June, mm
	Atrazine loss, g/ha				
1	18	9	27	1.6	170
2	1	2	3	0.2	30
3	51	61	113	6.6	255
4	13	32	45	2.6	165

Data abstracted from Gaynor et al. (1995).

of pesticides can raise serious health concerns (Box 15.1). Levels of pesticides in drinking water high enough to raise health concerns occur occasionally in the United States and Europe but commonly in newly industrializing counties such as China and India.

Upon contacting the soil, some pesticides undergo chemical modification independent of soil organisms. For example, iron cyanide compounds decompose within hours or days if exposed to bright sunlight. DDT, diquat, and the triazines are subject to slow photodecomposition in sunlight. The triazine herbicides (e.g., atrazine) and organophosphate insecticides (e.g., malathion) are subject to hydrolysis and subsequent degradation. While the complexities of molecular structure of the pesticides suggest different mechanisms of breakdown, it is important to realize that non-biological degradation can occur.

Figure 15.3
Herbicides in the main U.S. corn-growing region illustrate the direct and rapid connection between the use of a chemical on the land and its concentration in streams and rivers. The White River near Hazelton, Indiana, was monitored over a 5-year period. Note that the concentrations of the herbicide Alachlor peaked every year in June, about a month after most farmers in the watershed sprayed their corn and soybean fields. In year 3, a new compound, Acetochlor, partially replaced the older herbicide Alachlor. Within a year of the introduction of the newer compound, Acetochlor concentrations increased while Alachlor concentrations decreased.
[From Gilliom et al. (2006)]

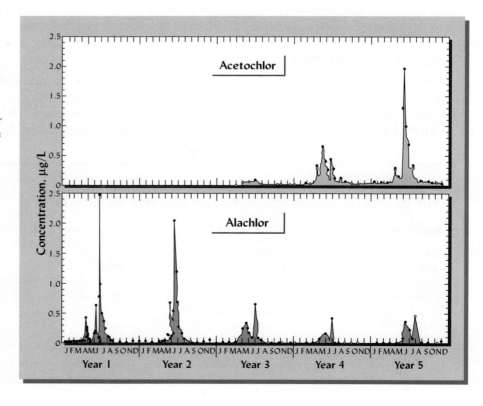

BOX 15.1
CONCENTRATIONS AND TOXICITY OF CONTAMINANTS IN THE ENVIRONMENT

As analytical instrumentation becomes more sophisticated, contaminants can be detected at much lower levels than was the case in the past. Since humans and other organisms can be harmed by almost any substance if large enough quantities are involved, the subject of toxicity and contamination must be looked at quantitatively. That is, we must ask *how much*, not simply *what*, is in the environment. Many highly toxic (meaning harmful in very small amounts) compounds are produced by natural processes and can be detected in the air, soil, and water—quite apart from any activities of humans.

The mere presence of a natural toxin or a synthetic contaminant may not be a problem. Toxicity depends on (1) the concentration of the contaminant, and (2) the level of exposure of the organism. Thus, low concentrations of certain chemicals that would cause no observable effect by a single exposure (e.g., one glass of drinking water) may cause harm (e.g., cancer, birth defects) to individuals exposed to these concentrations over a long period of time (e.g., three glasses of water a day for many years).

Regulatory agencies attempt to estimate the effects of long-term exposure when they set standards for no-observable-effect levels (NOEL) or health-advisory levels. Some species and individuals within a species will be much more sensitive than others to any given chemical. Regulators attempt to consider the risk to the most susceptible individual in any particular case. For nitrate in groundwater, this individual might be a human infant whose entire diet consists of infant formula made with the contaminated water. For DDT, the individual at greatest risk might be a bird of prey that eats fish that eat worms that ingest lake sediment contaminated with DDT. For a pesticide taken up by plants from the soil, the individual at greatest risk might be an avid gardener who eats vegetables and fruits mainly from the treated garden over the course of a lifetime.

It is important to get a feel for the meaning of the very small numbers used to express the concentration of contaminants in the environment. For instance, concentrations are often given in parts per billion (ppb). This is equivalent to micrograms per kilogram or µg/kg. In water this would be µg/L. To comprehend the number 1 billion imagine a billion golf balls: lined up, they would stretch completely around the Earth. One bad ball out of a billion (1 ppb) seems like an extremely small number. On the other hand, 1 ppb can seem like a very large number. Consider water contaminated with 1 ppb of potassium cyanide, a very toxic substance consisting of a carbon, a potassium, and a nitrogen atom linked together (KCN). If you drank just one drop of this water, you would be ingesting almost 1 trillion molecules of potassium cyanide:

$$\frac{6.023 \times 10^{23} \text{ molecles}}{1 \text{ mol}} \times \frac{1 \text{ mol}}{65 \text{ g KCN}} \times \frac{1 \text{ g KCN}}{10^6 \mu\text{g KCN}} \times \frac{1 \mu\text{g KCN}}{L} \times \frac{L}{10^3 \text{cm}^3} \times \frac{\text{cm}^3}{10 \text{ drops}} = \frac{9.3 \times 10^{11} \text{ molecules}}{\text{drop}}$$

In the case of potassium cyanide, the molecules in this drop of water would probably not cause any observable effect. However, for other compounds, this many molecules may be enough to trigger DNA mutations or the beginning of cancerous growth. Assessing these risks is still an uncertain business.

Biochemical degradation by soil organisms is the single most important method by which pesticides are removed from soils. Certain polar groups on the pesticide molecules, such as —OH, —COO⁻, and —NH₂, provide points of attack for the organisms.

DDT and other chlorinated hydrocarbons, such as aldrin, dieldrin, and heptachlor, are very slowly broken down, persisting in soils for 20 or more years. In contrast, the organophosphate insecticides, such as parathion, are degraded quite rapidly in soils, apparently by a variety of organisms. Likewise, most herbicides (e.g., 2,4-D, the phenylureas, the aliphatic acids, and the carbamates) are readily attacked by a host of organisms. Exceptions are the triazines, which are slowly degraded, primarily by chemical action. Most organic fungicides are also subject to microbial decomposition, although the rate of breakdown of some is slow, causing troublesome residue problems.

Pesticides are commonly absorbed by higher plants. This is especially true for those pesticides (e.g., systemic insecticides and most herbicides) that must be taken up in order to perform their intended function. The absorbed chemicals may remain intact inside the plant, or they may be degraded. Some degradation products are harmless, but others are even more toxic to humans than the original chemical that was absorbed. Understandably, society is quite concerned about pesticide residues found in the parts of plants that people eat, whether as fresh fruits and vegetables or as processed foods. The use of pesticides and the amount of pesticide residues in food are strictly regulated by law. Despite widespread concerns, there is little evidence that the small amounts of residues permissible in foods by law have had any ill effects on public health. Routine testing by regulatory agencies has shown that about 1 to 2% of food samples tested contain pesticide residues above the levels permissible, but only a tiny percentage of the food supply is actually tested.

The **persistence** of chemicals in the soil is the net result of all their reactions, movements, and degradations. Marked differences in persistence are the rule. For example, organophosphate insecticides may last only a few days in soils. The widely used herbicide 2,4-D persists in soils for only two to four weeks. The PCBs, DDT, and other chlorinated hydrocarbons may persist for 3 to 20 years or longer (Table 15.4). The persistence times of other pesticides and industrial organics fall generally between the extremes cited. Compounds that resist degradation have a greater potential to cause environmental damage.

Continued use of the same pesticide on the same land can increase the rate of microbial breakdown of that pesticide. Apparently, having a constant food source allows a population buildup of those microbes equipped with the enzymes needed to break down the compound. This is an advantage with respect to environmental quality and is a principle sometimes applied in environmental cleanup of toxic organic compounds. On the other hand, the breakdown may become sufficiently rapid to reduce a pesticide's effectiveness.

Groundwater Vulnerability   The vulnerability of groundwater to contamination by pesticide leaching varies greatly from one area to another. Highest vulnerability occurs in regions with high rainfall, an abundance of sandy soils, and intensive cropping systems that involve high usage of those types of pesticides that are most soluble and least strongly adsorbed by the soil colloids. However, pesticide hazards are site specific, and these regional generalizations might mask localized areas of vulnerability. For example, in arid regions irrigated areas of intensive vegetable crop production may experience considerable leaching of both pesticides and nitrates. Likewise, application of certain water-soluble pesticides may result in groundwater contamination even where the soil may not be coarse in texture.

Table 15.4
COMMON RANGE OF PERSISTENCE OF A NUMBER OF ORGANIC COMPOUNDS
Risks of environmental pollution are higher with those chemicals with greater persistence.

Organic chemical	Persistence in soils
Chlorinated hydrocarbon insecticides (e.g., DDT, chlordane, dieldrin)	3–20 yr
PCBs	2–10 yr
Triazine herbicides (e.g., atrazine, simazine)	1–2 yr
Glyphosate herbicide	6–20 mo
Benzoic acid herbicides (e.g., amiben, dicamba)	2–12 mo
Urea herbicides (e.g., monuron, diuron)	2–10 mo
Vinyl chloride	1–5 mo
Phenoxy herbicides (2,4-D, 2,4,5-T)	1–5 mo
Organophosphate insecticides (e.g., malathion, diazinon)	1–12 wk
Carbamate insecticides	1–8 wk
Carbamate herbicides (e.g., barban, CIPC)	2–8 wk

15.4   REMEDIATION OF SOILS CONTAMINATED WITH ORGANIC CHEMICALS

Soils contaminated with organic pollutants are found throughout the world. The wide areas contaminated with organic pesticides are best addressed by modifying the agro-ecosystems to enable the use of pesticides to be reduced or eliminated, or by using less toxic, less mobile, and more rapidly degradable pesticide compounds. In many cases, the soil ecosystem should be able to recover its function and diversity through **natural attenuation** over a reasonable period. Natural attenuation may involve any or all of the chemical, physical, and biological processes that were illustrated in Figure 15.2.

U.S. Geological Survey bioremediation projects: http://water.usgs.gov/wid/html/bioremed.html

Perhaps more problematic, however, are sites where accidental spills of toxic organic materials have occurred or where, through the decades, organic wastes from industrial and domestic processes have been dumped on soils. The levels of such *acute contamination* are often sufficiently high that plant growth is restrained or even prevented. Pollutants can move into the groundwater, making it unfit for human consumption. Fish and wildlife may be decimated. Because of public concerns, businesses and government are spending billions of dollars annually to clean up (**remediate**) these contaminated soils. We shall consider a few of the methods in use and under development in the rapidly evolving soil remediation industry. In general, efforts to remediate polluted soils face the need to compromise between speed and certainty that cleanup standards will be met on one hand, and expense and disruption of the site on the other (Figure 15.4).

Physical and Chemical Methods

The earliest and still most widely used methods of soil remediation involve physical and/or chemical treatment of the soil, either in place (*in situ*) or by moving the soil to a treatment site (*ex situ*). The latter may involve excavating the soil to treatment bins

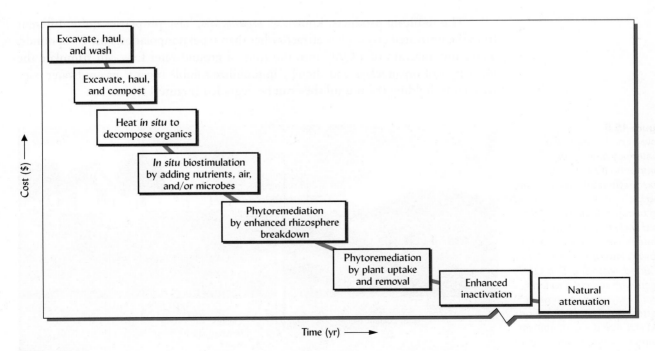

Figure 15.4

A wide range of methods is available to remediate (clean up) polluted soils. At one extreme are remediation techniques that are very expensive and disruptive, but usually quite rapid. Technologies at the other extreme may be quite inexpensive and nondisruptive, but usually take much more time to accomplish the cleanup. [Modified from Reynolds et al. (1999)]

where it may be incinerated to drive off volatile chemicals and to destroy other pollutants by high-temperature chemical decomposition. Water-soluble and volatile chemicals may also be removed by pushing or pulling air or water through the soil by vacuum extraction or leaching. Such treatments are usually quite effective in removing or destroying the contaminants, but are expensive, especially if large quantities of soil must be excavated and treated. And, of course, the treated soil is also destroyed as a living system and must be either replaced on the site or deposited in a landfill.

Alternative Cleanup
Technologies for Underground Storage Tank Sites:
www.epa.gov/swerust1/
pubs/tums.htm

In situ treatments are usually preferred if viable technologies are available. The soil is left in place, thereby reducing excavation, transport, treatment, and disposal costs and providing greater flexibility in future land use. The contaminants are either removed from the soil (*decontamination*) or are sequestered (*bound up*) in the soil matrix (*stabilized*). Decontamination *in situ* involves some of the same techniques of water flushing, leaching, vacuum extraction, and heating used in *ex situ* processes. Water treatment is not effective, however, with nonpolar compounds that are repelled by water. To help remove such compounds, scientists and engineers have sprayed onto the soil surface or have injected into the soil compounds called *surfactants*. As these move downward in the soil, they dissolve organic contaminants, which can then be pumped out of the soil as in the water-washing systems.

Certain surfactants may also be used to immobilize or stabilize soil contaminants. They are positively charged and through cation exchange can replace metal cations on soil clays. For example, one group of such surfactants, quaternary ammonium compounds (QACs), has the general formula $(CH_3)_3NR^+$, where R is an organic alkyl or aromatic group. The positive charges on QACs stimulate cation exchange by reactions such as the following, using a monovalent exchangeable cation such as K^+ as an example:

$$\boxed{\text{Colloid}}\ K^+ + (CH_3)_3NR^+ \longrightarrow \boxed{\text{Colloid}}\ (CH_3)_3NR^+ + K^+ \qquad (15.1)$$

Untreated clay   QAC   Organoclay

The resulting products, known as *organoclays*, have properties quite different from the untreated clays. They attract rather than repel nonpolar organic compounds. Thus, the injection of a QAC into the zone of groundwater flow can stimulate the formation of organoclays and thereby immobilize soluble organic groundwater contaminants, holding them until they can be degraded (Figure 15.5).

Figure 15.5

How a combination of a quaternary ammonium compound (QAC), hexadecyltrimethylammonium, and bioremediation by degrading bacteria could be used to hold and remove an organic contaminant. The pollutant is moving into groundwater from a buried waste site. The QAC reacts with soil clays to form organoclays and soil organic matter complexes that adsorb and stabilize the contaminant, giving microorganisms time to degrade or destroy it. [Redrawn from Xu et al. (1997)]

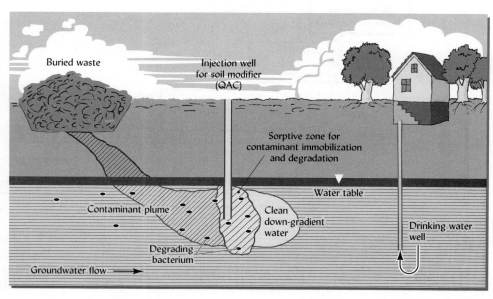

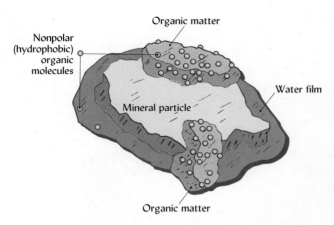

Figure 15.6

Because mineral colloids in soils are nearly always surrounded by at least a thin film of water, hydrophobic organic molecules tend to sorb onto humus more readily than clay. The nonpolar organic molecules cannot compete with the polar water molecule for a place on the charged mineral surfaces. This is one reason why, for certain organic contaminants, the soil organic matter content is more consistent than the clay content in characterizing the tendency to be held by various soils.

(Diagram courtesy of R. Weil)

The distribution coefficient K_d for the adsorption of many nonpolar organic compounds on untreated clays is very low because the clays are hydrophilic (water-loving) and their adhering water films repel the hydrophobic, nonpolar organic compounds (Figure 15.6). Surface soil horizons containing significant quantities of humus often exhibit a much higher K_d because of the sorption of the organic contaminant into the organic matter coatings. This is the reason that the organic carbon distribution coefficient K_{oc} is often a better measure of a compound's tendency to become immobilized in various surface soils (see Section 8.12 for an explanation of K_d and K_{oc}). Deeper soil layers, especially near and below the water table, generally contain little humus and so have limited capacity to immobilize organic contaminants.

In contrast, organoclays effectively sorb organic contaminants, leaving little in the soil solution, thereby reducing their movement into the groundwater and eventually into streams or drinking water. Consequently, the K_d values of organic contaminants on organoclays are commonly 100 to 200 times those measured on the untreated clays. Organoclays thus offer promising mechanisms for holding organic soil pollutants until they can be destroyed by biological or physicochemical processes.

Bioremediation[1]

For many heavily contaminated soils, there is a biological alternative to incineration, soil washing, and landfilling—namely, **bioremediation**. Simply put, this technology uses enhanced plant and/or microbial action to degrade organic contaminants into harmless metabolic products. Analysis of microbial DNA has shown that degradation of contaminants in soils is almost always the work of genetically diverse *consortia* of many organisms, rather than just one or two bacterial species. Petroleum constituents, including the more resistant polyacrylic aromatic hydrocarbons (PAHs), as well as several synthetic compounds, such as pentachlorophenol (PCP) and trichloroethylene (TCE), can be broken down, primarily by soil bacteria archaea and so-called white-rot fungi. Bioremediation is usually accomplished *in situ*, but polluted soil may also be excavated and treated *ex-situ*—that is, hauled to a treatment site where such techniques as high-temperature composting (see Section 11.10) may be used to destroy the organic contaminants in the soil (Figure 15.7).

---

[1]For reviews of the theories and technologies regarding this topic, see Wise et al. (2000) and Eccles (2007).

Figure 15.7

Hot water vapor rises in cold winter air as windrows of high-temperature compost are mixed and aerated by a special compost-turning machine in order to accelerate the breakdown of organic compounds. The method can hasten the degradation of organic pollutants in soil material excavated from a contaminated site, mixed with decomposable organic materials, and made into windrows. (Photo courtesy of R. Weil)

Bioaugmentation   In some cases, the remediation process depends on organisms native to the soil. In others, microbes specifically selected for their ability to remove the contaminants are introduced into the polluted soil zone to *augment* the natural microbial populations. This approach is called **bioaugmentation**. Some success has been achieved by inoculating contaminated soils with improved organisms that can degrade the pollutant more readily than can the native population. Although genetic engineering may prove useful in making "superbacteria" in the future, most inoculation has been achieved with naturally occurring organisms. Organisms isolated from sites with a long history of the specific contamination or grown in laboratory culture on a diet rich in the pollutant in question tend to become acclimated to metabolizing the target chemical.

For example, certain bacteria have been identified that can detoxify perchloroethene (PCE), a common, highly toxic groundwater pollutant that is suspected of being a carcinogen. Scientists can inoculate with these organisms to expedite the step by step removal of the four chlorines from the PCE, producing ethylene, a gas that is relatively harmless to humans:

$$\underset{\substack{\text{PCE}\\\text{(Suspected carcinogen)}}}{\mathrm{Cl_2C{=}CCl_2}} \xrightarrow[\;4HCl\;]{\;8H\;} \underset{\substack{\text{Ethylene}\\\text{(Harmless gas)}}}{\mathrm{H_2C{=}CH_2}}$$

Composting to clean up munitions-contaminated soils:
www.epa.gov/osw/
conserve/rrr/composting/
pubs/explos.pdf

Global interest group for bioremediation (click "BioLinks"):
www.bioremediationgroup
.org/AboutUs/Home.htm

Biostimulation   Technology that assists the naturally occurring microbial populations in breaking down chemicals is called **biostimulation**. Usually, soils naturally contain some microorganisms that can degrade the specific contaminant. But the rate of natural degradation may be far too slow to be very effective. The growth rate of the organisms capable of using the contaminant as a carbon source may be limited by insufficient mineral nutrients, especially N and P (see Section 11.3 for a discussion of the C/N ratio in organic decomposition). Special fertilizers have been formulated and used successfully to greatly speed up the degradation process. One such fertilizer of French manufacture is an oil-in-water microemulsion of urea, lauryl phosphate, and an emulsion stabilizer. It acts not only as a supplier of nutrients, but also as a surfactant that can enhance interaction between microbes and the organic contaminants. It received its first major test in Alaska with the Exxon Valdez oil spill (Figure 15.8).

Figure 15.8

Bioremediation of crude oil from the Exxon Valdez oil spill off the coast of Alaska. The oil contaminating the beach soils was degraded by indigenous bacteria when an oil-soluble fertilizer containing nitrogen and phosphorus was sprayed on the beach (o data points in graph on the left). The control sections of the beach (+ data points) were left unfertilized for 70 days. By then the effect of the fertilization was so dramatic that a decision was made to treat the control sections as well. The index of oil remaining is based on natural logarithms, so each whole number indicates more than doubling of the oil remaining. The photo (right) shows the clear delineation between the oil-covered control section and the fertilized parts of the beach. [Data from Bragg et al. (1994); reprinted with permission from *Nature*, © 1994 Macmillan Magazines Limited; photo courtesy of P. H. Pritchard, U.S. EPA, Gulf Breeze, FL; from Pritchard et al. (1992); reprinted by permission of Kluwer Academic Publishers]

In some cases, low soil porosity causes oxygen deficiency that limits microbial activity. Techniques are being developed that use *in situ* bioremediation to clean up oxygen-deficient soils and associated groundwater contamination. For example, organic-solvent-contaminated soils have been bioremediated (Figure 15.9) by piping in a mixture of air (for oxygen), methane (to act as a carbon source to stimulate specific bacteria), and phosphorus (a nutrient that is needed for bacteria growth).

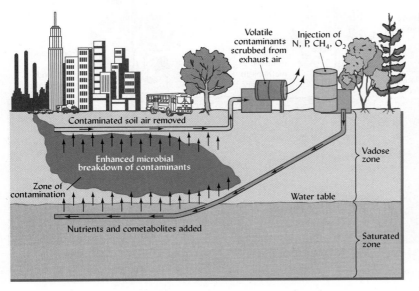

Figure 15.9

In situ bioremediation of soil and groundwater contaminated with volatile organic solvents. The scheme illustrated is typical of the biostimulation approach to soil remediation. Breakdown of the organic contaminant is stimulated by adding such components as nutrients, oxygen, and cometabolites that improve the soil environment for the growth of native bacteria capable of metabolizing the contaminant. In this instance, methane (CH_4) is added intermittently as a substrate for certain methane-oxidizing bacteria which multiply rapidly and turn to the solvent as a carbon source whenever methane is not available. The nutrients are added and the contaminated soil air is removed by pumps connected to perforated pipes that are inserted by horizontal well-drilling techniques. Such biostimulation schemes can significantly cut the time and cost for cleanup of contaminated soils. [Based on Hazen (1995)]

Phytoremediation

Links to many sites on phytoremediation: www.dsa.unipr.it/phytonet/ links.htm

Plants can also participate in bioremediation, a process termed **phytoremediation**. For years, plant-based systems have been used for the removal of municipal wastewater contaminants (Section 13.2). More recently, this concept has been extended to industrial pollutants and to the removal from shallow groundwater pollutants of all kinds, both organic and inorganic.

Phytoremediation uses plants in two fundamentally different ways (Figure 15.10). In the first, plant roots take up the pollutant from the soil. The plant may then either accumulate large amounts of the contaminant in aboveground biomass, or it may metabolize the contaminant into harmless by-products. The accumulation of unusually high concentrations of a contaminant in the plant biomass is called **hyperaccumulation**. Hyperaccumulating plants take up and tolerate very high concentrations of a contaminant, most commonly a toxic metal such as zinc or nickel, but also certain organics such as trinitrotoluene (TNT). Hyperaccumulation allows the contaminant to be removed by harvesting the plant tissue.

The second type of cleanup using plants is called **enhanced rhizosphere phytoremediation**. In this process, the plants do not take up the contaminant. Instead, the plant roots excrete into the soil carbon compounds that serve as microbial substrates and growth regulators (see Section 10.7). These compounds stimulate the growth of the rhizosphere bacteria that, in turn, degrade the organic contaminant. The transpiration of water by the plant causes soil water, with its load of dissolved contaminant molecules, to move toward the roots, thus increasing the efficiency of the rhizosphere reactions.

Research has shown that some plant species are better than others at stimulating the degradation of specific compounds in their rhizospheres. Many plant species, domesticated and wild, have been used in phytoremediation. Prairie grasses can stimulate the degradation of petroleum products, including PAHs, and spring wildflower plants in Kuwait have been found to degrade the hydrocarbons in oil spills. Fast-growing hybrid poplars can remove compounds such as TNT, as well as some pesticides and excess nitrates.

Figure 15.10

Two approaches to phytoremediation—the use of plants to help clean up contaminated soils. (Left) Hyperaccumulating plants take up and tolerate very high concentrations of an inorganic or organic contaminant. In the case of metal contaminants, the addition of chelating agents may increase the rate of metal uptake, but can add a major expense and may allow metals to migrate below the root zone. (Right) In enhanced rhizosphere phytoremediation, the plants do not take up the contaminant. Instead, the plant roots excrete substances that stimulate the microbes in the rhizosphere soil, speeding their degradation of organic contaminants. Transpiration-driven movement of water and dissolved contaminants to the enhanced rhizosphere zone improves the system's effectiveness.

(Diagram courtesy of R. Weil)

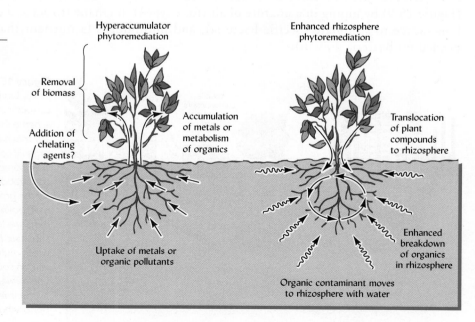

Phytoremediation is particularly advantageous where large areas of soil are contaminated with only moderate concentrations of organic pollutants. However, phytoremediation also commonly takes a longer time to remove large quantities of contaminants than do the more costly engineering procedures.

15.5 CONTAMINATION WITH TOXIC INORGANIC SUBSTANCES[2]

The toxicity of inorganic contaminants released into the environment every year is now estimated to exceed that from organic and radioactive sources combined. A fair share of these inorganic substances ends up contaminating soils. The greatest problems most likely involve mercury, cadmium, lead, arsenic, nickel, copper, zinc, chromium, molybdenum, manganese, selenium, fluorine, and boron. To a greater or lesser degree, all of these elements are toxic to humans and other animals. Cadmium and arsenic are extremely poisonous; mercury, lead, nickel, and fluorine are moderately so; boron, copper, manganese, and zinc are relatively lower in mammalian toxicity. Although the metallic elements (see periodic table, Appendix B) are not all, strictly speaking, "heavy" metals, for the sake of simplicity this term is often used in referring to them.

There are many sources of the inorganic chemical contaminants that can accumulate in soils. The burning of fossil fuels, smelting, and other processing techniques release into the atmosphere tons of these elements, which can be carried for miles and later deposited on the vegetation and soil. Lead, nickel, and boron are gasoline additives that may be released into the atmosphere and carried to the soil through rain and snow. Arsenic was for many years used as a wood preservative as well as an insecticide on cotton, tobacco, fruit crops, lawns, and as a defoliant or vine killer. Some of the toxic metals are being released to the environment in increasing amounts, while others (most notably lead, because of changes in gasoline formulation) are decreasing. All are daily ingested by humans, either through the air or through food, water, and—yes—soil (see Box 15.2).

Irrespective of their sources, toxic elements can and do reach the soil, where they become part of the food chain: soil→plant→animal→human (Figure 15.11). Unfortunately, once the elements become part of this cycle, they may accumulate to toxic levels. This situation is especially critical for animals (fish, other wildlife, and humans) at the top of the food chain. Governments must closely regulate the release of these toxic elements in the form of industrial wastes. Because of the globalization of our

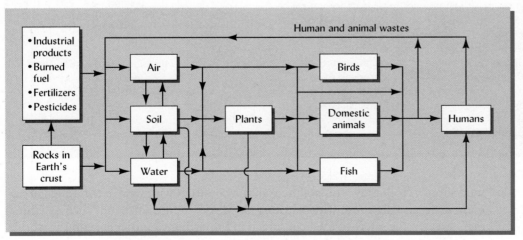

Figure 15.11
Sources of heavy metals and their cycling in the soil–water–air–organism ecosystem. It should be noted that the content of metals in tissue generally builds up from left to right, indicating the vulnerability of humans to heavy metal toxicity.

[2]For an excellent in-depth review of all aspects of this topic, see Adriano (2001). Additional information is available in Ahmad et al. (2006). Heavy metal contamination of food grown in China is discussed in Zamiska and Spencer (2007).

BOX 15.2
LEAD CONTAMINATION AND POISONING

Lead contamination is a serious and widespread form of inorganic soil pollution. Long-term exposure to low levels of lead can contribute to reduced mental capacities, poor academic performance, and juvenile delinquency. In the past, much of the lead exposure came from burning leaded fuels. The content of lead in soils commonly increases with proximity to major highways and as one approaches the center of a major city. Residents of inner cities generally live surrounded by lead-contaminated soils. A second reason for high lead concentrations in urban soils is related to the lead-based pigments used in paint prior to 1970. Paint chips, flakes, and dust from sanding painted surfaces spread the lead around, and eventually much of it ends up in the soil. During dry weather, soil particles blow about, spreading the lead and contributing to the dust that settles on floors, windowsills, plant foliage, and fruits.

Eating these garden products and breathing in lead-contaminated dust are two pathways for human lead exposure (see Figure 15.11). However, the most serious pathway for young children is hand-to-mouth activity—basically, eating dirt (see also Box 1.1). Anyone who has observed a toddler knows that a child's hands are continually in his or her mouth (Figure 15.12). Lead-contaminated dust on surfaces in the home can therefore be an important source of lead exposure for young children; so, too, can lead-contaminated soil in outdoor play areas. Having children wash their hands frequently can significantly cut down their exposure to this insidious toxin. The U.S. EPA has set standards for the cleanup of lead in soil around homes: 400 parts per million (ppm) of lead in bare soil in children's play areas or 1200 ppm average for bare soil in the rest of the yard. Soils with lead levels higher than these standards require remediation.

Current measures designed to protect children from lead in soil around the home include (1) excavation and removal of the soil, (2) dilution by mixing in large amounts of noncontaminated soil, or (3) stabilization of the lead away from the reach of children and dust-creating winds. Contaminated soil areas may be covered with a thick layer of uncontaminated topsoil, a wooden deck, or pavement. Well-maintained turfgrass will prevent most dust formation and soil ingestion.

Orthophosphate ions (PO_4^{3-}) react with Pb in the soil solution to precipitate very insoluble minerals, mainly lead pyromorphite [$Pb_5(PO_4)_3OH,Cl,F$]. Highly contaminated soil was amended with phosphorus (500 to 1000 times normal fertilization rates). After 32 months, P-treated and untreated soils were fed to young pigs (whose digestive systems are similar to those of humans). Lead levels in the blood of pigs fed the P-treated soil were much lower than those fed the control soil (Figure 15.13), suggesting that although P-treatment leaves the lead in the soil, it could substantially reduce the hazard to children.

Figure 15.12

Lead poisoning in young children. (Photo courtesy of R. Weil)

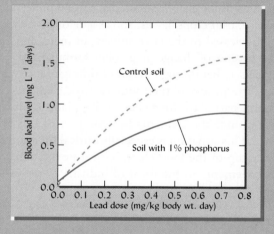

Figure 15.13

Adding 1% P reduced the bioavailability of soil lead, as evidenced by lower lead levels in the blood of pigs fed the soil. [Graphed from data in Ryan et al. (2004)]

food supply, crops grown on polluted soils in countries with weak environmental regulations (especially those with a history of "dirty" industrialization, such as China) may threaten the safety of food consumed in importing countries around the world.

Direct ingestion of soils and sludge is also an important pathway for human and animal exposure. Animals should not be allowed to graze on sludge-treated pastures until rain or irrigation has washed the sludge from the forage. Children may eat soil while they play, and a considerable amount of soil eventually becomes dust in many households. Direct ingestion of soil and dust is particularly harmful in lead toxicity.

15.6 REACTIONS OF INORGANIC CONTAMINANTS IN SOILS

Heavy Metals in Sewage Sludge

The domestic and industrial sewage sludges considered as nutrient sources in Chapter 13 can be important sources of potentially toxic chemicals. Nearly half of the municipal sewage sludge produced in the United States is being applied to the soil, either on agricultural land or to remediate land disturbed by mining and industrial activities. Industrial sludges commonly carry significant quantities of inorganic as well as organic chemicals that can have harmful environmental effects.

Farmers must be assured that the levels of inorganic chemicals in sludge are not sufficiently high to be toxic to plants (a possibility mainly for zinc and copper) or to humans and other animals who consume the plants (a serious consideration for Cd, Cr, and Pb). For relatively low-metal municipal sludges, application at rates just high enough to supply needed nitrogen seems to be quite safe (Table 15.5).

Concern over the possible buildup of heavy metals in soils resulting from large land applications of sewage sludges has prompted research on the fate of these chemicals in soils. Most attention has been given to zinc, copper, nickel, cadmium, and lead, which are commonly present in significant levels in these sludges. Many studies have suggested that if only moderate amounts of sludge are added, and the soil is not very acid (pH > 6.5), these elements are generally bound by soil constituents; they do not then easily leach from the soil, nor are they readily available to plants. Only in moderately to strongly acid soils have most studies shown significant movement down the profile from the layer of application of the sludge. Monitoring soil acidity and using judicious applications of lime have been widely recommended to prevent leaching into groundwaters and to minimize uptake by plants.

Forms Found in Soils Treated with Sludge   A very small proportion of heavy metals in sludge-treated soil is held in *soluble* or *exchangeable forms*, which are available for plant uptake. Another portion of the metals is bound by the *soil organic matter* and by

Virginia guidelines on land application of sludge: www.ext.vt.edu/pubs/compost/452-303/452-303.html

Sewage sludge—the case for caution: http://cwmi.css.cornell.edu/sewagesludge.htm

Table 15.5
UPTAKE OF METALS BY CORN AFTER 19 YEARS OF FERTILIZING A MINNESOTA SOIL (TYPIC HAPLUDOLLS) WITH LIME-STABILIZED MUNICIPAL SEWAGE SLUDGE
Note that the metals show the typical pattern of less accumulation in the grain than in the leaves and stalks (stover). The annual sludge rate of about 10.5 Mg was designed to supply the nitrogen needs of the corn. The sludge had little effect on the metal content of the plants, except in the case of zinc (which increased, but not beyond the normal range for corn).

	Zn	Cu	Cd	Pb	Ni	Cr
Cumulative metal applied in sludge, kg/ha	175	135	1.2	49	4.9	1045
Treatment	Uptake in stover, mg/kg					
Fertilizer	18	8.4	0.16	0.9	0.7	0.9
Sludge	46.5	7.0	0.18	0.8	0.6	1.4
	Uptake in grain, mg/kg					
Fertilizer	20	3.2	0.29	0.4	0.4	0.2
Sludge	26	3.2	0.31	0.5	0.3	0.2

Data abstracted from Dowdy et al. (1994).

the *organic materials* in the sludge. High proportions of the copper and chromium are commonly found in this form, while lead is not so highly attracted. Organically bound elements are not readily available to plants, but may be released over a period of time.

Heavy metals in soils are also associated with *carbonates* and with *oxides of iron and manganese.* These forms are less available to plants than either the exchangeable or the organically bound forms, especially if the soils are not allowed to become too acid. The remaining or *residual form* of the heavy metals consists of sulfides and other very insoluble compounds that are less available to plants than any of the other forms.

Most soil-applied heavy metals are not readily absorbed by plants and are not easily leached from the soil. However, this immobility means that metals will accumulate in soils if repeated sludge applications are made. Care must be taken not to add such large quantities that the capacity of the soil to react with a given element is exceeded. It is for this reason that regulations set maximum cumulative loading limits for each metal (see Table 15.6).

Other Inorganic Pollutants

Arsenic has accumulated in certain orchard soils following years of application of arsenic-containing pesticides. Being present in an anionic form (e.g., $H_2AsO_4^-$), this element is absorbed (as are phosphates) by hydrous iron and aluminum oxides, especially in acid soils. In spite of the capacity of most soils to tie up arsenates, long-term additions of arsenical sprays can lead to toxicities for sensitive plants and earthworms. The arsenic toxicity can be reduced by applications of sulfates of zinc, iron, and aluminum, which tie up the arsenic in insoluble forms.

The presence of arsenic in soils, groundwater, and well water is of concern to people around the world, but especially in Bangladesh, India, China, Chile, and Slovakia. In Bangladesh, for example, more than 20 million of the country's 126 million people are believed to be drinking arsenic-contaminated water. Thousands are suffering from skin cancer that is caused by naturally occurring arsenic toxicity. The well water in the United States is generally safe, but the arsenic contents of some wells exceed the current maximum contaminant level (MCL), which has recently become even more stringent.

Arsenic is found as a minor constituent of many minerals (especially sulfides). Upon their breakdown it becomes associated with the soil in two major forms, arsenite

Arsenic in drinking water—
USEPA rule:
www.epa.gov/safewater/
arsenic/index.html/

Table 15.6
REGULATORY LIMITS ON INORGANIC POLLUTANTS (HEAVY METALS) IN SEWAGE SLUDGE APPLIED TO AGRICULTURAL LAND

Element	Maximum concentration in sludge, U.S. EPA,[a] mg/kg	Annual pollutant loading rates, U.S. EPA, kg/ha/yr	Cumulative allowable pollutant loading, kg/ha U.S. EPA	Germany	Ontario
As	75	2.0	41	—	28
Cd	85	1.9	39	3.2	3.2
Cr	3000	150.0	3000	200	240
Cu	4300	75.0	1500	120	200
Hg	57	0.85	17	2	1.0
Mo	75	—	—	—	8
Ni	420	21	420	100	64
Pb	840	15	300	200	120
Se	100	5.0	100	—	3.2
Zn	7500	140	2800	400	440

[a]U.S. Environmental Protection Agency (1993).

[AsO_3^{3-}, or three-valent As(III)] and arsenate (AsO_4^{3-}, or five-valent As(V)]. Both forms are sorbed by oxides and hydroxides of iron, but As(V) is generally more strongly sorbed, especially in acid soils. Consequently, arsenites [As(III)] are generally more mobile and move more easily into groundwater, the source of drinking water in many parts of the world. For this reason, wet and reduction-prone conditions are to be avoided to minimize dissolution and movement of the most toxic forms of arsenic.

Scientists are evolving methods for the remediation of arsenic-contaminated waters. For example, they are trying to use hydrous oxides of iron as sorbing agents for the arsenic in drinking water. Also, they have discovered that certain plants are *hyper-accumulators* of arsenic.

Lead contaminates soils primarily from vehicle exhaust and from old lead-pigmented paints (paint chips and dust from painted woodwork). Most of the lead is tied up in the soil as low solubility carbonates, sulfides, and in combination with iron, aluminum, and manganese oxides. Consequently, the lead is largely unavailable to plants and not mobile enough to readily leach to groundwater. However, it can be absorbed by children who put contaminated soil in their mouths (Box 15.2).

Boron can contaminate soil via high-boron irrigation water, by excessive fertilizer application, or by the use of power plant fly ash as a soil amendment. Boron may be adsorbed by organic matter and clays but may still be available to plants, except at high soil pH. Boron is relatively soluble in soils, toxic quantities being leachable, especially from acid sandy soils. Boron toxicity in plants is usually a localized problem and is probably much less important than boron deficiency.

Fluorine toxicity is also generally localized. Drinking water for animals and fluoride fumes from industrial processes sometimes contain toxic amounts of fluorine. The fumes can be ingested directly by animals or deposited on nearby plants. If the fluorides are adsorbed by the soil, their uptake by plants is restricted. The fluorides formed in soils are highly insoluble, the solubility being least if the soil is well supplied with lime.

Mercury is released mainly from burning coal to generate electricity. When it contaminates lake beds and swampy areas, the result is toxic levels of mercury among certain species of fish. Insoluble forms of mercury in soils, not normally available to plants or, in turn, to animals, are converted by microorganisms to an organic form, methylmercury, in which it is more soluble and available for plant and animal absorption. The methylmercury is concentrated in fatty tissue as it moves up the food chain, until it accumulates in some fish to levels that may be toxic to humans. This series of transformations illustrates how reactions in soil can influence human toxicities.

Chromium in trace amounts is essential for human life, but, like arsenic, it is a carcinogen when absorbed in larger doses. This element is widely used in steel, alloys, and paint pigments. Chromium is found in two major oxidation states in ordinary soils: a trivalent form [Cr(III)] and a hexavalent form [Cr(VI)]. In contrast to most metals, the more highly oxidized state [Cr(VI)] is the more soluble, and its solubility increases above pH 5.5. This behavior is opposite that of Cr(III), which forms insoluble oxides and hydroxides above that pH level.

To remediate Cr(VI)-contaminated soil and water, it is useful to reduce the chromium to Cr(III) (see also Section 7.5). This reduction process is enhanced by anaerobic conditions [wet soil with an abundance of decomposable organic material to provide a large, biological oxygen demand (BOD)]. The organic matter serves as an electron donor and thereby hastens the reduction of Cr(VI) to the trivalent state [Cr(III)]. Provided the pH is maintained above 5.5, chromium in this reduced state will remain relatively stable, immobile, and nontoxic.

Selenium, which derives mainly from certain soil parent material, can accumulate in soils and plants to toxic levels, especially in arid regions.

Selenium is found in nature in four major solid forms and several volatile forms. The particular forms present determine the degree of toxicity much more than does the total amount of selenium in the soil. The relationship among these forms may be shown by the following reactions, which illustrate the microbiological reduction of the soluble and highly oxidized selenates to reduced and less soluble forms.

$$SeO_4^{2-} \rightleftharpoons SeO_3^{2-} \rightleftharpoons Se \rightleftharpoons Se^{2-} \rightleftharpoons (CH_3)_2Se \qquad (15.2)$$

[Se(VI)]	[Se(IV)]	[Se(0)]	[Se(-II)]	[Se(-II)organic]
Selenate	Selenite	Elemental selenium	Selenide	Dimethyl selenide

Selenates are most soluble and are prominent in well-aerated soils, especially if the pH is high (above 7). They seem to be responsible for most environmental selenium toxicity. Selenites are commonly dominant under acid (pH 4.5–6.5), poorly drained conditions, but are only slowly available since they are adsorbed by iron oxides. If added to soils to reduce selenium deficiencies they will not likely induce selenium toxicities.

Elemental selenium and selenides are quite insoluble and accumulate in wetland sediments, as do some Se-organic compounds. Some plants, in association with fungi and bacteria, absorb both organic and inorganic forms of selenium and produce volatile organics such as dimethyl selenide and dimethyl diselenide that can be released as gases to the atmosphere. These are relatively nontoxic compounds. As explained in Box 15.3, these reactions are used in a promising means of **bioremediation** to remove toxic levels of soluble selenium from soils and water.

BOX 15.3
SELENIUM—BOUND AND VAPORIZED

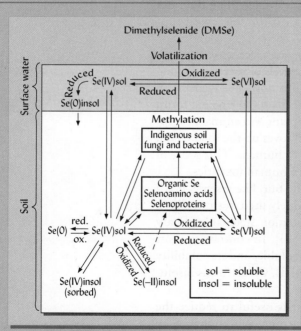

Irrigation waters carry two relatively soluble forms of selenium, selenates [Se(VI)O_4^{2-}] and selenites [Se(IV)O_3^{2-}]. When selenium first moves into the soil, some of it is reduced quickly to very insoluble elemental selenium (Se^0), which is largely unavailable to plants and is nontoxic. Further transformations take place as both soluble forms move downward into the soil (Figure 15.14). Reducing conditions favor formation of selenites, which tend to be tightly sorbed by iron oxides. Further reduction induced by microbes leads to the formation of not only elemental selenium [Se^0] but to selenides [Se^{2-}], both of which are quite insoluble. Thus, reducing conditions encourage the formation of insoluble forms, thereby lowering the toxicity of the selenium present.

As microbes and plants metabolize selenium, it is assimilated into organic forms such as selenoamino acids and selenoproteins, most of which are also quite insoluble. Certain plant species such as rice and members of the *Brassica family* (generally in association with soil fungi and bacteria) are able to attach methyl groups (methylation) to organoselenium compounds, thereby forming volatile gases such as dimethylselenide (DMSe). DMSe is 700 times less toxic than the selenates and can be dispersed into the atmosphere without any environmental damage. The process seems to work best in soils that are moist but not flooded and that are well supplied with organic materials to provide metabolic energy for these reactions. To allow continued irrigated crop production without damaging the environment, soil scientists are working to harness both pathways for selenium detoxification—the process that changes selenium to insoluble forms and the process that releases the selenium into the atmosphere.

Figure 15.14
Selenium transformation in wetland soils.

15.7 PREVENTION AND ELIMINATION OF INORGANIC CHEMICAL CONTAMINATION

Three primary methods of alleviating soil contamination by toxic inorganic compounds are (1) eliminate or drastically reduce the soil application of the toxins; (2) immobilize the toxin by means of soil management, to prevent it from moving into food or water supplies; and (3) in the case of severe contamination, remove the toxin by chemical, physical, or biological remediation.

The first method requires action to reduce unintentional aerial contamination from industrial operations and from automobile, truck, and bus exhausts. Decision makers must recognize the soil as an important natural resource that can be seriously damaged if its contamination by unintended addition of inorganic toxins is not curtailed. Also, there must be judicious reductions in intended applications to soil of the toxins through pesticides, fertilizers, irrigation water, and solid wastes.

Soil management can help reduce the continued cycling of these inorganic chemicals. This is done primarily by keeping the chemicals in the soil rather than encouraging their uptake by plants. The soil becomes a sink for the toxins, and thereby breaks the soil–plant–animal (humans) cycle through which the toxin exerts its effect. The soil breaks the cycle by immobilizing the toxins. For example, most of these elements are rendered less mobile and less available if the pH is kept near neutral or above (Figure 15.15). Liming of acid soils reduces metal mobility; hence, regulations require that the pH of sludge-treated land be maintained at 6.5 or higher.

Draining wet soils should be beneficial because the oxidized forms of the several toxic elements are generally less soluble and less available for plant uptake than are the reduced forms. However, the opposite is true for chromium. The oxidized Cr(VI) is mobile and highly toxic to humans.

Heavy phosphate applications reduce the availability of some metal cations (see Box 15.2) but may have the opposite effect on arsenic, which is found in the anionic form. Leaching may be effective in removing excess boron, although moving the toxin from the soil to water may not be of any real benefit.

Care should be taken in selecting plants to be grown on metal-contaminated soil. Generally, plants translocate much larger quantities of metals to their leaves than to their fruits or seeds. The greatest risk for food-chain contamination with metals is

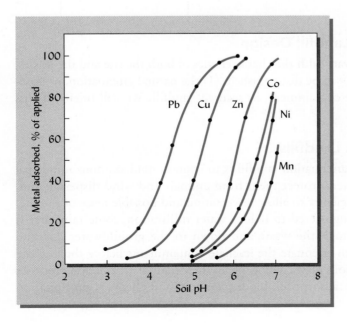

Figure 15.15

The effect of pH on the adsorption of six heavy metals. The metals were adsorbed by clay-sized goethite (an iron oxide mineral) that forms coatings on many soil particles. Maintaining the soil pH near 7 (neutral) is expected to maximize the sorption and thereby minimize the solution concentration of most heavy metals, especially of copper and lead. [Modified from Basta et al. (2005)]

therefore through leafy vegetables, such as lettuce and spinach, or through forage crops eaten by livestock.

Bioremediation by Metal Hyperaccumulating Plants

Certain plants that have evolved in soils naturally very high in metals are able to take up and accumulate extremely high concentrations of metals without suffering from toxicity. Plants have been found that accumulate more than 20,000 mg/kg nickel, 40,000 mg/kg zinc, and 1000 mg/kg cadmium. While such **hyperaccumulator** plants would pose a serious health hazard if eaten by animals or people, they may facilitate a new kind of bioremediation for metal-contaminated soils.

If sufficiently vigorously growing genotypes of such plants can be found, it may be possible to use them to remove metals from contaminated soils. For example, several plants in the genus *Thlaspi* have been grown in soils contaminated by smelter fumes. These soils are so contaminated that they are virtually barren. Accumulating nearly 40,000 mg/kg (about 4%) zinc in their tissues, the *Thlaspi* plants grown on such sites could be harvested to remove large quantities of the metals from the soil. The plant tissue is so concentrated that it could be used as an "ore" for smelting new metal. This and other bioremediation technologies for metals (e.g., the bioreduction of chromium and selenium discussed earlier) hold promise for cleaning up badly contaminated soils without resorting to expensive and destructive excavation and soil-washing methods.

A combination of chelates and phytoremediation has been used to remove lead from contaminated soil. This element is sparingly available to plants, being strongly bound by both mineral and organic matter. The chelates solubilize the lead, and plants such as Indian mustard are used to remove it.

15.8   LANDFILLS

A visit to a landfill would convince anyone of the wastefulness of modern societies. Roughly 300 million Mg of municipal wastes are generated each year by people in the United States. Most (about 70%) of this waste material is organic in nature, largely paper, cardboard, and yard wastes (e.g., grass clippings, leaves, and tree prunings). The other 30% consists mainly of such nonbiodegradeables as glass, metals, and plastic. Currently, despite an upsurge in recycling efforts, the great majority of these materials are buried in the ground.

Two Basic Types of Landfill Design

Although landfill designs vary with the characteristics of both the site and the wastes, two basic types of landfills can be distinguished: (1) the natural attenuation, or unsecured, landfill and (2) the containment, or secured, landfill. We will briefly discuss the main features of each.

Natural Attenuation Landfills

Leaking landfills (USGS): http://pubs.usgs.gov/fs/fs-040-03/

The purpose of a natural attenuation landfill is to contain nonhazardous municipal wastes in a sanitary manner, protect them from animals and wind dispersal, and, finally, to cover them sufficiently to allow revegetation and possible reuse of the site. Although the landfill is engineered to reduce water infiltration, some rainwater is allowed to percolate through the waste and down to the groundwater. Natural processes are relied upon to attenuate the leachate contaminants before the leachate reaches the groundwater. Soils play a major role in these natural attenuation processes through physical filtering, adsorption, biodegradation, and chemical precipitation (Table 15.7).

Table 15.7
SOME ORGANIC AND INORGANIC CONTAMINANTS IN UNTREATED LEACHATE FROM MUNICIPAL LANDFILLS

Range of concentrations and typical sources of the contaminants and mechanisms by which soils can attenuate the contaminants are also given. The ranges show that leachates vary greatly among landfills.

Chemical	Concentration, μg/L	Common sources	Mechanisms of attenuation
		Organics	
Dissolved organic matter, as chemical oxygen demand (COD)	140,000–150,000,000	Rotting yard wastes, paper, and garbage	Biological degradation
Benzene	0.2–1630	Adhesives, deodorants, oven cleaner, solvents, paint thinner, and medicines	Filtration, biodegradation, and methanogenesis
Trans 1,2-Dichloroethane	1.6–6500	Adhesives and degreasers	Biodegradation and dilution
Toluene	1–12,300	Glues, paint cleaners and strippers, adhesives, paints, dandruff shampoo, and carburetor cleaners	Biodegradation and dilution
Xylene	0.8–3500	Oil and fuel additives, paints, and carburetor cleaners	Biodegradation and dilution
		Metals	
Nickel	15–1300	Batteries, electrodes, and spark plugs	Adsorption and precipitation
Chromium	20–1500	Cleaners, paint, linoleum, and batteries	Precipitation, adsorption, and exchange
Cadmium	0.1–40	Paint, batteries, and plastics	Precipitation and adsorption

Leachate concentration ranges from a review of hundreds of landfills built since 1965, in Kjeldsen et al. (2002).

Soil Requirements   Finding a site with suitable soil characteristics is critical for a natural attenuation landfill. There must be at least 1.5 m of soil material between the bottom of the landfill and the highest groundwater level. This layer of soil should be only moderately permeable. If too permeable (sandy, gravelly, or highly structured), it will allow the leachate to pass through so quickly that little attenuation of contaminants will take place. The soil must have sufficient cation exchange capacity to adsorb NH_4^+, K^+, Na^+, Cd^{2+}, Ni^{2+}, and other metallic cations that the wastes are expected to release. The soil should also adsorb and retard organic contaminants long enough to allow a high degree of microbial degradation. On the other hand, if the soil is too impermeable, the leachate will build up, flood the landfill, and seep out laterally.

Daily and Final Soil Cover   The site for a natural attenuation landfill should also provide soils suitable for daily and final cover materials. At the end of every workday, the waste must be covered by a layer of relatively impermeable soil material. The final cover for the landfill is much thicker than the daily covers and includes a 60- to 100-cm-thick layer of low-permeability, clay soil material designed to minimize percolation of water into the landfill. This impermeable layer of compacted clay is usually then covered with a 30 to 45 cm layer of highly permeable medium to coarse sand. This sand layer is designed to allow water to drain laterally off the landfill to a collection area. On top

of the sand, a thinner layer of loamy "topsoil" is installed. The moderately permeable topsoil layer is meant to support a vigorous plant cover that will prevent erosion and use up water by evapotranspiration. The whole system is designed to limit the amount of water percolating through the waste, so that the amounts of contaminated leachate generated will not overwhelm the attenuating capacity of the soil between the landfill bottom and the groundwater (Plate 64).

Containment, or Secured, Landfills

The second main type of landfill is much more complex and expensive to construct, but its construction and function are much less dependent on the nature of the soils at the site. The design is intended to contain, pump, and treat all leachate from the landfill, rather than to depend on soil processes for cleansing the leachate on its way to the groundwater. To accomplish the containment, one or more impermeable liners are set in place around the sides and bottom of the landfill. These are often made of expanding clays (e.g., bentonite) that swell to a very low permeability when wet. Plastic, watertight geomembranes are also used in making the liners. The membranes are covered with a tough, nonwoven, synthetic fabric (geotextiles) and then covered with a thick layer of fine gravel or sand to protect the liner from accidental punctures. A system of slotted pipes and pumps is installed to collect all the leachate from the bottom of the landfill (Figure 15.16, *left*). The collected leachate is then treated on or off the site. The principal soil-related concerns are the requirement for suitable sources of sand and gravel, of soil for daily cover, for clayey material to form the final cover, and for topsoil to support protective vegetation.

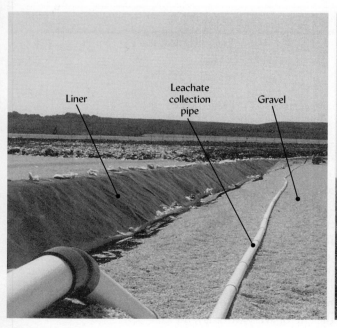

Figure 15.16
Engineered systems for the collection of leachate and gas emissions in a containment landfill. (Left) A black geomembrane liner covered with white pea gravel and a leachate collection pipe in a new cell being prepared in a containment-type landfill. The low hills in the background are completed cells blanketed with a vegetated final cover. The pollutant-laden leachate will be piped to a treatment facility. (Right) Gas wells collecting landfill gas (a mixture of mainly methane and carbon dioxide) from anaerobic decomposition in a completed landfill cell. The methane is used to fuel turbines that generate electricity that is used for the waste-disposal operation or sold to the local electric utility company. (Photos courtesy of R. Weil)

Environmental Impacts of Landfills

Today, regulations require that wastes be buried in carefully located and designed sanitary landfills. As a result, the number of landfill sites in the United States was reduced from about 16,000 in 1970 to fewer than 1700 in 2006. The remaining landfills are mostly very large, highly engineered containment-type systems. A major concern with regard to landfills is the potential water pollution from the rainwater that percolates through the wastes, dissolving and carrying away all manner of organic and inorganic contaminants (see Table 15.6). In addition to the general load of oxygen-demanding dissolved organic carbon compounds, many of the contaminants in landfill leachate are highly toxic and would create a serious pollution problem if they reached the groundwater under the landfill.

In addition to efficiency of resource use, avoidance of particular landfill management problems is another reason that the organic components of refuse (mainly paper, yard trimmings, and food waste) should be composted to produce a soil amendment rather than landfilled. First, as these materials decompose in a finished landfill, they lose volume and cause the landfill to settle and the landfill surface to subside. This physical instability severely limits the uses that can be made of the land once a landfill is completed.

Second, decomposition of the organic refuse produces undesirable liquid and gaseous products. Within a few weeks, decomposition uses up the oxygen in the landfill, and the processes of anaerobic metabolism take over, changing the cellulose in paper wastes into butyric, propionic, and other volatile organic acids, as well as hydrogen and carbon dioxide. After a month or so, methane-producing bacteria become dominant, and for several years (or even decades) a gaseous mixture of about one-third carbon dioxide and two-thirds methane (known as *landfill gas*) is generated in quantity.

The production of methane gas by the anaerobic decomposition of organic wastes in a landfill can present a very serious explosion hazard if this gas is not collected (and possibly burned as an energy source; see Figure 15.16, *right*). Where the soil is rather permeable, the gas may diffuse into basements up to several hundred meters away from the landfill. A number of fatal explosions have occurred by this process. Anaerobic decomposition in landfills also emits other harmful gases, the effects of which are less well known.

Gas problems in apartments built on a former landfill: www.eti-geochemistry.com/walnut/index.html

U.S. Environmental Protection Agency division of radiation: www.epa.gov/radiation/

15.9   RADON GAS FROM SOILS[3]

The Health Hazard

The soil is the primary source of the colorless, odorless, tasteless radioactive gas **radon**, which has been shown to cause lung cancer. Although landfills containing radioactive wastes have been known to emit radon gas at elevated concentrations, most concern regarding this potential toxin is directed toward radon that occurs naturally in soils. Therefore, radon is not usually considered a soil pollutant because it has not been introduced into the soil by human activity. Nonetheless, radon gas is thought to be a serious environmental health hazard when it moves from the soils and accumulates inside buildings. Deaths from breathing in radon are estimated at about 20,000 per year in the United States, some 10 to 50 times more numerous than deaths caused by contaminants in drinking water. The gas may be a causal factor in about 10% of lung cancer cases.

---

[3]For a straightforward discussion of the hazards of indoor radon and what you can do to protect yourself, see U.S. EPA (2005). For a review of the radon in soils and geologic material with a focus on mapping radon hazard areas, see Appleton (2007).

The health hazard from this gas stems from its transformation to radioactive polonium isotopes, which are solids that tend to attach to dust particles. The principal concern is with radon accumulation in homes, offices, and schools where people breathe the air in basement or ground-level rooms for extended periods.

How Radon Accumulates in Buildings

Geologic Factors   Radon originates from uranium (^{238}U) found in minerals, sorbed on soil colloids, or dissolved in groundwater. Over billions of years, the uranium undergoes radioactive decay forming radium, which in turn gives off radiation over thousands of years and transforms into radon. Both uranium and radium are solids; however, radon is a gas that can diffuse through pores and cracks and emerge into the atmosphere. Soils and rocks that contain high concentrations of uranium will likely produce large amounts of radon gas. Soils formed from certain highly deformed metamorphic rocks and from marine sediments, limestones, and coal or oil-bearing shales tend to have the highest potential levels for radon production. However, nearby houses built on soils formed from the same parent material may differ widely in their indoor radon concentrations (Figure 15.17), the difference being due to variations in soil properties and/or house construction.

Geology of radon and its indoor risk potential: http://energy.cr.usgs.gov/radon/georadon/4.html

Soil Properties   To become a hazard, radon must travel from its source in the underlying rock or soil, up through overlying soil layers, and finally into an enclosed building where it might accumulate to unhealthful concentrations. It must also make this trip quite rapidly, because the half-life of radon is only 3.8 days. Within several weeks, radon completely decays to polonium, lead, and bismuth, radioactive solids that last for only minutes. Whether significant quantities of radon reach a building foundation depends mainly on two factors: (1) the distance that the radon must travel from its source and (2) the permeability of the soil through which it travels. Since radon is an inert gas, the soil does not react with it, but merely serves as a channel through which the gas moves.

As explained in Sections 7.1 and 7.2, gases move through soil both by diffusion and by mass flow (convection). The rate of radon diffusion through a soil depends on the total soil porosity (more so than on pore size) and the degree to which the pores are filled with water. Radon diffuses through air-filled pores about 10,000 times faster than through water-filled pores. Movement is most rapid through sandy or gravelly

Figure 15.17

The soil that underlies a house plays a role in the movement of radon gas from its source in rock and soil minerals to the air inside the house. Dry, coarse-textured permeable soil layers allow much faster diffusion of radon gas than does a wet or fine-textured soil. If soils underlying a house are relatively impermeable, radon movement will be so slow that nearly all of the radon emitted will have decayed before it can reach the house foundation. Once it arrives at a house foundation, radon may enter through a variety of openings, such as cracks in the foundation blocks, joints between the walls and concrete floor, and gaps where utility pipes enter the house. (Diagrams courtesy of R. Weil)

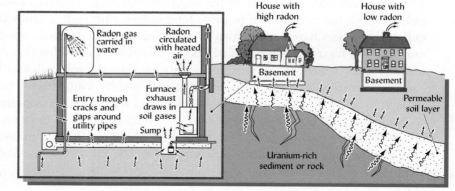

soil layers that tend to hold little water. Therefore, some of the highest indoor radon concentrations have been found in houses where only a thin, well-drained, gravelly soil separates the house foundation from uranium-rich rock. At the other extreme, a thick, wet clay layer would provide an excellent barrier against radon diffusion. Convective airflow, which is mainly stimulated by rainwater entering the soil and by changes in atmospheric pressure, may play a significant role in radon movement during stormy weather.

Radon Testing and Remediation

Testing   Since the occurrence of high radon levels cannot be accurately predicted, the only sure way to determine the risk of radon is to test for its presence. Testing is usually carried out in two stages. The first uses an inexpensive charcoal canister, which is placed in the test area, unsealed, and left to absorb radon for the specified period (usually 3 days). If the results suggest a radon level above 4 piCu/L (148 Bq/m^3), the U.S. EPA advises that a long-term test be conducted, using a somewhat more expensive alpha-track detector for a period of 3 to 12 months. If the long-term test also suggests levels above 4 piCi/L, modifications should be made to the building to reduce the accumulation of radon inside.

Remediation   Depending on the levels of radon and the condition of the building, the modification may be as simple as caulking cracks in the floor and walls and filling gaps around utility-pipe entrances. Remediation of higher radon levels may require alterations that are more extensive. Ventilation of the room with outside air can prevent unhealthful radon buildup, but a more energy-efficient solution is a sub-slab ventilation system. For the latter, perforated pipes are installed in a layer of gravel under the foundation slab, and the air pressure there is lowered either by a mechanical fan or by convective draw from a special chimney. In this way, gas coming from the soil is intercepted and redirected to the atmosphere before it can enter the building. Installation of a subslab ventilation system is much less expensive during new construction than as a retrofit, and is now standard practice in many areas with high-uranium soils.

15.10   CONCLUSION

Three major conclusions may be drawn about soils in relation to environmental quality. First, since soils are valuable resources, they should be protected from environmental contamination, especially that which does permanent damage. Second, because of their vastness and remarkable capacities to absorb, bind, and break down added materials, soils offer promising mechanisms for the disposal and utilization of many wastes that otherwise may contaminate the environment. Third, soil contaminants and the products of their breakdown in soil reactions can be toxic to humans and other animals if the soil is ingested or the contaminants move from the soil into plants, soil fauna, the air, and—particularly—into water supplies.

To gain a better understanding of how soils might be used and yet protected in waste-management efforts, soil scientists devote a considerable share of their research efforts to environmental-quality problems. Furthermore, soil scientists have much to contribute to the research teams that search for better ways to clean up environmental contamination. Some of the most promising technological advances have been in the field of bioremediation, in which the biological processes of the soil are harnessed to effect soil cleanup. Soil mapping and taxonomy (as discussed in Chapters 2 and 3) make essential contributions to finding appropriate sites where soils can be safely used to clean up or store hazardous materials.

STUDY QUESTIONS

1. What agricultural practices contribute to soil and water pollution, and what steps must be taken to reduce or eliminate such pollution?

2. Discuss the types of reactions pesticides undergo in soils, and indicate what we can do to encourage or prevent such reactions.

3. Discuss the environmental problems associated with the disposal of large quantities of sewage sludge on agricultural lands, and indicate how the problems could be alleviated.

4. What is *bioremediation*, and what are its advantages and disadvantages compared with physical and chemical methods of handling organic wastes?

5. Even though large quantities of the so-called heavy metals are applied to soils each year, relatively small quantities find their way into human food. Why?

6. Compare the design, management, and role of soil in today's containment landfills with the natural attenuation type most common 30 years ago, and indicate how the changes affect soil and water pollution.

7. What are *organoclays*, and how can they be used to help remediate soils polluted with nonpolar organic compounds?

8. Soil organic matter and some silicate clays chemically sorb some organic pollutants and protect them from microbial attack and leaching from the soil. What are the implications (positive and negative) of such protection for efforts to reduce soil and water pollution?

9. What are the comparative advantages and disadvantages of *in situ* and *ex situ* means of remediating soils polluted with organic compounds?

10. What are two approaches to *phytoremediation*, and for what kinds of pollutants are they useful? Explain.

11. Suppose a nickel-contaminated soil 15 cm deep contained 800 mg/kg Ni. Vegetation was planted to remove the nickel by phytoremediation. The above-ground plant parts average 1% Ni on a dry-weight basis and produce 4000 kg/ha of harvestable dry matter. If two harvests are possible per year, how many years will it take to reduce the Ni level in the soil to a target of 80 mg/kg?

12. Suppose you moved into a townhouse in an old inner-city neighborhood. How would you minimize lead toxicity to your baby?

REFERENCES

Adriano, D. C. 2001. *Trace Elements in Terrestrial Environments: Biogeochemistry, Bioavailability, and Risks of Metals* (New York: Springer).

Ahmad, I., S. Hayat, and J. Pichtel (eds.). 2006. *Heavy Metal Contamination of Soil: Problems and Remedies* (Enfield, NH: Science Publishers, Inc.).

Appleton, J. D. 2007. "Radon: Sources, health risks, and hazard mapping," *AMBIO: A Journal of the Human Environment,* **36**:85–89.

Basta, N. T., J. A. Ryan, and R. L. Chaney. 2005. "Trace element chemistry in residual-treated soil: Key concepts and metal bioavailability," *J. Environ. Qual.,* **34**:49–63.

Bragg, J. R., R. C. Prince, E. J. Harner, and R. M. Atlas. 1994. "Effectiveness of bioremediation for the *Exxon Valdez* oil spill," *Nature,* **368**:413–418.

Dowdy, R. H., C. E. Clapp, D. R. Linden, W. E. Larson, T. R. Halbach, and R. C. Polta. 1994. "Twenty years of trace metal partitioning on the Rosemount sewage sludge watershed," pp. 149–155, in C. E. Clapp, W. E. Larson, and R. H. Dowdy (eds.), *Sewage Sludge: Land Utilization and the Environment* (Madison, WI: Soil Science Society of America).

Eccles, H. 2007. *Bioremediation* (New York: Taylor & Francis).

Gaynor, J. D., D. C. MacTavish, and W. I. Findlay. 1995. "Atrazine and metolachlor loss in surface and subsurface runoff from three tillage treatments in corn," *J. Environ. Qual.,* **24**:246–256.

Gilliom, R. J., J. E. Barbash, C. G. Crawford, P. A. Hamilton, J. D. Martin, N. Nakagaki, L. H. Nowell, J. C. Scott, P. E. Stackelberg, G. P. Thelin, and D. M. Wolock. 2006. *The Quality of Our Nation's Waters: Pesticides in the Nation's Streams and Ground Water, 1992–2001,* USGS Circular 1291 (Reston, VA: U.S. Geological Survey). http://pubs.usgs.gov/circ/2005/1291/

Hazen, Terry C. 1995. "Savannah river site—A test bed for cleanup technologies," *Environ. Protection* (April):10–16.

Kabata-Pendias, A., and H. Pendias. 1992. *Trace Elements in Soils and Plants* (Boca Raton, FL: CRC Press).

Kjeldsen, P., M. Barlaz, A. Rooker, A. Baun, A. Ledin, and T. Christensen. 2002. "Present and long-term composition of MSW landfill leachate: A review," *Critical Reviews in Environmental Science and Technology*, **32**:297–336.

Pimental, D., H. Acquay, M. Biltonen, P. Rice, M. Silva, J. Nelson, V. Lipner, S. Giordano, A. Horowitz, and M. D'Amore. 1992. "Environmental and economic costs of pesticide use," *Bioscience*, **42**:750–760.

Pritchard, P. H., J. G. Mueller, J. C. Rogers, F. V. Kremer, and J. A. Glaser. 1992. "Oil spill bioremediation: Experiences, lessons and results from the *Exxon Valdez* oil spill in Alaska," *Biodegradation*, **3**:315–335.

Reynolds, C. M., D. C. Wolf, T. J. Gentry, L. B. Perry, C. S. Pidgeon, B. A. Koenen, H. B. Rogers, and C. A. Beyrouty. 1999. "Plant enhancement of indigenous soil microorganisms: A low cost treatment of contaminated soils," *Polar Record*, **35**(192):33–40.

Ryan, J. A., K. G. Scheckel, W. R. Berti, S. L. Brown, S. W. Casteel, R. L. Chaney, J. Hallfrisch, M. Doolan, P. Grevatt, M. Maddaloni, and D. Mosby. 2004. "Reducing children's risk from lead in soil," *Environ. Sci. Technol.*, **38**:18A–24A.

U.S. EPA. 1993. *Clean Water Act*, sec. 503, vol. 58, no. 32 (Washington, DC: U.S. Environmental Protection Agency).

U.S. EPA. 2005. *A Citizen's Guide to Radon: The Guide to Protecting Yourself and Your Family from Radon*. U.S. EPA 402-K-02-006, Revised. (Washington, DC: U.S. Environmental Protection Agency). www.epa.gov/radon/pubs/citguide.html#howdoes

Weber, J. B., and C. T. Miller. 1989. "Organic chemical movement over and through soil," in B. L. Sawhney and K. Prown (eds.), *Reactions and Movement of Organic Chemicals in Soils*. SSSA Special Publication no. 22 (Madison, WI: Soil Science Society of America).

Wise, D. L., D. J. Trantolo, E. J. Cichon, H. I. Inyang, and U. Stottmeister (eds.). 2000. *Bioremediation of Contaminated Soils* (New York: Marcel Dekker).

Xu, S., G. Sheng, and S. A. Boyd. 1997. "Use of organoclays in pollution abatement," *Advances in Agronomy*, **59**:25–62.

Zamiska, N., and J. Spencer. 2007. "China faces new worry: Heavy metals in the food," *Wall Street Journal*, July 2, 2007, p. A1.

appendix A

World Reference Base, Canadian, and Australian Soil Classification Systems

Table A.1

SOIL REFERENCE GROUPS IN THE WORLD REFERENCE BASE (WRB) FOR SOIL RESOURCES[a]

The World Reference Base provides a global vocabulary for communicating about different kinds of soils and a reference by which various national soil classification systems (such as U.S. Soil Taxonomy and the Canadian Soil Classification System discussed in this text) can be compared and correlated. The 32 Reference Soil Groups are differentiated mainly according to the primary pedogenesis process that has produced the characteristic soil features, except where "special" soil parent materials are of overriding importance. Each Reference Soil Group can be subdivided using a unique list of possible prefix and suffix qualifiers (not shown here[b]). These qualifiers indicate secondary soil-forming processes that significantly affected the primary soil characteristics especially important to soil use. To avoid making the classification of soils dependent on the availability of climatic data, separations are not based on specific climatic characteristics (as is the case in U.S. Soil Taxonomy).

Reference Soil Group[c]	Major Soil Characteristics	Approximate Equivalents in U.S. Soil Taxonomy[d]
Organic Soils		
Histosols (HS)	Composed of organic materials	Most Histosols and Histels
Mineral Soils Dominantly Influenced by Human Activity		
Anthrosols (AT)	Soils with long and intensive agriculture use	Anthrepts and Anthropic, Plaggic great groups and subgroups
Technosols (TC)	Soils containing many artifacts	Entisols, such *proposed* suborders as Urbents, Garbents
Soils with Limited Rooting Due to Shallow Permafrost or Stoniness		
Cryosols (CR)	Ice-affected soils: Cryosols	Gelisols
Leptosols (LP)	Shallow or extremely gravelly soils	Lithic subgroups of Inceptisols and Entisols
Soils Influenced by Water		
Vertisols (VR)	Alternating wet-dry conditions, rich in swelling clays	Vertisols
Fluvisols (FL)	Young soils in alluvial deposits	Fluvents and Fluvaquents
Solonchaks (SC)	Strongly saline soils	Salids and salic or halic great groups of other orders
Solonetz (SN)	Soils with subsurface clay accumulation, rich in sodium	Natric great groups of Alfisols, Aridisols, and Mollisols
Gleysols (GL)	Groundwater-affected soils	Endoaquic great groups (e.g., Endoaqualfs, Endoaquolls, Endoaquults, Endoaquents, Endoaquepts)
Soils in Which Aluminium (Al) Chemistry Plays a Major Role in Their Formation		
Andosols (AN)	Young soils from volcanic ash and tuff deposits	Andisols
Podzols (PZ)	Acid soils with a subsurface accumulation of iron-aluminium-organic compounds	Spodosols
Plinthosols (PT)	Wet soils with an irreversibly hardening mixture of iron, clay, and quartz in the subsoil	Plinthic great groups of Aqualfs, Aquox, and Ultisols

Table A.1   (Continued)

Reference Soil Group[c]	Major Soil Characteristics	Approximate Equivalents in U.S. Soil Taxonomy[d]
Soils in Which Aluminium (Al) Chemistry Plays a Major Role in Their Formation (Continued)		
Nitisols (NT)	Deep, dark red, brown, or yellow clayey soils having a pronounced shiny, nut-shaped structure	Parasesquic Inceptisols and some Oxisols and Ultisols
Ferralsols (FR)	Deep, strongly weathered soils with a chemically poor, but physically stable subsoil	Oxisols
Soils with Stagnant Water		
Planosols (PL)	Soils with a bleached, temporarily water-saturated topsoil on a slowly permeable subsoil	Albaqualfs and Albaquults and some albaquic subgroups of Alfisols and Ultisols
Stagnosols (ST)	Soil with temporarily water-saturated topsoil on structural or moderate textural discontinuity	Epiaquic great groups
Mineral Soils with Humus-Rich Topsoils and a High Base Saturation Typically in Grasslands		
Chernozems (CH)	Soils with a thick, dark topsoil, rich in organic matter with a calcareous subsoil	Calciudolls
Kastanozems (KS)	Soils with a thick, dark brown topsoil, rich in organic matter and a calcareous or gypsum-rich subsoil	Many Calciustolls and Calcixerolls
Phaeozems (PH)	Soils with a thick, dark topsoil, rich in organic matter and evidence of removal of carbonates	Many Cryolls, Udolls, and Albolls
Soils with Accumulation of Nonsaline Substances and Influenced by Aridity		
Gypsisols (GY)	Soils with accumulation of secondary gypsum	Gypsids and some Gypsic great groups of other orders
Durisols (DU)	Soils with accumulation of secondary silica	Durids and some Duric great groups of other orders
Calcisols (CL)	Soils with accumulation of secondary calcium carbonates	Calcids and Calcic great groups of Inceptisols
Mineral Soils with a Clay-Enriched Subsoil		
Albeluvisols (AB)	Acid soils with a bleached horizon penetrating into a clay-rich subsurface horizon	Some Glossudalfs
Alisols (AL)	Soils with subsurface accumulation of high activity clays, and low base saturation	Ultisols and Ultic Alfisols
Acrisols (AC)	Soils with subsurface accumulation of low activity clays and low base saturation	Kandic great groups of Alfisols and Ultisols
Luvisols (LV)	Soils with subsurface accumulation of high activity clays and high base saturation	Haplo and pale great groups of Alfisols
Lixisols (LX)	Soils with subsurface accumulation of low activity clays and high base saturation	Kandic great groups of Alfisols with high base saturation
Relatively Young Soils or Soils with Little or No Profile Development		
Umbrisols (UM)	Acid soils with a thick, dark topsoil, rich in organic matter	Many umbric great groups of Inceptisols
Arenosols (AR)	Very sandy soils featuring no or only very weak B horizon development	Psamments, grossarenic subgroups of other orders
Cambisols (CM)	Soils with only weakly to moderately developed B horizons	Cambids and many Inceptisols
Regosols (RG)	Soils with very limited soil development, often shallow to rock	Orthents, some Psamments and other Entisols

[a]Based on FAO (2006). World reference base for soil resources 2006: A framework for international classification correlation and communication. World Soil Resources Reports 103. Food and Agriculture Organization of the United Nations and United Nations Environmental Program, Rome. 128 pp. and on personal communication from Bob Engel (USDA/NRCS) and Michéli Erika (Univ. Agric. Sci., Hungary).
[b]As one example, the Kastanozems reference group can be subdivided using the prefix modifiers Vertic, Gypsic, Calcic, Luvic, Hyposodic, Siltic, Chromic, Anthric, and Haplic.
[c]Abbreviations (often used as symbols on soil maps) in parentheses.
[d]As discussed in Chapter 3 of this text.

Table A.2
THE AUSTRALIAN SOIL CLASSIFICATION SYSTEM AND APPROXIMATE CORRELATIONS WITH U.S. SOIL TAXONOMY

Order	Main Characteristics	Soil Taxonomy Order and Suborders
Anthroposols	"Human-made" soils	Some are Entisols (e.g., proposed Urbents, Spoilents)
Calcarosols	B horizon calcareous and lacking a marked clay accumulation	Aridisols, Alfisols (Ustalfs, Xeralfs)
Chromosols	Strong clay accumulation and pH > 5.5 in B horizon	Alfisols, some Aridisols
Dermosols	B horizon well structured but lacking a marked clay accumulation	Mollisols, Alfisols, Ultisols
Ferrosols	B horizon high in Fe and lacking a marked clay accumulation	Oxisols, some Alfisols
Hydrosols	Prolonged seasonal water saturation	Aquic subgroups of Alfisols, Ultisols, Inceptisols, Salic Aridisols, and some Histosols
Kandosols	B horizon massive and lacking a marked clay accumulation	Alfisols, Ultisols, and Aridisols with massive B horizon structure
Kurosols	Strong clay accumulation and pH < 5.5 in B horizon	Ultisols, some Alfisols
Organosols	Organic materials	Histosols
Podosols	Acid soils with subsurface accumulation of Fe, Al-organic compounds	Spodosols, some Entisols
Rudosols	Negligible (rudimentary) horizon differentiation	Entisols, Salic Aridisols
Sodosols	Strong clay accumulation in B horizon, with high sodium saturation	Natric subgroups of Alfisols, Aridisols
Tenosols	Weak horizon differentiation	Inceptisols, Aridisols, Entisols
Vertosols	High clay (>35%), deep cracks, slickensides	Vertisols

Modified from CSIRO Land and Water (2003): www.clw.csiro.au/aclep/asc_re_on_line/soilhome.htm

Table A.3
SUMMARY WITH BRIEF DESCRIPTIONS OF THE SOIL ORDERS IN THE SOIL CLASSIFICATION SYSTEM OF CANADA

The Canadian Soil Classification System is one of many national soil classification systems used in various countries around the world. Of these, it is perhaps the most closely aligned with the U.S. Soil Taxonomy. It includes five hierarchical categories: order, great group, subgroup, family, and series. The system is designed to apply principally to the soils of Canada. The soil orders of the Canadian System of Soil Classification are described in this Table and soil orders and some great groups are compared to the U.S. Soil Taxonomy in Table A.4. Videos with field scenes and discussions for each order in the Canadian Soil Classification system are available at: http://projects.oltubc.com/SOIL/HS.htm

Brunisolic	Soils with sufficient development to exclude them from the Regosolic order, but lacking the degree or kind of horizon development specified for other soil orders.
Chernozemic	Soils with high base saturation and surface horizons darkened by the accumulation of organic matter from the decomposition of plants from grassland or grassland-forest ecosystems.
Cryosolic	Soils formed in either mineral or organic materials that have permafrost either within 1 m of the surface or within 2 m if more than one-third of the pedon has been strongly cryoturbated, as indicated by disrupted, mixed, or broken horizons.
Gleysolic	Gleysolic soils have features indicative of periodic or prolonged saturation (i.e., gleying, mottling) with water and reducing conditions.

Table A.3 (CONTINUED)

Luvisolic
Soils with light-colored, eluvial horizons that have illuvial B horizons in which silicate clay has accumulated.

Organic
Organic soils developed on well- to undecomposed peat or leaf litter.

Podzolic
Soils with a B horizon in which the dominant accumulation product is amorphous material composed mainly of humified organic matter combined in varying degrees with Al and Fe.

Regosolic
Weakly developed soils that lack development of genetic horizons.

Solonetzic
Soils that occur on saline (often high in sodium) parent materials, which have B horizons that are very hard when dry and swell to a sticky mass of very low permeability when wet. Typically the solonetzic B horizon has prismatic or columnar macrostructure that breaks to hard to extremely hard, blocky peds with dark coatings.

Vertisolic
Soils with high contents of expanding clays that have large cracks during the dry parts of the year and show evidence of swelling, such as gilgae and slickensides.

Table A.4
COMPARISON OF U.S. SOIL TAXONOMY AND THE CANADIAN SOIL CLASSIFICATION SYSTEM
Note that because the boundary criteria differ between the two systems, certain U.S. Soil Taxonomy soil orders have equivalent members in more than one Canadian Soil Classification System soil order.[a]

U.S. Soil Taxonomy soil order	Canadian system soil order	Canadian system great group	Equivalent lower-level taxa in U.S. Soil Taxonomy
Alfisols	Luvisolic	Gray Brown Luvisols	Hapludalfs
		Gray Luvisols	Haplocryalfs, Eutrocryalfs, Fragudalfs, Glossocryalfs, Palecryalfs, and some subgroups of Ustalfs and Udalfs
	Solonetzic	Solonetz	Natrudalfs and Natrustalfs
		Solod	Glossic subgroups of Natraqualfs, Natrudalfs, and Natrustalfs
Andisols	Components of Brunisolic and Cryosolic		
Aridisols	Solonetzic		Frigid families of Natrargids
Entisols	Regosolic		Cryic great groups and frigid families of Entisols, except Aquents
		Regosol	Cryic great groups and frigid families of Folists, Fluvents, Orthents, and Psamments
Gelisols	Cryosolic	Turbic Cryosol	Turbels
		Organic Cryosol	Histels
		Stagnic Cryosol	Orthels
Histosols	Organic	Fibrisol	Cryofibrists, Sphagnofibrists
		Mesisol	Cryohemists
		Humisol	Cryosaprists
Inceptisols	Brunisolic	Melanic Brunisol	Some Eutrustepts
		Eutric Brunisol	Subgroups of Cryepts; frigid and mesic families of Haplustepts
		Sombric Brunisol	Frigid and mesic families of Udepts, and Ustept and Humic Dystrudepts
		Dystric Brunisol	Frigid families of Dystrudepts and Dystrocryepts
	Gleysolic		Cryic subgroups and frigid families of Aqualfs, Aquolls, Aquepts, Aquents, and Aquods
		Humic Gleysol	Humaquepts
		Gleysol	Cryaquepts and frigid families of Fragaquepts, Epiaquepts, and Endoaquepts

(continued)

Table A.4   (CONTINUED)

U.S. Soil Taxonomy soil order	Canadian system soil order	Canadian system great group	Equivalent lower-level taxa in U.S. Soil Taxonomy
Mollisols	Chernozemic	Brown	Xeric and Ustic subgroups of Argicryolls and Haplocryolls
		Dark Brown	Subgroups of Argicryolls and Haplocryolls
		Black	Typic subgroups of Argicryolls and Haplocryolls
		Dark Gray	Alfic subgroups of Argicryolls
	Solonetzic	Solonetz	Natricryolls and frigid families of Natraquolls and Natralbolls
		Solod	Glossic subgroups of Natricryolls
Oxisols	Not relevant in Canada		
Spodosols	Podzolic	Humic Podzol	Humicryods, Humic Placocryods, Placohumods, and frigid families of other Humods
		Ferro-Humic Podzol	Humic Haplocryods, some Placorthods, and frigid families of humic subgroups of other Orthods
		Humo-Ferric Podzol	Haplorthods, Placorthods, and frigid families of other Orthods and Cryods except humic subgroups
Ultisols	Not relevant in Canada		
Vertisols	Vertisolic	Vertisol	Haplocryerts
		Humic Vertisol	Humicryerts

[a]Based on information in Soil Classification Working Group, 1998, *The Canadian System of Soil Classification*, 3rd ed. (Ottawa: Agriculture and Agri-Food Canada). Publication No. A53-1646/1997E.

appendix B
SI Units, Conversion Factors, Periodic Table of the Elements, and Plant Names

---

BASIC SI UNITS OF MEASUREMENT

Parameter	Basic unit	Symbol
Amount of substance	mole	mol
Electrical current	ampere	A
Length	meter	m
Luminous intensity	candela	cd
Mass	gram (kilogram)	g (kg)
Temperature	kelvin	K
Time	second	s

PREFIXES USED TO INDICATE ORDER OF MAGNITUDE

Prefix	Multiple	Abbreviation	Multiplication factor
exa	10^{18}	E	1,000,000,000,000,000,000
peta	10^{15}	P	1,000,000,000,000,000
tera	10^{12}	T	1,000,000,000,000
giga	10^{9}	G	1,000,000,000
mega	10^{6}	M	1,000,000
kilo	10^{3}	k	1,000
hecto	10^{2}	h	100
deca	10	da	10
deci	10^{-1}	d	0.1
centi	10^{-2}	c	0.01
milli	10^{-3}	m	0.001
micro	10^{-6}	μ	0.000 001
nano	10^{-9}	n	0.000 000 001
pico	10^{-12}	P	0.000 000 000 001
femto	10^{-15}	f	0.000 000 000 000 001
atto	10^{-18}	a	0.000 000 000 000 000 001

FACTORS FOR CONVERTING NON-SI UNITS TO SI UNITS

Non-SI Unit	Multiply by[a]	To obtain SI Unit
Length		
inch, in.	2.54	centimeter, cm (10^{-2} m)
foot, ft	0.304	meter, m
mile,	1.609	kilometer, km (10^3 m)
micron, µ	1.0	micrometer, µm (10^{-6} m)
Ångstrom unit, Å	0.1	nanometer, nm (10^{-9} m)
Area		
acre, ac	0.405	hectare, ha (10^4 m^2)
square foot, ft^2	9.29×10^{-2}	square meter, m^2
square inch, in^2	645	square millimeter, mm^2
square mile, mi^2	2.59	square kilometer, km^2
Volume		
bushel, bu	35.24	liter, L
cubic foot, ft^3	2.83×10^{-2}	cubic meter, m^3
cubic inch, in^3	1.64×10^{-5}	cubic meter, m^3
gallon (U.S.), gal	3.78	liter, L
quart, qt	0.946	liter, L
acre-foot, ac-ft	12.33	hectare-centimeter, ha-cm
acre-inch, ac-in	1.03×10^{-2}	hectare-meter, ha-m
ounce (fluid), oz	2.96×10^{-2}	liter, L
pint, pt	0.473	liter, L
Mass		
ounce (avdp), oz	28.4	gram, g
pound, lb	0.454	kilogram, kg (10^3 g)
ton (2000 lb)	0.907	megagram, Mg (10^6 g)
tonne (metric), t	1000	kilogram, kg
Radioactivity		
curie, Ci	3.7×10^{10}	becquerel, Bq
picocurie per gram, pCi/g	37	becquerel per kilogram, Bq/kg
Yield and Rate		
pound per acre, lb/ac	1.121	kilogram per hectare, kg/ha
pounds per 1000 ft^2	48.8	kilogram per hectare, kg/ha
bushel per acre (60 lb), bu/ac	67.19	kilogram per hectare, kg/ha
bushel per acre (56 lb), bu/ac	62.71	kilogram per hectare, kg/ha
bushel per acre (48 lb), bu/ac	53.75	kilogram per hectare, kg/ha
gallon per acre (U.S.), gal/ac	9.35	liter per hectare, L/ha
ton (2000 lb) per acre	2.24	megagram per hectare, Mg/ha
miles per hour, mph	0.447	meter per second, m/s
gallon per minute (U.S.), gpm	0.227	cubic meter per hour, m^3/h
cubic feet per second, cfs	101.9	cubic meter per hour, m^3/h
Pressure		
atmosphere, atm	0.101	megapascal, MPa (10^6 Pa)
bar	0.1	megapascal, MPa
pound per square foot, lb/ft^2	47.9	Pascal, Pa
pound per square inch, lb/in^2	6.9×10^3	Pascal, Pa
Temperature		
degrees Fahrenheit (°F − 32)	0.556	degrees, °C
degrees Celsius (°C + 273)	1	Kelvin, K
Energy		
British thermal unit, BTU	1.05×10^3	joule, J
calorie, cal	4.19	joule, J
dyne, dyn	10^{-5}	Newton, N
erg	10^{-7}	joule, J
foot-pound, ft-lb	1.36	joule, J
Concentrations		
percent, %	10	gram per kilogram, g/kg
part per million, ppm	1	milligram per kilogram, mg/kg
milliequivalents per 100 grams	1	centimole per kilogram, cmol/kg

[a]To convert from SI to non-SI units, *divide* by the factor given.

PERIODIC TABLE OF THE ELEMENTS WITH NOTES CONCERNING RELEVANCE TO SOIL SCIENCE

Based on atomic mass of $^{12}C = 12.0$. Numbers in parentheses are the mass numbers of the most stable isotopes of radioactive elements.

Group IA	Group IIA	Group IIIB	Group IVB	Group VB	Group VIB	Group VIIB	Group VIIIB			Group IB	Group IIB	Group IIIA	Group IVA	Group VA	Group VIA	Group VIIA	Group VIIIA
1 H 1.01 Hydrogen																	2 He 4.00 Helium
3 Li 6.94 Lithium	4 Be 9.01 Beryllium											5 B 10.81 Boron	6 C 12.01 Carbon	7 N 14.01 Nitrogen	8 O 16.00 Oxygen	9 F 19.00 Fluorine	10 Ne 20.18 Neon
11 Na 22.99 Sodium	12 Mg 24.30 Magnesium											13 Al 26.98 Aluminum	14 Si 28.09 Silicon	15 P 30.97 Phosphorus	16 S 32.07 Sulfur	17 Cl 35.45 Chlorine	18 Ar 39.95 Argon
19 K 39.10 Potassium	20 Ca 40.08 Calcium	21 Sc 44.96 Scandium	22 Ti 47.88 Titanium	23 V 50.94 Vanadium	24 Cr 52.00 Chromium	25 Mn 54.94 Manganese	26 Fe 55.85 Iron	27 Co 58.93 Cobalt	28 Ni 58.69 Nickel	29 Cu 63.55 Copper	30 Zn 65.38 Zinc	31 Ga 69.72 Gallium	32 Ge 72.59 Germanium	33 As 74.92 Arsenic	34 Se 78.96 Selenium	35 Br 79.90 Bromine	36 Kr 83.80 Krypton
37 Rb 85.47 Rubidium	38 Sr 87.62 Strontium	39 Y 88.91 Yttrium	40 Zr 91.22 Zirconium	41 Nb 92.91 Niobium	42 Mo 95.94 Molybdenum	43 Tc (98) Technetium	44 Ru 101.07 Ruthenium	45 Rh 102.91 Rhodium	46 Pd 106.42 Palladium	47 Ag 107.87 Silver	48 Cd 112.41 Cadmium	49 In 114.82 Indium	50 Sn 118.71 Tin	51 Sb 121.75 Antimony	52 Te 127.60 Tellurium	53 I 126.90 Iodine	54 Xe 131.29 Xenon
55 Cs 132.91 Cesium	56 Ba 137.33 Barium	57 La 138.91 Lanthanum	72 Hf 178.49 Hafnium	73 Ta 180.95 Tantalum	74 W 183.85 Tungsten	75 Re 186.21 Rhenium	76 Os 190.2 Osmium	77 Ir 192.22 Iridium	78 Pt 195.08 Platinum	79 Au 196.97 Gold	80 Hg 200.59 Mercury	81 Tl 204.38 Thallium	82 Pb 207.2 Lead	83 Bi 208.98 Bismuth	84 Po (209) Polonium	85 At (210) Astatine	86 Rn (222) Radon
87 Fr (223) Francium	88 Ra (226) Radium	89 Ac (227) Actinium	104 Unq (261) Unnilquadium	105 Unp (262) Unnilpentium	106 Unh (263) Unnilhexium	107 Uns (262) Unnilseptium	108 Uno (265) Unniloctium										

58 Ce 140.12 Cerium	59 Pr 140.91 Praseodymium	60 Nd 144.24 Neodymium	61 Pm (145) Promethium	62 Sm 150.36 Samarium	63 Eu 151.96 Europium	64 Gd 157.25 Gadolinium	65 Tb 158.93 Terbium	66 Dy 162.50 Dysprosium	67 Ho 164.93 Holmium	68 Er 167.26 Erbium	69 Tm 168.93 Thulium	70 Yb 173.04 Ytterbium	71 Lu 174.97 Lutetium
90 Th (232) Thorium	91 Pa (231) Protactinium	92 U (238) Uranium	93 Np (237) Neptunium	94 Pu (244) Plutonium	95 Am (243) Americium	96 Cm (247) Curium	97 Bk (247) Berkelium	98 Cf (251) Californium	99 Es (252) Einsteinium	100 Fm (257) Fermium	101 Md (258) Mendelevium	102 No (259) Nobelium	103 Lr (260) Lawrencium

Metals ↔ Nonmetals

Atomic number — Symbol — Atomic mass
87 Fr (223) Francium

Elements known to be nutrients for animals or plants. Some are also toxic in excessive amounts.

Elements toxic to organisms in small amounts, and not known to serve as nutrients.

Other elements commonly studied in Soil Science because of soil-environmental impacts or because of their use as tracers or electrodes. (Br is used to trace anionic solutes such as nitrate. Isotopes of Rb and Sr are used to trace K and Ca in plants and soils. Cs and Ti are used to trace geological processes such as soil erosion. Pt and Ag are used in electrodes for measuring soil redox potential and pH, respectively.)

These 22 elements are needed as essential mineral nutrients by humans: macronutrients (calcium, chloride, magnesium, phosphorus, potassium, sodium, and sulfur) and micronutrients (chromium, cobalt, copper, fluoride, iodine, iron, manganese, molybdenum, nickel, selenium, silicon, tin, vanadium, and zinc).

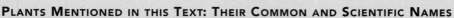

PLANTS MENTIONED IN THIS TEXT: THEIR COMMON AND SCIENTIFIC NAMES

acacia, apple ring	*Faidherbia albida* (Del.) A. Chev. [syn. *Acacia albida*]	carrot	*Daucus carota* L. ssp. *sativus* (Hoffm.) Arcang.
acacia, catclaw	*Acacia greggii* Gray	cassava	*Manihot esculenta* Crantz
alder	*Alnus spp.* P. Mill.	casuarina (sheoak)	*Casuarina spp.* Rumph. ex L.
alder, red	*Alnus rubra* Bong.	cattail, common	*Typa latifolia* L.
alfalfa	*Medicago sativa* L.	cauliflower	*Brassica oleracea* L. (Botrytis group)
alkali grass, Nutall's	*Puccinellia nuttalliana* (J.A. Schultes) A.S. Hitchc.	ceanothus	*Ceanothus spp.* L.
alkali sacaton	*Sporobolus airoides* (Torr.) Torr.	celery	*Apium graveolens* L. var. dulce (Mill.) Pers.
almond	*Prunus dulcis* (P. Mill.) D.A. Webber	cherry, flowering	*Prunus serrulata* Lindl.
andromeda (bog rosemary)	*Andromeda polifolia* L.	citrus	*Citrus spp.* L.
apple	*Malus spp.* P. Mill.	clover, alsike	*Trifolium hybridum* L.
apricot	*Prunus armeniaca* L.	clover, berseem	*Trifolium alexandrinum* L.
arborvitae	*Thuja occidentalis* L.	clover, crimson	*Trifolium incarnatum* L.
ash	*Fraxinus spp.* L.	clover, ladino	*Trifolium repens* L.
ash, white	*Fraxinus americana* L.	clover, red	*Trifolium pratense* L.
asparagus	*Asparagus officinalis* L.	clover, strawberry	*Trifolium fragiferum* L.
aspen	*Populus spp.* L.	clover, sweet	*Melilotus indica* All
aspen, quaking	*Populus tremuloides* Michx.	clover, white	*Trifolium repens* L.
autumn olive	*Elaeagnus umbellata* Thunb.	coffee	*Coffea spp.* L.
azalea	*Rhododendron spp.* L.	corn	*Zea mays* L.
azolla	*Azolla spp.* L.	cotton	*Gossypium hirsutum* L.
bahia grass	*Paspalum notatum* Flueggé	cottonwood	*Populus deltoidies* Bartr. Ex. Marsh
banana	*Musa acuminata* Colla	cowpea	*Vigna unguiculata* (L.) Walp.
barley, forage	*Hordeum vulgare* L.	cranberry	*Vaccinium macrocarpon* Ait.
bean, broad (faba)	*Vicia faba* L.	cranberry, small	*Vaccinium oxycoccos* L.
bean, common	*Phaseolus vulgaris* L.	cucumber	*Cucumis sativus* L.
bean, winged	*Psophocarpus tetragonobus* L. D.C.	currant	*Ribes spp.* L.
beech, American	*Fagus grandifolia* Ehrh.	cypress, bald	*Taxodium distichum* (L.) L.C. Rich.
beet, garden	*Beta procumbens* L.	dallisgrass	*Paspalum dilatatum* Poir
beet, sugar	*Beta vulgaris* L.	dogwood	*Cornus spp.* L.
bentgrass	*Agrostis stolonifera* L.	dogwood, grey	*Cornus racemosa* Lam.
bermudagrass	*Cynodon dactylon* (L.) Pers.	elaeagnus	*Elaeagnus spp.* L.
birch	*Betula spp.* L.	elm	*Ulmus spp.* L.
birch, black	*Betula lenta* L.	elm, American	*Ulmus americana* L.
black cherry	*Prunus serotina* Ehrh.	eucalyptus	*Eucalyptus spp.*
black locust	*Robinia pseudoacacia* L.	eucalyptus (jarrah)	*Eucalyptus marginata* Donn ex Sm.
blackberry	*Rubus spp.* L.	fescue	*Festuca spp.* L.
blueberry	*Vaccinium spp.* L.	fescue, meadow	*Festuca pratensis* Huds.
bluegrass, Kentucky	*Poa pratensis* L. ssp. *pratensis*	fescue, red	*Festuca rubra* L.
		fescue, sheep	*Festuca ovina* L.
bog rosemary	*Andromeda polifolia* L.	fescue, tall	*Festuca elatior* L.
bougainvillea	*Bougainvillea spp.* Comm. ex Juss.	fig	*Ficus carica* L.
		filbert	*Corylus spp.* L.
boxwood	*Buxus spp.* L.	fir, Douglas	*Pseudotsuga menziesii* (Mirbel) Franco
broccoli	*Brassica oleracea* L. var. *botrytis* L.	gamagrass, eastern	*Tripsacum dactyloides* (L.) L.
brome grass	*Bromus spp.* L.	gliricidia (quickstick)	*Gliricidia sepium* (Jacq.) Kunth ex Walp.
buckwheat	*Eriogonum spp.* Michx.	grape	*Vitus spp.* L.
buffalo grass	*Buchloe dactyloides* (Nutt.) Engelm.	grapefruit	*Citrus paradisi* Macfad. (pro sp.) [*maxima sinensis*]
cabbage	*Brassica oleracea* L.	grevillea	*Grevillea spp.* R. Br. ex Knight
canola (rapeseed)	*Brassica napus* L.	groundnut (peanut)	*Arachis hypogaea* L.
cantaloupe	*Cucumis melo* L.	guayule	*Parthenium argentatum* Gray

Plants Mentioned in this Text: Their Common and Scientific Names (Continued)

gunnera	*Gunnera spp.* L.
harding grass	*Phalaris tuberosa* L. var. *stenoptera* (Hack) A.S. Hitchc.
hemlock, Canadian	*Tsuga canadensis* (L.) Carr.
hemlock, Carolina	*Tsuga caroliniana* Engelm.
hibiscus	*Hibiscus spp.* L.
hickory, bitternut	*Carya cordiformis* (Wangenh.) K. Koch
hickory, shagbark	*Carya ovata* (P. Mill.) K. Koch
holly, American	*Ilex opaca* Ait.
holly, burford	*Ilex cornuta* Lindl. & Paxton
honeysuckle	*Lonicera spp.* L.
hydrangea	*Hydrangea spp.* L.
ipil ipil tree	*Leucaena leucocephala* Benth.
jarrah	*Eucalyptus marginata* Donn ex Sm.
Johnsongrass	*Sorghum halepense* (L.) Pers.
jojoba	*Simmondsia chinensis* (Link) Schneid.
juniper	*Juniperus spp.* L.
kale	*Brassica oleracea* L. (Acephala group)
kallargrass	*Leptochloa fusca* (L.) Kunth [syn. *Diplachne fusca* Beauv.]
kenaf	*Hibiscus cannabinus* L.
kochia, prostrate	*Kochia prostrata* (L.) Schrad.
kudzu	*Pueraria montana* (Lour.) Merr. Var. lobata (Wild.)
larch	*Larix spp.* P. Mill.
lemon	*Citrus limon* (L.) Burm. F.
lespedeza	*Lespedeza spp.* Michx.
lettuce	*Lactuca sativa* L.
leucaena (lead tree)	*Leucaena spp.* Benth.
lilac	*Syringa spp.* L.
linden	*Tillia spp.* L.
locust, black	*Robinia pseudoacacia* L.
locust, honey	*Gleditsia triacanthos* L.
lovegrass, weeping	*Eragrostis curvula* (Schrad.) Nees
lupine	*Lupinus spp.* L.
magnolia	*Magnolia spp.* L.
maiden cane	*Panicum hemitomon* J.A. Schultes
maize (corn)	*Zea mays* L.
mandarin orange	*Citrus reticulata* Blanco
maple	*Acer spp.* L.
maple, red	*Acer rubrum* L.
maple, sugar	*Acer saccharum* Marsh.
mosquito fern	*Azolla spp.* L.
mountain laurel	*Kalmia latifolia* L.
mulberry	*Morus spp.* L.
myrica	*Myrica spp.* L.
nut trees (e.g., almonds, hazelnuts)	*Prunus dulcis* (P. Mill.) D.A. Webber *Corylus spp.* L.
oak	*Quercus spp.* L.
oak, blackjack	*Quercus marilandica* Muenchh.
oak, chestnut	*Quercus prinus* L.
oak, northern red	*Quercus rubra* L.
oak, pin	*Quercus palustris* Muenchh.
oak, southern red	*Quercus falcata* Michx.
oak, swamp white	*Quercus bicolor* Wild.
oak, white	*Quercus alba* L.
oak, willow	*Quercus pellos* L.
oats	*Avena sativa* L.
olive	*Olea europaea* L.
onion	*Allium cepa* L.
orange	*Citrus sinensis* (L.) Osbeck
orchard grass	*Dactylis glomerata* L.
pangola grass	*Digitaria eriantha* Steud.
pea	*Pisum sativa* L.
pea, pigeon	*Cajanus cajan* (L.) Millsp.
peach	*Prunus persica* (L.) Batsch
peanut	*Arachis hypogaea* L.
pear	*Pyrus communis* L.
pecan	*Carya illinoinensis* (Wangenh.) K. Koch
phragmities reed	*Phragmities australis* (Cav.) Trin. Ex Steud.
pine, loblolly	*Pinus taeda* L.
pine, Monterey	*Pinus radiata* D. Don
pine, ponderosa	*Pinus ponderosa* Dougl. Ex P. & C. Laws.
pine, red	*Pinus resinosa* Ait.
pine, white	*Pinus strobus* L.
pine, white Scotch	*Pinus sylvestris* L.
pineapple	*Ananas comosus* (L.) Merrill
pitcher plant	*Sarracenia spp.* L.
plum (prune)	*Prunus domestica* L.
poinsettia	*Euphorbia pulcherrima* Willd. ex Klotzsch
pomegranate	*Punica granatum* L.
poplar	*Populus spp.* L.
potato	*Solanum tuberosum* L.
potato, sweet	*Ipomoea batatas* (L.) Lam.
povertygrass	*Danthonia spicata* (L.) Beauv. Ex Roem. & Schult.
privet	*Ligustrum spp.* L.
pueraria, kudzu	*Pueraria phaseoloides* (Roxb.) Benth.
quickstick	*Gliricidia sepium* (Jacq.) Kunth ex Walp.
radish	*Raphanus sativus* L.
rapeseed (see also canola)	*Brassica campestris* L. [syn. *B. rapa* L.]
raspberry	*Rubus idaeus* L.
red top	*Agrostis alba* L.
red top	*Agrostis gigantea* Roth
reed canarygrass	*Phalaris arundinacea* L.
rescuegrass	*Bromus catharticus* Vahl
rhododendron	*Rhododendron spp.* L.
rice	*Oryza spp.* L.
rice (paddy)	*Oryza sativa* L.
riverhemp, Egyptian	*Sesbania sesban* (L.) Merr.
rose-mallow, swamp	*Hibiscus moscheutos* L.
rosemary	*Rosmarinus officinalis* L.
roses	*Rosa spp.* L.
rye (grain, forage)	*Secale cereale* L.

(continued)

PLANTS MENTIONED IN THIS TEXT: THEIR COMMON AND SCIENTIFIC NAMES (CONTINUED)

rye, wild	*Elymus spp.*	tobacco	*Nicotiana spp.* L.
ryegrass, perennial	*Lolium perenne* L.	tomato	*Solanum lycopersicum* L.
safflower	*Carthamus tinctorius* L.	tree marigold	*Tithonia diversifolia* (Hemsl.) Gray
saltgrass, desert	*Distichlis spicta* L. var. *stricta* (Torr.) Bettle	trefoil, birdsfoot	*Lotus corniculatus* L.
sesbania	*Sesbania sesban* (L.) Merr.	tulip poplar	*Liriodendron tulipifera* L.
skunk cabbage	*Symplocarpus foetidus* (L.) Salisb. Ex Nutt.	turnip	*Brassica rapa* L. (Rapifera group)
sorghum	*Sorghum bicolor* (L.) Moench	velvetleaf	*Abutilon theophrasti* Medik.
soy beans	*Glycine max* (L.) Merr.	vetch	*Vicia spp.* L.
spartina (cordgrass)	*Spartina spp.* Schreb.	vetch, common	*Vicia angustifolia* L.
spinach	*Spinacia oleracea* L.	vetch, hairy	*Vicia villosa* Roth
spruce, black	*Picea mariana* (Mill.) B.S.P.	vetiver grass	*Vetiveria zizanioides* (L.) Nash ex Small
spruce, Norway	*Picea abies* (L.) Karst.		
spruce, red	*Picea rubens* Sarg.	viburnum	*Viburnum spp.* L.
spruce, white	*Picea glauca* (Moench) Voss	walnut	*Juglans spp.* L.
squash	*Cucurbita pepo* L.	water melon	*Citrullus lanatus* (Thunb.) Matsumura & Nakai
squash (zucchini)	*Cucurbita pepo* L. var. *melopepo* (L.) Alef.	water tupelo	*Nyssa aquatica* L.
star jasmine	*Jasminum multiflorum* (Burm. f.) Andr	wheat	*Triticum aestivum* L.
strawberry	*Fragaria x ananassa* Duch.	wheatgrass, crested	*Agropyron sibiricum* (Willd.) Beauvois
sudan grass	*Sorghum sudanense* (Piper) Stapf	wheatgrass, fairway	*Agropyron cristatum* (L.) Gaertn.
sugar beet	*Beta vulgaris* L.	wheatgrass, tall	*Agropyron elongatum* (Hort) Beauvois
sugar cane	*Saccharum officinarum* L.		
sumac	*Rhus spp.* L.	wheatgrass, western	*Pascopyrum smithii* (Rydb.) A. Löve
sunflower	*Helianthus annuus* L.		
sycamore	*Plantus occidentalis* L.	wild rye, altai	*Leymus angustus* (Trin.) Pilger
tamarix (tamarisk)	*Tamarix gallica* L.	wild rye, Russian	*Psathyrostachys juncea* (Fisch.) Nevski
tea	*Camellia sinensis* (L.) O. Kuntze	willow	*Salix spp.* L.
teaberry	*Gaultheria procumbens* L.	willow, black	*Salix nigra* L.
timothy	*Phleum pratense* L.	winged bean	*Psophocarpus tetragonolobus* (L.) DC
tithonia	*Tithonia diversifolia* (Hemsl.) Gray	yew	*Taxus spp.* L.

Glossary of Soil Science Terms[1]

A horizon The surface horizon of a mineral soil having maximum organic matter accumulation, maximum biological activity, and/or eluviation of materials such as iron and aluminum oxides and silicate clays.

abiotic Nonliving basic elements of the environment, such as rainfall, temperature, wind, and minerals.

accelerated erosion Erosion much more rapid than normal, natural, geological erosion; primarily as a result of the activities of humans or, in some cases, of animals.

acid cations Cations, principally Al^{3+}, Fe^{3+}, and H^+, that contribute to H^+ ion activity either directly or through hydrolysis reactions with water. *See also* non-acid cations.

acid rain Atmospheric precipitation with pH values less than about 5.6, the acidity being due to inorganic acids (such as nitric and sulfuric) that are formed when oxides of nitrogen and sulfur are emitted into the atmosphere.

acid saturation The proportion or percentage of a cation-exchange site occupied by acid cations.

acid soil A soil with a pH value <7.0. Usually applied to surface layer or root zone, but may be used to characterize any horizon. *See also* reaction, soil.

acid sulfate soils Soils that are potentially extremely acid (pH <3.5) because of the presence of large amounts of reduced forms of sulfur that are oxidized to sulfuric acid if the soils are exposed to oxygen when they are drained or excavated. A sulfuric horizon containing the yellow mineral jarosite is often present. *See also* cat clays.

acidity, active The activity of hydrogen ions in the aqueous phase of a soil. It is measured and expressed as a pH value.

acidity, residual Soil acidity that can be neutralized by lime or other alkaline materials but cannot be replaced by an unbuffered salt solution.

acidity, salt replaceable Exchangeable hydrogen and aluminum that can be replaced from an acid soil by an unbuffered salt solution such as KCl or NaCl.

acidity, total The total acidity in a soil. It is approximated by the sum of the salt-replaceable acidity plus the residual acidity.

Actinomycetes A group of bacteria that form branched mycelia that are thinner, but somewhat similar in appearance, to fungal hyphae. Includes many members of the order Actinomycetales.

active layer The upper portion of a Gelisol that is subject to freezing and thawing and is underlain by permafrost.

active organic matter A portion of the soil organic matter that is relatively easily metabolized by microorganisms and cycles with a half-life in the soil of a few days to a few years.

adhesion Molecular attraction that holds the surfaces of two substances (e.g., water and sand particles) in contact.

adsorption The attraction of ions or compounds to the surface of a solid. Soil colloids adsorb large amounts of ions and water.

adsorption complex The group of organic and inorganic substances in soil capable of adsorbing ions and molecules.

aeration, soil The process by which air in the soil is replaced by air from the atmosphere. In a well-aerated soil, the soil air is similar in composition to the atmosphere above the soil. Poorly aerated soils usually contain more carbon dioxide and correspondingly less oxygen than the atmosphere above the soil.

aerobic (1) Having molecular oxygen as a part of the environment. (2) Growing only in the presence of molecular oxygen, as aerobic organisms. (3) Occurring only in the presence of molecular oxygen (said of certain chemical or biochemical processes, such as aerobic decomposition).

aerosolic dust A type of eolian material that is very fine (about 1 to 10 µm) and may remain suspended in the air over distances of thousands of kilometers. Finer than most *loess*.

aggregate (soil) Many soil particles held in a single mass or cluster, such as a clod, crumb, block, or prism.

agric horizon A diagnostic subsurface horizon in which clay, silt, and humus derived from an overlying cultivated and fertilized layer have accumulated. Wormholes and illuvial clay, silt, and humus occupy at least 5% of the horizon by volume.

agroforestry Any type of multiple cropping land-use that entails complementary relations between trees and agricultural crops.

agronomy A specialization of agriculture concerned with the theory and practice of field-crop production and soil management. The scientific management of land.

air porosity The proportion of the bulk volume of soil that is filled with air at any given time or under a given condition, such as a specified moisture potential; usually the large pores.

albic horizon A diagnostic subsurface horizon from which clay and free iron oxides have been removed or in which the oxides have been segregated to the extent that the color of the horizon is determined primarily by the color of the primary sand and silt particles rather than by coatings on these particles.

Alfisols An order in *Soil Taxonomy*. Soils with gray to brown surface horizons, medium to high supply of bases, and B horizons of illuvial clay accumulation. These soils form mostly under forest or savanna vegetation in climates with slight to pronounced seasonal moisture deficit.

algal bloom A population explosion of algae in surface waters, such as lakes and streams, often resulting in high turbidity and green- or red-colored water, and commonly stimulated by nutrient enrichment with phosphorus and nitrogen.

alkaline soil Any soil that has pH >7. Usually applied to the surface layer or root zone but may be used to characterize any horizon or a sample thereof. *See also* reaction, soil.

allelochemical An organic chemical by which one plant can influence another. *See* allelopathy.

allelopathy The process by which one plant may affect other plants by biologically active chemicals introduced into the soil, either directly by leaching or exudation from the source plant, or as a result of the decay of the plant residues. The effects, though usually negative, may also be positive.

allophane A poorly defined aluminosilicate mineral whose structural framework consists of short runs of three-dimensional crystals interspersed with amorphous noncrystalline materials. Along with its more weathered companion, it is prevalent in volcanic ash materials.

alluvial fan Fan-shaped alluvium deposited at the mouth of a canyon or ravine where debris-laden waters fan out, slow down, and deposit their burden.

---

[1]This glossary was compiled and modified from several sources, including *Glossary of Soil Science Terms* [Madison, WI: Soil Science Society of America (1997)], *Resource Conservation Glossary* [Ankeny, IA: Soil Conservation Society of America (1982)], and *Soil Taxonomy* [Washington, DC: U.S. Department of Agriculture (1999)].

alluvium   A general term for all detrital material deposited or in transit by streams, including gravel, sand, silt, clay, and all variations and mixtures of these. Unless otherwise noted, alluvium is unconsolidated.

aluminosilicates   Compounds containing aluminum, silicon, and oxygen as main constituents. An example is microcline, $KAlSi_3O_8$.

amendment, soil   Any substance other than fertilizers, such as lime, sulfur, gypsum, and sawdust, used to alter the chemical or physical properties of a soil, generally to make it more productive.

amino acids   Nitrogen-containing organic acids that couple together to form proteins. Each acid molecule contains one or more amino groups ($—NH_2$) and at least one carboxyl group ($—COOH$). In addition, some amino acids contain sulfur.

Ammanox   A biochemical process in the N cycle by which certain anaerobic bacteria or archaea oxidize ammonium ions using nitrite ions as the electron acceptor, the main product being N_2 gas.

ammonification   The biochemical process whereby ammoniacal nitrogen is released from nitrogen-containing organic compounds.

ammonium fixation   The entrapment of ammonium ions by the mineral or organic fractions of the soil in forms that are insoluble in water and are at least temporarily nonexchangeable.

amorphous material   Noncrystalline constituents of soils.

anaerobic   (1) The absence of molecular oxygen. (2) Growing or occurring in the absence of molecular oxygen (e.g., anaerobic bacteria or biochemical reduction reaction).

anaerobic respiration   The metabolic process whereby electrons are transferred from a reduced compound (usually organic) to an inorganic acceptor molecule other than oxygen.

andic properties   Soil properties related to volcanic origin of materials, including high organic carbon content, low bulk density, high phosphate retention, and extractable iron and aluminum.

Andisols   An order in *Soil Taxonomy*. Soils developed from volcanic ejecta. The colloidal fraction is dominated by allophane and/or Al-humus compounds.

angle of repose   The maximum slope steepness at which loose, cohesionless material will come to rest.

anion   Negatively charged ion; during electrolysis it is attracted to the positively charged anode.

anion exchange   Exchange of anions in the soil solution for anions adsorbed on the surface of clay and humus particles.

anion exchange capacity   The sum total of exchangeable anions that a soil can adsorb. Expressed as centimoles of charge per kilogram ($cmol_c/kg$) of soil (or of other adsorbing material, such as clay).

anoxic   *See* anaerobic.

anthropic epipedon   A diagnostic surface horizon of mineral soil that has the same requirements as the mollic epipedon but that has more than 250 mg/kg of P_2O_5 soluble in 1% citric acid, or is dry more than 10 months (cumulative) during the period when not irrigated. The anthropic epipedon forms under long-continued cultivation and fertilization.

antibiotic   A substance produced by one species of organism that, in low concentrations, will kill or inhibit growth of certain other organisms.

Ap   The surface layer of a soil disturbed by cultivation or pasturing.

apatite   A naturally occurring complex calcium phosphate that is the original source of most of the phosphate fertilizers. Formulas such as $[3Ca_3(PO_4)_2] \cdot CaF_2$ illustrate the complex compounds that make up apatite.

aquic conditions   Continuous or periodic saturation (with water) and reduction, commonly indicated by redoximorphic features.

aquiclude   A saturated body of rock or sediment that is incapable of transmitting significant quantities of water under ordinary water pressures.

aquifer   A saturated, permeable layer of sediment or rock that can transmit significant quantities of water under normal pressure conditions.

arbuscule   Specialized branched structure formed within a root cortical cell by endotrophic mycorrhizal fungi.

arbuscular mycorrhiza   A common endo-mycorrhizal association produced by phy-comycetous fungi and characterized by the development, within root cells, of small structures known as *arbuscules*. Some also form, between root cells, storage organs known as *vesicles*. Host range includes many agricultural and horticultural crops. Formerly called vesicular arbuscular mycorrhiza (VAM). *See also* endotrophic mycorrhiza.

Archaea   One of the two domains of single-celled prokaryote microorganisms. Includes organisms adapted to extremes of salinity and heat, and those that subsist on methane. Similar appearing, but evolutionarily distinct from bacteria.

argillan   A thin coating of well-oriented clay particles on the surface of a soil aggregate, particle, or pore. A clay film.

argillic horizon   A diagnostic subsurface horizon characterized by the illuvial accumulation of layer-lattice silicate clays.

arid climate   Climate in regions that lack sufficient moisture for crop production without irrigation. In cool regions annual precipitation is usually less than 25 cm. It may be as high as 50 cm in tropical regions. Natural vegetation is desert shrubs.

Aridisols   An order in *Soil Taxonomy*. Soils of dry climates. They have pedogenic horizons, low in organic matter, that are never moist for as long as three consecutive months. They have an ochric epipedon and one or more of the following diagnostic horizons: argillic, natric, cambic, calcic, petrocalcic, gypsic, petrogypsic, salic, or a duripan.

aspect (of slopes)   The direction (e.g., south or north) that a slope faces with respect to the sun.

association, soil   *See* soil association.

Atterberg limits   Water contents of fine-grained soils at different states of consistency.
   liquid limit (LL)   The water content corresponding to the arbitrary limit between the liquid and plastic states of consistency of a soil.
   plastic limit (PL)   The water content corresponding to an arbitrary limit between the plastic and semisolid states of consistency of a soil.

autochthonous organisms   Those microorganisms thought to subsist on the more resistant soil organic matter and little affected by the addition of fresh organic materials. *Contrast with* zymogenous organisms. *See also* k-strategist.

autotroph   An organism capable of utilizing carbon dioxide or carbonates as the sole source of carbon and obtaining energy for life processes from the oxidation of inorganic elements or compounds such as iron, sulfur, hydrogen, ammonium, and nitrites, or from radiant energy. *Contrast with* heterotroph.

available nutrient   That portion of any element or compound in the soil that can be readily absorbed and assimilated by growing plants. ("Available" should not be confused with "exchangeable.")

available water   The portion of water in a soil that can be readily absorbed by plant roots. The amount of water released between the field capacity and the permanent wilting point.

B horizon   A soil horizon, usually beneath the A or E horizon, that is characterized by one or more of the following: (1) a concentration of soluble salts, silicate clays, iron and aluminum oxides, and humus, alone or in combination; (2) a blocky or prismatic structure; and (3) coatings of iron and aluminum oxides that give darker, stronger, or redder color.

Bacteria   One of two domains of single-celled prokaryote microorganisms. Includes all that are not Archaea.

bar A unit of pressure equal to 1 million dynes per square centimeter (10^6 dynes/cm^2). It approximates the pressure of a standard atmosphere.

base-forming cations (obsolete) Cations that form strong (strongly dissociated) bases by reaction with hydroxyl; e.g., K$^+$ forms potassium hydroxide (K$^+$ + OH). *See* non-acid cations.

base saturation percentage The extent to which the adsorption complex of a soil is saturated with exchangeable cations other than hydrogen and aluminum. It is expressed as a percentage of the total cation exchange capacity. *See* non-acid saturation.

bedrock The solid rock underlying soils and the regolith in depths ranging from zero (where exposed by erosion) to several hundred feet.

bench terrace An embankment constructed across sloping fields with a steep drop on the downslope side.

bioaccumulation A buildup within an organism of specific compounds due to biological processes. Commonly applied to heavy metals, pesticides, or metabolites.

bioaugmentation The cleanup of contaminated soils by adding exotic microorganisms that are especially efficient at breaking down an organic contaminant. A form of *bioremediation*.

biodegradable Subject to degradation by biochemical processes.

biological nitrogen fixation Occurs at ordinary temperatures and pressures. It is commonly carried out by certain bacteria, algae, and actinomycetes, which may or may not be associated with higher plants.

biomass The total mass of living material of a specified type (e.g., microbial biomass) in a given environment (e.g., in a cubic meter of soil).

biopores Soil pores, usually of relatively large diameter, created by plant roots, earthworms, or other soil organisms.

bioremediation The decontamination or restoration of polluted or degraded soils by means of enhancing the chemical degradation or other activities of soil organisms.

biosequence A group of related soils that differ, one from the other, primarily because of differences in kinds and numbers of plants and soil organisms as a soil-forming factor.

biosolids Sewage sludge that meets certain regulatory standards, making it suitable for land application. *See* sewage sludge.

biostimulation The cleanup of contaminated soils through the manipulation of nutrients or other soil environmental factors to enhance the activity of naturally occurring soil microorganisms. A form of *bioremediation*.

blocky soil structure Soil aggregates with blocklike shapes; common in B horizons of soils in humid regions.

broad-base terrace A low embankment with such gentle slopes that it can be farmed, constructed across sloping fields to reduce erosion and runoff.

broadcast Scatter seed or fertilizer on the surface of the soil.

brownfields Abandoned, idled, or underused industrial and commercial facilities where expansion or redevelopment is complicated by real or perceived environmental contamination.

buffering capacity The ability of a soil to resist changes in pH. Commonly determined by presence of clay, humus, and other colloidal materials.

bulk density, soil The mass of dry soil per unit of bulk volume, including the air space. The bulk volume is determined before drying to constant weight at 105 °C.

buried soil Soil covered by an alluvial, loessal, or other deposit, usually to a depth greater than the thickness of the solum.

by-pass flow *See* preferential flow.

C horizon A mineral horizon, generally beneath the solum, that is relatively unaffected by biological activity and pedogenesis and is lacking properties diagnostic of an A or B horizon. It may or may not be like the material from which the A and B have formed.

calcareous soil Soil containing sufficient calcium carbonate (often with magnesium carbonate) to effervesce visibly when treated with cold 0.1 N hydrochloric acid.

calcic horizon A diagnostic subsurface horizon of secondary carbonate enrichment that is more than 15 cm thick, has a calcium carbonate equivalent of more than 15%, and has at least 5% more calcium carbonate equivalent than the underlying C horizon.

caliche A layer near the surface, more or less cemented by secondary carbonates of calcium or magnesium precipitated from the soil solution. It may occur as a soft, thin soil horizon; as a hard, thick bed just beneath the solum; or as a surface layer exposed by erosion.

cambic horizon A diagnostic subsurface horizon that has a texture of loamy very fine sand or finer, contains some weatherable minerals, and is characterized by the alteration or removal of mineral material. The cambic horizon lacks cementation or induration and has too few evidences of illuviation to meet the requirements of the argillic or spodic horizon.

capillary fringe A zone in the soil just above the plane of zero water pressure (water table) that remains saturated or almost saturated with water.

capillary water The water held in the capillary or *small* pores of a soil, usually with a tension >60 cm of water. *See also* soil water potential.

carbon cycle The sequence of transformations whereby carbon dioxide is fixed in living organisms by photosynthesis or by chemosynthesis, liberated by respiration and by the death and decomposition of the fixing organism, used by heterotrophic species, and ultimately returned to its original state.

carbon/nitrogen (C/N) ratio The ratio of the weight of organic carbon (C) to the weight of total nitrogen (N) in a soil or in organic material.

carnivore An organism that feeds on animals.

casts, earthworm Rounded, water-stable aggregates of soil that have passed through the gut of an earthworm.

catena A group of soils that commonly occur together in a landscape, each characterized by a different slope position and resulting set of drainage-related proprieties. *See also* toposequence.

cation A positively charged ion; during electrolysis it is attracted to the negatively charged cathode.

cation exchange The interchange between a cation in solution and another cation on the surface of any surface-active material, such as clay or organic matter.

cation exchange capacity The sum total of exchangeable cations that a soil can adsorb. Sometimes called *total exchange capacity*, *base exchange capacity*, or *cation adsorption capacity*. Expressed in centimoles of charge per kilogram (cmol$_c$/kg) of soil (or of other adsorbing material, such as clay).

cemented Indurated; having a hard, brittle consistency because the particles are held together by cementing substances, such as humus, calcium carbonate, or the oxides of silicon, iron, and aluminum.

channery Thin, flat fragments of limestone, sandstone, or schist up to 15 cm (6 in.) in major diameter.

chelate (Greek "claw") A type of chemical compound in which a metallic ion is firmly combined with an organic molecule by means of multiple chemical bonds.

chert A structureless form of silica, closely related to flint, that breaks into angular fragments.

chisel, subsoil A tillage implement with one or more cultivator-type feet to which are attached strong knifelike units used to shatter or loosen hard, compact layers, usually in the subsoil, to depths below normal plow depth. *See also* subsoiling.

chlorite　A 2:1:1-type layer-structured silicate mineral having 2:1 layers alternating with a magnesium-dominated octahedral sheet.

chlorosis　A condition in plants relating to the failure of chlorophyll (the green coloring matter) to develop. Chlorotic leaves range from light green through yellow to almost white.

chroma (color)　*See* Munsell color system.

chronosequence　A sequence of related soils that differ, one from the other, in certain properties primarily as a result of time as a soil-forming factor.

classification, soil　*See* soil classification.

clay　(1) A soil separate consisting of particles <0.002 mm in equivalent diameter. (2) A soil textural class containing >40% clay, <45% sand, and <40% silt.

clay mineral　Naturally occurring inorganic material (usually crystalline) found in soils and other earthy deposits, the particles being of clay size, that is, <0.002 mm in diameter.

claypan　A dense, compact, slowly permeable layer in the subsoil having a much higher clay content than the overlying material, from which it is separated by a sharply defined boundary. Claypans are usually hard when dry and plastic and sticky when wet. *See also* hardpan.

climosequence　A group of related soils that differ, one from another, primarily because of differences in climate as a soil-forming factor.

clod　A compact, coherent mass of soil produced artificially, usually by such human activities as plowing and digging, especially when these operations are performed on soils that are either too wet or too dry for normal tillage operations.

coarse fragments　Mineral (rock) soil particles larger than 2 mm in diameter. *Compare to* fine earth fraction.

coarse texture　The texture exhibited by sands, loamy sands, and sandy loams (except very fine sandy loam).

cobblestone　Rounded or partially rounded rock or mineral fragments 7.5 to 25 cm (3 to 10 in.) in diameter.

co-composting　A method of composting in which two materials of differing but complementary nature are mingled together and enhance each other's decomposition in a compost system.

cohesion　Holding together: Force holding a solid or liquid together, owing to attraction between like molecules. Decreases with rise in temperature.

collapsible soil　Certain soil that may undergo a sudden loss in strength when wetted.

colloid, soil　(Greek "gluelike") Organic and inorganic matter with very small particle size and a correspondingly large surface area per unit of mass.

colluvium　A deposit of rock fragments and soil material accumulated at the base of steep slopes as a result of gravitational action.

color　The property of an object that depends on the wavelength of light it reflects or emits.

columnar soil structure　*See* soil structure types.

companion planting　The practice of growing certain species of plants in close proximity because one species has the effect of improving the growth of the other, sometimes by positive *allelopathic* effects.

compost　Organic residues, or a mixture of organic residues and soil, that have been piled, moistened, and allowed to undergo biological decomposition. Mineral fertilizers are sometimes added. Usually managed so as to reach thermophilic temperatures.

concretion　A local concentration of a chemical compound, such as calcium carbonate or iron oxide, in the form of grains or nodules of varying size, shape, hardness, and color.

conduction　The transfer of heat by physical contact between two or more objects.

conductivity, hydraulic　*See* hydraulic conductivity.

conservation tillage　*See* tillage, conservation.

consistence　The combination of properties of soil material that determine its resistance to crushing and its ability to be molded or changed in shape. Such terms as *loose*, *friable*, *firm*, *soft*, *plastic*, and *sticky* describe soil consistence.

consistency　The interaction of adhesive and cohesive forces within a soil at various moisture contents as expressed by the relative ease with which the soil can be deformed or ruptured.

consociation　*See* soil consociation.

consolidation test　A laboratory test in which a soil mass is laterally confined within a ring and is compressed with a known force between two porous plates.

constant charge　The net surface charge of mineral particles, the magnitude of which depends only on the chemical and structural composition of the mineral. The charge arises from isomorphous substitution and is not affected by soil pH.

consumptive use　The water used by plants in transpiration and growth, plus water vapor loss from adjacent soil or snow, or from intercepted precipitation in any specified time. Usually expressed as equivalent depth of free water per unit of time.

contour strip-cropping　Layout of crops in comparatively narrow strips in which the farming operations are performed approximately on the contour. Usually strips of grass, close-growing crops, or fallow are alternated with those of cultivated crops.

controlled traffic　A farming system in which all wheeled traffic is confined to fixed paths so that repeated compaction of the soil does not occur outside the selected paths.

convection　The transfer of heat through a gas or solution because of molecular movement.

cover crop　A close-growing crop grown primarily for the purpose of protecting and improving soil between periods of regular crop production or between trees and vines in orchards and vineyards.

creep　Slow mass movement of soil and soil material down relatively steep slopes, primarily under the influence of gravity, but facilitated by saturation with water and by alternate freezing and thawing.

crop rotation　A planned sequence of crops growing in a regularly recurring succession on the same area of land, as contrasted to continuous culture of one crop or growing different crops in haphazard order.

crotovina　A former animal burrow in one soil horizon that has been filled with organic matter or material from another horizon (also spelled *krotovina*).

crumb　A soft, porous, more or less rounded natural unit of structure from 1 to 5 mm in diameter. *See also* soil structure types.

crushing strength　The force required to crush a mass of dry soil or, conversely, the resistance of the dry soil mass to crushing. Expressed in units of force per unit area (pressure).

crust (soil)

　(1) **physical**　A surface layer on soils, ranging in thickness from a few millimeters to as much as 3 cm, that physical-chemical processes have caused to be much more compact, hard, and brittle when dry than the material immediately beneath it.

　(2) **microbiotic**　An assemblage of cyanobacteria, algae, lichens, liverworts, and mosses that commonly forms an irregular crust on the soil surface, especially on otherwise barren, arid-region soils. Also referred to as cryptogamic, cryptobiotic, or biological crusts.

cryophilic　Pertaining to low temperatures in the range of 5 to 15 °C, the range in which cryophilic organisms grow best.

cryoturbation　Physical disruption and displacement of soil material within the profile by the forces of freezing and thawing. Sometimes called *frost churning*, it results in irregular, broken horizons, involutions, oriented rock fragments, and accumulation of organic matter on the permafrost table.

cryptogam　*See* crust (2) microbiotic.

crystal A homogeneous inorganic substance of definite chemical composition bounded by planar surfaces that form definite angles with each other, thus giving the substance a regular geometrical form.

crystal structure The orderly arrangement of atoms in a crystalline material.

cultivation A tillage operation used in preparing land for seeding or transplanting or later for weed control and for loosening the soil.

cutans A modification of the texture, structure, or fabric at natural surfaces in soil materials due to concentration of particular soil constituents; e.g. "clay skins."

cyanobacteria Chlorophyll-containing bacteria that accommodate both photosynthesis and nitrogen fixation. Formerly called blue-green algae.

deciduous plant A plant that sheds all its leaves every year at a certain season.

decomposition Chemical breakdown of a compound (e.g., a mineral or organic compound) into simpler compounds, often accomplished with the aid of microorganisms.

deflocculate (1) To separate the individual components of compound particles by chemical and/or physical means. (2) To cause the particles of the *disperse phase* of a colloidal system to become suspended in the *dispersion medium*.

delineation An individual polygon shown by a closed boundary on a soil map that defines the area, shape, and location of a map unit within a landscape.

delivery ratio The ratio of eroded sediment carried out of a drainage basin to the total amount of sediment moved within the basin by erosion processes.

delta An alluvial deposit formed where a stream or river drops its sediment load upon entering a quieter body of water.

denitrification The biochemical reduction of nitrate or nitrite to gaseous nitrogen, either as molecular nitrogen or as an oxide of nitrogen.

density *See* particle density; bulk density.

desalinization Removal of salts from saline soil, usually by leaching.

desert crust A hard layer, containing calcium carbonate, gypsum, or other binding material, exposed at the surface in desert regions.

desert pavement A natural residual concentration of closely packed pebbles, boulders, and other rock fragments on a desert surface where wind and water action has removed all smaller particles.

desert varnish A thin, dark, shiny film or coating of iron oxide and lesser amounts of manganese oxide and silica formed on the surfaces of pebbles, boulders, rock fragments, and rock outcrops in arid regions.

desorption The removal of sorbed material from surfaces.

detritivore An organism that subsists on detritus.

detritus Debris from dead plants and animals.

diagnostic horizons (As used in *Soil Taxonomy*) Horizons having specific soil characteristics that are indicative of certain classes of soils. Horizons that occur at the soil surface are called *epipedons*; those below the surface, *diagnostic subsurface horizons*.

diatomaceous earth A geologic deposit of fine, grayish, siliceous material composed chiefly or wholly of the remains of diatoms. It may occur as a powder or as a porous, rigid material.

diatoms Algae having siliceous cell walls that persist as a skeleton after death; any of the microscopic unicellular or colonial algae constituting the class Bacillariaceae. They occur abundantly in fresh and salt waters and their remains are widely distributed in soils.

diffusion The movement of atoms in a gaseous mixture or of ions in a solution, primarily as a result of their own random motion.

dioctahedral sheet An octahedral sheet of silicate clays in which the sites for the six-coordinated metallic atoms are mostly filled with trivalent atoms, such as Al^{3+}.

disintegration Physical or mechanical breakup or separation of a substance into its component parts (e.g., a rock breaking into its mineral components).

disperse (1) To break up compound particles, such as aggregates, into the individual component particles. (2) To distribute or suspend fine particles, such as clay, in or throughout a dispersion medium, such as water.

dissimilatory nitrate reduction to ammonium (DNRA) A bacterial process by which nitrate is converted to ammonium under a wide range of oxygen and carbon levels. Compare to dentrification (a different type of dissimilatory nitrate reduction) which is strictly anaerobic and requires an energy source.

dissolution Process by which molecules of a gas, solid, or another liquid dissolve in a liquid, thereby becoming completely and uniformly dispersed throughout the liquid's volume.

distribution coefficient (K_d) The distribution of a chemical between soil and water.

diversion terrace *See* terrace.

drain (1) To provide channels, such as open ditches or drain tile, so that excess water can be removed by surface or by internal flow. (2) To lose water (from the soil) by percolation.

drain field, septic tank An area of soil into which the effluent from a septic tank is piped so that it will drain through the lower part of the soil profile for disposal and purification.

drainage, soil The frequency and duration of periods when the soil is free from saturation with water.

drift Material of any sort deposited by geological processes in one place after having been removed from another. Glacial drift includes material moved by the glaciers and by the streams and lakes associated with them.

drumlin Long, smooth, cigar-shaped low hills of glacial till, with their long axes parallel to the direction of ice movement.

dryland farming The practice of crop production in low-rainfall areas without irrigation.

duff The matted, partly decomposed organic surface layer of forest soils.

duripan A diagnostic subsurface horizon that is cemented by silica, to the point that air-dry fragments will not slake in water or HCl. Hardpan.

dust mulch A loose, finely granular or powdery condition on the surface of the soil, usually produced by shallow cultivation.

E horizon Horizon characterized by maximum illuviation (washing out) of silicate clays and iron and aluminum oxides; commonly occurs above the B horizon and below the A horizon.

earthworms Animals of the Lumbricidae family that burrow into and live in the soil. They mix plant residues into the soil and improve soil aeration.

ecosystem A dynamic and interacting combination of all the living organisms and non-living elements (matter and energy) of an area.

ecosystem services Products of natural ecosystems that support and fulfill the needs of human beings. Provision of clean water and unpolluted air are examples.

ectotrophic mycorrhiza (ectomycorrhiza) A symbiotic association of the mycelium of fungi and the roots of certain plants in which the fungal hyphae form a compact mantle on the surface of the roots and extend into the surrounding soil and inward between cortical cells, but not into these cells. Associated primarily with certain trees. *See also* endotrophic mycorrhiza.

edaphology The science that deals with the influence of soils on living things, particularly plants, including human use of land for plant growth.

effective cation exchange capacity The amount of cation charges that a material (usually soil or soil colloids) can hold at the pH of the material, measured as the sum of the exchangeable Al^{3+}, Ca^{2+}, Mg^{2+}, K^+, and Na^+, and expressed as moles or cmol of charge per kg of material. *See also* cation exchange capacity.

effective precipitation   That portion of the total precipitation that becomes available for plant growth or for the promotion of soil formation.

E_h   In soils, the potential created by oxidation-reduction reactions that take place on the surface of a platinum electrode measured against a reference electrode, minus the Eh of the reference electrode. This is a measure of the oxidation-reduction potential of electrode-reactive components in the soil. *See also* pe.

electrical conductivity (EC)   The capacity of a substance to conduct or transmit electrical current. In soils or water, measured in siemens/meter (or often dS/m), and related to dissolved solutes.

eluviation   The removal of soil material in suspension (or in solution) from a layer or layers of a soil. Usually, the loss of material in solution is described by the term "leaching." *See also* illuviation and leaching.

endoaquic (endosaturation)   A condition or moisture regime in which the soil is saturated with water in all layers from the upper boundary of saturation (water table) to a depth of 200 cm or more from the mineral soil surface. *See also* epiaquic.

endotrophic mycorrhiza (endomycorrhiza)   A symbiotic association of the mycelium of fungi and roots of a variety of plants in which the fungal hyphae penetrate directly into root hairs, other epidermal cells, and occasionally into cortical cells. Individual hyphae also extend from the root surface outward into the surrounding soil. *See also* arbuscular mycorrhiza.

enrichment ratio   The concentration of a substance (e.g., phosphorus) in eroded sediment divided by its concentration in the source soil prior to being eroded.

Entisols   An order in *Soil Taxonomy*. Soils that have no diagnostic pedogenic horizons. They may be found in virtually any climate on very recent geomorphic surfaces.

eolian soil material   Soil material accumulated through wind action. The most extensive areas in the United States are silty deposits (loess), but large areas of sandy deposits also occur.

epiaquic (episaturation)   A condition in which the soil is saturated with water due to a perched water table in one or more layers within 200 cm of the mineral soil surface, implying that there are also one or more unsaturated layers within 200 cm below the saturated layer. *See also* endoaquic.

epipedon   (As used in *Soil Taxonomy*:) A diagnostic surface horizon that includes the upper part of the soil that is darkened by organic matter, or the upper eluvial horizons, or both.

equilibrium phosphorus concentration   The concentration of phosphorus in a solution in equilibrium with a soil, the EPC_0 being the concentration of phosphorus achieved by desorption of phosphorus from a soil to phosphorus-free distilled water.

erosion   (1) The wearing away of the land surface by running water, wind, ice, or other geological agents, including such processes as gravitational creep. (2) Detachment and movement of soil or rock by water, wind, ice, or gravity.

esker   A narrow ridge of gravelly or sandy glacial material deposited by a stream in an ice-walled valley or tunnel in a receding glacier.

essential element   A chemical element required for the normal growth of plants.

eukaryote   An organism whose cells each have a visibly evident nucleus.

eutrophic   Having concentrations of nutrients optimal (or nearly so) for plant or animal growth. (Said of algal-enriched bodies of water)

eutrophication   Nutrient enrichment of lakes, ponds, and other such waters that stimulates the growth of aquatic organisms, which leads to a deficiency of oxygen in the water body.

evapotranspiration   The combined loss of water from a given area, and during a specified period of time, by evaporation from the soil surface and by transpiration from plants.

exchange capacity   The total ionic charge of the adsorption complex active in the adsorption of ions. *See also* anion exchange capacity; cation exchange capacity.

exchangeable ions   Positively or negatively charged atoms or groups of atoms that are held on or near the surface of a solid particle by attraction to charges of the opposite sign, and which may be replaced by other like-charged ions in the soil solution.

exchangeable sodium percentage   The extent to which the adsorption complex of a soil is occupied by sodium. It is expressed as follows:

$$ESP = \frac{\text{exchangeable sodium (cmol}_c/\text{kg soil)}}{\text{cation exchange capacity (cmol}_c/\text{kg soil)}} \times 100$$

exfoliation   Peeling away of layers of a rock from the surface inward, usually as the result of expansion and contraction that accompany changes in temperature.

expansive soil   Soil that undergoes significant volume change upon wetting and drying, usually because of a high content of swelling-type clay minerals.

external surface   The area of surface exposed on the top, bottom, and sides of a clay crystal.

facultative organism   An organism capable of both aerobic and anaerobic metabolism.

fallow   Cropland left idle in order to restore productivity, mainly through accumulation of nutrients, water, and/or organic matter. Preceding a cereal grain crop in semiarid regions, land may be left in *summer fallow* for a period during which weeds are controlled by chemicals or tillage and water is allowed to accumulate in the soil profile. In humid regions, fallow land may be allowed to grow up in natural vegetation for a period ranging from a few months to many years. *Improved fallow* involves the purposeful establishment of plant species capable of restoring soil productivity more rapidly than a natural plant succession.

family, soil   In *Soil Taxonomy*, one of the categories intermediate between the great group and the soil series. Families are defined largely on the basis of physical and mineralogical properties of importance to plant growth.

fauna   The animal life of a region or ecosystem.

fen   A calcium-rich, peat-accumulating wetland with relatively stagnant water.

ferrihydrite, $Fe_5HO_8 \cdot 4H_2O$ A dark reddish brown poorly crystalline iron oxide that forms in wet soils.

fertigation   The application of fertilizers in irrigation waters, commonly through sprinkler systems.

fertility, soil   The quality of a soil that enables it to provide essential chemical elements in quantities and proportions for the growth of specified plants.

fertilizer   Any organic or inorganic material of natural or synthetic origin added to a soil to supply certain elements essential to the growth of plants.

fibric materials   *See* organic soil materials.

field capacity (field moisture capacity)   The percentage of water remaining in a soil two or three days after its having been saturated and after free drainage has practically ceased.

fine earth fraction   That portion of the soil that passes through a 2 mm diameter sieve opening. *Compare to* coarse fragments.

fine texture   Consisting of or containing large quantities of the fine fractions, particularly of silt and clay. (Includes clay loam, sandy clay loam, silty clay loam, sandy clay, silty clay, and clay textural classes.)

fine-grained mica   A silicate clay having a 2:1-type lattice structure with much of the silicon in the tetrahedral sheet having been replaced by aluminum and with considerable interlayer potassium, which binds the layers together, prevents interlayer expansion and swelling, and limits interlayer cation exchange capacity.

fixation   (1) For other than elemental nitrogen: the process or processes in a soil by which certain chemical elements are converted from a soluble or exchangeable form to a much less soluble or to a nonexchangeable form; for example, potassium, ammonium, and phosphorus fixation. (2) For elemental nitrogen: process by which gaseous elemental nitrogen is chemically combined with hydrogen to form ammonia. *See* biological nitrogen fixation.

flagstone A relatively thin rock or mineral fragment 15 to 38 cm in length commonly composed of shale, slate, limestone, or sandstone.

flocculate To aggregate or clump together individual, tiny soil particles, especially fine clay, into small clumps or floccules. Opposite of *deflocculate* or *disperse*.

floodplain The land bordering a stream, built up of sediments from overflow of the stream and subject to inundation when the stream is at flood stage. Sometimes called *bottomland*.

flora The sum total of the kinds of plants in an area at one time. The organisms loosely considered to be of the plant kingdom.

fluorapatite A member of the apatite group of minerals containing fluorine. Most common mineral in phosphate rock.

fluvial deposits Deposits of parent materials laid down by rivers or streams.

fluvioglacial *See* glaciofluvial deposits.

foliar diagnosis An estimation of mineral nutrient deficiencies (excesses) of plants based on examination of the chemical composition of selected plant parts, and the color and growth characteristics of the foliage of the plants.

food web The community of organisms that relate to one another by sharing and passing on food substances. They are organized into trophic levels such as producers that create organic substances from sunlight and inorganic matter, to consumers and predators that eat the producers, dead organisms, waste products, and each other.

forest floor The forest soil O horizons, including litter and unincorporated humus, on the mineral soil surface.

fraction A portion of a larger store of a substance operationally defined by a particular analysis or separation method. For example, the fulvic acid fraction of soil organic matter is defined by a series of laboratory procedures by which it is solubilized. *Compare to* pool.

fragipan Dense and brittle pan or subsurface layer in soils that owes its hardness mainly to extreme density or compactness rather than high clay content or cementation. Removed fragments are friable, but the material in place is so dense that roots penetrate and water moves through it very slowly.

friable A soil consistency term pertaining to soils that crumble with ease.

frigid A soil temperature class with mean annual temperature below 8 °C.

fritted micronutrients Sintered silicates having total guaranteed analyses of micronutrients with controlled (relatively slow) release characteristics.

fulvic acid A term of varied usage but usually referring to the mixture of organic substances remaining in solution upon acidification of a dilute alkali extract from the soil.

functional diversity The characteristic of an ecosystem exemplified by the capacity to carry out a large number of biochemical transformations and other functions.

functional group An atom, or group of atoms, attached to a large molecule. Each functional group (e.g., —OH, —CH$_3$, —COOH) has a characteristic chemical reactivity.

fungi Eukaryote microorganisms with a rigid cell wall. Some form long filaments of cells called *hyphae* that may grow together to form a visible body.

furrow slice The uppermost layer of an arable soil to the depth of primary tillage; the layer of soil sliced away from the rest of the profile and inverted by a moldboard plow.

gabion A partitioned, wire fabric container, filled with stone at the site of use, to form flexible, permeable, and monolithic structures for earth retention.

gamma ray A high-energy ray (photon) emitted during radioactive decay of certain elements.

Gelisols An order in *Soil Taxonomy*. Soils that have permafrost within the upper 1 m, or upper 2 m if cryoturbation is also present. They may have an ochric, histic, mollic, or other epipedon.

gellic materials Mineral or organic soil materials that have *cryoturbation* and/or ice in the form of lenses, veins, or wedges and the like.

genesis, soil The mode of origin of the soil, with special reference to the processes responsible for the development of the solum, or true soil, from the unconsolidated parent material.

genetic horizon Soil layers that resulted from soil-forming (pedogenic) processes, as opposed to sedimentation or other geologic processes.

geographic information system (GIS) A method of overlaying, statistically analyzing, and integrating large volumes of spatial data of different kinds. The data are referenced to geographical coordinates and encoded in a form suitable for handling by computer.

geological erosion Wearing away of the Earth's surface by water, ice, or other natural agents under natural environmental conditions of climate, vegetation, and so on, undisturbed by man. Synonymous with *natural erosion*.

gibbsite, Al(OH)$_3$ An aluminum trihydroxide mineral most common in highly weathered soils, such as Oxisols.

gilgai The microrelief of soils produced by expansion and contraction with changes in moisture. Found in soils that contain large amounts of clay that swells and shrinks considerably with wetting and drying. Usually a succession of microbasins and microknolls in nearly level areas or of microvalleys and microridges parallel to the direction of the slope.

glacial drift Rock debris that has been transported by glaciers and deposited, either directly from the ice or from the meltwater. The debris may or may not be heterogeneous.

glacial till *See* till.

glaciofluvial deposits Material moved by glaciers and subsequently sorted and deposited by streams flowing from the melting ice. The deposits are stratified and may occur in the form of outwash plains, deltas, kames, eskers, and kame terraces.

gleyed A soil condition resulting from prolonged saturation with water and reducing conditions that manifest themselves in greenish or bluish colors throughout the soil mass or in mottles.

glomalin A protein-sugar group of molecules secreted by certain fungi resulting in a sticky hyphal surface thought to contribute to aggregate stability.

goethite, FeOOH A yellow-brown iron oxide mineral that accounts for the brown color in many soils.

granular structure Soil structure in which the individual grains are grouped into spherical aggregates with indistinct sides. Highly porous granules are commonly called *crumbs*. A well-granulated soil has the best structure for most ordinary crop plants. *See also* soil structure types.

granulation The process of producing granular materials. Commonly used to refer to the formation of soil structural granules, but also used to refer to the processing of powdery fertilizer materials into granules.

grassed waterway Broad and shallow channel, planted with grass (usually perennial species) that is designed to move surface water downslope without causing soil erosion.

gravitational potential That portion of the total *soil water potential* due to differences in elevation of the reference pool of pure water and that of the soil water. Since the soil water elevation is usually chosen to be higher than that of the reference pool, the gravitational potential is usually positive.

gravitational water Water that moves into, through, or out of the soil under the influence of gravity.

great group   A category in *Soil Taxonomy*. The classes in this category contain soils that have the same kind of horizons in the same sequence and have similar moisture and temperature regimes.

green manure   Plant material incorporated with the soil while green, or soon after maturity, for improving the soil.

greenhouse effect   The entrapment of heat by upper atmosphere gases, such as carbon dioxide, water vapor, and methane, just as glass traps heat for a greenhouse. Increases in the quantities of these gases in the atmosphere will likely result in global warming that may have serious consequences for humankind.

groundwater   Subsurface water in the zone of saturation that is free to move under the influence of gravity, often horizontally to stream channels.

grus   A sediment or soil material comprised of loose grains of coarse sand and fine gravel size composed of quartz, feldspar and rock fragments. Produced from rocks by physical weathering or selectively transported by burrowing insects.

gully erosion   The erosion process whereby water accumulates in narrow channels and, over short periods, removes the soil from this narrow area to considerable depths, ranging from 1 to 2 ft to as much as 23 to 30 m (75 to 100 ft).

gypsic horizon   A diagnostic subsurface horizon of secondary calcium sulfate enrichment that is more than 15 cm thick.

gypsum requirement   The quantity of gypsum required to reduce the exchangeable sodium percentage in a soil to an acceptable level.

halophyte   A plant that requires or tolerates a saline (high salt) environment.

hard armor   Pertains to the use of hard materials (such as large stones or concrete) to prevent soil and stream bank erosion by reducing the erosive force of flowing water. *See* soft armor.

hardpan   A hardened soil layer, in the lower A or in the B horizon, caused by cementation of soil particles with organic matter or with such materials as silica, sesquioxides, or calcium carbonate. The hardness does not change appreciably with changes in moisture content and pieces of the hard layer do not slake in water. *See also* caliche; claypan.

harrowing   A secondary broadcast tillage operation that pulverizes, smoothes, and firms the soil in seedbed preparation, controls weeds, or incorporates material spread on the surface.

heaving   The partial lifting of plants, buildings, roadways, fence posts, etc., out of the ground, as a result of freezing and thawing of the surface soil during the winter.

heavy metals   Those metals that have densities of 5.0 Mg/m or greater. Elements in soils include Cd, Co, Cr, Cu, Fe, Hg, Mn, Mo, Pb, and Zn.

heavy soil   (Obsolete in scientific use.) A soil with a high content of clay, and a high drawbar pull, hence difficult to cultivate.

hematite, Fe_2O_3   A red iron oxide mineral that contributes red color to many soils.

hemic material   *See* organic materials.

herbicide   A chemical that kills plants or inhibits their growth; intended for weed control.

herbivore   A plant-eating animal.

heterotroph   An organism capable of deriving energy for life processes only from the decomposition of organic compounds and incapable of using inorganic compounds as sole sources of energy or for organic synthesis. *Contrast with* autotroph.

histic epipedon   A diagnostic surface horizon consisting of a thin layer of organic soil material that is saturated with water at some period of the year unless artificially drained and that is at or near the surface of a mineral soil.

Histosols   An order in *Soil Taxonomy*. Soils formed from materials high in organic matter. Histosols with essentially no clay must have at least 20% organic matter by weight (about 78% by volume). This minimum organic matter content rises with increasing clay content to 30% (85% by volume) in soils with at least 60% clay.

horizon, soil   A layer of soil, approximately parallel to the soil surface, differing in properties and characteristics from adjacent layers below or above it. *See also* diagnostic horizons.

horticulture   The art and science of growing fruits, vegetables, and ornamental plants.

hue (color)   *See* Munsell color system.

humic acid   A mixture of variable or indefinite composition of dark organic substances, precipitated upon acidification of a dilute alkali extract from soil.

humic substances   A series of complex, relatively high molecular weight, brown- to black-colored organic substances that make up 60 to 80% of the soil organic matter and are generally quite resistant to ready microbial attack.

humid climate   Climate in regions where moisture, when distributed normally throughout the year, should not limit crop production. In cool climates annual precipitation may be as little as 25 cm; in hot climates, 150 cm or even more. Natural vegetation in uncultivated areas is forests.

humification   The processes involved in the decomposition of organic matter and leading to the formation of humus.

humin   The fraction of the soil organic matter that is not dissolved upon extraction of the soil with dilute alkali.

humus   That more or less stable fraction of the soil organic matter remaining after the major portions of added plant and animal residues have decomposed. Usually it is dark in color.

hydration   Chemical union between an ion or compound and one or more water molecules, the reaction being stimulated by the attraction of the ion or compound for either the hydrogen or the unshared electrons of the oxygen in the water.

hydraulic conductivity   An expression of the readiness with which a liquid, such as water, flows through a solid, such as soil, in response to a given potential gradient.

hydric soils   Soils that are water-saturated for long enough periods to produce reduced conditions and affect the growth of plants.

hydrogen bonding   Relatively low energy bonding exhibited by a hydrogen atom located between two highly electronegative atoms, such as nitrogen or oxygen.

hydrologic cycle   The circuit of water movement from the atmosphere to the Earth and back to the atmosphere through various stages or processes, as precipitation, interception, runoff, infiltration, percolation, storage, evaporation, and transpiration.

hydrolysis   A reaction with water that splits the water molecule into H^+ and OH^- ions. Molecules or atoms participating in such reactions are said to *hydrolyze*.

hydronium   A hydrated hydrogen ion (H_3O^+), the form of the hydrogen ion usually found in an aqueous system.

hydroperiod   The duration of the presence of surface water in seasonal wetlands.

hydroponics   Plant-production systems that use nutrient solutions and no solid medium to grow plants.

hydrostatic potential   *See* submergence potential.

hydrous mica   *See* fine-grained mica.

hydroxyapatite   A member of the apatite group of minerals rich in hydroxyl groups. A nearly insoluble calcium phosphate.

hygroscopic coefficient   The amount of moisture in a dry soil when it is in equilibrium with some standard relative humidity near a saturated atmosphere (about 98%), expressed in terms of percentage on the basis of oven-dry soil.

hyperaccumulator   A plant with unusually high capacity to take up certain elements from soil resulting in very high concentrations of these elements in the plant's tissues. Often pertaining to concentrations of heavy metals to 1% or more of the tissue dry matter.

hyperthermic A soil temperature class with mean annual temperatures >22 °C.

hypha (pl. hyphae) Filament of fungal cells. Actinomycetes also produce similar, but thinner, filaments of cells.

hypoxia State of oxygen deficiency in an environment so low as to restrict biological respiration (in water, typically less than 2 to 3 mg O_2/L).

hysteresis A relationship between two variables that changes depending on the sequences or starting point. An example is the relationship between soil water content and water potential, for which different curves describe the relationship when a soil is gaining water or losing it.

igneous rock Rock formed from the cooling and solidification of magma that has not been changed appreciably since its formation.

illite *See* fine-grained mica.

illuvial horizon A soil layer or horizon in which material carried from an overlying layer has been precipitated from solution or deposited from suspension. The layer of accumulation.

illuviation The process of deposition of soil material removed from one horizon to another in the soil; usually from an upper to a lower horizon in the soil profile. *See also* eluviation.

immature soil A soil with indistinct or only slightly developed horizons because of the relatively short time it has been subjected to the various soil-forming processes. A soil that has not reached equilibrium with its environment.

immobilization The conversion of an element from the inorganic to the organic form in microbial tissues or in plant tissues, thus rendering the element not readily available to other organisms or to plants.

imogolite A poorly crystalline aluminosilicate mineral with an approximate formula $SiO_2Al_2O_3 \cdot 2.5H_2O$; occurs mostly in soils formed from volcanic ash.

impervious Resistant to penetration by fluids or by roots.

improved fallow *See* fallow.

Inceptisols An order in *Soil Taxonomy*. Soils that are usually moist with pedogenic horizons of alteration of parent materials but not of illuviation. Generally, the direction of soil development is not yet evident from the marks left by various soil-forming processes or the marks are too weak to classify in another order.

induced systemic resistance Plant defense mechanisms activated by a chemical signal produced by a rhizosphere bacteria. Although the process begins in the soil, it may confer disease resistance to leaves or other aboveground tissues.

indurated (soil) Soil material cemented into a hard mass that will not soften on wetting. *See also* consistence; hardpan.

infiltration The downward entry of water into the soil.

infiltration capacity A soil characteristic determining or describing the *maximum* rate at which water *can* enter the soil under specified conditions, including the presence of an excess of water.

inner-sphere complex A relatively strong (not easily reversed) chemical association or bonding directly between a specific ion and specific atoms or groups of atoms in the surface structure of a soil colloid.

inoculation The process of introducing pure or mixed cultures of microorganisms into natural or artificial culture media.

inorganic compounds All chemical compounds in nature except compounds of carbon other than carbon monoxide, carbon dioxide, and carbonates.

insecticide A chemical that kills insects.

intergrade A soil that possesses moderately well-developed distinguishing characteristics of two or more genetically related great soil groups.

interlayer (mineralogy) Materials between layers within a given crystal, including cations, hydrated cations, organic molecules, and hydroxide groups or sheets.

internal surface The area of surface exposed within a clay crystal between the individual crystal layers. *Compare with* external surface.

interstratification Mixing of silicate layers within the structural framework of a given silicate clay.

ionic double layer The distribution of cations in the soil solution resulting from the simultaneous attraction toward colloid particles by the particle's negative charge and the tendency of diffusion and thermal forces to move the cations away from the colloid surfaces. Also described as a diffuse double layer or a diffuse electrical double layer.

ions Atoms, groups of atoms, or compounds that are electrically charged as a result of the loss of electrons (cations) or the gain of electrons (anions).

iron-pan An indurated soil horizon in which iron oxide is the principal cementing agent.

irrigation efficiency The ratio of the water actually consumed by crops on an irrigated area to the amount of water diverted from the source onto the area.

isomorphous substitution The replacement of one atom by another of similar size in a crystal lattice without disrupting or changing the crystal structure of the mineral.

isotopes Two or more atoms of the same element that have different atomic masses because of different numbers of neutrons in the nucleus.

joule The SI energy unit defined as a force of 1 Newton applied over a distance of 1 meter; 1 joule = 0.239 calorie.

K_d *See* distribution coefficient, K_d.

K_{oc} The distribution coefficient, K_d, calculated based on organic carbon content. $K_{oc} = K_d$/foc, where foc is the fraction of organic carbon.

kame A conical hill or ridge of sand or gravel deposited in contact with glacial ice.

kandic horizon A subsurface diagnostic horizon having a sharp clay increase relative to overlying horizons and having low-activity clays.

kaolinite An aluminosilicate mineral of the 1:1 crystal lattice group; that is, consisting of single silicon tetrahedral sheets alternating with single aluminum octahedral sheets.

K_{sat} Hydraulic conductivity when the soil is water saturated. *See also* hydraulic conductivity.

k-strategist An organism that maintains a relatively stable population by specializing in metabolism of resistant compounds that most other organisms cannot utilize. *Contrast with* r-strategist. *See also* autochthonous organisms.

labile A substance that is readily transformed by microorganisms or is readily available for uptake by plants.

lacustrine deposit Material deposited in lake water and later exposed either by lowering of the water level or by the elevation of the land.

land A broad term embodying the total natural environment of the areas of the Earth not covered by water. In addition to soil, its attributes include other physical conditions, such as mineral deposits and water supply; location in relation to centers of commerce, populations, and other land; the size of the individual tracts or holdings; and existing plant cover, works of improvement, and the like.

land capability classification A grouping of kinds of soil into special units, subclasses, and classes according to their capability for intensive use and the treatments required for sustained use. One such system has been prepared by the USDA Natural Resources Conservation Service.

land-use planning The development of plans for the uses of land that, over long periods, will best serve the general welfare, together with the formulation of ways and means for achieving such uses.

laterite An iron-rich subsoil layer found in some highly weathered humid tropical soils that, when exposed and allowed to dry, becomes very hard and will not soften when rewetted. When erosion removes the overlying layers, the laterite is exposed and a virtual pavement results. *See also* plinthite.

layer   (Clay mineralogy.) A combination in silicate clays of (tetrahedral and octahedral) sheets in a 1:1, 2:1, or 2:1:1 combination.

leaching   The removal of materials in solution from the soil by percolating waters. *See also* eluviation.

leaching requirement   The leaching fraction of irrigation water necessary to keep soil salinity from exceeding a tolerance level of the crop to be grown.

leaf area index   The ratio of the area of the total upper leaf surface of a plant canopy and the unit area on which the canopy is grown.

legume   A pod-bearing member of the Leguminosae family, one of the most important and widely distributed plant families. Includes many valuable food and forage species, such as peas, beans, peanuts, clovers, alfalfas, sweet clovers, lespedezas, vetches, and kudzu. Nearly all legumes are associated with nitrogen-fixing organisms.

lichen   A symbiotic relationship between fungi and cyanobacteria (blue-green algae) that enhances colonization of bare minerals and rocks. The fungi supply water and nutrients, the cyanobacteria the fixed nitrogen and carbohydrates from photosynthesis.

Liebig's law   The growth and reproduction of an organism are determined by the nutrient substance (oxygen, carbon dioxide, calcium, etc.) that is available in minimum quantity with respect to organic needs; the *limiting factor*. Also attributed to Sprengel.

light soil   (Obsolete in scientific use.) A coarse-textured soil; a soil with a low drawbar pull and hence easy to cultivate. *See also* coarse texture; soil texture.

lignin   The complex organic constituent of woody fibers in plant tissue that, along with cellulose, cements the cells together and provides strength. Lignins resist microbial attack and after some modification may become part of the soil organic matter.

lime (agricultural)   In strict chemical terms, calcium oxide. In practical terms, a material containing the carbonates, oxides, and/or hydroxides of calcium and/or magnesium used to neutralize soil acidity.

lime requirement   The mass of agricultural limestone, or the equivalent of other specified liming material, required to raise the pH of the soil to a desired value under field conditions.

limestone   A sedimentary rock composed primarily of calcite ($CaCO_3$). If dolomite ($CaCO_3 \cdot MgCO_3$) is present in appreciable quantities, it is called a *dolomitic limestone*.

limiting factor   *See* Liebig's law.

liquid limit (LL)   *See* Atterberg limits.

lithosequence   A group of related soils that differ, one from the other, in certain properties primarily as a result of parent material as a soil-forming factor.

loam   The textural-class name for soil having a moderate amount of sand, silt, and clay. Loam soils contain 7 to 27% clay, 28 to 50% silt, and 23 to 52% sand.

loamy   Intermediate in texture and properties between fine-textured and coarse-textured soils. Includes all textural classes with the words *loam* or *loamy* as a part of the class name, such as clay loam or loamy sand. *See also* loam; soil texture.

lodging   Falling over of plants, either by uprooting or stem breakage.

loess   Material transported and deposited by wind and consisting of predominantly silt-sized particles.

luxury consumption   The intake by a plant of an essential nutrient in amounts exceeding what it needs. For example, if potassium is abundant in the soil, alfalfa may take in more than it requires.

lysimeter   A device for measuring percolation (leaching) and evapotranspiration losses from a column of soil under controlled conditions.

macronutrient   A chemical element necessary in large amounts (usually 50 mg/kg in the plant) for the growth of plants. Includes C, H, O, N, P, K, Ca, Mg, and S. (*Macro* refers to quantity and not to the essentiality of the element.) *See also* micronutrient.

macropores   Larger soil pores, generally having a diameter greater than 0.06 mm, from which water drains readily by gravity.

map unit (mapping unit), soil   A conceptual group of one to many component soils, delineated or identified by the same name in a soil survey, that represent similar landscape areas. *See also* delineation, soil consociation, soil complex, soil association, and undifferentiated group.

marl   Soft and unconsolidated calcium carbonate, usually mixed with varying amounts of clay or other impurities.

marsh   Periodically wet or continually flooded area with the surface not deeply submerged. Covered dominantly with sedges, cattails, rushes, or other hydrophytic plants. Subclasses include freshwater and saltwater marshes.

mass flow   Movement of nutrients with the flow of water to plant roots.

matric potential   That portion of the total *soil water potential* due to the attractive forces between water and soil solids as represented through adsorption and capillarity. It will always be negative.

mature soil   A soil with well-developed soil horizons produced by the natural processes of soil formation and essentially in equilibrium with its present environment.

maximum retentive capacity   The average moisture content of a disturbed sample of soil, 1 cm high, which is at equilibrium with a water table at its lower surface.

mechanical analysis   (Obsolete term.) *See* particle size analysis; particle size distribution.

medium texture   Intermediate between fine-textured and coarse-textured (soils). It includes the following textural classes: very fine sandy loam, loam, silt loam, and silt.

melanic epipedon   A diagnostic surface horizon formed in volcanic parent material that contains more than 6% organic carbon, is dark in color, and has a very low bulk density and high anion adsorption capacity.

mellow soil   A very soft, very friable, porous soil without any tendency toward hardness or harshness. *See also* consistence.

mesic   A soil temperature class with mean annual temperature 8 to 15 °C.

mesofauna   Animals of medium size, between approximately 2 and 0.2 mm in diameter.

mesophilic   Pertaining to moderate temperatures in the range of 15 to 35 °C, the range in which mesophilic organisms grow best and in which mesophilic composting takes place.

metamorphic rock   A rock that has been greatly altered from its previous condition through the combined action of heat and pressure. For example, marble is a metamorphic rock produced from limestone, gneiss is produced from granite, and slate is produced from shale.

methane, CH_4   An odorless, colorless gas commonly produced under anaerobic conditions. When released to the upper atmosphere, methane contributes to global warming. *See also* greenhouse effect.

micas   Primary aluminosilicate minerals in which two silica tetrahedral sheets alternate with one alumina/magnesia octahedral sheet with entrapped potassium atoms fitting between sheets. They separate readily into visible sheets or flakes.

microfauna   That part of the animal population which consists of individuals too small to be clearly distinguished without the use of a microscope. Includes protozoans and nematodes.

microflora   That part of the plant population which consists of individuals too small to be clearly distinguished without the use of a microscope. Includes actinomycetes, algae, bacteria, and fungi.

micronutrient   A chemical element necessary in only extremely small amounts (<50 mg/kg in the plant) for the growth of plants. Examples are B, Cl, Cu, Fe, Mn, and Zn. (*Micro* refers to the amount used rather than to its essentiality.) *See also* macronutrient.

micropores Relatively small soil pores, generally found within structural aggregates and having a diameter less than 0.06 mm. *Contrast to* macropores.

microrelief Small-scale local differences in topography, including mounds, swales, or pits that are only 1 m or so in diameter and with elevation differences of up to 2 m. *See also* gilgai.

mineral (1) An inorganic compound of defined composition found in rocks. (2) An adjective meaning inorganic.

mineral nutrient An element in inorganic form used by plants or animals.

mineral soil A soil consisting predominantly of, and having its properties determined predominantly by, mineral matter. Usually contains <20% organic matter, but may contain an organic surface layer up to 30 cm thick.

mineralization The conversion of an element from an organic form to an inorganic state as a result of microbial decomposition.

minimum tillage *See* tillage, conservation.

minor element (Obsolete term.) *See* micronutrient.

moderately coarse texture Consisting predominantly of coarse particles. In soil textural classification, it includes all the sandy loams except the very fine sandy loam. *See also* coarse texture.

moderately fine texture Consisting predominantly of intermediate-sized (soil) particles or with relatively small amounts of fine or coarse particles. In soil textural classification, it includes clay loam, sandy loam, sandy clay loam, and silty clay loam. *See also* fine texture.

moisture potential *See* soil water potential.

mollic epipedon A diagnostic surface horizon of mineral soil that is dark colored and relatively thick, contains at least 0.6% organic carbon, is not massive and hard when dry, has a base saturation of more than 50%, has less than 250 mg/kg P_2O_5 soluble in 1% citric acid, and is dominantly saturated with bivalent cations.

Mollisols An order in *Soil Taxonomy*. Soils with nearly black, organic-rich surface horizons and high supply of bases. They have mollic epipedons and base saturation greater than 50% in any cambic or argillic horizon. They lack the characteristics of Vertisols and must not have oxic or spodic horizons.

molybdenosis A nutritional disease of ruminant animals in which high Mo in the forage interferes with copper absorption.

montmorillonite An aluminosilicate clay mineral in the smectite group with a 2:1 expanding crystal lattice, with two silicon tetrahedral sheets enclosing an aluminum octahedral sheet. Isomorphous substitution of magnesium for some of the aluminum has occurred in the octahedral sheet. Considerable

expansion may be caused by water moving between silica sheets of contiguous layers.

mor Raw humus; type of forest humus layer of unincorporated organic material, usually matted or compacted or both; distinct from the mineral soil, unless the latter has been blackened by washing in organic matter.

moraine An accumulation of drift, with an initial topographic expression of its own, built within a glaciated region chiefly by the direct action of glacial ice. Examples are ground, lateral, recessional, and terminal moraines.

morphology, soil The constitution of the soil, including the texture, structure, consistence, color, and other physical, chemical, and biological properties of the various soil horizons that make up the soil profile.

mottling Spots or blotches of different color or shades of color interspersed with the dominant color.

mucigel The gelatinous material at the surface of roots grown in unsterilized soil.

muck Highly decomposed organic material in which the original plant parts are not recognizable. Contains more mineral matter and is usually darker in color than peat. *See also* muck soil; peat.

muck soil (1) A soil containing 20 to 50% organic matter. (2) An organic soil in which the organic matter is well decomposed.

mulch Any material such as straw, sawdust, leaves, plastic film, and loose soil that is spread upon the surface of the soil to protect the soil and plant roots from the effects of raindrops, soil crusting, freezing, evaporation, etc.

mulch tillage *See* tillage, conservation.

mull A humus-rich layer of forested soils consisting of mixed organic and mineral matter. A mull blends into the upper mineral layers without an abrupt change in soil characteristics.

Munsell color system A color designation system that specifies the relative degrees of the three simple variables of color:

chroma The relative purity, strength, or saturation of a color.

hue The chromatic gradation (rainbow) of light that reaches the eye.

value The degree of lightness or darkness of the color.

mycelium A stringlike mass of individual fungal or actinomycetes hyphae.

myco Prefix designating an association or relationship with a fungus (e.g., mycotoxins are toxins produced by a fungus).

mycorrhiza The association, usually symbiotic, of fungi with the roots of seed plants. *See also* ectotrophic mycorrhiza; endotrophic mycorrhiza; arbuscular mycorrhiza.

natric horizon A diagnostic subsurface horizon that satisfies the requirements of an argillic horizon, but that also has prismatic, columnar, or blocky structure and a subhorizon having more than 15% saturation with exchangeable sodium.

necrosis Death associated with discoloration and dehydration of all or parts of plant organs, such as leaves.

nematodes Very small (most are microscopic) unsegmented round worms. In soils they are abundant and perform many important functions in the soil food web. Some are plant parasites and considered pests.

neutral soil A soil in which the surface layer, at least to normal plow depth, is neither acid nor alkaline in reaction. In practice this means the soil is within the pH range of 6.6 to 7.3. *See also* acid soil; alkaline soil; pH; reaction, soil.

nitrate depression period A period of time, beginning shortly after the addition of fresh, highly carbonaceous organic materials to a soil, during which decomposer microorganisms have removed most of the soluble nitrate from the soil solution.

nitrification The biochemical oxidation of ammonium to nitrate, predominantly by autotrophic bacteria.

nitrogen assimilation The incorporation of nitrogen into organic cell substances by living organisms.

nitrogen cycle The sequence of chemical and biological changes undergone by nitrogen as it moves from the atmosphere into water, soil, and living organisms, and upon death of these organisms (plants and animals) is recycled through a part or all of the entire process.

nitrogen fixation The biological conversion of elemental nitrogen (N_2) to organic combinations or to forms readily utilized in biological processes.

nodule bacteria *See* rhizobia.

non-acid cations Those cations that do not react with water by hydrolysis to release H^+ ions to the soil solution. These cations do not remove hydroxyl ions from solution, but form strongly dissociated bases such as potassium hydroxide ($K^+ + OH$). Formerly called *base cations* or *base-forming cations* in soil science literature.

non-acid saturation The proportion or percentage of a cation-exchange site occupied by non-acid cations. Formerly termed *base saturation*.

nonhumic substances The portion of soil organic matter comprised of relatively low molecular weight organic substances; mostly identifiable biomolecules.

nonlimiting water range The region bounded by the upper and lower soil water content over which water, oxygen, and mechanical resistance are not limiting to plant growth. *Compare with* available water.

nonpoint source   A pollution source that cannot be traced back to a single origin or source. Examples include water runoff from urban areas and leaching from croplands.

no-tillage   *See* tillage, conservation.

nucleic acids   Complex organic acids found in the nuclei of plant and animal cells; may be combined with proteins as nucleoproteins.

O horizon   Organic horizon of mineral soils.

ochric epipedon   A diagnostic surface horizon of mineral soil that is too light in color, too high in chroma, too low in organic carbon, or too thin to be a plaggen, mollic, umbric, anthropic, or histic epipedon, or that is both hard and massive when dry.

octahedral sheet   Sheet of horizontally linked, octahedral-shaped units that serve as the basic structural components of silicate (clay) minerals. Each unit consists of a central, six-coordinated metallic atom (e.g., Al, Mg, or Fe) surrounded by six hydroxyl groups that, in turn, are linked with other nearby metal atoms, thereby serving as interunit linkages that hold the sheet together.

oligotrophic   Environments, such as soils or lakes, which are poor in nutrients.

order, soil   The category at the highest level of generalization in *Soil Taxonomy*. The properties selected to distinguish the orders are reflections of the degree of horizon development and the kinds of horizons present.

organic farming   A system/philosophy of agriculture that does not allow the use of synthetic chemicals to produce plant and animal products, but instead emphasizes the management of soil organic matter and biological processes. In many countries, products are officially certified as being organic if inspections confirm that they were grown by these methods.

organic fertilizer   By-product from the processing of animal or vegetable substances that contain sufficient plant nutrients to be of value as fertilizers.

organic soil   A soil in which more than half of the profile thickness is comprised of organic soil materials.

organic soil materials   (As used in *Soil Taxonomy*.) (1) Saturated with water for prolonged periods unless artificially drained and having 18% or more organic carbon (by weight) if the mineral fraction is more than 60% clay, more than 12% organic carbon if the mineral fraction has no clay, or between 12 and 18% carbon if the clay content of the mineral fraction is between 0 and 60%. (2) Never saturated with water for more than a few days and having more than 20% organic carbon.

Histosols develop on these organic soil materials. There are three kinds of organic materials:

> **fibric materials**   The least decomposed of all the organic soil materials, containing very high amounts of fiber that are well preserved and readily identifiable as to botanical origin; with very low bulk density.

> **hemic materials**   Intermediate in degree of decomposition of organic materials between the less decomposed fibric and the more decomposed sapric materials.

> **sapric materials**   The most highly decomposed of the organic materials, having the highest bulk density, least amount of plant fiber, and lowest water content at saturation.

orographic   Influenced by mountains (Greek *oros*). Used in reference to increased precipitation on the windward side of a mountain range induced as clouds rise over the mountain, leaving a *rain shadow* of reduced precipitation on the leeward side.

ortstein   An indurated layer in the B horizon of Spodosols in which the cementing material consists of illuviated sesquioxides (mostly iron) and organic matter.

osmotic potential   That portion of the total *soil water potential* due to the presence of solutes in soil water. It will generally be negative.

osmotic pressure   Pressure exerted in living bodies as a result of unequal concentrations of salts on both sides of a cell wall or membrane. Water moves from the area having the lower salt concentration through the membrane into the area having the higher salt concentration and, therefore, exerts additional pressure on the side with higher salt concentration.

outer-sphere complex   A relatively weak (easily reversed) chemical association or general attraction between an ion and an oppositely charged soil colloid via mutual attraction for intervening water molecules.

outwash plain   A deposit of coarse-textured materials (e.g., sands and gravels) left by streams of meltwater flowing from receding glaciers.

oven-dry soil   Soil that has been dried at 105 °C until it reaches constant weight.

oxic horizon   A diagnostic subsurface horizon that is at least 30 cm thick and is characterized by the virtual *absence* of weatherable primary minerals or 2:1 lattice clays and the *presence* of 1:1 lattice clays and highly insoluble minerals, such as quartz sand, hydrated oxides of iron and aluminum, low cation exchange capacity, and small amounts of exchangeable bases.

oxidation   The loss of electrons by a substance; therefore, a gain in positive valence charge and, in some cases, the chemical combination with oxygen gas.

oxidation ditch   An artificial open channel for partial digestion of liquid organic wastes in which the wastes are circulated and aerated by a mechanical device.

oxidation-reduction potential   *See* E_h; pe.

Oxisols   An order in *Soil Taxonomy*. Soils with residual accumulations of low-activity clays, free oxides, kaolin, and quartz. They are mostly in tropical climates.

pans   Horizons or layers in soils that are strongly compacted, indurated, or very high in clay content. *See also* caliche; claypan; fragipan; hardpan.

parent material   The unconsolidated and more or less chemically weathered mineral or organic matter from which the solum of soils is developed by pedogenic processes.

particle density   The mass per unit volume of the soil particles. In technical work, usually expressed as metric tons per cubic meter (Mg/m^3) or grams per cubic centimeter (g/cm^3).

particle size   The effective diameter of a particle measured by sedimentation, sieving, or micrometric methods.

particle size analysis   Determination of the various amounts of the different separates in a soil sample, usually by sedimentation, sieving, micrometry, or combinations of these methods.

particle size distribution   The amounts of the various soil separates in a soil sample, usually expressed as weight percentages.

particulate organic matter   A microbially active fraction of soil organic matter consisting largely of fine particles of partially decomposed plant tissue.

partitioning   The distribution of organic chemicals (such as pollutants) into a portion that dissolves in the soil organic matter and a portion that remains undissolved in the soil solution.

pascal   An SI unit of pressure equal to 1 Newton per square meter.

pe   The negative logarithm of the electron activity, a unitless measure of redox potential. Low pe values signifiy high e^- activities and correspond to highly reducing chemical environment, while high pe values signify low e^- activities and correspond to a highly oxidizing chemical environment. At 25 °C, pe = $E_h/0.059$ Volt, where E_h is a similar measure of redox potential measured in volts (see also E_h).

peat   Unconsolidated soil material consisting largely of undecomposed, or only slightly decomposed, organic matter accumulated under conditions of excessive moisture. *See also* organic soil materials; peat soil.

peat soil   An organic soil containing more than 50% organic matter. Used in the United States to refer to the stage of decomposition of the organic matter, *peat* referring to the slightly decomposed or undecomposed deposits and *muck* to the highly decomposed materials. *See also* muck; muck soil; peat.

ped A unit of soil structure such as an aggregate, crumb, prism, block, or granule, formed by natural processes (in contrast to a *clod*, which is formed artificially).

pedology The science that deals with the formation, morphology, and classification of soil bodies as landscape components.

pedon The smallest volume that can be called *a soil*. It has three dimensions. It extends downward to the depth of plant roots or to the lower limit of the genetic soil horizons. Its lateral cross section is roughly hexagonal and ranges from 1 to 10 m^2 in size, depending on the variability in the horizons.

pedosphere The conceptual zone within the ecosystem consisting of soil bodies or directly influenced by them. A zone or sphere of activity in which mineral, water, air, and biological components come together to form soils. Usage is parallel to that for "atmosphere" or "biosphere."

pedoturbation Physical disturbance and mixing of soil horizons by such forces as burrowing animals (faunal pedoturbation) or frost churning (cryoturbation).

peneplain A once high, rugged area that has been reduced by erosion to a lower, gently rolling surface resembling a plain.

penetrability The ease with which a probe can be pushed into the soil. May be expressed in units of distance, speed, force, or work depending on the type of penetrometer used.

penetrometer An instrument consisting of a rod with a cone-shaped tip and a means of measuring the force required to push the rod into a specified increment of soil.

perc test *See* percolation test.

percolation, soil water The downward movement of water through soil. Especially, the downward flow of water in saturated or nearly saturated soil at hydraulic gradients of the order of 1.0 or less.

percolation test A measurement of the rate of percolation of water in a soil profile, usually to determine the suitability of a soil for use as a septic tank drain field.

perforated plastic pipe Pipe, sometimes flexible, with holes or slits in it that allow the entrance and exit of air and water. Used for soil drainage and for septic effluent spreading into soil.

permafrost (1) Permanently frozen material underlying the solum. (2) A perennially frozen soil horizon.

permanent charge *See* constant charge.

permanent wilting point *See* wilting point.

permeability, soil The ease with which gases, liquids, or plant roots penetrate or pass through a bulk mass of soil or a layer of soil.

petrocalcic horizon A diagnostic subsurface horizon that is a continuous, indurated calcic horizon cemented by calcium carbonate and, in some places, with magnesium carbonate. It cannot be penetrated with a spade or auger when dry; dry fragments do not slake in water; and it is impenetrable by roots.

petrogypsic horizon A diagnostic subsurface horizon that is a continuous, strongly cemented, massive gypsic horizon that is cemented by calcium sulfate. It can be chipped with a spade when dry. Dry fragments do not slake in water and it is impenetrable by roots.

pH, soil The negative logarithm of the hydrogen ion activity (concentration) in the soil solution. The degree of acidity (or alkalinity) of a soil as determined by means of a glass or other suitable electrode or indicator at a specified moisture content or soil-to-water ratio, and expressed in terms of the pH scale.

pH-dependent charge That portion of the total charge of the soil particles that is affected by, and varies with, changes in pH.

phase, soil A subdivision of a soil series or other unit of classification having characteristics that affect the use and management of the soil but do not vary sufficiently to differentiate it as a separate series. Included are such characteristics as degree of slope, degree of erosion, and content of stones.

photomap A mosaic map made from aerial photographs to which place names, marginal data, and other map information have been added.

phyllosphere The leaf surface.

physical properties (of soils) Those characteristics, processes, or reactions of a soil that are caused by physical forces and that can be described by, or expressed in, physical terms or equations. Examples of physical properties are bulk density, water-holding capacity, hydraulic conductivity, porosity, pore-size distribution, and so on.

physical weathering The breakdown of rock and mineral particles into smaller particles by physical forces such as frost action. *See also* weathering.

phytotoxic substances Chemicals that are toxic to plants.

placic horizon A diagnostic subsurface horizon of a black to dark reddish mineral soil that is usually thin but that may range from 1 to 25 mm in thickness. The placic horizon is commonly cemented with iron and is slowly permeable or impenetrable to water and roots.

plaggen epipedon A diagnostic surface horizon that is human-made and more than 50 cm thick. Formed by long-continued manuring and mixing.

plant nutrients *See* essential element.

plastic limit (PL) *See* Atterberg limits.

plastic soil A soil capable of being molded or deformed continuously and permanently, by relatively moderate pressure, into various shapes. *See also* consistence.

platy Consisting of soil aggregates that are developed predominantly along the horizontal axes; laminated; flaky.

plinthite (brick) A highly weathered mixture of sesquioxides of iron and aluminum with quartz and other diluents that occurs as red mottles and that changes irreversibly to hardpan upon alternate wetting and drying.

plow layer The soil ordinarily moved when land is plowed; equivalent to *surface soil*.

plow pan A subsurface soil layer having a higher bulk density and lower total porosity than layers above or below it, as a result of pressure applied by normal plowing and other tillage operations.

plowing A primary broad-base tillage operation that is performed to shatter soil uniformly with partial to complete inversion.

point of zero charge The pH value of a solution in equilibrium with a particle whose net charge, from all sources, is zero.

point source A pollution source that can be traced back to its origin, which is usually an effluent discharge pipe. Examples are a wastewater treatment plant or a factory. *Opposite of* nonpoint source.

polypedon (As used in *Soil Taxonomy*:) Two or more contiguous pedons, all of which are within the defined limits of a single soil series; commonly referred to as a *soil individual*.

pool A portion of a larger store of a substance defined by kinetic or theoretical properties. For example, the passive pool organic matter is defined by its very slow rate of microbial turnover. *Compare to* fraction.

pore size distribution The volume of the various sizes of pores in a soil. Expressed as percentages of the bulk volume (soil plus pore space).

porosity, soil The volume percentage of the total soil bulk not occupied by solid particles.

potential acidity The acidity that could potentially be formed if reduced sulfur compounds in a potential acid sulfate soil were to become oxidized.

precision farming The spatially variable management of a field or farm based on information specific to the soil or crop characteristics of many very small subunits of land. This technique commonly uses variable rate equipment, geo positioning systems, and computer controls.

preferential flow   Nonuniform movement of water and its solutes through a soil along certain pathways, which are often macropores.

primary consumer   An organism that subsists on plant material.

primary mineral   A mineral that has not been altered chemically since deposition and crystallization from molten lava.

primary producer   An organism (usually a photosynthetic plant) that creates organic, energy-rich material from inorganic chemicals, solar energy, and water.

primary tillage   *See* tillage, primary.

priming effect   The increased decomposition of relatively stable soil humus under the influence of much enhanced, generally biological, activity resulting from the addition of fresh organic materials to a soil.

prismatic soil structure   A soil structure type with prismlike aggregates that have a vertical axis much longer than the horizontal axes.

Proctor test   A laboratory procedure that indicates the maximum achievable bulk density for a soil and the optimum water content for compacting a soil.

productivity, soil   The capacity of a soil for producing a specified plant or sequence of plants under a specified system of management. Productivity emphasizes the capacity of soil to produce crops and should be expressed in terms of yields.

profile, soil   A vertical section of the soil through all its horizons and extending into the parent material.

prokaryote   An organism whose cells do not have a distinct nucleus.

protein   Any of a group of nitrogen-containing organic compounds formed by the polymerization of a large number of amino acid molecules and that, upon hydrolysis, yield these amino acids. They are essential parts of living matter and are one of the essential food substances of animals.

protonation   Attachment of protons (H$^+$ ions) to exposed OH groups on the surface of soil particles, resulting in an overall positive charge on the particle surface.

protozoa   One-celled eukaryotic organisms, such as amoeba.

puddled soil   Dense, massive soil artificially compacted when wet and having no aggregated structure. The condition commonly results from the tillage of a clayey soil when it is wet.

rain, acid   *See* acid rain.

reaction, soil   (No longer used in soil science.) The degree of acidity or alkalinity of a soil, usually expressed as a pH value or by terms ranging from extremely acid for pH values <4.5 to very strongly alkaline for pH values >9.0.

reactive nitrogen   All forms of nitrogen that are readily available to biota (mainly ammonia, ammonium, and nitrate with smaller quantities of other compounds including nitrogen oxide gases) as opposed to unreactive nitrogen that exists mostly as inert N$_2$ gas.

recharge area   A geographic area in which an otherwise confined aquifer is exposed to surficial percolation of water to recharge the groundwater in the aquifer.

redox concentrations   Zones of apparent accumulations of Fe-Mn oxides in soils.

redox depletions   Zones of low chroma (<2) where Fe-Mn oxides, and in some cases clay, have been stripped from the soil.

redox potential   The electrical potential (measured in volts or millivolts) of a system due to the tendency of the substances in it to give up or acquire electrons.

redoximorphic features   Soil properties associated with wetness that result from reduction and oxidation of iron and manganese compounds after saturation and desaturation with water. *See also* redox concentrations; redox depletions.

reduction   The gain of electrons, and therefore the loss of positive valence charge, by a substance. In some cases, a loss of oxygen or a gain of hydrogen is also involved.

regolith   The unconsolidated mantle of weathered rock and soil material on the Earth's surface; loose earth materials above solid rock. (Approximately equivalent to the term *soil* as used by many engineers.)

relief   The relative differences in elevation between the upland summits and the lowlands or valleys of a given region.

residual material   Unconsolidated and partly weathered mineral materials accumulated by disintegration of consolidated rock in place.

resilience   The capacity of a soil (or other ecosystem) to return to its original state after a disturbance.

rhizobacteria   Bacteria specially adapted to colonizing the surface of plant roots and the soil immediately around plant roots. Some have effects that promote plant growth, while others have effects that are deleterious to plants.

rhizobia   Bacteria capable of living symbiotically with higher plants, usually in nodules on the roots of legumes, from which they receive their energy, and capable of converting atmospheric nitrogen to combined organic forms; hence the term *symbiotic nitrogen-fixing bacteria.* (Derived from the generic name *Rhizobium.*)

rhizoplane   The root surface–soil interface. Used to describe the habitat of root-surface-dwelling microorganisms.

rhizosphere   That portion of the soil in the immediate vicinity of plant roots in which the abundance and composition of the microbial population are influenced by the presence of roots.

rill   A small, intermittent water course with steep sides; usually only a few centimeters deep and hence no obstacle to tillage operations.

rill erosion   An erosion process in which numerous small channels of only several centimeters in depth are formed; occurs mainly on recently cultivated soils. *See also* rill.

rip rap   Coarse rock fragments, stones, or boulders placed along a waterway or hillside to prevent erosion.

riparian zone   The area, both above and below the ground surface, that borders a river.

rock   The material that forms the essential part of the earth's solid crust, including loose incoherent masses such as sand and gravel, as well as solid masses of granite and limestone.

root interception   Acquisition of nutrients by a root as a result of the root growing into the vicinity of the nutrient source.

root nodules   Swollen growths on plant roots. Often in reference to those in which symbiotic microorganisms live.

rotary tillage   *See* tillage, rotary.

r-strategist   Opportunistic organisms with short reproductive times that allow them to respond rapidly to the presence of easily metabolized food sources. *Contrast with* k-strategist. *See also* zymogenous organisms.

runoff   The portion of the precipitation on an area that is discharged from the area through stream channels. That which is lost without entering the soil is called *surface runoff* and that which enters the soil before reaching the stream is called *groundwater runoff* or *seepage flow* from groundwater. (In soil science *runoff* usually refers to the water lost by surface flow; in geology and hydraulics *runoff* usually includes both surface and subsurface flow.)

salic horizon   A diagnostic subsurface horizon of enrichment with secondary salts more soluble in cold water than gypsum. A salic horizon is 15 cm or more in thickness.

saline seep   An area of land in which saline water seeps to the surface, leaving a high salt concentration behind as the water evaporates.

saline soil   A nonsodic soil containing sufficient soluble salts to impair its productivity. The conductivity of a saturated extract is >4 dS/m, the exchangeable sodium adsorption ratio is less than about 13, and the pH is <8.5.

saline–sodic soil   A soil containing sufficient exchangeable sodium to interfere with the growth of most crop plants and containing appreciable quantities of soluble salts. The exchangeable sodium adsorption ratio is >13, the conductivity of the saturation extract is >4 dS/m (at 25 °C), and the pH is usually 8.5 or less in the saturated soil.

salinization The process of accumulation of salts in soil.

saltation Particle movement in water or wind where particles skip or bounce along the stream bed or soil surface.

sand A soil particle between 0.05 and 2.0 mm in diameter; a soil textural class.

sapric materials *See* organic soil materials.

saprolite Soft, friable, weathered bedrock that retains the fabric and structure of the parent rock but is porous and can be dug with a spade.

saprophyte An organism that lives on dead organic material.

saturated paste extract The extract from a saturated soil paste, the electrical conductivity E_c of which gives an indirect measure of salt content in a soil.

saturation extract The solution extracted from a saturated soil paste.

saturation percentage The water content of a saturated soil paste, expressed as a dry weight percentage.

savanna (savannah) A grassland with scattered trees, either as individuals or clumps. Often a transitional type between true grassland and forest.

second bottom The first terrace above the normal floodplain of a stream.

secondary mineral A mineral resulting from the decomposition of a primary mineral or from the reprecipitation of the products of decomposition of a primary mineral. *See also* primary mineral.

sediment Transported and deposited particles or aggregates derived from soils, rocks, or biological materials.

sedimentary rock A rock formed from materials deposited from suspension or precipitated from solution and usually being more or less consolidated. The principal sedimentary rocks are sandstones, shales, limestones, and conglomerates.

seedbed The soil prepared to promote the germination of seed and the growth of seedlings.

self-mulching soil A soil in which the surface layer becomes so well aggregated that it does not crust and seal under the impact of rain but instead serves as a surface mulch upon drying.

semiarid Term applied to regions or climates where moisture is more plentiful than in arid regions but still definitely limits the growth of most crop plants. Natural vegetation in uncultivated areas is short grasses.

separate, soil One of the individual-sized groups of mineral soil particles—sand, silt, or clay.

septic tank An underground tank used in the deposition of domestic wastes. Organic matter decomposes in the tank, and the effluent is drained into the surrounding soil.

series, soil The soil series is a subdivision of a family in *Soil Taxonomy* and consists of soils that are similar in all major profile characteristics.

sewage effluent The liquid part of sewage or wastewater; it is usually treated to remove some portion of the dissolved organic compounds and nutrients present from the original sewage.

sewage sludge Settled sewage solids combined with varying amounts of water and dissolved materials, removed from sewage by screening, sedimentation, chemical precipitation, or bacterial digestion. Also called *biosolids* if certain quality standards are met.

shear Force, as of a tillage implement, acting at right angles to the direction of movement.

sheet (Mineralogy) A flat array of more than one atomic thickness and composed of one or more levels of linked coordination polyhedra. A sheet is thicker than a plane and thinner than a layer. Examples: tetrahedral sheet, octahedral sheet.

sheet erosion The removal of a fairly uniform layer of soil from the land surface by runoff water.

shelterbelt A wind barrier of living trees and shrubs established and maintained for protection of farm fields. Syn. *windbreak*.

shifting cultivation A farming system in which land is cleared, the debris burned, and crops grown for 2 to 3 years. When the farmer moves on to another plot, the land is then left idle for 5 to 15 years; then the burning and planting process is repeated.

short-range order minerals Minerals, such as allophane, whose structural framework consists of short distances of well-ordered crystalline structure interspersed with distances of noncrystalline amorphous materials.

shrinkage limit (SL) The water content above which a mass of soil material will swell in volume, but below which it will shrink no further.

side-dressing The application of fertilizer alongside row-crop plants, usually on the soil surface. Nitrogen materials are most commonly side-dressed.

siderophore A nonporphyrin metabolite secreted by certain microorganisms that forms a highly stable coordination compound with iron.

silica/alumina ratio The molecules of silicon dioxide (SiO_2) per molecule of aluminum oxide (Al_2O_3) in clay minerals or in soils.

silica/sesquioxide ratio The molecules of silicon dioxide (SiO_2) per molecule of aluminum oxide (Al_2O_3) plus ferric oxide (Fe_2O_3) in clay minerals or in soils.

silt (1) A soil separate consisting of particles between 0.05 and 0.002 mm in equivalent diameter. (2) A soil textural class.

silting The deposition of waterborne sediments in stream channels, lakes, reservoirs, or on floodplains, usually resulting from a decrease in the velocity of the water.

site index A quantitative evaluation of the productivity of a soil for forest growth under the existing or specified environment.

slash-and-burn *See* shifting cultivation.

slick spots Small areas in a field that are slick when wet because of a high content of alkali or exchangeable sodium.

slickensides Stress surfaces that are polished and striated and are produced by one mass sliding past another.

slope The degree of deviation of a surface from horizontal, measured in a numerical ratio, percent, or degrees.

slow fraction (of soil organic matter) That portion of soil organic matter that can be metabolized with great difficulty by the microorganisms in the soil and therefore has a slow turnover rate with a half-life in the soil ranging from a few years to a few decades. Often this fraction is the product of some previous decomposition.

smectite A group of silicate clays having a 2:1-type lattice structure with sufficient isomorphous substitution in either or both the tetrahedral and octahedral sheets to give a high interlayer negative charge and high cation exchange capacity and to permit significant interlayer expansion and consequent shrinking and swelling of the clay. Montmorillonite, beidellite, and saponite are in the smectite group.

sodic soil A soil that contains sufficient sodium to interfere with the growth of most crop plants, and in which the sodium adsorption ratio is 13 or greater.

sodium adsorption ratio (SAR)

$$SAR = \frac{[Na^+]}{\sqrt{1/2([Ca^{2+}] + [Mg^{2+}])}}$$

where the cation concentrations are in millimoles of charge per liter ($mmol_c/L$).

soft armor The bioengineering use of organic and/or inorganic materials combined with plants to create a living vegetation barrier of protection against erosion.

soil (1) A dynamic natural body composed of mineral and organic solids, gases, liquids, and living organisms which can serve as a medium for plant growth. (2) The collection of natural bodies occupying parts of the Earth's surface that is capable of supporting plant growth and that has properties resulting from the integrated effects of climate and living organisms acting upon parent material, as conditioned by topography, over periods of time.

soil air   The soil atmosphere; the gaseous phase of the soil, being that volume not occupied by soil or liquid.

soil alkalinity   The degree or intensity of alkalinity of a soil, expressed by a value >7.0 on the pH scale.

soil amendment   Any material, such as lime, gypsum, sawdust, or synthetic conditioner, that is worked into the soil to make it more amenable to plant growth.

soil association   A group of defined and named taxonomic soil units occurring together in an individual and characteristic pattern over a geographic region, comparable to plant associations in many ways.

soil auger   A tool used to bore small holes up to several meters deep in soils in order to bring up samples of material from various soil layers. It consists of a long T-handle attached to either a cylinder with twisted teeth or a screwlike bit.

soil classification   The taxonomy or systematic arrangement of soils into groups or categories on the basis of their characteristics. *See* taxonomy, soil; order; suborder; great group; subgroup; family; and series.

soil complex   A mapping unit used in detailed soil surveys where two or more defined taxonomic units are so intimately intermixed geographically that it is undesirable or impractical, because of the scale being used, to separate them. A more intimate mixing of smaller areas of individual taxonomic units than that described under *soil association*.

soil compressibility   The property of a soil pertaining to its capacity to decrease in bulk volume when subjected to a load.

soil conditioner   Any material added to a soil for the purpose of improving its physical condition.

soil conservation   A combination of all management and land-use methods that safeguard the soil against depletion or deterioration caused by nature and/or humans.

soil consociation   A kind of soil map unit that is named for the dominant soil taxon in the delineation, and in which at least half of the pedons are of the named soil taxon, and most of the remaining pedons are so similar as to not affect most interpretations.

soil correlation   The process of defining, mapping, naming, and classifying the kinds of soils in a specific soil survey area, the purpose being to ensure that soils are adequately defined, accurately mapped, and uniformly named.

soil erosion   *See* erosion.

soil fertility   *See* fertility, soil.

soil genesis   *See* genesis, soil.

soil geography   A subspecialization of physical geography concerned with the areal distributions of soil types.

soil horizon   *See* horizon, soil.

soil loss tolerance (T value)   (1) The maximum average annual soil loss that will allow continuous cropping and maintain soil productivity without requiring additional management inputs. (2) The maximum soil erosion loss that is offset by the theoretical maximum rate of soil development, which will maintain an equilibrium between soil losses and gains.

soil management   The sum total of all tillage operations, cropping practices, fertilizer, lime, and other treatments conducted on or applied to a soil for the production of plants.

soil map   A map showing the distribution of soil types or other soil mapping units in relation to the prominent physical and cultural features of the Earth's surface.

soil moisture potential   *See* soil water potential.

soil monolith   A vertical section of a soil profile removed from the soil and mounted for display or study.

soil morphology   The physical constitution, particularly the structural properties, of a soil profile as exhibited by the kinds, thicknesses, and arrangement of the horizons in the profile, and by the texture, structure, consistence, and porosity of each horizon.

soil order   *See* order, soil.

soil organic matter   The organic fraction of the soil that includes plant and animal residues at various stages of decomposition, cells and tissues of soil organisms, and substances synthesized by the soil population. Commonly determined as the amount of organic material contained in a soil sample passed through a 2-mm sieve.

soil porosity   *See* porosity, soil.

soil productivity   *See* productivity, soil.

soil profile   *See* profile, soil.

soil quality   The capacity of a specific kind of soil to function, within natural or managed ecosystem boundaries, to sustain plant and animal productivity, maintain or enhance water and air quality, and support human health and habitation. Sometimes considered in relation to this capacity in the undisturbed, natural state.

soil reaction   *See* reaction, soil; pH, soil.

soil salinity   The amount of soluble salts in a soil, expressed in terms of percentage, milligrams per kilogram, parts per million (ppm), or other convenient ratios.

soil separates   *See* separate, soil.

soil series   *See* series, soil.

soil solution   The aqueous liquid phase of the soil and its solutes, consisting of ions dissociated from the surfaces of the soil particles and of other soluble materials.

soil strength   A transient soil property related to the soil's solid phase cohesion and adhesion.

soil structure   The combination or arrangement of primary soil particles into secondary particles, units, or peds. These secondary units may be, but usually are not, arranged in the profile in such a manner as to give a distinctive characteristic pattern. The secondary units are characterized and classified on the basis of size, shape, and degree of distinctness into classes, types, and grades, respectively.

soil structure classes   A grouping of soil structural units or peds on the basis of size from the very fine to very coarse.

soil structure grades   A grouping or classification of soil structure on the basis of inter- and intraaggregate adhesion, cohesion, or stability within the profile. Four grades of structure, designated from 0 to 3, are recognized: *structureless*, *weak*, *moderate*, and *strong*.

soil structure types   A classification of soil structure based on the shape of the aggregates or peds and their arrangement in the profile, including platy, prismatic, columnar, blocky, subangular blocky, granulated, and crumb.

soil survey   The systematic examination, description, classification, and mapping of soils in an area. Soil surveys are classified according to the kind and intensity of field examination.

soil temperature classes   A criterion used to differentiate soils in *Soil Taxonomy*, mainly at the family level. Classes are based on mean annual soil temperature and on differences between summer and winter temperatures at a depth of 50 cm.

soil textural class   A grouping of soil textural units based on the relative proportions of the various soil separates (sand, silt, and clay). These textural classes, listed from the coarsest to the finest in texture, are sand, loamy sand, sandy loam, loam, silt loam, silt, sandy clay loam, clay loam, silty clay loam, sandy clay, silty clay, and clay. There are several subclasses of the sand, loamy sand, and sandy loam classes based on the dominant particle size of the sand fraction (e.g., loamy fine sand, coarse sandy loam).

soil texture   The relative proportions of the various soil separates in a soil.

soil water deficit   The difference between PET and ET, representing the gap between the amount of evapotranspiration water atmospheric conditions "demand" and the amount the soil can actually supply. A measure of the limitation that water supply places on plant productivity.

soil water potential (total) A measure of the difference between the free energy state of soil water and that of pure water. Technically it is defined as "that amount of work that must be done per unit quantity of pure water in order to transport reversibly and isothermically an infinitesimal quantity of water from a pool of pure water, at a specified elevation and at atmospheric pressure, to the soil water (at the point under consideration)." This *total* potential consists of *gravitational*, *matric*, and *osmotic* potentials.

solarization The process of heating a soil in the field by covering it with clear plastic sheeting during sunny conditions. The heat is meant to partially sterilize the upper 5 to 15 cm of soil to reduce pest and pathogen populations.

solum (pl. sola) The upper and most weathered part of the soil profile; the A, E, and B horizons.

sombric horizon A diagnostic subsurface horizon that contains illuvial humus but has a low cation exchange capacity and low percentage base saturation. Mostly restricted to cool, moist soils of high plateaus and mountainous areas of tropical and subtropical regions.

sorption The removal from the soil solution of an ion or molecule by adsorption and absorption. This term is often used when the exact mechanism of removal is not known.

species diversity The variety of different biological species present in an ecosystem. Generally, high diversity is marked by many species with few individuals in each.

species richness The number of different species present in an ecosystem, without regard to the distribution of individuals among those species.

specific gravity The ratio of the density of a mineral to the density of water at standard temperature and pressure.

specific heat capacity The amount of kinetic (heat) energy required to raise the temperature of 1 g of a substance (usually in reference to soil or soil components).

specific surface The solid particle surface area per unit mass or volume of the solid particles.

splash erosion The spattering of small soil particles caused by the impact of raindrops on very wet soils. The loosened and separated particles may or may not be subsequently removed by surface runoff.

spodic horizon A diagnostic subsurface horizon characterized by the illuvial accumulation of amorphous materials composed of aluminum and organic carbon with or without iron.

Spodosols An order in *Soil Taxonomy*. Soils with subsurface illuvial accumulations of organic matter and compounds of aluminum and usually iron. These soils are formed in acid, mainly coarse-textured materials in humid and mostly cool or temperate climates.

stem flow The process by which rain or irrigation water is directed by a plant canopy toward the plant stem so as to wet the soil unevenly under the plant canopy.

stratified Arranged in or composed of strata or layers.

strip-cropping The practice of growing crops that require different types of tillage, such as row and sod, in alternate strips along contours or across the prevailing direction of wind.

structure, soil See soil structure.

stubble mulch The stubble of crops or crop residues left essentially in place on the land as a surface cover before and during the preparation of the seedbed and at least partly during the growing of a succeeding crop.

subgroup, soil In *Soil Taxonomy*, subdivisions of the great groups into central concept subgroups that show the central properties of the great group, intergrade subgroups that show properties of more than one great group, and other subgroups for soils with atypical properties that are not characteristic of any great group.

submergence potential The positive hydrostatic pressure that occurs below the water table.

suborder, soil A category in *Soil Taxonomy* that narrows the ranges in soil moisture and temperature regimes, kinds of horizons, and composition, according to which of these is most important.

subsoil That part of the soil below the plow layer.

subsoiling Breaking of compact subsoils, without inverting them, with a special knife-like instrument (chisel), which is pulled through the soil at depths usually of 30 to 60 cm and at spacings usually of 1 to 2 m.

sulfidic Adjective used to describe sulfide-containing soil materials that initially have a pH >4.0 and exhibit a drop of at least 0.5 pH unit within 8 weeks of aerated, moist incubation. Found in potential acid sulfate soils.

sulfuric horizon A diagnostic subsurface horizon in either mineral or organic soils that has a pH <3.5 and fresh straw-colored mottles (called *jarosite mottles*). Forms by oxidation of sulfide-rich materials and is highly toxic to plants.

summer fallow See fallow.

surface runoff See runoff.

surface seal A thin layer of fine particles deposited on the surface of a soil that greatly reduces the permeability of the soil surface to water.

surface soil The uppermost part of the soil, ordinarily moved in tillage, or its equivalent in uncultivated soils. Ranges in depth from 7 to 25 cm. Frequently designated as the *plow layer*, the *Ap layer*, or the *Ap horizon*.

surface tension The elasticlike phenomenon resulting from the unbalanced attractions among liquid molecules (usually water) and between liquid and gaseous molecules (usually air) at the liquid–gas interface.

swamp An area of land that is usually wet or submerged under shallow fresh water and typically supports hydrophilic trees and shrubs.

symbiosis The living together in intimate association of two dissimilar organisms, the cohabitation being mutually beneficial.

synergism (1) The nonobligatory association between organisms that is mutually beneficial. Both populations can survive in their natural environment on their own, although, when formed, the association offers mutual advantages. (2) The simultaneous actions of two or more factors that have a greater total effect together than the sum of their individual effects.

talus Fragments of rock and other soil material accumulated by gravity at the foot of cliffs or steep slopes.

taxonomy, soil The science of classification of soils; laws and principles governing the classifying of soil. Also a specific *soil classification* system developed by the U.S. Department of Agriculture.

tensiometer A device for measuring the negative pressure (or tension) of water in soil *in situ*; a porous, permeable ceramic cup connected through a tube to a manometer or vacuum gauge.

tension, soil-moisture See soil water potential.

terrace (1) A level, usually narrow, plain bordering a river, lake, or the sea. Rivers sometimes are bordered by terraces at different levels. (2) A raised, more or less level or horizontal strip of earth usually constructed on or nearly on a contour and designed to make the land suitable for tillage and to prevent accelerated erosion by diverting water from undesirable channels of concentration; sometimes called *diversion terrace*.

tetrahedral sheet Sheet of horizontally linked, tetrahedron-shaped units that serve as one of the basic structural components of silicate (clay) minerals. Each unit consists of a central four-coordinated atom (e.g., Si, Al, Fe) surrounded by four oxygen atoms that, in turn, are linked with other nearby atoms (e.g., Si, Al, Fe), thereby serving as interunit linkages to hold the sheet together.

texture See soil texture.

thermal analysis (differential thermal analysis)   A method of analyzing a soil sample for constituents, based on a differential rate of heating of the unknown and standard samples when a uniform source of heat is applied.

thermic   A soil temperature class with mean annual temperature 15 to 22 °C.

thermophilic   Pertaining to temperatures in the range of 45 to 90 °C, the range in which thermophilic organisms grow best and in which thermophilic composting takes place.

thermophilic organisms   Organisms that grow readily at temperatures above 45 °C.

thixotrophy   The property of certain clay soils of becoming fluid when jarred or agitated and then setting again when at rest. Similar to *quick*, as in quick clays or quicksand.

tile, drain   Pipe made of burned clay, concrete, or ceramic material, in short lengths, usually laid with open joints to collect and carry excess water from the soil.

till   (1) Unstratified glacial drift deposited directly by the ice and consisting of clay, sand, gravel, and boulders intermingled in any proportion. (2) To plow and prepare for seeding; to seed or cultivate the soil.

tillage   The mechanical manipulation of soil for any purpose; but in agriculture it is usually restricted to the modifying of soil conditions for crop production.

tillage, conservation   Any tillage sequence that reduces loss of soil or water relative to conventional tillage, which generally leaves at least 30% of the soil surface covered by residues, including the following systems:

　minimum tillage   The minimum soil manipulation necessary for crop production or meeting tillage requirements under the existing soil and climatic conditions.

　mulch tillage   Tillage or preparation of the soil in such a way that plant residues or other materials are left to cover the surface; also called *mulch farming, trash farming, stubble mulch tillage*, and *plowless farming*.

　no-tillage system   A procedure whereby a crop is planted directly into a seedbed not tilled since harvest of the previous crop; also called *zero tillage*.

　ridge till   Planting on ridges formed by cultivation during the previous growing period.

　strip till   Planting is done in a narrow strip that has been tilled and mixed, leaving the remainder of the soil surface undisturbed.

tillage, conventional   The combined primary and secondary tillage operations normally performed in preparing a seedbed for a given crop grown in a given geographic area. Usually said of non-conservation tillage.

tillage, primary   Tillage that contributes to the major soil manipulation, commonly with a plow.

tillage, rotary   An operation using a power-driven rotary tillage tool to loosen and mix soil.

tillage, secondary   Any tillage operations following primary tillage designed to prepare a satisfactory seedbed for planting.

tilth   The physical condition of soil as related to its ease of tillage, fitness as a seedbed, and its impedance to seedling emergence and root penetration.

topdressing   An application of fertilizer to a soil after the crop stand has been established.

toposequence   A sequence of related soils that differ, one from the other, primarily because of *topography* as a soil-formation factor, with other factors constant.

topsoil   (1) The layer of soil moved in cultivation. *See also* surface soil. (2) Presumably fertile soil material used to top-dress roadbanks, gardens, and lawns.

trace elements   Elements present in the Earth's crust in concentrations less than 1000 mg/kg. When referring to plant nutrients, the term *micronutrients* is preferred.

trophic levels   Levels in a food chain that pass nutrients and energy from one group of organisms to another.

truncated   Having lost all or part of the upper soil horizon or horizons.

tuff   Volcanic ash usually more or less stratified and in various states of consolidation.

tundra   A level or undulating treeless plain characteristic of arctic regions.

Ultisols   An order in *Soil Taxonomy*. Soils that are low in bases and have subsurface horizons of illuvial clay accumulations. They are usually moist, but during the warm season of the year some are dry part of the time.

umbric epipedon   A diagnostic surface horizon of mineral soil that has the same requirements as the mollic epipedon with respect to color, thickness, organic carbon content, consistence, structure, and P_2O_5 content, but that has a base saturation of less than 50%.

universal soil loss equation (USLE)   An equation for predicting the average annual soil loss per unit area per year; $A = RKLSPC$, where R is the climatic erosivity factor (rainfall plus runoff), K is the soil erodibility factor, L is the length of slope, S is the percent slope, P is the soil erosion practice factor, and C is the cropping and management factor.

unsaturated flow   The movement of water in a soil that is not filled to capacity with water.

vadose zone   The aerated region of soil above the permanent water table.

value (color)   *See* Munsell color system.

variable charge   *See* pH-dependent charge.

varnish, desert   A glossy sheen or coating on stones and gravel in arid regions.

vermicompost   Compost made by earthworms eating raw organic materials in moist aerated piles, which are kept shallow to avoid heat buildup that could kill the worms.

vermiculite   A 2:1-type silicate clay, usually formed from mica, that has a high net negative charge stemming mostly from extensive isomorphous substitution of aluminum for silicon in the tetrahedral sheet.

Vertisols   An order in *Soil Taxonomy*. Clayey soils with high shrink–swell potential that have wide, deep cracks when dry. Most of these soils have distinct wet and dry periods throughout the year.

vesicles   (1) Unconnected voids with smooth walls. (2) Spherical structures formed inside root cortical cells by vesicular arbuscular mycorrhizal fungi.

virgin soil   A soil that has not been significantly disturbed from its natural condition.

water deficit (soil)   The amount of available water removed from the soil within the vegetation's active rooting depth, or the amount of water required to bring the soil to field capacity.

water potential, soil   *See* soil water potential.

water table   The upper surface of groundwater or that level below which the soil is saturated with water.

water table, perched   The surface of a local zone of saturation held above the main body of groundwater by an impermeable layer of stratum, usually clay, and separated from the main body of groundwater by an unsaturated zone.

waterlogged   Saturated with water.

watershed   All the land and water within the geographical confines of a drainage divide or surrounding ridges that separate the area from neighboring watersheds.

water-stable aggregate   A soil aggregate stable to the action of water, such as falling drops or agitation, as in wet-sieving analysis.

water-use efficiency   Dry matter or harvested portion of crop produced per unit of water consumed.

weathering   All physical and chemical changes produced in rocks, at or near the Earth's surface, by atmospheric agents.

wetland   An area of land that has hydric soil and hydrophytic vegetation, typically flooded for part of the year, and forming a transition zone between aquatic and terrestrial systems.

wetting front   The boundary between the wetted soil and dry soil during infiltration of water.

wilting point (permanent wilting point) The moisture content of soil, on an oven-dry basis, at which plants wilt and fail to recover their turgidity when placed in a dark, humid atmosphere.

windbreak Planting of trees, shrubs, or other vegetation perpendicular, or nearly so, to the principal wind direction to protect soils, crops, homesteads, etc., from wind and snow.

xenobiotic Compounds foreign to biological systems. Often refers to compounds resistant to decomposition.

xerophytes Plants that grow in or on extremely dry soils or soil materials.

zero tillage *See* tillage, conservation.

zymogenous organisms So-called opportunist organisms found in soils in large numbers immediately following addition of readily decomposable organic materials. *Contrast with* autochthonous organisms. *See also* r-strategist.

index

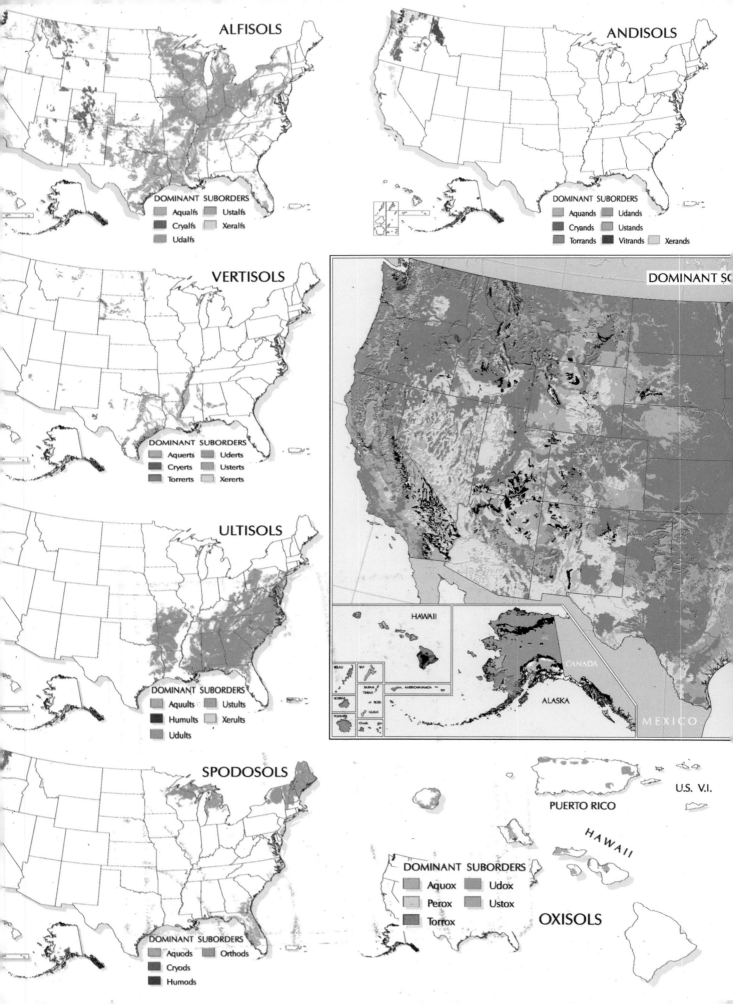

ALFISOLS

DOMINANT SUBORDERS

- Aqualfs
- Cryalfs
- Udalfs
- Ustalfs
- Xeralfs

ANDISOLS

DOMINANT SUBORDERS

- Aquands
- Cryands
- Torrands
- Udands
- Ustands
- Vitrands
- Xerands

VERTISOLS

DOMINANT SUBORDERS

- Aquerts
- Cryerts
- Torrerts
- Uderts
- Usterts
- Xererts

ULTISOLS

DOMINANT SUBORDERS

- Aquults
- Humults
- Udults
- Ustults
- Xerults

SPODOSOLS

DOMINANT SUBORDERS

- Aquods
- Cryods
- Humods
- Orthods

DOMINANT SO

HAWAII

CANADA

ALASKA

MEXICO

PALAU
YAP
SAIPAN
TINIAN
ROTA
KOSRAE
GUAM
POHNPEI
Chuuk
AMERICAN SAMOA

PUERTO RICO

U.S. V.I.

HAWAII

OXISOLS

DOMINANT SUBORDERS

- Aquox
- Perox
- Torrox
- Udox
- Ustox